Diseases of Edible Oilseed Crops

Diseases of Edible Oilseed Crops

C. Chattopadhyay
Director
ICAR: National Research Centre for Integrated Pest Management
New Delhi, India

S.J. Kolte
Professor of Plant Pathology (Retired)
G.B. Pant University of Agriculture and Technology
Pantnagar, India

F. Waliyar
Principal Plant Pathologist
International Crops Research Institute for the Semi-Arid Tropics (ICRISAT)
Bamako, Mali

CRC Press
Taylor & Francis Group
Boca Raton London New York

CRC Press is an imprint of the
Taylor & Francis Group, an **informa** business

CRC Press
Taylor & Francis Group
6000 Broken Sound Parkway NW, Suite 300
Boca Raton, FL 33487-2742

Printed on acid-free paper
Version Date: 20150831

International Standard Book Number-13: 978-1-4665-9565-1 (Hardback)

Visit the Taylor & Francis Web site at
http://www.taylorandfrancis.com

and the CRC Press Web site at
http://www.crcpress.com

Contents

SECTION IV Sunflower

SECTION V Sesame

SECTION VI Safflower

SECTION VII Soybean

Foreword

Annual edible field-grown oilseed crops have wide adaptability and are grown under varied agroclimatic conditions. They occupy a special place in agricultural economies all over the world. Almost all such crops have a great potential for the diversification of major cropping systems in developing countries. In the period 2009–2012, these crops recorded impressive compound growth rates in terms of area and production. However, the average yield per hectare is hardly 30%–50% of what is obtainable under real-farm situations in rain-fed areas as well as in areas with assured moisture supply in developing countries. One of the major reasons for this huge gap is the occurrence of diseases that adversely affect these crops.

In the war against hunger and malnutrition, it is necessary to enhance and update knowledge about these diseases, their occurrence, epidemiology, and disease management, including transgenic technology. To this end, the International Crops Research Institute for the Semi-Arid Tropics (ICRISAT) has developed and characterized transgenic peanut (groundnut) lines with aflatoxin resistance conferred by the rice chitinase gene and also lines with bud necrosis virus resistance imparted by expressing the viral coat protein gene. Elsewhere, similar approaches in the development of stem necrosis virus–resistant transgenic sunflower and *Alternaria*-resistant transgenic mustard have been cultivated. Enriched with these recent developments, this book is useful as an updated basic reference volume in the conduct of research and development activities toward obtaining increased productivity and sustainability of oilseed production in the world.

In view of this, I appreciate and compliment all three authors for bringing out this important book. I hope that readers of this book will contribute to pushing forward the frontiers in the war against hunger and malnutrition.

William D. Dar
Director General
International Crops Research Institute for the Semi-Arid Tropics
Patancheru, Telangana, India

Preface

The proposal for bringing out this book is the result of the pressing need and demand from edible oilseed crop researchers, university faculty members and students, and the interest taken by the publishers in the proposed authors based on the previously published book *Diseases of Annual Edible Oilseed Crops* by CRC Press in 1984–1985. The new volume is an improved one covering the latest developments in host–pathogen interactions and disease management, including molecular breeding for disease resistance and developments of transgenics, if any, in edible oilseed crops. Accordingly, the proposal was prepared and submitted for review along with the expert opinion of scientists as per CRC Press norms and regulations, and was approved after incorporating suggestions given by the experts.

This book deals primarily with the diseases of cultivated annual edible oilseeds, that is, peanut (groundnut), rapeseed–mustard, sunflower, sesame, safflower, and soybean. In recent years, soybean, though not a high-oil-containing seed, has been identified more as an oilseed crop than a bean crop by the United Nations Food and Agriculture Organisation (UN FAO). Diseases of other annual crops, for example, cotton, corn, rice bran, and perennial oil palm, which also contribute significantly to the world supply of edible vegetable oils and fats, have been excluded. Linseed oil is mostly useful for industrial purposes. However, a new edible grade oil crop termed *Linola* has been created (through collaborative research between Australia and the United States) out of conventional linseed varieties through mutation breeding, which is likely to expand as a good source of vegetable oils for human consumption.

A great deal of information has been accumulated on the diseases of peanut, rapeseed–mustard, sesame, sunflower, safflower, and soybean since the publication of the 1984–1985 edition. Besides a pressing need for such a comprehensive work, the experience of the authors in research in this pertinent field has prompted the attempt to bring together the scattered information on the subject in a comprehensive manner in order to present it in a useful form. An attempt has been made to present a broader view of the subject than that generally included in bulletins and manuals. Discussions on the development of a straightforward and also of a controversial nature have been included to stimulate thinking especially among graduate students. The information presented represents a careful synthesis of research articles. The survey of literature has been made as complete as possible up to the beginning of 2014. In most cases, original papers are consulted, and the temptation to use review articles or abstracts as a major source of information is avoided.

The "Introduction" deals with the uses and chemistry of vegetable oils and fats, trends in world production and consumption, production constraints, crop management, and disease problems. Depending upon the available literature, the treatment of all the previously mentioned crop diseases follows a uniform pattern under headings such as Symptoms, Geographical Distribution and Losses, Pathogen, Epidemiology, Disease Cycle, and Diseases Management covering Host Plant Resistance, Molecular Breeding, Cultural Control, Biological Control, etc., in each chapter. The aim has been to make the subject matter regarding each disease as complete and self-contained as possible. At first, the reader is introduced to the respective edible oilseed crop in each chapter with a brief botanical description of the crop and its genomics, origin, and distribution. The diseases are arranged under each crop on the basis of their global economic importance.

Acknowledgments

The authors first wish to express their appreciation to their respective work organization for providing them with opportunities to independently investigate the disease problems of oilseed crops: the Indian Council of Agricultural Research (C. Chattopadhyay), G.B. Pant University of Agriculture and Technology (S.J. Kolte), and the International Crops Research Institute for the Semi-Arid Tropics (ICRISAT) (F. Waliyar). It is basically because of this research work and their experience as the team leaders of the oilseed pathology discipline in their respective institutes that the authors could come together and make a genuinely convenient and desirable, firm, support base to write this book, while being involved at some time or an other in some active collaborative research on certain aspects of edible oilseed crop diseases.

During the process of preparing the manuscript, the authors sacrificed time they could spend with their family members and friends. They were constantly encouraged by their near and dear ones to prioritize the writing of this book. C. Chattopadhyay acknowledges the encouragement given by his parents (Amiya Kumar and Arati Chattopadhyay), wife (Aparna), daughter (Nivedita), teachers (Dr. Chitreshwar Sen and Dr. Bineeta Sen) and research managers (Dr. S. Ayyappan [DG-ICAR], Dr. S.K. Datta [DDG-CS-ICAR], and all mentors, scientists, staff of ICAR: NCIPM, other scientist colleagues, all dear friends, and other relatives, especially Chaitali. S. J. Kolte expresses his appreciation to his wife, Rekha, for being the best critic and for being very supportive of him continuing his job even after retirement, and to his great teacher, Professor Y.L. Nene, for encouragement. F. Waliyar acknowledges the encouragement from his family by dedicating this book to his parents and to Marion, Simine, Edrice, Engela, Martin, Charline, and Alice.

Most of the illustrations, particularly the photographs of the signs and symptoms of different diseases, are the results of the authors' original work. A few photographs have been obtained from other scientists working in India, the United States, and Nigeria, whose help is acknowledged in the captions. The following people deserve particular mention: Dr. Tom Gulya, research plant pathologist, USDA-ARS, Northern Crop Science Lab, Fargo, North Dakota, USA; Dr. Shrishail Navi, associate scientist, Department of Plant Pathology and Microbiology, Iowa State University, Ames, USA; Dr. B.A. Tunwari, Department of Crop Production and Protection, Federal University, Wukari (Taraba), Nigeria; Dr. H. Nahunnaro, Department of Crop Protection, Modibbo Adama University of Technology, Yola, Nigeria; Dr. G.K. Gupta, head, Division of Crop Protection, ICAR-Directorate of Soybean Research, Indore, India; Dr. S. Chander Rao, principal scientist, and Dr. K.S. Varaprasad, project director, ICAR-Directorate of Oilseed Research, Hyderabad, India; Dr. Anil Kotasthane, head, Department of Plant Pathology, Indira Gandhi Krishi Vishwavidyalaya, Raipur, India; and Dr. A.K. Tewari, professor of plant pathology, G.B. Pant University of Agriculture and Technology, Pantnagar, India.

Some of the figures are partly adapted from previous published records, and the sources of illustrations have been included as needed. The authors are grateful to Manish M. Patil for his help in sending the finished manuscript to CRC Press. The courteous cooperation in all matters relevant to the development of this book that the authors received from the publisher, especially Randy Brehm, acquiring editor and main contact person, and Kari Budyk, senior project coordinator, CRC Press, is gratefully acknowledged.

The authors are grateful to Dr. William D. Dar, director general, ICRISAT, Patancheru, Telangana, India, for writing the foreword for this book.

Authors

Chirantan Chattopadhyay is currently the director of the National Research Centre for Integrated Pest Management (NCIPM) in New Delhi under the Indian Council of Agricultural Research (ICAR) system. He earned a PhD in mycology and plant pathology from the Indian Agricultural Research Institute (IARI), New Delhi, and is a fellow of the Indian National Academy of Agricultural Sciences. He has more than 22 years of research experience in oilseed crop disease management and has significantly contributed in the areas of understanding the host–pathogen interactions, epidemiology, and integrated disease management in sunflower, safflower, sesame, peanut, and rapeseed–mustard crops. He has published more than 70 peer-reviewed research papers and 4 books. Dr. Chattopadhyay has to his credit several academic awards and one patent. His significant research achievements include the development of *Fusarium* wilt–resistant genotypes in safflower, the identification of genotypes of *Brassica juncea* resistant to a mixture of isolates of *Albugo candida*, the development of a biocontrol agent formulation to reduce the use of fungicides in the management of major diseases of oilseeds (*Brassica*), and seed treatment (and foliar application) of aqueous bulb extract of *Allium sativum* for the management of major foliar diseases (*Alternaria* blight, white rust, powdery mildew, and *Sclerotinia* stem rot) of mustard.

S.J. Kolte, professor of plant pathology (Retd.), has 36 years of experience in teaching both undergraduate and postgraduate courses in the Department of Plant Pathology, G.B. Pant University of Agriculture and Technology, Pantnagar, India. He has contributed considerably in the areas of oilseed pathology and plant pathology by being principal investigator of three international collaborative research projects funded by the Department for International Development (United Kingdom) and the International Development Research Centre (Canada) on disease and drought resistance in oilseed crops for more than two decades. He has contributed outstanding research work in the area of oilseed crop pathology, written 8 books and guided 26 students for their PhD and MS programs. He has also published 125 research papers in reputed national and international journals and has contributed several review articles and book chapters. In recognition of his contributions in plant pathology, Dr. Kolte has been bestowed with more than half a dozen awards, notable among them being the Lifetime Achievement Award by the Society for Rapeseed–Mustard Research, Y.L. Nene Outstanding Best Teacher award by the Indian Society of Mycology and Plant Pathology, and the IPS Recognition Award by the Indian Phytopathological Society. Induced host resistance, genetics of host resistance, and development of a new concept of the Divya mustard ideotype are some of his significant contributions.

Farid Waliyar is a principal scientist of plant pathology at the International Crops Research Institute for the Semi-Arid Tropics (ICRISAT). He is currently ICRISAT director for West and Central Africa with regional office in Bamako. He earned a PhD in plant pathology (1984) from the University of Paris M&C, where he specialized in research on the role of *Aspergillus flavus* and safety measures for aflatoxin contamination and other soil fungi on peanut and their evolution during the cropping season in Senegal. Dr. Waliyar has a total of 30 years experience in strategic basic research related to the development of integrated crop disease management particularly in the peanut crop. He has published widely (more than 200 publications) as well as supervised and examined several undergraduate, MS, and PhD students. He has expertise in administration and management of research projects involving international collaboration. Dr. Waliyar has received several awards, the most recent from the Malian Ministry of Agriculture in recognition of his contribution to the development of agriculture in Mali.

Section I

Introduction

1 Edible Oilseed Crops

Edible oilseed plants are those whose seeds bear fixed nonvolatile oil, and oilseed crops are grown primarily for the oil contained in the seeds. The oil content of small grains (e.g., wheat) is only 1%–2%, and that of oilseeds ranges from about 20% for soybeans to over 40% for sunflowers and canola rapeseed. Crops like rapeseed–mustard, peanut, and sunflower have oil recovery ratio of 45%, 40%, and 30%, respectively, whereas cottonseed and soybean have oil recovery ratio of 11.5% and 17% only (Kumar 2014). Some of the oilseeds like peanut, sesame, sunflower could be consumed directly or may be eaten fried, roasted, or pounded and mixed with sugar; or the oil may be extracted from such seeds and directly used for cooking food or for confectionery purposes. Usually, refining of the oil is done before it is used as food. Edible vegetable oils may, however, be used occasionally for industrial purposes, for example, manufacturing of soaps, varnishes, hair oils, and lubricants. The residues left, that is, the oil cakes, serve as excellent animal or poultry feed. Oil cakes may also be used as manure to increase the fertility status of soils. The demand for edible oilseeds for human consumption in different parts of the world is principally derived from three categories of cultivated crop plants: (1) primarily cultivated annual oilseed crops, for example, peanut (*Arachis hypogaea* L.), rapeseed–mustard (*Brassica campestris* L., *Brassica napus* L., *Brassica juncea* [L.], Czern and Coss. *Eruca sativa* Lam.), sunflower (*Helianthus annuus* L. var. *macrocarpus* [DC] Ckll.), sesame (*Sesamum indicum* L.), safflower (*Carthamus tinctorius* L.), niger seed (*Guizotia abyssinica* Cass.), and soybean (*Glycine max* [L.] Merrill); (2) an annual fiber crop cotton (*Gossypium hirsutum* L.) through its seed by-products; and (3) perennial oilseed plants such as coconut palm (*Cocos nucifera* L.) and oil palm (*Elaeis guineensis* Jacq). Corn (*Zea mays* L.) also contributes significantly to the world edible oil supply. Besides traditionally grown oilseed crops, technological innovations in refining, bleaching and deodorization, newer oils like cottonseed and rice bran oils have also become popular in the recent times. Thus, the range of plants that could be cultivated for edible oils is extensive, but only a few that are included in the first (1) category are suitable for large-scale commercial production or produce oil that is required in large quantities. In this chapter, only this category of primarily cultivated annual oilseed crops is considered with respect to diseases and their management.

CHEMICAL NATURE OF EDIBLE OILS AND FATS

Edible oils and fats of vegetable origin are composed of triglycerides that are esters of one molecule of glycerol and three molecules of fatty acids. A reaction leading to the formation of a triglyceride is shown in Figure 1.1.

Triglycerides that are solids at room temperature are termed as *fats*, whereas the liquid ones are termed as *oils*. The latter contain ester-bound unsaturated fatty acids.

Fatty acids by and large are straight-chain aliphatic monocarboxylic acids. Most of the members of this series contain an even number of carbon atoms in the molecule. Individual fatty acids are distinguished from one another by the nature of the hydrocarbon chain. This chain can vary in length from 4 to 24 carbon atoms. When fatty acid contains one or more double bonds in the molecule, it is said to be unsaturated. Thus, the fatty acid may be saturated (no double bond) as stearic acid, monounsaturated (one double bond) as oleic acid, or polyunsaturated (with two or more double bonds) as linoleic acid. The fatty acids are abbreviated according to the number of carbon atoms in the molecule and degree of unsaturation (number of double bonds). The common names,

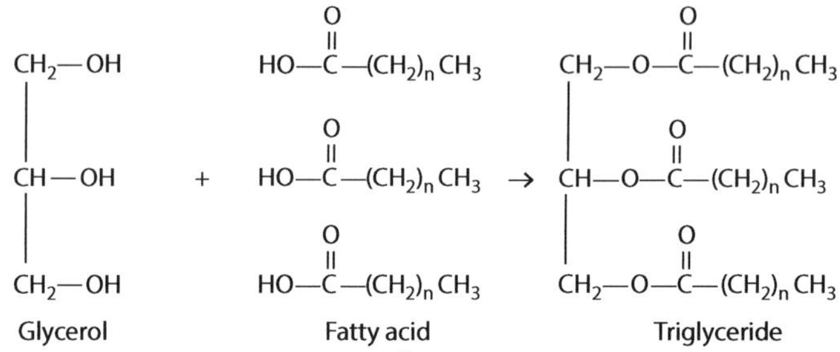

FIGURE 1.1 A chemical reaction leading to the formation of triglyceride.

TABLE 1.1
Common Names, Symbols, Systematics, and Structural Formulae of Certain Important Fatty Acids Found in Vegetable Oils

Common Name	Symbol	Systematic	Structural Formula
Saturated fatty acids (1–4)			
1. Myristic	$C_{14.0}$	Tetradecanoic	$C_{13}H_{27}COOH$
2. Palmitic	$C_{16.0}$	Hexadecanoic	$C_{15}H_{31}COOH$
3. Stearic	$C_{18.0}$	Octadecanoic	$C_{17}H_{35}COOH$
4. Arachidic	$C_{20.0}$	Eicosanoic	$C_{19}H_{39}COOH$
Unsaturated fatty acids (5–10)			
5. Palmitoleic	$C_{16.1(9)}$	9-Hexadecenoic	$C_{15}H_{29}COOH$
6. Oleic	$C_{18.1(9)}$	9-Octadecenoic	$C_{17}H_{33}COOH$
7. Linoleic	$C_{18.2(9,12)}$	9,12-Octadecadienoic	$C_{17}H_{31}COOH$
8. Linolenic	$C_{18.3(9,12,15)}$	9,12-Octadecatrienoic	$C_{17}H_{29}COOH$
9. Gadoleic	$C_{20.1(9)}$	Eicosenoic	$C_{19}H_{37}COOH$
10. Erucic	$C_{22.1(13)}$	13-Docosenoic	$C_{21}H_{41}COOH$

Sources: Dutcher, R.A. et al., *Introduction to Agricultural Biochemistry*, John Wiley & Sons, New York, 1951, p. 72; Vaisey-Genser, M. and Eskin, N.A.M., Canadian rapeseed oil—Properties, processes and food quality, Publication No. 54, Rapeseed Association of Canada, Winnipeg, Manitoba, Canada, 1978, p. 13.

Note: Figures in the parenthesis indicate the position of double bonds (=) in the fatty acid chain at carbon numbers starting from carboxyl group.

abbreviated symbols, systematic, and structural formulae of certain important fatty acids found in vegetable oils are given in Table 1.1. The natural configuration of fatty acids is the *cis* configuration, which is considered to be nutritionally more desirable.

TRENDS IN WORLD PRODUCTION AND CONSUMPTION OF VEGETABLE OILS AND FATS

Commonly Cultivated Annual Edible Oilseed Crops

The oilseeds sector has remained vibrant globally with 4.1% growth per annum in the last three decades. The production of annual oilseed field crops has increased considerably since 1960, and now constitutes over 50% of the total production of fats and oils in the world. However, the supply of vegetable oils from annual field crops tends to remain quite flexible from year to year in relation to the total world supply of vegetable oils and fats (Sharma et al. 2012). The present average

TABLE 1.2
Average World Oilseed Production (2010–2012) and Projected Oilseed Crop Production (2022)

	Production (kilotons)		Growth (%)		Per Capita Food Use (kg/annum)	Per Capita Food Use (kg/annum)
World/Country	Average (2010–2012 Estimated)	Projected (2022)	2003–2012	2013–2022	Average (2010–2012 Estimated)	Projected (2022)
World	300,414	400,460	3.13	2.07	18.3	20.0
Developed countries	165,474	203,242	3.70	1.63	24.0	24.4
North America	107,682	128,468	2.35	1.13	37.9	32.2
Canada	18,184	23,367	8.05	1.62	23.9	20.8
United States	89,497	105,101	1.44	1.03	39.5	33.5
Europe	52,349	66,678			21.8	23.3
Oceania developed	2,861	4,130	9.33	2.53	26.4	27.0
Developing countries	224,040	28,728	2.71	2.39	16.7	19.0
Africa	10,043	12,910	1.11	2.65	11.4	12.4
Latin America and Caribbean	139,470	189,415	4.17	2.84	19.2	22.3
Asia and Pacific	175,427	84,893	0.62	1.42	17.7	20.5
China	44,380	47,951	0.01	1.20	21.9	26.3
India	23,222	27,165	1.29	1.71	13.3	16.0

Source: OECD-FAO, OECD-FAO agricultural outlook 2013, Chapter 5—Oilseeds and oilseed products, OECD/FAO Secretariats, pp. 139–282.

per capita consumption of edible oils and fats is 39.5 kg/annum (highest) in the United States, 13.3 kg/annum (low) in India and other south Asian countries, and 11.4 kg/annum (lowest) in Africa. Thus, the consumption of fats and oils in Asia and Africa and in other developing countries is much less as against the required minimum consumption level of 30 kg/annum (Table 1.2). This is a serious situation, particularly when it comes to meeting the requirement of an essential fatty acid, linoleic ($C_{18.2}$), and the energy supply for body functions under a balanced diet pattern. Considering the global minimum per capita consumption as required for keeping human health, the increasing world population by about 2% every year, the present rate of oilseed production on a global scale is not and will not be satisfactory. However, developed countries such as the United States, Canada, and the Russian Federation have been and should continue to be the major producing areas. Population growth and rising per capita income are expected to lead to an average 2.1%/annum growth of food vegetable oil use in developing countries. Annual food vegetable oil use per capita is expected to average 19 kg/annum across developing countries, but no more than 9.5 kg/annum in least developed countries by 2022. As a group, developed countries are showing a stable consumption level of 24–25 kg/annum, but individual countries differ based on tastes and preferences (OEDC-FAO 2013). Biotechnology offers a number of solutions to meet the growing need for affordable vegetable oils with improved fatty acid composition for food and industrial uses (Lu et al. 2011). The six annual edible oilseed crops, as considered in this chapter, are grown in different parts of the world, covering a wide range of geographical areas. Total world's oilseed production from major oil crops has been 423.55 million tons from 205.08 million hectares during 2009–2010. The leading countries in oilseed production are the United States, Brazil, Argentina, China, and India (Yadav et al. 2012). The yield of these crops is of higher magnitude in the developed countries as compared with the developing ones (Table 1.2). For example, the average yield of peanuts in the developed countries, particularly in the

TABLE 1.3
Edible Oilseed Crop Productivity (q/ha) in India vis-à-vis World (2012)

Crop	India	World	Country with Highest Productivity[a]
Peanut	11.7	16.7	46.9 (United States)
Rapeseed–mustard	11.4	18.7	36.9 (Germany)
Soybean	12.0	23.7	27.8 (Paraguay)
Sunflower	07.6	14.8	24.9 (China)
Sesame	04.26	05.1	13.1 (Egypt)
Safflower	06.5	09.6	14.8 (Mexico)

Sources: FAO, FAOSTAT world oilseed production, 2012, available at: http://faostat.fao.org; Paroda, R.S., The Indian oilseeds scenario: Challenges and opportunities, in: *The First Dr. M.V. Rao Lecture*, Indian Society of Oilseeds Research, Hyderabad, India, August 24, 2013, p. 26.

[a] Among the countries with >80% global contribution.

United States, is 46.9 q (quintals)/ha; whereas in India and in other semiarid countries, it is only about 11 q/ha as given in Table 1.3 (FAO 2012, Paroda 2013). A similar situation appears to be true with respect to the high production of rapeseed (now canola) in Canada and sunflower in Russian Federation, compared with the yield performance of these crops in developing countries. Safflower production is about 16.4 q/ha in the United States, 16.8 q/ha in Mexico, and only 6.3 q/ha in India (FAO 2011, Padmavati and Virmani 2012). The average yield of sesame varies from a high of 11.75 q/ha in Egypt to a low of 1.52 q/ha in Sudan (Ranganatha et al. 2012).

Linola: A New Annual Edible Oilseed Crop

There are reports showing that in certain linseed species extent and degree of polyunsaturated fatty acids are so low that the oil extracted is perfect for edible purposes. For example, *Linum strictum* L. is largely cultivated for edible oil and fodder purposes in Afghanistan (Richaria 1962). The edible oils are characterized by rather having higher content of oleic, palmitoleic, and linoleic acids. The linseed oil obtained from the seeds of *Linum usitatissimum* cannot normally be used for edible purposes directly or in the edible products because of its high linolenic acid contents; though in certain regions of Chhattisgarh (formerly a part of Madhya Pradesh) and adjoining eastern part of Vidarbha Region of Maharashtra State in India, linseed oil is used for cooking food. It is noteworthy that Green (1986a,b) has been successful in obtaining two low-linolenic acid (28%–30%) mutants *M1589* and *M1722* by treating the seeds of the linseed (*Linum usitatissimum* L.) cultivar *Glenelg* with ethyl methanesulfonate. The two mutants through crossing together have been further combined within a single genotype that has only 1% linolenic acid and increase in linoleic acid to 50%–70% depending on the temperature during seed maturation. Consequently, through traditional plant-breeding procedures, a joint venture between Commonwealth Scientific and Industrial Research Organization (CSIRO) of Australia and United Grain Growers Ltd. of Winnipeg, Canada, has led to the development of edible linseed oil (Gunstone 2011). The fatty acid composition of the new oilseed crop named *linola* (a registered trademark of CSIRO) has been changed, and the level of linolenic acid substantially reduced from 50%–60% to 2%. This greatly increases the oxidative stability of the oil that is a polyunsaturated oil identical to sunflower, safflower, or corn oil in fatty acid composition. The oxidative stability of oil of this newly created linseed genotype is equivalent to that of sunflower oil and much better than high-linolenic common linseed oil (Green 1986a–c). The color of the linola seed is also changed to pale yellow, which allows it to be distinguished from brownish traditional flaxseed/linseed.

The new oilseed crop can be grown wherever flax and linseed varieties are currently cultivated. The climate in northern Europe is highly suitable for the production of linola where sunflower and corn cannot be produced. Linola seed can be processed in existing crushing plants using standard procedures, and linola meal can also be used in ruminant feed in the same way as linseed meal. Refining of crude linola oil by conventional steps produces a pale-colored oil with good oxidative stability. The Food and Drug Administration has given GRAS approval to linola (Solin: the common generic name) oil for use as a general-purpose cooking oil, frying, and salad oil. Thus, linola oil and seed of the new oilseed crop appear to have a promising future. Anticipating the adoption and likely expansion of acreage of linola as a new edible oilseed crop, the linola crop is likely to be affected by the same diseases that affect the traditional nonedible grade linseed/flax (Kolte and Fitt 1997).

PRODUCTION CONSTRAINTS

Basic Constraints

It is true that high-yielding varieties of oilseeds do not have the genetic potential to yield at par with cereals, even at the optimum management level. Besides, it could be observed that production of a unit quantity of fats and oils by a plant requires more energy than production of carbohydrates by cereals. In making comparisons, one should always keep in view the differential energy requirements for the plants to produce a quintal of oil. If, for example, a plant produces 1 g of glucose, the conversion of this results in the formation of 0.83 g carbohydrate, while if glucose is converted into lipid, only 0.38 g is formed (Swaminathan 1979). It is because of this high-differential energy requirement that oil yield from oilseeds has continued to be restricted.

Other Constraints in Developing Countries

Poor plant population arising from poor-quality seed, particularly in the case of soybean, peanut, and sunflower, inadequate nutrient status of soil and nutrient supply, no rhizobial inoculation or use of inefficient rhizobial cultures in the case of soybean and peanuts, poor plant protection measures, and poor postharvest technology have been some other constraints for poor yields of oilseeds in developing countries. Besides, much of the oilseed acreage in developing countries—particularly in India—is rainfed, and therefore, a certain degree of instability is inherent in the production process. Absence of rain or lack of irrigation water at critical stages of the crop growth before maturity causes significant loss in yield. Thus, productivity in developing countries is still low compared to other oilseed-producing countries in the world. The main cause is low cultivation of oilseeds on account of switchover to other profitable crops and dependence on rainfall rather than on irrigation (Narayan et al. 2011).

CROP MANAGEMENT

Oilseed crop management must be seriously considered in view of the very low yield of these crops in developing countries. Considerable advancement in research has led to an increase in the productivity of the oilseed crops, both in developed and in developing countries particularly in China and India. In some crops, like and safflower, it is now possible to plan on the exploitation of hybrid vigor. Higher productivity of the sunflower in Canada and other developed countries is attributed to the cultivation of hybrid cultivars. In developing countries, adoption of a package approach (technological package) supported by package of services (seed, fertilizer, chemical supplies, etc.) constitutes an important major thrust to intensify oilseed production. There is still considerable scope for introduction of short-duration varieties of oilseeds in irrigated as well as in dry-farming systems favoring multiple cropping pattern all over the world.

It becomes necessary to obtain a thorough updated knowledge of a particular crop in terms of land preparation, techniques of sowing, varieties, fertilizer requirements, and intensive care during crop-growing season. Preparation of seedbeds, with sufficient conservation of soil moisture, is necessary for the most oilseed crops with special reference to peanut, sunflower, and rapeseed crops. Seed treatment with most recently recommended fungicides (thiram, carbendazim, or with a mixture formulation of such fungicides) at the rate of 2–3 g/kg of seed may be necessary for the soybean, peanuts, and sunflowers to get good seed germination and plant stand, directly increasing yields through such treatment. Some crops like rapeseed and mustard are still sown by broadcast method in India. It has now been demonstrated through planned field experiments and on-farm farmers participatory research that the yield of rapeseed–mustard crop can be increased considerably by line sowing. It is, therefore, considered best that rapeseed–mustard crop be sown in lines through seed drills. The requirements for fertilizers will be determined by the fertility status of the soil, the nature of oilseed crop to be grown, and time of sowing. For the peanut crop, application of calcium through gypsum may be quite important for better pod and seed development. Some other nutritional problems with respect to deficiency of boron, zinc, and iron have been encountered in oilseed crops in different geographic areas. A direct yield loss of U.S. $1.5 billion/annum is estimated due to low crop yields besides huge loss due to disease concerns arising out of Zn malnutrition in the country (Singh 2010, Suresh et al. 2013). Timely steps should be taken to correct the aforementioned deficiencies. Other management practices include spraying of suitable insecticides and fungicides at the appropriate time for the management of insect pests and diseases. In the case of rapeseed–mustard, the crop must be essentially protected from aphid attack under Indian conditions.

DISEASE PROBLEMS

Peanut, rapeseed–mustard, sunflower, sesame, safflower, and soybean are subject to attack by several infectious and noninfectious diseases. The loss in yield of the crop may vary, depending upon the nature of the pathogen and the severity of the attack. Considering all the vegetable oil–producing crops, the quantity lost, on a world basis, is estimated to be more than about 14.00 million tons/year—amounting to a monetary loss of about U.S. $16 million. This excludes the newly developed diseases for which loss estimates have not yet been determined. Thus, the overall losses may be of a higher magnitude. With an increasing emphasis on oilseed production, it is expected that limited land resources through intensive farming, higher cropping intensity, better seeds, and greater use of fertilizers and herbicides, the production of oilseeds will increase; however, this might create new disease problems under the changed environments, in addition to the already existing diseases. Such a shift in the disease situation, as discussed in the following chapters, has already taken place in the case of peanuts due to the use of benomyl for early and late leaf spot management, consequently favoring more peanut rust and Sclerotium rot development in the United States, and with respect to rapeseed–mustard due to the use of Barban® herbicide favoring development of *Sclerotinia* rot in Canada. Use of dalapon herbicide has increased the susceptibility of rapeseed to light leaf spot (*Pyrenopeziza brassicae* Sutton and Rawlinson) in the United Kingdom (Kolte 1985). A similar situation appears to be true with respect to nutrient status and susceptibility of rapeseed and sunflowers to fungal diseases at a lower concentration of erucic acid and glucosinolates. Derivatives of glucosinolates have been known to be fungitoxic. Some volatile derivatives of glucosinolates are reported to be more abundant in light leaf spot–resistant varieties than in susceptible types of rapeseed. So, the consequences of this trend, that is, breeding for low glucosinolates and for other quality characters, must be thoroughly examined in the general context of rapeseed diseases. Although climate change and variability is considered an altering situation and a big challenge to oilseed production, there is sometimes a positive impact of it regarding the disappearance of sesame phyllody disease (caused by phytoplasma) in an unusually cool and rainy growing season in the west Mediterranean region of Turkey. This is a unique case of influence of climate variability characterized by higher and frequent rainfalls and consequently causing lower temperatures but higher humidity on the *nonoccurrence* of phyllody disease transmitted and spread by leaf hopper vectors (Cagirgan et al. 2013). In contrast to this insect- and vector-borne phytoplasma, the fungal pathogen, *Sclerotinia sclerotiorum*, and several other pathogens find such weather conditions with higher rainfalls and lower temperatures most congenial to cause epidemics in rapeseed–mustard and sunflowers (Boomiraj et al. 2010, Evans et al. 2010).

Oilseed crops are affected by foliage diseases such as the rusts, downy mildews, leaf spots and blights. The management of these diseases through the use of chemical sprays and host resistance has been achieved in a satisfactory manner, but the situation with respect to control of a number of soil-borne root diseases, for example, charcoal rot, *Sclerotinia* rots, *Verticillium* wilts, and *Fusarium* wilts, is not satisfactory. Oilseed crops have a rather low-yield genetic potential. Therefore, the least expensive management measures, such as use of host resistance and cultural control, will find favor with farmers and others concerned with more oilseed production. In recent years, the gains in productivity of oilseed crops have been achieved primarily through exploitation of genetic variability (Anjani 2012, Azeez and Morakinyo 2011, Zhang and Johnson 1999). Conventional breeding coupled with modern tools such as biotechnology should now be the primary focus in crop improvement programs. Investigations to develop disease-resistant transgenics are underway all over the world. In India, for example, *Alternaria*-resistant mustard transgenics using antifungal chitinase and glucanase genes have been successfully developed. Similarly, transgenic peanuts have been developed with coat protein genes for resistance to peanut bud necrosis and peanut stem necrosis viruses (Paroda 2013).

REFERENCES

Anjani, K. 2012. Genetic improvement in safflower: Possible avenues. In: *Safflower Research and Development in the World: Status and Strategies*, ed., I.Y.L.N. Murthy et al. Indian Society of Oilseeds Research, Hyderabad, India, pp. 1–26.

Azeez, M.A. and J.A. Morakinyo. 2011. Genetic diversity of the seed physical dimensions in cultivated and wild relatives of sesame (Genera *Sesamum* and *Ceratotheca*). *Int. J. Plant Breed. Genet.* 5: 369–378.

Boomiraj, K., B. Chakrabarti, P.K. Aggarwal, R. Choudhary, and S. Chander. 2010. Assessing the vulnerability of Indian mustard to climate change. *Agric. Ecosyst. Environ.* 138: 265–273.

Cagirgan, M.I., N. Mbaye, R.S. Silme, N. Ouedraogo, and H. Topuz. 2013. The impact of climate variability on occurrence of sesame phyllody and symptomatology of the disease in a Mediterranean environment. *Turkish J. Field Crops* 18: 101–108.

Dutcher, R.A., C.O. Jensen, and P.M. Althouse. 1951. *Introduction to Agricultural Biochemistry*. John Wiley & Sons, New York, p. 72.

Evans, N., M.H. Butterworth, A. Baierl et al. 2010. The impact of climate change on disease constraints on production of oilseed rape. *Food Sec.* 2: 143–156.

FAO. 2011. FAOSTAT world oilseed production. http://faostat.fao.org.

FAO. 2012. FAOSTAT world oilseed production. http://faostat.fao.org.

Green, A.G. 1986a. A mutant genotype of flax (*Linum usitatissimum* L.) containing very low levels of linolenic acid in its seed oil. *Can. J. Plant Sci.* 66: 499.

Green, A.G. 1986b. Effect of temperature during seed maturation on the oil composition of low-linolenic genotypes of flax. *Crop Sci.* 26: 961.

Green, A.G. 1986c. Genetic conservation of linseed oil from industrial to edible quality. *J. Am. Oil Chem. Soc.* 63: 464.

Gunstone, F.D. 2011. *Vegetable Oils in Food Technology: Composition, Properties and Uses*, 2nd edn. Wiley-Blackwell, Chichester, U.K.

Kolte, S.J. 1985. *Diseases of Annual Edible Oilseed Crops*, Vol. II: Rapeseed-Mustard and Sesame Diseases. CRC Press, Boca Raton, FL.

Kolte, S.J. and B.D.L. Fitt. 1997. *Diseases of Linseed and Fibre Flax*. Shipra Publications, New Delhi, India.

Kumar, A. 2014. Challenges of edible oils: Can Brassicas deliver? In: *Second National Brassica Conference on "Brassicas for Addressing Edible oil and National Security"*, PAU, Ludhiana, February 14–16, 2014, p. 8.

Lu, C., J.A. Napier, T.E. Clemente, and E.B. Cahoon. 2011. New frontiers in oilseed biotechnology: Meeting the global demand for vegetable oils for food, feed, biofuel, and industrial applications. *Curr. Opin. Biotechnol.* 22: 252–259.

Narayan, P., M.S. Chauhan, and S. Chauhan. 2011. Oilseeds scenario in India. *Agriculture Today*, December 2011, pp. 40–43.

OECD/FAO. 2013. OECD-FAO agricultural outlook 2013, Chapter 5—Oilseeds and oilseed products. OECD/FAO Secretariats, pp. 139–282.

Padmavathi, P. and S.M. Virmani. 2012. Climate change: Impact on safflower (*Carthamus tinctorius* L.) productivity and production. In: *Safflower Research and Development in the World: Status and Strategies*, eds., I.Y.L.N. Murthy et al. Indian Society of Oilseeds Research, Hyderabad, India, pp. 45–64.

Paroda, R.S. 2013. The Indian oilseeds scenario: Challenges and opportunities. In: *The First Dr. M.V. Rao Lecture*, Indian Society of Oilseeds Research, Hyderabad, India, August 24, 2013, p. 26.

Ranganatha, A.R.G., R. Lokesha, A. Tripathi, A. Tabassum, S. Paroha, and M.K. Shrivastava. 2012. Sesame improvement—Present status and future strategies. *J. Oilseeds Res.* 29: 1–26.

Richaria, R.H. 1962. *Linseed.* Indian Central Oilseeds Committee, Hyderabad, India.

Sharma, M., S.K. Gupta, and A.K. Mondal. 2012. Production and trade of major world oil crops. In: *Technological Innovations in Major World Oil Crops*, Vol. I, ed., S.K. Gupta. Springer Science + Business Media, LLC, New York, pp. 1–15.

Singh, M.V. 2010. Detrimental effect of zinc deficiency on crops productivity and human health. Presented at the *First Global Conference on Biofortification*, Harvest Plus, Washington, DC.

Suresh,G., I.Y.L.N. Murthy, S.N. Babu and K.S. Varaprasad. 2013. An overview of zinc use and its management in oilseed crops. *J. SAT Agric. Res.* 11: 1–11.

Swaminathan, M.S. 1979. A integrated strategy for increasing the production and consumption of oilseeds and oils in India. 6th D.C.M. Chemical Works, S.S. Ramaswamy Memorial Lecture 1979, presented at the *International Congress on Oilseeds and Oilsand 34th Convention, Oil Technologists Association of India*, Ashoka Hotel, New Delhi, India, February 9–13, 1979.

Vaisey-Genser, M. and N.A.M. Eskin. 1978. Canadian rapeseed oil—Properties, processes and food quality. Publication No. 54. Rapeseed Association of Canada, Winnipeg, Manitoba, Canada, p. 13.

Yadava, D.K., S. Vasudev, N. Singh, T. Mohapatra, and K.V. Prabhu. 2012. Breeding major oil crops: Present status and future research needs. In: *Technological Innovations in Major World Oil Crops*, Vol. 1: Breeding, ed., S.K. Gupta. Springer Science + Business Media, LLC, Berlin, Germany, p. 17.

Zhang, Z. and R.C. Johnson. 1999. Safflower germplasm collection directory. IPGRI Office for East Asia, Beijing, China.

Section II

Peanut

Peanut (groundnut) belongs to the family Fabaceae (= Leguminosae) and the genus *Arachis*, derived from the Greek *a-rachis*, meaning without spine, and refers to the absence of erect branches. The species name *hypogaea* is derived from *hupo-ge*, which in Greek means below the earth. It is an allotetraploid having South American origin (Nigam 2000, Wang et al. 2011). Recent studies reveal that peanut originated in northern Argentina or southern Bolivia from hybridization between the diploid wild species *Arachis duranensis* and *Arachis ipaensis*. Cultivated peanut thus has an allotetraploid genome (AABB, $2n + 4x = 40$). It is an annual herbaceous plant growing to a height of 30–60 cm with an angular hairy stem and spreading branches. Leaves occur alternately, one at each node; they are pinnate with two pairs of ovate leaflets. The peanut flowers are perfect, and self-pollination is the general rule although natural cross-pollination may occur at times. After pollination, the perianth withers and at the base of the ovary a meristematic region grows into a stalklike peg that pushes the ovary into the soil. Groundnut has a taproot system that is often covered with root nodules resulting from a symbiosis with nitrogen-fixing bacteria (collectively called rhizobia). Among the rhizobia identified on groundnut, *Bradyrhizobium* species are the most prominent ones. The rhizobia penetrate the root tissue, induce cell division, and settle inside root cells where they convert atmospheric nitrogen (N_2) into ammonia, which in turn is used by the plant.

Because its ancestors were two different species, today's peanut is a polyploid, meaning this species can carry two separate genomes, designated as A and B subgenomes. *A. duranensis* serves as a model for the A subgenome of the cultivated peanut, while *A. ipaensis* represents the B subgenome. Very recently in April 2014, peanut genome has been successfully sequenced as a result of the collaborative research done by the International Peanut Genome Initiative (IPGI)—a group of multinational crop geneticists from the United States, China, Brazil, India, and Israel. The peanut genome sequences will now provide researchers access to 96% of all peanut genes in their genomic context and provide the molecular map needed to more quickly breed for peanut disease resistance and other economically important traits producing more improved high-yielding peanut cultivars all over the world (UGA Today—University of Georgia News Services—October 20, 2014; Mallikarjuna and Varshney 2014).

The varieties in cultivation fall into two main groups: the bunch or erect and the runner or spreading types. The basic chromosome number of *Arachis hypogaea* is 20 pairs ($2n = 40$) with large genome size (2.82 Gb DNA, 2800 Mb/1C). The oil content of the seed varies from 44% to 55%, and protein content of the seed is about 25%–28% in different varieties. About two-thirds of the world's production is crushed for oil and the remaining one-third is consumed as food. Today, peanut is widely distributed and is cultivated in more than 80 countries in tropical and subtropical regions of the world. Peanut requires warm, sunny climate with a well-distributed rainfall of at least 500 mm and temperatures ranging from 25°C to 30°C. It thrives best in well-drained sandy-loam soils with a pH ranging from 5.5 to 7.0. Asia with 63.4% area produces 71.7% of the world's peanut production followed by Africa with 31.3% area and 18.6% production and North–Central America with 3.7% area and 7.5% production. Important peanut-producing countries are China, India, Indonesia, Myanmar, Thailand, and Vietnam in Asia; Nigeria, Senegal, Sudan, Zaire, Chad, Uganda, Republic of Ivory Coast, Mali, Burkina Faso, Guinea, Mozambique, and Cameroon in Africa; Argentina and Brazil in South America; and the United States and Mexico in North America (Hegde 2009). The most favorable conditions for peanuts are moderate rainfall during the growing season, an abundance of sunshine, and relatively high temperature. The plants need ample soil moisture from the beginning of blooming up to 2 weeks before harvest. The crop thrives best on sandy-loam, loam, and well-drained black soils. The crop is affected by several diseases, causing large losses in both yield and quality of seeds. The peanut diseases are described in Chapters 2 through 4 as follows.

2 Fungal Diseases

SEED ROT AND SEEDLING DISEASE COMPLEX

Symptoms

A wide range of fungi, acting synergistically or in succession, attack the plant from the time seed is planted, until a few weeks after emergence to cause the seedling disease syndrome. The symptoms may be divided into four categories according to the development stage of the plant when the damage occurs. These categories are seed rot, preemergence damping-off, postemergence damping-off, and seedling blight. The rotted seeds become soft and mushy, turn brown, shrink, and finally disintegrate, often showing the presence of the fungal growth. This gives a patchy stand of the crop. The patchy stand of the crop may also be due to infection of seedlings after the seed has germinated but before the seedling has emerged above the soil level, which is called as preemergence damping-off. They may be killed even before the hypocotyls have broken the seed coat. The radicle and the plumule, when they come out of the seed, undergo complete rotting. Since this happens under the soil surface, the disease is often not visible except for the resulting patchy stand. In both cases, infection takes place before emergence of the seedlings above the soil level.

The postemergence seedling blight is characterized by the toppling over of infected seedlings any time after they emerge from the soil until the stem of the peanut plant soon becomes lignified and resistant to postemergence damping-off. The fungi that cause seedling blight symptoms infect the cotyledons and leaves as they emerge, either from inoculum carried on the seed or when the cotyledons come into contact with contaminated crop residues. Seedlings may also be attacked by certain species of fungi at the roots and sometimes at or below the soil line. Such seedlings usually show collar rot, root rot, brown root rot (caused by *Fusarium solani* [Martius] Sacc. in Argentina), and wilt-like and damping-off symptoms. In the case of collar rot caused by *Lasiodiplodia theobromae* (Pat.) Griff. & Maubl., the leaflets and stem remain green until the seedlings die and black pycnidia are found on the collar region at the soil level (Chi Mai Thi et al. 2006).

Geographical Distribution and Losses

These diseases occur all over the peanut-growing countries in the world and can cause serious reduction in yield through reduced plant stand. Losses recorded in any one field for a given inoculum level will vary from one season to the next depending on crop residues, soil conditions, seed quality, and climatic factors during the critical 3–4 weeks after planting. Extensive losses have been reported in the range of 25%–50% in Malawi, Senegal, Sudan, Niger, Nigeria, and other West African countries (Subrahmanyam et al. 1991, Kolte 1997, Thiessen and Woodward 2012), Southern African countries (Subrahmanyam et al. 1997), Egypt (Wakil and Ghonim 2000), India (Kolte 1984), and Pakistan (Riaz et al. 2002). Under conducive conditions like drought stress, losses due to brown root diseases in peanuts could reach 95% in some fields in Argentina (Rojo et al. 2007). Patchy stands of peanuts due to these diseases are the single most important factor for low production of the crop in almost all peanut-growing states of India.

Pathogens

The causal fungi associated with the preemergence seed rot, preemergence and postemergence damping-off of peanut seedlings, and seedling blights are *Rhizoctonia solani* Kuhn, *F. solani* (Mart) Apple and Wr., *Pythium ultimum* Trow., *P. myriotylum* Drechs, *P. debaryanum* Hesse emend. Middleton, *P. aphanidermatum* (Edson) Fitzp., and *P. butleri* Subram (Kolte 1984, Rashid et al. 2004, Cavallo et al. 2005, Thiessen and Woodward 2012).

Both the preemergence and postemergence damping-off of peanut seedlings can also be the result of infection of *Sclerotium rolfsii* Sacc., *Rhizoctonia bataticola* (Taub.) Butl., a sclerotial stage of *Macrophomina phaseolina* (Tassi) Goid (Chakrabarty et al. 2005), *Aspergillus niger* van Tieghem, and *A. flavus* (Link) ex Fries. The latter four fungal pathogens have the capacity to cause root rots, collar rot diseases beyond the stages of crop growth that could reasonably be considered as seedling. Diseases caused by these are described separately in "*Sclerotium* Stem Rot," "*Aspergillus* Collar Rot," "Yellow Mold and Aflarroot," and "Charcoal Rot" Sections under *F. solani* causing the brown root disease is one of the most serious diseases of peanuts in the southern region of Argentina in recent years (Oddino et al. 2008). Collar rot of peanut seedlings caused by *L. theobromae* (Pat.) Griff. & Maubl. (syn *Botryodiplodia theobromae* Pat.) has been reported to become a major disease problem in peanut production in North Vietnam (Chi Mai Thi et al. 2006). Two or more of the aforementioned pathogens can act together as a complex of the cause of seedling diseases.

Factors Affecting Infection

Normally, the peanut seed contains a tannin-like substance that acts as an antioxidant to retard the breakdown of the oil and entrance of the fungi that cause decay of seeds. Fungi associated with the peanut seed may decay it, particularly if it is damaged. In the mechanically damaged seeds, the seed coat is scratched and broken. Unless done with extreme care, machine-shelled seeds show reduction in germination by 25%–75%, as compared with hand-shelled seeds. The scratches on the seed surface provide points of entry of fungi. Delayed germination because of deep sowing or lack of moisture in the soil or waterlogging conditions may all influence the development of seed rot and seedling blight diseases within the first week after planting.

Soil temperature influences the involvement of the kind of pathogens in seedling disease complex and the severity of incidence of seedling diseases. For example, at higher soil temperature (>35°C), *R. solani* isolate from peanuts (a warm weather crop) is more virulent than the isolate from wheat (a cool weather crop) (Sreedharan et al. 2010). Thus, soil temperature as affected by planting date should be considered in areas where wheat is planted following peanut, if root disease caused by *Rhizoctonia* is a concern. There appear to be biotypes of *R. solani* capable of causing the disease over a wide range of optimum temperature. For seedling disease caused by *S. rolfsii* and *R. bataticola*, the optimum temperature is 25°C or above. There is a definite wide range of temperature optima for different species of *Pythium*. *P. ultimum* requires low soil temperatures, while *P. aphanidermatum* is more damaging at higher temperatures. Interestingly, *R. bataticola*, on the other hand, can survive and its incidence can substantially increase in infected peanut seed even at −18°C, a temperature recorded for long-term storage. The implication of these results is that for ensuring a high level of germination in peanut seeds under such situations, only dry and pathogen (*R. bataticola*) free or seeds with very low infection grade be used for long-term storage (Singh et al. 2003b).

Disease Management

Though there is a low to high range variation in the degree of susceptibility of peanut genotypes to seedling diseases caused by various pathogens, for example, *S. rolfsii*, there is the least possibility of obtaining acceptable level of seedling blight disease resistance in peanut seeds (Gour and

Sharma 2009). Therefore, a combined management strategy utilizing cultural, biological and chemical management practices is important in reducing losses caused by seed rot and seedling diseases in peanut crop production.

Cultural Control

Only sound seed without any evidence of injury should be used for sowing. Care should be taken to avoid injury to seeds during shelling and while the seeds are sown through seed drills. Hand-shelled seeds, if sown, without injury to the seed coat, give a higher stand of the crop even without fungicidal treatment. The loss in the crop stand because of seed rot and seedling blights can also be compensated by increasing the seed rate from 60 to 75 kg/ha. The effect appears to be similar to that of the fungicidal seed treatment. Crop rotation and tillage practices greatly influence the seedling disease complex in peanuts. For example, reduction of brown root rot of peanut seedlings caused by *F. solani* is greater in a 2-year rotation including corn–soybean or soybean–corn prior to peanuts than in a 1-year crop rotation, and the tillage system using the paratill subsoiler before seeding peanuts in a no-till system is a suitable strategy to improve peanut root growth and reduce the disease incidence, which provides a promising alternative in the control of peanut seedling brown root rot in Argentina (Oddino et al. 2008). In Egypt, seedling diseases of peanuts caused by *F. solani* and *R. solani* could be managed by amending the soil with gypsum (500 kg/acre) and by balanced application of nitrogen (100 kg/acre) and potassium (50 kg/acre) fertilizers as well as by soil moisture (55%–70% of field capacity) (El-Korashy 2001).

Chemical Control: Fungicidal Seed Treatment

The shelled seeds should be immediately treated with the fungicide rather than treating them after lapse of time. It should also be noted that no amount of seed treatment will change poor seeds into good ones when they are stored under conditions detrimental to their keeping quality. Therefore, seed treatment is not a corrective for improper storage. The seed should be treated with effective fungicide, usually a mixture of thiram and carbendazim (2:1) or thiram and carboxin (2:1) at 3% that takes care of the variety of seed-borne pathogens (Kolte 1994, 1997, Akgul et al. 2011). Seed treatment with a mixture of thiram and carbendazim at 2 g ai/kg seed is recommended as a routine treatment in plant quarantine labs in India to prevent the seed transmission of *R. bataticola* infection from one region to another (Chakrabarty et al. 2005). A mixture of the aforementioned three fungicides (thiram + carboxin + carbendazim) is known to be the most efficient seed treatment for the control of most seedling diseases of peanuts in Argentina (Cavallo et al. 2005), whereas a mixture of thiram and thiophanate methyl is reported to be the best in comparison to other fungicides (Meena and Chattopadhyay 2002). A number of other fungicides are recently investigated to be of potential use in the management of seedling disease complex of peanuts caused by Basidiomycetes such as *R. solani* (= *Thanatephorus cucumeris* (Frank) Donk.) and *S. rolfsii* (= *Athelia rolfsii* (Curzi) Tu & Kimbrough). These are triazole fungicides, like tebuconazole and propiconazole, and flutolanil and strobilurins, such as azoxystrobin. But these are not effective in controlling the infection caused by *Pythium*, which can, however, be controlled by seed treatment with phenylamides such as metalaxyl. Hence, if *Pythium* species is involved in the seedling disease complex, a component of metalaxyl be included in the mixture of fungicides for seed treatment (Thiessen and Woodward 2012). A new seed treatment fungicide *Stamina* (Headline) provides broader spectrum of control of seed-borne and seedling diseases caused by species of *Rhizoctonia*, *Pythium*, and *Fusarium*, which needs to be investigated for peanut crop.

Presowing treatment of peanuts with antioxidant hydroquinone (in 20 mM water solution) for 12 h is reported to be useful in completely inhibiting seed-borne pathogenic fungi and enhancing the plant growth parameters producing a 50% increase in yield (El-Wakil 2003). Use of properly cleaned seeds and integrating mancozeb seed treatment (3 g/kg seed) is reported to be beneficial in getting significantly higher seedling emergence and higher dry pod yield in eastern Ethiopia (Tarekegn et al. 2007).

Biological Control

Peanut seed pelleting with *Trichoderma harzianum* Rifai strain Th-5 is reported to be effective in protecting seeds from *M. phaseolina* infection by 79.6%, resulting in improved seedling vigor (Malathi and Doraisamy 2004). Similarly, seed treatment with another isolate of *T. harzianum* at 10 g/kg provides maximum protection to peanut seedlings, which is reported to be superior to fungicidal seed treatment under similar conditions (Rakholiya and Jadeja 2010). Rojo et al. (2007) have also reported the control of brown root rot of peanuts (*F. solani*) with *T. harzianum* strain ITEM 3636 through seed treatment in Argentina. The mass inoculum of the antagonist (*T. harzianum*) could be produced on low-cost agricultural waste products (rice bran, wheat bran, mustard cake) for the control of seedling collar rot of peanuts caused by *S. rolfsii* by seed treatment followed by soil application of *T. harzianum* along with farmyard manure (FYM) (Bhagat and Sitansu 2007). A combination of *Trichoderma* strain 5 MI with Rovral 50 WP fungicide and garlic extract gives the best control of seedling diseases caused by *R. solani* and *S. rolfsii* (Islam et al. 2005). The presence of *Bacillus subtilis* (Ehrenberg) Cohn has been naturally observed on peanut kernels. It is thus possible that the seeds that show the presence of this bacterium remain free from attack due to seed-borne fungi (Kolte 1984). El-Shehaby and Morsy (2005) from Egypt have demonstrated the usefulness of soil treatment of four isolates of an antagonist bacterium, *Bacillus sphaericus*, for the management of seedling diseases of peanuts caused by *R. solani*, *F. solani*, and *S. rolfsii*.

Extracts of leaves of *Azadirachta indica* A. Juss. (Kadam et al. 2008b) and *Moringa* seed (Donli and Dauda 2003) can be used for treating peanut seed, as alternative methods to fungicidal seed treatments for the control of seed and seedling diseases of peanuts.

EARLY AND LATE LEAF SPOTS

Symptoms

Peanut leaf spot diseases, the early leaf spot (ELS) and late leaf spot (LLS), are caused by two distinct, but closely related fungal pathogens. These are *Cercospora arachidicola* Hori (causing ELS) and *Phaeoisariopsis personata* (Berk. & Curt.) V. Arx. (= *Cercosporidium personatum* (Berk. & M.A. Curtis) Deighton) (causing LLS); both may occur simultaneously on the same leaf. (Other diseases too may cause spots on leaves, but they are not referred to as *leaf spots*). The ELS appears earlier about 10–18 days after emergence than the LLS, which appears 28–35 days after emergence or may appear at the time of harvest. In both cases, symptoms become visible as pale areas on the upper surface of the leaves. As the spot develops, it becomes yellow; necrosis occurs from the center of the lesion, and later the entire spot becomes necrotic. Infection with either leaf spot fungus produces hormonal changes in the leaf that cause leaf drop. Defoliation usually starts at the base of the central or lateral stem and then progresses upward. Initially, the ELS and LLS are indistinguishable. The distinguishing features, as the spots are fully developed, become quite evident as described under ELS and LLS.

ELS: Circular to irregular, larger, measures 1–10 mm in diameter; spots are characterized by a yellow halo of visible width (Figure 2.1). At maturity, the spots are reddish brown to black on upper surface but the lower surface of spot is distinctly orange in color. Cushions of conidiophores are formed at first on the upper surface (epiphyllous), but sometimes these are found on the lower surface of older spots and during periods of heavy cloud cover and frequent showers particularly when defoliation is at the peak (Figure 2.2); masses of clear to olive colored spores may be seen with the use of a hand lens on the upper surface of the spot; thus, the intensity of sporulation usually on the upper surface of the spot becomes a visible sign of the causal fungus.

LLS: Tends to remain distinctly round, 1.5–5.00 mm in diameter; yellow halos are not visible around the newly formed spots but are found only with mature spots. Spots are almost black on both surfaces, but lower surface of the spot is distinctly carbon black (Figure 2.3). Conidiophores are always found confined to the lower leaf surface (hypophyllous),

FIGURE 2.1 ELS of peanuts.

FIGURE 2.2 Defoliation of the peanut crop at harvest due to ELS.

and these are usually in the plainly visible concentric circles. Because the LLS pathogen produces many spores especially on the lower surfaces, the lesions usually have a raised or tufted appearance. Spots on stem, petiole, and pegs are similar and are irregular or elliptical in shape, when defoliation occurs (Figure 2.4).

Geographical Distribution and Losses

Berkeley (1875) from the United States was the first to describe peanut leaf spot. Later, Woodroof (1933) gave a clear account of the existence of two distinct spots: *ELS* and *LLS*. Presently, the two diseases have been reported throughout the world wherever peanuts are grown. The two diseases,

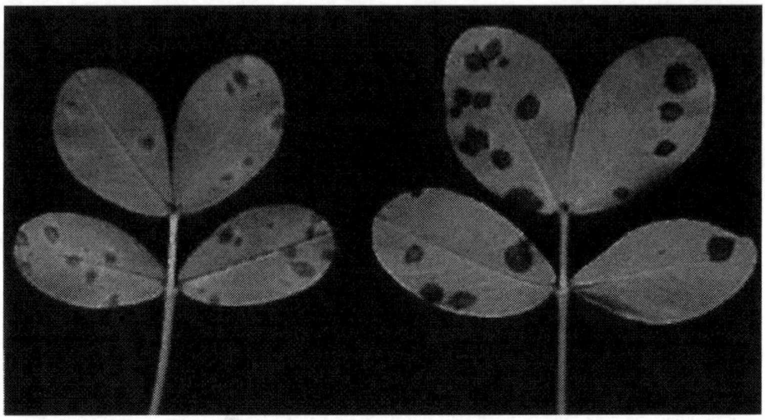

FIGURE 2.3 LLS of peanuts. Note the carbon black color of the undersurface of the spot.

FIGURE 2.4 Defoliation of the peanut crop at harvest due to LLS.

though occur simultaneously in the same area, differ quite considerably in their relative preponderance from one region to another, depending upon the prevailing weather conditions and type of peanut varieties under cultivation (Kolte 1984, Das and Roy 1995, Sawargaonkar et al. 2010). Reduction in yield is largely due to loss in photosynthetic tissue and defoliation (Naab et al. 2005, Singh et al. 2011a,b). Spots on the pegs also tend to decrease the yield by restricting translocation of food to the seeds. Annual crop losses in the range of 10%–50% or more are common worldwide particularly in the peanut-growing regions in Australia (Kelley et al. 2012), the southeastern United States (Nutter and Shokes 1995), and savanna zones in Nigeria (Izge et al. 2007, Iwo and Olorunju 2009) and Ghana (Nutsugah et al. 2007a,b); in other areas of West Africa (Waliyar 1991, Waliyar et al. 2000), DR Congo and Central Africa (Tshilenge 2010, Tshilenge-Lukanda et al. 2012), sub-Saharan Africa (Hamasselbe et al. 2007), and Malawi (Kisyombe et al. 2001); and in South Asia covering India (Sarkar and Chowdhury 2005, Gopal et al. 2006a), Pakistan (Ijaz et al. 2008), Bangladesh (Hossain et al. 2010), and Nepal (Thakur et al. 2013). The leaf spots have been a serious problem in case of early-season rains and near maturity of the peanut crop in the northern parts of Vietnam, while in the southern parts of that country, it causes damage from the beginning of the crop (Mehan and Hong 1994). Losses of great magnitude have been documented by comparing yields in plots in which the leaf spot diseases have been controlled with fungicides to those in untreated check plots in which the diseases have been allowed to progress. Plants not treated with fungicides may shed most of their leaves (90% in severe cases) prematurely, causing significant yield losses. Since there are established negative relationships between biomass and disease rating or defoliation and between pod yield and disease rating or defoliation, the visual rating of disease and measured defoliation could be used as a rapid and inexpensive tool to initially assess yield losses caused by the ELS and LLS diseases (Naab et al. 2005). Of several disease assessment methods, the best is proved to be percent main-stem defoliation above the fourth node and percent diseased leaf area estimated from visual leaf spot score (Adomou et al. 2005). Besides the loss in yield of kernels, the value of the hay that may be used as fodder for cattle is also adversely affected. When the leaf spots occur in combination with rust (caused by *Puccinia arachidis*), the losses involved are still more and need more attention (Gughe et al. 1981, Vidyasekaran 1981). The relative importance of each of the leaf spot disease varies from place to place and from season to season, depending on the cropping system and the environmental conditions.

The ELS and LLS lower the peanut haulm yield and quality (Pande et al. 2003). Crude fiber, crude protein, fat, and dry matter content of haulm are significantly lower in severely infected haulm, whereas ash, moisture content, and nitrogen-free extracts get increased with increasing leaf spot severity (Bdliya 2006, 2007).

Pathogens: *C. arachidicola* Hori (perfect stage, *Mycosphaerella arachidis* Deighton) and *P. personata* (Berk. & M.A. Curtis) Arx (perfect stage, *Mycosphaerella berkeleyi* W. Jenkins)

Classification
Kingdom: Fungi
Phylum: Ascomycota
Class: Dothideomycetes
Subclass: Dothideomycetidae
Order: Capnodiales
Family: Mycosphaerellaceae
Genus: *Mycosphaerella*
Species: *arachidis* or *berkeleyi*

C. arachidicola: The perfect stage of this fungus is *M. arachidis*. Its mycelium is septate. In plant tissue, initially, it is intercellular and then it becomes intracellular. Mycelium penetrates directly in plant cell and does not form haustoria. Conidiophores are supported by dark-brown stromata of 25–100 μ in diameter. In the early stages of development, conidiophores mostly are epiphyllous, but in the later stages of disease development, these become amphigenous. Conidiophores arise from stroma and are fasciculate and geniculate, of yellowish-brown color. Conidiophores are usually continuous but may also be with several septations. These measure 20–45 μ × 3–6 μ. Conidiospores are obclavate to clavate, mostly curved, and subhyaline to olivaceous in color. These measure 35–108 μ × 2–5.4 μ, having 4–12 septa with rounded to distinctly truncate base and subacute tips, and germinate by producing germ tubes from different cells. Secondary conidiospores and conidia are seen on slide made from host tissue kept under extremely favorable environmental conditions (Kolte 1984). In case of teleomorphic state, perithecia are scattered mostly along margins of lesions produced by spores of imperfect state. These are amphigenous, somewhat embedded in leaf tissue, erumpent, ovate to nearly globose, and black in color. These are of a size 47.6–84 μ × 44.4–74 μ. Ostiolem is slightly papillate. Asci are cylindrical, club-shaped short stipitate, fasciculate, aparaphysate, and bitunicate with eight ascospores. Asci measure 27–37 μ × 7–8.4 μ. Ascospores are uniseriate to biseriate in the ascus. These are bicellular with upper cell slightly curved and hyaline. Their sizes range from 7 to 15.4 μ × 3 to 4 μ (Kolte 1984).

P. personata (= *C. personatum*): The mycelium of *P. personata* is septate and exclusively intercellular. Its haustoria puncture into the palisade and mesophyll tissue. Dense, globular, brown to black stromata measuring a diameter of 20–30 μ are produced. Conidiophores are mostly hypophyllous, but sometimes amphigenous. In the later stages of disease development, conidiophores arise in clearly concentric tufts from heavy stromatic base. These are fasciculate, geniculate, and reddish brown in color with mostly hyaline tips and nonseptate or severally septate. Conidiophore sizes range from 24 to 54 μ × 2 to 8.2 μ. Conidia or conidiospores of the fungus are obclavate with attenuated tips and pale-brown dilutely olivaceous color measuring 18–60 μ × 5–11 μ with one to nine septa and bluntly rounded top cells. In *P. personata*, secondary conidia and conidiophores are not reported. The perithecia, asci, and ascospores of teleomorphic stage of *P. personata* only differ from *C. arachidicola* in size. The sizes of perithecia, asci, and ascospores are 84–140 μ × 70–112 μ, 30–40 μ × 4–6 μ, and 19.6 μ × 2.9–3.83 μ, respectively (Kolte 1984). The teleomorphic stage of the LLS pathogen, *M. berkeleyi* Jenkins, is rarely seen on peanuts (Shokes and Culbreath 1997). Molecular variation in the pathogen has been studied (Kumari et al. 2009a,b, 2012).

Epidemiology and Disease Cycle

The fungi that cause ELS and LLS reproduce and infect by means of asexual spores conidia. Both fungi are capable of producing tremendous numbers of conidia on infected plant parts. Conidial production is favored by high humidity. The primary inocula that cause the initial leaf spot infections during the growing season are spores produced on infested peanut residues in the soil. Visible spots develop 10–14 days after infection. New conidia are produced in spots on infected leaves. These conidia will subsequently infect plants and produce secondary infections. Conidia are spread by wind, splashing rain, and insects. Leaf spot can increase rapidly under favorable conditions as several secondary cycles may occur per season. The ELS/LLS stage characterized by the higher level of symptom expression is found not associated with the plant phase of highest emerged leaves, but the disease symptoms reach the peak only after the phase of intense leaf development in peanut crop (Tshilenge-Lukanda et al. 2012).

When enough precipitation of monsoon rains makes a film of water on leaves or a relative humidity more than 90% prevails with a temperature of 20°C–29°C for 6–7 days, peanut crop is severely affected by the ELS and LLS. Disease incidence and severity may vary depending on prevailing climatic conditions. Intermittent rains from flowering to pod development stage of the crop favor the infection and development of leaf spots (Pande et al. 2000). The maximum temperature range of 31°C–35°C and minimum temperature range of 18°C–33°C favor leaf spot outbreak on peanuts. The influence of climatic elements like temperature and relative humidity on the development of ELS and LLS of peanuts has extensively been studied (Dubey 2005, Kadam et al. 2008a, Ijaz et al. 2011). Abundant peanut residue in fields where peanuts are cropped continuously often results in early and rapid development of leaf spot. The first appearance of leaf spot and its continuous progress throughout the growing season are heavily dependent upon weather conditions. Environmental conditions required for both types of leaf spots are warm temperatures and long periods of high humidity or leaf wetness (Pande et al. 2004). Wet periods of sufficient duration to support infection usually consist of dew periods at night or extended rainy periods. When adequate moisture is present, leaf spot infections may occur in a relatively short period when temperatures are warm, but a longer wet period is required when temperatures are cool. For these reasons, potential for damage from leaf spot is greater where levels of humidity and rainfall are high (Muhammad et al. 2008). Frequent irrigation with small amounts of water can also create prolonged periods of high humidity and leaf wetness favorable for infection. A model has been developed by taking into consideration the relative humidity of more than 95% and the minimum temperature of 22°C and maximum

30°C. The model is used or compared with a calendar-based schedule in the United States (Smith 1986), Argentina (Pezzopane et al. 1998), and Brazil (Moraes et al. 2002). Based on prediction of favorable weather conditions and application of high-resolution Weather Research and Forecasting model, the model for ELS control in peanut crop has been developed. The short-term prediction of weather parameters and their use in management of the leaf spot diseases is a viable and promising technique, which could help the growers make accurate management decisions through optimum timing of fungicide applications (Olatinwo et al. 2012).

There appears to be a reduction in lesion size and intensity of sporulation (inoculum potential) of ELS pathogen due to infection (Subhalakshmi and Chowdhury 2008).

Disease Management

Host Plant Resistance

In Cultivated Arachis hypogaea

Host resistance to both ELS and LLS diseases is reported in *A. hypogaea*, and it is quite variable (Varman 2001, Izge et al. 2007, Padi 2008, Giri et al. 2009, Tallury et al. 2009, Dolma et al. 2010, Visnuvardan et al.2011). Generally, late-maturing alternately branched cultivars are either runner or spreading bunch type with dark-green foliage possessing very high degree of resistance to the leaf spot pathogens, whereas early forms with sequential branching are usually erect bunch type with light-green foliage showing a high degree of susceptibility to the disease (Kolte 1984). Sequentially, branched early cultivars are more susceptible to *Cercospora*, possibly because there are a greater proportion of stomata of penetrable size (more than 13.4 μ) on the upper leaf surface. In alternately branched forms, there is thicker palisade tissue in the leaf, which may partly explain the slower rate of lesion growth on them as well as account for their dark-green foliage (Hemingway 1957, Gibbons and Bailey 1967). Because of the reduced infection rate characterized in part due to lower infection frequencies, smaller lesions, lower sporulation index, longer latent periods, and lesser leaf area damage and disease score in resistant varieties, the resistance in these varieties is considered to be a partial one (Hossain and Ilag 2000, Dwivedi et al. 2002, Pande et al. 2002, Cantonwine et al. 2008b). Selection based on components of resistance to LSS may not lead to plants with higher retained green leaf area. The remaining green leaf area on the plant should, therefore, be the major selection criteria for resistance to LLS in breeding programs (Dwivedi et al. 2002). Improving levels of resistance along with foliar application of fungicides to manage the disease in locally adapted varieties would substantially increase peanut yields in developing countries (Waliyar et al. 1993, 1995, 1998). Currently, there are only a few varieties possessing tolerance to foliar diseases. However, one to two sprays depending upon the suitable time of application increase the pod yield significantly (Waliyar et al. 1998). A Bolivian land race cultivar *Bayo Grande* and several other Bolivian-derived genotypes show promise for use in a reduced fungicide and/or conservative tillage system with a potential to lessen fungicides compared to standard production practices (Gremillion et al. 2011a). Some of the most promising genotypes resistant or moderately resistant (tolerant) to either or both of the leaf spots are given in Table 2.1. High heritability coupled with high genetic variation is usually noticed for LLS and rust and pod yield indicating that additive gene effects are functional for these characters (Venkataravana and Injeti 2008).

Two genotypes of peanut, namely, cv 850 and cv 909, are reported to be resistant to *P. personata* (*C. personatum*) and show symptoms similar to hypersensitive response (HR) lesions, and the production of HR lesions is due to a novel *O′*-methyltransferase gene (Nobile et al. 2008). Higher amount and greater accumulation rate of total free phenol and stilbene phytoalexin production are the most possible biochemical mechanisms of resistance in peanuts against the two leaf spot pathogens (Motagi et al. 2004, Sobolev et al. 2007, Bhaskar and Parakhia 2010). The isoenzyme variability, for example, the presence of phosphatase band and two esterase bands, and stilbene phytoalexins are found to be specific to resistant cultivars, which can possibly be used as biochemical

TABLE 2.1
Peanut Genotypes Resistant (R) or Moderately Resistant (MR) to ELS and/or LLS as Reported from Different Countries in the World

Genotype	Country	R/MR	Reference(s)
ICGV 98369	South Africa	R	Mathews et al. (2007)
ICGV SM99529	Malawi		
ICGV 91225	Sub-Saharan Africa	R to ELS	Hamasselbe et al. (2007)
Samnut II	Sub-Saharan Africa	R to LLS	
Golden Mutant 96 C	Pakistan	R to CLS	Naeem-ud-Din et al. (2009)
INS-1-2006, AIS-2006-11	India	R to LLS	Sheela (2008)
ICGV-IS-96805		R to ELS + LLS	Iwo and Olorunju (2009)
C689-2, Georgia-01R, C12-3-114-58, C11-154-6, Tifguard, and Georganic	Southern United States	R to ELS	Li et al. (2012)
Charmwon, HyQ(CG)S-10 (*A. hypogaea* ssp. *fastigiata*)	Korea	R to ELS	Pae et al. (2008)
ICGV 05033, ICGV 03037, ICGV 05099, ICGX020063-P11, ICGV 04093, ICGV 03016, ICGV 04071, ICGV 86031, ICGV 03157, PAFRGVT60	India	R to LLS + rust	Venkataravana and Injeti (2008)
ICGV 99057, ICGV 00228, ICGV 99068, ICGV 99057, ICGV 00169	India	MR or R to LLS	Venkataravana et al. (2008)
259/88, 262/88	Bangladesh	High R to LLS + ELS	Hossain et al. (2007)
269/89	Bangladesh	MR to ELS	Hossain et al. (2007)
DP-1, Georganic	United States	R to ELS + LLS	Cantonwine et al. (2008)
Georgia-01R, Georgia-05E	United States	R to ELS + LLS + TSW	Branch and Culbreath (2008)
CV100, PI648033	United States	R to ELS + LLS + TSW	Holbrook and Culbreath (2008)
ICGV-IS-96808	Nigeria	R to ELS/LLS	Izge et al. (2007)
CV 850, CV 909		High R to LLS	Nobile et al. (2008)
PI 390590	India	R to LLS	Suryawanshi et al. (2006)
R8972	India	R to LLS + rust	Gopal et al. (2006b)
ICGV 92099, ICGV 90084	Ghana	R to ELS + LLS	Frimpong et al. (2006)
TFDRG1, TFDRG2, TFDRG3, TFDRG4, TFDRG5, VG9514	India	R to LLS + rust	Badigannavar et al. (2005)
N96076L (GP-125, PI641950)	United States	Multiple disease resistance including LLS	Isleib et al. (2006)
Nkatiesari	Ghana	R to ELS + LLS	Padi et al. (2006)
SP 8638	Korea	R to LLS	Pae et al. (2005)
Huayu 22	China	R to LLS + web blotch	Chen et al. (2005)
Kokwang	Korea	R to ELS	Park et al. (2004)
Jakwang	Korea	R to ELS	Pae et al. (2004)
FDRS-10	India	R to LLS	Jyosthna et al. (2004)
Zhonghua 9	Hubei Province of China	R to LLS + rust	Liao et al. (2004)
GPBD-4	India	R to LLS + rust	Gowda et al. (2002a)
Mutant 28-2	India	R to LLS	Gowda et al. (2002b)
C-99R	United States	R to LLS + stem rot + TSW	Gorbet and Shokes (2002a)
Florida MDR 98	United States	MR-R to LLS + stem rot + TSW	Gorbet and Shokes (2002b)

(Continued)

TABLE 2.1 (*Continued*)
Peanut Genotypes Resistant (R) or Moderately Resistant (MR) to ELS and/or LLS as Reported from Different Countries in the World

Genotype	Country	R/MR	Reference(s)
ICGV 92267	ICRISAT, India	MR to LLS + rust	Upadhyaya et al. (2002)
Georgia-01R	United States	R to ELS + LLS + rust	Branch (2002)
GP-NCWS11, GP-NCWS12, GP-NCWS-13, GP-NCWS14, GP-NCWS15	United States	R to ELS + LLS	Stalker et al. (2002)
VRI Gn 5	India	R to LLS + rust	Vindhiyavarman and Mohammed (2001)
ICGV 92080, ICGV 92093	India	R to LLS	Mohammed et al. (2001)
Huayu 17	Shandong Province, China	High R to LLS	Yu et al. (2000)

ELS, early leaf spot; LLS, late leaf spot; TSW, tomato spotted wilt.

markers in the identification of the leaf spot pathogen–resistant cultures in peanuts (Jyosthana et al. 2004, Sobolev et al. 2007). In order to be effective in improving the efficiency of identifying genes of interest of ELS/LLS resistance in the entire germplasm collection, core collection as developed in the United States could be of immense value (Holbrook and Dong 2005, Gremillion et al. 2011b).

As resistance to ELS and LLS is inherited independently (Higgins 1935) and in some genotypes, duplicate complementary recessive genes are reported to control the LLS resistance (Motagi et al. 2000), and these diseases vary in their relative preponderance in different regions based on prevalent cropping system and spectrum of locally prevalent races, a separate breeding program may have to be used, if survey work reveals that only one of these is only important in a particular area.

In Wild Arachis *Species*

The cultivated peanut (*A. hypogaea* L.) is an allotetraploid with an AABB genome and low genetic diversity. Because of its limited genetic diversity, this species lacks resistance to a number of important pests and diseases. In contrast, wild species of *Arachis* are genetically diverse and are rich sources of disease-resistant genes (Varman et al. 2000, Fávero et al. 2009). The genus *Arachis* is native to South America and consists of 22 described species and possibly more than 40 undescribed. Collections are maintained in Brazil, the United States, and India. Many of these accessions have been screened for resistance to pathogens and insect pests, and it is demonstrated that sources of resistance are available in the wild species belonging to the taxonomic section *Arachis* with either A or B (or *non-A*) genomes and these can be used for introgression of resistance genes against two leaf spot diseases and other diseases (Mallikarjuna et al. 2004, 2012, Yadav et al. 2007, Fávero et al. 2009). International Crops Research Institute for the Semi-Arid Tropics (ICRISAT) is currently using diploid species of section *Arachis*, which are cross compatible with the tetraploid-cultivated peanut to transfer useful genes through interspecific hybridization. Thus, stable, tetraploid interspecific hybrids with resistance to the LLS and rust showing high yield potential have been developed (Yadav et al. 2007). Twenty-nine percent (29%) of interspecific derivatives from the cross *A. hypogaea* ($2n = 40$) × *A. kempff-mercado* ($2n = 20$) at the BC2 F2 generation have been established to show resistance to both ELS and LLS (Mallikarjuna et al. 2004). Hybrids formed between cultivated and wild species are generally alternately branched, giving low yield, but these may provide the basis for selection in a breeding program.

Resistance to *P. personata* is found in accessions of different species of *Arachis*, from A genome species, namely, *A. stenosperma*, *A. kuhlmannii*, *A. helodes*, *A. simpsonii*, *A. diogoi*, *A. aff. diogoi*, *A. microsperma*, *A. linearifolia*, and *A. cardenasii*, and *non-A* genome species, namely, *A. cruziana*, *A. hoehnei*, *A. magna*, *A. valida*, *A. batizocoi*, and *A. williamsii*. Differential gene expression in *A. diogoi* upon interaction with *P. personata* reveals that the pathogen induces cyclophilin-like proteins (Kumar and Kirti 2011). Similarly, resistance to *C. arachidicola* is found in accessions of different species of *Arachis*, from A genome species, namely, *A. kuhlmannii*, *A. helodes*, *A. cardenasii*, *A. kempff-mercado*, *A. linearifolia*, and *A. stenosperma*, and from *non-A* genome species, namely, *A. hoehnei*, *A. magna*, and *A. batizocoi* (Fávero et al. 2009). The resistance to LLS and rust (*P. arachidis*) studied by Pande and Rao (2001) in the 74 accessions of wild species of *Arachis* revealed the accession *KG3006* of *A. hoehnei* immune to either of the two leaf spot diseases. Thus, there is resistance to *P. personata* and *C. arachidicola* in many accessions of wild species, and these accessions may be different among accessions of the same species. Interestingly, molecular markers for resistance genes (R genes) that encode a putative nucleotide-binding site (NBS) domain and a leucine-rich repeat (NBS-LRR genes) have been developed to generate resistance gene analogues (RGAs) in wild *Arachis* spp. in Brazil (Guimarães et al. 2005, Garcia et al. 2006). Inclusion of these RGAs in *Arachis* genetic map will be of paramount importance in breeding for disease resistance not only to ELS and LLS diseases but also to rust and other economically important diseases of peanuts.

Molecular Breeding and Transgenic Peanuts for ELS and LLS Resistance

Genetic transformation has launched a new era in peanut breeding and germplasm creativity through transformational methods including *Agrobacterium tumefaciens*–mediated, particle bombardment, and nontissue culture techniques (Wei et al. 2008). Comparison of molecular profile among peanut cultivars and breeding lines with differential reaction against LLS and other diseases is facilitated by the use of random amplified polymorphic DNA (RAPD) and intersimple sequence repeat (ISSR) markers (Dwivedi and Gurtu 2002, Dwivedi et al. 2003, Mondal et al. 2008a, 2009, Khedikar et al. 2010). For example, transgenic peanut plants constitutively expressing the mustard defensin gene have been generated by cloning the complete cDNA containing an open reading frame (ORF) of 243 bp of a defensin of mustard (defensins are small positively charged antimicrobial peptides, 5 kDa in size, and show potent antifungal activity). Such transgenic peanut plants show enhanced resistance against both the leaf spot diseases (Anuradha et al. 2008). The ISSR marker UBC810 (540) has been found associated with LLS + rust resistance but UBC810 (500) only with LLS resistance (Mondal et al. 2009). Leal-Bertioli et al. (2009) have identified candidate genome regions that control disease resistance and placed candidate disease-resistant genes and quantitative trait loci (QTLs) against LLS disease on the genetic map of the A genome of *Arachis*, which is based on microsatellite markers and legume anchor markers enabling mapping of a total of 34 sequence confirmed candidate disease-resistant genes and 5 QTLs. Among the polymorphic sclerotium stem rot (SSR) markers developed through crossing LLS-susceptible cultivar TMV-2 and LLS-resistant genotype COG-0437, the primer PM 384(100) has association with resistance and could therefore be utilized in the marker-assisted breeding program over a wide range of genetic background (Shoba et al. 2012). A double-gene construct with *Solanum nigrum* osmotin-like protein (SniOLP) and *Raphanus sativus* antifungal protein 2 (Rs-AF2) genes under separate constitutive 35S promoters has been developed to transform peanut plants. Such transgenic peanut plants expressing the SniOLP and Rs-AF2 genes show enhanced resistance to LLS based on reduction of number and size of lesions on leaves and delay of the onset of the LLS (Gowda et al. 2010, Vasavirama and Kirti 2012). The molecular diversity analysis using SSR reveals high level of genetic polymorphism for resistance to LLS and rust diseases, which provides valuable information for peanut breeders designing strategies for incorporating and pyramiding LLS and rust resistances and creating inbred line populations to map these traits (Mondal et al. 2005, Mace et al. 2006). Similarly, RAPD assays using 10 oligonucleotide primers have revealed existence of DNA-level variation within the LLS and rust-resistant genotypes. The susceptible lines can be clustered distinctly away from the resistant group, and clustering of genotypes based on phenotyping of LLS and rust can serve as basis for tagging resistant genes (Reddy et al. 2004). Marker-assisted backcross breeding should be able to minimize the linkage drag as it greatly facilitates monitoring of introgressed chromosome segments carrying disease-resistant genes from wild species to cultivated peanuts. Transgenic peanuts with resistance to biotic stresses as ELS and LLS and others have been reported to be produced and in various stages of characterization under containment and/or controlled field conditions (Dwivedi et al. 2003, Luo et al. 2005). For example, fertile transgenic plants of peanut cv TMV-2 expressing tobacco (*Nicotiana* spp.) chitinase and neomycin phosphotransferase (nptII) genes have been generated using *A. tumefaciens*–modified transformation system, and peanut plants containing transgenically increased activity of chitinase are resistant to attack by *C. arachidicola* to different degrees (Rohini and Rao 2001). Similarly, transgenic peanut plants possessing rice chitinase gene expressing resistance to ELS pathogen have been obtained through *Agrobacterium* mediation (Iqbal et al. 2012). Such strategies are of great potential for the control of the leaf spot diseases.

Chemical Control

Fungicides

The economic benefit of using fungicides to control the leaf spot diseases depends on the climatic conditions, the variety of peanut grown, the relative importance of the fungus species affecting the crop, and the general pattern of farming in the area concerned (Johnson et al. 2007a, Hoque et al. 2008, Naab et al. 2009, Muhammad and Bdliya 2011, Wann et al. 2011). Better results with increase in yield have been reported on early-maturing sequentially branched varieties compared with alternately branched late varieties. The reason is that early-maturing sequential type cannot produce branches and leaves at the end of growing season; consequently, all the assimilates produced at this stage are available for the growth of the nuts so that when the disease is controlled, there is considerable increase in the kernel yield. Conversely, the alternate forms continue to produce many new branches and leaves even up to the end of the season, and therefore, the effect on kernel yield as a result of disease control is proportionately less. The time of the first application may be dependent on the susceptibility of the variety. In long-duration varieties like 28-206 and 47-16, it is better to apply fungicides at the later stages of growth. Both of these lines produced 3.16 and 2.94 tons/ha pod yield when fungicide is applied at 70 days after sowing (DAS) (Waliyar et al. 1993, 1995, 1998). A new Australian cultivar Sutherland has significantly higher resistance to LLS and would need reduced or fewer number of fungicide spray in the management of the disease (Kelley et al. 2012). Research results obtained by Carley et al. (2009) reinforce the value of controlling ELS and LLS (and web blotch) with timely fungicide applications and importance of digging at optimum pod maturation of peanuts contradicting the general recommendation or belief that the percentage of canopy defoliation justifies early digging to prevent the yield loss. Similarly, such experimental evidence has been obtained in relation to fungicide application in refinement harvest timing guidelines based on the distribution of pod maturity as defined by mesocarp color categories (Chapin and Thomas 2005). An interaction of number and properly timed fungicide sprays for the control of ELS and LLS can result in substantial monetary gains for peanut farmers in West Africa (Waliyar et al. 2000). Several fungicides that have been effective in the control of the ELS and LLS are given in the following paragraphs.

During the initial stages of the development of fungicides, elemental sulfur (sulfur dust, wettable sulfur) and copper compounds (Bordeaux mixture, copper oxide, copper oxychloride) were in use for the control of the leaf spot diseases of peanuts (Kolte 1984). Later as many modern compounds like dithiocarbamates (maneb, zineb, mancozeb) and chlorothalonil became available, these then largely replaced copper-based compounds as protective fungicides. However, currently, both copper-based and elemental sulfur fungicides are considered to be organically acceptable fungicides on cultivars with partial resistance to one or both the ELS and LLS pathogens (Cantonwine et al. 2006, 2007a, 2008). Chlorothalonil spray at 0.2% has been found better than mancozeb and copper oxychloride, which is proved to be an economical fungicide (Culbreath et al. 1992a,b). Organotin compounds, though performing better than dithiocarbamates, could not become a commercially successful class of fungicides for the control of ELS and LLS. In 1968, a systemic fungicide, benomyl, was introduced and registered for better and efficient control of both the leaf spots through its protective and curative effects. Consequently, similar other two systemic fungicides, for example, carbendazim (0.05%) and thiophanate methyl (0.05%), found a commercially successful use all over the world for the control of the leaf spot diseases (Kolte 1984, Noriega-Cantú et al. 2000, Biswas and Singh 2005). All these systemic fungicides are reported to give better control of the leaf spots than mancozeb or copper oxychloride (Vidyasekaran 1981, Bolonhezi et al. 2004, Biswas and Singh 2005). However, curative action of systemic fungicide, that is, carbendazim when combined with a dithiocarbamate (mancozeb), has given good control of the leaf spots at longer intervals at various locations in the peanut-growing regions in different countries (Joshi et al. 2000, Srinivas et al. 2002, Biswas and Singh 2005, Satish et al. 2007, Johnson et al. 2007a, Lokesh et al. 2008). The first spray of mancozeb on initiation of the leaf spots followed by the carbendazim spray 10 days after mancozeb and again the spray of mancozeb 10 days after carbendazim give maximum control of the leaf spot disease, avoiding the possibility of development of fungicide resistance (Sawant 2000). The continued use of sprays of benomyl or benomyl-like compound carbendazim resulted in the development of resistant strains of *C. arachidicola* and *P. personata* in the mid-1970s. Consequently, use of benomyl on peanuts in the southeastern United States was discontinued, but chlorothalonil still proved to be an effective standard fungicide and is used extensively for the control of ELS and LLS diseases not only in the United States but also in Malawi (Kisyombe et al. 2001, Culbreath et al. 2002). Both mixtures and alternate applications of chlorothalonil and benomyl are effective for the management of the leaf spot in fields where benomyl alone did not provide season-long leaf spot control due to fungicide resistance (Culbreath et al. 2002). Additional options became available for the leaf spot control in the 1990s with registration of the demethylation inhibitors (DMIs), that is, sterol (ergosterol) biosynthesis–inhibitor (SBI) fungicides, which are triazole fungicides, namely, difenoconazole, propiconazole, hexaconazole, cyproconazole, tebuconazole, and epoxiconazole for use on peanut crops. These fungicides improved the management of one or both leaf spot diseases compared to chlorothalonil alone (Moraes et al. 2001). They control not only the leaf spots but peanut rust also (Culbreath et al. 1992b, Dahmen and Staub 1992, Jadeja et al. 1999). Similarly, quinol oxidation (Qo site) inhibitor (QoI) fungicides, also referred to as strobilurin fungicides, first launched in 1996, are very effective for the control of the leaf spots. The three most commonly used strobilurin fungicides used for the ELS and LLS on peanuts are azoxystrobin, pyraclostrobin, and trifloxystrobin. Recently, the new pyrazole carboxamide fungicide penthiopyrad at 0.20 kg ai/ha is reported to have excellent potential for the management of the LLS and may complement current SBI and QoI fungicides (Culbreath et al. 2009). Prothioconazole applied alone (0.18–0.20 kg/ha) or in combination, that is, 0.085 kg ai/ha of prothioconazole + 0.17 kg ai/ha of tebuconazole or mixtures of prothioconazole at 0.063 kg ai/ha with trifloxystrobin at 0.063 kg ai/ha, gives similar or better leaf spot control than chlorothalonil (Culbreath et al. 2008). Pyraclostrobin (Chapin and Thomas 2005, Culbreath et al. 2006), azoxystrobin (Bowen et al. 2006, Hagan et al. 2006), tebuconazole (Hossain et al. 2005, Culbreath et al. 2006, Nutsugah et al. 2007), and a new systemic fungicide Nativo (Khan et al. 2014) have been found as effective fungicides for the control of leaf spot diseases of peanuts in different countries around the world. Weather-based fungicide advisory sprays of these fungicides in the United States reflect the improved disease control, consequently improving the yield of the peanut crop (Grichar et al. 2005). The cost–benefit analysis of fungicidal control of the leaf spots of peanuts in the Sudan savanna region of Nigeria revealed positive returns per hectare from the use of fungicides. Application of Bentex T (benomyl + thiram), for instance, has been found to give 78.13% seed yield increase over untreated plants, which could be translated into a mean net profit in the range of U.S. $387–$909/ha depending on the number of effective efficacies of sprays (Bdliya and Gwio-Kura 2007a,b). Similarly, high incremental cost–benefit ratio has been obtained with hexaconazole (1:4.72) followed by propiconazole (1:2.05), difenoconazole (1:1.29), and chlorothalonil (1:1.13) in the control of leaf spot diseases in India (Gururaj et al. 2002, 2005a). In some other studies, however, the highest cost–benefit ratios (1:6.9–1:675) with 0.1% difenoconazole spraying at 30, 50, and 70 DAS (Gopal et al. 2003) and with hexaconazole at 60 and 75 DAS (Johnson and Subramanyam 2003) and effective control with single spray (Chandra et al. 2007) have been obtained in India. Rainfall-based advisories for the chemical control of LLS in Brazil reveal that tebuconazole sprays at an interval of 15 days show higher efficiency in controlling the LLS in comparison to a scheme of control with chlorothalonil sprayings at five fixed dates (the first at 49–50 days after planting and the following ones at intervals of 14–15 days) besides promoting an average reduction of one to three sprayings (Moraes et al. 2002). When the proportion of ELS- and LLS-affected leaves exceeds 10% and the wet index total exceeds the threshold value (<2.3), the first application of fungicide is advised, followed by at least two successive sprays at a 14-day interval (Butler et al. 2000).

Nonconventional Chemicals

Among nonconventional chemicals using inorganic and metal salts, catechol at 10(-3M) (Maiti et al. 2005), nickel chloride, and cupric sulfate at 10(-3M) (Kishore et al. 2001b) have been found significantly effective in controlling the ELS and LLS infection through induction of host resistance in peanuts. Resistance interaction is correlated with early and rapid induction of phenylalanine ammonia lyase (PAL) enzyme regulating the biosynthesis of antifungal phytoalexin medicarpin (Kale and Choudhary 2001). Similarly, the effect of salicylic acid (SA) on induction of resistance in peanuts against LLS through foliar application of SA at a concentration of 1 mM is observed to be due to the enhanced activity of PAL, chitinase, β-1,3-glucanase, peroxidase (PO), and polyphenol oxidase (PPO) (catechol oxidase) enzymes and in total phenolic contents in peanuts after application of SA and inoculation with *P. personata* (Meena et al. 2001a).

Cultural Control

Since the ELS and LLS pathogens have a very restricted host range and largely to *Arachis* species, effective crop rotation of a 2–3-year duration of peanuts with nonsusceptible crops, such as cotton or corn or soybean or maize alternating with peanut or not following peanut with peanut, has been found useful in delaying the initial infections by both leaf spot fungi and reducing the incidence of the disease early in the season by 88%–93% (Kucharek 1975). This level of reduction permits growers to take some of the *pressure off* of a fungicide spray and delay the first fungicide application where crop rotation is followed requiring fewer or at least less expensive fungicide application. Though there can be regional differences based on different cropping schemes, rotation into nonhost crops is essential to sustainable, long-term production of peanuts. About 72%–86% more yield has been reported by following crop rotation. Eliminating volunteer peanut plants in the fields that follow peanut main or forage crops immediately after harvest of the peanut is a critical component of successful crop rotation program. Keeping the peanut crop free from weed (weeding once at either 4 or 6 weeks after planting) is also helpful in minimizing the severity of the leaf spots (Abudulai et al. 2007) in Ghana. Clewis et al. (2001) from the state of North Carolina in the United States provided evidence as to how high density of common ragweed (*Ambrosia artemisiifolia*) in peanut fields increases the incidence and severity of LLS on the peanut crop.

The relationship between disease severity and yield increase at different plant spacings is reported. Lower incidence and severity of the leaf spots is reported at a 50 × 30 cm intrarow spacing in comparison to 50 × 20 cm in Nigeria (Garba et al. 2005). Reduced tillage, the strip-tilled peanut crop with the presence of cover crop residue at the soil surface of strip-tilled fields in the United States, is less severely affected by ELS epidemics than conventionally tilled fields. The strip tillage delays ELS epidemics due to a fewer initial infections, most likely due to cover crop residue interfering with the dispersal of primary inoculum from overwintering stroma in the soil to the peanut plant (Cantonwine et al. 2007b). Consequently, the number of fungicide applications, if any, could be reduced without compromising the control of ELS when reduced tillage is used especially when combined with moderately resistant cultivars (Monfort et al. 2004). Intercropping of millet (Bdliya and Muhammad 2006) and that of maize (Ihejirika 2007) in Nigeria and of pearl millet in India (Srinivas et al. 2002) with peanut show significantly lower ELS/LLS severity on peanut in comparison with sole peanut crop without adversely affecting the yield of inter crops. NPK fertilizer rates and peanut plant population per hectare significantly influence the severity of ELS, the least severity of the ELS being at the density of 250,000 peanut plants/ha in comparison to 1,000,000 plants/ha (Ihejirika et al. 2006a). Among several methods of irrigation studied (Jordan and Johnson 2007, Woodward et al. 2008), subsurface drip irrigation is useful in avoiding the increase and spread of the ELS compared with overhead sprinkler irrigation (Lanier et al. 2004).

Depending on the more frequent occurrence of the disease in a particular area, a choice of suitable planting dates (either early or late) and a choice of using early- or late-maturing variety may be important in reducing the severity of the disease (Reddy and Reddy 2000, Naidu and Vasanthi 2002). The least incidence and severity of the leaf spots is noted when the crop is planted in the first week of May in the state of Gujarat in India (Hazarika et al. 2000). Planting date and maturity of the variety then also significantly influence the effectiveness of fungicide spray application. For example, fungicide spraying in early-sown crop in the state of Chhattisgarh in India is reported to be uneconomical because reduction in pod yield due to the leaf spots in the early-sown crop is not significant. In contrast, two sprays (or even only one spray) are economical for late-sown crops under similar conditions (Tiwari et al. 2005). Interestingly, long-duration (120 days) variety (F-mix) of peanuts when sown early and treated with fungicide for the leaf spot control under optimum and timely crop management practices is reported to have produced greater yield than short-duration (90 days) variety (Chinese) under both with and without fungicide-treated environments and more than three- to fourfold increase over the average peanut yields in Ghana (Naab et al. 2005). Deep burying of crop residues in the soil by mold board plow and destruction of crop residues by burning have been recommended as additional aids for the control of the leaf spot diseases of peanuts.

Biological Control

Biological control of leaf spot diseases using antagonistic fungi or bacteria may be an alternative approach to use of chemical fungicides. This appears to be desirable in view of the development of resistance against the most effective systemic chemicals and side effects from frequent use of chemicals, as is evident from the increase in population of phylloplane mycoflora and foliar mites as a result of the spray of chemicals. In India, a mycoparasite, *Hansfordia pulvinata*, has been found parasitizing *C. arachidicola* and *C. personatum* (Krishna and Singh 1980, Siddaramaiah and Jayaramaiah 1981). *Hansfordia* sp. (exact species not mentioned) has also been reported from the United States to be parasitizing only *C. personatum* and not *C. arachidicola* (Taber and Pettit 1981). It is observed that the leaf spots, which are parasitized by the aforementioned fungus, do not show conidiophores and conidia of the respective causal fungus species. Among several isolates of *Pseudomonas fluorescens* screened for their efficacy for the control of the LLS, the *strain Pfl* has been found to be effective in reducing the LLS disease index through seed treatment (10 g/kg seed) combined with soil application of *P. fluorescens* strain Pf1 (2.5 kg/ha at 30 and 45 DAS) or foliar spray application of talc-based powder formulation (1–2.5 kg/ha) of the strain Pf1 (Meena et al. 2000, 2002, 2006, Zhang et al. 2001, Johnson and Subramanyam 2009, Meena 2010, Meena and Marimuthu 2012). Following the seed treatment, the antagonist colonizes soil in the peanut rhizosphere. All such treatments also show increase in plant height with enhanced yield of pod possibly due to the production of IAA and induction of *P. fluorescens*–mediated systemic resistance against the leaf spot pathogens. *P. fluorescens*–treated peanut plants show increase in the activity of PAL, phenolic content, and lytic enzymes. Chitin-supplemented application of antifungal and chitinolytic bacteria *Bacillus circulans* GRS 243 and *Serratia marcescens* GPS 5 effectively results in the control of LLS (Kishore et al. 2005a) through enhanced activity of four defense-related enzymes such as chitinase, β-1,3-glucanase, PO, and PAL (Kishore et al. 2005b). The nonchitinolytic *Pseudomonas aeruginosa* GSE 18 (Kishore et al. 2005b) and chlorothalonil-tolerant *P. aeruginosa* have also given effective control of the LLS (Kishore et al. 2005b). Kondreddy and Podile (2012) reported a new integrated approach, where both direct antagonism and induced resistance got combined to reduce the incidence of the LLS in peanuts. Chlorothalonil-tolerant chitinolytic bacterium has been genetically engineered to secrete elicitor protein harpin Pss of *Pseudomonas syringae* pv. *syringae* for the dual benefit of growth promotion of peanut plants and the control of LLS.

Mycorrhizal (*Glomus* sp. and *Gigaspora* sp.) symbiosis with peanut roots increases the resistance of plants to leaf spot pathogens reducing the severity of the leaf spot diseases by 54% (Zachée et al. 2008).

Effect of Plant Extracts

Seed extract of *A. indica* (Srinivas et al. 2000, Alabi and Olorunju 2004, Nandgopal and Ghewande 2004, Ambang et al. 2007, 2011, Badliya and Alkali 2010a,b); aqueous leaf extracts of *A. indica* (Aage et al. 2003, Ihejirika et al. 2006b); Hemi Fern, that is, *Hemionitis arifolia* (Sahayaraj et al. 2009), *Prosopis juliflora* at 2% (Kishore and Pande 2005b), *Polyalthia longifolia* at 10% (Adiver 2004), *Lawsonia inermis* at 5%, and *Datura metel* at 2% (Kishore et al. 2001a, 2002, Kishore and Pande 2005a); seed extract of *Thevetia peruviana* (Ambang et al. 2007, 2011); and *Mahogany* bark extract (Salaudeen and Salako 2009), when sprayed on peanut plants, give reductions in the leaf spot disease index on peanuts. Aqueous leaf extracts of *A. indica* and that of *D. metel* and neem (*A. indica*) seed could be of potential economical and eco-friendly alternative usages for the control of the leaf spots particularly in areas where farmers cannot afford to use fungicides particularly in the Sudan savanna region of Nigeria.

RUST DISEASE OF PEANUTS

Symptoms

Orange-red to chestnut-brown elliptical raised uredo pustules appear on the abaxial surface of the leaves (Figure 2.5). The pustules are 0.3–2.00 mm in diameter and usually surrounded by a yellow halo. The adaxial surface of the leaf might present a gray appearance due to the formation of flecks that correspond to the position of the uredo pustules below. The uredo pustules are either isolated or in groups; they are formed subepidermally on compact stomata but soon burst through the epidermis and become exposed. Consequently, a reddish-brown mass of spores becomes visible on the

FIGURE 2.5 Peanut rust symptoms on the leaflet. Note the numerous small-sized uredo pustules.

FIGURE 2.6 Field view of peanut rust–affected crop.

surface of the leaves. The uredo pustules are also formed on the adaxial surface at advanced stages of infection. They may also be formed on stipules, petioles, and stem. As the infection advances, the pustules turn dark brown and frequently coalesce to cover larger areas. Eventually, the leaflets may curl and drop off resulting in defoliation. Severely affected plants appear as light-brown patches in normal green plants in the field and can be easily seen from a longer distance (Figure 2.6). Pods of severely affected plants are low in number and mature 2–3 weeks early. The seeds of such plants remain small in size. The sequence of the development of symptoms has been studied on artificially inoculated plants (Mallaiah and Rao 1979).

Geographical Distribution and Losses

Peanut rust has been known since 1884 from specimens on cultivated peanut plants collected in Caaguazu, Paraguay (Spegazzini 1884). It has now become a disease of major economic importance in almost all peanut-growing areas in North, Central, and South America, the Caribbean Islands, the West Indies, Asian and West and East African countries, and Australia (Kolte 1984, Subrahmanyam et al. 1985). In recent years, peanut rust has spread to and become established in South Africa (Mathews et al. 2007). In the United States, rust causes considerable economic losses in South Texas, and earlier rust, as such, had not been considered a serious problem in peanut production (Hammons 1977), but in recent years, the disease has been recorded to occur and impact yields in the southeastern United States (Gremillion 2007).

Since the appearance of the rust coincides with the appearance of ELS and LLS, several workers have adopted the method of estimating the loss in yield due to the disease by controlling the leaf spots by spraying plants first with benomyl or carbendazim and then superimposing this treatment with another fungicide (tridemorph or chlorothalonil or hexaconazole) effective against the rust. Thus, information on loss in yield due to rust or the leaf spots alone or due to both has been obtained. The loss in pod yield due to rust alone has been reported in the range of 14%–70%, depending on the variety and geographical region and climatic conditions (Kolte 1984, Gururaj and Kulkarni 2006). It is reported that early establishment of the disease in Australia may advance harvesting of the crop by up to 28 days. Significant yield loss to the extent of 100% is observed in most rust-affected peanut crops in Northern Australia in every season (Middleton and Shorter 1987). A serious outbreak of peanut rust appeared in 1973 in northern Territory in Australia (O'Brien 1977) and in the Maharashtra State of India in 1976–1977, and the crop yield then declined by 35% (Mayee 2009, Tashildar et al. 2012).

In Nicaragua, the commercial crop cost for peanuts has increased by 48% because of measures to control peanut rust and leaf spots in nonrotated fields. In the P.R. of China, losses due to rust are estimated to be 49%, 41%, 31%, and 18% at flowering, pegging, pre-pod-forming, and mid-pod-forming stages, respectively (Zhou et al. 1980).

In India, the combined infections of rust and leaf spots cause losses to the extent of 29%–70% in pod yield and 27% in kernel weight depending on the variety (Ghuge et al. 1981, Tashildar et al. 2012). In addition to direct yield losses, rust can lower down seed quality by reducing seed size and seed viability and oil content (Subrahmanyam et al. 1991, 1997). The loss in oil content due to rust infection alone has been estimated to be about 7%–10% (Kenjale et al. 1981).

Pathogen: *P. arachidis* Speg.

Classification
Kingdom: Fungi
Phylum: Asidiomycota
Class: Urediniomycetes
Subclass: Incertae Sedis
Order: Uredinales
Family: Pucciniaceae
Genus: *Puccinia*
Species: *arachidis* Speg.

Usually, the pathogen is observed in the uredial stage on peanut plants. Uredosori are subepidermal, amphigenous, and scattered. Each sorus contains numerous uredospores. The uredospores are round to oval and pedicellate. The pedicel of the uredospore is short, fragile, and hyaline. The epispore is thin walled, echinulate, and cinnamon colored. The germ pores are 2–3 or 3–4 and are located equatorially in the spore. The spores measure in the range of 18.56–33.00 × 17.47–26.48 μm, with an average size of 24.96 × 21.22 μm.

The telial stage is reported on the peanuts in Brazil (Spegazzini 1884) and India (Chahal and Chohan 1971), but its presence has been found to be quite rare. It is interesting that Chahal and Chohan (1971) have reported only the telial stage and not the uredial stage of peanut rust while noting that it is the first report of peanut rust occurrence in India. Such a situation could be attributed to possible differences in variability in the isolates of *P. arachidis*, for example, out of several isolates collected from the state of Karnataka in India, only one *Gadag isolate* shows a rare phenomenon of teliospore formation (Tashildar et al. 2012). Aecia and pycnia are not known. An alternate host, if any, is also not known.

Epidemiology and Disease Cycle

Little new information on epidemiology and disease cycle of peanut rust has become available in the last three decennia. The sexual stage (teleutospore formation) is rare in the main areas of peanut cultivation and is epidemiologically insignificant, and the absence of an alternate host or a collateral host indicates that in the endemic areas, the fungus perennates in the uredinial stage on either volunteer self-sown plants or autumn-sown crops. There has been a large swing to autumn and spring crops in many peanut-growing countries including India, and this combined with the increased number of peanut crops has favored the pathogen survival. Uredospores, as such, without the living host have a very short life except at very low temperature. Uredospores can be stored at low temperature without loss of viability, but at 40°C temperature, they rapidly loose viability. For instance, uredospores in the exposed crop debris lose all viability within 4 weeks under postharvest conditions at Hyderabad under Indian conditions. It thus appears that in tropical and subtropical countries, the pathogen does not survive uredospores per se from year to year. Hence, survival and development of the disease are limited by both temperature and survival of host tissues. Such epidemiological studies on peanut rust have also been carried out in detail in Mexico (Noriega-Cantú et al. 2000). Logistic model for prediction of the rust of peanuts under Karnataka conditions has been given by Gururaj et al. (2006). Similar prediction model has been developed by Narayana et al. (2006).

The possibility of the pathogen being carried through seed as a surface contaminant has been indicated, as introduction of rust on the peanuts in the United States, Brunei, Australia, and Papua New Guinea is reported to be the result of seed transmission of the pathogen through imported seed (Kolte 1984). Usually, the primary infection is attributed to uredospores transported by wind from short or long distances where the volunteer or self-sown rust-affected peanut plants are located or where the unexposed affected plant debris is present (Gururaj and Kulkarni 2007). In the southern part of India, there is an extensive and continuous cropping of peanuts at all times of the year, and thus, there is an easy availability of uredospores as primary inoculum from one season to another. It is clear that peanut rust is now well established in India and that it shows a clear pattern of spread from one or two origins where the crop circumstances permit inoculum buildup. Thus, the infection appears in July–August in southern India, in September and October in central India, and in November and December in the northeast. It appears, therefore, that the initial inoculum infecting peanuts in northern Andhra Pradesh and Maharashtra is derived from southern states and that, in turn, central Indian crops serve to provide the initial inoculum infecting those in West Bengal and Assam (Mayee et al. 1977). In the United States, the fungus does not overwinter but blows in from subtropical areas (Van Arsdel 1974).

After deposition of the uredospores on the leaves of the peanut plants, they germinate by giving rise to a germ tube at the temperature range of 15°C–30°C, with an optimum range being 20°C–25°C (Kono 1977). For spores to germinate on a leaf surface, the presence of free water is a must. The uredospores do not germinate at a relative humidity below 100%. Relative humidity above 80% supports the germination when spores are placed in a thin film of water (Cook 1980b). After germination, the germ tube grows and forms an appressorium over stomata. Cook (1980a) has studied the infection process in detail. According to him, close adhesion of the germ tube leaf surface is essential for appressorium formation. From the appressorium, the infection hypha arises and penetrates the tissue through stomata. After transversing the length of the stomatal passage, the infection peg swells and forms a vesicle in the substomatal chamber. Several infection hyphae then arise from such a vesicle, and the subsequently formed mycelia become intercellular by producing knob-shaped haustoria into the cells. Most infections take place successfully at 22°C–26°C. About 8–10 days after incubation, symptoms become visible with the production of a new crop of spores that become wind borne and cause secondary infection.

The factors that directly affect the spore germination have an indirect effect on disease development. Light inhibits uredospore germination and germ tube elongation (Subrahmanyam et al. 1980). Thus, it appears that there are more chances of getting a crop infected with the pathogen during evening or night hours than through the day. The density of spore concentration also affects the spore germination and subsequent infection process. Spores do not germinate in clumps and dense patches because of a high concentration of self-inhibitor within them and in surrounding water. The self-inhibitor has been isolated by Foudin and Macko (1974) and identified as methyl *cis*-3,4-dimethoxy cinnamate. It is found that *P. arachidis* is more sensitive to self-inhibitor chemicals than any other rust fungus. Leaf surface influences the infection, probably because of the differences in wettability of the leaf surfaces. It is usually seen that the abaxial surface of the leaf is more wettable compared with adaxial surface, and therefore, more infection sites develop on abaxial surface.

Under conditions of high rainfall and humidity in the *postrera* planting season in Honduras and Nicaragua (Central America), the disease becomes devastating, and it then becomes difficult to control (Arneson 1970). In Venezuela, the rust becomes severe when the rainy season is nearly over or when dew is abundant (Hammons 1977). In India, a continuous dry period characterized by high temperature (>26°C) and low relative humidity (<70%) is reported to delay rust occurrence and severity, whereas intermittent rain, high relative humidity, and 20°C–26°C temperature favor disease development (Siddaramaiah et al. 1980). Aerial dissemination of uredospores has been studied, and diurnal periodicity with peak occurrences around noon was reported under Indian conditions (Mallaiah and Rao 1982), and currently changing scenario of prevalence of rust in the late 1990s from the three important southern states Andhra Pradesh, Karnataka, and Tamil Nadu has been described by Pande and Rao (2000).

In India, during summer months (May/June) when incidence of the rust is low, the incubation period is long (18 days), while in winter, when the rust is abundant, the incubation period is only 7 days (Mallaiah 1976). More or less similar observations, with respect to incubation in summer and winter seasons, have been reported from Taiwan (Fang 1977).

Susceptibility of the plants appears to be related to age, with plants becoming susceptible to disease at 5–6 weeks of age (McVey 1965).

There appears to be a relationship between altitude and appearance of the rust on peanuts. In lowveld areas (altitude 430 m) in Zimbabwe, the crop is severely affected, whereas in highveld areas of that country, rust appears quite late and in less severe form. A close relationship in the climatic requirements of rust and LLS has also been noted, as both are favored at lower elevations (Rothwell 1975). A similar situation appears to be true with the development of rust and LLS in Malawi (Sibale and Kisyombe 1980) and perhaps in southern parts of India. Though the pathogen is mainly restricted in its host range to *A. hypogaea*, it has recently been found to naturally infect *Arachis repens* with the formation of both uredospores and otherwise rarely formed teliospores in Brazil (Rodrigues et al. 2006).

Disease Management

Host Plant Resistance

In Cultivated Arachis *Species*

Screening has been conducted on large scale under both natural conditions and artificial inoculation of plants (4–5 weeks old) with uredospores to locate the sources of resistance especially at ICRISAT where a range of techniques have been developed (Subrahmanyam et al. 1982, Waliyar et al. 1993, Sudhagar et al. 2009).

On the basis of the percentage of leaves infected due to rust or percentage of defoliation due to rust, rating scales for discriminating resistant and susceptible genotypes have been used. No symptoms of rust, leaves showing slight infection, and less than 25% of leaves showing severe symptoms or defoliation due to rust disease have been taken as criteria for determining resistance.

It is found that highly and moderately rust-resistant genotypes are characterized by higher cuticular and epidermal cell thickness, lesser and smaller epidermal cells, lower number of stomata, and more wax content at the later stages of crop growth (Gururaj and Kulkarni 2006). Differences in the degree of and development of rust mycelium in the substomatal cavities are manifested in resistant and susceptible reactions. In highly resistant and immune types, the germ tube dies after penetration through stomata without the further development (Nevill 1980). In the nonphysiological resistant types, chloronemic flecks are formed without the formation of uredia and uredospore release, but in the physiological type of resistance, less than half of the chloronemic flecks are developed into uredia as in NCI3 cultivar (Cook 1972, 1980a). In nonphysiological resistant cultivars, resistance is related to leaf wettability, which in turn determines the spore retention capability of the cultivar. Cultivars with thin and less waxy leaves are generally affected earlier than those with thick and waxy leaves (Chen et al. 1981). Sudhagar et al. (2009) reported that the activity of PO and PPO is maximum at 80 DAS and that ascorbic acid oxidase and chitinase enzymes exhibit their maximum activity at 80 DAS

TABLE 2.2
Peanut Genotypes Resistant (R) or Moderately Resistant (MR) to Rust Disease of Peanuts as Reported from Different Countries in the World

Genotype	Country	R/MR	Reference(s)
DH22 (red), DH22 (tan), GPBD-4, K-134, R8808, R9214, R9227, R2001-1, R2001-2, and R2001-3	India	MR (partial resistance)	Gururaj and Kulkarni (2008)
ICGV 93207 (named as Sylvia)	Released for Mauritius (ICRISAT)	R	Reddy et al. (2001a)
ICGV 94361	India (ICRISAT)	MR (early)	Upadhyaya et al. (2001a)
ICGV 87354	India (ICRISAT)	R	Reddy et al. (2001b)
ICG 8954 (*A. kuhlmannii*)	India (ICRISAT)	Immune (asymptomatic)	Pande and Rao (2001)
ICGV 87853 (Venus)	India (ICRISAT)	R	Reddy et al. (2000)
M-5, 255/88	Bangladesh	MR	Hossain et al. (2007)
JALW-20, JL-501	India	MR to rust + LLS	Deshmukh et al. (2009)
VL-1	India	R	Kumar et al. (2012)
ICGV 98383	India	R	Patil et al. (2010)
ICGV 99003, ICGV 99005, ICGV 99012, ICGV 99015	India (ICRISAT)	R	Singh et al. (2003a), Dwivedi et al. (2002)
NC17090	—	R	Pensuk et al. (2003)

with prominent expression of a 56 kDa protein in rust-resistant genotypes of peanut. The potential amount and activity of these are genetically determined, and such changes in the quantity of isozyme and protein can be relied for screening rust-resistant/rust-tolerant genotypes. Induction and accumulation of phenols and PPO enzyme have been found to be at a faster rate in rust-resistant genotypes ICG 1697 and ICG 10053 on inoculation with uredospores (Kumar and Balasubramanian 2000).

Marked sources of resistance in the cultivated peanut have been reported by several workers (Tables 2.1 and 2.2). Some genotypes are resistant to both rust and LLS. Some peanut genotypes such as GPBD-4, DH22 (red and tan), and R9214 (Gururaj and Kulkarni 2008) and genotypes such as ICGVs 99003, 99005, 99012, and 99015 (Dwivedi et al. 2002) show the slow-rusting ability characterized by longer incubation period, low sporulation index, lesser number of pustules per unit area, and smaller pustule size. In such cases, the resistance is controlled by several genes. A new Australian peanut cultivar *Sutherland* has significantly higher level of resistance to rust (and also to LLS) under the Queensland conditions in Australia (Kelly et al. 2012).

In Wild Arachis *Species*

The genus *Arachis* is native to South America and also consists of 22 described species and more than 40 undescribed. Gene banks are maintained in Brazil, the United States, and India. Many germplasm accessions have been screened and several peanut genotypes with immunity or high level of resistance to peanut rust have been identified, with sources for resistance mainly originating from Peru, Bolivia, and India (Wynne et al. 1991, Varman et al. 2000, Yadav et al. 2007). A very high degree of rust resistance in wild species of the genus *Arachis* has been reported. Such species are *A. duranensis* (PI 219823, section *Arachis*), *A. correntia* (PI 331194, section *Arachis*), *A. cardenasii* (PI 262141, section *Arachis*), *A. chacoense* (PI 276235, section *Arachis*), *A. chacoense* × *A. cardenasii* (F1 hybrid), *A. pusilla* (PI 338448, section *Triseminalae*), *A.* sp. 9667 (PI 262848, section *Rhizomatosae*), *A.* sp. 10596 (PI 276233, section *Rhizomatosae*), *A. glabrata* (PI 118457, 231318, 262287, 262801, section *Rhizomatosae*), *A. villosulicarpa* (PI 336985, section *Extranervosae*), and *A. villosa* (Subrahmanyam et al. 1982). *A. stenosperma* accession V 10309 is resistant to rust and LLS and the experimental evidence reveals that in *A. stenosperma*, infection is hampered at the stage of penetration (Leal-Bertioli et al. 2010).

Resistance to rust in some wild species has been found to be mostly recessive and governed by monogenic (3:1), digenic (15:1), and trigenic (63:1) F2 segregation ratios (Joel et al. 2006). Susceptibility is dominant to resistance and F2 population segregates in the ratio of 3:1 with resistance governed by a single recessive gene (Paramasivam et al. 1990). However, rust resistance in peanuts is also reported to be controlled by both additive and nonadditive gene actions and additive gene effects are predominant (Ghewande 2009). Thus, wild germplasm accessions of both A and B genome types are available to be used for the introgression of resistance genes against rust fungal pathogen (Fávero et al. 2009).

Molecular Breeding and Transgenic Peanuts for Rust Disease Resistance

Comparison of molecular profile among different peanut cultivars, genotypes, and breeding lines with differential disease reaction against rust has been carried out using RAPD (Mondal et al. 2005, 2008a,b) and ISSR marker analysis (Varma et al. 2005, Mace et al. 2006, Mondal et al. 2008a, 2009). Of the two markers, ISSR reveals higher polymorphism (74.5%) than RAPD (47.1%) with the average number of polymorphic bands per assay unit being 5.4 in ISSR and 3.3 in RAPD (Mondal et al. 2008b). Mondal et al. (2008b) have been the first to report on the identification of RAPD markers linked to rust resistance in peanuts in India and that RAPD marker $J7_{1300}$ is reported to be applicable for marker-assisted selection in the peanut rust resistance breeding program. The ISSR primer UBC 810_{540} is found to be associated with both rust and LLS resistances (Mondal et al. 2009). A study has been conducted by Mace et al. (2006) using ISSR analysis to identify diverse rust disease–resistant germplasm for the development of mapping populations and for their introduction into breeding programs. Twenty-three SSRs have been screened across 22 groundnut genotypes with differing levels of resistance to rust and LLS. Rust resistance in peanuts is associated with two SSR alleles ($pPGPseq3A1_{271}$ and $pPGPseq3A1_{390}$) in ICGV 99003 × TMV-2 and seven SSR alleles ($pPGPseq5D5_{270}$, $pPGPseq5D5_{295}$, $pPGPseq5D5_{325}$, $pPGPseq16F1_{315}$, $pPGPseq16F1_{424}$, $pPGPseq7F6_{128}$, and $pPGPseq13A7_{292}$) in ICGV 99005 × TMV-2. SSR markers associated with rust resistance should facilitate the rapid identification and transfer of chromosomal region(s) into elite breeding lines by using marker-assisted backcross breeding in peanuts (Varma et al. 2005).

QTL analysis using inbred lines of a mapping population TAG 24 × GPBD-4 segregating for rust reaction reveals 12 QTLs for rust. Interestingly, a major QTL associated with rust ($QTL_{rust}01$), contributing 6.90%–55.20% variation, has been identified by both composite interval mapping and single-marker analysis. A candidate SSR marker (IPAHM 103) linked with this QTL has been validated using a wide range of resistant/susceptible breeding lines as well as progeny lines of another mapping population (TG 26 × GPBD-4). Therefore, this marker is considered to be useful for introgressing the major QTL for rust in desired lines/varieties of peanut through marker-assisted backcrossing (Khedikar et al. 2010). Similarly, the two more SSR markers pPGpseq4A05 and gi56931710 have been found to show significant association with the rust reaction, and these flank the rust resistance genes at map distances of 4.7 and 4.3 cM, respectively, in linkage Group 2 (Mondal and Badigannavar 2010, Mondal et al. 2012). Thus, tagging of the rust resistance locus with linked SSR markers can be useful in selecting the rust-resistant genotypes from segregating populations and in introgressing the rust resistance genes from diploid wild species.

Chemical Control

In order to ensure elimination of infection from the surface-contaminated seeds, treatment of seed with thiram or captan at 3–4 g/kg of seed or with any effective seed treatment fungicide is advisable.

Mancozeb and Calixin (chlorothalonil, fentin hydroxide, tridemorph, triadimefon, and benodanil) have had been the fungicides of choice for peanut rust control through foliar sprays in the past (Kolte 1984). Chlorothalonil appears to be more effective than mancozeb. In recent years, some newer fungicides such as hexaconazole (Johnson and Subramanyam 2003, Hossain et al. 2010), difenoconazole (Gopal et al. 2003, Kalaskar et al. 2012), tebuconazole (Besler et al. 2006, Hagan et al. 2006), and azoxystrobin (Hagan et al. 2006) have been found to be more effective in controlling the peanut rust in comparison to mancozeb and chlorothalonil. In case the rust-resistant or partially resistant peanut cultivar is used, for example, the use of peanut cultivar *Sutherland* in Australia or in years with low rust disease pressure, few or reduced number of fungicide application will be needed to manage the disease (Kelly et al. 2012).

In order to control both rust and leaf spots, a mixture of two chemicals has been found effective. A mixture of systemic fungicide benomyl or carbendazim (0.05%) and mancozeb (0.2%) or a mixture of benomyl (0.05%) and mancozeb (0.2%) (Kolte 1984, Ghewande 2009) or a mixture of chlorothalonil (0.2%) and hexaconazole or a combination of tridemorph plus carbendazim (Mathur and Doshi 1990) has been reported to be superior in bringing about the control of rust and leaf spot diseases of peanuts with increase in peanut yield by 30%–40%.

Among the inorganic and metal salts, ammonium dihydrogen orthophosphate (monoammonium phosphate) and cobalt chloride have been found to be effective in controlling rust infection caused by *P. arachidis* (Kishore et al. 2001b).

Cultural Control

Cultural practices that destroy volunteer peanut plants or crop debris are an important measure to limit primary sources of inoculum. This is particularly important in Caribbean and Central American countries. In Australia, in the Atherton Tableland region, growers are encouraged to eliminate volunteer plants within 2 months before sowing to reduce the amount of inoculum early in the season. It has also been possible to introduce a degree of uniformity in planting time in Australia to minimize the probability of late plants being close to an early planted rust-affected crop. Under Indian conditions, early planting (15 days earlier than normal), plant spacing of 45 × 10 cm, and intercropping with red gram (*Cajanus cajan*) and castor (*Ricinus communis*) (Kodmelwar and Ingle 1989, Ghewande 2009) and with sorghum and pearl millet (Reddy et al. 1991) have been effective in reducing the peanut rust incidence and severity. Care should be taken to use seeds for sowing from healthy plants and noninfested regions.

Biological Control

The possibility of biological control of the disease by the use of antagonistic microorganisms has been indicated. Uredosori on peanut leaves have been found to be parasitized by mycoparasites such as *Darluca filum* (Biv.) Bem. ex. Fr, *Eudarluca caricis* (Fr.) O. Erik, *Daluea phylum* Byv and *Tuberculina* cos-*traricana* Sy, *Verticillium lecanii* (Zimm.) Viégas, and *Penicillium islandicum*. Mycophagous thrips, *Euphysothrips minozzi* Bagnall and *Dipteron* maggots (Patil et al. 2000) have also been reported to feed on uredospores of *P. arachidis* (Kolte 1984). However, no serious attempt has been made to use mycoparasites in the control of rust in peanuts, though sprays of culture filtrate of *V. lecanii* and *P. islandicum* were demonstrated to be effective in reducing the rust severity under field conditions (Ghewande 1993, 2009). A new fungal antagonist, *Fusarium chlamydosporum* Wollenw. & Reinking, has been isolated from the pustules of peanut rust significantly reducing the rust infection by *P. arachidis* using conidia and culture filtrate of the antagonist in artificial infection studies (Mathivanan and Murugesan 2000). Similarly, antagonistic isolate of *F. solani* has also been isolated from peanut rust pustules, and its antagonistic activity against *P. arachidis* is found to be due to chitinase and β-1,3-glucanase enzymes (Mathivanan 2000).

Effect of Plant Extracts

Foliar application of neem (*Azadirachta indica*) seed kernel extract (NSKE) at 3% (Gururaj et al. 2005b, Ghewande 2009) and aqueous leaf extracts of *Azadirachta indica* (Zade et al. 2005, Ghewande 2009) at 2% spray and that of *Prosopis juliflora* at 2% (Kishore and Pande 2005b), *Lawsonia inermis* at 5%, and *Datura metel* at 2% (Kishore et al. 2001, 2002, Kishore and Pande 2005, Zade et al. 2005) give reduction in the severity of rust disease index by 65%–74%. Aqueous plant extracts (30–40 g/L) of *Lippia multiflora*, *Boscia senegalensis*, and *Ziziphus mucronata*, the three local plants from the vicinity of Burkina Faso, show very strong antifungal activity against *P. arachidis* in vitro, and these have been found to be as good as or superior to fungicides in controlling the *P. arachidis* infection on peanut leaves (Koïta et al. 2012). Integrating neem leaf extract (2%) with potassium (1.0% K_2O) (Hossain and Rahman 2007) and the NSKE (5%) or leaf extract (2%) of *D. metel* with fungicide chlorothalonil or difenoconazole is more effective in disease control (Kishore and Pande 2005, Kalaskar et al. 2012). More recently, flaxseed oil has been found to be effective in the control of peanut rust by adversely affecting the uredospore germ tube length and by completely suppressing appressorium formation, which is essential for pathogen (*P. arachidis*) to form an infection peg to pass through the stomatal aperture and infect the host tissue (Chen and Ko 2014). All such measures may be more economical, eco-friendly, and useful in improved control of the rust with lesser dependence on fungicides particularly for the resource-poor farmers in less developed countries.

Sclerotium STEM ROT

SYMPTOMS

All the aboveground and underground plant parts can be affected. Several kinds of symptoms of the disease become visible depending upon the stage of plant growth, but the stem rot is more common. Usually, the disease appears more frequently as the plant approaches maturity. Infection may take place on the stem just above the soil surface or at the foot of the plant 1–2.5 cm below the ground level. In the beginning, the symptoms become visible in the form of a deep-brown lesion around the main stem at the soil level. It may occur on the stem below the soil surface under dry conditions or above the ground in wet weather. Soon after, the lesion becomes covered with white radiating mycelium that encircles the affected portion of the stem. The distinct rot occurs beneath the fungal weft leading to wilt like symptoms characterized by yellowing and browning of the foliage that show drooping while remaining attached on the plants. Such plants remain upright in the row under field conditions. The entire plant or one or two branches may be killed. Death of the aboveground portion of plant sometimes takes place very rapidly, particularly in extremely hot weather. Coarse white strands of the pathogenic fungus growing in a fan-shaped pattern may be present on the surface of the affected plant parts or on the soil surface or leaf litter adjacent to affected plants. Later, brown-colored mustard seed–like sclerotia are often noticed intermingled with the fungal strands on affected plant parts or around the affected plants on soil surface (Figure 2.7). This facilitates spread of the disease on plants sown in rows, and thus, *the row effect* becomes evident under field conditions.

GEOGRAPHICAL DISTRIBUTION AND LOSSES

The *Sclerotium* stem rot or SSR of peanuts is variously named as southern blight, sclerotium wilt, sclerotium blight, white mold, crown rot, foot rot, sclerotial disease, and sclerosteosis. The disease occurs throughout peanut-growing areas of the world in the tropics and in warmer parts of the temperate zones (Kolte 1984, Subrahmanyam et al. 1991, Momotaz et al. 2009, Shakil and Noor 2012). It is the most important disease on peanuts in India, Israel, and the southeastern United States, and the average annual peanut crop losses due to SSR are usually in the range of 25%–27% in India (Kolte 1984), Israel (Bowen et al. 1992), the United States (Damicone and Melouk 2009, Thiessen and Woodward 2012), and Australia (Middleton 1980). The disease incidence in the farmer's fields may be in the range of 0%–60%, and yield losses usually do not exceed 25% but may be as great as 80%. The losses are usually greater than what are apparent from field observations, as peg decay severs many nuts from the plant and they are left in the soil at harvest. The disease occurs in distinct foci as it spreads to adjacent plants in the row, and the number of disease foci is reported to be linearly related to yield loss in peanuts as reported by Rodriguez-Kabana et al. (1975). However, yield loss models from data from selected fungicide treatments indicate that the loss caused by SSR at low disease incidence may be proportionately greater than the yield loss at higher disease incidence and indicate that the relationship between SSR incidence and peanut yields may be nonlinear (Bowen et al. 1992). The SSR is, however, becoming a greater threat to irrigated peanut crop

FIGURE 2.7 SSR of peanuts. Note the mycelial growth and small mustard seedlike bodies of the fungus on rotted parts of the peanut plant.

in tropical countries. The temperature relation in the development of the disease and growth of the fungus seems to be the limiting factor in geographical distribution of the disease. More recently, the disease has spread in a more severe form over 80% in the heavily infested fields in the main peanut production regions of Henan Province in the P.R. of China (Xu et al. 2011) and to the extent of 25% in central Vietnam (Le 2011), and its prevalence is increasing in Indonesia (Taufiq et al. 2007).

Pathogen: Asexual Anamorph Stage (*S. rolfsii* Sacc); Sexual Stage (Teleomorph, *A. rolfsii* (Curzi) Tu & Kimbrough)

Classification
Kingdom: Fungi
Phylum: Basidiomycota
Subphylum: Agaricomycotina
Class: Agaricomycetes
Order: Atheliales
Family: Atheliaceae
Genus: *Athelia*
Species: *rolfsii* (Curzi) C.C. Tu & Kimbr.

S. rolfsii does not produce conidia and is a deuteromycete in the group mycelia sterilia. The mycelium is septate and hyaline with conspicuous branching at right angles. A bud-like structure forms at the growing tip. It continues to grow and gives rise to a branch. The main branch hyphae are relatively large (5–9 μ in diameter) compared to many fungi that are more typically with hyphal diameter of 2–4 μ. The well-developed mycelium, which forms the main vegetative body of the fungus, is in cord-like strands and grows as a creeping mycelium. The hyphae show the presence of clamps. Many a time, a number of hyphae are found to anastomose among them. Clamp connections in the form of forks and hooks or H-like in shape can be noticed. The young growing mycelial mass on the host surface, as well as on the medium, is generally snow-white with a silky luster.

Smaller-diameter (2–4 μ) hyphal cells called *feeding branches* arise from the main branches and penetrate the plant tissue. The well-developed mycelium growing in strands at its tips initiates the sclerotial formation in about 6–12 days, either on a host surface or on a medium. Sclerotia arise singly or quite close to each other, so as to assume a cluster or group at the tips of the growing mycelium. Initially, the sclerotia are whitish, but later they become yellowish white. A water drop is given out from the sclerotium during the course of its development, which on drying becomes dark brown and forms an outer coating. As the sclerotia age, they become dark brown to chocolate brown in color and are like mustard seed in appearance and size. The outer dark-brown ring of sclerotium consists of thickened cells followed by the formation of pseudoparenchymatous tissue.

The development of the perfect (sexual) stage is very rarely found in nature, and it has been the common practice to use the name of the sclerotial stage. Interestingly, the fungus is known to occur in its perfect stage as *A. rolfsii* causing the collar rot phase of the disease in the Marathwada Region of Maharashtra in India (Kadam et al. 2011). The development of perfect stage consists of the formation of a structure called a basidium in which meiosis occurs. Four haploid basidiospores are produced at the tips of small structures on the basidium called as sterigmata. *A. rolfsii* produces basidia in an unprotected layer (hymenium) that develops under humid conditions at the margins of lesions. Hymenial production consists of aerial, button-like loosely formed white growth, and it measures 0.1–0.75 mm. Basidia are club shaped and slender. They measure 20–25 μ × 4–6 μ. The sterigmata are shorter but more stout and measure 1.5–3 μ and vary from 1 to 4 in number. Basidiospores are hyaline, smooth, obclavate and measure 4–4.8 μ × 2–3.5 μ; when mature, the basidiospores are forcibly discharged.

Epidemiology and Disease Cycle

The pathogen is a facultative parasite found on a wide range of soils. The fungus can affect plants in about 100 families representing about 500 plant species. Most of these plants are dicotyledons comprising mainly composites and legumes. Members of the Gramineae family are regarded less susceptible.

Several investigators have compared the morphology and host range of different isolates from various hosts and concluded that the differences are small. Geographical variability among *S. rolfsii* populations is demonstrated. The importance of variability has been realized in the light of the discovery of pentachloronitobenzene (PCNB) tolerant strains of *S. rolfsii* isolated from a Texas peanut field in 1985 (Nelin 1992, Shim et al. 1998, Sharma and Singh 2002). There exists a variability among isolates of *S. rolfsii* in their ability to produce oxalic acid, which is correlated with the mortality of the seedlings; the lower the oxalic acid production by the isolate, less pathogenic it is in causing the seedling mortality (Palaiah and Adiver 2004). The variability in population structure of *S. rolfsii* isolates from peanut fields in Japan reveals the presence of four mycelial compatibility groups (MCGs). These are MCGs A, B, C, and D, and most isolates of the same MCG are proven to be clonal (Okabe and Matsumoto 2000). Similarly, in Vietnam, three distinct groups among isolates of *S. rolfsii* have been identified to display their diversity of genetic and phenotypic traits in mycelial compatibility, growth rate, and sclerotial characteristics (Le et al. 2012a). Based on internal transcribed spacer (ITS) ribosomal DNA sequence analyses, three distinct groups have been identified among field isolates of *S. rolfsii* from peanuts, tomatoes, and *taros* in India, and these show differences in aggressiveness, suggesting thereby that the most aggressive isolates be used in a consortium for the development of sick field plots for screening of resistance (Bagwan 2011a).

S. rolfsii survives through sclerotia or mycelium on plant debris in the top 6 cm of soil. It has a high oxygen demand and soil aeration affects survival; hence, viable sclerotia are found at greater depths in lighter sandy soil compared to heavier soils. Mycelia from germinated sclerotia first colonize dead or senescent plant tissue on the soil surface before infecting healthy plants. Colonized food bases supply energy for invasion of living plant organs and bridge distances between germinated sclerotia and the host. The fungus is most active at the soil surface and a mat of hyphae is formed over the basal portion of peanut. It clings to the epidermis but does not apparently penetrate living cells. It grows into host cells killed in advance by metabolites such as oxalic acid and/or pectolytic or cellulolytic enzymes produced by the fungus as it grows over the host tissues (Kolte 1997). Warm (30°C–35°C) aerated conditions in the upper few centimeters of the soil favor the development of the fungus. Generally, temperatures remaining at 29°C–35°C during most of the day and seldom dropping below 23°C during the night are more favorable for disease development. A soil moisture content of 40%–50% of water-holding capacity is optimum for disease development.

Sclerotial germination throughout the growing season is favored by low humidity and particularly by alternate drying and wetting (Porter et al. 1984). The accumulation of dried leaves around the peanut plants following defoliation due to leaf spots, insect damage, and drought creates optimum conditions for disease development by providing a food base for *S. rolfsii*. Phosphorus fertilization is reported to increase disease incidence but this is decreased by potassium. Peanut plants treated with benomyl often exhibit greater SSR problems, primarily because benomyl reduces the soil population of antagonistic *Trichoderma*. Rodriguez-Kabana et al. (1979) reported that 1,2-dibromo-3-chloropropane (DBCP) is stimulatory for the germination of dormant sclerotia in soil by producing mycelium that can use available organic debris to produce new sclerotia. This therefore leads to the production of more sclerotia leading to a higher inoculum density in DBCP-treated soil, subsequently resulting in higher incidence of the disease. Peanut plants infected by *S. rolfsii* emit a blend of organic compounds such as methyl salicylate and linalool, which appear to be inhibitory to the growth of *S. rolfsii* in vitro, suggesting thereby that the emission of these compounds by infected plants may constitute a direct defense against *S. rolfsii* (Cardoza et al. 2002). It has been shown that *S. rolfsii* can be seed borne in the peanut. But the possibility of survival and subsequent infection from this source is greatly overshadowed by soilborne inoculum.

Disease Management

Host Plant Resistance

Significant negative correlation between SSR incidence and yield per unit area reveals the importance of the development of SSR-resistant cultivars in obtaining higher yield under pathogen stress conditions (Krishnakanth et al. 2005). Several workers have attempted to spot out resistant varieties of peanuts to SSR. Peanuts of the bunch type are killed outright, but plants of runner peanuts are not killed by the disease, since portions away from the attack on runner peanuts are usually supported by the adventitious root system. It is probably because of this behavior that runner peanuts have been reported as less susceptible (Garren 1966).

Although the development of resistant cultivars to *S. rolfsii* is rather difficult, in recent years, a few peanut cultivars that have been registered to be resistant to tomato spotted wilt disease (caused by tomato spotted wilt virus [TSWV]) are also known to be resistant to SSR (Branch and Brenneman 2009). Such cultivars are Georgia-07W, Georgia-03L, and AP-3 (Gorbet 2007, Branch and Brenneman 2008, 2009), Florida-07 (Gorbet and Tillman 2009), DP-1 (Gorbet and Tillman 2008), Andru II (Gorbet et al. 2006a), and Carver (Gorbet et al. 2006b). Ashok et al. (2004a,b) from India confirmed through artificial infection studies that 11 genotypes, 3 belonging to Virginia bunch (*A. hypogaea* var. *hypogaea*; TCG 1525, PI 269710, NCAc 38), 2 belonging to Virginia runner (*A. hypogaea* var. *hypogaea*; Haryanawadi, ND 8-2), 4 belonging to *vulgaris* (*A. hypogaea* var.

fastigiata; SS 34, VRR 472, Tai son, and PI 1268559), and 2 belonging to Spanish (*A. hypogaea* var. *vulgaris*; NCAc18019 and RR5290), are highly resistant to SSR.

Moderate resistance to SSR is reported in registered peanut cultivars such as Tamrun OL 02 (Simpson et al. 2006), Phule Unap (Patil et al. 2005), Dh 8 (Krishnakanth et al. 2003, 2005), C-99R (Gorbet and Shokes 2002a), Florida MDR 98 (Gorbet and Shokes 2002b), GG-11 and GG-13 (Rakholiya and Jadeja 2010), R9227 (Pujar et al. 2011), JL-365 (Thakare et al. 2007), and Local 235 (Abd-El-Moneem et al. 2003). Root exudates of the moderately resistant *Local 235* cultivar from Egypt are characterized by the presence of arabinose, lysine, and tryptophan (Abd-El-Moneem et al. 2003). Host genotype and biocontrol treatment combinations have established that the biological control using *T. harzianum* is more convenient, less costly, eco-friendly, and more effective in partially resistant genotypes (Krishnakanth et al. 2003). The same conclusion should be valid when chemical control using propiconazole is combined with the partial resistance of the peanut cultivars.

Chemical Control

The application of sterol biosynthesis inhibitors (SBI) also referred to as DMIs and quinone outside inhibitor (QoI) fungicides are now known to be more effective than previously recommended fungicides such as quintozene and carboxin for the control of SSR (Johnson and Subramanyam 2000, Besler et al. 2006, Grichar et al. 2010, Augusto and Brenneman 2011, 2012). The SBI triazole systemic fungicides, namely, diniconazole, propiconazole, tebuconazole (Minton et al. 1990, 1991, Culbreath et al. 1992, 2009, Adiver and Anahosur 1995), cyproconazole (Culbreath et al. 1992, Adiver and Anahosur 1995), hexaconazole (Johnson et al. 2007b), difenoconazole (Cilliers et al. 2003), and prothioconazole (Augusto and Brenneman 2012), and the QoI strobilurin fungicides, that is, azoxystrobin (Johnson and Subramanyam 2000, Rideout et al. 2002, Bowen et al. 2006, Sconyears et al. 2007, Hagan et al. 2010) and pyraclostrobin (Hagan et al. 2007, Grichar et al. 2010), have been proved to give very efficient and effective control of SSR of peanuts. The QoI fungicides should be applied preventively or as early as possible in the disease cycle as these are effective in inhibiting early mycelial growth. Once the fungus is growing inside the plant tissues, QoI fungicides have little or no effect. The strobilurins act at one specific site in the cytochrome system in the fungus and inhibit the mitochondrial respiration and are in the same cross-resistance group (same mode of action) and belong to Fungicide Resistance Action Committee (FRAC) Code 11. Similarly, SBI fungicides belong to FRAC Code 3, which include triazoles. Although these fungicides pose less risk to human health and/or environment than alternative pesticides, they appear to be vulnerable to rapid buildup of resistance in fungal population, and hence, their use must be managed carefully to avoid appearance of fungicide resistance (Vincelli 2002). Mixing triazole (SBI fungicides) or strobilurins (QoI fungicides) with other fungicides that have different mode of actions is desirable in minimizing the development of fungicide resistance. Fontelis (penthiopyrad), the new pyrazole carboxamide, is recently established to show excellent control of SSR of peanuts in the United States and may complement current SBI and QoI fungicides. Fontelis would be an acceptable rotation partner for resistance management purposes with SBI triazole (Group 3) and strobilurin (Group 11) fungicides (Culbreath et al. 2009, Hagan 2012) in the management of SSR of peanuts. These chemicals are most frequently applied as granules at the pegging stage but may also be directed as sprays (Hagan et al. 1991). Flutolanil, a benzanilide systemic compound, has been found to show protective and curative effect for the control of SSR of peanuts in the United States (Csinos 1987, Timper et al. 2001) and Nicaragua (Augusto et al. 2010c). Seed treatment with triadimenol, another systemic triazole-type fungicide, or insecticide chlorpyrifos or carboxin, or ipconazole or azoxystrobin is also reported to be effective in controlling the SSR in the seedling stage of peanuts (Bowen et al. 1992, Rodriguez-Kabana and Kokalis-Burelle 1997, Rakholiya and Jadeja 2010, Akgul et al. 2011, Rakholiya et al. 2012).

Presowing treatment of susceptible peanut seeds for 24 h using 10^{-4} to 10^{-7} M dilute solutions of four growth-regulating chemicals, that is, indoleacetic acid, Cycocel (chlormequat), 2,4-dichloroacetic acid, and 2,4,5-trichloroacetic acid, significantly inhibits the development of *sclerotium* blight symptoms and reduces mortality in 2-week-old plants inoculated with the

S. rolfsii (Chowdhury 2003). Chitosan, a deacylated product of chitin, when used as seed treatment at 0.05%–1%, gives effective control of collar rot phase of the disease caused by *S. rolfsii* (Chowdhury 2002). Peanut plants raised from seeds treated with the aforementioned growth-regulating chemicals and chitosan show increased activity of PO and PPO enzymes producing more phenols and proteins. A combination of soil applications of insecticide and fungicide such as aldicarb + flutolanil (Minton et al. 1991) and chlorpyrifos + quintozene (Hagan et al. 1988) has been found to be more effective than either alone in the control of SSR.

Cultural Control

Deep burial of surface organic matter and crop debris by plowing it to a depth of 8–10 in. in soil during land preparation can improve the yield of peanuts by more than 50%, largely due to elimination of the food base of the pathogen and subsequent reduction of disease incidence (Kolte 1984, Desai and Bagwan 2005). The nondirting of peanuts during cultivation (as an interculture operation) likewise has been shown to reduce losses from SSR in the peanut crop, and a similar effect can be achieved by planting the peanut on slightly raised beds.

Since the pathogen is omnivorous, there is little chance of any control of the disease by following crop rotation. However, inoculum buildup of the pathogen in soil can be brought under control by rotating peanuts with less susceptible crop plants belonging to Gramineae family. For example, when maize precedes the peanut crops and the crops are under conservation tillage, population of biocontrol agents becomes higher, which then appears to be an important practice in the SSR management strategies as per the investigations carried out in Argentina (Vargas Gil et al. 2008). Crop rotation of maize, pearl millet, sorghum, garlic, and onion with peanut may be useful for the management of SSR of peanuts and also for the reduction of soil population of *S. rolfsii* due to the presence of certain antifungal compounds in the root exudates of these crops (Bagwan 2010, Vinod Kumar et al. 2012). The SSR incidence can be reduced to 62% with increase in pod yield by 15.5% when onions precede peanuts in the crop rotation (Zeidan et al. 1986).

Nondirting cultivation in combination with minimizing defoliation due to the leaf spot control can bring about significant control of SSR of peanuts in the subsequently planted crop. Soil amendment with basal application of gypsum at 500 kg/ha (additional to the normal practice of gypsum application at flowering) followed by neem cake at 150 kg/ha has been found to be useful in reducing the SSR of peanuts by 31%–39%, resulting in increase in pod yield by 200–260 kg/ha (Johnson et al. 2003). Among different forms of nitrogen and potassium fertilizers, application of calcium ammonium nitrate results in minimizing the incidence of SSR (Johnson et al. 2007b). Interaction of fungicide application timing and postspray irrigation is significant for SSR control and peanut yield. Applying fungicides at night when the leaves are folded and using irrigation water after fungicidal spray have both been shown to increase deposition of fungicides in the lower plant canopy, which subsequently improves control of SSR (Augusto and Brenneman 2011). Mulching the soil surface with wheat straw (80%–90% soil coverage) helps in restricting the increase in the SSR incidence, despite the reported increase in the density of *S. rolfsii* inoculum in such soils (Ferguson and Shew 2001).

Soil solarization is a nonchemical method for controlling the disease by means of solar heating of the soil. In this method, naturally infested soil when mulched for 6 weeks during July–August with transparent polyethylene (TPE) sheets raises the temperature of soil to 40°C–53°C, enabling killing or inactivation of the fungus. When such a field area is sown with peanuts, the following spring, the disease in the spring-planted crop is kept to minimum level, giving 52% more yield compared with the crop raised on untreated plots. This method has become more useful in Israel (Grinstein et al. 1979) and can be used in most tropical and subtropical countries including India (Reddy et al. 2007a,b).

Biological Control

Biocontrol agents particularly the fungi (*T. harzianum*, *Trichoderma viride*) and bacteria (*B. subtilis*, *Ps. fluorescens*, *Ps. aeruginosa*, *Ps. chlororaphis*) are seen to be an alternative and viable option for the management of SSR of peanuts as these have been found antagonistic to the growth of

S. rolfsii reducing its inoculum potential (Biswas and Sen 2000, Ray and Mukherjee 2002, Abd-Alla et al. 2003, Sahu and Senapati 2003, Desai et al. 2004, Pal et al. 2004, Saralamma and Reddy 2004, Abd-Allah 2005, Kishore et al. 2005c, Saralamma and Reddy 2005, Bagwan 2011b, Sharma et al. 2012). But the newly introduced biocontrol agents should be able to survive in the new ecological niche (Podile et al. 2002). For this purpose, low-cost local agricultural waste products such as wheat bran; oil cakes like mustard cake, castor cake, and neem cake; and FYM or compost can be used as the substrate for supporting the growth of effective antagonists as well as for retention of appropriate population ($\times 10^8$ cfu/g) of antagonists under field conditions (Vikram and Hamzehzarghani 2001, Nandagopal and Ghewande 2004, Bhagat and Pan 2007, Thiruvudainambi et al. 2010). Thus, combined soil application of 2.5 kg *T. viride* + 6 kg compost + 500 kg neem cake/ha gives effective control of SSR (Dandnaik et al. 2006), and oil cakes (mahua cake, neem cake, pungam cake) in combination with *T. viride* or *Ps. fluorescens*, each at 5 kg/ha of soil, give best degree of SSR control in peanuts (Varadharajan et al. 2006). Soil application of *T. harzianum* or *T. viride* in combination with thiophanate methyl + neem cake or with wheat bran saw dust + carboxin results in sustaining the antagonist population in soil giving better control of SSR (Patibanda et al. 2002, Saralamma and Reddy 2005). Similarly, lowest SSR incidence is reported when soil application of *Ps. fluorescens* is combined with FYM and tryptophan (Johnson et al. 2008a,b). A combined application of *Rhizobium* and *T. harzianum* is also beneficial in reducing the incidence of SSR (Ganesan et al. 2007). A diatomaceous earth with granules impregnated in a 10% molasses solution has been found suitable for the growth and delivery of *T. harzianum* to peanut fields. Granules coated with the growth of *T. harzianum* are applied 70 and 100 days after planting to the infested soil; this brings about significant disease control (Backman and Rodriguez-Kabana 1975, 1977). Besides antagonistic effects of *T. harzianum* or *T. viride* and that of *Ps. fluorescens*, plants treated with these biocontrol agents do show additional enhanced activity of plant defense-related enzymes, that is, *PO* and *polyphenol oxidase* (Varadharajan et al. 2006), whereas *Ps. aeruginosa* inhibits the plant cell wall–degrading enzymes (polygalacturonase and cellulose) of *S. rolfsii* and reduces the severity of peanut SSR (Kishore et al. 2005c). Some other strains of *Ps. chlororaphis* are, however, known to produce phenazines and lipopeptide surfactants—the thanamycin inhibiting the hyphal growth of *S. rolfsii* and suppressing the incidence of SSR (Le 2011, Le et al. 2012a,b). Some other strains of *Pseudomonas* sp. (strain BREN6) and *Bacillus* sp. (strain CHEP5) are capable of mobilizing infection-induced cellular defense responses (priming) in peanut plant. Inoculation of these strains increases the activity of PAL and PO enzymes, after challenge inoculation of peanut plants with *S. rolfsii*, and reduces the severity of the disease (Tonelli et al. 2011), indicating the induction of induced systemic resistance (ISR). Similar report on ISR against *S. rolfsii* infection has been made using fungal components of *S. rolfsii* in the form of fungal culture filtrate and the mycelial cell wall (Durgesh et al. 2010). Talc-based bioformulation mixture consisting of *Beauveria bassiana* (B2 strain) + *Ps. fluorescens* (strain TDK 1) + *Ps. fluorescens* (strain Pf1) amended with chitosan when applied through seed, soil, and foliar spray has been found to give effective control of *S. rolfsii* infection in peanuts (Senthilraja et al. 2010).

ISR in peanuts also becomes functional in response to use of certain arbuscular mycorrhizal fungi such as *Glomus caledonium* or *G. fasciculatum* when used alone or in combination with *Trichoderma* species for the control of SSR of peanuts (Ozgonen et al. 2010, Doley and Jite 2012).

In case of the seed treatment, *T. harzianum* or *T. viride* can be combined with compatible fungicides iprodione (Raihan et al. 2003, Manjula et al. 2004, Saralamma et al. 2004, Islam et al. 2005), thiophanate-methyl (Saralamma and Reddy 2005), difenoconazole (Cilliers et al. 2003), and chlorpyrifos (Rakholiya and Jadeja 2010) for the control of SSR in peanuts. Peanut seed treatment with *Ps.* cf. *monteilii* has been found to decrease the SSR incidence by 45%–66% (Rakh et al. 2011).

Water-soluble substances that occur naturally in the oat suppress the growth of *S. rolfsii*. Soil microorganisms that decompose oat residue also become more numerous and active during the decomposition process and suppress or destroy *S. rolfsii*. Based on this information, rotation of peanut with oat or rye to reduce the incidence of the disease is suggested (Webb 1971).

Effect of Plant Extract

Aqueous leaf and seed kernel extracts of neem (*A. indica*) (Ume-Kulsoom et al. 2001), aqueous leaf extract of *P. juliflora* and *Agave americana* (Kiran et al. 2006), and garlic, onion, pearl millet, sunflower, and sorghum at 5% (Vinod Kumar et al. 2009) show inhibitory effect on mycelial growth and sclerotial formation of *S. rolfsii*, indicating their potential uses in the management of SSR.

Aspergillus COLLAR ROT OR CROWN ROT

Symptoms

Collar rot or crown rot of peanut seedling is essentially a postemergence disease, but the preemergence phase where the seeds may rot and become covered with sooty black masses of spores can occur. On germination, the emerging hypocotyl is rapidly killed by the lesion below ground, resulting in rotting of the seedlings before their emergence from the soil. In the postemergence phase, crown rot is characterized in the field by wilting and death of seedlings accompanied by rotting of the hypocotyl. Under relatively moist conditions, accompanied by the high atmospheric humidity and high temperature prevailing during the monsoon period, hypocotyl rot itself is seen first as a yellowish-brown lesion that extends into the plant tissue, and affected collar region becomes shredded with a lapse of time and shows profusely sporulating, black growth of the causal fungus. Eventually, the hypocotyl becomes blackened and rotten. Most affected plants die within 30 days of planting, which leads to *patchy* crop stand. As plants develop woody stems and taproots, the disease is less likely to occur. However, later in the season, individual branches or entire plants may develop similar symptoms. Splitting the crown and taproot of affected plants reveals an internal discoloration of the vascular system that is dark gray in color. The dried branches are readily detached from the disintegrated collar region and are blown away by wind. Under dry conditions, the lesion in the collar region remains restricted, bringing about slow wilting and death of shoots in the proximity of lesions, and the rest of the shoots survive. If plants escape early infection, that is, immediate postemergence phase, the plants reach maturity, and crown rot symptom may develop. Occasionally, rotting is continued to the lower portion of the main root, in which case the plants produce adventitious roots above the diseased area. Such plants seldom thrive and usually die during dry weather (Kolte 1997).

Geographical Distribution and Losses

Collar rot or crown of peanuts caused by *A. niger* van Tieghem was perhaps first reported from Sumatra in 1925 (Jochem 1926). It was then reported from all over the peanut-growing countries in the world (Kolte 1984, Cantonwine et al. 2011). Since the disease causes considerable seedling mortality in the early stages of crop growth, crop yield is directly affected by reduction in the stand of the crop. The disease may cause an average 5% loss in yield but in some areas it may cause as high as a 40% loss. Collar rot is a more serious problem in sandy soil (Gibson 1953, Chohan 1965). In Punjab (India), the mortality losses of plants due to the disease may amount to 40%–50% (Aulakh and Sandhu 1970). Similarly, Ghewande et al. (2002) reported that losses in terms of mortality of plants due to collar rot range from 28% to 50%. Plant stand losses as high as two plants due to the disease per meter of planted row have been recorded in the North Territory (Queensland) in Australia.

A serious incidence of the disease was reported in Australia in 1951 (Morwood 1953) and in the United States, resulting in serious stand deterioration, in Georgia in 1961 (Jackson 1962), in Texas in 1962 and 1963 (Ashworth et al. 1964), and in New Mexico in 1965 (Hsi 1966). With increasing interest in producing organic peanuts in traditional peanut production areas in the southeastern United States, postemergence phase of collar rot (*A. niger*) has become the major obstacle regardless of the peanut cultivar being used in organic peanut production, though the disease along with other seedling diseases could be controlled by standard chemical seed treatments (Ruark and Shew 2010, Cantonwine et al. 2011). Infection occurring within 50 days of sowing from untreated seed causes serious losses and can kill up to 40% of the plant stand.

Pathogen

The disease is known to be caused by the fungus *A. niger* van Tieghem. (In some instances, *Aspergillus pulverulentus* (McAlpine) Thom has been found as the cause of the disease.)

Classification
Domain: Eukaryota
Kingdom: Fungi
Phylum: Ascomycota
Subphylum: Pezizomycotina
Class: Eurotiomycetes
Order: Eurotiales
Family: Trichocomaceae
Genus: *Aspergillus*
Species: *niger* van Tieghem

A. niger is a member of the genus *Aspergillus* that includes a set of fungi that are generally considered asexual, although perfect forms (forms that reproduce sexually) have been found. Conidial heads are globose and black and measure 700–800 μm in diameter; conidiophores have thick smooth walls and measure 1.5–3.00 mm × 15–20 μ and show colorless to brownish shades in the upper half, with sterigmata in two series; primary sterigmata vary with the strain and with the age of the conidial heads and measure 20–30 μ × 5–6 μ at the beginning of sporulation, but often reach 60–70 μ × 8–10 μ at maturity; secondary sterigmata are more uniform, ranging usually from 7 to 10 μ × 3 to 3.5 μ; conidia are globose at maturity, echinulated, somewhat variable in size, and mostly measure 4–5 μ in diameter. Sclerotia are produced in some strains and may dominate the colony character; sclerotia are globose to subglobose, 0.8–1.2 mm in diameter (Kolte 1984).

A. niger has a total genome size that ranges from 35.5 to 38.5 Mb and is composed of about 13,000 genes. Of these genes, about 8000–8500 genes have functional assignments. In addition, about 14,000 ORFs have been identified in the genome that could potentially encode a protein. The DNA sequence of *A. niger* consists approximately of 33.9 million base pairs. The possible function of 6500 genes could be established, which is only about 45% of its total gene count (Debets et al. 1990).

Epidemiology and Disease Cycle

A. niger is a saprophyte found in almost every type of tropical soil. It can tolerate low soil moisture and develops best at temperature between 30°C and 35°C. It can survive on seed or in the soil. It may be carried in or under the seed testa (Kolte 1984, Desai and Bagwan 2005). Seeds become infected during the last days of maturation in the soil and also during harvesting, shelling, and handling. Both soilborne and seed-borne inocula serve as primary sources of infection and adversely affect the seed germination (Mohapatra 2011). The fungus enters the host through a wound on the seed coat (testa) or through the stem, and the cotyledons usually act as the site for primary infection. It is interesting that conidia alone are not capable of causing infection of uninjured tissues. The presence of mycelium is essential for the infection of the uninjured tissue (Nema et al. 1955). Conidia can cause infection only when the testa of the seed is broken, and infection through conidia as the inoculum source occurs only to the extent of 25%, whereas the combination of mycelium and conidia can cause 100% infection through injured seeds. If the injury is extended to the cotyledons, the infection is so rapid that seedlings die before emergence. No spread of the disease in a particular crop season is seen. This is probably because of the absence of secondary infection and critical stages of infection period. A seed carrying infection gives rise to cotyledons with the development of the lesion, which in turn affects the hypocotyl or stem of the seedlings. The pathogen present in the soil infects either the cotyledons or the hypocotyls directly. What is usually observed is that peanut seedlings grow out of soil in such a way that their cotyledons remain slightly covered with a thin layer of soil. Even if they are not covered with the soil just after germination, they get covered immediately thereafter by the wind-blown soil. The soil-covered cotyledons of the peanut seedlings form a good substrate and are in proper environment for the growth of *A. niger* already present in the soil. Once the fungus is established on cotyledons, it grows into the collar region and causes collar rot of seedlings. In most of the cases, infection takes place within 30–35 days after planting. This period corresponds to maturity of the hypocotyl and shedding of the cotyledons.

The incidence of the disease is positively correlated with high soil inoculum levels and is more prevalent in fields continuously cropped with peanuts than in fields grown with nonhost crops. The main carryover from season to season is through plant debris in soil. Predisposition is a major factor in the development of *Aspergillus* crown rot. Adverse weather conditions, extreme fluctuations in soil moisture, poor seed quality, seedling damage from pesticides or fertilizer application, and any other factor that delays emergence are associated with the disease (Kolte 1997). Young plants are particularly susceptible especially if the seeds are planted too shallow (<3 cm) or too deep (>8 cm) and the soil is exceptionally wet or dry. High soil temperatures also increase the risk of infection. As plants mature and the soil cools, plants become less susceptible and the mortality rate declines. *A. niger* produces oxalic acid and pectinase (polygalacturonase) enzyme *in vitro* and *in vivo*. This indicates that oxalic acid and pectinase enzymes are involved in the pathogenesis, which is further substantiated by the fact that only virulent isolates of the fungus produce oxalic acid and avirulent ones are not capable of doing so. At a soil moisture level of 13%–16%, peanut seedlings are affected most by collar rot. Sandy or sandy-loam soil and amendment of soil with sulfur, FYM, and gypsum have been found to increase the incidence of collar rot.

Etiolated plants are more susceptible to the disease, and high light intensity has been found to render seedlings almost immune to infection (Ashworth et al. 1964). Lesions induced by drought stress and high temperatures predispose the plant to infection. This type of predisposition can occur in the field during the early period of crop growth by partially covering seedlings with hot sand (Ashworth et al. 1964, Kokalis-Burelle et al. 1997).

Disease Management

Host Resistance

Breeders have not actively selected and bred for peanut resistance to *A. niger* in cultivar development perhaps because the disease is successfully brought under control due to fungicidal seed treatment and only partial resistance has been reported in cultivated peanut germplasm. For example, genotypes, namely, EC 21115 (U-4-47-7), B-4, B-21, B-60, B-76, B-101, B-18L, Asiriya Mwitunde (Kolte 1984), N03081T (Bailey), Perry5 (Ruark and Shew 2010), J-11, and GG-2 (Gajera et al. 2013), have been found to be moderately or partially resistant to the disease. The runner types of peanut genotypes are less susceptible to collar or crown rot in comparison to bunch types (Wynne et al. 1991).

Chemical Control

Effects of Fungicides

Many seed-dressing fungicides are reported to be effective against collar rot of groundnut (Gangopadhyay et al. 1996, Karthikeyan 1996). Seed treatment with thiram at 0.5% or captan at 0.25% or Vitavax 200 (carboxin 37.5% + thiram 37.5%) at 0.4% has been found effective for the control of both the phases of the disease under field conditions almost all over the world (Purss 1960, Jackson 1964, Sidhu and Chohan 1969, Agnihotri and Sharma 1972, Rakholiya et al. 2012).

The better efficacy of thiram and carboxin is related to its fungitoxic ability and its influence in increasing the population of *T. viride* antagonistic to *A. niger* around the treated seed. Among the newer fungicides, azoxystrobin alone (Rideout et al. 2002, 2008) or Dynasty PD (a mixture of azoxystrobin + fludioxonil + metalaxyl-M) at 2.5 g/kg of peanut seed in dry form can be used, using equipment specifically designed to apply a dust seed treatment to peanut seed (Australian Govt: Australian Pesticides and Veterinary Medicine Authority Permit No.: Per 13513).

Effects of Nonconventional Chemicals

Copper hydroxide at 0.25 g/kg of peanut seeds could provide an option for growers in organic peanut production for the control of the disease (Tarekegn et al. 2007, Ruark and Shew 2010). Seed treatment with nonconventional inorganic salt chemicals such as barium sulfate, zinc sulfate, and zinc chloride has been found to be effective in the control of collar rot of peanuts (Dasgupta et al. 2000). Supplementing zinc ions as an antioxidant treatment during peanut seed germination has been found to be effective in controlling the hyper increase of reactive oxygen species (ROS) in certain peanut varieties (as GG-11 and GG-24), which then imparts host resistance to *A. niger* infection due to the formation of oligomeric protein of 110 kDa. It, therefore, reveals that control of ROS could control the *A. niger* infection in peanuts (Jajda and Thakkar 2012).

Cultural Control

Only sound, undamaged healthy seeds should be selected and treated with thiram or captan before sowing. The seeds should be sown in good soil moisture conditions, avoiding deep sowing (preferably not more than 2 in.) so that the emergence of seedlings is hastened and cotyledons come above the soil soon, and the pathogen is less liable to cause infection, thereby escaping the disease (Chohan and Kapoor 1967). During interculture operations, care should be taken to avoid injury to seedlings and deposition of soil particles on cotyledons. If cotyledons and the collar region remain exposed to aeration and light under field conditions, the symptoms of collar rot do not develop on peanut seedlings. This constitutes another important measure to control the disease, which can be achieved by ordinary hand hoe during the hoeing operation.

In India under the Punjab conditions, mixed cropping with short-stature crops like moth (*Phaseolus aconitifolius*) in alternate rows has been found useful in decreasing the incidence of collar rot caused by *A. niger*. Moth is a leguminous short-stature crop, and it does not compete with peanut, with regard to both nutrients and light (Chohan and Kapoor 1967).

A. niger fungal populations are rich in soils where continuous cultivation of peanut is a regular practice or peanut is included in the cropping system (Emmanuel et al. 2011). Therefore, it is appropriate that crop sequence of chickpea peanut or wheat peanut has been suggested for reducing the intensity of collar rot under Indian conditions (Chohan and Kapoor 1967). Planting of peanuts on land that has been kept fallow or cropped with grain sorghum the year before (or for longer periods) has been suggested under the U.S. conditions in New Mexico area (Hsi 1966). Irrigation of the fields within 28 days of sowing has been reported to be useful in protecting plants from severe damage in China (Lin 1982). Soil solarization with TPE tarping at 0.05 or 0.10 mm during April for 30 days is useful in reducing the crown rot incidence by 70%–95% (Reddy et al. 2007a,b).

Biological Control

Treatment of premoist peanut seed with talc and 0.5% carboxymethyl cellulose (CMC)-based formulation of antagonistic fungi *T. viride*, particularly strain 60, and *T. harzianum* at 3–4 g/kg seed alone and/or soil application of *T. harzianum* or *T. viride* at 25–62 kg/ha preferably in conjunction with organic amendment such as castor cake or neem cake or mustard cake at 500 kg/ha has been found to be effective in the management of collar rot disease (Raju and Murthy 2000, Rao and Sitaramaiah 2000, Sheela and Packiaraj 2000, Kishore et al. 2001c, 2006, Devi and Prasad 2009,

Mohapatra and Sahoo 2011, Gajera et al. 2011, 2013, Bagwan 2011b, Gajera and Vakharia 2012). *T. viride* is also reported to be tolerant or compatible with seed treatment fungicides like thiram and captan, and seed treatment with *T. viride* can be combined with reduced half dose of thiram or captan, which gives better control of the disease (Kishore et al. 2001c, Devi and Prasad 2009) than either of the *Trichoderma* or fungicidal seed treatment.

At least six plant growth-promoting rhizobacteria (PGPR) have been successfully investigated as biocontrol agents for the control of *Aspergillus* crown rot disease of peanuts. These are (1) fluorescent pseudomonads like *P. fluorescens* (Dilip et al. 1999, Haggag and Abo-Sadera 2000, Sheela and Packiaraj 2000, Dey et al. 2004, Anand and Kulothungan 2010), (2) *P. aeruginosa* strain GSE 18 (Achira et al. 2002, Kishore et al. 2005c, 2006), (3) *B. subtilis* strain G303 formulated for commercial seed treatment use as Kodiak FL (Ruark and Shew 2010, Cantonwine et al. 2011), (4) biofilm-producing *Paenibacillus polymyxa* (Haggag 2007, Haggag and Timmusk 2008), (5) other *Bacillus* species (Prabakaran and Ravimycin 2012, Yuttavanichakul et al. 2012), and (6) *Methylobacterium* sp. (Madhaiyan et al. 2006).

Out of those, Kodiak, a flowable formulation (Bayer Crop Science) that contains a select strain of *B. subtilis* G303 (not less than 5.5×10^{10} viable endospores), is worth mentioning. It is designed to use in combination with other registered seed-applied fungicides such as thiram, captan, and carboxin to extend window of protection. Within 4–8 h of planting, the bacterial endospores in Kodiak begin to reproduce, reaching a population of up to 1 million cells/g of root, and the actively growing bacteria surround the growing roots blocking the intrusion of pathogen, *A. niger*, into the plants. They also produce a chemical inhibitor that can slow the growth of the pathogen. Thus, *B. subtilis* or Kodiak is often effective and can be considered as standard bioagent seed treatment at 2.5 g/kg of seed for the control of crown rot and peanut stand establishment not only in intensive peanut crop production but also in organic peanut crop production (Ruark and Shew 2010, Cantonwine et al. 2011).

Postulated mechanism for better crown rot control due to the aforementioned biocontrol agents (*Trichoderma*, *Pseudomonas*, *Bacillus*, and *Methylobacterium* species) includes inhibition of growth of the pathogen by lytic enzymes (β-1,3-glucanase, chitinase, protease) produced by the antagonists and induction of systemic host resistance followed by promoted plant growth in terms of plant height, increased plant vigor, and efficient rhizosphere colonization and biofilm formation (Sailaja et al. 1998, Lashin et al. 1989, Haggag and Abo-Sadera 2000, Kishore et al. 2005d, 2006, Haggag et al. 2007, Haggag and Timmusk 2008, Devi and Prasad 2009, Anand and Kulothungam 2010, Gajera and Vakharia 2012, Yuttavanichakul et al. 2012). Interestingly, biocontrol of *Aspergillus* crown rot (dry rot) disease has been reported by using transconjugants obtained by the horizontal gene transfer from *P. fluorescens* to *Rhizobium*, and the percentage control efficacy has been found to be better due to the application of transconjugants (Ade and Gangawane 2010).

Effects of Plant Extracts

Seed treatment with *Calotropis procera* leaf extract at 10 mL/kg seed alone or in combination with *T. viride* at 4 g/kg seed has been reported to give significant control of *Aspergillus* collar rot of peanuts (Srinivas et al. 2005). Essential oils (Kishore et al. 2007) and Xenorhabdus metabolites (Vyas et al. 2005) are also reported to be inhibitory to *A.niger* pathogen causing the collar rot disease.

YELLOW MOLD AND AFLAROOT

Symptoms

Yellow Mold Phase

Because of the fungal growth and its secretion, the peanut seed disintegrates within 4–8 DAS, and the seed becomes yellowish brown in color. The testa loses its natural color, turns dark purple to black, and becomes brittle. The seeds become rancid, shrivelled, and turn leathery in texture (Figure 2.8). When the seeds are split open, the mycelium and sporulation of the fungus are clearly visible in the cavity

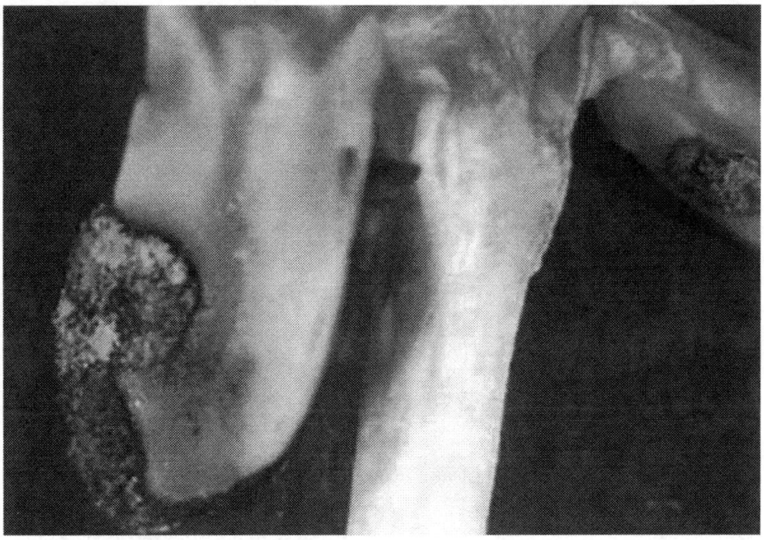

FIGURE 2.8 Yellow mold of peanuts caused by *A. flavus*.

between the cotyledons of the seeds. Under low soil moisture conditions, the decay is very rapid, since the activity of the fungus gets prolonged, owing to the delay in emergence of seedlings. Seedlings and ungerminated seeds shrivel to become a dried brown to black mass covered by yellow or green spores.

Aflaroot Phase

The germinating seeds, which escape the yellow mold phase, may show symptoms on the cotyledons. It is interesting that the hypocotyl is not affected. The pathogen is first seen on cotyledons, and from there it inhibits the growth of the plumule and the root. Cotyledon surfaces are covered with masses of yellow-green spores. The affected cotyledons show necrosis of the central tissues by forming reddish-brown lesions. Necrosis of the cotyledons terminates at or near the cotyledonary axis. The true leaves, which emerge from the affected seedlings, become reduced in size, with pointed tips, and show much variation in shape (Figure 2.9). The color of the affected

FIGURE 2.9 Aflaroot disease of peanuts caused by *A. flavus*.

leaves is yellowish green in comparison with the deep-green color of the leaves of healthy plants. The leaflets also show vein-clearing symptoms. The auxiliary branch arising from the side of affected cotyledons does not grow normally, but remains quiescent. If both the cotyledons are affected, plants remain quite stunted and show great variation in shape and size of the leaflets. The leaves remain thin with a shortened petiole. Leaves are rough to the touch and leathery and appear to be deficient in chlorophyll. When the affected plants are pulled out and examined, it becomes evident that the radicle is without secondary root development denoting the condition described as *aflaroot* by Chohan and Gupta (1968). Under field conditions, the diseased plants can be easily spotted because of their reduced growth and general chlorosis. Since the necrosis of the seedlings does not proceed to the hypocotyl, the aflaroot disease-affected seedlings continue to live till maturity.

Geographical Distribution and Losses

These diseases occur in most peanut-growing countries of the world particularly in Egypt, India, and Sudan (Mehan et al. 1991, Kolte 1997). Incidence of *Aspergillus* (*A. flavus*) contamination of peanut seeds favoring potential occurrence of seedling infection (aflaroot) is recently highlighted from Ethiopia (Mohamed and Chala 2014). Yellow mold was first observed in 1984 in a commercial peanut farm in South Texas in the United States (Subrahmanyam et al. 1987). Losses of 10%–20% in seedling emergence due to yellow mold phase and 5%–11% due to the aflaroot phase have been reported from India (Chohan and Gupta 1968).

Pathogen: *A. flavus* (Link) ex. Fries

Classification

Domain: Eukaryota
Kingdom: Fungi
Phylum: Ascomycota
Subphylum: Pezizomycotina
Class: Eurotiomycetes
Order: Eurotiales
Family: Trichocomaceae
Genus: *Aspergillus*
Species: *flavus* (Link) ex. Fries

The fungus is ubiquitous and saprophytic, and it grows rapidly on a variety of media. Conidial heads typically radiate measuring 500–600 μ in diameter, with the conidiophore coarsely rough, heavy walled, and usually less than 1 mm in length—but it may measure up to 2–2.5 mm—sterigmata are either uniseriate or biseriate with the two conditions rarely occurring in the same heads. Primary sterigmata measure 6–10 μ × 4–5.5 μ. Secondary sterigmata measure 6.5–10 μ × 3–5 μ. Uniseriate sterigmata are variable in size from 6.5 to 14 μ × 3 to 5.5 μ with conidium forming tips that are usually phialiform; conidia are globose to subglobose and inconspicuously echinulate and vary in diameter size from 3 to 6 μ, but most measure 3.5 to 4.5 μ. Sclerotia are produced in many strains and sometimes these dominate the colony characters. Sclerotia are globose to subglobose and red brown in color and measure about 400–700 μ. The toxigenic isolates produce more aflatoxin B and show stronger pathogenicity, whereas nontoxigenic isolates are poor producers of the aflatoxin (Tang et al. 2002). Maximum production of the aflatoxin is produced by toxigenic isolate at 25°C. It is reported that highly toxigenic isolates produce abundant sclerotia in culture media.

Epidemiology and Disease Cycle

The pathogen is seed borne as well as soilborne and survives in crop debris, and its ability to cause disease is related to strong saprophytic competitive ability. The pathogen is very active when soil moisture is below the field capacity and the atmospheric relative humidity is high. It can grow over a temperature range of 17°C–42°C and the optimum temperature for aflatoxin production is 25°C–35°C. The population of *A. flavus* propagules remains high in the top 5 cm of soil and gradually decreases with soil depth. Vertisol soils support a smaller population of *A. flavus* than the alfisols. Damage to pods by the lesser cornstalk borer, *Elasmopalpus lignosellus* (Zeller), exacerbates the disease (Kolte 1997). The development of infection is related to the production of aflatoxin. Nontoxigenic isolates of the fungus are not pathogenic to seedlings. However, no correlation between mycelial growth and aflatoxin production is observed (Lisker et al. 1993). The extent of yellow mold damage and aflatoxin is dependent on the environmental conditions, production, harvesting, and storage practices. Following harvest, further infections may develop with fungal growth covering the seed surface and invading the seed itself. A yellow or brown discoloration of pods and weight loss of pods result in subsequent losses (Satish Kumar and Popat 2010).

Disease Management

Since the fungus is a weak parasite, agronomic practices that favor rapid germination and vigorous growth of the seedlings will reduce the chances of *A. flavus* infection. Peanut germplasm lines possessing thick seed testa structure and protein characteristics can resist the infection by *A.flavus*. (Shan et al. 2006, Upadhyay et al. 2001a,b, Wang et al. 2010). Seed treatment with some of the fungicides, for example, carbendazim, captafol, mancozeb, or thiram, at 3–3.5 g/kg peanut seed has been found to manage the disease significantly under field conditions (Kolte 1984). The yellow mold and the possibility of controlling the aflaroot by the use of antagonistic isolates of *T. viride* and *T. harzianum* have been studied (Desai et al. 2000, Anjaiah et al. 2006, Bagwan 2011b); besides, there exists a potential of using certain isolates of plant growth-promoting fluorescent pseudomonads for the control of the infection caused by *A. flavus* and yield enhancement attributes in peanuts (Dey et al. 2004), whereas biocontrol of aflaroot disease has been reported by using transconjugants obtained by the horizontal gene transfer from *P. fluorescens* to *Rhizobium*, and the percentage control efficacy has been found to be better due to application of transconjugants (Ade and Gangawane 2010) and transgene over-expressing a tobacco β-1,3-glucanase in peanut (Sundaresha et al. 2010). Gamma-ray peanut mutant resistant to *A. flavus* infection has been reported (Azza et al. 2003).

CHARCOAL ROT

Symptoms

The symptoms of the disease appear in different phases. The development of water-soaked lesions on the hypocotyl near the soil surface is a characteristic symptom of this disease. After the hypocotyl is girdled, the seedling dies. Usually, the plants show typical symptoms when they approach maturity. Initial symptoms become visible in the form of the development of a red-brown water-soaked lesion on the stem just above the soil surface. Gradually, the lesion then becomes dark and spreads upward as well as downward, covering larger areas of the stem and root. The stem rot symptoms develop partially or completely by girdling of the stem by the lesion. If the stem is completely girdled by the lesion, the affected plants show wilting, followed by rapid colonization of the branches, which might result in defoliation. The plant then turns brown and subsequently dies.

Usually, rotting of the stem is associated with the rotting of the roots also. The root rot symptoms, independent of stem rot symptoms, appear rarely. If the root alone is affected, the taproot shows rotting, which becomes visible by a shredded appearance of the tissue. The dead tissue is covered with abundant minute black sclerotia, giving a charcoal or ashy-gray appearance to the tissue. Although not normally classified as a foliar pathogen, *M. phaseolina* does cause leaf spots (Gupta and Kolte 1982).

Geographical Distribution and Losses

Charcoal rot of peanuts has a wide geographical distribution and is especially found in tropical and subtropical countries with arid to semiarid climates in Africa, Asia, and North and South America. It is particularly important in Burma, Gambia, India, Israel, Kenya, Malawi, Mauritius, Nigeria, Senegal, Sudan, Syria, and Venezuela, but is of minor importance in the United States (Kolte 1997). All phases of the disease are economically important. Reduction in yield of the crop may be because of poor stand due to seedling rot or by killing of the plant at maturity. The percent incidence of mortality due to the disease may be as high as 65%–72% as reported from India and from Argentina. About 10% loss in yield due to the root rot phase of the disease has been reported from India and Palestine (Kolte 1997).

Pathogen: *M. phaseolina* (Tassi) Goid (syns. *M. phaseoli* (Maubl.) Ashby, *R. bataticola* (Taub.) Briton-Jones, *Sclerotium bataticola* (Taub.), and *Botryodiplodia phaseoli* (Maubl.) Thrium)

Classification
Kingdom: Fungi
Phylum: Ascomycota
Class: Dothideomycetes
Subclass: Incertae Sedis
Order: Botryosphaeriales
Family: Botryosphaeriaceae
Genus: *Macrophomina*
Species: *phaseolina* (Tassi) Goid

M. phaseolina is an anamorphic fungus in the ascomycete family Botryosphaeriaceae. It is normally being without an observable sexual stage, but many isolates have been shown to be anamorphs with sexual teleomorphs so they are generally grouped as mitosporic fungi (Coelomycetes) that have enclosed conidia. It is highly variable, with isolates differing in microsclerotial size and the presence or absence of pycnidia. Microsclerotia are of uniform texture and jet black in color and appear smooth and round to oblong or irregular. Across isolates, microsclerotia vary on size and shape and on different substrates. They are made up of anastomosing mycelial cells that are thick walled and dark brown, and several such cells compose an individual microsclerotium. The pycnidial stage is common on peanuts. Pycnidia are initially embedded in host tissue, then erumpent at maturity. They are 100–200 μm in diameter; dark to grayish, becoming black with age; globose or flattened globose; and membranous to subcarbonaceous with an inconspicuous or definite truncate ostiole. The pycnidia bear simple, rod-shaped conidiophores, 10–15 μm long. Conidia (14–33 × 6–12 μm) are single-celled hyaline, elliptical, or oval. Despite its wide host range and wide phenotypic diversity among isolates, the genus *Macrophomina* contains only one species *M. phaseolina* (Edraki and Banihashemi 2010). Molecular studies using species-specific probes and primers and RFLP and RAPD techniques show no variation among isolates from different hosts in restriction pattern of DNA fragments amplified by PCR of ITS region and also confirm that *M. phaseolina* constitutes a single species (Su et al. 2001, Babu et al. 2007). Sequencing of 92.3% of the genome of *M. phaseolina* has been done, and about 14,249 ORFs, that is, protein-encoding genes, are predicted and 9,934 validated by transcriptome (Islam et al. 2012). It has an abundance of oxidases, POs, and hydrolytic enzymes for degrading cell wall polysaccharides and lignocelluloses to penetrate into the host tissues.

Epidemiology and Disease Cycle

The causal fungus is both seed borne and soilborne. Mycelium in seed and mycelium and microsclerotia or pycnidia in plant debris in soil are primary sources of inoculum. The pathogen can be detected in the seed coat, cotyledons, and embryo of peanuts (Chakrabarty et al. 2005). Peanut germplasm stored for various durations at 4°C commonly shows 10%–29% infection of this pathogen. Infected seeds do not germinate or produce seedlings that die soon after emergence (Singh et al. 2003b). The microsclerotia can remain viable in dry soils for many years, but rapidly loose viability in wet soils. The microsclerotia are black and spherical to oblong in structure that are produced in the host tissue and released into the soil as the infected plant decays. These multicelled structures allow the persistence of the fungus under adverse conditions such as low soil nutrient levels and temperature above 30°C. Microsclerotial survival is greatly reduced in wet soils surviving no more than 7–8 weeks and mycelium no more than 7 days. Germination of the microsclerotia occurs throughout the growing season when temperatures are between 28°C and 35°C. Sclerotia are then stimulated to germinate by giving rise to the mycelium that grows in the direction of the host surface. A certain amount of vegetative growth on the host surface appears to be essential prior to penetration, and then penetration occurs through cotyledons during emergence and through rootlets. Peanut seedlings are infected more rapidly and severely at 29°C and 35°C than at 18°C or 24°C, with plumules being invaded more frequently than roots. The intact pods are invaded most rapidly at 26°C–32°C. Mature pods are invaded, whereas actively growing pods remain free of the disease. When dry pods are allowed to hydrate over a 6-day period at 26°C, 32°C, or 39°C, pods are penetrated by the fungus quite extensively. Microsclerotia germinate on the root surface and germ tubes form appressoria that penetrate the host epidermal cell walls by mechanical pressure and enzymatic digestion or through natural openings. The hyphae grow first intercellularly in the cortex and then intracellularly through the xylem colonizing the vascular tissue (Okwulehie and Okpara 2002). Once in the vascular tissue, *M. phaseolina* spreads through the taproot and lower stem of the plant producing microsclerotia that plug the vessels. The rate of infection increases with higher soil temperatures, and low soil moisture will further enhance disease severity. Hot, dry weather promotes infection and development of charcoal rot. The charcoal rot is a greater problem when the plant is under drought stress. It is reported that incidence of the disease is higher in shallow cultivated fields than in those planted at a depth of 9 in. Damaged pods and kernels during harvesting and shelling are liable to be affected more by *M. phaseolina*. The fungal growth in pods is increased by rain after harvest. The mechanical plugging of the xylem vessels by microsclerotia, phaseolinone toxin production, hydrolytic and lignocellulose enzymatic action, and mechanical pressure during penetration lead to disease development (Islam et al. 2012). The population of *M. phaseolina* in soil will increase when susceptible hosts are cropped in successive years and can be redistributed by tillage practices.

Disease Management

Host Plant Resistance

Some of the peanut lines possessing less susceptibility or tolerance to the disease have been identified. For example, Spanish bunch peanut (*A. hypogaea* subsp. *fastigiata* var. *vulgaris*) cultivar TG 38 developed by irradiating with 300 Gy gamma-ray F1 seeds of the cross Girnar 1 × TG 26 has been released as resistant cultivar to charcoal rot in India (Kale et al. 2007). Drought-, waterlogging-, and PStV-tolerant peanut cultivar Huayu 16 is also resistant to *M. phaseolina*, and the same is released for wider adaptability in northern China (Li and Qiu 2000).

Chemical Control

Seed treatment with captafol or captan (0.38%) or thiram (0.45%) or Rizolex T (50%) has been found effective in improving seed germination and seedling stand. This results in the reduction of the disease incidence, increasing the yield by 13%–23%. However, the effect of seed treatment does not persist for so long period as to check the development of the disease at maturity or on developing pods. Soil treatment with benomyl drench (0.1%) followed by quintozene drench (0.5%) or with Rizolex T (50%) during the growth period is helpful to control the disease at a later period of crop growth (Kolte 1997, El-Wakil and Ghonim 2000). If the seed colonization is <20%, seed treatment with a mixture of carbendazim and thiram (1:1) at 2 g ai/kg is used, routinely in plant quarantine laboratories to eliminate *M. phaseolina* (= *R. bataticola*) from peanut seeds, and seed samples having >20% colonization are rejected (El-Habbaa et al. 2002, Chakrabarty et al. 2005).

Cultural Control

Late sowing of the crop is suggested to reduce the incidence of the disease, but late sowing should not be done in areas where rosette is an important disease problem. Deficiency in soil moisture affects the physiology of peanut plants influencing the increase in the incidence of the disease that can, however, be reduced by irrigating the crop field (Okwulehie 2000, 2004, Chougule and Kore 2004). Balanced fertilization inclusive of trace elements such as copper, manganese, and zinc and insect pest control ensure good vigorous growth of the plants and help in the reduction of incidence of charcoal in peanuts. Application of gypsum at 150 kg/ha in severe cases results in a significant reduction of the disease (Kolte 1997).

Biological Control

The use of fungal antagonists *Trichoderma virens/Gliocladium virens* (Maheshwari et al. 2001, Christopher et al. 2008), *T. viride* (Malathi and Doraisamy 2004), *T. harzianum* (Malathi and Doraisamy 2004), *T. hamatum* (Vimala et al. 2000), and bacterial antagonist *B. sphaericus* (El-Shehaby and Morsy 2005) has been investigated to be useful in the control of charcoal of peanuts. Generally, seed treatment with the fungal antagonist at 4 g/kg of peanut seed in conjunction with soil application of the antagonist at 100 g/m^2 at the time of sowing and 30 DAS combined with FYM shows maximum reduction in the charcoal rot incidence (Christopher et al. 2008). Seed bacterization with talc-based formulation of PGPR such as *P. fluorescens* strain Pf1 or GRC strain (Meena et al. 2001b, Gupta et al. 2002, Shanmugam et al. 2002, 2003, Ramesh and Korikanthimath 2010) and *Bradyrhizobium* sp. (Deshwal et al. 2003) results in reduction of incidence of charcoal rot, making the PGPR as potential biocontrol agents for the control of the disease. With a view to understanding the broader aspects of control of the disease, the qualitative and quantitative aspects of rhizosphere mycoflora of peanuts as influenced by seed bacterization or seed-dressing fungicides, sprays of chemicals on the crop, and inorganic and organic soil amendments have been studied.

Effects of Plant Extracts

Cold-water extracts of *Allium sativum*, *Polyalthia longifolia*, and thyme have been proved to be effective in significantly reducing the incidence of charcoal of peanuts (El-Habbaa et al. 2002, Udhayakumar et al. 2008).

Sclerotinia BLIGHT

Symptoms

Sclerotinia species are causing the disease often referred to as stem rot, but the disease is truly a blight, characterized by sudden and serious damage to all aerial parts of the plant. Usually, the pegs are invaded first at the soil level, facilitating the colonization of the lateral branches. Light-tan to brown lesions demarcating the healthy and affected tissue appear on the lateral branches. The lesions then become dark brown, and shredding of the tissue becomes evident from an affected branch, and the fungus moves into and colonizes the main branch also. Leaves of such plants become chlorotic, turn brown, and wither, resulting in defoliation and death of the lateral branches or of the whole plant.

Pods of severely affected plants also show rotting. The taproot becomes necrotic and turns black in color. Abundant white fluffy mycelium appears on the soil surface in close proximity of the affected parts or debris in the field. Sclerotia of the fungus are also produced on the surface and within the affected branches, in the central portion of the taproot, on the pegs, on the surface of the pods, on the interface of the shell, and inside the seed.

Geographical Distribution and Losses

Occurrence of the disease is likely to be restricted only to such areas in temperate regions and under cool tropical conditions where the mean temperature drops below 25°C during the growing season of the crop often at intermediate altitude. The first report of occurrence of the disease caused by *Sclerotinia* sp. on peanuts was from Argentina in 1922 by Marchionatto (1922). Subsequently, the disease was reported from Australia, China, India, Israel, Japan, Australia, Mauritius, Taiwan, and the United States (Kolte 1984, Thiessen and Woodward 2012). The first reports of *Sclerotinia minor* causing the disease in the United States were made in Virginia in 1971 and then in North Carolina in 1972. Since its inception in the United States, the fungal disease has become widespread, having moved to Oklahoma, Texas, and New Mexico as well as becoming severe in Virginia and North Carolina. *S. minor* is more prevalent in most peanut-growing states in the United States, but *Sclerotinia sclerotiorum* has also been reported to be associated with the disease in Oklahoma (Wadsworth 1979), Georgia (Woodward et al. 2006), New Mexico (Sanogo and Puppala 2007), and Texas (Woodward et al. 2008) in the United States and in Argentina (Kolte 1984). Epidemic outbreaks of *Sclerotinia* blight of peanuts caused by *S. minor* have been of regular occurrence from 1994 to 1999 in Argentina (Marinelli et al. 2001).

Under low-temperature conditions, the disease can be quite destructive. Yield losses of up to 50% for peanut producers in the southwestern United States, particularly in Oklahoma, North Carolina, and Virginia, have been reported (Chenault et al. 2006). Pod yield losses have been correlated with aggravated disease incidence resulting from damage done to the plant. Midseason onset of a crown infection has the greatest impact on plant productivity, causing a severe decrease in pod yield and seed quality. Yield losses of near 80% in some areas have been reported in some areas in the United States. The requirement of calcium and intensive use of fungicides for disease control cause producers in the Virginia–North Carolina regions to have a higher production cost than most other growers in the United States (Partridge et al. 2006). Consequently, peanut crop production acreage in Virginia reduced from 23,473 ha in 2002 to only 6,880 ha in 2006.

Pathogen(s): *S. minor* Jagger and *S. sclerotiorum* (Lib.) de Bary

Classification

Kingdom: Fungi
Phylum: Ascomycotina
Class: Leotiomycetes
Subclass: Leotiomycetidae
Order: Helotiales
Family: Sclerotiniaceae
Genus: Sclerotinia
Species: *mino*r Jagger and *sclerotiorum* (Lib.) de Bary

Small sclerotia type, for example, *S. minor* Jagger, and large sclerotia type, for example, *S. sclerotiorum* (Lib.) de Bary, have been reported to cause the disease. Both these species are necrotrophic fungal pathogens in the phylum Ascomycota, the order Helotiales, and the family Sclerotiniaceae.

Both produce dense mat of mycelium on the surface of the host and on adjacent soil surfaces; dense white bodies then form within this fluffy white mass of mycelium. *S. minor* produces small (0.5–2 mm in diam.), rough angular sclerotia, while *S. sclerotiorum* produces large (2–10 mm diam.) smooth rounded sclerotia. Identification should be made based on a group of sclerotia from the same colony rather a single sclerotium. In general, *S. minor* sclerotia are more numerous, smaller, and more angular than the sclerotia of *S. sclerotiorum*. *S. minor* sclerotia usually germinate by producing a mass of fungal threads and seldom produce ascospores, but sclerotia of *S. sclerotiorum* can produce ascospores from apothecia.

Epidemiology and Disease Cycle

Overwintering sclerotia of commonly occurring *S. minor* in the soil provide the primary inoculum for the disease. They germinate myceliogenically, producing a mass of mycelium that directly penetrates peanuts and causes infection on the pegs or lateral branches near the soil initiating the disease (Faske et al. 2006, Smith et al. 2006). Low temperatures (10°C–25°C) and high soil moisture favor infection, and the disease becomes severe when there are more cold days in a growing season. Myceliogenic sclerotial germination of *S. minor* and growth as well as infection and colonization of peanut tissues are optimum at soil matric potential of −7.2 kPa and optimum temperature of 30°C with 95%–100% relative humidity (Smith et al. 2006). *Sclerotinia* blight becomes more severe as soil pH increases from 6.0 to 6.5. The presence of volatile substances (aldehydes, esters, and halogenated hydrocarbons) from moist undecomposed peanut plant tissues has been shown to initiate sclerotial germination of *S. minor.* Field indices of available moisture in the form of rainfall or relative humidity >95%, air or soil temperatures <29°C, plant growth, and density of foliar canopy are used for predicting outbreaks of *Sclerotinia* blight and the need for fungicide sprays in the United States. As the *Sclerotinia* blight develops in the peanut canopy, numerous sclerotia are produced in and on diseased tissues. Sclerotia are then shed into the soil and increase the inoculum density of the pathogen, which subsequently becomes available in future growing seasons. Sclerotia of *S. minor* are able to remain viable for as many as 8 years in the soil. Since *S. minor* seldom produces ascocarps and ascospores, the latter are unimportant in the epidemiology of *S. minor*–caused disease. It usually attacks its host root and stem at or near the soil line in contrast to the most common formation of apothecia and ascospores in case of *S. sclerotiorum.* Hence, it is not uncommon to see *S. sclerotiorum* infections on the foliar parts of the plants because ascospores become airborne to spread the fungus and cause disease throughout the field. Senescent flower parts are an ideal site for ascospores of *S. sclerotiorum* to colonize. From this tissue, the pathogen can quickly invade healthy leaves, stems, and nuts. It can be a more serious pathogen at the flowering stage of the crop growth (Thiessen and Woodward 2012). Interestingly, unlike *S. sclerotiorum*, *S. minor* does not need a food base of dead or senescing tissues to infect. It has the ability to infect stems and branches in contact with the soil with no food bases (Shew 2011). Pathogenic *Sclerotinia* species produce oxalic acid, termed as pathogenicity factor, which predisposes plants to infection. Godoy et al. (1990) showed that *S. sclerotiorum* mutants that are unable to produce oxalic acid cannot infect susceptible plants. This indicates that oxalic acid is a necessary pathogenicity factor for the development of the disease. Plants injured during interculture operation are predisposed to infection. Peanut plants sprayed with captafol or chlorothalonil (0.56–2.24 kg/ha) are affected more severely by the disease, possibly because plants sprayed with these chemicals favor more production of oxalic acid by *S. minor.*

Disease Management

Host Plant Resistance

In Cultivated Arachis

It is difficult to find a high degree of resistance to the disease, but differences in the degree of host reaction through screening for resistance can be noticed. In general, Spanish and Valencia genotypes because of their upright plant canopy tend to exhibit greater resistance to the disease than the dense spreading Virginia and runner genotypes. The increasing level of resistance exhibited by these erect genotypes appears to be a plant developmental trait escape and early-maturity mechanism (Faske et al. 2006, Damicone et al. 2010). Some of the peanut varieties/lines, for example, Tamspan 98 (Simpson et al. 2000), Tamrun (Damicone et al. 2010), TxAG-4, VA 93B (Cruickshank et al. 2002), GP-NC-WS 12 (Hollowell et al. 2003b), Tamrun 98 (Tx 901417), and Perry (N93112C), and advanced breeding line N92056C (Lemay et al. 2002) have been found to show moderate resistance or tolerance or less susceptibility to the disease, but none have allowed or have a potential for doing away with the use of fungicide sprays.

Molecular Breeding and Transgenic Peanuts for Sclerotinia *Blight Resistance*

Genetically modified peanut lines (1) N70-8-B, P53-28-B, and W73-27-B (Partridge-Telenko et al. 2011) and (2) N70, P39, and W171 (Chriscoe 2008) expressing oxalate oxidase (Oxox) gene from barley have been developed successfully to enhance the resistance to *Sclerotinia* blight through degradation of oxalic acid by Oxox enzyme to produce carbon dioxide and hydrogen peroxide preventing predisposition of peanut cells to infection caused by the pathogen. It is demonstrated that the transgene Oxox in the peanuts has the potential to eliminate costs of fungicide use, increasing profits and promoting a more environmentally sound strategy for *Sclerotinia* blight management (Livingstone et al. 2005). Similarly, peanut transgenic lines No. 654 and No. 487 transformed with chitinase and glucanase genes (Douglas 2004), and about 32 some other peanut transgenic peanut lines possessing a rice chitinase and/or an alfalfa glucanase antifungal gene (Chenault et al. 2005) have been developed for resistance to *Sclerotinia* blight that may be useful in traditional breeding and disease management strategies. Simple sequence repeat (SSR) primer has been identified as a molecular marker associated with resistance to *Sclerotinia* blight in peanuts. Thus, identification of the marker and development of PCR-based screening method is extremely useful to peanut breeders in screening germplasm collections and segregating populations as well as in pyramiding *S. minor* resistance with other desirable traits into superior peanut lines (Chenault et al. 2009).

Chemical Control

While the use of resistant cultivars remains a viable option for the management of *Sclerotinia* blight, the use of fungicides may be necessary in cases of severe infestation. Presently, a number of fungicides have been tested for use in the control of this disease (Bowen et al. 2000, Ryley et al. 2000, Smith et al. 2008). The two most preferred fungicides for use in the control of this disease are fluazinam (Omega 500 F from Syngenta) and boscalid (Endura 70 WG from BASF). These fungicides are effective against *Sclerotinia* blight when applied as preventive measures. Timing of the first spray is critical. Fields with a history of serious problems should be scouted carefully beginning when plants are within 6 in. of touching. Spray the crop when *Sclerotinia* blight is first observed or 60–70 days after planting (calendar program) or according to a *Sclerotinia* blight advisory. A weather-based *Sclerotinia* blight advisory can be used to time applications and prevent unnecessary fungicide applications. If the disease continues to spread, one or two more applications may be made at 3–4-week intervals or according to the advisory (Shew 2011). In general, two or three sprays of fluazinam at 0.75–1.00 kg ai/ha can reduce the disease incidence in the range of 56%–80% on partially resistant cultivars. Boscalid performs marginally better than fluazinam (Smith et al. 2008). Foliar application of procymidone (0.68–0.75 kg ai/ha) has been found effective in the control of the disease.

Cultural Control

Damaging levels of the disease can be prevented by rotating the peanuts with nonhost cereal crops such as corn, sorghum, or cotton, avoiding cool-season vegetables. However, this practice has shown limited effectiveness in controlling the *Sclerotinia* blight because the sclerotia are able to remain viable for as many as 4 years in the soil (Partridge et al. 2006, Shew 2011). In addition, many winter annual weed species serve as hosts for *S. minor* during winter fallow, potentially reducing the benefits of crop rotation (Hollowell et al. 2003a). Hence, it is imperative that sanitation by way of weed control be practiced as one of the disease management strategies. Planting density influences the disease incidence and severity and could become a useful aid in disease management depending on the cultivars used (Maas et al. 2006). To reduce the spread of sclerotia of infested farm from one location to another by contaminated soil and plant debris lodged in farm equipment and shoes, it is recommended that the infested items be removed from these sites and be immersed in 6% sodium hypochlorite (NaOCl) solution for 60 min. Sodium hypochlorite is proved to be lethal to bare sclerotia when immersed in its solution of 6% for 5 min or 3% for 10 min (Wilson et al. 2010).

Biological Control

Among the antagonistic fungi parasitizing sclerotia of *Sclerotinia* spp., *Coniothyrium minitans* has been identified to have great potential in the management of *Sclerotinia* blight of peanuts caused by apothecium-producing *S. sclerotiorum*. The effects have been shown to be long term (Partridge et al. 2006, Whipps et al. 2008). A spore suspension of *C. minitans* applied across field crops infected by *S. sclerotiorum* at the first appearance of apothecia results in a reduced population of sclerotia in soil at the end of a 7-year period even when susceptible crops are planted. However, less information is known about the effectiveness of *C. minitans* on *S. minor*. If *C. minitans* is active on *S. minor*, as has been found in some studies, then it may provide a way to reduce sclerotial numbers in peanut fields heavily infested by *S. minor*, thereby reducing losses. However, more studies are needed to confirm the effectiveness of *C. minitans* against *S. minor* causing *Sclerotinia* blight of peanuts. The possibility of some endophytic and epiphytic bacteria as potential bacterial antagonists against some soilborne pathogens of peanuts including *S. sclerotiorum* has been evaluated (Tonelli et al. 2010).

CYLINDROCLADIUM BLACK ROT

Symptoms

Under field conditions, the affected plants show chlorosis and wilting. Such plants also exhibit blighting of the leaf tips and margins. The lateral foliage is usually less affected than the erect primary branches. On artificial inoculation, circular brown spots of 0.5–1 mm diameter appear on leaves. The spots are surrounded by chlorotic halos that measure up to 2 mm in diameter. The hypocotyls and taproots become necrotic and black, with necrosis terminating at the soil level. The tips of the lateral roots and taproots are sloughed off, leaving short dark-brown to black fragmented stubs. Adventitious roots often develop on diseased plants at the soil level. Dark-brown to black and slightly sunken lesions appear on the pegs and pods, but the size of the lesions on pods is larger. The distinguishing feature of the disease is the presence of numerous orange-red perithecia at the base of the stems at the soil level or on the pegs and pods under the soil.

Geographical Distribution and Losses

The causal fungus is found in tropical and subtropical regions, and it is thought the fungus was introduced from Asia during the establishment of a tea plantation in coastal Georgia in the 1950s. Bell and Sobers (1966) first observed the occurrence of *Cylindrocladium* black rot (CBR) of peanuts in Georgia in 1965. Since then, this disease has been reported in all peanut-producing areas of the southeastern United States (Kucharek et al. 2000, Wheeler and Black 2005) and in Japan, India, and Australia (Kolte 1984). The disease has been a cause of major concern, particularly in Virginia and North Carolina because of its widespread occurrence and chronic threat to peanut production (Branch and Brenneman 2003). In other parts of the United States, outbreaks of CBR have remained static, and yield losses are generally of secondary importance to other soilborne diseases. The first report of its occurrence with 50% incidence from the Guangdong Province in China has been made in 2008 (Pan et al. 2009) followed by another first report of its occurrence from the Jiangxi Province in China in 2012 (Gai et al. 2012). Depending on the age of plant at the time of infection, the effect of the disease can be devastating. When the underground parts of the plants are destroyed due to the disease, the loss in pod yield and mature kernels may occur up to 50% and 30%, respectively.

Pathogen: Anamorph, *Cylindrocladium parasiticum* Crous, Wingfield & Alfenas (teleomorph, *Calonectria ilicicola* Boedijn & Reitsma)

Anamorph: *C. parasiticum* Crous, Wingfield & Alfenas (teleomorph, *C. ilicicola* Boedijn & Reitsma). It is an ascomycete in the Sordariomycetes group. The fungus is homothallic and produces orange-red subglobose to oval or obviate perithecia, 300–500 × 290–370 µm; asci are hyaline, clavate, long-stalked, and thin walled, contain eight ascospores, and measure 95–138 × 13–19 µm; ascospores are hyaline, granular, and fusoid to falcate, with one to three septa, slightly constricted at the medium septum, and measure 34–58 × 6.3–7.8 µm; conidia are cylindrical and hyaline, with one to three septa, are produced by apical budding, and measure 58–107 × 4–7 µm; conidiophores bearing stipes appear at right angles from the host.

The fungus grows well on potato dextrose agar and produces light, gray-to-white web-like aerial mycelium. It can produce conidia, ascospores, and microsclerotia in culture and infected plants. The microsclerotia vary in size with an average size of 52.7 × 88.4 µm. A selective medium named as sucrose-QT medium has been developed for isolation and inoculum quantification of *C. parasiticum* (Griffin 1977).

Epidemiology and Disease Cycle

The pathogen survives through microsclerotia produced in the roots of affected plants. They are produced abundantly in cortical tissues and *Rhizobium* nodules of infected peanut roots. After disintegration of the affected roots, the microsclerotia are released into soil and spread locally during cultivation and aqueous runoff. Soil movement on equipment can redistribute microsclerotia within and between fields. Plant debris blown into the air during harvesting and combining can be carried by wind to other fields. Birds have been implicated in disseminating the pathogen by ingesting infected peanuts and depositing the microsclerotia in droppings. Conidia and ascospores are quite susceptible to desiccation, and these cannot remain viable to play any effective role in the survival. The primary infection mainly appears through surviving microsclerotia in soil. Inoculum levels as low as 10 microsclerotia/g of soil have been reported to initiate epidemics. The number of observed infections on roots and the level of symptom expression by plants are directly proportional to microsclerotial densities in soil. Soil temperatures of 20°C–25°C and moisture levels near field capacity are most conducive to infection and rot of peanut roots by *C. parasiticum*. Randall-Schadel et al. (2001) confirmed seed transmission of the pathogen and CBR-infected seed could also be a source of primary inoculum to initiate disease epidemics. Infection cushions are formed on the epidermis and complete colonization of the cortex by the mycelium occurs. Epidermal cells beneath the infection cushion die and the necrosis of the surrounding cortical cell appears, suggesting the involvement of phytotoxins in pathogenesis. Fibrous roots emerging from the taproot of the peanut plant disrupt the protective periderm, which subsequently provides favorable courts of infection. Evidence for the occurrence of multiple root infection is reported, but it appears that only a portion of the multiple infection contributes appreciably to disease development. Since conidia are formed infrequently in nature, besides being less viable, the possibility of secondary infection through conidia is eliminated. Perithecia are formed on peanut stems after the establishment of infection in large quantities if adequate moisture is available. Mature ascospores subsequently develop and are discharged between 20°C and 30°C and maximally at 25°C more or less coinciding with vegetative growth. Ascospore formation and discharge appear to be controlled by day–night relative humidity fluctuations and can be dispersed by rain splash and runoff and appear to play a significant role in secondary disease spread within a growing season. All legumes are susceptible to *C. parasiticum*. The incidence of CBR on peanuts increases when soybean is included in the crop rotation sequence (Jordan et al. 2008). Nonleguminous crops such as cotton and tobacco are also known to be susceptible, but infection of these plants does not increase the inoculum in soil. The root-knot nematode species, *Meloidogyne hapla* and *M. arenaria*, have been reported to interact with the fungus enhancing the disease severity and incidence on both susceptible and resistant varieties of peanuts. For example, root-knot (*M. arenaria*)-resistant peanut genotype C724-19-15 can withstand the interaction effect of infection due to *C. parasiticum* and *M. arenaria* when both pathogens are inoculated simultaneously, but other root-knot-resistant peanut genotypes such as C724-19-25 and Georgia-02C show mortality under similar conditions (Dong et al. 2009). Because of its sexual stage and wide host range, there is a possibility of the development of physiological races of *C. parasiticum*. But currently *C. parasiticum* appears to consist of a genetically homogeneous population with mainly clonal reproduction or inbreeding contributing to the population genetic structure. This is because of absence of random mating due to homothallic nature of *C. parasiticum* as well as the clonality of the population (Wright et al. 2010).

Disease Management

Host Plant Resistance

Use of resistant varieties appears to be more useful. In general, Spanish and runner-type varieties are most resistant to CBR (Dong et al. 2008). Some of the most resistant runner-type peanut genotypes are Georgia-06G, Georgia-07W, Georgia-02C, and Carver (Branch and Brenneman 2012). A breeding peanut line, 90x7-1-5-1-b2-B, has been developed to be resistant to the CBR and tomato-spotted wilt virus (Kucharek et al. 2000). Argentine and NC 3033 varieties grow and survive in soil having as high as 1000 microsclerotia/g soil, whereas only 0.5 microsclerotia/g soil is sufficient to cause severe disease in susceptible varieties. Histological study of the nature of resistance in peanut genotypes suggests that in response to infection in resistant genotypes, additional effective periderm is formed, which sloughs an entire quadrant of infected taproot.

Chemical Control

Several broad spectrum chemical fumigants such as Vapam and Vorlex 201 have been evaluated as preplant (4 weeks before planting) treatments, and other chemicals such as captafol, quintozene, chloroneb, tebuconazole (Kucharek et al. 2000), and gypsum have been evaluated for the management of the disease, which appears to be less useful. The soil fumigant, metam sodium, was first recognized in 1981 to have commercial value for the control of CBR in Virginia. Following applications into soil, metam sodium converts rapidly to methyl isothiocyanate (MIT), which is the active ingredient. MIT is a highly toxic, broad spectrum biocide. Because of its highly toxic and nonspecific nature, MIT should be applied at least 2 weeks prior to planting to avoid crop injury. The severity of CBR increases when roots are parasitized by the northern root-knot nematode (*M. hapla*) and the ring nematode (*Criconemella ornata*). Metam sodium treatment along with the application of Temik 15G at 5–7 lb/A in the seed furrow provides good control of these nematodes as well as CBR.

Cultural Control

Very little is known about methods of control of the disease through cultural practices. It is, however, reported that corn and small grains are highly resistant to the disease, and these should be useful as rotational crops to check the incidence of the disease in peanuts. But cotton can be a better rotation crop than corn with respect to peanut yield and gross economic return (Jordan et al. 2002). Well-drained fields should be used for peanut production, and to prevent soil movement, farm equipment should be thoroughly cleaned prior to field-to-field transport. The harvest of peanut seeds in areas of fields with high incidence of CBR should be avoided. This measure is expected to lower the number of speckled seeds entering commercial seed lots and reduce the risk for the spread of CBR (Glenn et al. 2003).

PEG AND POD ROTS

Symptoms

No apparent symptoms of pod rot appear above-ground except that plants with severe pod rot may flower profusely and appear abnormally dark green in color late in the season. Below ground, symptoms of pod rot caused by *Rhizoctonia* and *Pythium* are difficult to distinguish because both pathogens are often present. Light-brown areas develop on pods that later turn dark brown or black (Figure 2.10). A few to nearly all of the pods on one plant may be affected. *Pythium* usually causes a black, watery rot. Pods rotted by *Rhizoctonia* have a firm brown decay, and the seeds and inner pod wall may be lined with a cream-colored fungus. The seeds within rotted pods are usually completely decayed or severely damaged.

Geographical Distribution and Losses

Peanut plants with a few rotted pods are found in most fields all over the peanut-growing countries in the world. Severe outbreaks of pod rot are less common, but can be devastating where they occur. Pod rot tends to be severe in sandy soils and in fields intensively cropped to peanuts. Peanut pod rot primarily caused by *P. myriotylum* is one of the most important diseases in the pacific coast region of Cosiguina in Nicaragua (Augusto et al. 2010a,b). *Neocosmospora striata* Udagawa & Y. Horie causing peanut pod rot to the extent of 90% incidence in the Old Yellow River drainage area (Sun et al. 2012a) and *Neocosmospora vasinfecta* E. F. Smith causing similar pod rot disease in peanuts to the extent of 30%–100% incidence in the Sha River drainage area (Sun et al. 2012b) have been reported as first reports of occurrences in China. Depending on the geographical location, the losses in yield due to rotting of the pods might vary in the range of 5%–50%.

FIGURE 2.10 Pod rot complex of peanuts caused by fungi.

Pathogens: Complex of Fungi in Preharvest Peg and Pod Rots

There are several types of pod-rotting fungi. The specific pod rots could be caused by fungi such as *Pythium* spp., especially *P. myriotylum*; *R. solani*; *Fusarium* spp., especially *F. solani*; *Verticillium albo-atrum*; and *Botrytis cinerea*. Recently, two species of fungi, namely, *N. striata* and *N. vasinfecta*, have been reported as the causes of peanut pod rots (Sun et al. 2012a,b). Rotting of pegs and pods caused by *S. rolfsii*, *M. phaseolina*, *S. sclerotiorum*, and *C. parasiticum* has been described in the preceding pages. The fungi that cause pod rot are normally found at some level in most peanut soils. Most grow on the above-ground parts of the plant also. *Pythium* is the exception as it only grows below ground. Advanced stages of *Pythium* and *Rhizoctonia* pod rots result in complete decay of the pod and kernels. In most peanut-growing areas, *Rhizoctonia* is the main problem in terms of total acreage. On a field-by-field basis, *Sclerotinia* is the most devastating, but it is not so wide spread. Nematode (*Pratylenchus brachyurus*) and black hull fungus (*Thielaviopsis basicola*) also cause pod discoloration, but the decay is usually superficial.

Epidemiology and Disease Cycle

A complex of factors in addition to the fungi is probably responsible for these severe outbreaks of pod rots. These factors include excessive soil moisture, wide fluctuations in soil moisture, calcium deficiency, insect and nematode feeding, and irrigation with poor-quality (salty) water (Choppakatla et al. 2008). One very important factor among these is low level of available calcium in soil, especially in large-seeded cultivars. Pod surface area and surface to volume ratio are important in determining the quantity of calcium in seeds, and this explains why large-seeded cultivars are more sensitive to calcium deficiency–induced susceptibility to pod rots caused by a complex of fungi (Augusto et al. 2010a). Sudden increase in maturity of kernels and concomitant development of pale testa kernels influence more pod rots. The increase in pod rot due to *Fusarium oxysporum* and *F. solani* later in the season is associated with pale testa kernels found in pods, which to the unaided eye appear healthy. Early defoliation of plants hastens maturity and causes peg breaking. Thus, any disease or insect injury resulting in leaf shedding will increase the incidence of rotting of pegs and pods. For example, under Malawi conditions, the increase in pod rot is associated with higher incidence of ELS and LLS and senescence (Kolte 1984). A long wet season and wetness of the top soil increase the incidence of pod rot caused by *P. myriotylum* particularly in sandy well-aerated soil under Israel conditions. Frequent sprinkling of sandy Israeli soil encourages infection of pods by *P. myriotylum*. Application of fertilizers in midseason increases the amount of succulent tissue, favoring more pod breakdown by *P. myriotylum*. While nitrogen amendments with nitrogenous compounds may improve plant health, an overabundance of nitrogen may cause an increase in disease such as *Rhizoctonia* pod rot (Thiessen and Woodward 2012). Several instances of definite interactions taking place between insect or nematode injury and invasion of pods by *P. myriotylum* have been reported. Feeding sites of corn root worm, *Diabrotica undecimpunctata*, and of a mite, *Caloglyphus michaeli*, on peanut pods favor the entry of *P. myriotylum* resulting in the development of rotting of pods. Interaction studies between *P. myriotylum*, *F. solani*, and *M. arenaria* revealed that *P. myriotylum* interacts synergistically with *F. solani* and *M. arenaria*, but not with *R. solani* in causing pod rots of peanuts. *F. solani* alone has been found to predispose the pods to pathogenic activity of *P. myriotylum*.

Disease Management

Host Resistance

Pod rot management through host resistance may be effective but is dependent on the identification of the causal agent. Spanish cultivars, especially Toalson, may provide resistance to both *Pythium* spp. and *R. solani*. Partial resistance to *R. solani* has also been identified in the runner peanut Georgia Browne. Resistance to *Sclerotinia* has been shown in the varieties Virginia 81B, Virginia 93B,

Tamspan 90 and Southwest Runner, and Tamrun OL 07 (Thiessen and Woodward 2012). Some of the peanut varieties have been found to possess field resistance to pod rots (Gopal 2003, Krishnakanth et al. 2005, Gopal et al. 2006c). Resistance in *Sehwar Z-21* and *TMV-2* is attributed to the shortness of the gynophore (inherited independently) enabling the development of pods in a shallow layer of soil and subsequent escaping of the attack by *P. myriotylum*. The phenotype of pod is positively correlated with pod rot severity (Yang et al. 2002). For example, in the *Amani* variety, resistance to pod rot is related to the thicker shell of the pods. It is thus seen that efforts to breed varieties with a shorter gynophore and thick pod shell be encouraged in order to combat losses due to pod rots.

Chemical Control

Some of the fungicides such as Vitavax, thiram, Rizolex T and Topsin M 70 (El-Deeb et al. 2002), mefenoxam, azoxystrobin (Augusto et al. 2010b), and quintozene when applied to the soil before planting or at the fruiting zone at the flowering time give good control of pod rots. The use of metalaxyl + quintozene (PCNB) or metalaxyl + tolclofos-methyl is useful in controlling pod rot diseases (*Pythium* and *Rhizoctonia* spp.) of peanuts (Filonow and Jackson 1989).

Cultural Control

Since pod rot is caused by a complex of several fungi, cultural control measures assume special significance in the management of the pod rot diseases of peanuts. Planting of peanuts in infested soil should be avoided. Care should be taken to avoid mechanical injury to the developing peg and pods during interculture operations. Defoliation of plants due to leaf spot diseases and insect pests should be controlled to check rotting of the pods. Infrequent irrigation resulting in drying out of the top soil brings about reduction in the incidence of pod rot under Israel conditions. The pod rots can be brought under control by mixing gypsum with the upper 15 cm soil layer at the peak of flowering period. The gypsum reduces the soil inoculum and increases calcium contents of pods resisting invasion of pods by fungi (Kolte 1997).

Biological Control

Experimental evidence indicates that infection of fungi on peanut pods can be checked by certain formulation (e.g., plant guard) of antagonistic microorganisms particularly *T. harzianum* (Kanth et al. 2000, El-Deeb et al. 2002). Endomycorrhizal fungus, *Glomus mosseae*, has been proved to be of potential usefulness in protecting the peanut plants from infection by pod rot fungal pathogens (Abdalla and Abdel-Fattah 2000). But the possibility of actual use of such method over large area has not been investigated.

PROBLEM OF AFLATOXIN CAUSES

Geographical Distribution and Losses

In 1960, in Britain, widespread fatalities occurred in poultry fed on peanut meal that had been imported from Brazil, India, and African countries to Britain, the source of toxin being traced to *A. flavus*. Attempts to isolate any microorganism from the peanut meal resulted in isolation of the fungus, *A. flavus*, which was capable of producing four closely related toxins. The toxins were named as aflatoxins (*A. flavus* toxins) identifying their generic origin. Such toxins were isolated from the imported peanut meal also. Thus, aflatoxins were established as the cause of *Turkey X* disease. Species of *Aspergillus* are normal components of soil microflora, and the principal producers of aflatoxins are the species of *A. flavus* group including *A. parasiticus*. *A. flavus* is the most common species in Africa and Asia, while *A. parasiticus* is predominant in America. In South Africa, *A. parasiticus* is predominant and is associated at twice the frequency of *A. flavus* (Ncube et al. 2010). Aflatoxins are known to cause liver cirrhosis in livestock and humans and they possess potent carcinogenic properties. Aflatoxin contamination in peanuts is a serious and worldwide problem especially in developing countries concerning food safety and human health (Williams et al. 2004, Alwakeel and Nasser 2011). Aflatoxins can result in serious economic hardships to producers and adverse health impacts in both humans and domestic animals (Duran et al. 2009). Aflatoxin contamination costs the U.S. peanut industry over $20 million annually (Holbrook et al. 2009). It is reported to be more serious in the southern parts of China and considered to be a crucial factor affecting sustainable development of peanut industry in that country (Liao et al. 2009). Similar is the situation in India where aflatoxins are found in a very high range of 1400–3600 μg/kg of peanut cake. Therefore, aflatoxin contamination is also of significance in relation to public health and export. Aflatoxin levels are higher than the minimum parts per billion (ppb) permitted level set by the Food and Drug Administration in the United States (20 ppb), the European Union (6 ppb), the Department of Health in South Africa (10 ppb), and the Brazilian Regulatory Authorities in Brazil (20 μg/kg) for peanuts that are intended for direct human consumption (Oliveira et al. 2009). Interestingly, aflatoxin levels of up to 131–160 ppb are reported to be present in peanuts produced in the three northern provinces KwaZulu-Natal, Mpumalanga, and Limpopo in South Africa (Ncube et al. 2010). Peanuts stored and consumed in rural areas in Mali (West Africa) have been observed to be highly contaminated by *A. flavus* and aflatoxin B1, with average rates of aflatoxin B1 significantly above the accepted international standards (Passone et al. 2005, Soler et al. 2010). *A. flavus* population in soils from the peanut-growing regions in Argentina (South America) indicates prevalence of three strains (sclerotium-forming S and L and nonsclerotial strains). The S strains produce higher mycotoxin levels than the L and nonsclerotial strains, and about 10% of the S strains simultaneously produce aflatoxins B and G and cyclopiazonic acid. These strains are of great concern in food safety as there is a higher probability of aflatoxin contamination in peanuts in Argentina (Barros et al. 2005, 2006, Nesci et al. 2011). Prevalence of such similar strains of *A. flavus* is also reported from peanut soils in Iran (Amani et al. 2012) and India (Raina and Desai 2006). Consequently, aflatoxin contamination of peanuts becomes one of the most important constraints to peanut production in many countries and becomes a crucial factor in restricting the export of peanuts from one country to another (Asis et al. 2009, Xie et al. 2009). Not only animals but also plants are susceptible to aflatoxins (El-Khadem 1968). Most countries/institutions in Asia and Africa give high priority to research on the peanut aflatoxin problem (Wynne et al. 1991, Waliyar 1997, Zobia et al. 2012), and there appears to be the need for aflatoxin awareness campaigns and management programs to be implemented in rural areas in most countries of Asia and Africa.

Chemical Nature of Aflatoxins

Hartley et al. (1963) were the first to successfully isolate and characterize four closely related furanocoumarin compounds, which have been designated as aflatoxins B_1, B_2, G_1, and G_2. Four more aflatoxins are also known, and these are designated as M_1, M_2, B_2a, and G_2a. The structural formulae of different aflatoxins are given in Figure 2.11. The aflatoxins B and G give a characteristic blue and green fluorescence, respectively, under ultraviolet light.

Invasion of Peanut Pods and Kernels and Conditions for Aflatoxin Production.

As a rule, the seeds are free of *A. flavus* at maturity. However, invasion of peanut seeds by aflatoxigenic strains of *A. flavus* and subsequent aflatoxin contamination can occur both before and after harvest especially during drought stress situation. Colonization of seeds by *A. flavus* increases after maturity and during the period between removal from soil and drying. It is observed that conidia of *A. flavus* do not germinate in peg geocarposphere and germination of conidia occurs only in traces in the fruit geocarposphere. This is possibly because of the presence of volatile and nonvolatile fungistatic substances and increased competition from other soil microorganisms in the peanut geocarposphere. What has been observed is that injury of peanut pods, due to growth cracks, mechanical agents, or biological agents such as root-knot nematode (*M. arenaria*) (Timper et al. 2001, 2004), pod burrower bug (*Pangaeus bilineatus*) (Chapin et al. 2004), and insect pest particularly *Tribolium confusum* in storage (Mohale et al. 2010), predisposes the peanut pods to colonization by *A. flavus* or the development of

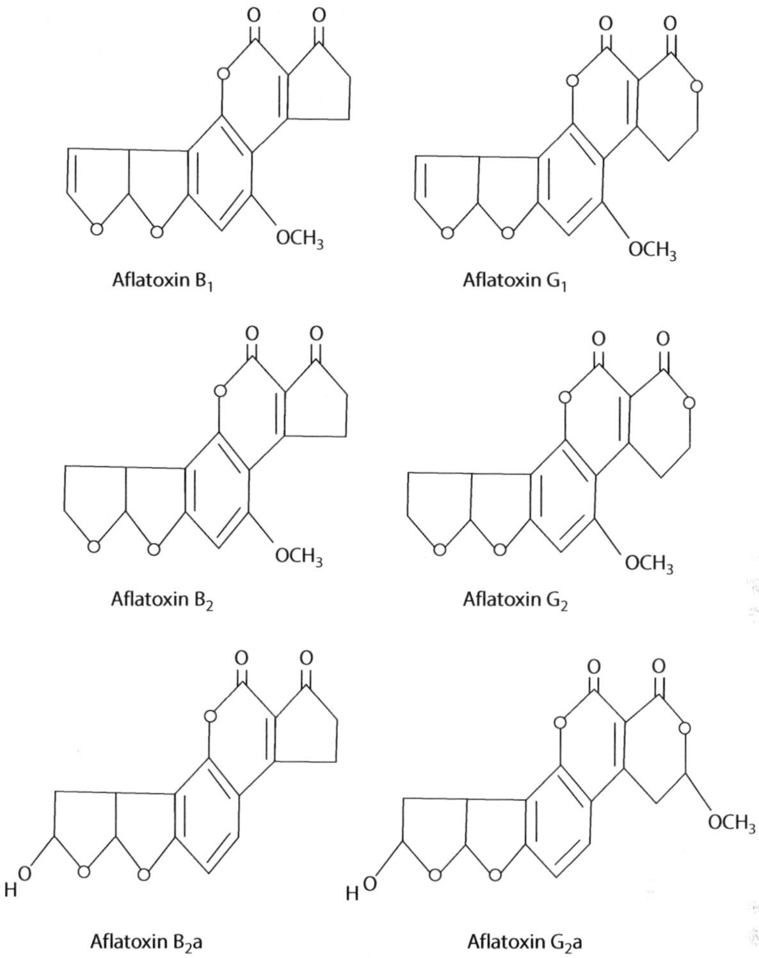

FIGURE 2.11 The structural formulae of different aflatoxins. (From Kolte, S.J., *Diseases of Annual Edible Oilseed Crops*, Vol. I, Peanut Diseases, CRC Press, Boca Raton, FL, 1984.)

aflatoxin in kernels. Following injury to pods or kernels, certain carbon (glucose) and nitrogen substrates (amino acids) are exuded that support the conidial germination of *A. flavus*, enabling it to invade the damaged pods and kernels. Low levels of colonization of peanut fruits by *A. flavus* via flower and aerial peg colonization appear to be possible under field conditions (Kolte 1984). Cropping system appears to have an influence on *A. flavus* infection and accumulation of aflatoxin in the peanut (Kumar et al. 2008). Peanuts harvested from lands planted with peanuts during the previous season are infested with more fungi and contain more aflatoxin than in the case of peanuts raised on lands planted with rye, oats, etc., as the previous crops. Population dynamics of *A. flavus* in soil is also reported to be dependent on the varieties of peanut used in cultivation (Yang and Ma 2003), and its density and genetic diversity based on analysis of vegetative compatibility group (VCG) can vary greatly among regions and fields used for cultivation (McAlpin et al. 2002, Barros et al. 2003, 2006, Horn 2006). Molecular characterization reveals that each VCG represents a single isolate that produces unique DNA fingerprints. This has, therefore, been useful to identify isolates of toxigenic potentials and/or VCG affiliations (McAlpin et al. 2002, Victoria Novas and Cabral 2002, Chen et al. 2002, 2005, Pildain et al. 2004, 2005, Barros et al. 2005, 2007, Reis et al. 2012). Population of *A. flavus* is low in summer crop than in rainy season crop, and its population increases toward pod development stage, and aflatoxin production is negatively related with relative water content, pod wall integrity, and moisture content at harvest (Thakur et al. 2003a,b, Sudhakar et al. 2007). Infection and aflatoxin concentration can be related to the soil types, alfisol soils being more conducive than vertisols, and to the occurrence of soil moisture stress during pod filling when soil temperatures are near optimal for *A. flavus*. Average soil temperatures of 28°C–34°C are favorable for aflatoxin contamination (Craufurd et al. 2006). Thus, late-season drought stress is the most important factor for *A. flavus* invasion and aflatoxin contamination. Peanuts invaded by aflatoxigenic strains of *A. flavus* in the soil before harvest can lead to a serious contamination during drying and storage, and the postharvest drying conditions tend to influence the degree of seed infection and aflatoxin contamination. Delayed harvesting also leads to increased fungal invasion. *A. flavus* grows best between 10°C and 45°C temperature and at relative humidity of 75% or more. The optimum aflatoxin production, however, takes place at 25°C–30°C temperature at 80%–85% relative humidity. Peanut seeds containing more than 9% moisture are likely to be affected by *A. flavus*, and moisture content of seeds only up to 30% is favorable for aflatoxin production. Prematurely dried pods and rainfall at the time of harvesting favor production of aflatoxin during storage (Kolte 1984).

Aflatoxin Management

Host Plant Resistance

a. In cultivated *Arachis* species: Efforts on the development of screening techniques especially at ICRISAT Center for resistance to *A. flavus* infection and aflatoxin production have led to the foundations for conventional resistance breeding program that has resulted in the identification of accessions (genotypes) and development of breeding lines that have resistance to seed infection by *A. flavus* and low aflatoxin production relative to standard susceptible control cultivars in several peanut-growing countries (Jiang et al. 2005, Ntare et al. 2006, Holbrook et al. 2000, 2008, Nigam et al. 2009). Table 2.3 gives the list of peanut genotypes/lines from different countries that show seed infection and colonization equal to or less than most commonly used standard resistant control peanut genotypes *J 11* from India. Some such genotypes show stability of resistance and high yield across seasons and in multilocation environments. For example, three lines (ICGVs 87084, 87094, and 87110), bred at ICRISAT Center in India for resistance to seed infection, had also been found to be resistant in Niger, Senegal, and Burkina Faso in West Africa (Waliyar et al. 1994). Several land races (local germplasm lines) in China have been identified as resistant to seed infection by *A. flavus* and aflatoxin production (Liao et al. 2009). The resistance of peanut seeds to *A. flavus* and aflatoxin production is associated with certain morphological and biochemical characteristics, namely, structure of seed coat, size of wax layer, junction between epidermal cells, thickness of cell wall, and presence of cracks. Resistance depends upon the intact and undamaged testa. So, protective role of seed testa has been emphasized in case of selection for resistance to seed colonization by aflatoxigenic isolates of *A. flavus* (Asis et al. 2005, Lei et al. 2006, Liang et al. 2009). Resistance to peanut seed infection by *A. flavus* is associated with higher content or higher activity of some biochemical constituents such as resveratrol (Liang et al. 2006a,b, Wang et al. 2012), lipoxygenase (LOX) (Liang et al. 2002, Tsitsigiannis et al. 2005, Kumari et al. 2012), β-1,3-glucanase (Liang et al. 2005), oleic acid (Jiang et al. 2006, Ebrahimi et al. 2009), trypsin inhibitor (Liang et al. 2003), superoxide radical generation (Liu et al. 2012, Zhou et al. 2012), storage protein including proteinase inhibitor (Yan et al. 2012b), hydrogen peroxide (H_2O_2), malondialdehyde accumulation (Zhou et al. 2012), and high total phenols (Kumar et al. 2002, Latha et al. 2007, Kumari et al. 2011). In contrast to the preceding report on oleic acid, high-oleic peanut lines are reported to have nearly twice as much aflatoxin as normal lines (Xue et al. 2003). Some accessions have been identified that exhibit low preharvest aflatoxin contamination (PAC) in multiple environments that are tolerant to drought stress conditions (Cleveland et al. 2003, Wang et al. 2004, Liang et al. 2006b, Arunyanark et al. 2010, 2012, Girdthai et al. 2010a,b). Some such genotypes are J 11, 55-437, and PI 337394F, HY 22 (Liu et al. 2012). Traits related to efficient ability to nitrogen fixation under drought conditions may be used as indirect selection criteria for

TABLE 2.3
Peanut Genotypes Resistant (R) or Moderately Resistant (MR) to *A. flavus* Infection and Aflatoxin Contamination as Reported from Different Countries in the World

Genotype	Country	R/MR	Reference(s)
Tifguard	United States	R (also possesses resistance to root-knot nematode and TSWV)	Holbrook et al. (2009)
Minhua 6, Kanghuang 1, Yueyou 9 (released as *A. flavus*–resistant cultivars)	China	R	Zhuang et al. (2007)
ICGVs 86590, 89104, 94350, 99029; IC 48, ICGS-76	India	R (also shows resistance to drought)	Sudhakar et al. (2007)
J 11	India	R	Kumari et al. (2012)
J 11, PI 337394, PI 337409; breeding lines Manfredi 68, Colorado Irradiado, and Florman INTA	Argentina	R/MR	Asis et al. (2005)
9843-26-2, 9817-36-2	China	R	Chen et al. (2005a)
S230, R8808, ICGV 86590, Spanish improved, mutant 28-2	India	R	Harish Babu et al. (2004a)
Significantly superior to J 11 genotypes, ICGV 86155, ICGV 86699, and ICGV 96266; comparable to J 11 genotypes, ICGV 96262, ICG 1697, and R 9227	India	R	Harish Babu et al. (2004b)
ICGV 86590, ICGV 93280, ICGV 95322 (tolerant to end-of-season drought)	Indonesia	R/MR	Rahmianna et al. (2004)
H2030, H2060, H2063, H2095 (tolerant to end-of-season drought)	China	R	Wang et al. (2004)
S206, KRG1, GPBD-4	India	MR	Varma et al. (2001)
KB 153 (high content of storage protein + proteinase inhibitor)	China	R	Yan et al. (2012a)
ICG 12625	China	R (to aflatoxin production)	Jiang et al. (2010)
ICG 4750	China	R (to invasion)	Jiang et al. (2010)
G 845, G 8	China	R	Jiang et al. (2006)
GT-YY 9, GT-YY 20	United States	R (to *A. flavus*)	Liang et al. (2005)
EF 7284	China	MR	Jiang et al. (2002)
J 11, HY 22	China	R	Liu et al. (2012)
J 11, IC 48, ICGV 89104, ICGS-76	India	R	Latha et al. (2007)
Xiaohongmao (high oleic acid content and small seed size; bacterial wilt resistant)	China	R	Liao et al. (2003)
Taishan Zhenzhu, 93-76	China	R (to aflatoxin production)	Liao et al. (2003)

resistance to aflatoxin production in peanuts (Arunyanark et al. 2012). Resistance to *A. flavus* infection in peanuts is independently attributed to three genes: (1) ARAhPR10 (Xie et al. 2009, 2013), (2) PnLOX2 (Yan et al. 2012a), and (3) $PnAG_3$ (Liu et al. 2012). The $PnAG_3$ gene has been found to be expressed more prominently in *A. flavus*–resistant genotypes than susceptible ones under drought stress conditions (Liu et al. 2012). LOXs are nonheme, nonsulfur iron dioxygenases and are encoded by a multigene family and widely distributed in higher plants. Its metabolic products as jasmonic acid, SA, etc., are anti-insect or antibiotic active substances in which active oxygen radicals can destroy cytomembrane and inhibit fungus and aflatoxin generation. Certain agronomic traits as associated with resistance to *A. flavus* in peanuts are valuable. About 13 common loci (63 alleles having increasing effect) are found to be associated with both agronomic traits and resistance to *A. flavus* (Huang et al. 2012). Microanalysis of resistant and susceptible peanut cultivars infected with *A. flavus* (or with *A. parasiticus*) has resulted in the identification of 62 genes in resistant cultivars that are upexpressed in response to *A. flavus* infection, whereas 22 putative *Aspergillus* resistance genes have been identified to be constitutively upexpressed in resistant cultivars in comparison to the susceptible ones (Guo et al. 2009, 2011). These sources among others have been used in breeding programs, and several lines have been derived to possess resistance and produce high yield. The most promising breeding lines developed at ICRISAT reported to be resistant to seed infection and colonization are ICGVs 87084, 87094, 87110, 91278, and 91284. These sources of resistance to PAC have been crossed with cultivars and breeding lines that have high yield and acceptable grade and resistance to TSWV and root-knot nematode (*M. arenaria*). One such first exemplary high-yielding PAC-resistant peanut cultivar is *Tifguard*, which is also resistant to TSWV and root-knot nematode, has been released in the United States (Holbrook et al. 2009). More such resistant cultivars adapted to different production systems need to be developed to meet the requirements of producers and users. The levels of resistance could be improved further by pyramiding resistance genes from different and diverse sources. Liang et al. (2009) have reviewed the peanut host resistance mechanisms to aflatoxin contamination and suggested functional genomics approach as a valuable tool to understand the comprehensive mechanism of the resistance pathways.

b. In wild *Arachis* species: *Arachis chiquitana* has been identified as one of the few wild species of *Arachis* showing resistance to *A. flavus* colonization, and initial seed screening of *A. chiquitana* for *A. flavus* has shown promise of obtaining hybrids resistant to *A. flavus* colonization, though further studies remain to be done whether interspecific hybrid so obtained is also resistant to aflatoxin production (Mallikarjuna 2005).
c. Molecular breeding and transgenic peanuts for resistance to aflatoxin production: It is expected that transgenic resistance against *A. flavus* (or *A. parasiticus*) infection and aflatoxin production in combination with conventional breeding may lead to the development of agronomically superior peanuts that are free of aflatoxin contamination (Ozias-Akins et al. 2002, Nigam et al. 2009). For example, a nonheme chloroperoxidase gene (cpo-p) from *Pseudomonas pyrrocinia*, a growth inhibitor of mycotoxin-producing fungi, has been introduced into peanuts by particle bombardment method. Such transgenic peanuts show inhibition of *A. flavus* hyphal growth that could be translated to a reduction in aflatoxin contamination (Niu et al. 2009). Transgenic peanut lines developed by Xie et al. (2013) overexpressing the effects of ARAhPR10 gene have been established to play an important role in peanut host resistance to *A. flavus* infection and alleviation of aflatoxin production in peanuts. Besides this, the recourse to biotechnology, through modification of the aflatoxin biosynthesis pathway, or the use of variants of hydrolytic enzymes (chitinases and glucanases) to provide transgenic protection to peanuts against infection by aflatoxin-producing fungi may help in obtaining peanuts free from aflatoxin. For example, at ICRISAT, transgenic peanut lines with aflatoxin resistance conferred by rice chitinase gene have been developed and characterized (Sharma et al. 2006).

Chemical Control

Aflatoxin production during storage can be prevented by treating peanut pods with aureofungin. Penetration and invasion by *A. flavus* on pods and kernels can also be prevented by spraying chemicals like propionic acid (5%), sorbic acid (0.1%), and chlorothalonil (0.15%) on pods of peanut plants (harvested at proper stage of maturity) inverted in windrows for drying under field conditions (Kolte 1984). Postharvest control of *A. flavus* infection and aflatoxin contamination can be obtained by the use of formulations of food-grade antioxidants such as butylated hydroxyanisole, butylated hydroxytoluene, and propylparaben (Passone et al. 2007, 2008, 2009). Methyl jasmonate treatment to peanut seeds promotes resistance of peanut plants to *A. flavus* infection and production of aflatoxin B1 (He et al. 2004).

Cultural Control

Several workers (Waliyar 1997, Kasno 2004, Rahmianna et al. 2007) have suggested measures to prevent damage to peanut pods and kernels during cultivation, harvesting, and storage conditions for the prevention of infection due to *A. flavus* and aflatoxin production. Some of the important precautionary measures are (1) avoiding damage to plants and pods from soilborne diseases and during cultivation; (2) avoiding late-season drought stress by the manipulation of crop duration and supplementary irrigation; (3) harvesting and lifting the crop at optimum maturity; (4) discarding damaged pods; (5) drying pods to below 8% moisture content; (6) storing under clean, dry, and insect-free conditions; and (7) avoiding rewetting of pods/seeds during storage. Peanut seeds washed with table salt solution (Sumartini and Yusnawan 2005) and liming the soil have been reported to be useful in reducing the seed infection of *A. flavus* and aflatoxin production (Pereira and Rossetto 2008). Cultural control treatments such as soil application of FYM at 15 tons/ha + gypsum at 500 kg/ha when combined with *Trichoderma* soil application + soil drenching with mancozeb at 1000 ppm result in significant reduction of aflatoxin content (6.6 ppb) with reduced seed infection (13.3%) and maximum pod yield (Ramanaiah et al. 2008). The possibility of inhibition of aflatoxin B_1 production by irradiating the peanut seeds with gamma irradiation has been studied (Prado et al. 2005, Borges et al. 2007). *Musa paradisiaca* fruit peel can be used to suppress aflatoxin production (Sharma and Sharma 2011)

Biological Control

Application of Atoxigenic Strains of A. flavus *or* A. parasiticus

Aflatoxin contamination of peanuts in the field can be reduced by 77%–98% with biological control through the application of nontoxigenic strains of these species, which competitively exclude native aflatoxin-producing strains from developing peanuts (Cotty 1990, Horn et al. 2001, Barros et al. 2003, 2006, Horn and Dorner 2009, Yin et al. 2009). These technologies rely on application of highly competitive atoxigenic strains on solid nutritive substrates. They must be applied at a time and in a manner that allows successful competition with aflatoxin producers. Application timing and placement greatly influence the efficacy of these formulations in preventing aflatoxin contamination. The retention of conidia of atoxigenic strains in the upper soil layers is important in reducing aflatoxin contamination of peanuts (Horn et al. 2001). Proven atoxigenic strains can be routinely applied once per growing season at 11.2–22.4 kg/ha (10–20 lb/acre) of formulation. Reduction of aflatoxins in peanut seeds depends on both the density and the aflatoxin-producing potential of native populations and on the fungal strain used for biological control and nitrate-nonutilizing mutants, which can be used for evaluating the efficacy of biocontrol strains (Horn and Dorner 2009).

Application of Native Antagonistic Fungi and Bacteria

Antagonistic fungi, namely, *T. viride* (Thakur et al. 2003a,b), *T. harzianum* (Thakur et al. 2003a,b, El-Moneim et al. 2010), *Saccharomycopsis schoenii* and *S. crataegensis* (Prado et al. 2008), and *S. cerevisiae* (Prado et al. 2011), and antagonistic bacteria, namely, *B. subtilis* (El-Moneim et al. 2010), a strain of marine bacterium, *Bacillus megaterium* isolated from the Yellow Sea of East China (Kong et al. 2010), *Streptomyces* sp. strain ASBV-1 (Zucchi et al. 2008), and *Burkholderia* sp. strain TNAU-1 (Ayyathurai et al. 2010), have been found to be significantly effective in reducing the aflatoxin production in peanuts. The dose of 20 g of *T. harzianum* or *B. subtilis* formulation per kg of peanut pods and the seed treatment at 10 g/kg or soil application at 2.5 kg/ha at 30, 45, and 60 DAS with the formulation of *Burkholderia* sp. result in significant reduction in the infection by *A. flavus* and aflatoxin B_1 contamination in peanut kernels.

Effect of Plant Extracts

Garlic bulb extract (Sumartini and Yusnawan 2005, Bora et al. 2010), aqueous extract of leaves of lemon (Tewari et al. 2004), aqueous *Moringa* seed extract (Donli and Dauda 2003), and onion and neem extracts (Bora et al. 2010) have shown promise through peanut seed treatment in the control of peanut seed infection by *A. flavus* and aflatoxin production.

TABLE 2.4
Other Fungal Diseases of Peanuts

Disease	Pathogen	Geographical Distribution	Reference(s)
Aerial blight	Rhizoctonia solani (= *Thanatephorus cucumeris*)	India, Malaya, United States	Kolte (1984)
Alternaria leaf spot/ blight	*Alternaria tenuissima (Kunze ex. Pers.) Wilts, A. alternata (Fr.) Keissler, A. arachidis*	India, United States	Kolte (1984), Kumar et al. (2012)
Anthracnose	*Colletotrichum arachidis, C. dematium (Pers. and Fries) Grov. Sensuvona*	Argentina, India, Panama, Taiwan, Uganda, United States	Kolte (1984)
Black root rot	*Thielaviopsis basicola* (Berk and Berk) Ferr.	United States	Kolte (1984)
Botrytis blight	*Botrytis cinerea* (= *Botryotinia fuckeliana*)	Australia, Japan, Romania, Taiwan, Tanzania, United States, Venezuela, Zambia	Kolte (1984), Thiessen and Woodward (2012)
Brown blotch (of Bambara groundnut)	*Colletotrichum capsici*	Nigeria	Obagwu (2003a,b)
Bud blight	*Phoma glomerata*	India	Kolte (1984)
Chlorosis and tip spot	*Psedoplea trifolii*	Mauritius	Kolte (1984)
Choanephora leaf spot	*Choanephora* spp.	United States	Porter (1993)
Collar rot	*Lasiodiplodia theobromae* (= *Diplodia gossypina*)	United States	Porter (1993)
Diaporthe blight	*Diaporthe phaseolorum* var. *sojae* (Lehm.) Wehm.	United States	Kolte (1984)
Fusarium disease (wilt, dry stem rot)	*Fusarium oxysporum, F. oxysporum* var. *orthoceras*	Bulgaria	Vitanova (2003)
Fusarium wilt	*Fusarium martii phaseoli* Burkh	United States	Kolte (1984)
Leaf scorch	*Leptosphaerulina trifolii (Rost.) Pert.*	India, Malawi	Kolte (1984), Desai and Bagwan (2005)
Leaf spot	*Cristulariella pyramidalis Wat. and Marshall*	India, United States	Kolte (1984)
Leaf spot	*Macrophomina phaseolina*	India	Gupta and Kolte (1982)
Leaf blight	*Drechslera spicifera*	India	Jat et al. (2004)
Limb rot (foliar blight phase)	*Rhizoctonia solani*	United States	Thiessen and Woodward (2012)
Melanosis	*Stemphylium botryosum*		Porter (1993)
Myrothecium leaf blight	*Myrothecium roridum* Tode ex Fr.	India	Kolte (1984)
Muddy spot	*Ascochyta* sp.	Brazil	Kolte (1984)
Passalora leaf spot	*Passalora arachidicola* (= *Mycosphaerella arachidis*)	China	Zhang et al. (2010)
Pepper spot and scorch	*Leptosphaerulina crassiasca* (Sechet) Jackson and Bell (= *L. arachidicola* Ye, Chen, and Huang)	Argentina, India, United States	Kolte (1984)
Phomopsis stem blight	*Phomopsis* sp., *P. longicolla*	India, New Mexico (United States)	Kolte (1984), Sanogo and Etarock (2009)
Phyllosticta leaf spot	*Phyllosticta arachidis hypogaea* Vasant Rao	Argentina, India, Israel, Myanmar, Sudan, Taiwan, United States	Desai and Bagwan (2005), Kolte (1984)
Powdery mildew	*Erysiphe polygoni* DC	Mauritius	Kolte (1984)
	Oidium arachidis Chorin	Bulgaria, Israel	Kolte (1984)

(*Continued*)

TABLE 2.4 (*Continued*)
Other Fungal Diseases of Peanuts

Disease	Pathogen	Geographical Distribution	Reference(s)
Olpidium root rot	*Olpidium brassicae*	India	Porter (1993)
Scab	*Sphaceloma arachidis* Bit. and Jenk.	Argentina, Brazil, China	Desai and Bagwan (2005), Fang et al. (2007), Kolte (1984), de Godoy et al. (2001), Moraes et al. (2006), Wang et al. (2009)
Smut (peanut smut)	*Thecaphora frezii* Carranza and Lindquist	Cordoba (Argentina)	Marinelli et al. (2008)
Texas root rot	*Phymatotrichum omnivorum*	United States	Kolte (1984)
Verticillium wilt	*Verticillium dahliae*	Argentina, Australia, Bulgaria, Israel, Southern High Plains of the United States	Kolte (1984), Nakova et al. (2003a,b), Thiessen and Woodward (2012)
Web blotch (net blotch)	*Phoma arachidicola* Marasas, Pauer, and Boerema (= *Mycosphaerella arachidicola* Khokhr)	Australia, South Africa, United States, Zimbabwe	Kolte (1984), Mozingo et al. (2004)
Web blotch	*Ascochyta arachidis Woron.* (= *Mycosphaerella argentinensis*)	Argentina, Russia, United States	Kolte (1984)
Web blight	*Rhizoctonia solani*	India	Dubey (2000)
Zonate leaf spot	*Cristulariella moricola* (*Grovesinia pyramidalis*)	United States	Porter (1993)

OTHER FUNGAL DISEASES

Other less important fungal diseases affecting the peanut crop in different peanut-growing regions of the world are given in Table 2.4.

REFERENCES

Aage, V.E., S.J. Gaikwad, G.T. Behere, and V.S. Tajame. 2003. Efficacy of extracts of certain indigenous medicinal plant against *Cercospora* leaf spot of groundnut. *J. Soils Crops* 13: 140–144.

Abd-Allah, E.F. 2005. Effect of a *Bacillus subtilis* isolate on southern blight (*Sclerotium rolfsii*) and lipid composition of peanut seeds. *Phytoparasitica* 33: 460–466.

Abd-Alla, E.F., S.M. Ezzat, M.M. Sarhan, and A.A. Abd-El-Mottleb. 2003. Biocontrol of peanut southern blight (*Sclerotium rolfsii*) by *Bacillus subtilis*. *Egypt. J. Microbiol.* 38: 207–216.

Abdalla, M.E. and G.M. Abdel-Fattah. 2000. Influence of the endomycorrhizal fungus *Glomus mosseae* on the development of peanut pod rot disease in Egypt. *Mycorrhiza* 10: 29–35.

Abd-El-Moneem, K.M.H., M.H.A. Moharam, and M.M.E. El-Sherif. 2003. Role of root exudates of certain peanut cultivars in resistance to root-rot disease. *Assiut J. Agric. Sci.* 34: 193–209.

Abudulai, M., I.K. Dzomeku, A.B. Salifu, S.K. Nutsugah, R.L. Brandenburg, and D.L. Jordan. 2007. Influence of cultural practices on soil arthropods, leaf spots, pod damage and yield of peanut in northern Ghana. *Peanut Sci.* 34: 72–78.

Achira, C., K. Satyaprasad, and A. Nagamani. 2002. Population dynamics of *Trichoderma viride* and *Pseudomonas aeruginosa* in rhizosphere of groundnut. In: *Frontiers in Microbial Biotechnology and Plant Pathology*, eds., C. Manoharachary, D.K. Purohit, S. Reddy, and M.A. Singaracharya. Scientific Publishers, Jodhpur, India, pp. 247–252.

Ade A.B. and L.V. Gangawane. 2010. Biocontrol of aflaroot and dry root rot of *groundnut* by transgenic *Rhizobium*. *Indian Phytopathol.* 63: 42–44.

Adiver, S.S. 2004. Field efficacy of botanical preparations on late leaf spot *Phaeoisariopsis personata* (Berk. and A.M. Curt) Van Arx. of groundnut (*Arachis hypogaea* L.). *J. Oilseeds Res.* 21: 369–370.

Adiver, S.S. and K.H. Anahosur. 1995. Efficacy of some triazole fungicides against late leaf spot of groundnut and their subsequent effects on *Sclerotium rolfsii*. *Indian Phytopathol.* 48: 459–462.

Adomou, M., P.V.V. Prasad, K.J. Boote, and J. Detongnon. 2005. Disease assessment methods and their use in simulating growth and yield of peanut crops affected by leafspot disease. *Ann. Appl. Biol.* 146: 469–479.

Agnihotri, J.P. and M.P. Sharma. 1972. Efficacy of fungicides in controlling collar rot of groundnut (*Arachis hypogaea* L.). *Acta Agron. Acad. Sci. Hung.* 21: 222.

Akgul, D.S., H. Ozgonen, and A. Erkilic. 2011. The effects of seed treatments with fungicides on stem rot caused by *Sclerotium rolfsii* Sacc. in peanut. *Pak. J. Bot.* 43: 2991–2996.

Alabi, O. and P.E. Olorunju. 2004. Evaluation of neem seed extract, black soap and cow dung for the control of groundnut leaf spot at Samaru, Nigeria. *Arch. Phytopathol. Plant Prot.* 37: 123–127.

Alwakeel, S.S. and L.A. Nasser. 2011. Microbial contamination and mycotoxins from nuts in Riyadh, Saudi Arabia. *Am. J. Food Technol.* 6: 613–630.

Amani, S., M. Shams-Ghahfarokhi, M. Banasaz, and M. Razzaghi-Abyaneh. 2012. Mycotoxin-producing ability and chemotype diversity of *Aspergillus* section flavi from soils of peanut-growing regions in Iran. *Indian J. Microbiol.* 52: 551–556.

Ambang, Z., N. Bekolo, E. Petga, J.P.N. Dooh, A. Asanga, and K.Z. Ahmed. 2007. Effect of seed extracts of *Thevetia peruviana* weeds on development of leaf spot disease of groundnut (*Arachis hypogaea* L.) caused by *Cercospora* sp. *Eighth African Crop Science Society Conference*, El-Minia, Egypt, October 27–31, 2007, pp. 797–800.

Ambang, Z., B. Ndongo, G. Essono et al. 2011. Control of leaf spot disease caused by *Cercospora* sp. on groundnut (*Arachis hypogaea*) using methanolic extracts of yellow oleander (*Thevetia peruviana*) seeds. *Aust. J. Crop Sci.* 5: 227–232.

Anand, R. and S. Kulothungan. 2010. Antifungal metabolites of *Pseudomonas fluorescens* against crown rot pathogen of *Arachis hypogaea*. *Ann. Biol. Res.* 1: 199–207.

Anjaiah, V., R.P. Thakur, and N. Koedam. 2006. Evaluation of bacteria and *Trichoderma* for biological control of pre-harvest seed infection by *Aspergillus flavus* in groundnut. *Biocontrol Sci. Technol.* 16: 431–436.

Anuradha, T.S., K. Divya, S.K. Jami, and P.B. Kirti. 2008. Transgenic tobacco and peanut plants expressing a mustard defensin show resistance to fungal pathogens. *Plant Cell Rep.* 27: 1777–1786.

Arneson, P.A. 1970. Chemical control of rust and *Cercospora* leaf spots on peanuts in Honduras and Nicaragua. *Phytopathology* 60: 1539.

Arunyanark, A., S. Jogloy, S. Wongkaew et al. 2010. Heritability of aflatoxin resistance traits and correlation with drought tolerance traits in peanut. *Field Crop Res.* 117: 258–264.

Arunyanark, A., S. Pimratch, S. Jogloy et al. 2012. Association between aflatoxin contamination and N_2 fixation in peanut under drought conditions. *Int. J. Plant Prot.* 6: 161–172.

Ashok, J., B. Fakrudin, M.S. Kuruvinashetti, and B.Y. Kullaiswami. 2004a. Regeneration of stem and pod rot resistant groundnut plants from *Sclerotium rolfsii* Saac. culture filtrate treated callus. *Indian J. Genet. Plant Breed.* 64: 221–224.

Ashok, J., B. Fakrudin, H. Paramesh, P.V. Kenchanagoudar, and B.Y. Kullaiswamy. 2004b. Identification of groundnut (*Arachis hypogaea* L.) germplasm resistant to stem and pod rot caused by *Sclerotium rolfsii* Saac. *Indian J. Genet. Plant Breed.* 64: 247–248.

Ashworth, Jr. I.J., B.C. Langley, M.A. Mian, and C.J. Wrenn. 1964. Epidemiology of a seedling disease of Spanish peanut caused by *Aspergillus niger*. *Phytopathology* 54: 1161.

Asis, R., D.L. Barrionuevo, L.M. Giorda, M.L. Nores, and M.A. Aldao. 2005. Aflatoxin production in six peanut (*Arachis hypogaea* L.). *J. Agric. Food Chem.* 53: 9274–9280.

Asis, R., V. Muller, D.L. Barrionuevo, S.A. Araujo, and M.A. Aldao. 2009. Analysis of protease activity in *Aspergillus flavus* and *A. parasiticus* on peanut seed infection and aflatoxin contamination. *Eur. J. Plant Pathol.* 124: 391–403.

Augusto, J. and T.B. Brenneman. 2011. Implications of fungicide application timing and post-spray irrigation on disease control and peanut yield. *Peanut Sci.* 38: 48–56.

Augusto, J. and T.B. Brenneman. 2012. Interactive effects of fungicide, application timing and spray volume on peanut diseases and yield. *Plant Health Prog.*, doi:10.1094/PHP-2010-0420-01-Rs.

Augusto, J., T.B. Brenneman, J.A. Baldwin, and N.B. Smith. 2010c. Maximizing economic returns and minimizing stem rot incidence with optimum plant stands of peanut in Nicaragua. *Peanut Sci.* 37: 137–143.

Augusto, J., T.B. Brenneman, and A.S. Casino. 2010a. Etiology of peanut pod rot in Nicaragua: I. The effect of pod size, calcium, fungicide, and nematicide. Online. *Plant Health Prog*. doi: 10.1094/PHP-2010-0215-01-Rs.

Augusto, J., T.B. Brenneman, and A.S. Casino. 2010b. Etiology of peanut pod rot in Nicaragua: II. The role of *Pythium myriotylum* as defined by application of gypsum and fungicides. Online. *Plant Health Prog*. doi: 10.1094/PHP-2010-0215-02-Rs.

Aulakh, K.S. and R.S. Sandhu. 1970. Reaction of groundnut varieties against *Aspergillus niger*. *Plant Dis. Res*. 54: 337.

Ayyathurai, V., V. Selvaraj, P. Vaikuntavasan, S. Ramasamy, and V. Rethinasamy. 2010. Integrated management of aflatoxin B_1 contamination of groundnut (*Arachis hypogaea* L.) with *Burkholderia* sp. and zimmu (*Allium sativum* L. × *Allium cepa* L.). *J. Plant Interact*. 5: 59–68.

Azza, A.M. Shahin, and S.A. Azer. 2003. Evaluation of the productivity of some gamma-rays peanut mutants-resistant to *Aspergillus flavus* infection. *Ann. Agric. Sci*. (Cairo) 48: 147–158.

Babu, B.K., A.K. Saxena, A.K. Srivastava, and D.K. Arora. 2007. Identification and detection of *Macrophomina phaseolina* by using species-specific oligonucleotide primers and probes. *Mycologia* 99: 797–803.

Backman, P.A. and R. Rodriguez-Kabana. 1975. A system for the growth and delivery of biological control agents to the soil. *Phytopathology* 65: 819.

Backman, P.A. and R. Rodriguez-Kabana. 1977. Predisposition of peanuts to disease and suppression of *Sclerotium rolfsii* by pesticides, the role of antagonism. In: Current Trends in Plant Pathology, ed., Z. Kiraly, Accad. Kiado, Budapest, p. 209.

Badigannavar, A.M., D.M. Kale, S. Mondal, and G.S. Murty. 2005. Trombay groundnut recombinants resistant to foliar diseases. *Mutat. Breed. Newsl. Rev*. 1: 11–12.

Bagwan, N.B. 2010. In vitro antifungal effect of crop root exudates against *Sclerotium rolfsii*, causing stem rot in groundnut. *Int. J. Plant Prot*. 3: 298–305.

Bagwan, N.B. 2011a. Morphological variation in *Sclerotium rolfsii* Sacc. isolates causing stem rot in groundnut (*Arachis hypogaea* L.). *Int. J. Plant Prot*. 4: 68–73.

Bagwan, N.B. 2011b. Evaluation of biocontrol potential of *Trichoderma* species against *Aspergillus niger* and *Aspergillus flavus* and *Sclerotium rolfsii*. *Int. J. Plant Prot*. 4: 107–111.

Barros, G., A. Torres, and S. Chulze. 2005. *Aspergillus flavus* population isolated from soil of Argentina's peanut-growing region. Sclerotia production and toxigenic profile. *J. Sci. Food Agric*. 85: 2349–2353.

Barros, G., A. Torres, G. Palacio, and S. Chulze. 2003. *Aspergillus* species from section Flavi isolated from soil at planting and harvest time in peanut-growing regions of Argentina. *J. Sci. Food Agric*. 83: 1303–1307.

Barros, G.G., M.L. Chiotta, M.M. Reynoso, A.M. Torres, and S.N. Chulze. 2007. Molecular characterization of *Aspergillus* section Flavi isolates collected from peanut fields in Argentina using AFLPs. *J. Appl. Microbiol*. 103: 900–909.

Barros, G.G., A.M. Torres, M.I. Rodriguez, and S.N. Chulze. 2006. Genetic diversity within *Aspergillus flavus* strains isolated from peanut-cropped soils in Argentina. *Soil Biol. Biochem*. 38: 145–152.

Bdliya, B.S. 2006. Effect of *Cercospora* leaf spot on the quality of groundnut haulm in northeastern Nigeria. *Trop. Sci*. 46: 126–128.

Bdliya, B.S. 2007. Groundnut haulm quality as affected by *Cercospora* leaf spot severity. *J. Plant Prot. Res*. 47: 231–241.

Bdliya, B.S. and G. Alkali. 2010a. Cost-benefit of using some plant extracts in controlling *Cercospora* leaf spot of groundnut in the Sudan savanna of Nigeria. *Arch. Phytopathol. Plant Prot*. 43: 95–104.

Bdliya, B.S. and G. Alkali. 2010b. Efficacy of some plant extracts in the management of *Cercospora* leaf spot of groundnut in the Sudan savanna of Nigeria. *Arch. Phytopathol. Plant Prot*. 43: 507–518.

Bdliya, B.S. and K.K. Gwio-Kura. 2007a. Efficacy of some fungicides in the management of *Cercospora* leaf spot of groundnut in the Sudan savanna of Nigeria. *J. Plant Prot. Res*. 47: 243–253.

Bdliya, B.S. and K.K. Gwio-Kura. 2007b. Cost-benefit of fungicidal control of *Cercospora* leaf spot of groundnut in the Sudan savanna of Nigeria. *J. Plant Prot. Res*. 47: 329–338.

Bdliya, B.S. and A.S. Muhammad. 2006. Effect of intercropping millet with groundnut on the control of *Cercospora* leaf spot of groundnut in the Sudan Savanna of northeastern Nigeria. *J. Sustain. Agric*. 29: 19–41.

Bell, D.K. and E.K. Sobers. 1966. A peg, pod, and root necrosis of peanuts caused by a species of *Calonectria*. *Phytopathology* 56: 1361–1364.

Berkeley, M.J. 1875. Notices on North American fungi. *J. Grev*. 3: 97.

Besler, B.A., W.J. Grichar, J.A. Starr, S.A. Senseman, R.G. Lemon, and A.J. Jaks. 2006. Effects of peanut row pattern, cultivar and fungicides on control of southern stem rot, early leaf spot and rust of peanut. *Peanut Sci*. 33: 1–6.

Bhagat, S. and P. Sitansu. 2007. Mass multiplication of *Trichoderma harzianum* on agricultural by-products and their evaluation against seedling blight (*Rhizoctonia solani*) of mungbean and collar rot (*Sclerotium rolfsii*) of groundnut. *Indian J. Agric. Sci.* 77: 583–588.

Bhaskar, A.V. and A.M. Parakhia. 2010. Biochemical changes in resistant and susceptible varieties of peanut (*Arachis hypogaea*) in relation to early and late leaf spot disease. *J. Oilseeds Res.* 27: 195–196.

Biswas, K.K. and C. Sen. 2000. Management of stem rot of groundnut caused by *Sclerotium rolfsii* through *Trichoderma harzianum. Indian Phytopathol.* 53(3): 290–295.

Biswas, S. and N.P. Singh. 2005. Fungicidal management of foliar diseases of groundnut in Tripura. *Indian Phytopathol.* 58: 500–502.

Bolonhezi, D., S.B. Verdade, and J.A. Marques. 2004. Efficiency of fungicides for the control of leaf spot and scab in peanut cv. IAC-Tatu ST. *Arquivos do Instituto Biologico* (*Sao Paulo*), 71(Supplemento, RAIB): 358–360.

Bora, M.V., R.T. Sapkal, and S.B. Latake. 2010. Anti-fungal properties of plant extracts against *Aspergillus flavus* inciting groundnut. *J. Maharashtra Agric. Univ.* 35: 337–339.

Borges, V.B., Z.B. Miranda, H.C. de Vital and C.A.R. da Rocha. 2007. Elimination or inactivation of fungi of the genus *Aspergillus* in groundnuts (*Arachis hypogea* L.) and increase in storage life when treated with gamma radiation. *Higiene Alimentar* 21: 61–68.

Bowen, C., H.A. Melouk, K.E. Jackson, and M.E. Payton. 2000. Effect of a select group of seed protectant fungicides on growth of *Sclerotinia minor* in vitro and its recovery from infested peanut seed. *Plant Dis.* 84: 1217–1220.

Bowen, K.L., A.K. Hagan, H.L. Campbell, and L. Wells. 2006. Comparison of calendar and AU-Pnut advisory programs with azoxystrobin for the control of leaf spot and southern stem rot on peanut. *Peanut Sci.* 33: 76–82.

Bowen, K.L., A.K. Hagan and R. Weeks. 1992. Seven years of *Sclerotium rolfsii* in peanut fields: yield losses and means of minimization. *Plant Dis.* 76: 982–985.

Branch, W.D. 2002. Registration of 'Georgia-01R' peanut. *Crop Sci.* 42:1750–51.

Branch, W.D. and T.B. Brenneman. 2003. Field resistance to Cylindrocladium black rot and tomato spotted wilt virus among advanced runner-type peanut breeding lines. *Crop Prot.* 22: 729–734.

Branch, W.D. and T.B. Brenneman. 2008. Registration of 'Georgia-07W' peanut. *J. Plant Regist.* 2(2): 88–91.

Branch, W.D. and T.B. Brenneman. 2009. Field evaluation for the combination of white mould and tomato spotted wilt disease resistance among peanut genotypes. *Crop Prot.* 28: 595–598.

Branch, W.D. and T.B. Brenneman. 2012. New sources of *Cylindrocladium* black rot resistance among runner type peanut cultivars. *Peanut Sci.* 39: 38–42.

Branch, W.D. and A.K. Culbreath. 2008. Disease and insect assessment of candidate cultivars for potential use in organic peanut production. *Peanut Sci.* 35(1): 61–66.

Butler, D.R., K.D.R. Wadia, R.K. Reddy et al. 2000. A weather-based scheme to advise on limited chemical control of groundnut leaf spot diseases in India. *Experimental Agriculture* 36: 469–478.

Cantonwine, E.G., A.K. Culbreath, C.C. Holbrook, and D.W. Gorbet. 2008b. Disease progress of early leaf spot and components of resistance to *Cercospora arachidicola* and *Cercosporidium personatum* in runner-type peanut cultivars. *Peanut Sci.* 35: 1–10.

Cantonwine, E.G., A.K. Culbreath, B.B. Shew, and M.A. Boudreau. March 2008a. Efficacy of organically acceptable fungicides for management of early and late leaf spots diseases on partially resistant peanut cultivars. *Plant Health Prog.* doi:10.1094/PHP-2008-0317-03-RS.

Cantonwine, E.G., A.K. Culbreath, K.L. Stevenson et al. 2006. Integrated disease management of leaf spot and spotted wilt of peanut. *Plant Dis.* 90: 493–500.

Cantonwine, E.G., A.K. Culbreath, and K.L. Stevenson. 2007a. Effects of cover crop residue and preplant herbicide on early leaf spot of peanut. *Plant Dis.* 9: 822–827.

Cantonwine, E.G., A.K. Culbreath, and K.L. Stevenson. 2007b. Characterization of early leaf spot suppression by strip tillage in peanut. *Phytopathology* 97: 187–194.

Cantonwine, G., C.C. Holbrook, A.K. Culbreath, R.S. Tubbs, and M.A. Boudreau. 2011. Genetic and seed treatment effects in organic peanut. *Peanut Sci.* 38: 115–121.

Cardoza, Y.J., H.T. Alborn, and J.H. Tumlinson. 2002. In vivo volatile emissions from peanut plants induced by simultaneous fungal infection and insect damage. *J. Chem. Ecol.* 28: 161–174.

Carley, D.S., D.L. Jordan, B.B. Shew, T.B. Sutton, L.C. Dharmasri, and R. Brandenburg. 2009. Influence of digging date and fungicide program on canopy defoliation and pod yield of peanut (*Arachis hypogaea* L.). *Peanut Sci.* 36: 77–84.

Cavallo, A.R., R.J. Novo, and M.A. Perez. 2005. Fungicide efficiency in the control of seedborne fungi in peanut (*Arachis hypogaea* L.) in Argentina. *AgriScientia* 22(1/2): 9–16.

Chahal, C.S. and J.S. Chohan. 1971. *Puccinia* rust on groundnut. *FAO Plant Prot. Bull.* 19: 90.
Chakrabarty, S.K., A.G. Girish, K. Anitha et al. 2005. Detection, seedborne nature, disease transmission and eradication of seedborne infection of *Rhizoctonia bataticola* (Taub.) Butler in groundnut. *Indian J. Plant Prot.* 33: 85–89.
Chandra, S., A.K. Singh, and P. Baiswar. 2007. Efficacy of single spray of fungicides on early leaf spot (*Cercospora arachidicola*) of groundnut (*Arachis hypogaea* L.). *Indian J. Agric. Sci.* 77: 201–203.
Chapin, J.W., J.W. Dorner, and J.S. Thomas. 2004. Association of a burrower bug (*Heteroptera: Cydnidae*) with aflatoxin contamination of peanut kernels. *J. Entomol. Sci.* 39: 71–83.
Chapin, J.W. and J.S. Thomas. 2005. Effect of fungicide treatments, pod maturity, and pod health on peanut peg strength. *Peanut Sci.* 32: 119–125.
Chen, J., L.R. Wu, H.R. Miao, and W.G. Hu. 2005b. Huayu 22—A high-yielding large-seeded groundnut variety with improved seed quality. *Int. Arachis Newsl.* 25: 26.
Chen, R.S., J.G. Tsay, Y.F. Huang, and R.Y.Y. Chiou. 2002. Polymerase chain reaction-mediated characterization of molds belonging to the *Aspergillus flavus* group and detection of *Aspergillus parasiticus* in peanut kernels by a multiplex polymerase chain reaction. *J. Food Prot.* 65: 840–844.
Chen, S.L., Z.X. Tang, and G.X. Li. 2005a. A study on breeding and screening of groundnut varieties with resistance to *Aspergillus flavus*. *Acta Agriculturae Universitatis Jiangxiensis* 27: 77–80.
Chen, W.L., D. Xing, and W.G. Chen. 2005c. Rapid detection of *Aspergillus flavus* contamination in peanut with novel delayed luminescence spectra. *Photochem. Photobiol.* 81: 1361–1365.
Chen, W.Y. and W.H. Ko. 2014. Flaxseed oil for control of peanut rust and its disease control mechanism. *J. Phytopathol.* 2014. doi:10.1111/jph.121316.
Chenault, K.D., A.L. Maas, J.P. Damicone, M.E. Payton, and H.A. Melouk. 2009. Discovery and characterization of a molecular marker for *Sclerotinia minor* Jagger resistance in peanut. *Euphytica* 166: 357–365.
Chenault, K.D., H.A. Melouk, and M.E. Payton. 2005. Field reaction to *Sclerotinia* blight among transgenic peanut lines containing antifungal genes. *Crop Sci.* 45: 511–515.
Chenault, K.D., H.A. Melouk, and M.E. Payton. 2006. Effect of *Sclerotinia minor* infection loci on peanut production parameters. *Peanut Sci.* 33(1): 36–40.
Chi Mai Thi, N., T.V.D. Thanh, X.N. Hong et al. 2006. Collar rot of groundnut caused by *Lasiodiplodia theobromae* in North Vietnam. *Int. Arachis Newsl.* 26: 25–27.
Chohan, J.S. 1965. Collar rot of groundnut (*Arachis hypogaea* L.) caused by *Aspergillus niger* Van Tiegham and *Aspergillus pulverulentus* (McAlpine) Thom. in the Punjab. *J. Res.* Ludhiana (India) 2: 25–33.
Chohan, J.S. and V.K. Gupta. 1968. Aflaroot, a new disease of groundnut caused by *Aspergillus flavus* Link. *Indian J. Agric. Sci.* 38: 568–570.
Chohan, J.S. and S.P. Kapoor. 1967. Cultural factors in relation to collar rot of groundnut. *J. Res.* Ludhiana (India) 4: 536.
Choppakatla, V., T.A. Wheeler, G.L. Schuster, C. Robinson, and D.O. Porter. 2008. Relationship of soil moisture with the incidence of pod rot in peanut in West Texas. *Peanut Sci.* 35: 116–122.
Chougule, R.L. and S.S. Kore. 2004. Management of soil and seed borne groundnut diseases through cultural practices, irrigation levels and varieties. *J. Maharashtra Agric. Univ.* 29: 297–299.
Chowdhury, A.K. 2002. Effect of chitosan on collar rot of peanut caused by *Sclerotium rolfsii* Sacc. *Res. Crops* 3: 667–669.
Chowdhury, A.K. 2003. Control of *sclerotium* blight of groundnut by plant growth substances. *Crop Res.* (Hisar) 25: 355–359.
Chriscoe, S.M. 2008. Characterization of transgenic peanuts expressing oxalate oxidase for governmental approval of their release for control of *Sclerotinia* blight. M.Sc. thesis, the Virginia Polytechnic Institute and State University, Blacksburg, VA, p. 124.
Christopher, D.J., S. Usharani, and R. Udhayakumar. 2008. Management of dry root rot (*Macrophomina phaseolina* (Tassi.) Goid) of peanut (*Arachis hypogaea* L.) by the integration of antagonistic (*Trichoderma virens*) and organic amendments. *Adv. Plant Sci.* 21: 389–392.
Cilliers, A.J., Z.A. Pretorius, and P.S. van Wyk. 2003. Integrated control of *Sclerotium rolfsii* on groundnut in South Africa. *J. Phytopathol.* 151: 249–258.
Cleveland, T.E., P.F. Dowd, A.E. Desjardnis, D. Bhatnagar, and P.J. Cotty. 2003. United States Department of Agriculture Research Service: Research on pre-harvest contamination of mycotoxins and mycotoxigenic fungi in U.S. *Pest Manag. Sci.* 59: 629–642.
Clewis, S.B., S.D. Askew, and J.W. Wilcut. 2001. Common ragweed interference in peanut. *Weed Sci.* 49: 768–772.
Cook, M. 1972. Screening of peanut for resistance to peanut rust in the greenhouse and field. *Plant Dis. Rep.* 56: 382.

Cook, M. 1980a. Host-parasite relations in uredial infections of peanut by *Puccinia arachidis*. *Phytopathology* 70: 822.

Cook, M. 1980b. Peanut leaf wettability and susceptibility to infection by *Puccinia arachidis*. *Phytopathology* 70: 826.

Cotty, P.J. 1990. Effect of atoxigenic strains of *Aspergillus flavus* on aflatoxin contamination of developing cottonseed. *Plant Dis.* 74: 233–235.

Craufurd, P.Q., P.V.V. Prasad, F. Waliyar, and A. Taheri. 2006. Drought, pod yield, pre-harvest *Aspergillus* infection and aflatoxin contamination on peanut in Niger. *Field Crops Res.* 98: 20–29.

Cruickshank, A.W., M. Cooper, and M.J. Ryley. 2002. Peanut resistance to *Sclerotinia minor* and *S. sclerotiorum*. *Aust. J. Agric. Res.* 53: 1105–1110.

Csinos, A.S.1987. Control of Southern stem rot and Rhizoctonia limb rot of peanut with flutolanil. *Peanut Sci.* 14: 55–58.

Culbreath, A.K., T.B. Brenneman, R.C. Kemerait Jr., and G.G. Hammes. 2009. Effect of new pyrazole carboxamide fungicide penthiopyrad on late leaf spot and stem rot of peanut. *Pest Manag. Sci.* 65: 66–73.

Culbreath, A.K., T.B. Brenneman, F.M. Shokes, A.C. Casinos, and H.S. McLean. 1992b. Tank-mix application of cyproconazole and chlorothalonil for control of foliar and soil-borne diseases of peanut. *Plant Dis.* 76: 1241.

Culbreath, A.K., R.C. Kemerait Jr., and T.B. Brenneman. February 1–11, 2006. Management of early leaf spot of peanut as affected by fungicide and date of spray program initiation. *Plant Health Prog.* doi:10.1094/PHP-2006-0214-01-RS.

Culbreath, A.K., R.C. Kemerait Jr., and T.B. Brenneman. 2008. Management of leaf spot disease of peanut with prothioconazole applied alone or in combination with tebuconazole or trifloxystrobin. *Peanut Sci.* 35(2): 149–158.

Culbreath, A.K., N.A. Minton, T.B. Brenneman, and B.G. Mullinix. 1992a. Response of Florunner and Southern runner cultivars to chemical management of late leaf spot, southern stem rot and nematodes. *Plant Dis.* 76: 1199.

Culbreath, A.K., K.L. Stevenson, and T.B. Brenneman. 2002. Management of late leaf spot of peanut with benomyl and chlorothalonil: A study in preserving fungicide utility. *Plant Dis.* 86: 349–355.

Dahmen, H. and T. Staub. 1992. Protective, curative and eradicant activity of difenoconazole against *Venturia inaequalis*, *Cercospora arachidicola* and *Alternaria solani*. *Plant Dis.* 76: 774.

Damicone, J.P., C.C. Holbrook, D.L. Smith, H.A. Melouk, and K.D. Chamberlin. 2010. Reaction of core collection of peanut germplasm to *Sclerotinia* blight and pepper spot. *Peanut Sci.* 37: 1–11.

Damicone, J.P. and H.A. Melouk. 2009. Soilborne disease of peanut. Tech-Rep., EPP-7664, Oklahoma Cooperative Extension Service, Oklahoma City, OK.

Dandnaik, B.P., P.K. Dhoke, and S.N. Pensalwar. 2006. Integrated management of stem rot (*Sclerotium rolfsii*) in groundnut. *J. Plant Dis. Sci.* 1: 234–235.

Das, S. and T.K. Roy. 1995. Assessment of losses in groundnut due to early and late leaf spots. *IAN* 15: 34–36.

Dasgupta, S., S.K. Raj, and S. Das. 2000. Control of collar rot disease of groundnut by *Aspergillus niger* by seed soaking with non-conventional chemicals. *Indian Phytopathol.* 53: 227–29.

Debets, A., E. Holub, K. Swart, H. van den Broek, and C. Bos. 1990. An electrophoretic karyotype of *Aspergillus niger*. *Mol. Gen. Genet.* 224: 264–268.

de Godoy, I.J., S.A. de Moraes, A.R.A. de Moraes et al. 2001. Yield potential of upright early maturing peanut lines with and without control of foliar diseases. *Bragantia* 60: 101–110.

Desai, S. and N.B. Bagwan. 2005. Fungal and bacterial diseases of groundnut. In *Diseases of Oilseed Crops*, eds., G.S. Saharan, N. Mehta, and M.S. Sangwan. Indus Publishing Company, New Delhi, India, pp. 108–151.

Desai, S., N.B. Bagwan, and R.D. Yoele. 2004. Effect of mustard cake extract on *Sclerotium rolfsii* Sacc. causing stem rot in groundnut and *Trichoderma harzianum* Rifai, a common biocontrol agent. *Indian J. Plant Prot.* 32: 158–159.

Desai, S., R.P. Thakur, V.P. Rao, and V. Anjaiah. 2000. Characterization of isolates of *Trichoderma* for biocontrol potential against *Aspergillus flavus* infection in groundnut. *Int. Arachis Newsl.* 20: 57–59.

Deshmukh, M.P., D.A. Shambharkar, D.A. Patil, T.R. Patil, and B.P. Kurundkar. 2009. Evaluation of water use efficient genotypes of groundnut against late leaf spot and rust. *J. Plant Dis. Sci.* 4: 156–159.

Deshwal, V.K., R.C. Dubey, and D.K. Maheshwari. 2003. Isolation of plant growth-promoting strains of *Bradyrhizobium* (*Arachis*) sp. with biocontrol potential against *Macrophomina phaseolina* causing charcoal rot of peanut. *Curr. Sci.* 84: 443–448.

Devi, M.C. and R.D. Prasad. 2009. Biointensive management of collar rot of groundnut caused by *Aspergillus niger*. *J. Biol. Control* 23: 21–24.

Dey, R., K.K. Pal, D.M. Bhatt, and S.M. Chauhan. 2004. Growth promotion and yield enhancement of peanut (*Arachis hypogaea* L.) by application of plant growth-promoting rhizobacteria. *Microbiol. Res.* 159: 371–394.

Dilip, C., B.S. Kumar, and H.C. Dube. 1999. Disease suppression and plant growth promotion of peanut by fluorescent pseudomonad strains, FPSC 32 and FP04, in soil infested with collar rot fungi, *Aspergillus niger*. *Indian Phytopathol.* 52: 415–416.

Doley, K. and P.K. Jite. 2012. Efficacy of *Glomus fasciculatum* and *Trichoderma viride* for biological control of stem rot caused by pathogen *Sclerotium rolfsii* in groundnut. *Adv. Plant Sci.* 25: 647–652.

Dolma, T., M.R. Sekhar, and K.R. Reddy. 2010. Genetic variability, correlation and path analysis for yield, its components and late leaf spot resistance in groundnut (*Arachis hypogaea*). *J. Oilseeds Res.* 27: 154–157.

Dong, W.B., T.B. Brenneman, C.C. Holbrook, and A.K. Culbreath. 2008. Evaluation of resistance to *Cylindrocladium parasiticum* of runner-type peanut in the greenhouse and field. *Peanut Sci.* 35: 139–148.

Dong, W.B., T.B. Brenneman, C.C. Holbrook, P. Timper, and A.K. Culbreath. 2009. The interaction between *Meloidogyne arenaria* and *Cylindrocladium parasiticum* in runner peanut. *Plant Pathol.* 58: 71–79.

Donli, P.O. and H. Dauda. 2003. Evaluation of aqueous *Moringa* seed extract as a seed treatment biofungicide for groundnut. *Pest Manag. Sci.* 59: 1060–1062.

Douglas, T.N. 2004. Field evaluation of transgenic peanut lines for resistance to *Sclerotinia* blight and yield. M.Sc. thesis, Oklahoma State University, Stillwater, OK, p. 62.

Dubey, S.C. 2000. Biological management of web blight of groundnut (*Rhizoctonia solani*). *J. Mycol. Plant Pathol.* 30: 89–90.

Dubey, S.C. 2005. Role of weather on development of *cercospora leaf* spot (*Cercospora arachidicola*) on groundnut (*Arachis hypogaea*). *Indian J. Agric. Sci.* 75: 232–234.

Duran, R.M., J.W. Cary, and A.M. Calvo. 2009. The role of veA in *Aspergillus flavus* infection of peanut, corn and cotton. *Open Mycol. J.* 3: 27–36.

Durgesh, N., J.S.S. Mohan, and G. Singh. 2010. Induction of systemic acquired resistance in *Arachis hypogaea* L. by *Sclerotium rolfsii* derived elicitors. *J. Phytopathol.* 158: 594–600.

Dwivedi, S.L., J.H. Crouch, S.N. Nigam, M.E. Ferguson, and A.H. Paterson. 2003. Molecular breeding of groundnut for enhanced productivity and food security in the semi-arid tropics: Opportunities and challenges. *Adv. Agron.* 80: 153–221.

Dwivedi, S.L. and S. Gurtu. 2002. Molecular diversity among accessions possessing varying levels of resistance to early leaf spot in groundnut. *Int. Arachis Newsl.* 22: 36–37.

Dwivedi, S.L., S. Pande, J.N. Rao, and S.N. Nigam. 2002. Components of resistance to late leaf spot and rust among interspecific derivatives and their significance in a foliar disease resistance breeding in groundnut (*Arachis hypogaea* L.). *Euphytica* 125: 81–88.

Ebrahimi, T., M. Khomeiri, Y. Maghsoudlou, and G.M. Ahmadi. 2009. Relation between chemical compositions of peanut with its resistance to *Aspergillus flavus*. *J. Agric. Sci. Nat. Res.* 16(1-A): unpaginated.

Edraki, V. and Z. Banihashemi. 2010. Phenotypic diversity among isolates of *Macrophomina phaseolina* and its relation to pathogenicity. *Iran. J. Plant Pathol.* 46: 93–100.

El-Deeb, A.A., S.M. Abdel-Momen, and A.A. Hanafi. 2002. Effect of some fungicides and alternative compounds on root and pod rots in peanut. *Egypt. J. Agric. Res.* 80: 71–82.

El-Habbaa, G.M., M.S. Felaifel, A.M. Zahra, and R.E. Abdel-Ghany. 2002. In vitro evaluation of some fungicides, commercial biocontrol formulations and natural plant extracts on peanut root rot pathogens. *Egypt. J. Agric. Res.* 80: 1017–1031.

El-Khadem, M. 1968. The role of aflatoxin in the seedling disease of groundnut caused by *Aspergillus flavus*. *Phytopathol. Z* 61: 218.

El-Korashy, M.A. 2001. Disease incidence and yield of peanut as affected by gypsum amendment, soil moisture and fertilization. *Ann. Agric. Sci.* (Moshtohor) 39(1): 211–224.

El-Moneim, M.L.A., A.F. Tolba, and A.M. Gomaa. 2010. Effect of some bioagents on inhibiting toxins produced by *Aspergillus flavus* in peanut pods. *Ann. Agric. Sci.* (Cairo) 55(1): 109–119.

El-Shehaby, A.I. and S.M.A. Morsy. 2005. Biological control of damping-off disease by *Bacillus sphaericus* soil treatment. *Egypt. J. Agric. Res.* 83: 1–9.

El-Wakil, A.A. and M.I. Ghonim. 2000. Survey of seed borne mycoflora of peanut and their control. *Egypt. J. Agric. Res.* 78: 47–61.

El-Wakil, M.A. 2003. Use of antioxidant hydroquinone in the control of seed-borne fungi of peanut with special reference to the production of good quality seed. *Pak. J. Plant Pathol.* 2: 75–79.

Emmanuel, E.S.C., S. Rathnavel, S. Maruthamuthu, and R.S. Yadav. 2011. A survey of collar rot disease of *Arachis hypogaea* by *Aspergillus niger* in certain locations of Madurai district (Tamil Nadu), India. *Plant Arch.* 11: 645–647.

Fang, H.C. 1977. Studies on peanut rust in Taiwan. *Plant Prot. Bull. Taiwan* 19(4): 218.

Fang, S.M., Z.R. Wang, Y.Q. Ke, Y.S. Chen, C.M. Huang, and J.X. Yu. 2007. The evaluation of resistance and resistant mechanisms of peanut varieties to scab disease. *Sci. Agric. Sin.* 40: 291–297.

Faske, T., H.A. Melouk, and M.E. Payton. 2006. Comparison of Sclerotinia minor inocula for differentiating the reaction of peanut genotypes to *sclerotinia blight*. *Peanut Sci.* 33(1): 7–11.

Fávero, A.P., S.A. de Moraes, A.A.F. Garcia, J.F.M. Valls, and N.A. Vello. 2009. Characterization of rust, early and late leaf spot resistance in wild and cultivated peanut germplasm. *Sci. Agricola* 66: 110–117.

Ferguson, L.M. and B.B. Shew. 2001. Wheat straw mulch and its impacts on three soilborne pathogens of peanut in microplots. *Plant Dis.* 85: 661–667.

Filonow, A.B. and K.E. Jackson. 1989. Effect of metalaxyl plus PCNB or metalaxyl plus tolclofos-methyl on peanut pod rot and soil populations of *Pythium* spp. and *Rhizoctonia solani*. *Peanut Sci.*16: 25–32.

Foudin, A.S. and V. Macko. 1974. Identification of self-inhibitor and some germination characteristics of peanut rust uredospores. *Phytopathology* 64: 990.

Frimpong, A., F.K. Padi, and J. Kombiok. 2006. Registration of foliar-disease-resistant and high-yielding groundnut varieties ICGV 92099 and ICGV 90084. *Int. Arachis Newsl.* 26: 22–24.

Gai, Y., Q. Deng, R. Pan, and X. Chen. 2012. First report of *Cylindrocladium* black rot of peanut caused by *Cylindrocladium parasiticum* (*Teleomorph calonectria ilicicola*) in Jiangxi province, Guangzhou, China. *APSnet* 96(4): 586.

Gajera, H., K. Rakholiya, and D. Vakharia. 2011. Bioefficacy of *Trichoderma* isolates against *Aspergillus niger* Van Tieghem inciting collar rot in groundnut (*Arachis hypogaea* L.). *J. Plant Prot. Res.* 51: 240–247.

Gajera, H.P., K. Rathod, S.V. Patel, and B.A. Golakiya. 2013. Biochemical markers induced by *Trichoderma* against rot infection in groundnut (*Arachis hypogaea* L.) and their appraisal with DNA fingerprinting. *Can. J. Plant Prot.* 1: 49–63.

Gajera, H.P. and D.N. Vakharia. 2012. Production of lytic enzymes by *Trichoderma* isolates during in vitro antagonism with *Aspergillus niger*, the causal agent of collar rot of peanut. *Braz. J. Microbiol.* 43: 43–52.

Ganesan, S., R.G. Kuppusamy, and R. Sekar. 2007. Integrated management of stem rot disease (*Sclerotium rolfsii*) of groundnut (*Arachis hypogaea* L.) using Rhizobium and *Trichoderma harzianum* (ITCC-4572). *Turk. J. Agric. Forest.* 31: 103–108.

Gangopadhyay, S., J.V. Bhatia, and S.L. Godara. 1996. Evaluation of fungicides for the control of collar rot of groundnut. *Indian J. Mycol. Plant Pathol.* 26: 278–279.

Garba, A., S.D. Abdul, G.N. Udom, and B.M. Auwal. 2005. Influence of variety and intra-row spacing on Cercospora leaf spot disease of groundnut in Bauchi Nigeria. *Global J. Agric. Sci.* 4: 177–182.

Garcia, G.M., S.P. Tallury, H.T. Stalker, and G. Kochert. 2006. Molecular analysis of Arachis interspecific hybrids. *Theor. Appl. Genet.* 112: 1342–1348.

Garren, K.K. 1966. Peanut (Groundnut) microflora and pathogenesis in peanut pod. *Phytopathol. Z* 55: 359.

Ghewande, M.P. 1993. Biological control of groundnut (*Arachis hypogaea* L.) rust (*Puccinia arachidis*) in India. *Trop. Pest Manag.* 36: 17–20.

Ghewande, M.P. 2009. Rust of groundnut—An overview. In: *Integrated Pest and Disease Management*, eds., R.V. Upadhyay, K.G. Mukerji, B.P. Chamola, and O.P. Dubey. APH Publishing Corporation, New Delhi, India, pp. 326–334.

Ghewande, M.P., S. Desai, and M.S. Basu. 2002. Diagnosis and management of major diseases of groundnut. *NRCG Bull.* 8–9.

Gibbons, R.W. and B.E. Bailey. 1967. Resistance to *Cercospora arachidicola* in some species of *Arachis*. *Rhod. Zamb. Malawi J. Agric. Res.* 5: 57.

Gibson, I.A.S. 1953. Crown rot seedling diseases of groundnut caused by *A. niger* II. Anomalous effect of orange mercurial seed dressings. *Trans. Br. Mycol. Soc.* 36: 324–334.

Girdthai, T., S. Jogloy, N. Vorasoot, C. Akkasaeng, S. Wongkaew, and C.C. Holbrook. 2010a. Heritability of, and genotypic correlations between, aflatoxin traits and physiological traits for drought tolerance under end of season drought in peanut (*Arachis hypogaea* L.). *Field Crops Res.* 118: 169–176.

Girdthai, T., S. Jogloy, N. Vorasoot, C. Akkasaeng, S. Wongkaew, and C.C. Holbrook. 2010b. Associations between physiological traits for drought tolerance and aflatoxin contamination in peanut genotypes under terminal drought. *Plant Breed.* 129: 693–699.

Giri, R.R., V.N. Toprope, and P.K. Jagtap. 2009. Genetic variability, character association and path analysis for yield, its component traits and late leaf spot, *Phaeoisariopsis personata* (Berk and curt), in groundnut. *Int. J. Plant Sci.* 4: 551–555.

Glenn, D.L., P.M. Phipps, and R.J. Stipes. 2003. Incidence and survival of *Cylindrocladium parasiticum* in peanut seed. *Plant Dis.* 87: 867–871.

Godoy, G., J.R. Steadman, M.B. Dickman, and R. Dam. 1990. Use of mutants to demonstrate the role of oxalic acid in pathogenicity of *Sclerotinia sclerotiorum* on *Phaseolus vulgaris*. *Physiol. Mol. Plant Pathol.* 37: 179–191.

Gopal, K. 2003. Reaction of certain advanced Spanish groundnut genotypes to pod rot disease. *Ann. Agric. Res.* 24: 184–186.

Gopal, K., S.K. Ahamed, and G.P. Babu. 2006a. Estimation of losses due to tikka leaf spot in groundnut (*Arachis hypogaea* L.). *Legume Res.* 29: 289–291.

Gopal, K., S.K. Ahamed, and G.P. Babu. 2006b. Evaluation of advanced Spanish bunch groundnut cultures under kharif seasons of Northern Telangana region of Andhra Pradesh. *Legume Res.* 29: 292–294.

Gopal, K., S.K. Ahamed, and G.P. Babu. 2006c. Relative resistance in groundnut genotypes to pod rot disease. *Legume Res.* 29: 205–208.

Gopal, K., R. Jagadeswar, and G.P. Babu. 2003. Efficacy of new systemic fungicide in controlling late leaf spot and rust in groundnut. *Indian J. Plant Prot.* 31: 76–79.

Gorbet, D.W. 2007. Registration of 'AP-3' peanut. *J. Plant Regist.* 1(2): 126–127.

Gorbet, D.W., S. Machado, and S.E. Petrie. 2006a. Registration of 'Andru II' peanut. *Crop Sci.* 46: 2712–2713.

Gorbet, D.W., S. Machado, and S.E. Petrie. 2006b. Registration of 'Carver' peanut. *Crop Sci.* 46: 2713–2714.

Gorbet, D.W. and F.M. Shokes. 2002a. Registration of 'C-99R' peanut. *Crop Sci.* 42(6): 2207.

Gorbet, D.W. and F.M. Shokes. 2002b. Registration of 'Florida MDR98' peanut. *Crop Sci.* 42(6): 2207.

Gorbet, D.W. and B.L. Tillman. 2008. Registration of 'DP-1' peanut. *J. Plant Regist.* 2(3): 200–204.

Gorbet, D.W. and B.L. Tillman. 2009. Registration of 'Florida-07' peanut. *J. Plant Regist.* 3(1): 14–18.

Gour, H.N. and P. Sharma. 2009. Effect of *Sclerotium rolfsii* Sacc. on seed germination and seedling vigor of groundnut germplasm. *Indian Phytopathol.* 62: 522–525.

Gowda, M.V.C., R.S. Bhat, B.N. Motagi, V. Sujay, V. Kumari, and B. Sujatha. 2010. Association of high-frequency origin of late leaf spot resistant mutants with AhMITE1 transposition in peanut. *Plant Breed.* 129: 567–569.

Gowda, M.V.C., B.N. Motagi, G.K. Naidu, S.B. Diddimani, and R. Sheshagiri. 2002a. GPBD 4: A spanish bunch groundnut genotype resistant to rust and late leaf spot. *Int. Arachis Newsl.* 22: 29–32.

Gowda, M.V.C., B.N. Motagi, R. Sheshagiri, G.K. Naidu, and M.N. Rajendraprasad. 2002b. Mutant 28-2: A bold-seeded disease and pest resistant groundnut genotype for Karnataka, India. *Int. Arachis Newsl.* 22: 32–34.

Gremillion, S.K. 2007. Contributions to management of diseases of peanut (*Arachis hypogaea*) through Bolivan-derived host resistance, integrated disease management and knowledge of pathogen variability. Dissertation Univ of Georgia, Athens, GA, p. 158.

Gremillion, S., A. Culbreath, D.W. Gorbet et al. 2011a. Response of progeny bred from Bolivian and North American cultivars in integrated management systems for leaf spot of peanut (*Arachis hypogaea*). *Crop Prot.* 30: 698–704.

Gremillion, S.K., A.K. Culbreath, D.W. Gorbet et al. 2011b. Field evaluations of leaf spot resistance and yield in peanut genotypes in the United States and Bolivia. *Plant Dis.* 95: 263–268.

Grichar, W.J., A.J. Jaks, and B.A. Besler. 2005. Response of peanuts (*Arachis hypogaea*) to weather-based fungicide advisory sprays. *Crop Prot.* 24: 349–354.

Grichar, W.J., A.J. Jaks, and J. Woodward. 2010. Using prothioconazole plus tebuconazole for foliar and soil-borne disease control in Texas peanut. *Crop Manag.* doi:10.1094/CM-2010-0405-02-RS.

Griffin, G.J. 1977. Improved selective medium for isolating *Cylindrocladium crotalariae* microsclerotia from naturally infested soil. *Can J. Microbiol.* 23: 680.

Grinstein, A., J. Katan, A.A. Rajik, O. Zeydan, and Y. Elad. 1979. Control of *Sclerotium rolfsii* and weed in peanut by solar heating of the soil. *Plant Dis. Rep.* 63: 1056–1059.

Gughe, S.S., C.D. Mayee, and G.M. Godbole. 1981. Assessment of losses in peanut due to rust and tikka leaf spots. *Indian Phytopathol.* 34: 179.

Guimarães, P.M., A.C.F.V. José, K. Proite, D.J. Bertioli, and S.C.M. Leal-Bertioli. 2005. Development of molecular markers for resistance gene analogs in wild *Arachis* spp. *Fitopatol. Bras.* 30(6): 663–667.

Gupta, C.P., R.C. Dubey, and D.K. Maheshwari. 2002. Plant growth enhancement and suppression of *Macrophomina phaseolina* causing charcoal rot of peanut by fluorescent *Pseudomonas*. *Biol. Fertil. Soils* 35(6): 399–405.

Gupta, S.C. and S.J. Kolte. 1982. A comparative study of two isolates of *Macrophomina phaseolina* from leaf and root of groundnut. *Indian Phytopathol.* 35: 222–225.

Guo, B.Z., N.D. Fedorova, X.P. Chen et al. 2011. Gene expression profiling and identification of resistance genes to Aspergillus flavus infection in peanut through EST and microarray strategies. *Toxins* 3(7): 737–753.

Guo, B.Z., J. Yu, and C.C. Holbrook. 2009. Studies in prevention of preharvest aflatoxin contamination in peanuts: Aflatoxin biosynthesis, genetics and genomics. *Plant Sci.* 36: 11–20.

Gururaj, S., P.V. Kenchanagoudar, and M. Reddy. 2002. Bio-efficacy of fungicides for control of leaf spots of groundnut in northeastern Dry Zone of Karnataka, India. *Int. Arachis Newsl.* 22: 44–46.
Gururaj, S. and S. Kulkarni. 2006. Studies on structural and biochemical mechanism of resistance in groundnut to *Puccinia arachidis*. *Indian Phytopathol.* 59: 323–328.
Gururaj, S. and S. Kulkarni. 2007. Studies on perpetuation and carry over of groundnut rust (*Puccinia arachidis* Speg.) in Northern Karnataka. *Karnataka J. Agric. Sci.* 20: 297–300.
Gururaj, S. and S. Kulkarni. 2008. Slow rusting in groundnut genotypes against *Puccinia arachidis*. *Indian Phytopathol.* 61: 456–460.
Gururaj, S., S. Kulkarni, and V.I. Benangi. 2005b. Efficacy of neem seed kernel extract in combination with selected fungicides for groundnut rust management. *Int. Arachis Newsl.* 25: 39–42.
Gururaj, S., S. Kulkarni, V.I. Benagi, and P.V. Kenchanagoudar. 2006. Logistic model for prediction of groundnut rust caused by *Puccinia arachnids* in Northern Eastern Dry Zone of Karnataka. *Karnataka J. Agric. Sci.* 19: 553–557.
Gururaj, S., R.K. Mesta, and M. Reddy. 2005a. Field efficacy of some fungicides for effective and economical control major foliar diseases of groundnut. *Karnataka J. Agric. Sci.* 18: 995–997.
Hagan, A.K. 2012. *Peanut Fungicide Update*. Timely information (Agriculture & Natural Resources), Plant Pathology Series. Alabama Cooperative Extension System, Auburn University, Auburn, AL, p. 714.
Hagan, A.K., K.L. Bowen, H.L. Campbell, and L. Wells. 2007. Calendar-based and AU-Pnuts advisory programs with pyraclostrobin and chlorothalonil for the control of early leaf spot and stem rot of peanut. *Peanut Sci.* 34: 114–121.
Hagan, A.K., H.L. Campbell, K.L. Bowen, and M. Pegues. November 1–9, 2006. Calendar and AU-Pnuts advisory programs compared for leaf spot diseases and rust control on peanut in southwest Alabama. *Plant Health Prog.* doi:10.1094/PHP-2006-1110-01-RS.
Hagan, A.K., H.L. Campbell, K.L. Bowen, L. Wells, and R. Goodman. 2010. Managing early leaf spot and stem rot with reduced fungicide inputs on disease-resistant peanut cultivars. *Peanut Sci.* 37: 129–136.
Hagan, A.K., J.R. Weeks, and K. Bowen. 1991. Effects of application of timing and method on control of Southern stem rot of peanut with foliar-applied fungicides. *Peanut Sci.* 18: 47–50.
Hagan, A.K., J.R. Weeks, and J.A. McGuire. 1988. Comparison of soil insecticides alone and in combination with PCNB for suppression of Southern stem rot of peanut. *Peanut Sci.* 15: 35–38.
Haggag, W.M. 2007. Colonization of exopolysaccharide-producing *Paenibacillus polymyxa* on peanut roots for enhancing resistance against crown rot disease. *Afr. J. Biotechnol.* 6: 1568–1577.
Haggag, W.M. and S.A. Abo-Sadera. 2000. Influence of iron sources and siderophore producing *Pseudomonas* on crown-rot disease incidence and seed contamination of peanut with pathogenic Aspergilli. *Ann. Agric. Sci.* (Moshtohor, Egypt) 38: 1933–1948.
Haggag, W.M. and S. Timmusk. 2008. Colonization of peanut roots by biofilm-forming *Paenibacillus polymyxa* initiates biocontrol against crown rot disease. *J. Appl. Microbiol.* 104: 961–969.
Hamasselbe, A., M. Ishuyaku, I. Kuresh, and G. Sanusi. 2007. Evaluation of groundnut (*Arachis hypogaea* L.) genotypes for resistance to *Cercospora* leaf spots in north Cameroon. *Cameroon J. Agric. Sci.* 3(2): http://dx.doi.org/10.4314/cjas.v3i2.48368.
Hammons, R.O. 1977. Groundnut rust in the United States and the Caribbean. *PANS* 23: 300.
Harish Babu, B.N., M.V.C. Gowda, and G.K. Naidu. 2004b. Screening advanced breeding lines of groundnut for resistance to in vitro seed colonization by *Aspergillus flavus*. *Int. Arachis Newsl.* 24: 10–12.
Harish Babu, B.N., M.V.C. Gowda, and G. Shirnalli. 2004a. Evaluation of groundnut varieties of Karnataka for resistance to *Aspergillus flavus*. *Karnataka J. Agric. Sci.* 17: 566–567.
Hartley, R.D., B.E. Nesbill, and I. Okelly. 1963. Toxic metabolites of *Aspergillus flavus*. *Nature* 198: 1056.
Hazarika, D.K., L.N. Dubey, and K.K. Das. 2000. Effect of sowing dates and weather factors on development of leaf spot and rust of groundnut. *J. Mycol. Plant Pathol.* 30(1): 27–30.
Hegde, D.M. 2009. Can India achieve self-reliance in vegetable oils? In *National Symposium on Vegetable Oils Scenario: Approaches to Meet the Growing Demands*, Indian Society of Oilsseed Research, Hyderabad, India, January 29–31, 2009, pp. 1–15.
He, L.H., H.W. He, and J.H. Bin. 2004. Effects of methyl jasmonate on the resistance of peanut seeds to *Aspergillus flavus*. *J. South China Agric. Univ.* 25(3): 10–13.
Hemingway, J.S. 1957. The resistance of groundnuts to *Cercospora* leaf spots. *Emp. J. Exp. Agric.* 25: 60.
Higgins, B.B. 1935. Breeding peanuts for disease resistance Abstr. *Phytopathology* 25: 971.
Holbrook, C.C. and A.K. Culbreath. 2008. Registration of "Georganic" peanut. *J. Plant Regist.* 2(1): 17.
Holbrook, C.C. and W.B. Dong. 2005. Development and evaluation of a mini core collection for the U.S. peanut germplasm collection. *Crop Sci.* 45: 1540–1544.

Holbrook, C.C., B.Z. Guo, D.M. Wilson, and P. Timper. 2009. The U.S. breeding program to develop peanut with drought tolerance and reduced aflatoxin contamination. *Peanut Sci.* 36: 50–53.

Holbrook, C.C., P. Ozias-Akins, P. Timper et al. 2008. Research from the Coastal Plain Experiment Station, Tifton, Georgia, to minimize aflatoxin contamination in peanut. *Toxin Rev.* 27(3/4): 391–410.

Holbrook, C.C., D.M. Wilson, M.E. Matheron, J.E. Hunter, D.A. Knauft, and D. Gorbet. 2000. *Aspergillus* colonization and aflatoxin contamination in peanut genotypes with reduced linoleic acid composition. *Plant Dis.* 84: 148–150.

Hollowell, J.E., B.B. Shew, M.A. Cubeta, and J.W. Wilcut. 2003a. Weed species as hosts of *Sclerotinia minor* in peanut fields. *Plant Dis.* 87: 197–199.

Hollowell, J.E., B.B. Shew, and T.G. Isleib. 2003b. Evaluating isolate aggressiveness and host resistance from peanut leaflet inoculations with *Sclerotinia minor*. *Plant Dis.* 87: 402–406.

Hoque, M.Z., F. Begum, M.M. Alam, M.M.A. Patwary, and M.A. Sadat. 2008. Efficacy of some fungicides against tikka and rust diseases of groundnut (*Arachis hypogaea* L.). *Ann. Bangladesh Agric.* 12: 21–28.

Horn, B.W. 2006. Relationship between soil densities of *Aspergillus* species and colonization of wounded peanut seeds. *Can. J. Microbiol.* 52: 951–960.

Horn, B.W. and J.W. Dorner. 2009. Effect of nontoxigenic *Aspergillus flavus* and *A. parasiticus* on aflatoxin contamination of wounded peanut seeds inoculated with agricultural soil containing natural fungal populations. *Biocontrol Sci. Technol.* 19: 249–262.

Horn, B.W., R.L. Greene, R.B. Sorensen, P.D. Blankenship, and J.W. Dorner. 2001. Conidial movement of nontoxigenic *Aspergillus flavus* and *A. parasiticus* in peanut fields following application to soil. *Mycopathologia* 151: 81–92.

Hossain, M.B. and M.Z. Rahman. 2007. Efficacy of foliar spray with potash, neem leaf extract and Bavistin to manage leaf spot and rust of groundnut. *Bangladesh J. Plant Pathol.* 23: 85–88.

Hossain, M.D. and L.L. Ilag. 2000. Pathogenic variability of *Phaeoisariopsis personata* (Berk. & Curt.) V. Arx. in groundnut (*Arachis hypogaea* L.). *Bangladesh J. Bot.* 29: 61–65.

Hossain, M.D., M.Z. Rahman, K. Abeda, and M.M. Rahman. 2007. Screening of groundnut genotypes for leaf spots and rust resistance. *Int. J. Sustain. Crop Prod.* 2(1): 7–10.

Hossain, M.S., K.M. Khalequzzaman, M.A. Hossain, M. Hossain, and M.R.A. Mollah. 2005. Effect of variety and fungicides in controlling tikka and rust diseases of groundnut. *J. Subtrop. Agric. Res. Dev.* 3(1): 570.

Hossain, M.S., M.M. Karim, and M.D. Hossain. 2010. Minimum number of fungicidal spray for economic control of leaf spot and rust of groundnut. *Bangladesh J. Plant Pathol.* 26: 59–63.

Hsi, D.C. 1966. Observations on an outbreak of *Aspergillus* crown rot of Valencia peanuts in New Mexico. *Plant Dis. Rep.* 50: 175.

Huang, L., X.P. Ren, X.J. Zhang, Y.N. Chen, and H.F. Jiang. 2012. Association analysis of agronomic traits and resistance to *Aspergillus flavus* in the ICRISAT peanut mini-core collection. *Acta Agron. Sin.* 38: 935–946.

Ihejirika, G.O. 2007. Effects of maize intercropping on incidence and severity of leafspot disease (*Cercospora arachidicola* Hori) and nodulation of groundnut. *Int. J. Agric. Res.* 2: 504–507.

Ihejirika, G.O., M.I. Nwufo, C.I. Duruigbo, E.U. Onwerenadu, O.P. Obilo, E.R Onuoha, and K.O. Ogbede. 2006b. Effects of plant extracts and plant density on the severity of leafspot disease of groundnut (*Cercospora arachidicola* Hori) in Imo State. *J. Plant Sci.* 1(4): 374–377.

Ihejirika, G.O., M.I. Nwufo, E. Oputa, O.P. Obilo, K.O. Ogbede, and V.N. Onyia. 2006a. Effects of NPK fertilizer rates and plant population on foliar diseases, insect damage and yield of groundnut. *J. Plant Sci.* 1: 362–367.

Ijaz, M., M.I. Haque, C.A. Rauf, Fayyaz-ul-Hassan, A. Riaz, and S.M. Mughal. 2011. Correlation between humid thermal ratio and epidemics of *Cercospora* leaf spot of peanut in Pothwar. *Pak. J. Bot.* 43: 2011 2016.

Ijaz, M., C.A. Rauf, I.U. Haque, F.U. Hussan, and A. Mahmood. 2008. Distribution and severity of *Cercospora* leaf spots of peanut in rainfed regions of Punjab. *Pak. J. Phytopathol.* 20: 165–172.

Iqbal, M.M., N. Farhat, A. Shaukat et al. 2012. Over expression of rice chitinase gene in transgenic peanut (*Arachis hypogaea* L.) improves resistance against leaf spot. *Mol. Biotechnol.* 50: 129–136.

Islam, Md.S., M.M. Islam, E.M. Emdad et al. 2012. Tools to kill: Genome of one of the most destructive plant pathogenic fungi *Macrophomina phaseolina*. *BMC Genomics* 13: 493.

Islam, M.N., M.G. Raihan, and Z.A. Rafiq. 2005. In vitro evaluation of *Trichoderma*, fungicides and plant extracts against *Rhizoctonia solani* and *Sclerotium rolfsii* of peanut. *Int. J. Sustain. Agric. Technol.* 1(1): 14–23.

Isleib, T.G., Rice, P.W., Mozingo, R.W. et al. 2006. Registration of N96076L peanut germplasm line. *Crop Sci.* 46(5): 2329–2330.

Iwo, G.A. and P.E. Olorunju. 2009. Yield stability and resistance to leaf spot diseases and rosette in groundnut. *Czech. J. Genet. Plant Breed.* 45: 18–25.

Izge, A.U., Z.H. Mohammed, and A. Goni. 2007. Levels of variability in groundnut (*Arachis hypogaea* L.) to *Cercospora* leaf spot disease—Implications for selection. *Afr. J. Agric. Res.* 2: 182–186.

Jackson, C.R. 1962. *Aspergillus* crown rot of peanut in Georgia, seed treatment fungicides for control of seed-borne fungi in peanut. *Plant Dis. Rep.* 46: 888.

Jackson, C.R. 1964. Field comparisons of seed treatment fungicides for the control of *Aspergillus* crown rot and other seedling diseases of peanut. *Plant Dis. Rep.* 48: 264.

Jadeja, K.B., D.M. Nandolia, I.U. Dhruj, and R.R. Khandar. 1999. Efficacy of four triazole fungicides in the control of the leaf spots and rust of groundnut. *Indian Phytopathol.* 52: 421.

Jajda, H.M. and V.R. Thakkar. 2012. Control of *Aspergillus niger* infection in varieties of *Arachis hypogaea* L. by supplementation of zinc ions during seed germination. *Arch. Phytopathol. Plant Prot.* 45: 1464–1478.

Jat, R.G., J.P. Goyal, and S.C. Jain. 2004. Effect of fertilizers on leaf blight disease and pod yield of groundnut. *J. Mycol. Plant Pathol.* 34: 73–74.

Jiang, H.F., X.P. Ren, and S.U. Wang. 2005. Evaluation of resistance to *Aspergillus flavus* invasion and aflatoxins production in groundnut. *Chin. J. Oil Crop Sci.* 27: 21–25.

Jiang, H.F., X.P. Ren, S.U. Wang et al. 2010. Development and evaluation of peanut germplasm with resistance to *Aspergillus flavus* from core collection. *Acta Agron. Sin.* 36: 428–434.

Jiang, H.F., X.P. Ren, S.Y. Wang, and B.S. Liao. 2006. Durability of resistance to *Aspergillus flavus* infection and effect of intact testa without injury on aflatoxin production in peanut. *Acta Agron. Sin.* 32: 851–855.

Jiang, H.F., S.Y. Wang, and X.P. Ren. 2002. Responses of groundnut germplasms to *Aspergillus flavus* invasion. *Chin. J. Oil Crop Sci.* 24: 23–25.

Jochem, S.C.J. 1926. *Aspergillus niger* on groundnut. *Indisch Culturen* (Teysmannia) 11: 325–326.

Joel, A.J., P. Sumathi, T.S. Raveendran, and R.S. Yadav. 2006. Genetics of rust (*Puccinia arachidis* Speg.) and its association with rust related traits in groundnut (*Arachis hypogaea* L.). *Plant Arch.* 6: 553–556.

Johnson, M. and K. Subramanyam. 2003. Management of groundnut late leaf spot and rust through triazole fungicides. *Ann. Plant Prot. Sci.* 11: 395–397.

Johnson, M., and K. Subramanyam. 2009. *Pseudomonas fluorescens* for the management of late leaf spot in groundnut. *Indian J. Plant Prot.* 37: 161–165.

Johnson, M., K. Subramanyam, D. Balaguravaiah, and M.J. Sudheer. 2003. Management of stem rot in groundnut through soil amendments. *Ann. Plant Prot. Sci.* 11: 83–85.

Johnson, M. and K.A. Subramanyam. 2000. In-vitro efficacy of fungicides against stem rot pathogen (*Sclerotium rolfsii*) of groundnut. *Ann. Plant Prot. Sci.* 8: 255–257.

Johnson, M., P.N. Reddy, and D.R. Reddy. 2007a. Management of late leaf spot and stem rot in groundnut. *Indian J. Plant Prot.* 35: 81–85.

Johnson, M., P.N. Reddy, and D.R. Reddy. 2007b. Soil amendments with different forms of nitrogen and potassium fertilizers on stem rot and late leaf spot of groundnut. *Indian J. Plant Prot.* 35: 304–308.

Johnson, M., P.N. Reddy, and D.R. Reddy. 2008a. Comparative efficacy of rhizosphere mycoflora, fungicides, insecticides and herbicides against groundnut stem rot caused by *Sclerotium rolfsii*. *Ann. Plant Prot. Sci.* 16: 414–418.

Johnson, M., P.N. Reddy, and D.R. Reddy. 2008b. Effective management of stem rot of groundnut through application of *Pseudomonas fluorescens*. *Ann. Plant Prot. Sci.* 16: 428–432.

Jordan, D.L., J.E. Bailey, J.S. Barnes et al. 2002. Yield and economic return of ten groundnut-based cropping systems. *Agron. J.* 94: 1289–1294.

Jordan, D.L., J.S. Barness, T. Corbett et al. 2008. Crop response to rotation and tillage in peanut-based cropping systems. *Agron. J.* 100: 1580–1566.

Jordan, D.L. and P.D. Johnson. September 2007. Comparison of irrigation systems and fungicide programs in Virginia market-type peanut. *Crop Manag.* doi:10.1094/CM-2007-0921-01-RS.

Joshi, M.S., P.G. Borkar, and A.M. Mandokhot. 2000. Bio-efficacy of carbendazim and mancozeb-based fungicide in control of early and late leaf spots of groundnut. *Int. Arachis Newsl.* 20: 53–54.

Jyosthna, M.K., N.P.E. Reddy, T.V. Chalam, and G.L.K. Reddy. 2004. Morphological and biochemical characterization of *Phaeoisariopsis personata* resistant and susceptible cultivars of groundnut (*Arachis hypogaea*). *Plant Pathol. Bull.* (Taiwan) 13(4): 243–250.

Kadam, R.M., S.D. Dhavle, R.B. Allapure, and V.S. Nagpurne. 2008b. Protection of pathogenic seedborne fungi of groundnut by using leaf extract of *Azadirachta indica* A. Juss. *Int. J. Plant Prot.* 1(2): 110–111.

Kadam, R.M., N.J.M. Reddy, B.S. Jadhav, and B.S. Nagpurne. 2008a. Aerobiological approach to leaf spot and rust diseases of groundnut (*Arachis hypogaea* L.). *Int. J. Plant Prot.* 1: 63–65.

Kadam, T., S.P.V. Khalikar Nikam, and P. S Nikam. 2011. Survey and surveillance of collar rot of groundnut caused by *Sclerotium rolfsii* in Marathwada Region. *J. Plant Dis. Sci.* 6: 204–05.

Kalaskar, R.R., R.L. Parate, R.R. Rathod, P.S. More, and S. Yeturi. 2012. Management of groundnut rust (*Puccinia arachidis* Speg.) by fungicides, plant extracts and bioagents. *J. Soils Crops* 22: 420–423.

Kale, D.M., G.S.S. Murty, and A.M. Badigannavar. 2007. New Trombay groundnut variety TG 38 suitable for the residual moisture situation in India. *J. SAT Agric. Res.* 3(1): 1–2.

Kale, M.C. and A.D. Choudhary. 2001. Induction of phenylalanine ammonia-lyase in groundnut cultivars in response to biotic and abiotic stress. *Indian Phytopathol.* 54: 288–292.

Kanth, A.K., M.V.C. Gowda, and S. Lingaraju. 2000. Role of partial resistance in integrated management of stem and pod rots in groundnut. *Karnataka J. Agric. Sci.* 133: 726–728.

Karthikeyan, A. 1996. Effect of organic amendments, antagonist *Trichoderma viride* and fungicides on seed and collar rot of groundnut. *Plant Dis. Res.* 11: 72–74.

Kasno, A. 2004. Prevention of *Aspergillus flavus* infection and aflatoxin contamination in groundnut. *J. Penelitian dan Pengembangan Pertanian* (Indonesia) 23(3): 75–81.

Kelly, L.A., M.J. Ryley, P.R. Trevorrow, J.R. Tatnell, C. Nastasi, and Y. Chauhan. 2012. Reduced fungicide use on a new Australian peanut cultivar, highly resistant to the late leaf spot and rust pathogens. *Aust. Plant Pathol.* 41: 359–373.

Kenjale, L.D., P.A. Shinde, and P.G. Utikar. 1981. Loss in groundnut yield due to rust. *Indian Phytopathol.* 38: 374.

Khan, A.R., M. Ijaz, L.U. Haq, A. Farz, and M. Tariqzaved. 2014. Management of *Cercospora* leaf spots of groundnut (*Cercospora arachidicola* and *Cercosporidium personatum*) through the use of systemic fungicides. *Cercetări Agronomice* (Moldova) 67: 97–102.

Khedikar, Y.P., M.V.C. Gowda, C. Sarvamangala, K.V. Patgar, H.D. Upadhyaya, and R.K. Varshney. 2010. A QTL study on late leaf spot and rust revealed one major QTL for molecular breeding for rust resistance in groundnut (*Arachis hypogaea* L.). *Theor. Appl. Genet.* 121: 971–984.

Kiran, K.S., S. Lingaraju, and S.S. Adiver. 2006. Effect of plant extracts on *Sclerotium rolfsii*, the incitant of stem rot of groundnut. *J. Mycol. Plant Pathol.* 36: 77–79.

Kishore, G.K. and S. Pande. 2005a. Integrated applications of aqueous leaf extract of *Datura metel* and chlorothalonil improved control of late leaf spot and rust of groundnut. *Aust. Plant Pathol.* 34: 261–264.

Kishore, G.K. and S. Pande. 2005b. Integrated management of late leaf spot and rust diseases of groundnut (*Arachis hypogaea* L.) with *Prosopis juliflora* leaf extract and chlorothalonil. *Int. J. Pest Manag.* 51: 327–334.

Kishore, G.K., S. Pande, and G. Harish. 2007. Evaluation of essential oils and their components for broad-spectrum antifungal activity and control of late leaf spot and crown rot diseases in peanut. *Plant Dis.* 91: 375–379.

Kishore, G.K., S. Pande, and A.R. Podile. 2005a. Biological control of late leaf spot of peanut (*Arachis hypogaea*) with chitinolytic bacteria. *Phytopathology* 95: 1157–1165.

Kishore, G.K., S. Pande, and A.R. Podile. 2005b. Management of late leaf spot of groundnut (*Arachis hypogaea*) with chlorothalonil-tolerant isolates of *Pseudomonas aeruginosa*. *Plant Pathol.* 54: 401–408.

Kishore, G.K., S. Pande, J.N. Rao, and A.R. Podile. 2005c. *Pseudomonas aeruginosa* inhibits the plant cell wall degrading enzymes of *Sclerotium rolfsii* and reduces the severity of groundnut stem rot. *Eur. J. Plant Pathol.* 113: 315–320.

Kishore, G.K., S. Pande, and A.R. Podile. 2006. *Pseudomonas aeruginosa* GSE 18 inhibits the cell wall degrading enzymes of *Aspergillus niger* and activates defence-related enzymes of groundnut in control of collar rot disease. *Aust. Plant Pathol.* 35: 259–263.

Kishore, G.K., S. Pande, and J.N. Rao. 2001a. Control of late leaf spot of groundnut (*Arachis hypogaea*) by extracts from non-host plant species. *Plant Pathol. J.* 17: 264–270.

Kishore, G.K., S. Pande, and J.N. Rao. 2001b. Control of foliar diseases of groundnut using inorganic and metal salts. *Int. Arachis Newsl.* 21: 33–35.

Kishore, G.K., S. Pande, and J.N. Rao. 2002. Field evaluation of plant extracts for the control of late leaf spot in groundnut. *Int. Arachis Newsl.* 22: 46–48.

Kishore, G.K., S. Pande, J.N. Rao, and A.R. Podile. 2001c. Biological control of crown rot of groundnut by *Trichoderma harzianum* and *T. viride*. *Int. Arachis Newsl.* 21: 39–40.

Kisyombi, C.T., I.M.C. Phiri, A.R. Saka, and E.H.C. Chilembwe. 2001. An assessment of grain yield losses caused by foliar fungal diseases in some groundnut (*Arachis hypogaea* L.) genotypes in Malawi. *Proceedings of the First Annual Science Conference*, Malawi Institute of Management, Lilongwe, Malawi, November 6–10, 2000, pp. 303–318.

Kodmelwar, R.V. and A.Y. Ingle. 1989. Effect of sowing dates, spacing and meteorological factors on the development of tikka and rust of groundnut. *Indian Phytopathol.* 42: 274.

Koïta, K., F.B. Neya, A.T. Nana, and P. Sankara. 2012. Antifungal activity of plant extracts against Burkina Faso local *Puccinia arachidis* Speg., rust pathogen of groundnut (*Arachis hypogaea* L.). *Kenya J. Appl. Biosci.* 57: 4142–4150.

Kokalis-Burelle, N., D.M. Porter, R. Rodriguez-Kabana, D.H. Smith, and P. Subrahmanyam. 1997. *Compendium of Peanut Diseases*, 2nd edn. APS Press, St. Paul, MN, 94pp.

Kolte, S.J. 1984. *Diseases of Annual Edible Oilseed Crops*, Vol. I: Peanut Diseases. CRC Press, Boca Raton, FL.

Kolte, S.J. 1994. Disease problems in seed crop of oilseeds and methods to tackle them. *Indian Farming* (ICAR, India) 44: 7–12.

Kolte, S.J. 1997. Annual oilseed crops. In: *Soilborne Diseases of Tropical Crops*, eds., R.J. Hillocks and J.M. Waller. CAB International, Wallingford, U.K., pp. 253–276.

Kondreddy, A. and A.R. Podile. 2012. HarpinPss-mediated enhancement in growth and biological control of late leaf spot in groundnut by a chlorothalonil-tolerant *Bacillus thuringiensis* SFC24. *Microbiol. Res.* 167: 194–198.

Kong, Q., S.H. Shan, Q.Z. Liu, X.D. Wang, and F.T. Yu. 2010. Biocontrol of *Aspergillus flavus* on peanut kernels by use of a strain of marine *Bacillus megaterium*. *Int. J. Food Microbiol.* 139: 31–35.

Kono, A. 1977. On the uredospore germination of peanut leaf rust (*Puccinia arachidis* Speg) on peanut leaves. *Bull. Fac. Agric. Miyazaki Univ.* 24: 225.

Krishna, A. and R.A. Singh. 1980. *Hansfordia pulvinata* mycoparasite on *Cercospora* species causing tikka disease of groundnut. *Indian Phytopathol.* 32: 318.

Krishnakanth, A., M.V.C. Gowda, and G.K. Naidu. 2003. Evaluation of Spanish type groundnuts for resistance to stem-and pod-rot caused by *Sclerotium rolfsii* Sacc. *J. Oilseeds* 20: 254–256.

Krishnakanth, A., G.K. Naidu, and M.V. Gowda. 2005. Selection for resistance to stem and pod rot in groundnut, *Arachis hypogaea* L. *J. Oilseeds Res.* 22: 226–227.

Kucharek, T.A. 1975. Reduction of *Cercospora* leaf spots of peanut with crop rotation. *Plant Dis. Rep.* 59: 822.

Kucharek, T.A., J.D. Atkins, D.W. Gorbet, and R.C. Kemerait. 2000. The performance of select genotypes of peanut to natural inocula of *Cylindrocladium* black rot and tomato spotted wilt virus in Florida. *Soil Crop Sci. Soc. Florida* 59: 72–76.

Kumar, A.L.R. and P. Balasubramanian. 2000. Induction of phenols in groundnut rust resistance. *Int. Arachis Newsl.* 20: 55–57.

Kumar, K.V.K., S. Desai, V.P. Rao, H.A. Nur, P. Srilakshmi, and R.P. Thakur. 2002. Evaluation of an integrated management package to reduced preharvest seed infection by *Aspergillus flavus* in groundnut. *Int. Arachis Newsl.* 22: 42–44.

Kumar, K.R.R. and P.B. Kirti. 2011. Differential gene expression in *Arachis diogoi* upon interaction with peanut late leaf spot pathogen, *Phaeoisariopsis personata* and characterization of a pathogen induced cyclophilin. *Plant Mol. Biol.* 75: 497–513.

Kumar, V., N.B. Bagwan, U.M. Vyas, and D. Singh. 2008. Dynamics of soil population of *Aspergillus flavus* and aflatoxin contamination in groundnut based production system in Gujarat. *Indian Phytopathol.* 61: 343–347.

Kumar, V., C. Lukose, N.B. Bagwan, V.G. Koradia, and R.D. Padavi. 2012. Occurrence of *Alternaria* blight of groundnut in Gujarat and reaction of some genotypes against the disease. *Indian Phytopathol.* 65: 25–30.

Kumar, V.M., N. Harisatyanarayana, and K.M. Kumar. 2012. Evaluation of promising ground genotypes for yield and their reaction to leaf spot diseases in North Central Zones of Andhra Pradesh. *Int. J. Plant Prot.* 5: 319–323.

Kumari, A.M., U. Maheswari Devi, and A. Sucharitha. 2011. Differential effect of lipoxygenase on aflatoxin production by *Aspergillus* spp. *Int. J. Plant Pathol.* 2: 153–164.

Kumari, A.M., A. Sucharitha, and P.U.M. Devi. 2012. Impact of 13-lipoxygenase on induction of resistance in peanut against aflatoxigenic *Aspergillus*. *Continental J. Biol. Sci.* 5: 4–13.

Kumari, S.S. Adiver, V.I. Benangi, A.S. Byadgi, and H.L. Nadaf. 2009a. Molecular variation in *Phaeoisariopsis personata* (Berk. and M.A. Curtis) van Arx causing late leaf spot of groundnut (*Arachis hypogaea* L.). *Karnataka J. Agric. Sci.* 22: 336–339.

Kumari, S.S. Adiver, S.B. Mallesh, M.A. Pasha, and R.K. Singh. 2009b. Isozyme variability in *Phaeoisariopsis personata* (Berk, and Curt.) von Arx causing late leaf spot of groundnut (*Arachis hypogaea* L.). *Int. J. Plant Prot.* 2: 219–223.

Kumari, V., N. Jaiswal, M.V.C. Gowda, and M.K. Meena. 2012. Evaluation of rust and late leaf spot mapping RILs population in groundnut using AhMITE 1 specific PCR. *Int. J. Plant Prot.* 5: 167–171.

Lanier, J.E., D.L. Jordan, J.S. Barnes et al. 2004. Disease management in overhead sprinkler and subsurface drip irrigation systems for peanut. *Agron. J.* 96: 1058–1065.

Lashin, S.M., H.I.S. El Nasr, M.A.A. El Nagar, and M.A. Nofal. 1989. Biological control of *Aspergillus niger* the causal organism of peanut crown rot by Trichoderma harzianum. *Ann. Agric. Sci.* 34: 795–803.

Latha, P., P. Sudhakar, Y. Sreenivasulu, P.H. Naidu, and P.V. Reddy. 2007. Relationship between total phenols and aflatoxin production of peanut genotypes under end-of-season drought conditions. *Acta Physiol. Plantarum* 29: 563–566.

Le, C.N. 2011. *Diversity and Biological Control of Sclerotium rolfsii, Causal Agent of Stem Rot of Groundnut*. Wageningen University, Wageningen, the Netherlands, 152pp.

Le, C.N., M. Kruijt, and J.M. Raaijimakers. 2012b. Involvement of phenazines and lipopeptides in interactions between *Pseudomonas* species and *Sclerotium rolfsii*, causal agent of stem rot disease on groundnut. *J. Appl. Microbiol.* 112: 390–403.

Le, C.N., R. Mendes, M. Kruijt, and J.M. Raaijmakers. 2012a. Genetic and phenotypic diversity of *Sclerotium rolfsii* in groundnut fields in central Vietnam. *Plant Dis.* 96: 389–397.

Leal-Bertioli, S.C.D.M., A.C.V.F. Jose, and D.M.T. Alves-Freitas. 2009. Identification of candidate genome regions controlling disease resistance in Arachis. *BMC Plant Biol.* 9: 112.

Leal-Bertioli, S.C.D.M., M.P.D. Farias, P.Í.T. Silva et al. 2010. Ultra-structure of the initial interaction of *Puccinia arachidis* and *Cercosporidium personatum* with leaves of *Arachis hypogaea* and *Arachis stenosperma. J. Phytopathol.* 158: 792–796.

Lei, Y., B.S. Liao, S.Y. Wang, Y.B. Zhang, D. Li, and H.F. Jiang. 2006. A SCAR marker for resistance to *Aspergillus flavus* in peanut (*Arachis hypogaea* L.). *Hereditas* (Beijing) 28: 1107–1111.

Lemay, A.V., J.E. Bailey, and B.B. Shew. 2002. Resistance of peanut to *Sclerotinia* blight and the effect of acibenzolar-*S*-methyl and fluazinam on disease incidence. *Plant Dis.* 86: 1315–1317.

Li, Y., C.Y. Chen, A.K. Culbreath, C.C. Holbrook and B. Guo. 2012. Variability in field response of peanut genotypes from the U.S. and China to Tomato spotted wilt virus and leaf spots. *Peanut Sci.* 39: 30–37.

Li, Z.C. and Q.S. Qiu. 2000. Huayu 16: A new high-yielding, improved quality groundnut cultivar with wide adaptability for northern China. *Int. Arachis Newsl.* 20: 31–32.

Liao, B.S., Y. Lei, S.Y. Wang, D. Li, H.F. Jiang, and X.P. Ren. 2003. Aflatoxin resistance in bacterial wilt resistant groundnut germplasm. *Int. Arachis Newsl.* 23: 24.

Liao, B.S., W.J. Zhuang, R.H. Tang et al. 2009. Peanut aflatoxin and genomics research in China: Progress and perspectives. *Peanut Sci.* 36: 21–28.

Liang, X.Q., C.C. Holbrook, R.E. Lynch, and B.Z. Guo. 2005. β-1,3-Glucanase activity in peanut (*Arachis hypogaea*) seed is induced by inoculation with *Aspergillus flavus* and copurifies with a conglutin-like protein. *Phytopathology* 95: 506–511.

Liang, X.Q., M. Luo, and B.Z. Guo. 2006. Resistance mechanisms to *Aspergillus flavus* infection and aflatoxin contamination in peanut (*Arachis hypogaea*). *Plant Pathol. J.* (Faisalabad) 5: 115–124.

Liang, X.Q., R.C. Pan, and G.Y. Zhou. 2002. Active oxygen generation and lipid peroxidation as related to *Aspergillus flavus* resistance/susceptibility in peanut cultivars. *Chin. J. Oil Crop Sci.* 24: 19–23.

Liang, X.Q., R.C. Pan, and G.Y. Zhou. 2003. Relationship of trypsin inhibitor in peanut seed and resistance to *Aspergillus flavus* invasion. *Acta Agron. Sin.* 29: 295–299.

Liang, X.Q., G.U. Zhou, Y.B. Hong, X.P. Chen, H.Y. Liu, and S.X. Li. 2009. Overview of research progress on peanut (*Arachis hypogaea* L.) host resistance to aflatoxin contamination and genomics at the Guangdong Academy of Agricultural Sciences. *Peanut Sci.* 36: 29–34.

Liang, X.Q., G.Y. Zhou, and S.C. Zou. 2006a. Differential induction of resveratrol in susceptible and resistant peanut seeds infected by *Aspergillus flavus. Chin. J. Oil Crop Sci.* 28: 59–62.

Lin, Y.S. 1982. Factors influencing the development of peanut crown rot caused by *Aspergillus niger* and the measures for disease management. *J. Agric. Res. China* 31: 144.

Lisker, N., R. Michaeli, and Z.R. Frank. 1993. Mycotoxigenic potential of *Aspergillus flavus* isolated from groundnuts growing in Israel. *Mycopathologia* 122: 177–183.

Liu, Y., C.J. Li., Y.X. Zheng, C.X. Yan, T.T. Zhang, and S.H. Shan. 2012. Effects of drought during peanut (*Arachis hypogaea* L.) seeding stage on disease resistance gene PnAG3 expression and physiological index related to drought resistance. *J. Agric. Biotechnol.* 20: 642–648.

Livingstone, D.M., J.L. Hampton, P.M. Phipps, and E.A. Grabau. 2005. Enhancing resistance to *Sclerotinia minor* in peanut by expressing a barley oxalate oxidase gene. *Plant Physiol.* 137: 1354–1362.

Lokesh, B.K., S.N. Upperi, G.N. Maraddi, Amaresh, and V.B. Nargund. 2008. Evaluation of chemical fungicides and genotypes for tikka, rust and bud necrosis virus disease of groundnut. *Environ. Ecol.* 26: 287–289.

Luo, M., P. Dang, M.G. Bausher, C.C. Holbrook et al. 2005. Identification of transcripts involved in resistance responses to leaf spot disease caused by *Cercosporidium personatum* in peanut (*Arachis hypogaea*). *Phytopathology* 95: 381–387.

Maas, A.L., K.E. Dashiell, and H.A. Melouk. 2006. Planting density influences disease incidence and severity of *Sclerotinia* blight in peanut. *Crop Sci.* 46: 1341–1345.

Mace, E.S., D.T. Phong, H.D. Upadhyaya, S. Chandra, and J.H. Crouch. 2006. SSR analysis of cultivated (*Arachis hypogaea* L.) germplasm resistant to rust and late leaf spot diseases. *Euphytica* 152: 317–330.

Madhaiyan, M., B.V.S. Reddy, R. Anandham et al. 2006. Plant growth-promoting *Methylobacterium* induces defense responses in groundnut (*Arachis hypogaea* L.) compared with rot pathogens. *Curr. Microbiol.* 53: 270–276.

Maheshwari, D.K., R.C. Dubey, and V.K. Sharma. 2001. Biocontrol effects of *Trichoderma virens* on *Macrophomina phaseolina* causing charcoal rot of peanut. *Indian J. Microbiol.* 41: 251–256.

Maiti, M.K., S.K. Raj, and S. Das. 2005. Management of leaf spot (*Cercospora arachidicola* and *Phaeoisariopsis personata*) of groundnut (*Arachis hypogaea*) by seed treatment with non-conventional chemicals. *Indian J. Agric. Sci.* 75: 452–453.

Malathi, P. and S. Doraisamy. 2004. Effect of seed priming with *Trichoderma* on seedborne infection by *Macrophomina phaseolina* and seed quality of groundnut. *Ann. Plant Prot. Sci.* 12: 87–91.

Mallaiah, K.V. 1976. A note on the seasonal changes in the incubation time of groundnut rust. *Curr. Sci.* (India) 45: 737.

Mallaiah, K.V. and A.S. Rao. 1979. Groundnut rust: Factors influencing disease development, sporulation and germination of uredospores. *Indian Phytopathol.* 32: 382.

Mallaiah, K.V. and A.S. Rao. 1982. Aerial dissemination of uredospores of groundnut rust. *Trans. Br. Mycol. Soc.* 78: 21.

Mallikarjuna, N. 2005. Production of hybrids between *Arachis hypogaea* and *A. chiquitana* (section *Procumbentes*). *Peanut Sci.* 32: 148–152.

Mallikarjuna, N., D.R. Jadhav, K. Reddy, F. Husain, and K. Das. 2012. Screening new *Arachis amphidiploids*, and autotetraploids for resistance to late leaf spot by detached leaf technique. *Eur. J. Plant Pathol.* 132: 17–21.

Mallikarjuna, N., S. Pande, D.R. Jadhav, D.C. Sastri, and J.N. Rao. 2004. Introgression of disease resistance genes from *Arachis kempff-mercado* into cultivated groundnut. *Plant Breed.* 123: 573–576.

Mallikarjuna, N. and R.K. Varshney, eds. 2014. *Genetics, Genomics and Breeding of Peanuts*. CRC Press, Boca Raton, FL.

Manjula, K., A.G. Girish, S.D. Singh, and G.K. Kishore. 2004. Combined application of *Pseudomonas fluorescens* and *Trichoderma viride* has an improved biocontrol activity against stem rot in groundnut. *Plant Pathol. J.* 20: 75–80.

Marchionatto, J.B. 1922. Peanut wilt in Argentina. *Revista de la Facultad de Agronomia, Universidad Nacional de La Plata* 3: 65.

Marinelli, A., G.J. March, and C. Oddino. 2008. Biological and epidemiological aspects of peanut (*Arachis hypogaea* L.) smut caused by *Thecaphora frezii* Carranza & Lindquist. *AgriScientia* 25(1/2): 1–5.

Marinelli, A., G.J. March, A. Rago, J. Giuggia, and M. Kearney. 2001. Outbreaks of *Sclerotinia* blight of peanut caused by *Sclerotinia minor* in Argentina. *Boletín de Sanidad Vegetal, Plagas* 27(1): 75–84.

Mathews, C., M.D. Lengwati, M.F. Smith, S.N. Nigam, and K.Z. Ahmed. 2007. New groundnut varieties for smallholder farmers in Mpumalanga, South Africa. *Eighth African Crop Science Society Conference*, El-Minia, Egypt, October 27–31, 2007, pp. 251–257.

Mathivanan, N. 2000. Chitinase and β-1,3 glucanase of *Fusarium solani*: Effect of crude enzymes on *Puccinia arachidis*, groundnut rust. *J. Mycol. Plant Pathol.* 30: 327–330.

Mathivanan, N. and K. Murugesan. 2000. *Fusarium chlamydosporum*, a potent biocontrol agent to groundnut rust, *Puccinia arachidis*. *Zeitschrift für Pflanzenkrankheiten und Pflanzenschutz* 107: 225–234.

Mathur, S. and A. Doshi. 1990. Field evaluation of some fungicides for the control of leaf spots and rust diseases of groundnut. *Summa Phytopathol.* 16: 162–165.

Mayee, C.D. 2009. Enhancing productivity of groundnut in India through IDM. In: *Integrated Pest and Disease Management*, eds., R.V. Upadhyay, K.G. Mukerji, B.P. Chamola, and O.P. Dubey. APH Publishing Corporation, New Delhi, India, pp. 344–357.

Mayee, C.D., G.M. Godbole, and F.S. Patil. 1977. Appraisal of groundnut rust in India: Problems and approach. *PANS* 23: 162.

McAlpin, C.E., D.T. Wicklow, and B.W. Horn. 2002. DNA fingerprinting analysis of vegetative compatibility groups in *Aspergillus flavus* from a peanut field in Georgia. *Plant Dis.* 86: 254–258.

McVey, D.V. 1965. Inoculation and development of rust on peanut grown in the greenhouse. *Plant Dis. Rep.* 49: 191–192.

Mehan, V.K., D. Mc Donald, L.J. Haravu, and S. Jayanthi. 1991. *The Groundnut Aflatoxin: Review and Literature Database*. International Crops Research Institute for the Semi-Arid Tropics (ICRISAT), Patancheru, India.

Meena, B. 2010. Survival and effect of *Pseudomonas fluorescens* formulation developed with various carrier materials in the management of late leaf spot of groundnut. *Int. J. Plant Prot.* 3: 200–202.
Meena, B. and Marimuthu, T. 2012. Effect of application methods of *Pseudomonas fluorescens* for the late leaf spot of groundnut management. *J. Biopest.* 5: 14–17.
Meena, B., T. Marimuthu, and R. Velazhahan. 2001a. Salicylic acid induces systemic resistance in groundnut against late leaf spot caused by *Cercosporidium personatum. J. Mycol. Plant Pathol.* 31: 139–145.
Meena, B., T. Marimuthu, and R. Velazhahan. 2006. Role of fluorescent pseudomonads in plant growth promotion and biological control of late leaf spot of groundnut. *Acta Phytopathologica et Entomologica Hungarica* 41: 203–212.
Meena, B., T. Marimuthu, P. Vidhyasekaran, and R. Velazhahan. 2001b. Biological control of root rot of groundnut with antagonistic *Pseudomonas fluorescens* strains. *Zeitschrift für Pflanzenkrankheiten und Pflanzenschutz* 108: 369–381.
Meena, B., R. Radhajeyalakshmi, T. Marimuthu, P. Vidhyasekaran, and R. Velazhahan. 2002. Biological control of groundnut late leaf spot and rust by seed and foliar applications of a powder formulation of *Pseudomonas fluorescens. Biocontrol Sci. Technol.* 12: 195–204.
Meena, B., V. Ramamoorthy, T. Marimuthu, and R. Velazhahan. 2000. *Pseudomonas fluorescens* mediated systemic resistance against late leaf spot of groundnut. *J. Mycol. Plant Pathol.* 30: 151–158.
Meena, P.D. and C. Chattopadhyay. 2002. Effect of some physical factors, fungicides on growth of *Rhizoctonia solani* Kuhn and fungicidal seed treatment on groundnut seed germination. *Indian J. Plant Prot.* 30: 172–176.
Mehan, V.K. and N.X. Hong. 1994. Disease constraints to groundnut production in Vietnam: Research and management strategies. *Int. Arachis Newsl.* 14: 8–11.
Middleton, K. and R. Shorter. 1987. Occurrence and management of groundnut rust in Australia. Groundnut rust disease. *Proceedings of a Discussion of Group Meeting*, ICRISAT Centre, Patancheru, India, September 24–28, 1984.
Middleton, K.J. 1980. Groundnut production utilization research problems and further needs in Australia. In: *Proceedings of International Workshop on Groundnuts*, eds., W. Gibbons and J.V. Mertin. International Crops Research Institute for the Semi-Arid Tropics (ICRISAT), Patancheru, India, pp. 223–225.
Minton, N.A., A.S. Csinos, R.E. Lynch, and T.B. Brenneman. 1991. Effects of two cropping and two tillage systems and pesticides on peanut pest management. *Peanut Sci.* 18: 41–46.
Minton, N.A., A.S. Csinos, and L.W. Morgan. 1990. Relationships between tillage and nematodes, fungicides and insecticide treatments on pests and yield and peanuts double-cropped with wheat. *Plant Dis.* 74: 1025–1029.
Mohale, S., J. Allotey, and B.A. Siame. 2010. Control of *Tribolium confusum* J. du Val by diatomaceous earth (Protect-It™) on stored groundnut (*Arachis hypogaea*) and *Aspergillus flavus* link spore dispersal. *Afr. J. Food Agric. Nutr. Dev.* 6: 2678–2694.
Mohamed, A. and A. Chala. 2014. Incidence of *Aspergillus* contamination of groundnut (*Arachis hypogaea* L.) in Eastern Ethiopia. *Afr. J. Microbiol.* 8: 759–765.
Mohammed, S.E.N., P. Vindhiyavarman and K. Sachithanandam. 2001. Evaluation of foliar disease resistant groundnut varieties of ICRISAT at Vriddhachalam, India. *Intl. Arachis News L.* 21: 12–13.
Mohapatra, K.B. 2011. Effect of *Aspergillus niger* and its culture filtrate on seed germination and seedling vigour of groundnut. *J. Plant Prot. Environ.* 8: 61–64.
Mohapatra, K.B. and M.K. Sahoo. 2011. Pathogenicity and biocontrol of *Aspergillus niger* the causal agent of seed and collar rot of groundnut. *J. Plant Prot. Environ.* 8: 78–81.
Momotaz, R., M.M. Islam, M.M. Karim, M.N. Islam, and M.S. Ali. 2009. Control of foot rot disease of groundnut caused by *Sclerotium rolfsii. Bangladesh J. Plant Pathol.* 25(1/2): 61–65.
Mondal, S. and A.M. Badigannavar. 2010. Molecular diversity and association of SSR markers to rust and late leaf spot resistance in cultivated groundnut (*Arachis hypogaea* L.). *Plant Breed.* 129: 68–71.
Mondal, S., A.M. Badigannavar, and S.F. D'Souza. 2012. Molecular tagging of a rust resistance gene in cultivated groundnut (*Arachis hypogaea* L.) introgressed from *Arachis cardenasii. Mol. Breed.* 29: 467–476.
Mondal, S., A.M. Badigannavar, and G.S.S. Murty. 2008b. RAPD markers linked to a rust resistance gene in cultivated groundnut (*Arachis hypogaea* L.). *Euphytica* 159: 233–239.
Mondal, S., S. Ghosh, and A.M. Badigannavar. 2005. RAPD polymorphism among groundnut genotypes differing in disease reaction to late leaf spot and rust. *Int. Arachis Newsl.* 25: 27–30.
Mondal, S., S.R. Sutar, and A.M. Badigannavar. 2008a. Comparison of RAPD and ISSR marker profiles of cultivated peanut genotypes susceptible or resistant to foliar diseases. *J. Food Agric. Environ.* 6: 181–187.
Mondal, S., S.R. Sutar, and A.M. Badigannavar. 2009. Assessment of genetic diversity in cultivated groundnut (*Arachis hypogaea* L.) with differential responses to rust and late leaf spot using ISSR markers. *Indian J. Genet. Plant Breed.* 69: 219–224.

Monfort, W.S., A.K. Culbreath, K.L. Stevenson, T.B. Brenneman, D.W. Gorbet, and S.C. Phatak. 2004. Effects of reduced tillage, resistant cultivars, and reduced fungicides inputs on progress of early leaf spot of peanut (*Arachis hypogaea*). *Plant Dis.* 88: 858–864.

Moraes, A.R.A., S.A. Moraes, A.L. Lourenção, I.J. Godoy, and A.L.M. Martins. 2006. Effect of thiamethoxam application to thrips control on the reduction of peanut scab severity. *Fitopatol. Bras.* 31: 164–170.

Moraes, S.A., I.J. Godoy, J.C.V.N.A. Pereira, and A.L.M. Martins. 2002. Rainfall-based advisories for chemical control of peanut late leaf spot on IAC-Caiapó cultivar. *Summa Phytopathol.* 28: 229–235.

Moraes, S.A., I.J. Godoy, J.R.M. Pezzopane, J.C.V.N.A. Pereira, and L.C.P. Silveira. 2001. Efficiency of fungicides in the control of peanut late leaf spot and scab by monitoring method. *Fitopatol. Bras.* 26: 134–140.

Morwood, R.B. 1953. Peanut pre-emergence and crown rot investigations. *Qld. J. Agric. Sci.* 10.

Motagi, B.N., M.V.C. Gowda, and G.K. Naidu. 2000. Inheritance of late leafspot resistance in groundnut mutants. *Indian J. Genet. Plant Breed.* 60: 347–352.

Motagi, B.N., M.V.C. Gowda, and N. Prabhakar. 2004. Biochemical basis of resistance to late leaf spot in groundnut. *Plant Pathol. Newsl.* 22: 19–20.

Mozingo, R.W., T.A. Coffelt, C.W. Swannand, and P.M. Phipps. 2004. Registration of 'Wilson' peanut. *Crop Sci.* 44: 1017–1018.

Muhammad, A.S. and B.S. Bdliya. 2011. Effects of variety and fungicidal rate on *Cercospora* leaf spots disease of groundnut in the Sudan Savanna. *J. Basic Appl. Sci.* 19: 135–141.

Muhammad, I., C.A. Rauf, I.U. Haque, F.U. Hussan, and A. Mahmood. 2008. Distribution and severity of *Cercospora* leaf spot of peanut in rainfed region of Punjab. *Pak. J. Phytopathol.* 20: 165–172.

Naab, J.B., S.S. Seini, K.O. Gyasi et al. 2009. Groundnut yield response and economic benefits of fungicides and phosphorus application in farmer-management trials in Northern Ghana. *Exp. Agric.* 45: 385–399.

Naab, J.B., F.K. Tsigbey, P.V.V. Prasad, K.J. Boote, J.E. Bailey, and R.L. Brandenburg. 2005. Effects of sowing date and fungicide application on yield of early and late maturing peanut cultivars grown under rainfed conditions in Ghana. *Crop Prot.* 24: 325–332.

Naeem, U.D., M. Abid, G.S.S. Khatak et al. 2009. High yielding groundnut (*Arachis hypogaea* L.) variety "Golden". *Pak. J. Bot.* 41: 2217–2222.

Naidu, P.H. and R.P. Vasanthi. 2002. Influence of sowing dates on tikka late leaf spot (*Phaeoisariopsis personata*) in rabi and summer groundnut. *Legume Res.* 25: 279–281.

Nakova, M., J. Eniola, and B. Nakov. 2003a. *Verticillium* wilt on groundnuts in Bulgaria. *Rasteniev'dni Nauki* 40: 260–265.

Nakova, M., S. Falireas, and H. Kaciulis. 2003b. Integrated control of *Verticillium* wilt on groundnuts in Bulgaria. *Rasteniev'dni Nauki* 40: 569–574.

Nandagopal, V. and M.P. Ghewande. 2004. Use of neem products in groundnut pest management in India. *Nat. Prod. Radiance* 3: 150–155.

Narayana, L., S.P. Raut, and U.A. Gadre. 2006. Linear disease prediction model in groundnut rust epidemics. *Ann. Plant Prot. Sci.* 14: 173–176.

Ncube, E., B.C. Flett, C. Waalwijk, and A. Viljoen. 2010. Occurrence of aflatoxins and aflatoxin-producing *Aspergillus* spp. associated with groundnut production in subsistence farming systems in South Africa. *South Afr. J. Plant Soil* 27: 195–198.

Nelin, M. June 1992. *White Mould in Peanuts*. Peanut Grower, pp. 18–20.

Nema, K.G., A.C. Jain, and R.P. Asthana. 1955. Further studies on *Aspergillus* blight of groundnut seedlings: Its occurrence and control. *Indian Phytopathol.* 8: 13–21.

Nesci, A., A. Montemarani, and M. Etcheverry. 2011. Assessment of mycoflora and infestation of insects, vector of *Aspergillus* section *Flavi*, in stored peanut from Argentina. *Mycotoxin Res.* 27: 5–12.

Nevill, D.J. 1980. Studies of resistance to foliar diseases. In: *Proceedings of the International Workshop on Groundnuts*, eds., R.W. Gibbons and J.V. Mertin. ICRISAT, Patancheru, India, October 13–17, 1980, p. 199.

Nigam, S.N. 2000. Some strategic issues in breeding for high and stable yield in groundnut in India. *J. Oilseed Res.* 17: 1–10.

Nigam, S.N., F. Waliyar, R. Aruna et al. 2009. Breeding peanut for resistance to aflatoxin contamination at ICRISAT. *Peanut Sci.* 36: 42–49.

Niu, C., Y. Akasaka-Kennedy, P. Faustinelli et al. 2009. Antifungal activity in transgenic peanut (*Arachis hypogaea* L.) conferred by a nonheme chloroperoxidase gene. *Peanut Sci.* 36: 126–132.

Nobile, P.M., C.R. Lopes, C. Basalobres-Cavallari et al. 2008. Peanut genes identified during initial phase of *Cercosporidium personatum* infection. *Plant Sci.* 174: 78–87.

Noriega-Cantú, D.H., J. Pereyra-Hernández, I.C. Joaquín-Torres et al. 2000. Epidemiology of late leaf spot and rust of groundnut in Guerrero, Mexico. *Int. Arachis Newsl.* 20: 40–42.

Ntare, B.R., F. Waliyar, A.H. Mayeux, and H.Y. Bissala. 2006. Strengthening conservation and utilization of ground-nut (*Arachis hypogaea* L.) genetic resources in West Africa. *Plant Genet. Resour. Newsl.* (Rome, Italy) 147: 18–24.

Nutsugah, S.K., M. Abudulai, C. Oti-Boateng, R.L. Brandenburg, and D.L. Jordan. 2007a. Management of leaf spot diseases of peanut with fungicides and local detergents in Ghana. *Pak. Plant Pathol. J.* 6: 248–253.

Nutsugah, S.K., C. Oti-Boateng, F.K. Tsigbey, and R.L. Brandenburg. 2007b. Assessment of yield losses due to early and late leaf spots of groundnut (*Arachis hypogaea* L.). *Ghana J. Agric. Sci.* 40: 21–27.

Nutter, F.W. Jr. and F.M. Shokes. 1995. Management of foliar diseases caused by fungi. In: *Peanut Health Management*, eds., H.A. Melouse and F.M. Shokes. American Phytopathological Society, St. Paul, MN, pp. 65–74.

Obagwu, J. 2003a. Evaluation of bambara groundnut (*Vigna subterranea* (L.) Verdc) lines for reaction to *Cercospora* leaf spot. *J. Sustain. Agric.* 22: 93–100.

Obagwu, J. 2003b. Control of brown blotch of bambara groundnut with garlic extract and benomyl. *Phytoparasitica* 31: 207–209.

O'Brien, R.G. 1977. Observations on the development of groundnut rust in Australia. *PANS* 23: 297.

Oddino, C.M., A.D. Marinelli, M. Zuza, and G.J. March. 2008. Influence of crop rotation and tillage on incidence of brown root rot of peanut caused by *Fusarium solani* in Argentina. *Can. J. Plant Pathol.* 30: 575–580.

Okabe, I. and N. Matsumoto. 2000. Population structure of *Sclerotium rolfsii* in peanut fields. *Mycoscience* 41: 145–148.

Okwulehie, I.C. 2000. Translocation of ^{14}C-(labelled) photosynthates in groundnut (*Arachis hypogaea* L.) infected with *Macrophomina phaseoli* (Maub) Ashby. *Photosynthetica* 38: 473–476.

Okwulehie, I.C. 2004. Studies on *Macrophomina phaseoli* (Maub.) Ashby growth and some physiological aspect of groundnut (*Arachis hypogaea* L.) plant infected with the fungus. *Global J. Pure Appl. Sci.* 10: 23–29.

Okwulehie, I.C. and M.L. Okpara. 2002. Anatomical changes in groundnut due to infection with *Macrophomina phaseoli* (Maub) Ashby. *J. Sustain. Agric. Environ.* 4: 249–257.

Olatinwo, R.O., T.V. Prabha, J.O. Paz, and G. Hoogenboom. 2012. Predicting favorable conditions for early leaf spot of peanut using output from the Weather Research and Forecasting (WRF) model. *Int. J. Biometeorol.* 56: 259–268.

Oliveira, C.A.F., N.B. Goncalves, R.E. Rosim, and A.M. Fernandes. 2009. Determination of aflatoxins in peanut products in the northeast region of São Paulo, Brazil. *Int. J. Mol. Sci.* 10: 174–183.

Ozias-Akins, P., H. Yang, R. Gill, H. Fan, and R.E. Lynch. 2002. Reduction of aflatoxin contamination in peanut: A genetic engineering approach. In: *Crop Biotechnology*, eds., K. Rajasekaran, T.J. Jacks and J.W. Finley, American Chemical Society, Washington D.C. Chapter 12, pp. 151–160.

Ozgonen, H., D.S. Akgul, and A. Erkîlic. 2010. The effect of arbuscular mycorrhizal fungi on yield and stem rot caused by *Sclerotium rolfsii* Sacc. in peanut. *Afr. J. Agric. Res.* 5: 128–132.

Padi, F.K. 2008. Genotype × environment interaction for yield and reaction to leaf spot infections in groundnut in semiarid West Africa: Genotype × environment interaction and leaf spot resistance in groundnut. *Euphytica* 164: 143–161.

Padi, F.K., A. Frimpong, J. Kombiok, A.B. Salifu, and K.O. Marfo. 2006. Registration of 'Nkatiesari' peanut. *Crop Sci.* 46: 1397–1398.

Pae, S.B., Y.K. Cheong, J.T. Kim, K.H. Park, and D.Y. Suh. 2004. A new early maturing, leaf spot resistant and high quality peanut cultivar, "Jakwang". *Korean J. Breed.* 36: 375–376.

Pae, S.B., C.H. Park, Y.K. Cheong et al. 2005. A new peanut cultivar with purple testa, large grains and high yield, Danuri. *Korean J. Breed.* 37: 251–252.

Pae, S.B., Y.K. Cheong, C.S. Jung et al. 2008. A new red testa and high quality vegetable peanut cultivar "Charmwon". *Korean J. Breed. Sci.* 40: 58–62.

Pal, K.K., R. Dey, and D.M. Bhatt 2004. Groundnut, *Arachis hypogaea* L. growth, yield and nutrient uptake as influenced by inoculation of plant growth promoting rhizobacteria. *J. Oilseeds Res.* 21: 284–287.

Palaiah, P. and S.S. Adiver. 2004. Bio-chemical and pathogenic variation in isolates of *Sclerotium rolfsii* causing stem rot of groundnut. *Karnataka J. Agric. Sci.* 17. 843–845.

Pan, R., M. Guan, D. Xu, X. Gao, X. Yan, and H. Liao. 2009. *Cylindrocladium* black rot caused by *Cylindrocladium parasiticum* newly reported on peanut. *Plant Pathol.* 58: 1176.

Pande, S., R. Bandyopadhyay, M. Blümmel et al. 2003. Disease management factors influencing yield and quality of sorghum and groundnut crop residues. *Field Crop Res.* 84: 89–103.

Pande, S., T.R. Rajesh, K.C. Rao, and G.K. Kishore. 2004. Effect of temperature and leaf wetness period on the components of resistance to late leaf spot disease in ground nut. *Plant Pathol. J.* 20: 67–74.

Pande, S. and J.N. Rao. 2000. Changing scenario of groundnut diseases in Andhra Pradesh, Karnataka, and Tamil Nadu states of India. *Intl. Arachis News L.* 20: 42–44.

Pande, S. and J.N. Rao. 2001. Resistance of wild *Arachis* species to late leaf spot and rust in greenhouse trials. *Plant Dis.* 85: 851–855.

Pande, S., J.N. Rao, and S.L. Dwivedi. 2002. Components of resistance to late leaf spot caused by *Phaeoisariopsis personata* in inter-specific derivatives of groundnut. *Indian Phytopathol.* 55: 444–450.

Pande, S., J.N. Rao, and E. Kumar. 2000. Survey of groundnut diseases in India, Survey Report. ICRISAT, Patancheru, AP, India. www.icrisat.org/gt3/r3. Accessed on July 14, 2014.

Paramasivam, K., M. Jayasekhar, R. Rajasekharan, and P. Veerabadhiran. 1990. Inheritance of rust resistance in groundnut (*Arachis hypogaea* L.). *Madras Agric. J.* 77: 50–52.

Park, K.H., Y.K. Cheong, H.S. Doo, S.D. Kim, D.Y. Suh, and S.H. Lee. 2004. A new high-yielding peanut cultivar, "Kokwang" with leaf spot resistance, large seed, and wide adaptability. *Korean J. Breed.* 36: 373–374.

Partridge, D.E., T.B. Sutton, D.L. Jordan, V.L. Curtis, and J.E. Bailey. 2006. Management of *Sclerotinia* blight of peanut with the biological control agent. *Coniothyrium minitans. Plant Dis.* 90: 957–963.

Partridge-Telenko, D. E., J. Hu, D.M. Livingstone, B.B. Shew, P.M. Phipps and E. Grabau. 2011. Sclerotinia blight resistance in Virginia-type peanut transformed with a barley oxalate oxidase gene. *Phytopathology* 101: 786–793.

Passone, M.A., S.L. Resnik, and M.G. Etcheverry. 2005. In vitro effect of phenolic antioxidants on germination, growth and aflatoxin B_1 accumulation by peanut *Aspergillus section Flavi. J. Appl. Microbiol.* 99: 682–691.

Passone, M.A., S. Resnik, and M.G. Etcheverry. 2007. Potential use of phenolic antioxidants on peanut to control growth and aflatoxin B_1 accumulation by *Aspergillus flavus and Aspergillus parasiticus. J. Sci. Food Agric.* 87: 2121–2130.

Passone, M.A., S. Resnik, and M.G. Etcheverry. 2008. The potential of food grade antioxidants in the control of *Aspergillus section Flavi*, interrelated mycoflora and aflatoxin B_1 accumulation on peanut grains. *Food Control* 19: 364–371.

Passone, M.A., M. Ruffino, V. Ponzio, S. Resnik, and M.G. Etcheverry. 2009. Postharvest control of peanut *Aspergillus section Flavi* populations by a formulation of food-grade antioxidants. *Int. J. Food Microbiol.* 131: 211–217.

Patibanda, A.K., J.P. Upadhyay, and A.N. Mukhopadhyay. 2002. Efficacy of *Trichoderma harzianum* Rifai alone or in combination with fungicides against *Sclerotium* wilt of groundnut. *J. Biol. Control* 16: 57–63.

Patil, P.D., M.B. Dalvi, and S.A. Chavhan. 2010. Performance of groundnut genotypes for leaf spot and rust disease in Konkan region. *Ann. Plant Prot. Sci.* 18: 547–548.

Patil, R.B., S.S. Patil, M.P. Deshmukh, R.S. Bhadane, R.B. Jadhav, and T.R. Patil. 2005. Phule Unap—A new groundnut variety for Western Maharashtra, India. *Int. Arachis News* 25: 17–19.

Patil, R.K., Shekharappa, I.K. Kalappanavar, and K. Giriraj. 2000. *Dipteran* (*Cecidomyiid*) maggots feeding on rust spores of groundnut. *Insect Environ.* 6: 54–55.

Pensuk, V., A. Patanothai, S. Jogloy, S. Wongkaew, C. Akkasaeng, and N. Vorasoot. 2003. Reaction of peanut cultivars to late leafspot and rust. *Songklanakarin J. Sci. Technol.* 25: 289–295.

Pereira, E.L. and C.A.V. Rossetto. 2008. Fungal population on peanut cultivation soil as affected by liming, genotypes and sampling times. *Ciência Agrotecnol.* 32: 1176–1183.

Pezzopan, J.R.M., M.J. Jior, S.A. Moraes, I.J. Godoy, J.N.V. Paternal, and L.C. Silveira. 1998. Rain and pervis? Of ocaso of pulveriza? for control of spots foliares of the peanut. *Bargantia* 57(2): http://216.239.37.104/translate_c?hl=en&sl=pt&u=http://www.scielo.br/scielo.php%3Fpid%3.

Pildain, M.B., D. Cabral, and G. Vaamonde. 2005. *Aspergillus flavus* populations in cultivated peanut from different agroecological zones of Argentina, toxigenic and morphological characterisation. *Revista de Investigaciones Agropecuarias* 34: 3–19.

Pildain, M.B., G. Vaamonde, and D. Cabral. 2004. Analysis of population structure of *Aspergillus flavus* from peanut based on vegetative compatibility, geographic origin, mycotoxin and sclerotia production. *Int. J. Food Microbiol.* 93: 31–40.

Podile, A.R., G.K. Kishore, and S.S. Gnanamanickam. 2002. Biological control of peanut diseases. In: *Biological Control of Crop Diseases*., ed., Samuel S. Gnanamanickam, Marcel Dekker, Inc., New York, 131–161.

Porter, D.M. 1993. *Diseases of Peanut (Arachis hypogaea L.)—Update-APS net.* American Phytopathological Society, St. Paul, MN.

Porter, D.M., D.H. Smith, and R. Rodriguez-Kabana. 1984. *Compendium of Peanut Diseases*. The American Phytopathological Society, St. Paul, MN, p. 73.

Prabakaran, G. and T. Ravimycin. 2012. Screening of *Bacillus* isolates against *Aspergillus niger* causing collar rot of groundnut. *Int. J. Plant Prot.* 5: 111–115.

Prado, G., E.P. Carvalho, M.S. Oliveira et al. 2005. Gamma-irradiation effect on aflatoxin B_1 production and growth of toxigenic strain of *Aspergillus flavus* in peanut (*Arachis hypogaea*). *Revista do Instituto Adolfo Lutz* 64: 85–90.

Prado, G., J.E.G.C. Madeira, V.A.D. Morais et al. 2011. Reduction of aflatoxin B1 in stored peanuts (*Arachis hypogaea* L.). using *Saccharomyces cerevisiae*. *J. Food Prot.* 74: 1003–1006.

Prado, G., R. de A. Souza, V.A.D. Morais et al. 2008. *Saccharomycopsis schoenii* and *Saccharomycopsis crataegensis* effect on B_1 and G_1 aflatoxins production by *Aspergillus parasiticus* in peanut (*Arachis hypogaea* L.). *Revista do Instituto Adolfo Lutz* 67: 177–182.

Pujar, S.B., P.V. Kenchanagoudar, M.V.C. Gowda, K.G. Parameshwarappa, and S.S. Adiver. 2011. Isolation of superior segregants for different quantitative traits and *Sclerotium* wilt resistance in groundnut. *Karnataka J. Agric. Sci.* 24: 230–233.

Purss, G.S. 1960. Further studies on the control of pre-emergence rot and crown rot of peanuts. *Qld. J. Agric. Sci.* 17: 1–14.

Rahmianna, A.A., A. Taufiq, and E. Yusnawan. 2004. Evaluation of ICRISAT groundnut genotypes for end-of-season drought tolerance and aflatoxin contamination in Indonesia. *Int. Arachis Newsl.* 24: 14–17.

Rahmianna, A.A., A. Taufiq, and E. Yusnawan. 2007. Effect of harvest timing and postharvest storage conditions on aflatoxin contamination in groundnuts harvested from the Wonogiri regency in Indonesia. *J. SAT Agric. Res.* 5: 1–3.

Raihan, M.G., M.K.A. Bhuiyan, and N. Sultana. 2003. Efficacy of integration of an antagonist, fungicide and in vitro control of *Colletotrichum dematium* causing anthracnose of soybean by fungicides, plant extracts and *Trichoderma harzianum* garlic extract to suppress seedling mortality of peanut caused by *Rhizoctonia solani* and *Sclerotium rolfsii*. *Bangladesh J. Plant Pathol.* 19(1 and 2): 69–73.

Raina, P. and S. Desai. 2006. Variability among isolates of *Aspergillus flavus* from groundnut for aflatoxin and cyclopiazonic acid production. *Elec. J. Env. Agricult. Food Chem*, 5(4): 1458–1463.

Raju, M.R.B. and K.V.M.K. Murthy. 2000. Efficacy of *Trichoderma* spp. in the management of collar rot of groundnut caused by *Aspergillus niger* Van Tieghem. *Indian J. Plant Prot.* 28: 197–199.

Rakh, R.R., L.S. Raut, S.M. Dalvi, and A.V. Manwar. 2011. Biological control of *Sclerotium rolfsii*, causing stem rot of groundnut by *Pseudomonas cf. monteilii*. *Recent Res. Sci. Technol.* 3: 26–34.

Rakholiya, K.B. and K.B. Jadeja. 2010. Effect of seed treatment of biocontrol agents and chemicals for management of stem and pod rot of groundnut. *Int. J. Plant Prot.* 3: 276–278.

Rakholiya, K.B., K.B. Jadeja, and A.M. Parakhia. 2012. Management of collar rot of groundnut through seed treatment. *Int. J. Life Sci. Pharma Res.* 2: 62–66.

Ramanaiah, M., N.P.E. Reddy, B.V.B. Reddy, S.D. Prasad, and P. Sudhakar. 2008. Integrated management of *Aspergillus flavus* (Link ex Fries) in groundnut—A preliminary study. *Curr. Biotica* (India) 2(3): 308–322.

Ramesh, R. and V.S. Korikanthimath. 2010. Seed treatment with bacterial antagonists—A simple technology to manage groundnut root rot under residual moisture conditions. *J. Biol. Control* 24: 58–64.

Randall-Schadel, B.L., J.E. Bailey, and M.K. Beute. 2001. Seed transmission of *Cylindrocladium parasiticum* in peanut. *Plant Dis.* 85: 362–370.

Rao, S.K.T. and K. Sitaramaiah. 2000. Management of collar rot disease (*Aspergillus niger*) in groundnut with *Trichoderma* spp. *J. Mycol. Plant Pathol.* 30: 221–224.

Rashid, S., S. Dawar, and A. Ghaffar. 2004. Location of fungi in groundnut seed. *Pak. J. Bot.* 36: 663–668.

Ray, S.K. and N. Mukherjee. 2002. Suppression of *Sclerotium rolfsii* causing foot rot of groundnut by *Bacillus* sp. *J. Mycopathol. Res.* 40: 89–92.

Reddy, L.J., S.N. Nigam, R.C.N. Rao, and N.S. Reddy. 2001b. Registration of ICGV 87354 peanut germplasm with drought tolerance and rust resistance. *Crop Sci.* 41: 274–275.

Reddy, L.J., S.N. Nigam, P. Subrahmanyam, F.M. Ismael, N. Govinden, and P.J.A. van der Merwe. 2000. Registration of groundnut cultivar Venus (ICGV 87853). *Int. Arachis Newsl.* 20: 29–31.

Reddy, L.J., S.N. Nigam, P. Subrahmanyam et al. 2001a. Registration of groundnut cultivar Sylvia (ICGV 93207). *Int. Arachis Newsl.* 21: 20–22.

Reddy, P.M., D.H. Smith, and D. McDonald. 1991. Inhibitory effect of sorghum and pearl millet pollen on urediniospore germination and germ tube growth of groundnut rust. *Int. Arachis Newsl.* 10: 21–22.

Reddy, R.N., K.G. Parameshwarappa, and H.L. Nadaf. 2004. Molecular diversity for resistance to late leaf spot and rust in parents and segregating population of a cross in groundnut. *Int. Arachis Newsl.* 24: 31–33.

Reddy, T.B.M., M.R. Govindappa, A.S. Padmaja, K.S. Shankarappa, A.L. Siddaramaiah, and H.V. Nanjappa. 2007a. Influence of soil solarization on yields of groundnut and tomato. *Environ. Ecol.* 25: 734–738.

Reddy, T.B.M., M.R. Govindappa, A.S. Padmaja, K.S. Shankarappa, A.L. Siddaramaiah, and H.V. Nanjappa. 2007b. Soil-borne disease management in groundnut and tomato by soil solarization. *Environ. Ecol.* 25: 788–790.

Reddy, V.C. and N.S. Reddy. 2000. Performance of groundnut varieties at various sowing dates during kharif season. *Curr. Res.* (UAS Bangalore) 29: 107–109.

Reis, G.M., L. de O. Rocha, D.D. Atayde, M.J.M. Batatinha, and B. Corrêa. 2012. Molecular characterization by amplified fragment length polymorphism and aflatoxin production of *Aspergillus flavus* isolated from freshly harvested peanut in Brazil. *World Mycotoxin J.* 5: 187–194.

Riaz, A., A.S. Khan, and I.H. Muhammad. 2002. Frequency of seed-borne mycoflora of groundnut and their impact on seed germination. *Pak. J. Phytopathol.* 14: 36–39.

Rideout, S.L., T.B. Brenneman, and A.K. Culbreath. September 1–9, 2002. Peanut disease management utilizing an in-furrow treatment of azoxystrobin. *Plant Health Prog.* doi:10.1094/PHP-2002-0916-01-RS.

Rideout, S.L., T.B. Brenneman, A.K. Culbreath, and D.B. Jr. Langston. 2008. Evaluation of weather-based spray advisories for improved control of peanut stem rot. *Plant Dis.* 92: 392–400.

Rodrigues, A.A.C., G.S. Silva, F.H.R. Moraes, and C.L.P. Silva. 2006. Arachis repens: New host of *Puccinia arachidis*. *Fitopatol. Bras.* 31: 411.

Rodriguez-Kabana, R., P.A. Backman, and J.C. Williams. 1975. Determination of yield losses due to *Sclerotium rolfsii* in peanut fields. *Plant Dis. Rep.* 59: 855.

Rodriguez-Kabana, R., M.K. Beute, and P.A. Backman. 1979. Effect of dibromochloropropane fumigation on the growth of *Sclerotium rolfsii* and the incidence of southern blight in field-grown peanut. *Phytopathology* 69: 1219.

Rodriguez-Kabana, R. and N. Kokalis-Burelle. 1997. Chemical and biological control. In: *Soilborne Diseases of Tropical Crops*, eds., R.J. Hillocks and J.M. Waller. CAB International, Wallingford, U.K., pp. 397–418.

Rohini, V.K. and K.S. Rao. 2001. Transformation of peanut (*Arachis hypogaea* L.) with tobacco chitinase gene: Variable response of transformants to leaf spot disease. *Plant Sci.* 160: 889–898.

Rojo, F.G., M.M. Reynoso, M. Ferez, S.N. Chulze, and A.M. Torres. 2007. Biological control of *Trichoderma* species of *Fusarium solani* causing peanut brown root rot under field conditions. *Crop Prot.* 26: 549–555.

Rothwell, A. 1975. Peanut rust in Rhodesia. *Plant Dis. Rep.* 59: 802.

Ruark, S.J. and B.B. Shew. 2010. Evaluation of microbial, botanical, and organic treatments for control of peanut seedlings. *Plant Dis.* 94: 445–454.

Ryley, M.J., N.A. Kyei, and J.R. Tatnell. 2000. Evaluation of fungicides for the management of *Sclerotinia blight* of peanut. *Aust. J. Agric. Res.* 51: 917–924.

Sahayaraj, K., J.A.F. Borgio, and G. Raju. 2009. Antifungal activity of three fern extracts on causative agents of groundnut early leaf spot and rust diseases. *J. Plant Prot. Res.* 49: 141–144.

Sahu, K.C. and A.K. Senapati. 2003. Efficacy of *Trichoderma* spp. against *Sclerotium rolfsii causing* stem rot of groundnut. *J. Appl. Biol.* 13: 38–40.

Sailaja, P.R., A.R. Podile, and P. Reddanna. 1998. Biocontrol strain of *Bacillus subtilis* AF 1 rapidly induces lipoxygenase in groundnut (*Arachis hypogaea* L.) compared to crown rot pathogen *Aspergillus niger*. *Eur. J. Plant Pathol.* 104: 125–132.

Salaudeen, M.T. and E.A. Salako. 2009. Evaluation of plant extracts for the control of groundnut leaf spot. *Arch. Phytopathol. Plant Prot.* 42: 1096–1100.

Sanogo, S. and B.F. Etarock. 2009. First report of *Phomopsis longicolla* causing stem blight of Valencia peanut in New Mexico. *Plant Dis.* 93: 965.

Sanogo, S. and N. Puppala. 2007. Characterization of a darkly pigmented mycelial isolate of *Sclerotinia sclerotiorum* on Valencia peanut in New Mexico. *Plant Dis.* 91: 1077–1082.

Sarkar, S.C. and A.K. Chowdhury. 2005. Performance of metiram 70%WDG (dithiocarbamate) on the infection and yield loss associated with tikka disease of groundnut (*Arachis hypogaea* L.) in West Bengal. *J. Mycopathol. Res.* 43: 63–65.

Saralamma, S. and T.V. Reddy. 2004. In vitro mycoparasitism of *Trichoderma harzianum Rifai* on *Sclerotium rolfsii* Sacc. causing root rot of groundnut. *Indian J. Plant Prot.* 32: 102–104.

Saralamma, S. and T.V. Reddy. 2005. Integrated management of root rot of groundnut. *J. Plant Prot.* 33: 264–267.

Saralamma, S., T.V. Reddy, and C.P.C. Kumar. 2004. Biological control of root rot of groundnut incited by *Sclerotium rolfsii* Sacc. *J. Res. ANGRAU* 32: 103–104.

Satish, C., A.K. Singh, and P. Baiswar. 2007. Efficacy of single spray of fungicides on early leaf spot (*Cercospora arachidicola*) of groundnut (*Arachis hypogaea*). *Indian J. Agric. Sci.* 77: 201–202.

Satish Kumar, G.D. and M.N. Popat. 2010. Factors influencing the adoption of aflatoxin management practices in groundnut (*Arachis hypogaea* L.). *Int. J. Pest Manag.* 56(2): 165–171.

Sawargaonkar, S.L., R.R. Giri, and B.V. Hudge. 2010. Character association and path analysis of yield component traits and late leaf spot disease traits in groundnut (*Arachis hypogaea* L.). *Agric. Sci. Dig.* 30: 115–119.

Sawant, G.G. 2000. Relative efficacy of number of sprays of mancozeb 75 WP and carbendazim 50 WP against tikka/leaf spot disease of groundnut. *Indian J. Environ. Toxicol.* 10: 90–92.

Sconyers, L.E., T.B. Brenneman, K.L. Stevenson, and B.G. Mullinix. 2007. Effects of row pattern, seeding rate, and inoculation date on fungicide efficacy and development of peanut stem rot. *Plant Dis.* 91: 273–278.

Senthilraja, G., T. Anand, C. Durairaj et al. 2010. A new microbial consortia containing entomopathogenic fungus, *Beauveria bassiana* and plant growth promoting rhizobacteria *Pseudomonas fluorescens* for simultaneous management of leaf miners and collar rot disease in groundnut. *Biocontrol Sci. Technol.* 20: 449–464.

Shakil, A. and Z. Noor. 2012. Evaluation of groundnut germplasm against root rot disease in agro-ecological conditions of district Chakwal, Punjab-Pakistan. *Afr. J. Microbiol. Res.* 6: 5862–5865.

Shan, S.H., H.X. Wang, C.J. Li, S.B. Wan, H.T. Liu, and G.Y. Jiang. 2006. Research of seed testa structure and storage material of peanut germplasm with different resistance to *Aspergillus flavus*. *Agric. Sci. China* 5: 478–482.

Shanmugam, V., T. Raguchander, A. Ramanathan, and S. Samiyappan. 2003. Management of groundnut root rot disease caused by *Macrophomina phaseolina* with *Pseudomonas fluorescens*. *Ann. Plant Prot. Sci.* 11: 304–308.

Shanmugam, V., N. Senthil, T. Raguchander, A. Ramanathan, and R. Samiyappan. 2002. Interaction of *Pseudomonas fluorescens* with Rhizobium for their effect on the management of peanut root rot. *Phytoparasitica* 30: 169–176.

Sharma, A and K. Sharma. 2011. *Musa paradisiaca* var. *sapientum* fruit peel for suppression of aflatoxin producing *Aspergillus* spp. on groundnuts. *Indian Phytopathol.* 64: 275–279.

Sharma, B.K. and U.P. Singh. 2002. Variability in Indian isolates of *Sclerotium rolfsii*. *Mycologia* 94: 1051–1058.

Sharma, K.K., F. Waliyar, P. Lava Kumar et al. 2006. Development and evaluation of transgenic groundnut expressing the rice chitinase gene for resistance to *Aspergillus flavus*. Paper presented at *the International Conference on "Groundnut Aflatoxin Management and Genomics"*, Guangdong, China, November 5–9, 2006.

Sharma, P., M.K. Saini, D. Swati, and V. Kumar. 2012. Biological control of groundnut root rot in farmer's field. *J. Agric. Sci.* (Toronto) 4: 48–59.

Sheela, J. 2008. Screening of groundnut cultures against rust and late leaf spot diseases of groundnut. *Madras Agric. J.* 95: 237–239.

Sheela, J. and D. Packiaraj. 2000. Management of collar rot of groundnut by *Pseudomonas fluorescens*. *Int. Arachis Newsl.* 20: 50–51.

Shew, B.B. 2011. *Plant Disease Fact Sheets: Sclerotinia Blight of Peanut*. Updated June 11, 2011. North Carolina State University, Raleigh, NC.

Shim, M.Y., J.L. Starr, N.P. Keller, K.E. Woodard, and T.A. Lee Jr. 1998. Distribution of isolates of *Sclerotium rolfsii* tolerant to pentachloronitrobenzene in Texas peanut fields. *Plant Dis.* 82: 103–106.

Shoba, D., N. Manivannan, P. Vindhiyavaraman, and S.N. Nigam. 2012. SSR markers associated for late leaf spot disease resistance by bulked segregant analysis in groundnut (*Arachis hypogaea* L.). *Euphytica* 188: 265–272.

Shokes, F.M. and A.K. Culbreath. 1997. Early and late leaf spot. In: *Compendium of Peanut Diseases*, 2nd edn., eds., N.D. Kokalis-Burelle, M. Porter, R. Rodriguez-Kabana, D.H. Smith, and P. Subrahmanyam. APS Press, St. Paul, MN, pp. 17–20.

Sibale, P.K. and C.T. Kisyombe. 1980. Groundnut production, utilization, research, problems, and further research needs in Malawi. In: *Proceedings of the International Workshop on Groundnuts*, eds., R.W. Gibbons and J.V. Mertin. ICRISAT, Patancheru, India, October 13–17, 1980, p. 249.

Siddaramaiah, A.L, S.A. Desai, R.K. Hegde, and H. Jayaramaiah. 1980. Effect of different dates of sowing of groundnut on rust development in Karnataka. *Proc. Indian Natl. Sci. Acad. B* 46: 380.

Siddaramaiah, A.L. and S.A. Jayaramaiah. 1981. Occurrence and severity of a mycoparasite on tikka leaf spots of groundnut. *Curr. Res.* (Karnataka, India) 10: 14.

Sidhu, G.S. and J.S. Chohan. 1969. Field scale testing of different fungicides for the control of collar rot of groundnut in the Punjab. *J. Res.* Ludhiana (India) 8: 211.

Simpson, C.E., M.R. Baring, A.M. Schubert, M.C. Black, H.A. Melouk, and Y. Lopez. 2006. Registration of 'Tamrun OL 02' peanut. *Crop Sci.* 46: 1813–1814.

Simpson, C.S., O.D. Smith, and H.A. Meloux. 2000. Registration of "Tamspan 98" peanut. *Crop Sci.* 40: 859.

Singh, A.K., S.L. Dwivedi, S. Pande, J.P. Moss, S.N. Nigam, and D.C. Sastri. 2003a. Registration of rust and late leaf spot resistant peanut germplasm lines. *Crop Sci.* 43: 440–441.

Singh, D., A.G. Girish, N.K. Rao, P.J. Bramel, and S. Chandra. 2003b. Survival of *Rhizoctonia bataticola* in groundnut seed under different storage conditions. *Seed Sci. Technol.* 31: 169–175.

Singh, M.P., J.E. Erickson, K.J. Boote, B.L. Tillman, J.W. Jones, and A.H.C. van Bruggen. 2011b. Late leaf spot effects on growth, photosynthesis, and yield in peanut cultivars of differing resistance. *Agron. J.* 103: 85–91.

Singh, M.P., J.E. Erickson, K.J. Boote, B.L. Tillman, A.H.C. van Bruggen, and J.W. Jones. 2011a. Photosynthetic consequences of late leaf spot differ between two peanut cultivars with variable levels of resistance. *Crop Sci.* 51: 2741–2748.

Smith, D.H. 1986. Disease forecasting method for groundnut leaf spot disease. In: *Agro-Meteorology of Groundnut*: *Proceedings of International Symposium*, ICRISAT, Patancheru, India, pp. 229–242.

Smith, D.L., M.C. Garrison, J.E. Hollowell, T.G. Isleib, and B.B. Shew. 2008. Evaluation of application timing and efficacy of the fungicides fluazinam and boscalid for control of *Sclerotinia blight* of peanut. *Crop Prot.* 27: 823–833.

Smith, D.L., J.E. Hollowell, T.G. Isleib, and B.B. Shew. 2006. Analysis of factors that influence the epidemiology of *Sclerotinia minor* on peanut. *Plant Dis.* 90: 1425–1432.

Sobolev, V.S., B.Z. Guo, C.C. Holbrook, and R.E. Lynch. 2007. Interrelationship of phytoalexin production and disease resistance in selected peanut genotypes. *J. Agric. Food Chem.* 55: 2195–2200.

Soler, C.M.T., G. Hoogenboom, R. Olatinwo, B. Diarra, F. Waliyar, and S. Traore. 2010. Peanut contamination by *Aspergillus flavus* and aflatoxin B1 in granaries of villages and markets of Mali, West Africa. *J. Food Agric. Environ.* 8: 195–203.

Spegazzini, C.L. 1884. Fungi guaranitici. *Anales Sociedad Cientifica Argentina* 17: 69–96 and 119–134. *Puccinia arachidis* Speg. (n.sp.), 90.

Sreedharan, A., R.M. Hunger, L.L. Singleton, M.E. Payton, and H.A. Melouk. 2010. Pathogenicity of three isolates of *Rhizoctonia* sp. from wheat and peanut on hard red winter wheat. *Int. J. Agric. Res.* 5: 132–147.

Srinivas, T., M. Manjulatha, and D. Venkateswarlu. 2005. Biointensive management of collar rot, *Aspergillus niger* and stem rot, *Sclerotium rolfsii* Sacc. in groundnut (*Arachis hypogaea* L.). *J. Oilseeds Res.* 22: 103–104.

Srinivas, T., M.S. Rao, P.S. Reddy, and P.N. Reddy. 2000. Comparative effects of plant extracts and chemicals for the management of leaf spot of groundnut (*Arachis hypogaea* L.). *Trop. Agric.* 77: 58–60.

Srinivas, T., P.S. Reddy, and K.C. Rao. 2002. Sustainable management of late leaf spot of groundnut (*Arachis hypogaea* L.) in semi-arid tropics. *Indian J. Plant Prot.* 30: 164–166.

Stalker, H.T., M.K. Beute, B.B. Shew, and T.G. Isleib. 2002. Registration of five leaf spot-resistant peanut germplasm lines. *Crop Sci.* 42: 314–316.

Su, G., S.-O. Suh, R.W. Schneider, and J.S. Russin. 2001. Host specialization in the charcoal rot fungus, *Macrophomina phaseoli*. *Phytopathology* 91: 120–126.

Subhlakshmi, T. and A.K. Chowdhury. 2008. Inoculum potential of *Cercospora arachidicola* and *phaeoisariopsis personata* with infection of groundnut (*Arachis hypogaea* L.) viruses. *J. Mycopathol. Res.* 46: 113–115.

Subrahmanyam, P., D.C. Greenberg, S. Savary, and J.P. Bosc. 1991. Diseases of groundnut in West Africa and their management: Research priorities and strategies. *Trop. Pest Manag.* 37: 259–269.

Subrahmanyam, P., D. McDonald, R.W. Gibbons, S.N. Nigam, and D.J. Nevill. 1982. Resistance to rust and late leaf spot diseases in some genotypes of *Arachis hypogaea* L. *Peanut Sci.* 9: 6.

Subrahmanyam, P., V.K. Mehan, D.J. Nevill, and D. McDonald. 1980. Research on fungal diseases of groundnut at ICRISAT. In: *Proceedings of the International Workshop on Groundnuts*, eds., R.W. Gibbons and J.V. Mertin. ICRISAT, Patancheru, India, October 13–17, 1980, p. 193.

Subrahmanyam, P., L.J. Reddy, R.W. Gibbons, and D. Mc Donald. 1985. Peanut rust: A major threat to peanut production in the semiarid tropics. *Plant Dis.* 69: 813–819.

Subrahmanyam, P., D.H. Smith, R.A. Raber, and E. Sheperd, E. 1987. An outbreak of yellow mold of peanut seedlings in Texas. *Mycopathologia* 100: 97–102.

Subrahmanyam, P., P.S. VanWye, C.T. Kisyombe et al. 1997. Diseases of groundnut in the Southern African Development Community (SADC) region and their management. *Int. J. Pest Manag.* 43: 261–273.

Sudhagar, R., D. Sassikumar, V. Muralidharan, A. Gopalan, and R. Vivekananthan. 2009. Screening for rust (*Puccinia arachidis* Speg.) resistant genotypes in groundnut using biochemical markers. *Acta Phytopathologica et Entomologica Hungarica* 44: 225–238.

Sudhakar, P., P. Latha, M. Babitha, P.V. Reddy, and P.H. Naidu. 2007. Relationship of drought tolerance traits with aflatoxin contamination in groundnut. *Indian J. Plant Physiol.* 12: 261–265.

Sumartini and E. Yusnawan. 2005. An effort to inhibit *Aspergillus flavus* development in peanut. *J. Penelitian dan Pengembangan Pertanian* 24: 109–112.

Sun, W.M., L.N. Feng, W. Guo et al. 2012a. First report of *Neocosmospora striata* causing peanut pod rot in China. *Plant Dis.* 96(1): 146.

Sun, W.M., L.N. Feng, W. Guo et al. 2012b. First report of Neocosmospora vasinfecta in northern China. *Plant Dis.* 96(3): 455.

Sundaresha, S., A.M. Kumar, S. Rohini et al. 2010. Enhanced protection against two major fungal pathogens of groundnut, *Cercospora arachidicola* and *Aspergillus flavus* in transgenic groundnut over-expressing a tobacco β-1-3 glucanase. *Eur. J. Plant Pathol.* 126: 497–508.

Suryawanshi, A.P., C.D. Mayee, P.K. Dhoke, and B.S. Indulkar. 2006. Role of phenols and sugars in resistance of groundnut against late leaf spot caused by *Phaeoisariopsis personata. J. Plant Dis. Sci.* 1: 195–197.

Taber, R. and R.E. Pettit. 1981. Potential for biological control of Cercosporidium leaf spot of peanuts by Hansfordia, Abstr. *Phytopathology* 71: 260.

Tallury, S.P., T.G. Isleib, and H.T. Stalker. 2009. Comparison of Virginia-type peanut cultivars and interspecific hybrid derived breeding lines for leaf spot resistance, yield, and grade. *Peanut Sci.* 36: 144–149.

Tang, Y.X., R.C. Ji, G.X. Li., Y.M. Kang, and X.Y. Li. 2002. Physiological difference of *Aspergillus flavus* strains from groundnut. *Fujian Agric. Sci. Technol.* 4(1): 8–9.

Tarekegn, G., P.K. Sakhuja, W.J. Swart, and T. Tamado. 2007. Integrated management of groundnut root rot using seed quality and fungicide seed treatment. *Int. J. Pest Manag.* 53: 53–57.

Tashildar, C.B., S.S. Adiver, S.N. Chattannavar et al. 2012. Morphological and isozyme variations in *Puccinia arachidis* Speg. causing rust of peanut. *Karnataka J. Agric. Sci.* 25: 340–345.

Taufiq, A., A.A. Rahmianna, S. Hardaningsih, and F. Rozi. 2007. Increasing groundnut yield on dryland Alfisols in Indonesia. *J. SAT Agric. Res.* 5: 1–3.

Tewari, S.N., M. Mishra, and S.K. Das. 2004. Performance of lemon (*Citrus medica*) leaves extract against *Pyricularia grisea*, *Aspergillus niger* and *A. flavus* pathogens associated with-rice (*Oryza sativa* L.)/ groundnut (*Arachis hypogea* L.) diseases. *Proc. Natl. Acad. Sci. B: Biol. Sci.* 74: 59–64.

Thakare, C.S., D.A. Shambharkar, and R.B. Patil. 2007. Screening of groundnut germplasm against *Sclerotium* stem and pod rot. *J. Maharashtra Agric. Univ.* 32: 161–162.

Thakur, R.P., V.P. Rao, and K. Subramanyam. 2003a. Influence of biocontrol agents on population density of *Aspergillus flavus* and kernel infection in groundnut. *Indian Phytopathol.* 56: 408–412.

Thakur, R.P., V.P. Rao, H.D. Upadhyaya, S.N. Nigam, and H.S. Talwar. 2003b. Improved screening techniques for in-vitro seed colonization and pre-harvest seed infection by *Aspergillus flavus* in groundnut. *Indian J. Plant Prot.* 31: 54–60.

Thakur, S.B., S.K. Ghimire, N.K. Chaudhary, S.M. Shrestha, and B. Mishra. 2013. Determination of relationship and path-coefficient between pod yield and yield component traits of groundnut cultivars. *Nepal J. Sci. Technol.* 14: 1–8.

Thiessen, L.D. and J.E. Woodward. 2012. Diseases of peanut caused by soilborne pathogens in the Southern United States. *ISRN Agron.* 2012: 9pp. doi.10.5402/2012/517905.

Thiruvudainambi, S., G. Chandrasekar, G. Baradhan, and R.S. Yadav. 2010. Management of stem rot (*Sclerotium rolfsii* Sacc) of groundnut through non-chemical methods. *Plant Arch.* 10: 633–635.

Timper, P., N.A. Minton, A.W. Johnson et al. 2001. Influence of cropping systems on stem rot (*Sclerotium rolfsii*), *Meloidogyne arenaria*, and the nematode antagonist *Pasteuria penetrans* in peanut. *Plant Dis.* 85: 767–772.

Timper, P., D.M. Wilson, C.C. Holbrook, and B.W. Maw. 2004. Relationship between *Meloidogyne arenaria* and aflatoxin contamination in peanut. *J. Nematol.* 36: 167–170.

Tiwari, R.K.S., R.B. Tiwari, A. Singh, and M.L. Rajput. 2005. Influence of time of sowing on the incidence of leaf spot caused by *Cercospora arachidicola* Hori and *Cercosporidium personatum* (Berk. & Curt.) (Deighton) and economics of spray schedules in groundnut (*Arachis hypogea* L.). *Adv. Plant Sci.* 18: 591–595.

Tonelli, M.L., A. Furlan, T. Taurian, S. Castro, and A. Faba. 2011. Peanut priming by biocontrol agents. *Physiol. Mol. Plant Pathol.* 75: 100–105.

Tonelli, M.L., T. Taurin, F. Ibanez, J. Angelini, and A. Faba. 2010. Selection and in vitro characterization of biocontrol agents with potential to protect plants against fungal pathogens. *J. Plant Pathol.* 92: 73–82.

Tshilenge, L. 2010. Pathosystems groundnut (*Arachis hypogaea* L.), *Cercospora* spp. and environment in DR-Congo: Overtime Interrelations. In: *Contribution to Food Security and Malnutrition in DR-Congo*, ed., K.K.C. Nkongolo. Laurentian Press, Congo, Africa, pp. 195–221.

Tshilenge-Lukanda, L., K.K.C. Nkongolo, A. Kalonji-Mbuyi, and R.V. Kizungu. 2012. Epidemiology of the groundnut (*Arachis hypogaea* L.) leaf spot disease: Genetic analysis and developmental cycles. *Am. J. Plant Sci.* 3: 582–588. doi: 10.4236/ajps.2012.350720.

Tsitsigiannis, D.I., S. Kunze, D.K. Willis, I. Feussner, and N.P. Keller. 2005. *Aspergillus* infection inhibits the expression of peanut 13S-HPODE-forming seed lipoxygenases. *Mol. Plant Microbe Interact.* 18: 1081–1089.

Udhayakumar, R., S.U. Rani, and D.J. Christopher. 2008. Effect of cold water extracts of plant products on the *Macrophomina phaseolina* Tassi. (Goid) the causative agent of dry root rot of groundnut. *Adv. Plant Sci.* 21: 415–417.

Ume-Kulsoom, A. Iftikhar, S.A. Malik et al. 2001. Efficacy of neem products against *Sclerotium rolfsii* infecting groundnut. In: *Proceedings of the Meeting on the Science and Application of Neem*, Glasgow, U.K., pp. 33–37.

Upadhyaya, H.D., S.N. Nigam, V.K. Mehan, A.G.S. Reddy, and N. Yellaiah. 2001b. Registration of *Aspergillus flavus* seed infection resistant peanut germplasm ICGV 91278, ICGV 91283, and ICGV 91284. *Crop Sci.* 41: 559–600.

Upadhyaya, H.D., S.N. Nigam, S. Pande, A.G. Reddy, and N. Yellaiah. 2001a. Registration of early-maturing, moderately resistant to rust germplasm ICGV 94361. *Crop Sci.* 41: 598–599.

Upadhyaya, H.D., S.N. Nigam, A.G.S. Reddy, and N. Yellaiah. 2002. Registration of early-maturing, rust, late leaf spot, and low temperature tolerant peanut germplasm line ICGV92267. *Crop Sci.* 42: 2220–2221.

Van Arsdel, E.P. 1974. The spread of air-borne pathogens using peanut rust as an example. Presented at the *Symposium on Plant Diseases and the Affairs of Man: Proceedings*, Texas Academy of Science, Denton, TX, p. 40.

Varadharajan, K., S. Ambalavanan, and N. Sevugaperumal. 2006. Biological control of groundnut stem rot caused by *Sclerotium rolfsii* (Sacc). *Arch. Phytopathol. Plant Prot.* 39: 239–246.

Vargas Gil, S., R. Haro, C. Oddino et al. 2008. Crop management practices in the control of peanut diseases caused by soilborne fungi. *Crop Prot.* 27: 1–9.

Varma, T.S.N., S.L. Dwivedi, S. Pande, and M.V.C. Gowda. 2005. SSR markers associated with resistance to rust (*Puccinia arachidis* Speg.) in groundnut (*Arachis hypogaea* L.). *SABRO J. Breed. Genet.* 37(2): 107–109.

Varma, T.S.N., S. Geetha, G.K. Naidu, and M.V.C. Gowda. 2001. Screening groundnut varieties for in vitro colonization with *Aspergillus flavus*. *Int. Arachis Newsl.* 21: 13–14.

Varman, P.V. 2001. Genetic introgression from wild species into cultivated groundnut. *Madras Agric. J.* 87: 263–266.

Varman, P.V., K.N. Ganesan, and A. Mothilal. 2000. Wild germplasm: Potential source for resistance breeding in groundnut. *J. Ecobiol.* 12: 223–228.

Vasavirama, K. and P.B. Kirti. 2012. Increased resistance to late leaf spot disease in transgenic peanut using combination of PR genes. *Funct. Integr. Genomics* 12: 625–634.

Venkataravana, P. and S.K. Injeti. 2008. Assessment of groundnut germplasm and advanced breeding lines and isolation of elite foliar disease resistant genotypes for southern Karnataka. *J. Soils and Crops* 18: 282–286.

Venkataravana, P., N. Jankiraman, D. Nuthan, and V.L.M. Prasad. 2008. Genetic enhancement of multiple resistance to foliar diseases in groundnut. *Legume Res.* 31: 155–156.

Victoria Novas, M. and D. Cabral. 2002. Association of mycotoxin and sclerotia production with compatibility groups in *Aspergillus flavus* from peanut in Argentina. *Plant Dis.* 86: 215–219.

Vidyasekaran, P. 1981. Control of rust and Tikka disease of groundnut. *Indian Phytopathol.* 34: 20.

Vikram, A. and H. Hamzehzarghani. 2011. Integrated management of *Sclerotium rolfsii* (Sacc.) in groundnut (*Arachis hypogaea* L.) under pot culture conditions. *Pest Technol.* 5: 33–38.

Vimala, R., J. Sheela, and D. Packiaraj. 2000. Management of root rot disease of groundnut by bioagents. *Madras Agric. J.* 87: 352–354.

Vincelli, P. 2002. QoI (strobilurin) fungicides: Benefits and risks. *The Plant Health Instructor.* doi:10.1094/PHI-1-2002-0809-02.

Vindhiyavarman, P. and S.E.N. Mohammed. 2001. Release of foliar disease resistant groundnut cultivar VRI Gn 5 in Tamil Nadu, India. *Int. Arachis Newsl.* 21: 16–17.

Vinod Kumar, N.B. Bagwan, and A.L. Rathnakumar. 2009. Influence of organic amendments on mycelial growth, sclerotial formation and stem rot incidence in groundnut (*Arachis hypogaea* L.) by *sclerotium rolfsii* Sacc. Abstr. In: *Fifth International Conference on Plant Pathology in the Globalized Era*, New Delhi, November 10–13, 2009, p. 314.

Vinod Kumar, A.L. Rathnakumar, and N.B. Bagwan. 2012. Effect of crop residues and root exudates on mycelial growth, sclerotial formation, and *Sclerotium rolfsii*-induced stem rot disease of groundnut. *Indian Phytopathol.* 65: 238–243.

Vishnuvardhan, K.M., R.P. Vasanthi, and K.H. Reddy. 2011. Combining ability of yield, yield traits and resistance to late leaf spot and rust in ground. *J. SAT Agric. Res.* 9: http://ejournal.icrisat.org/Volume9/Groundnut/Combining.pdf.

Vitanova, M. 2003. Species from genus *Fusarium* responsible for peanut *Fusarium* disease. *Rasteniev'dni Nauki* 40: 560–564.

Vyas, R.V., A.B. Maghodia, B. Patel, and D.J. Patel. 2005. In vitro testing of *Xenorhabdus* metabolites against groundnut collar rot fungus *Aspergillus niger*. *Int. Arachis Newsl.* 25: 34–36.

Wadsworth, D.F. 1979. *Sclerotinia* blight of peanut in Oklahoma and occurrence of the sexual stage of the pathogen. *Peanut Sci.* 6: 77–79.

Wakil, E. and M.I. Ghonim. 2000. Survey of seed-borne mycoflora of peanut and their control. *Egypt. J. Agric. Res.* 78: 47–61.

Waliyar, F., 1991. Evaluation of yield losses due to groundnut leaf diseases in West Africa. Proceedings of the 2nd ICRISAT Regional Groundnut Meeting for West Africa, September 11–14, 1991, Niamey, Niger. ICRISAT, Patancheru, India, pp. 32–33.

Waliyar, F. 1997. An overview of research on the management of aflatoxin contamination of groundnut. In: *Aflatoxin Contamination Problems in Groundnut in Asia*: *Proceedings of the First Asia Working Group Meeting*, eds., V.K. Mehan and C.L.L. Gowda. Ministry of Agriculture and Rural Development, Hanoi, Vietnam, May 27–29, 1996. International Crops Research Institute for the Semi-Arid Tropics, Patancheru, India, pp. 13–17.

Waliyar, F., M. Adamou, and A. Traoré. 2000. Rational use of fungicide applications to maximize peanut yield under foliar disease pressure in west Africa. *Plant Dis.* 84: 1203–1211.

Waliyar, F., A. Ba, H. Hassan, S. Bonkoungou, and J.P. Bosc. 1994. Sources of resistance to *Aspergillus flavus* and aflatoxin contamination in groundnut in West Africa. *Plant Dis.* 78: 704–708.

Waliyar, F., J.P. Bosc, and S. Bonkoungou. 1993. Sources of resistance to foliar diseases of groundnut and their stability in West Africa. *Oléagineux* (Paris) 48: 283–287.

Waliyar, F., A. Moustapha, and A. Traore. 1998. Effect of fungicides application on yield of groundnut genotype under foliar disease pressure in West Africa. *IAN* 18: 51–53.

Waliyar, F., B.B. Shew, R. Sidahmed, and M.K. Beute. 1995. Effects of host resistance on germination of *Cercospora arachidicola* on peanut leaf surface. *Peanut Sci.* 22: 154–157.

Wang, C.T., X.Z. Wang, Y.Y. Tang et al. 2009. Cloning of the 18S rDNA sequence from *Sphaceloma arachidis*, the causal pathogen of groundnut scab. *J. SAT Agric. Res.* 7: 1–3.

Wang, H.M., J.Q. Huang, Y. Lei et al. 2012. Relationship of resveratrol content and resistance to aflatoxin accumulation caused by *Aspergillus flavus* in peanut seeds. *Acta Agron. Sin.* 38: 1875–1883.

Wang, S.Y., B.S. Liao, Y. Lei, D. Li, and D.R. Xiao. 2004. Role of in vitro resistance to aflatoxin contamination in reducing pre-harvest aflatoxin contamination under drought stress in groundnut. *Int. Arachis Newsl.* 24: 38–39.

Wang, T., E. Zhang, X. Chen, L. Li, and X. Liang. 2010. Identification of seed proteins associated with resistance to pre-harvested aflatoxin contamination in peanut (*Arachis hypogaea* L.). *BMC Dev. Biol.* 10: 267. doi:10.1186/1471-2229-10-267.

Wang, X.J., T.L. Shuan, X.A. Han, B.W. Shu, Z.Z. Chuan, and Q.L. Ai. 2011. Peanut (*Arachis hypogaea* L.) omics and biotechnology in China. *Plant Omics J.* 4: 339–349.

Wann, D.Q., R.S. Tubbs, and A.K. Culbreath. October 2011. Genotype and approved fungicide evaluation for reducing leaf spot diseases in organically-managed peanut. *Plant Health Prog.* doi:10.1094/PHP-2010-1027-01-RS.

Webb, D.M. 1971. Profitable practices for groundnut. *World Farming* 13: 14.

Wei, L.Q., X. Feng, L. Shan, and Y.P. Bi. 2008. Recent progress of transgenic research in peanut. *China Biotechnol.* 28: 124–129.

Wheeler, T.A. and M.C. Black. 2005. First report of cylindrocladium black rot caused by *Cylindrocladium parasiticum* on peanut in Texas. *Plant Dis.* 89: 1245.

Whipps, J.M., S. Sreenivasaprasad, S. Muthumeenakshi, C.W. Rogers, and M.P. Challen. 2008. Use of *Coniothyrium minitans* as a biocontrol agents and some molecular aspects of sclerotial parasitism. *Eur. J. Plant Pathol.* 121: 323–330.

Williams, J.H., T.D. Phillips, P.E. Stiles, C.M. Jolly, and D. Aggarwal. 2004. Human aflatoxicosis in developing countries: A review of toxicology, exposure, potential health consequences, and interventions. *Am. J. Clin. Nutr.* 80: 1106–1122.

Wilson, J.N., T.A. Wheeler, and B.G. Mullinix Jr. 2010. Effect of sodium hypochlorite on mortality of *Sclerotinia minor* sclerotia. *Peanut Sci.* 37: 92–94.

Woodroof, N.C. 1933. Two leaf spots of peanut (*Arachis hypogaea* L.). *Phytopathology* 23: 627.

Woodward, J.E., T.B. Brenneman, R.C. Kemerait Jr., A.K. Culbreath, and J.R. Clark. 2006. First report of *Sclerotinia* blight caused by *Sclerotinia sclerotiorum* on peanut in Georgia *Plant Dis.* 90(1): 111–112.

Woodward, J.E., T.B. Benneman, R.C. Kemerait Jr., N.B. Smith, A.K. Culbreath, and K.L. Stevenson. 2008. Use of resistant cultivars and reduced fungicide programs to manage peanut diseases in irrigated and non-irrigated fields. *Plant Dis.* 92: 896–902.

Wright, L.P., A.J. Davis, B.D. Wingfield, P.W. Crous, T. Brenneman, and M.J. Wingfield. 2010. Population structure of *Cylindrocladium parasiticum* infecting peanut (*Arachis hypogaea* L.) in Georgia, USA. *Eur. J. Plant Pathol.* 127: 199–206.

Wynne, J.C., M.K. Beute, and S.N. Nigam. 1991. Breeding for disease resistance in peanut (*Arachis hypogaea* L.). *Annu. Rev. Phytopathol.* 29: 279–303.

Xie, C.Z., X.Q. Liang, L. Li, and H.Y. Liu. 2009. Cloning and prokaryotic expression of ARAhPR10 gene with resistance to *Aspergillus flavus* in peanut. *Genomics Appl. Biol.* 28: 237–244.

Xie, C.Z., S. Wen, H. Liu et al. 2013. Over expression of ARAhPR10, a member of the PR 10 family decreases levels of *Aspergillus flavus* infection in peanut seeds. *Am. J. Plant Sci.* 4: 602–607.

Xu, X.N., X.U. Zhang, B.Y. Huang et al. 2011. The outbreak causes and control measures of Sclerotium blight in part regions of Henan province. *China J. Henan Agric. Sci.* 40: 99–101.

Xue, H.Q., T.G. Isleib, G.A. Payne, R.F. Wilson, W.P. Novitzky, and G. O'Brian. 2003. Comparison of aflatoxin production in normal- and high-oleic backcross-derived peanut lines. *Plant Dis.* 87: 1360–1365.

Yadav, V.P., R.V. Kumar, and V.K. Tripathi. 2007. Role of wild species in improvement of cultivated groundnut (*Arachis hypogaea* L.). *Asian J. Horticulture* 2: 204–210.

Yan, C.X., C.J. Li, S.B. Wan., T.T. Zhang, Y.X. Zheng, and S.H. Shan. 2012a Cloning, expression and characterization of $PnLOX_2$ gene related to Aspergillus flavus-resistance from peanut (*Arachis hypogaea* L.) seed coat. *J. Agric. Sci.* (Toronto) 4: 67–75.

Yan, H.Y., C.Z. Zong, G. Zhai, S.C. Ma, and S.H. Shan. 2012b. Relationship between SR, α-tubulin and resistance to *Aspergillus flavus* in peanut. *China Biotechnol.* 32: 32–38.

Yang, K.H., W.L. Tsaus, G.J. Shieh et al. 2002. The effect of crop season on pod rot disease of peanut (*Arachis hypogaea* L.) varieties. *J. Agric. Res. China* 51: 28–36.

Yang, W.L. and G.Z. Ma. 2003. Population and dynamic analysis of *Aspergillus flavus* and other fungi in groundnut. *J. Hebei Vocation-Tech. Teachers College* 17: 19–22.

Yin, Y., T. Lou, L. Yan, T.J. Michailides, and Z. Ma. 2009. Molecular characterization of toxigenic and atoxigenic *Aspergillus flavus* isolates, collected from peanut fields in China. *J. Appl. Microbiol.* 107: 1857–1865.

Yu, S.L., Y.L. Cao, S.Y. Gu, and P. Min. 2000. A new high-yielding low oil content groundnut variety. *Int. Arachis News l.* 20: 32–33.

Yuttavanichakul, W., P. Lawongsa, S. Wongkaew et al. 2012. Improvement of peanut rhizobial inoculant by incorporation of plant growth promoting rhizobacteria (PGPR) as biocontrol against the seed borne fungus, *Aspergillus niger*. *Biol. Control* 63: 87–97.

Zachée, A., N. Bekolo, N.D. Bime, M. Yalen, and N. Godswill. 2008. Effect of mycorrhizal inoculum and urea fertilizer on diseases development and yield of groundnut crops (*Arachis hypogaea* L.). *Afr. J. Biotechnol.* 7: 2823–2827.

Zade, S.R., A.N. Buldeo, P.W. Lanje, and V.G. Gulhane. 2005. Evaluation of plant extracts and culture filtrate of bioagents against *Puccinia arachidis* Speg. in groundnut. *J. Soils Crops* 15: 150–154.

Zeidan, O., Y. Elad, Y. Hadan and I. Chet. 1986. Integrating onion in crop to control *Sclerotium rolfsii*. *Plant Dis.* 70: 426–428.

Zhang, S., M.S. Reddy, N. Kokalis-Burelle, L.W. Wells, S.P. Nightengale, and J.W. Kloepper. 2001. Lack of induced systemic resistance in peanut to late leaf spot disease by plant growth-promoting rhizobacteria and chemical elicitors. *Plant Dis.* 85: 879–884.

Zhang, S.G., Y.N. Guo, and R.E. Bradshaw. 2010. Genetics of dothistromin biosynthesis in the peanut pathogen *Passalora arachidicola*. *Toxins* 2: 2738–2753.

Zhou, L., C. Huo, J. Liu, and Z. Liu 1980. Studies on peanut rust in Guangdong Province. *Acta Phytophylacica Sin.* 7: 65–74.

Zhou, Y.Q., Y.M. Chen, Y.H. Zhang et al. 2012. Advances in *Aspergillus flavus*-resistant substances and genes in peanut seeds. *Guizhou Agric. Sci.* 1: 36–40.

Zhuang W., S.M. Fang, Y. Li, Y.S. Chen, Z. Cheng, and Y.S. Chen. 2007. Screening and identification of resistant peanut varieties and lines to *Aspergillus flavus*. *Fujian J. Agric. Sci.* 22: 261–265.

Zobia, J., A. Riaz, K. Sultana et al. 2012. Incidence of *Aspergillus flavus* and extent of aflatoxin contamination in peanut samples of Pothwar region of Pakistan. *African J. Microbiol. Res.* 6: 1942–1946.

Zucchi, T.D., L.A.B. de Moraes, and I.S. de Melo. 2008. *Streptomyces* sp. ASBV-1 reduces aflatoxin accumulation by *Aspergillus parasiticus* in peanut grains. *J. Appl. Microbiol.* 105: 2153–2160.

3 Virus Diseases of Peanut

PEANUT ROSETTE DISEASE COMPLEX

Symptoms

In order to understand the symptomatology of this disease, it is at the outset important to understand that the disease is caused by a complex of three agents: peanut rosette virus (PRV) and its satellite RNA (SatRNA) and peanut rosette assistor virus (PRAV).

Peanut rosette disease (PRD) complex occurs as two symptom variants, chlorotic rosette and green rosette, with considerable variation within each type (Murant 1989, Naidu et al. 1999). Both forms of the disease cause plants to be severely stunted, with shortened internodes and reduced leaf size, resulting in a bushy appearance of plants. In chlorotic rosette, leaves are usually bright yellow with a few green islands, and leaf lamina is curled (Figure 3.1). In the green rosette, leaves appear dark green, with light green to dark green mosaic (Figure 3.2). Chlorotic rosette occurs throughout the sub-Saharan Africa (SSA), whereas green rosette has been reported from Angola, Kenya, Malawi, Swaziland, Uganda, and West Africa (Naidu et al. 1999). Variability in SatRNA is mainly responsible for symptom variations (Murant and Kumar 1990, Taliansky and Robinson 1997). In addition, differences in genotypes, plant stage at infection, variable climatic conditions, and mixed infections with other viruses also contribute to symptom variability under field conditions (Naidu and Kimmins 2007). Stunting is more severe in diseased peanut plants containing all the three agents than in diseased peanut plants containing only PRV and SatRNA (Ansa et al. 1990). Some reports have suggested that PRAV or PRV infection alone in peanut results in transient mottle symptoms with insignificant impact on the plant growth and yield (Taliansky et al. 2000). These results have, however, been contradicted by studies that provide evidence for the first time that PRAV infection alone, without PRV and Sat RNA, affects plant growth and contributes to significant yield losses in susceptible groundnut cultivars (Naidu and Kimmins 2007).

Geographical Distribution and Losses

PRD, first reported in 1907 from Tanganyika (presently Tanzania), is endemic in peanut-growing areas of SSA including its offshore islands such as Madagascar; it is limited to peanut crop and the African continent (Zimmermann 1907, Reddy 1991, Naidu et al. 1999). There is no evidence of PRD occurrence anywhere outside Africa. Earlier reports on its occurrence based on rosette-like symptoms in peanut in India, Java, and Australia were later confirmed as caused by other viruses such as Indian peanut clump virus (IPCV).

Yield losses due to PRD depend on the growth stage at which infection occurs (Olorunju et al. 1991). Infection due to chlorotic or green rosette disease occurring in young plants (prior to flowering) will result in 100% yield loss. In contrast, plants infected during later growth stages (between flowering and pod setting) may show symptoms only in some branches or parts of branches, and yield loss depends on the severity of infection. Infection after pod setting/maturation causes negligible effects on pod yield. An average annual yield loss due to PRD is estimated to be between 5% and 30% in nonepidemic years, and epidemics often result in 100% yield loss (Alegbejo and Abo 2002).

FIGURE 3.1 Chlorotic rosette of peanut.

FIGURE 3.2 Green rosette of peanut.

PRD usually occurs in small proportions every growing season, but its severity increases in groundnut crops sown late in the season. When epidemics do occur, peanut production is significantly reduced, and the disease has the potential to cripple rural economies in SSA (Naidu et al. 1999). An epidemic in northern Nigeria in 1975 destroyed approximately 0.7 million ha of groundnut, with an estimated loss of U.S. $250 million (Yayock et al. 1976). Similarly, an epidemic in 1995 in eastern Zambia affected approximately 43,000 ha causing an estimated loss of U.S. $4.89 million. In the following year in the central region of Malawi, peanut production was reduced by 23%. As per the estimates of ICRISAT, PRD causes an annual yield loss of U.S. $156 million in SSA (Waliyar et al. 2007).

Pathogen(s): The Causal Virus Complex

Three causal agents, as mentioned earlier, are involved in PRD etiology: PRAV, PRV, and SatRNA (Murant et al. 1988, Murant 1990, Taliansky et al. 2000). The three agents, PRAV, PRV, and SatRNA, synergistically interact.

PRAV: PRAV virions are nonenveloped, isometric shaped with 28 nm diameter particles of polyhedral symmetry. There are no reports on the occurrence of strains of PRAV. The genome is a nonsegmented, single molecule of linear positive-sense, single-stranded RNA of c. 6900 nucleotides which encodes for structural and nonstructural proteins (Murant et al. 1998). It was first recognized as a component of PRD by Hull and Adams (1968) who later identified it as a *Luteovirus* belonging to the family Luteoviridae (Casper et al. 1983, Reddy et al. 1985a). In general, any *Luteovirus* purification protocol can be applied for the purification of PRAV particles from infected peanut plants (Murant 1989, Waliyar et al. 2007). Like other members of the *Luteovirus*, PRAV is thought to encode for six open reading frames (ORFs). Only coat protein (CP) region of the genome has been sequenced (Scott et al. 1996, Murant et al. 1998). Virions are made of single CP subunits of size 24.5 kDa, and the virus is antigenically related to bean/pea leaf roll virus, beet western yellows virus, and potato leaf roll virus (Scott et al. 1996). The virus replicates autonomously in the cytoplasm of phloem tissue. PRAV is transmitted by *Aphis craccivora* in a persistent manner, and experimentally by grafting, but not by mechanical sap inoculation, seed, pollen, or contact between the plants. Peanut is the only known natural host of the PRAV. The virus is reported to occur wherever PRD has been reported. The virus on its own causes symptomless infection or transient mottle and can cause significant yield loss in susceptible peanut cultivars (Naidu and Kimmins 2007).

Groundnut rosette virus (*GRV*): The virus is restricted to SSA and its offshore islands. It is first isolated and characterized by Reddy et al. (1985b) and has no structural (coat) protein (Taliansky et al. 2003), and thus, no conventional virus particles of GRV are formed. Enveloped bullet-shaped structures that can be detected in the ultrathin sections of infected cells could be shown to be cytopathological structures due to GRV infection, as opposed to real virions (Taliansky et al. 2003). The virus genome is a nonsegmented, single linear molecule of single-stranded, positive-sense RNA of size c. 4019 nucleotides which encodes for four ORFs (Taliansky et al. 1996). The genome of an isolate has been completely sequenced (GenBAnk accession # Z 66910), and several partial sequences are available in the gene bank. The virus replicates autonomously in the cytoplasm of the infected tissues (Taliansky and Robinson 2003). GRV on its own causes transient symptoms, but a SatRNA associated with GRV is responsible for rosette disease symptoms. GRV depends on groundnut rosette assistor virus for encapsidation of its RNA and transmission by *A. craccivora* in a persistent mode (Robinson et al. 1999). The virus is transmitted by grafting and mechanical inoculation, but not through seed, pollen, or contact between the plants. Peanut is the only known natural host, but several experimental hosts in the families Chenopodiaceae and Solanaceae have been reported (Murant et al. 1998). GRV belongs to the genus *Umbravirus*. No strain of GRV has been reported. The virus is restricted to SSA and its offshore islands.

SatRNA: SatRNA is responsible for rosette symptoms and plays a critical role in helper virus. The SatRNA (subviral RNAs) of GRV belongs to the Subgroup-2 (small linear) SatRNAs. It is a single-stranded, linear, nonsegmented RNA of 895–903 nucleotides (Murant et al. 1988, Blok et al. 1994, Taliansky et al. 2000). It totally depends on GRV for its replication, encapsidation, and movement, both within and between the plants.

Diagnosis

Various diagnostic techniques based on biological, serological (protein based), and genomic properties (nucleic acid) of the PRD agents have been developed (Kumar and Waliyar 2007, Waliyar et al. 2007). PRD can be diagnosed in the field based on the characteristic symptoms on peanut. Mechanical inoculation on to *Chenopodium amaranticolor* indicates the presence of PRV (infected plants show minute necrotic lesions on inoculated leaves about 4 days after inoculation) (Murant et al. 1998). Serological and nucleic acid–based diagnostic methods can be used for the detection of PRAV, but only nucleic acid–based methods can be used for the detection of PRV and SatRNA. Triple-antibody sandwich-enzyme-linked immunosorbent assay (ELISA) has been developed for the detection of PRAV (Rajeshwari et al. 1987) and dot blot hybridization and reverse transcription-polymerase chain reaction (RT-PCR) to detect all the three PRD agents in plants and aphids (Blok et al. 1995, Naidu et al. 1998b).

Transmission

1. *Sap transmission*: Transmissibility of sap-transmissible component is best achieved by extracting the sap in potassium phosphate buffer (K_2HPO_4) of pH 7.3 containing Mg or Na bentonite 25 mg/mL and 0.01 M diethyldithiocarbamate. By artificial mechanical sap inoculations, experimental hosts of PRV and SatRNA have been identified in several species in Leguminosae, Chenopodiaceae, and Solanaceae (Murant et al. 1998, Waliyar et al. 2007). *C. amaranticolor* and *Chenopodium murale* are local lesion hosts; *C. amaranticolor, Glycine max*, *Phaseolus vulgaris*, *Nicotiana benthamiana*, and *Nicotiana clevelandii* are systemic hosts of PRV. Apart from peanut, experimental hosts of both PRAV and PRV and SatRNA are *Gomphrena globosa*, *Stylosanthes gracilis*, *Stylosanthes mucronata*, *Stylosanthes sundaica*, *Spinacia oleracea*, *Trifolium incarnatum*, and *Trifolium repens* (Murant et al. 1998).
2. *Aphid transmission*: *A. craccivora*, commonly known as the cowpea aphid, is the principal vector involved in the transmission of all the PRD agents in a persistent and circulative manner (Hull and Adams 1968). PRV and SatRNA must be packaged within the PRAV CP to be aphid transmissible. Studies have shown that all the PRAV particles whether they contain PRAV RNA or PRV RNA and SatRNA are acquired by the aphid vector from phloem sap in 4 and 8 h acquisition access feeding for chlorotic and green rosette, respectively (Misari et al. 1988). Then, there is a latent period of 26 h 40 min and 38 h 40 min for chlorotic and green rosette, respectively, and the inoculation access feeding period of 10 min for both forms (Misari et al. 1988). Once acquired, aphid can transmit virus particles for up to 2 weeks and beyond. All stages of the aphid can acquire and transmit the disease agents. Transmission rates of 26%–31% have been reported with one and two aphids per plant and 49% with five aphids per plant (Misari et al. 1988).

Aphid vector does not always transmit all the three agents together (Naidu et al. 1999). Under natural conditions, some PRD-affected plants (PRV and SatRNA positive) have been found to be free from PRAV, and PRAV can be detected in some nonsymptomatic plants (no PRV and SatRNA) (Naidu et al. 1999). This situation appears to be due to the difference in inoculation feeding behavior of the vector leading to the transmission of (1) all the three agents together, (2) only PRAV, or (3) PRV and SatRNA, as demonstrated by the electrical penetration graph studies of aphid stylet activities (Naidu et al. 1999). This reveals that during short inoculation feeding (test probe or stylet pathway phase) vector aphids probe peanut leaves without reaching the phloem, transmitting only PRV and SatRNA, which multiply in the epidermal and mesophyll cells. Even if PRAV particles are deposited in the mesophyll cells, they cannot replicate, as they can replicate only in the phloem cells (Naidu et al. 1999). However, vector aphids can transmit PRAV and PRV–SatRNA when the stylets penetrate sieve elements (salivation phase) of the phloem cells. Therefore, the success of transmitting all the three agents together is high when inoculation feeding period is longer or increasing the number of aphids per plant (Misari et al. 1988). Vector aphids fail to acquire or transmit PRV and SatRNA from diseased plants lacking PRAV, and such plants become dead-end sources. However, if such plants receive PRAV later due to vector feeding, the plants again serve as source of inoculum.

Epidemiology and Disease Cycle

The epidemiology of PRD is complex, involving interactions between and among two viruses and a SatRNA, the vector, the host plant and environment. Since none of the causal agents is seed borne, primary infection of crops depends on the survival of infected plants (virus sources) and vectors (aphids) (Naidu et al. 1998a). Possible source from which rosette could spread are infected peanut plants surviving between cropping seasons. In regions where there are no sources of infection, initial infection may depend on the influx of viruliferous aphids from other parts of Africa on prevailing wind currents (Bunting 1950, Adams 1967). The vector *A. craccivora* is polyphagous and can survive on as many as 142 plant species in addition to peanut. One or more of these 142 plant species could be a source of the rosette complex (Adams 1967, Naidu et al. 1998a). Efforts thus far have failed to identify any alternative natural hosts of the PRD agents (Waliyar et al. 2007). Kenyan isolates of the virus are closer to the Malawian than to the Nigerian isolates (Wangai et al. 2001).

PRD is a polycyclic disease because each infected plant serves as a source for initiating subsequent disease spread in the field. Winged aphids are responsible for primary spread of the disease. Secondary spread from the initial foci of disease within the fields also occurs by way of the movement of aphid vector, but largely apterae and nymphs (Naidu et al. 1998a). In general, primary infection at early stages of the crop growth provides a good opportunity for repeating cycles of infection to occur before crops mature and vector populations decline. The nature and pattern of disease spread is influenced by plant age, cultivar, crop density, time of infection, transmission efficiency of aphids, proximity to the source of infection, and climatic conditions (van der Merwe et al. 2001, Herselman et al. 2004, Waliyar et al. 2007).

Disease Management

Host Plant Resistance

Efforts in breeding for host plant resistance and evaluation of peanut germplasm collection held in ICRISAT genebank have contributed to the development of several peanut genotypes and identification of germplasm lines with acceptable levels of field resistance to rosette disease (Olorunju et al. 1991, 2001, van der Merwe and Subrahmanyam 1997, Subrahmanyam et al. 1998, 2001). Evaluation of 12,500 lines from ICRISAT's genebank collection of peanut germplasm has resulted in the identification of 150 resistant sources, of which 130 are long-duration Virginia types and 20 are short-duration Spanish types (Subrahmanyam et al. 1998, Olorunju et al. 2001). Evaluation of

116 wild *Arachis* accessions representing 28 species identified 25 accessions resistant to rosette disease (Subrahmanyam et al. 2001). Out of 2301 germplasm lines evaluated in Samaru in Nigeria, only 65 new sources of resistance to rosette could be identified, 55 of which are long-duration Virginia types and 10 are short-duration Spanish types (Ntare and Olorunju 2001). It is not known whether these resistant sources carry the same or different kinds of resistance genes. Generally, resistance to rosette disease in a genotype is assessed by the lack of symptom expression, and therefore, such resistance is largely against PRV and SatRNA (the two components responsible for rosette symptoms) (Bock et al. 1990, Subrahmanyam et al. 1998, Olorunju et al. 2001). This resistance has been shown to be controlled by two independent recessive genes and is effective against both chlorotic and green forms of rosette (Nigam and Bock 1990, Olorunju et al. 1992). This resistance is directed against PRV and consequently to SatRNA and is not effective against PRAV (Bock et al. 1990). This form of resistance has been transformed into early maturing cultivars that are useful for cultivation in regions that are often characterized by short length of growing periods. Some of the rosette-resistant early varieties released in the West and Central Africa region are ICGV-SM 90704, ICG 12991, ICGV-SM 99568, ICGV 93437, SAMNUT 23 (ICGV-IS 96894), SAMNUT 21 (UGA 2), and SAMNUT 22 (M572.80I) (Ntare et al. 2002, Waliyar et al. 2007).

Yield reduction in genotypes that are resistant to PRV and SatRNA is observed to be presumably due to their susceptibility to PRAV (Subrahmanyam et al. 1998, Olorunju et al. 2001). This is finally confirmed in a study that separated PRAV from PRV and SatRNA and demonstrated that PRAV infection alone can significantly reduce peanut seed yield (Naidu and Kimmins 2007).

There is a possibility for the development and deployment of transgenic forms of resistance using genes derived from the virus itself (pathogen-derived resistance) (Deom 1999, Deom et al. 2000). Resistance to PRV has been detected in plants transformed with constructs derived from a mild variant of the SatRNA in *N. benthamiana* (Taliansky et al. 1998). Research in this direction is in progress at the ICRISAT centers.

Chemical Control

Seed treatment with imidacloprid and followed by regular systemic insecticide spray in the early stages of the crop growth (from emergence to 40th day) will control vector aphids and consequent protection against PRD. Long acquisition access feeding period required by the vector provides an opportunity to control aphids with chemical sprays before they can spread the disease. Various insecticides have been used to control *A. craccivora* to minimize or prevent the spread of rosette disease in field trials (Waliyar et al. 2007). Dosage and type of insecticide utilized is critical for controlling aphids. However, insecticides are an unviable option in SSA due to high costs and scarcity, thus seldom preferred by the farmers. Furthermore, insecticide applications pose detrimental effects on health and environment, and their usage is being discouraged.

Cultural Control

Information on the control of PRD by cultural practices has been obtained in different parts of SSA (Naidu et al. 1998a, 1999, Waliyar et al. 2007). Early sowing (particularly in June) and high seed rate (80–120 kg/ha) have been recommended in much of Africa as a standard measure for the control of peanut rosette. Early sowing in the season is to take advantage of low aphid populations, and maintaining good plant density without any gaps (aphids prefer widely spaced plantings for landing) has been shown to reduce rosette disease incidence. However, early sowings may not be effective in areas where groundnut is grown continuously, as this allows perpetuation of virus and vector. Peanut crops intersown with field beans, maize, and sesame remain less affected with the rosette disease that is subsequently useful in preventing the spread of the disease (Alegbejo 1997). Roguing of voluntary sources and early-infected plants prevents the spread of the rosette (Kolte 1984, Waliyar et al. 2007). Overall, the combination of resistant genotypes with early sowing and optimum plant population is economically useful in the management of the disease even under high disease pressure (Subrahmanyam et al. 2002).

PEANUT STEM NECROSIS DISEASE

Symptoms

The first appearance of symptoms becomes visible in the form of the development of necrotic lesions on terminal leaflets and petioles and the death of top growing bud on the main stem followed by necrosis of all top buds on primaries. Complete stem necrosis and often total necrosis of the entire plant occur in early infection. The plants that survive the viral infection become stunted and show proliferation of axillary shoots and reduction in leaflet size and exhibit chlorosis as secondary symptoms in contrast to mosaic and mottling of leaf lamina in the case of peanut bud necrosis disease (PBND). Pods from affected plants show black necrotic lesions on the pod shells, and kernels become smaller adversely affecting the marketability of the pods and kernels.

Geographical Distribution and Losses

The first known case of the occurrence of epidemic of peanut stem necrosis disease (PSND) caused by tobacco streak virus (TSV) was recorded in the monsoon season crop in the year 2000 from the major peanut-growing district of Anantapur in the state of Andhra Pradesh in India (Reddy et al. 2002). The disease now occurs in almost all peanut-growing states of India particularly in those regions, where sunflower necrosis disease caused by the same virus (TSV) is endemically present (Ravi et al. 2001, Bhat et al. 2002). During the 2000 epidemic in Andhra Pradesh, crop losses have been estimated to exceed U.S. $64 million. The TSV, the causal virus, because it infects a wide range of crops like sunflower, safflower, cotton, cowpea, okra, urd bean, and mung bean, assumes a great significance of economic importance, and hence, the TSV is currently regarded as an emerging threat to crop production in India.

Pathogen: The Causal Virus

TSV belonging to the genus *Ilarvirus* in the family Bromoviridae is the cause of the PSND (Reddy et al. 2002). Purified virions of TSV are nonenveloped, isometric, measuring 25–35 nm in diameter. It consists of a single capsid protein of 28 kDa. The virus genome is a single-stranded RNA, has positive polarity, is linear, and is a tripartite of size 3.7, 3.1, and 2.2 with 0.9 kb subgenomic RNA. The TSV genome has been sequenced. Cowpea (*Vigna unguiculata* cv. C-152) and *P. vulgaris* (cv. Top Crop) are the suitable hosts for propagating the virus.

Diagnosis

Symptom-based identification can be misleading as similar symptoms can also be due to peanut bud necrosis virus (PBNV). Hence, diagnosis of the infection caused by TSV should be done using indicator host plants reaction test and ELISA.

The two most important diagnostic hosts are cowpea cv. C-152 and French bean cv. Top Crop. On sap-inoculated leaves of cowpea and French bean, TSV produces necrotic lesions and veinal necrosis as early as within 2–3 days, whereas PBNV produces concentric chlorotic/necrotic lesions on these indicator hosts as late as 4–5 days after inoculation.

ELISA and RT-PCR polyclonal antibodies to TSV have been produced, and the direct antibody-coated-ELISA-based virus detection technique has been developed for the reliable diagnosis of PSND. Oligonucleotide primers from CP gene have been designed for RT-PCR-based virus detection (Prasad Rao et al. 2004).

Transmission

The virus is transmitted through thrips *Frankliniella schultzei*, *Scirtothrips dorsalis*, and *Megalurothrips usitatus* in a very peculiar manner. The thrips fed on infected leaves alone and do not transmit the virus. But they do so through their wounding of plants during feeding only in the presence of infected pollens particularly from *Parthenium hysterophorus* and/or sunflower in the vicinity of these wounds rather than entering into specific virus–vector relationships. The virus in peanut is not transmissible through seed, but it can be transmitted through sap inoculation.

Epidemiology and Disease Cycle

Though TSV is pollen borne, the virus is not transmitted through seed in the case of peanut (Reddy et al. 2007). Thrips vectors, viz., *F. schultzei*, *S. dorsalis*, and *M. usitatus*, aid in passive transmission of the virus as carriers of pollens from infected plants. Among the three thrips species, *F. schultzei* plays a major role in transmission and spread of the virus particularly from the flowers of infected *P. hysterophorus* and other weed plants (*Abutilon indicum*, *Ageratum conyzoides*, *Croton sparsiflorus*, *Commelina benghalensis*, *Cleome viscosa*, *Euphorbia hirta*, *Lagascea mollis*, *Tridax procumbense*), sunflower, and marigolds. *F. schultzei* carries 8–10 pollen grains on its body from these ranges of weed hosts and 60–70 pollen grains from sunflower flowers, and when these thrips become wind borne and visit groundnut plants, the pollen grains then get dislodged from the insect's body, and during the feeding process, virus present in the pollen grains infects the peanut plants. Since peanut is self-pollinated and early-infected peanut plants do not flower, the peanut plants on its own do not contribute to spread of the virus inoculum and the PSND in a crop field. *Parthenium* is widely distributed and present all year-round in the vicinity of peanut crop fields producing several flushes of pollen grains during its life cycle ensuring continuous supply of virus-infected pollens for effective transmission through the thrips (Prasad Rao et al. 2004, Kumar et al. 2008).

Based on 2000 PSND epidemic in Anantapur district in Andhra Pradesh in India, the following factors are conducive for the occurrence of PSND epidemic: (1) early rains during late May or early June that encourage germination and growth of *Parthenium*, (2) sowing peanut during July by which time *Parthenium* is in full bloom, (3) normal rain that promotes good growth of peanut crop as well as *Parthenium*, and (4) one or two dry periods of 3 week duration that encourages thrips multiplication and movement for the spread of virus.

Disease Management

Host Plant Resistance

1. *In peanut germplasm*: Resistance to PSND has not been found in the germplasm of cultivated peanut (Kalyani et al. 2007). However, a considerable number of peanut genotypes have been reported to be resistant or moderately resistant or promisingly less susceptible to the disease. These genotypes are ICGV 99057, 00169, 99068, 86325, 92267, 94379, ICG 4983 (*Arachis chacoense*), ICGS 37, RP 251, S 206, DH 40, CSMG 84-1, and M-22. Among these, the multiple disease-resistant genotypes ICGV 99057 and ICGV 00169 besides being of high potential of PSND resistance also exhibit good shelling percentage and oil content. Hence, it is rewarding to incorporate resistance into good agronomic types by hybridization and selection (Kumar et al. 2008, Venkataravana et al. 2008).
2. *Transgenic peanut for resistance to PSND*: Transgenic resistance in peanut to PSND has been obtained by transferring CP gene of TSV through *Agrobacterium*-mediated transformation of deembryonated cotyledons and immature leaves of peanut cultivars Kadiri 6 and Kadiri 134. The transgenic lines are reported to remain symptomless throughout and show traces or no systemic accumulation of virus indicating the tolerance/resistance to the TSV infection. CP gene expression has been confirmed in transgenic lines by RT-PCR, real-time PCR, and ELISA. This is an effective strategy for developing peanut with resistance to PSND (Mehta et al. 2013).

Chemical Control

Seed treatment with imidacloprid (Gaucho 70 WS) at 2 mL/kg seed followed by spraying of monocrotophos at 800 mL or dimethoate at 1000 mL or imidacloprid at 200 mL/ha in 500 L of water at 25–30 days after sowing is recommended in epidemic regions. It is noteworthy that in the case of most of the thrips-transmitted viruses, the use of insecticides after the appearance of the disease has no effect on the control of the disease (Prasad Rao and Reddy 2005).

Cultural Control

Removal of weeds such as *P. hysterophorus* and other weeds around the peanut fields is helpful to reduce the disease incidence; however, roguing of early-infected peanut plants may not limit further spread of the disease in the field. Barrier crops like bajra, maize, and sorghum should be planted in four to eight rows around the peanut field. These will prevent thrips and wind-borne weed pollen-carrying virus. Practicing intercropping with bajra, maize, and sorghum in the ratio of 7:1 or 11:1 is also helpful in decreasing the incidence of PSND in the peanut crop. Maintenance of optimum plant density is important to discourage landing of the thrips (Prasad Rao et al. 2004, Kumar et al. 2008).

PEANUT BUD NECROSIS DISEASE

Symptoms

Under field conditions, initially quadrifoliate leaf immediately below the terminal bud shows distinct chlorotic ring spots or chlorotic speckling and becomes flaccid about 30–40 days after planting. These symptoms in peanut due to PBNV are difficult to distinguish from those caused by tomato spotted wilt virus (TSWV). The vascular tissue of the shoot just below the growing tip becomes necrotic and the terminal bud is killed, which later dries and becomes brown. This is a characteristic symptom that occurs on peanut plants of crops grown in the rainy and post-rainy seasons, when ambient temperatures are relatively high above 30°C. The necrosis may proceed downward, and the whole branch may become blighted. Necrosis may also be seen on petioles and along the stems. Proliferation of axillary shoots takes place, but the leaves of such shoots remain smaller than normal and show a wide range of symptoms, including distortion, mosaic mottling, and general chlorosis. Infection with PBNV reduces the concentration of chlorophyll *a* and *b* and increases the specific activity of chlorophyll oxidase and peroxidase enzymes (Hema and Sreenivasulu 2002). Affected plants remain stunted because of the reduction in the length of the internodes, and the whole plants may show a bushy appearance. If plants are infected early, they are stunted and bushy. If plants older than 1 month are infected, the symptoms may be restricted to a few branches or to the apical parts of the plants. Seeds from such plants are small, shriveled, mottled, and discolored. Such seeds show poor germination, or they may fail to germinate. Late-infected plants may produce seed of normal size. However, the testae on such seed are often mottled and cracked.

Geographical Distribution and Losses

PBND was first recorded in India in 1949 as per the reports made from Indian Agriculture Research Institute, New Delhi (Reddy et al. 1995). The economic importance of the disease was realized during the late 1960s when incidences up to 100% were recorded in many peanut-growing regions in India. The disease was described under different names such as ring spot of peanut, ring mosaic, spotted wilt, bud blight, and bud necrosis (Kolte 1984). It was shown to be economically important in parts of Tamil Nadu, Karnataka, Andhra Pradesh, Maharashtra, and Uttar Pradesh. Although it was earlier reported to be caused by TSWV, currently the causal virus of PBND in India has been shown to be a serologically distinct *Tospovirus*, now referred to as PBNV, transmitted by *Thrips palmi*. Surveys in many groundnut-growing countries indicate that PBNV is restricted to South and Southeast Asia (Reddy et al. 1995, Poledate et al. 2007, Damayanti and Naidu 2009). Reports of its occurrence have been made from the main peanut-growing province of Córdoba, in Argentina in South America (de Breuil et al. 2008), and from the Golestan Province of Iran (Golnaraghi et al. 2002). The disease poses a major threat to peanut production in Thailand in dry seasons (Poledate et al. 2007). In India, average yield loss caused by the PBND is more than 50% in peanut that amounts to estimation of about U.S. $89 million/annum (Reddy et al. 1995, Kendre et al. 2000). The causal virus, the PBNV, appears to be economically more significant because different isolates (strains) of PBNV have been proved to be the primary pathogens in causing a number of diseases in different crop plants such as chilli (Gopal et al. 2011a), cucumber (Gopal et al. 2011a), cowpea (Jain et al. 2002, Akram and Naimuddin 2009, Gopal et al. 2011a), mung bean (Jain et al. 2002, Thien et al. 2003, Sreekanth et al. 2006a, Saritha and Jain 2007), okra (Kunkalikar et al. 2012), sesame (Gopal et al. 2011a), soybean (Kumari et al. 2003), sunflower (Pranav et al. 2008), taro (Sivaprasad et al. 2011), tomato (Jain et al. 2002, Raja and Jain 2006, Venkat et al. 2008, Venkatesan et al. 2009, Manjunatha et al. 2010a, Ramana et al. 2011, Akhter et al. 2012), potato (Akram et al. 2003, Pundhir et al. 2012), urd bean (Prasad Rao et al.2003, Kumar et al. 2006), pea (Akram and Naimuddin 2010), and watermelon (Gopal et al. 2011a).

Causal Virus: Peanut Bud Necrosis Virus

The virus causing the PBND does not react with the antisera to TSWV obtained from different sources (Reddy et al. 1992), and based on serological cross reactions (Adam et al. 1993) and amino acid sequence homology of the nucleoprotein (de Avila et al. 1993), it has revealed the existence of a distinct virus, different from the TSWV and impatiens necrotic spot virus (INSV), and thus, the virus-causing PBND has been identified as a distinct *Tospovirus* and named as PBNV. With ELISA as well as Western blots, PBNV has been shown to be serologically distinct from TSWV and INSV (Reddy et al. 1992). PBNV contains three RNA species of about 9.0 kb (large, L RNA), 5.0 kb (medium, M RNA), and 3.0 kb (small, S RNA), and the nucleotide sequences are 8911 for L RNA, 4801 for M RNA, and 3057 for S RNA (Satyanarayana et al. 1996a,b). The virus protein consists of four polypeptides of molecular weights of 27, 52, 58, and 78 × 10^3 Da. The particles are 70–90 nm in diameter and are surrounded by a double membrane of protein and lipid and sediment at 520–530s. Nonstructural protein, NSs, of the PBNV encoded by the S RNA is a bifunctional enzyme, which could participate in viral movement, replication, or suppression of the host defense mechanism (Lokesh et al. 2010, Bhat and Savithri 2011). Typical of a *Tospovirus*, the PBNV has extremely low thermal inactivation point of 45°C for 10 min; the dilution end point is between 10^{-2} and 10^{-3} and short longevity in vitro of less than 5 h at room temperature. PBNV classified as a virus in serogroup IV of *Tospoviruses* (Bunyaviridae) (Akram et al. 2004).

Transmission

Though the PBNV is graft transmissible, it is not transmitted through aphids and seed. Transmission through sap and thrips transmission are most common that are as follows.

Sap transmission: PBNV can be transmitted by mechanical sap inoculations if care is taken to extract the virus only from young infected leaflets with primary symptoms. Extracts should be prepared in neutral phosphate buffer (0.05 M phosphate buffer, pH 7.0) containing an antioxidant such as mercaptoethanol (0.2%) or thioglycerol (0.075%) and must be kept cold throughout the inoculation process.

Thrips transmission: *T. palmi* transmits PBNV. Other thrips species *S. dorsalis* and *F. schultzei*, which are also present on the plants, do not transmit the PBNV. Adults of *T. palmi* cannot acquire the virus; however, their larvae can acquire the virus and such larvae and subsequently developed adults transmit the virus persistently (Reddy et al. 1991, Sreekanth et al. 2006a,b). A minimum of 15 min acquisition access period by larvae and a 45 min inoculation access period by adult thrips are required for successful transmission of the PBNV (Sreekanth et al. 2004a, 2006a). Maximum transmission (100%) can be obtained when there are 10 adults per plant. The majority of individual adult thrips transmit the virus for more than half of their life period. Cowpea has been found to be the best host for rearing and multiplying *T. palmi* under laboratory conditions.

Diagnosis

Sap inoculations of virus extracts on diagnostic hosts like cowpea (*Vigna unguiculata*) cv. C-152 and *Petunia* hybrid can be used to identify the PBNV. Cowpea produces concentric chlorotic and necrotic lesions on inoculated primary leaves 4–5 days after inoculations, and subsequently systemic infection develops on newer leaves (Akram and Naimuddin 2009), whereas *Petunia* produces only necrotic lesions on inoculated leaves 3 or 4 days after inoculation (Reddy et al. 1991). ELISA using polyclonal antibodies clearly distinguish PBNV from TSWV and INSV (German et al. 1992, Reddy et al. 1992, Jain et al. 2005, Nagaraja et al. 2005b, Raja and Jain 2006). If young tissues showing initial symptoms are used, PBNV particles can be observed in leaf extracts or even in leaf dip preparations. They are 80–100 nm in diameter and are surrounded by a double membrane of protein and lipid.

Epidemiology and Disease Cycle

There is a possibility of existence of different strains of PBNV, as different isolates of the virus collected from different regions show some differences in host range and reaction of susceptible hosts. Though the detailed study in strain differentiation has not been done, it is revealed through the nucleotide and amino acid sequences of the movement protein (NSm) genes of different isolates that the NSm genes of PBNV isolates are identical in length (924 bp encoding 307 amino acids) suggesting their common origin (Akram et al. 2003, 2004, 2012). The primary sources of inoculum may be different host plant members of Leguminosae and Solanaceae specially crop plants such as tomato, mung bean, urd bean, sunflower, and cucumber that are cultivated in summer and weed hosts such as *Ageratum conyzoides*, *Acanthospermum hispidum*, and *Cassia tora*, which are more commonly present in and around the peanut crop fields and which sustain virus infection and effective thrips vector population (Nagaraja et al. 2005a, Reddy et al. 2011, Gopal et al. 2011a). Thus, the incidence of the PBND in peanut depends on infection by viruliferous thrips that acquire the virus from such alternative hosts and their mass migration flights from alternative hosts to peanut crop fields. Most migrations occur when air temperature is in the range of 20°C–30°C. Warm and dry weather favor disease buildup and prevalence of thrips (Thiara et al. 2004, Pensuk et al. 2010). An optimum temperature of 25°C is the best for rearing *T. palmi* and the total number of larvae produced per female is greater at 25°C (Vijayalakshmi et al. 2000). A wind velocity of 10 km/h at 3 m above the crop canopy is more conducive to mass flights of thrips (Prasad Rao and Reddy 2005). Interestingly, secondary spread from infected peanut plants within a peanut field is considered to be negligible.

Disease Management

Host Plant Resistance

Three peanut germplasm lines, viz., IC 10, IC 34, and ICGV 86388, have been confirmed to be the best PBND-resistant parental lines and useful sources of resistance not only to PBNV (Reddy et al. 2000, Pensuk et al. 2002a–c, 2004, Kesmala et al. 2006) but also to TSWV (do Nascimento et al. 2006), a closely related species of PBNV. Heritability estimates for both incidence and severity of PBND are found to be favorably high enough in these three PBNV-resistant parental lines for further improvement of these characters. Both genotypic and phenotypic correlations between PBNV resistance parameters and desirable agronomic traits are also high (Kesmala et al. 2003, 2004, Tonsomros et al. 2006). However, susceptibility of peanut to PBNV is somewhat associated with large-seed size, and this, therefore, might be an interfering factor for breeding large-seeded peanut cultivars with resistance to PBNV (Puttha et al. 2008). Two peanut cultivars such as ICGS 11 and ICGS 44 possessing resistance to PBND have been released in India, and some other promising PBND-resistant peanut genotypes are ICGV 92269, 89/94-3-2, ICGV 91229, ICGV 91193, 89/94-7-3, 83/151-7, 85/203-6, ICGV 91248, ICGV 91117, and ICGV 86031 (Gopal et al. 2004). Three more peanut genotypes, viz., GPBD-4, JSSP-9, and DH-53, are reported to be least affected by PBND indicating their very high degree of tolerance to the disease (Nagaraja et al. 2005b). Among 83 wild *Arachis* germplasm screened for resistance to PBNV, one accession each of *Arachis benensis* and *Arachis cardenasii* and two accessions of *Arachis villosa* are confirmed to be resistant to PBNV (Reddy et al. 2000). In these wild *Arachis* PBNV-resistant accessions, the inoculated leaves do show infection, but subsequently developed leaves do not show the presence of virus in spite of repeated sap inoculation indicating the resistance in these accessions appears to be due to a block in systemic movement of the virus. Since *A. cardensii* and *A. villosa* are the progenitors of cultivated peanut and can be hybridized with the latter, the resistant wild *Arachis* accessions can be successfully utilized in conventional breeding program to transfer PBNV resistance to widely cultivated peanut cultivars (Reddy et al. 2000).

Cultural Control

The occurrence and severity of the disease depends on the migration of the thrips. Therefore, there exists a scope for choice of planting dates in reducing or avoiding the disease (Sreekanth et al. 2002, Gopal et al. 2007). Under Andhra Pradesh (India) conditions, early planting at the onset of the rainy season decreases disease incidence. Interrow and intrarow spacing of 20 × 7.5 or 10 cm gives a high density of plant population (2–3 million plants per hectare), which ensure close canopy, leading to reduction in incidence of PBND (Gopal et al. 2007). Roguing of infected peanut plants should not be practiced as this will reduce the density of crop canopy leading to increased incidence of PBND. The movement of the thrips vector is decreased when pearl millet, sorghum, pigeon pea, or maize is intercropped with peanut in the ratio 3:1 resulting in significant decrease in the incidence of PBND (Sreekanth et al. 2004b, Gururaj et al. 2005, Gopal et al. 2010).

Several weed species particularly *Achyranthes aspera*, *Ageratum conyzoides*, *Alysicarpus rugosus*, *Commelina benghalensis,* and *Vigna trilobata* have been found to be reservoirs of PBNV in and around the peanut fields in major peanut-growing regions in India. Removal and destruction of these weed species acting as primary sources of infection would be useful in reducing the incidence of PBND (Gopal et al. 2011a).

Effect of Botanicals

PBNV-susceptible peanut cultivar sprayed separately with sorghum leaf, coconut leaf, and neem kernel or neem cake extracts (10%) 20 and 35 days after planting alone or in combination with 1.25 mL of monocrotophos significantly reduce the incidence of PBND that subsequently results in increase in pod yield by 60%–100% (Kulkarni et al. 2003, Gopal et al. 2011b). Antivirus principle in sorghum and coconut leaf extracts appear to be proteinous compound, and its translocation mechanism might afford protection besides induction of its effect in the accumulation of high concentration of defense-related enzymes such as phenylalanine ammonia lyase, peroxidase, and polyphenol oxidases (Manjunatha et al. 2010b).

SPOTTED WILT

Symptoms

TSWV-infected peanuts first appear at random throughout a field as early as 21 days after the seedlings emerge. Earliest symptoms of the disease are brown speckles on the underside of the first leaf below one or more terminal buds along the leaf. The leaf below the terminal bud, showing typical yellowing and mottling, appears wilted, while the rest of the plant looks healthy. With time, clusters of diseased plants may be seen. The virus usually spreads within a field down the row from plants infected at the start of the growing season. Brown necrotic spots or streaks may also be seen on the leaf petiole and stem and at times on the terminal bud. These spots may develop into a shoot dieback, which may ultimately kill the plant. Any new leaves are about half their normal size, crinkled and display a range of symptoms including chlorosis, concentric chlorotic ring spots, ring spots with green centers, and chlorotic line patterns. Severe stunting is a common symptom of TSWV infection of susceptible peanut cultivars and is generally more severe when young plants are infected.

Geographical Distribution and Losses

Significant economic losses have been recorded by peanut growers in the southeastern United States since the peanut growers in one Texas County suffered an estimated U.S. $3 million loss in 1986 due to TSWV. By 1988, symptomatic plants could be seen quite commonly in peanut crop stand in Alabama, Florida, and Georgia. However, the numbers of TSWV-infected plants in most fields remained extremely low. In recent years, severe outbreaks of spotted wilt in peanuts have occurred in South Central Georgia. In some fields, an estimated 40% to nearly 100% of peanut plants have been found to be infected with the disease. The disease has been particularly damaging in mid-April planted peanut (Olatinwo et al. 2009). TSWV is now well established throughout the southeastern peanut belt and has become a serious problem in the Virginia/Carolina peanut-growing regions of the United States. During 2002, the disease was present in 47% of the North Carolina hectarage and caused a 5% yield reduction in Virginia (Herbert et al. 2007). There appears to be a significant correlation between spotted wilt intensity and peanut yield (Olatinwo et al. 2010). Recent epidemics in Alabama, Florida, Georgia, Mississippi, and Texas show that the virus is a serious threat to peanut production in the region (Culbreath et al. 2011). The disease is also reported to occur in the province of Cordoba in Argentina (de Breuil et al. 2008).

Pathogen

Spotted wilt disease of peanut (*Arachis hypogaea*) is caused by TSWV (genus *Tospovirus*, family Bunyaviridae). Virions of TSWV are complex compared to many plant viruses. There are three RNAs in the virus genome that are individually encapsidated and are collectively bound by a membrane envelope, that is, of host origin. This complex virion structure is a characteristic that distinguishes TSWV from most other plant viruses. TSWV virions are roughly spherical and are 80–110 nm in diameter. Two virus proteins processed during replication to contain sugars, that is, glycoproteins (GPs), are dispersed throughout the surface of the viral envelope. These proteins are called GP C and GP N and differ slightly in size. Inside the viral envelope is each of the three viral RNAs, individually bound by multiple copies of a nucleocapsid protein. The three RNAs differ in size and are called large, middle, and small. Also, inside the envelope are several copies of a virus-encoded *replicase* protein that is required to initiate virus replication in a new host. The GPs in the envelope function in the maturation and assembly of virions and appear to play a role in the acquisition of TSWV by thrips. Envelope-deficient isolates of TSWV, generated by serial mechanical passage in plants, are infectious in plants but are not transmitted by thrips. It is evident that the GPs are not required for replication in plants, but are required for virus infection of thrips leading to subsequent virus replication in and transmission by thrips.

In addition to the GPs, the M RNA segment encodes a nonstructural protein (NSm). The NSm is unique to *Tospoviruses* in the family Bunyaviridae and is thought to be an adaptation of *Tospoviruses* to plants to facilitate *Tospovirus* the movement from cell to cell through plant cell walls via the plasmodesmata. Because enveloped particles are too large to be transported through plasmodesmata, the role of NSm is to form tubules that facilitate the movement of nucleocapsids (RNA plus protein) from cell to cell. In addition to the nucleocapsid protein, the S RNA segment encodes a nonstructural protein (NSs). Crystalline-like structures of NSs are produced in infected insect cells and plant cells. The NSs protein has RNA-silencing suppressor activity and may play a role in posttranscriptional gene silencing or RNA metabolism. TSWV is one member of the dozen or so different viruses in the genus *Tospovirus*. One striking difference in these viruses is the variation in their host ranges. TSWV is renowned for having an extensive host range, whereas other members of the genus *Tospovirus* such as peanut yellow spot virus or *Iris* yellow spot virus have narrow host ranges.

Diagnosis

TSWV can be detected using ELISA in both leaf and root crown tissues throughout the peanut-growing season to determine the time and percentage of infected plants (Rowland et al. 2005, Murakami et al. 2006). Diagnosis of TSWV in peanut can also be accomplished by RT-PCR. ELISA and RT-PCR are comparable for detecting TSWV infection rate in field-grown peanuts. A delayed accumulation of TSWV in a cultivar is a reliable indicator of host plant resistance (Dang et al. 2009, 2010).

Transmission, Epidemiology, and Disease Cycle

Worldwide, seven species of thrips are known to be vectors of TSWV. Two of these thrips vectors, the tobacco thrips, *Frankliniella fusca*, which is by far the most abundant, followed by the western flower thrips, *Frankliniella occidentalis*, are known to be efficient vectors of TSWV (Riley et al. 2011). However, the western flower thrips are only a minor component of the total thrips population on peanuts. Thrips may be the primary source of TSWV. Adult thrips carrying TSWV overwinter in the soil and crop debris and transmit the virus at or shortly after seedling emergence. Possibly, weed and crop reservoirs of TSWV also determine whether virus-carrying thrips overwinter in these hosts. Newly emerged peanut seedlings are infested by adult thrips migrating into the field. Adult female thrips usually lay eggs between the young, folded leaflets. After 3–5 days, the first-stage larvae emerge and feed for about 2 days before changing into larger, second-stage larvae. These larvae feed for 3–5 days before changing into a nonfeeding, inactive prepupal stage. Adult thrips then emerge 3 days later. The average time required to complete the cycle from egg to adult is about 13 days for tobacco thrips. Thrips damage to peanut is characterized by scarring and deformation of new leaves, which often results in a stunted, slow-growing seedling. Adult female tobacco thrips are small (1.3 mm) and dark brown. Male tobacco thrips are smaller (1 mm) and pale yellow. Tobacco thrips of both sexes occur in winged or wingless forms. During the growing season, the ratio of females to males may be 6:1 or greater. Female western flower thrips are also small (1.5 mm), with a yellow to blotchy brown abdomen. Males are smaller (1 mm), with a pale yellow body. Larvae of both species range from pale to bright yellow and have bright red eyes. Thrips larvae acquire TSWV by feeding on virus-infected plants. However, the thrips are capable of transmitting the virus only as adults, and they can do so throughout the remainder of their lives. The average life span of an adult female tobacco thrips is about 33 days. TSWV must be acquired by thrips during the larval stage of their development to be transmitted. Thus, only immature thrips that acquire TSWV, or adults derived from such immatures, transmit the virus. The ability of thrips to acquire TSWV decreases as the thrips age. Although the time in development that thrips can acquire the virus is limited, the wide host range for both virus and thrips facilitates the development of epidemics (Culbreath et al. 2011). Once acquired by the larvae, the virus is passed transstadially, that is, TSWV persists through insect molts from larval to adult stages. The virus replicates in thrips, and the thrips can transmit the virus during their entire life. Some evidence indicates that the viral GPs bind to the midgut epithelium and have a role in the process of virus uptake in the midgut. The virus then moves to other cells and organs, becomes well established in the muscle cells epithelium, and have a role in the process of virus uptake in the midgut. Another perspective is that the temporary association between the midgut, visceral muscle, and salivary gland complex in the larval stage provides the avenue for the virus to become systemically established in the thrips. Eventually, the virus enters the salivary glands.

Indirect evidence indicates that virions are excreted with the saliva into host plants during thrips feeding. The TSWV can be transmitted mechanically, and transmission efficiency is improved by the use of two antioxidants (sodium sulfite and mercaptoethanol) and two abrasives (Celite and Carborundum) in extracting the sap inoculum and by application of the inoculum rubbing with a cotton swab dipped in the inoculum as well as pricking with an inoculation needle (Mandal et al. 2001, 2006, Al-Saleh et al. 2007).

Disease Management

Host Plant Resistance

1. *In peanut germplasm*: Although genes to confer resistance to TSWV have been found in some peanut germplasm lines and used to develop new cultivars, there has been rapid adaptation of new forms of the virus to cultivars that have been released. The genetic diversity in this virus group may be fostered by their replication in both different species of plants and different species of thrips. Biological diversity of TSWV may, however, be useful in developing more durable TSWV-resistant crop through induced systemic resistance (Mandal et al. 2006). Thus, currently there are virtually no cultivars of peanut with significant levels of resistance to TSWV that have remained resistant in the field for more than a few years. However, the development of tolerant cultivars has proven to be one of the most promising methods to manage the disease (Riniker et al. 2008). Peanut genotypes reported to be resistant/tolerant to TSWV are given in Table 3.1. High levels of field resistance to TSWV in peanut breeding lines have been derived from hypogaea and hirsuta botanical varieties (Culbreath et al. 2005).

TABLE 3.1
Peanut Genotypes Resistant (R) or Moderately Resistant (MR) to Tomato Spotted Wilt Virus (TSWV) as Reported from Different Countries in the World

Genotype	Country	R/MR	Reference(s)
Georgia-08V (PI 655573)	United States	R	Branch (2009)
Georgia-07W, Georgia-03L, AP-3	United States	R to both TSWV and SR	Branch and Brenneman (2009)
AP-3, York, Tifguard, Georgia-03L	United States	R/MR	Culbreath et al. (2008)
C724-19-15, Tifguard	United States	R to both TSWV and root-knot nematode	Holbrook et al. (2008a)
F NC94022-1-2-1-1-b3	United States	R	Culbreath et al. (2005)
Georgia-01, Georgia-05	United States	R (multiple pest resistance)	Branch and Culbreath (2008)
Geoorganic cv. 100 (PI 648033)	United States	R	Holbrook and Culbreath (2008)
TifGP-1 (PI 648354)	United States	R to both TSWV and root-knot nematode	Holbrook et al. (2008b)
Georgia-06G (CV 94, PI 644220)	United States	R	Branch (2007a)
Georgia Greener (CV 95, PI 644219)	United States	R	Branch (2007b)
ANorden (CV 97, PI 636442)	United States	R	Gorbet (2007a)
Tifrunner (CV 93, PI 644011)	United States	R	Holbrook and Culbreath (2007)
AP-3 (CV 99, PI 633912)	United States	R	Gorbet (2007b)
CHAMPS	United States	MR (less susceptible)	Mozingo et al. (2006)
Tamrun OLO7	United States	R	Baring et al. (2006)
IC 10, IC 34, ICGV 86388	Brazil	R (higher resistance than standard *Georgia Green* for TSWV resistance)	do Nascimento et al. (2006)
C-99R, C11-2-39	United States	R	Mandal et al. (2002)

2. *Transgenic peanut for resistance to TSWV*: Considerable effort has been expended to develop transgenic peanut plants that have virus-derived genes to confer resistance to TSWV. Transgenic peanut progenies that express antisense nucleocapsid (N) gene of TSWV when subjected to natural infection of the virus under field conditions or to challenge inoculation under controlled environmental conditions show significantly lower incidence of the spotted wilt disease. But these have not been used commercially, but could be used in a traditional breeding program to enhance host resistance (Magbanua et al. 2000, Schwach et al. 2004, Yang et al. 2004).

Vector Control through Insecticides

Use of insecticides alone to control thrips populations in the field is often ineffective. Contact insecticides generally do not reach where the thrips are located on the plant, and systemic insecticides do not act rapidly enough to prevent virus transmission. As more is learned about thrips feeding, treatments that deter feeding or induce host resistance to deter thrips feeding may be used. Some success, however, has been achieved as furrow application of aldicarb and phorate results in significant levels of thrips control with reduced incidence of TSWV and significant increase in peanut pod yield (Wiatrak et al. 2000, Herbert et al. 2007).

Cultural Control

In general, timing of planting to avoid major thrips migrations during critical early plant growth periods is feasible to reduce disease. An integrated approach that addresses many parameters that affect tomato spotted wilt development has been successful in mitigating tomato spotted wilt in peanut in the southeastern United States. Peanut variety, planting date, plant population, insecticide application, disease history, row pattern, and tillage have been identified as factors affecting disease development. These factors are weighted to determine a *risk index* for TSWV in the crop. The grower obtains a low, moderate, or high risk value that can be considered when implementing crop production practices. Establishment of high peanut populations of the most resistant cultivars through the use of high seed germination rate in a well-prepared seedbed in early to mid-May plantings, when soil temperatures and moisture conditions favor uniform germination and rapid seedling growth, helps suppress epidemics of spotted wilt (Tillman et al. 2007, Culbreath et al. 2008, 2012).

PEANUT STRIPE

Symptoms

Symptoms on peanut plants vary, depending on virus isolate and peanut cultivar. For most isolates, the initial symptoms appear as chlorotic flecks or rings on young quadrifoliates. The plants are slightly stunted. Subsequently, the older leaves show symptoms that are more specific to the isolate: mild mottle, blotch, stripe, chlorotic ring mottle, chlorotic line pattern, oak leaf pattern, or necrosis (Wongkaew and Dollet 1990). The name *peanut stripe* has been given to the disease on the basis of stripes and green banding symptoms along lateral veins (Demski et al. 1984), characteristic of infected peanut plants. Subsequently, research on peanut stripe virus (PStV) obtained from different regions of the world indicated the existence of specific strains of the virus producing distinct symptoms on peanut. The stripe isolate produces discontinuous stripes along the lateral veins on young quadrifoliates; older leaflets show striping, mosaic in the form of green islands, and an oak leaf pattern. For most other PStV isolates, the initial symptoms appear as chlorotic flecks followed by mild mottle, blotch, or chlorotic ring mottle symptoms. Some isolates have been reported to produce leaf necrosis (Wongkaew and Dollet 1990). PStV differs from peanut mottle virus (PeMoV) in that the stripe symptoms persist in older leaflets, and early-infected plants are stunted in the case of isolates from Asia.

Geographic Distribution and Losses

The PStV is a major cause of yield reductions in peanut crops in many countries. Naturally occurring infections have been reported in China, Japan, Thailand, Philippines, Malaysia, Indonesia, and Myanmar, and the virus has entered the United States from China in 1982 (Demski et al. 1984) and India in 1987 (Demski et al. 1993) with germplasm introductions. Yield losses due to infection under dry season of peanut production are frequently as high as 75%–80%.

In Gujarat, India, the disease incidence has been recorded up to 40%, and it is prevalent in all other four major peanut-growing states (Andhra Pradesh, Karnataka, Maharashtra, and Tamil Nadu) in India (Jain et al. 2000). In Southeast Asia, high incidences of up to 38% have been reported causing yield reduction of 30%–60% in the peanut-growing area in Indonesia (Saleh et al. 1989) and the Philippines (Adalla and Natural 1988) and up to 100% disease incidence in South Korea (Choi et al. 2001). In northern China, where more than 65% of the nation's peanuts are produced, an incidence of over 50% has been reported (Xu et al. 1991). The virus has also been detected in Senegal. The risk of accidental introduction of the virus in any peanut-growing country in imported raw peanuts is considered high, and aphids capable of transmitting it are widespread in peanut crops. PStV infection has a highly variable effect on peanut yield, depending on the geographical conditions, cultivar, and virus isolate.

Causal Virus: Peanut Stripe Virus

Other scientific names for the virus are groundnut stripe virus, peanut stripe potyvirus, groundnut mild mottle virus, groundnut mosaic virus, peanut chlorotic ring mottle virus, sesame yellow mottle virus, peanut mild mottle virus, peanut chlorotic ring virus, peanut mosaic virus, and sesame yellow mosaic virus. PStV is a member of the potyvirus group and consists of filamentous flexuous rods, approximately 752 nm long and 12 nm in diameter, which have a sedimentation coefficient of 150 S and buoyant density in cesium chloride of 1.31 g/cm. Each particle consists of a single protein species of 33,500 Da. Molecular sequencing and analysis of the viral genome of Ts strain of PStV has been done (Wang et al. 2005). The genome is a single-stranded positive-sense RNA molecule of about 9500 nucleotides. The CP contains 287 amino acid residues with a molecular weight of 32,175 Da (Mckern et al. 1991, Cassidy et al. 1993). The particles are relatively stable and can be stained with 2% phosphotungstate or ammonium molybdate pH 6.5 (Demski et al. 1993). PStV strains have a CP sequence variability of below 10% and can be defined according to geographic origin and symptom type (Higgins et al. 1999). A study of the biological and genetic variability of PStV isolates in Indonesia, Thailand, and China found geographically related groups with wide symptom diversity. Indonesian isolates of PStV have been identified as intraspecies recombinants. This information is significant for future diagnosis.

Transmission

The virus is transmitted by several species of aphids in a nonpersistent manner, which is also the only means of disease spread under field conditions. *A. craccivora* is the major vector for the transmission of PStV. Apart from *A. craccivora*, *Myzus persicae* and *Aphis gossypii*, *Hysteroneura setariae* have been shown to be highly efficient PStV vectors for the transmission of the disease.

PStV transmission through groundnut seed can be as high as 37% in artificially inoculated plants (Demski et al. 1984, Demski and Warwick 1986). Under natural conditions, however, the transmission frequency is up to 7%. PStV seed transmission frequency can be influenced by the virus isolate, groundnut cultivar, and environment. The virus can be detected in both the embryo and the cotyledon, but not in the seed testa. Most cultivars tested from natural infection show less than 4% seed transmission (Demski and Reddy 1988).

Diagnosis

It is a common experience that symptoms of the disease caused by PStV resemble those of PeMoV. However, PStV can be distinguished by infectivity assays using indicator hosts reaction. For example, *C. amaranticolor* on sap inoculation produces chlorotic or necrotic lesions in response to PStV, whereas PeMoV does not produce any infection. *Phaseolus vulgaris* cv. Top Crop, on the other hand, remain uninfected due to PStV, but it produces reddish local lesions in response to PeMoV infection. *Pisum sativum* remains uninfected due to PStV inoculation, but it produces systemic mosaic due to infection caused by PeMoV.

PStV particles react strongly with antisera of black-eye cowpea mosaic, clover yellow vein, and soybean mosaic viruses but not with PeMoV antiserum. An immunocapture (IC)-RT-PCR technique that detects the virus in seed lots, which is more sensitive than ELISA, has been developed (Gillaspie et al. 2000). A technique has been developed for the detection of PStV in individual seeds without affecting their germination. The virus can be detected in both the embryo axis and the cotyledon. ELISA can detect one PStV-infected seed in a pool of 25 healthy samples. A dot blot hybridization technique has also been applied to detect PStV in seeds. The sensitivity of this technique is about 10 times greater than ELISA (Bijaisoradat and Kuhn 1988). Further, the technique can differentiate between the presence of PStV and PeMoV in peanut seeds.

Host Range

Besides *A. hypogaea* (peanut), the PStV can infect *Calopogonium caeruleum, Centrosema pubescens* (centro), *Crotalaria pallida* (smooth crotalaria), *Desmodium* (tick clovers), *G. max* (soybean), *Indigofera* (indigo), *Lupinus albus* (white lupine), *Medicago sativa* (lucerne), *Pogostemon cablin* (patchouli), *Pueraria phaseoloides* (tropical kudzu), *Senna obtusifolia* (sicklepod), *Senna occidentalis* (coffee senna), *Senna tora* (sicklepod), *Sesamum indicul* (sesame), *Stylosanthes* (pencil flower), *Uraria crinita* (medicinal plant in Taiwan), *Vigna radiata* (mung bean), and *V. unguiculata* (cowpea) (Liao et al. 2004, Singh et al. 2009).

Epidemiology and Disease Cycle

Though the main source of primary infection in the new planting is PStV-infected peanut seed, several crop plants and weed hosts in peanut-cropping system that fall in the host range of the virus do also serve as effective source of primary infection. Consequently, many aphid species mentioned earlier transmit the PStV in a nonpersistent manner and do play the solitary means of spread of the disease under field conditions (Demski et al. 1993).

Disease Management

Host Plant Resistance

1. *Among the cultivated and wild Arachis germplasm*: Resistance to PStV among the cultivated peanut accessions is not available. However, several accessions of wild *Arachis* spp. are either immune or highly resistant to the virus. For example, *Arachis diogi* accessions PI 468141, PI 468142; *Arachis helodes* accession PI 468144; *A. cardenasii* accessions PI 475998, PI 476012, and PI 476013; *A. chacoense* accession PI 276235; and *Arachis paraguariensis* accession PI 468176 are highly resistant to PStV (Prasad Rao and Reddy 2005, Nigam et al. 2012). But efforts of making crosses to introduce this trait have not been successful, due to incompatibility between species.
2. *Transgenic peanut for resistance to PStV*: A practical and efficient genetic transformation and regeneration system for cultivars of peanut has been developed. Using particle bombardment technology in Australia and China, viral resistance genes have been introduced

into peanut. Also, an alternative *Agrobacterium*-mediated transformation system is investigated. RNA silencing, an intrinsic defense mechanism, has been successfully induced in transgenic peanut plants to specifically eliminate PStV RNA (Waterhouse et al. 2001, Dietzgen et al. 2004). These plants are highly resistant to PStV infection, and the resistance is stably inherited. An international collaborative research program funded by the Australian Centre for International Agricultural Research has now applied this technology to peanuts to control stripe disease in commercial peanut cultivars (Higgins et al. 2004). These plants will be particularly useful for Indonesian growers to combat a major constraint in production and may provide a source of resistance in peanut breeding programs. The genetic improvement of the major Indonesian cv. Gajah for PStV resistance is of particular significance, since this cultivar is also resistant to bacterial wilt, another economically important disease in Southeast Asia. Transgenic peanut lines carrying a CP gene of PStV and showing resistance to the virus have been developed in Indonesia (Hapsoro et al. 2005, 2007a,b, 2008). The transgene has been proven to be stabile up to seven generations of selfing (Hapsoro et al. 2007b). The transgenic peanut lines have been shown to carry the PStV CP transgene and inherited according to the Mendel law (Hapsoro et al. 2008). Therefore, these transgenic pure lines could be used as parents in a breeding program of pyramiding character of resistance to PStV and other novel characters in peanut plants. For example, peanut cultivar Gajah that is resistant to PStV obtained through genetic engineering could be combined with resistance to leaf spot disease, a nontransgenic high yielding character, through hybridization. This demonstrates that transgenic character can be treated just as nontransgenic character in a breeding program employing hybridization (Hapsoro et al. 2010). Resistance to PStV mediated by inverted repeat of the CP gene in transgenic tobacco plants had been developed (Yan et al. 2007, 2012).

Chemical Control

Since the PStV is transmitted by the aphid vectors in nonpersistent manner, managing the disease through vector control through insecticide sprays is not effective.

Cultural Control

It is advisable not to use seeds from crops infected with the PStV from the preceding crop season. Virus-free seed can be produced by growing healthy seed, tested by ELISA in areas where PStV is currently not known to occur. Moreover, in areas where aphid vectors are not likely to be active when peanut crop is young, the proportion of seed containing PStV is not likely to be high. However, if a small number of plants with PStV infection from seeds are observed, they can be rogued. Selection of large-size seeds could reduce the source of primary inoculum, and thus, this practice appears to be useful in decreasing the incidence of the peanut stripe disease. Peanut seed production should be done in the wet season when aphid populations are low and PStV incidence is negligible. Roguing of infected peanut is not effective in controlling PStV, because by the time the symptoms develop, aphids could already have transmitted the virus to other plants (Demski et al. 1993).

Regulatory Control

One way of controlling PStV is through early detection by using PStV antisera for virus detection as well as seed certification and follow-up of plant quarantine. For example, Australian peanut crops are free of the PStV, and it is important that quarantine remains effective in keeping it out. The risk of accidental introduction of the virus in imported raw peanuts is considered high, and aphids capable of transmitting it are widespread in Australian peanut crops (Persley et al. 2001). Therefore, under similar conditions in other countries too, strictly observing plant quarantine regulations to exchange only PStV-free peanut germplasm is of great significance. Seed lots for experimental purposes should be tested by nondestructive methods before their distribution to non-infested areas. Sowing near leguminous crops or other potential hosts of PStV should be avoided (Demski et al. 1993). Growing groundnuts for seed at a distance from commercial groundnut fields is important. For example, in the United States, 100 m is regarded as a safe distance (Demski et al. 1993), whereas in China, it is 200 m (Xu and Zhang 1986). In seed production fields, roguing diseased plants when they are noticed should reduce the chance of having contaminated seed in the lot. This method may not be practical in large-scale production, but is a common practice in countries such as Thailand.

PEANUT MOTTLE

SYMPTOMS

The symptoms of PeMoV infection can vary with cultivar, time of infection, and environment. The most common symptom, although it may not be readily noticed, is a mild mottle as irregular dark green island or mosaic on the youngest leaves of infected plants. The symptoms are not as clear on older leaves and thus can be easily missed even when the virus is in epidemic proportions in the field. The light and dark green areas of affected leaves can best be seen if leaves are held up to light. Margins of leaflets may curl up and depressions in the leaf tissue between the veins may become prominent. Plants are generally only mildly stunted, if at all. As plants mature, the symptom expression generally declines, particularly during hot and dry weather. Pods from infected plants may be reduced in size and have irregular gray to brown patches. The seed coat of affected seed may also be discolored.

GEOGRAPHICAL DISTRIBUTION AND LOSSES

Peanut mottle disease caused by PeMoV was first reported to occur in Georgia in the United States in 1965 (Kuhn 1965). Since then, it has been reported to occur in many southeastern states of the United States, East Africa, Northeast Australia, China, India, Indonesia, Japan, Malaysia, Philippines, Bulgaria, Sudan, and Venezuela (Kolte 1984). In 2008, reports of its occurrence have been made from the Gilan Province of Iran (Elahinia et al. 2008) and Israel (Spiegel et al. 2008). Five major strains of the PeMoV as mild mottle strains (PeMoV-M1 and PeMoV-M2), severe mosaic strain (PeMoV-S), necrosis strain (PeMoV-N), and chlorotic line pattern strain (PeMoV-CLP) are known to occur, and the losses in yield caused by the disease in a particular geographical area depend on the prevalence of the particular strain of the virus, the severe mosaic strain being more damaging as reported from North Carolina in the United States reducing the yield by 41%–72% (Sun and Herbert 1972). PeMoV causes substantial yield losses in many parts of the world. In some Southeast Asian countries, yield losses up to 30%–48% have been reported, and in the Indian subcontinent, the virus is a potential threat to peanut production (Reddy 1991, Prasada Rao and Reddy 2005).

CAUSAL VIRUS

PeMoV belongs to the genus *Potyvirus* in the family Potyviridae. As with other members of this virus family, PeMoV has flexuous, filamentous, nonenveloped particles ranging from 740 to 750 nm in length and 15 nm in diameter with one molecule of positive-sense, single-stranded RNA (3.0–3.5×10^6 Da) and one coat polypeptide species (32–36,000 Da). Virus infection is often associated with intracellular cytoplasmic and nuclear inclusions, pinwheels, bundles, and laminated aggregates. It is difficult to purify the virus due to its tendency to aggregate and to be inactivated by most clarifying agents. Purified virus shows typical absorption spectrum of nucleoproteins with minimum and maximum absorbance at 246 and 260 nm, respectively (Paguio and Kuhn 1973, Kolte 1984).

Although the virus infects several species within the Leguminosae, its host range outside this family is extremely limited. PeMoV is also known by other names such as peanut green mosaic virus, peanut chlorotic mottle virus, and peanut mild mosaic virus.

Transmission

In addition to being mechanically transmissible, PeMoV is also transmitted in a nonpersistent manner by several species of aphid, including *A. craccivora*, *A. gossypii*, *Hyperomyzus lactucae*, *M. persicae*, *Rhopalosiphum maidis*, and *Rhopalosiphum padi* (Pietersen and Garnett 1992). Out of the five strains, the PeMoV-N is not transmitted by aphid. PeMoV is seed borne up to 20% in peanuts (Bashir et al. 2000). Adams and Kuhn (1977) reported that seed transmission is due to the presence of the virus in the embryo. The seed transmission, however, varies depending on the virus strain, host cultivar, time of infection, and temperature.

Diagnosis

PeMov can be initially diagnosed by its sap inoculation on the leaves of *P. vulgaris* cv. Top Crop that produces its characteristics reddish-brown local lesions. ELISA is also commonly used to detect PeMov in leaves as well as seeds (Bharathan et al. 1984, Puttaraju et al. 2001). In infected plant cells, the virus makes characteristic potyvirus cylindrical inclusions that are visible in the light microscope with proper staining. Since 2000, a new procedure known as an IC-RT-PCR method has been in use for testing large number of seed lots of peanut germplasm to detect PeMoV, and the IC-RT-PCR could be adopted to test other plants and detect other plant viruses (Gillaspie et al. 2000). In this method, a small slice is removed from each seed distal to the radicle of a 100-seed sample, the slices are extracted in buffer and centrifuged, and a portion of the supernatant is incubated in a tube that has been coated with antiserum to the PeMoV. Following immunocapture of the virus (PeMoV), the tube is washed, the RT-PCR mix (with primers designed from conserved sequences within the capsid region of the virus) is placed in the same tubes, and the test then is said to be complete. Results indicate good correlation between the virus detected by the IC-RT-PCR method and the virus detected from the same seed lots by ELISA. But the IC-RT-PCR method is more sensitive and efficient than ELISA. This method has been used for the first time as molecular evidence for the occurrence of PeMoV in China, and the phylogenetic studies done in that country reveal that PeMoV can be clustered into three groups, America, Asia, and Australia, which are found to be consistent with their geographical origins (Liu et al. 2010).

Epidemiology and Disease Cycle

Besides peanuts, this virus is known to infect several legume crops particularly soybeans (*G. max*), French bean (*P. vulgaris*), peas (*P. sativum*), and various weeds that occur in peanut fields. All these can serve as the sources of primary inoculum. In the United States, the virus is known to affect peanuts as well as soybeans. Since soybeans and peanut production areas in the United States are contiguous and overlapping in several southeastern states, the virus infection chain is maintained from one season to another. The susceptibility of several *Cassia* species both in the East Africa and in the United States indicates the presence of a potential reservoir of infection. Since 2007, PeMoV has been reported to infect Rhizoma peanuts (*Arachis glabrata*) in Georgia (Nischwitz et al. 2007). This plant is propagated by cuttings and is a perennial crop. If this virus spreads in perennial peanuts in the southern United States, this plant could become a reservoir of the virus and increase its spread to field peanut and soybean via aphid transmission. Infected seeds are also considered as the source of survival of virus. Seed samples collected from the farmers and fields in India have been found to show 2%–7% PeMoV infection (Puttaraju et al. 2001). In Zimbabwe, the Bambara groundnut (*Vigna subterranea*) seeds have been detected to be infected by PeMoV (Sibiya et al. 2002). The infected seed thus can therefore be one very important source of primary inoculum in a newly planted peanut crop.

Disease Management

Host Plant Resistance

The most promising peanut germplasm lines with resistance to PeMoV are PI 261946, PI 261949, and ICG 504, but this resistance has not been incorporated into any commercial varieties. Some accessions of wild species of *Arachis* that are resistant to PeMoV are PI 172223, 262817, 262818, 262794, 421707, 468141, 468142, 468169, 468171, 468174, 468363, 468366, 468371, and AM 3867 (Prasad Rao et al. 1993, Prasad Rao and Reddy 2005). These accessions can be of potential use in breeding for resistance to PeMoV.

Chemical Control

The spread of the disease may also be controlled by the use of effective insecticide sprays through aphid control, though the control of nonpersistently aphid-transmitted viruses as PeMoV is difficult.

Cultural Control

The use of PeMoV-free seed is the most feasible approach for control, as this prevents the disease from becoming initially established in the field. Thus, cultivars with low or no seed transmission can be of immense use in eliminating the initial source of virus inoculum. It is noteworthy to note that at least there are two peanut genotypes, viz., EC 76466 and NCAC 17133 (PI 259747), which do not show any evidence of transmission of mottle strains of PeMoV through seed (Bharathan et al. 1984, Prasad Rao and Reddy 2005). If PeMoV-free seed is used, volunteer plants must be completely removed and the field situated so that PeMoV hosts, such as clovers, southern pea, and navy bean, are at least 100 yards away. Geographical locations in peanut-growing countries, where there is low or no incidence of occurrence of the disease, are reported to be identified for virus seed production.

PEANUT CLUMP

Symptoms

Plants affected by clump disease are conspicuous in the field because of their severe stunting and dark green appearance (Figure 3.3). Initial symptoms appear on young leaflet as mottling, mosaic, and chlorotic rings, but later turn dark green with or without faint mottling as the leaves mature. Early-infected plants become severely stunted. Late-infected plants may not show conspicuous stunting but appear dark green with faint mottling on younger leaflets. Clump symptoms are similar to those of green rosette, and it is likely that the two diseases are confused in some areas of Africa where they both occur (Reddy 1991). In late-infected plants, clumping may be restricted to few branches. Infected plants become bushy and produce several flowers, but the pegs do not develop pods of normal size. Early-infected plants may not produce any pods and late-infected plants may produce poorly developed pods. These plants often occur in patches, and the disease reoccurs in the same area of the groundnut field in successive years.

Geographical Distribution and Losses

The disease first reported from Senegal in West Africa now affects peanut in several countries in Africa including Benin, Burkina Faso, Chad Congo, Gabon, Ivory Coast, Mali, Niger, Sudan, and also in South Africa and in Asia (the states of Punjab Haryana, Andhra Pradesh, Gujarat, Rajasthan and Uttar Pradesh of India, and in Pakistan) (CABI 2006, Dieryck et al. 2009).

Peanut clump virus (PCV)-infected plants do not produce pods, and yield losses in peanut grown in light sandy soils are as high as 60% even in late-infected crops (Reddy et al. 1988). Peanut clump disease is known to cause losses exceeding U.S. $40 million to peanut alone on a global scale.

(a)

(b)

FIGURE 3.3 Peanut clump caused by the peanut clump virus. Note the severe stunting with dark green leaves (a) and the mosaic mottling on young quadrifoliate leaves (b).

The virus has also been shown to infect a range of monocotyledonous plants, which include wheat, barley, maize, sorghum, foxtail millet, pearl millet, and various grassy weeds (Delfosse et al. 2002). In addition to peanut, the virus has the potential to cause crop losses at least in sugarcane, wheat, barley, maize, sorghum, chillies, and Bambara groundnut. PCV also infects such important crops as cowpea (niebe) and forage legumes (e.g., *Stylosanthes* spp.) and highlights a correlation between the countries cultivating these crops and the virus distribution, but crop losses in these hosts have not been investigated experimentally.

Causal Virus: Peanut Clump Virus

PCV that causes clump disease is characterized by straight tubular particles of two lengths, containing single-stranded RNA, precisely termed as RNA-1 and RNA-2 (Thouvenel et al. 1976, Thouvenel and Fauquet 1981a, Hemmer et al. 2003). However, isolates of the virus (PCV) causing the clump disease in West Africa are serologically distinct from the isolates that cause the clump disease in India and are referred to as isolates of IPCV. However, IPCV has been shown to resemble PCV in symptomatology, in particle morphology, as well as in other aspects. Both PCV and IPCV belong to the genus *Pecluvirus* whose members are typified by a bipartite S RNA genome and by having a fungus vector, now better known as Plasmodiophorid protozoa, the *Polymyxa graminis*. The nucleotide sequence of IPCV RNA-2 and the amino acid sequence of the CP of IPCV have been found to be 61% identical to PCV supporting the contention that the Indian and West African diseases are caused by distinct but related viruses (Naidu et al. 2000, 2003). Both viruses are known to exist as a range of strains: IPCV isolates from India have been grouped into three distinct serotypes—IPCV-H (Hyderabad), IPCV-D (Durgapura), and IPCV-L (Ludhiana)—whereas the common and yellowing strains of PCV have been recognized in West Africa (Thouvenel and Fauquet 1981b). The variability in both PCV and IPCV has been elucidated by studying the respective virus genome, and the complete nucleotide sequence of one of the two RNA species in each virus has been determined (Naidu et al. 2003). Since a polyclonal antiserum produced from one serotype usually does not detect others, a DNA probe derived from RNA has been used to detect all the known serotypes of IPCV (Reddy et al. 1994).

Diagnosis

RT-PCR technique can be used to detect PCV in the host tissue and to detect virus acquisition and transmission of PCV by the vector (Dieryck et al. 2011). *Phaseolus vulgaris* cv. Local and *Chenopodium quinoa* are found to be good diagnostic hosts. Although existence of diversity among various isolates of PCV has been reported, no attempts have been made to determine the distribution and biological characteristics of different isolates. For example, rice stripe necrosis virus (RSNV) is also transmitted by *Polymyxa* sp. and occurs in Côte d'Ivoire, Nigeria, Liberia, and Sierra Leone. It is a virus very similar to PCV, but it is not fully described; its serological and genomic properties need to be clarified, and it is possible that RSNV is a member of the *Pecluvirus*, the genus to which PCV and IPCV belong. This information is vital for the diagnosis of these viruses and for implementing the management practices. Additionally, very little is known about the diversity among the isolates of *Polymyxa* spp. in West Africa. The *Polymyxa* sp. transmitting RSNV infects rice roots, whereas this crop is not a host for *Polymyxa* sp. transmitting IPCV. The studies on the host range of both *Polymyxa* and the viruses it transmits are crucial before developing strategies for the management of this group of viruses in West Africa.

Transmission, Epidemiology, and Disease Cycle

Both PCV and IPCV are seed and soil transmitted and are vectored by the persistent, soil-inhabiting root parasite, Plasmodiophorid protozoa, the *Polymyxa graminis* f. sp. *tropicalis* (Delfosse et al. 2005, Dieryck et al. 2005, 2008, Otto et al. 2005). This fungus-like parasite survives for many years in the soil in the form of highly resistant resting spores, and clump disease occurs in patches in fields. The disease recurs when groundnut and certain IPCV-susceptible cereal hosts like pearl millet (*Pennisetum glaucum*), sorghum (*Sorghum bicolor*), wheat (*Triticum aestivum*), and barley (*Hordeum vulgare*) are grown regularly.

PCV is seed transmitted in peanut and suspected to be transmitting through the seed of a range of monocotyledonous and dicotyledonous hosts (Dieryck et al. 2005). Evidence has been obtained to show that IPCV can establish in disease-free areas if virus-containing seed from monocots is planted in soils containing *Polymyxa* species. IPCV is seed transmitted up to 11% in groundnut and also through the seeds of finger millet, pearl millet, fox tail millet, wheat, and maize (Ratna et al. 1991). Seed transmission frequency to the extent of 4.29% has been recorded in the case of pearl millet accessions, and the virus can be detected in 96% of the root tips of pearl millet seedlings infected through seed, raising concern regarding their role in the spread of the disease. It is revealed that capsidial proteins of CPV are localized in several parts of the root apexes, notably in the root caps of pearl millet, indicating that the virus is perhaps able to multiply in parts of the apical meristems, which is uncommon to the general characteristics of plant virus infection (Otto et al. 2005). Accumulation of PCV during infection is accompanied by specific association of PCV RNA-1 encoded proteins with membranes of the endoplasmic reticulum and other organelles (Dunoyer et al. 2002).

There is some evidence that quantity and distribution of rainfall influences the incidences of IPCV-H and *P. graminis*; that is, high rainfall with temperatures ranging from 23°C to 30°C results in high incidences of the virus and *P. graminis*, and a weekly rainfall of 14 mm is sufficient enough for *P. graminis* to initiate infection for natural virus transmission (Delfosse et al. 2002).

Disease Management

Host Plant Resistance

The host resistance to PCV and IPCV could not be identified in any of nearly 10,000 *Arachis* germplasm lines. The variation in resistance/tolerance reaction in genotypes in the sick plots has been found to be due to uneven distribution of virus inoculum in the fields, which depends on the germination of resting spores of the vector *P. graminis* and environmental conditions. A reliable virus inoculation procedure is therefore essential to accurately evaluate peanut germplasm for resistance to IPCV/PCV. One such method is mechanical sap inoculation reported by Reddy et al. (2005), where infected peanut seed stored at −70°C is used as initial virus inoculum source, that is, infected seed material (1:10 w/v) is macerated in chilled inoculum buffer and immediately inoculated to French bean (*P. vulgaris* cv. Top Crop) to get IPCV/PCV-infected French bean that then should serve as the source of inoculum for further efficient transmission of the virus to peanut by mechanical sap inoculation. This method is convenient and allows reliable screening of elite peanut germplasm for resistance to various PCV/IPCV isolates in a relatively short period in comparison to soil-borne inoculum that depends on the germination of resting spores of the vector *P. graminis* (Reddy et al. 2005). On biotechnological front, molecular work is in progress to obtain one or more virus CP genes that could be possibly used to transform peanut plants to induce transgenic resistance (Reddy et al. 1994).

Chemical Control

Soil application of biocides such as Nemagon and Temik or furrow application of a systemic insecticide, carbofuran at 5 kg ai/ha, 1 week before planting, although effective, is found to be either hazardous or uneconomical (Reddy 1991, Delfosse et al. 2005).

Cultural Control

Extreme caution is essential in selecting the virus-free seed. Based on epidemiological studies, early sowing of the peanut crop, before the onset of monsoon rains and using judicious irrigation, has been shown to be simple and the most effective cultural method of reducing disease incidence in irrigated areas (Delfosse 2000). As a result of the baiting technique used to monitor IPCV and *P. graminis* infection, a trap-cropping method using pearl millet has been developed and tested successful at different sites in India. Pear millet is planted soon after the monsoon rains and uprooted in 2 weeks after germination. This permits the infection by *P. graminis* but not the development into sporosori. This is useful in reducing the inoculum load and peanut crop sown subsequently shows lower incidence of the clump disease. As *P. graminis* multiplies intensively in monocots, these should be avoided in cropping systems without peanut (Delfosse 2000). Hence, crop rotation with these crops is not to be recommended.

Soil solarization has been found to reduce the incidence of clump disease in India where well-cultivated soils are profusely irrigated before being covered with layers of transparent polyethylene sheeting for at least 70 days during summer (Reddy 1991). But the economic benefit of this method needs to be determined.

PEANUT STUNT

Symptoms

Symptoms vary depending on the host plant and host plant cultivar and the strain of the virus. In peanuts, there are various degrees of stunting, shortening of the petioles, reduced leaf size, mild mottling, and malformation of pods. Severe reduction in leaf size, especially in width, occasionally results in the complete absence of leaflet lamina. Seeds from infected plants appear deformed, frequently with a split pericarp wall, and have poor viability.

Geographical Distribution and Losses

Though the disease is considered to be of less economic importance, there have been records of epidemic occurrence of peanut stunt in the 1960s in Virginia, North Carolina, and Georgia in the United States (Miller and Troutman 1966, Herbert 1967). In 1966, about 50%–90% loss in yield due to the disease was reported in Virginia. The causal peanut stunt virus (PSV) sporadically causes a high incidence of peanut stunt disease in Hebei, Henan, and Liaoning provinces in China (Xu et al. 1992). The PSV, otherwise, is an economically important pathogen of legumes worldwide including several countries in Europe (France, Hungary, Italy, Poland, and Spain), in Asia (China, Georgia, Japan, Korea, and India), in Africa (Morocco and Sudan), and also in the United States (Subrahmanyam et al. 1992). In Kentucky as well as in the southeastern United States, PSV is widespread in forage legumes and is considered a major constraint to production and stand longevity.

Causal Virus: Peanut Stunt Virus

The PSV belongs to the family Bromoviridae. It is a member of the genus *Cucumovirus*, the type member of which is *Cucumber mosaic virus*. PSV particles are isometric or polyhedral, with a diameter of ca. 25–30 nm. The CP of PSV contains a single polypeptide with an apparent molecular weight of about 26 kDa. PSV has a positive-sense tripartite genome (designated RNA-1, -2, and -3 in order of decreasing size), largest genome 3.355 kb (RNA-1), the second largest 2.946 kb (RNA-2), and the third largest 2.186 kb (RNA-3), and has base composition being 24% G, 26% A, 21% C, and 29% U. In addition to the genomic RNAs, the virions also encapsulate a fourth RNA (called RNA-4), which is a subgenomic RNA that functions as mRNA for the viral CP (Naidu et al. 1995, Suzuki et al. 2003). Three types of native particle exist, each consisting of the same protein shell, yet containing different RNA species. One type of particle contains genomic RNA-1, another contains RNA-2, and the third contains genomic RNA-3 and subgenomic RNA-4. However, all the particles have the same sedimentation coefficient (S20w). Nongenomic nuclei found in virions are subgenomic mRNA that encodes the CPs, are named RNA-4, and are of 0.986 kb. Naturally occurring virions of PSV may also package a fifth RNA designated as SatRNA along with their genomic and subgenomic RNAs. PSV-associated SatRNAs are linear, single-stranded RNA molecules, ranging in size from 391 to 393 nucleotides. PSV SatRNA has essentially no sequence homology with its helper virus (i.e., PSV) genomic RNAs. Depending on the PSV strain and host species involved, SatRNAs can modulate the symptoms caused by PSV. PSV supports the replication of its SatRNAs but not those associated with *Cucumber mosaic virus*.

Infectivity of the virus in sap is lost between 51°C and 56°C temperature when exposed to 10 min, between 10^{-4} and 10^{-5} dilutions and between 4 and 24 h at room temperature (Kolte 1984).

Several strains of the PSV-infecting peanut have been described. PSV-V from Virginia and PSV-W from Washington have been described from the United States, and these have been referred to as eastern (PSV-E) and western (PSV-W) American strains of PSV, respectively (Naidu et al. 1995). Some other strains such as PSV-P (Polish) from Poland (Obrepalska-Steplowska et al. 2008), PSV-RP (Robina) from Hungary (Kiss et al. 2009), and PSV-Mi from China (Xu et al. 2004) have been reported.

Diagnosis

Plants suspected of a viral infection should be sent to a plant diagnostic laboratory where the presence or absence of the virus can be confirmed by serological double-antibody sandwich-ELISA and an indirect ELISA, RT-PCR, or host range tests (Dai et al. 2011). Antiserum and sequence data are available for this virus. The PSV produces chlorotic lesions followed by systemic mottle on *Cucumis sativus* and *P. vulgaris* cv. Top Crop, Kentucky wonder, and Bountiful. The virus also produces chlorotic lesions on *C. amaranticolor* and *Cyamopsis tetragonoloba* (Kolte 1984).

Transmission, Epidemiology, and Disease Cycle

All three genomic RNAs, but not subgenomic RNA 4, are essential for infection, and its CP, as in other viruses, plays an important role in many processes during viral life cycle and has great impact on the infectivity (Obrepalska-Steplowska et al. 2008). The CP gene is essential and sufficient for the production of unusual cytoplasmic ribbonlike inclusions that is a strain-specific trait of the virus (Bashir et al. 2004). PSV is transmitted from plant to plant by several species of aphids (*A. craccivora, A. spiraecola*, and *M. persicae*) in a nonpersistent manner. It can also be transmitted by mechanical inoculation. It has been shown to be transmitted by seeds in peanuts at a very low level, but this is not considered to be very important to the spread of this virus. The virus can be introduced into a susceptible field crop by aphids from a nearby reservoir (infected perennial hosts like clover, alfalfa, or perennial peanuts) and then is spread further into the field by aphids (Blount et al. 2002). It can be spread in perennial crops by harvesting (mechanical transmission) and possibly by the root grafts.

Disease Management

In the absence of any satisfactory source of resistance among peanut germplasm or PSV-resistant cultivar, good control of the disease appears to be only through vector control by means of insecticides. Incidence of the disease can be brought under control if peanut fields are kept away from clover fields.

It is worth mentioning that there is an occurrence of attenuating PSV SatRNAs that, when coinoculated with PSV, elicit the suppression of virus replication and spread. The symptom-attenuating properties of SatRNAs have been successfully exploited in the development of SatRNA-mediated transgenic protection against *Cucumber mosaic virus* and tobacco ring spot virus, and there is a possibility of developing this technology-based transgenic peanut for the management of the stunt disease (Naidu et al. 1995).

REFERENCES

Adalla, C.B. and M.P. Natural. 1988. Peanut stripe virus disease in the Philippines. In: *First Coordinator's Meeting on Peanut Stripe Virus*, Malang, Indonesia, June 9–12, 1987 (Patancheru, India: ICRISAT), p. 9.

Adam, G., S.D. Yeh, D.V. Reddy, and S.K. Green. 1993. The serological comparison of tospoviruses isolates from Taiwan and India with impatiens necrotic spot virus and different tomato spotted wilt virus isolates. *Arch. Virol.* 130: 237–250.

Adams, A.N. 1967. The vectors and alternative hosts of groundnut rosette virus in Central Province, Malawi. Rhodesia, Zambia. *Malawi J. Agric. Res.* 5: 145–151.

Adams, D.B. and C.W. Kuhn. 1977. Seed transmission of peanut mottle virus in peanuts. *Phytopathology* 67: 1126–1129.

Akhter, M.S., S.K. Holkar, A.M. Akanda, B. Mandal, and R.K. Jain. 2012. First report of groundnut bud necrosis virus in tomato in Bangladesh. *Plant Dis*. 96: 917–918.

Akram, M., R.K. Jain, V. Chaudhary, Y.S. Ahlawat, and S.M.P. Khurana. 2003. Characterization of the movement protein (NSm) gene of groundnut bud necrosis virus from cowpea and potato. *Indian Phytopathol.* 56: 237–238.

Akram, M., R.K. Jain, V. Chaudhary, Y.S. Ahlawat, and S.M.P. Khurana. 2004. Comparison of groundnut bud necrosis virus isolates based on movement protein (NSm) gene sequences. *Ann. Appl. Biol.* 145: 285–289.

Akram, M. and K. Naimuddin. 2009. Movement protein gene sequence based characterization of groundnut bud necrosis virus infecting cowpea. *J. Food Legumes* 22: 82–85.

Akram, M. and K. Naimuddin. 2010. First report of bud necrosis infecting pea (*Pisum sativum*) in India. *New Dis. Rep*. 21: Article 10.

Akram, M., K. Naimuddin, and R.K. Jain. 2012. Sequence diversity in the NSm gene of groundnut bud necrosis virus isolates originating from different hosts and locations in India. *J. Phytopathol.* 160: 424–427.

Alegbejo, M.D. 1997. Survey of the effect of intercropping of groundnut with cereals on the incidence of groundnut rosette virus disease in Northern Nigeria. *Int. Arachis Newsl.* 17: 39–40.

Alegbejo, M.D. and M.E. Abo. 2002. Etiology, ecology, epidemiology and control of groundnut rosette disease in Africa. *J. Sustain. Agric*. 20: 17–29.

Al-Saleh, M.A., H.A. Melouk and P. Mulder. 2007. Reaction of peanut cultivars to tomato spotted wilt virus (TSWV) under field conditions and their response to mechanical inoculation by TSWV under greenhouse conditions. *Peanut Sci.* 34: 44–52.

Ansa, O.A., C.W. Kuhn, S.M. Misari, J.W. Demski, R. Casper, and E. Breyel. 1990. Single and mixed infections of groundnut (peanut) with groundnut rosette virus and groundnut rosette assistor virus (abstr.). *Proc. Am. Peanut Res. Educ. Soc.* 22: 40.

Baring, M.R., Y. Lopez, and M.D. Burow. 2006. Registration of "Tamrun OLO7" peanut. *Crop Sci.* 46: 2721–2722.

Bashir, M., Z. Ahmad, and N. Murata. 2000. *Seed-Borne Viruses: Detection, Identification and Control.* Pakistan Agricultural Research Council, National Agricultural Research Center, Islamabad, Pakistan, 156pp.

Bashir, N.S., M. Sanger, U. Järlfors, and S.A. Ghabrial. 2004. Expression of the peanut stunt virus coat protein gene is essential and sufficient for production of host-dependent ribbon-like inclusions in infected plants. *Phytopathology* 94: 722–729.

Bharathan, N., D.V.R. Reddy, R. Rajeshwari, V.K. Murthy, V.R. Rao, and R.M. Lister. 1984. Screening groundnut germplasm lines by enzyme-linked immunosorbent assay for seed transmission of peanut mottle virus. *Plant Dis.* 68: 757–758.

Bhat, A.I., R.K. Jain, and M. Ramiah. 2002. Detection of tobacco streak virus from sunflower and other crops by reverse transcription polymerase chain reaction. *Indian Phytopathol.* 55: 2118.

Bhat, A.S. and H.S. Savithri. 2011. Investigations on the RNA binding and phosphorylation of groundnut bud necrosis virus nucleocapsid protein. *Arch. Virol.* 156: 2163–2172.

Bijaisoradat, M. and C.W. Kuhn. 1988. Detection of two viruses in peanut seeds by complementary DNA hybridization tests. *Plant Dis.* 72: 1042–1046.

Blok, V.C., A. Ziegler, D.J. Robinson, and A.F. Murant. 1994. Sequences of 10 variants of the satellite-like RNA-3 of groundnut rosette virus. *Virology* 202: 25–32.

Blok, V.C., A. Ziegler, K. Scott, D.B. Dangora, D.J. Robinson, and A.F. Murant. 1995. Detection of groundnut rosette umbravirus infections with radioactive and non-radioactive probes to its satellite RNA. *Ann. Appl. Biol.* 127: 321–328.

Blount, A.R., R.N. Pittman, B.A. Smith et al. 2002. First report of peanut stunt virus in perennial peanut in North Florida and Southern Georgia. *Plant Dis*. 86: 326.

Bock, K.R., A.F. Murant, and R. Rajeshwari. 1990. The nature of the resistance in groundnut to rosette disease. *Ann. Appl. Biol.* 117: 379–384.

Branch, W.D. 2007a. Registration of 'Georgia-06G' peanut. *J. Plant Registrations* 1: 120

Branch, W.D. 2007b. Registration of "Georgia Greener" peanut. *J. Plant Registrations* 1: 121.

Branch, W.D. 2009. Registration of 'Georgia-08V' peanut. *J. Plant Registrations* 3: 143–145.

Branch, W.D. and T.B. Brenneman. 2009. Field evaluation for the combination of white mould and tomato spotted wilt disease resistance among peanut genotypes. *Crop Prot.* 28: 595–598.

Branch, W.D. and A.K. Culbreath. 2008. Disease and insect assessment of candidate cultivars for potential use in organic peanut production. *Peanut Sci.* 35: 61–66.

Bunting, A.H. 1950. Review of groundnut rosette disease and its causal agents. *Ann. Appl. Biol.* 37: 699.

CABI. 2006. Peanut clump virus—Distribution map. In: *Distribution Maps of Plant Diseases*, 1st edn., Map 988. CABI, Wallingford, UK.

Casper, R., S.M. Meyer, D.E. Lesemann et al. 1983. Detection of a luteovirus in groundnut rosette diseased groundnuts (*Arachis hypogaea*) by enzyme-linked immunosorbent assay and immunoelectron microscopy. *Phytopathol. Z.* 108: 12–17.

Cassidy, B., J.L. Sherwood, and R.S. Nelson. 1993. Cloning of the capsid protein gene from a blotch isolate of peanut stripe virus. *Arch. Virol.* 128: 287–297.

Choi, H.S., J.S. Kim, J.U. Cheon, J.K. Choi, S.S. Pappu, and H.R. Pappu. 2001. First report of peanut stripe virus (Family *Potyviridae*) in South Korea. *Plant Dis.* 85: 679.

Culbreath, A.K., W.D. Branch, J.P. Beasley Jr., R.S. Tubbs, and C.C. Holbrook. 2012. Peanut genotype and seeding rate effects on spotted wilt. *Plant Health Progress.* doi:10.1094/PHP-2012-0227-03-RS.

Culbreath, A.K., D.W. Gorbet, N. Martinez-Ochoa et al. 2005. High levels of field resistance to tomato spotted wilt virus in peanut breeding lines derived from hypogaea and hirsuta botanical varieties. *Peanut Sci.* 32: 20–24.

Culbreath, A.K., R. Srinivasan, and J.M. Thresh et al. 2011. Epidemiology of spotted wilt disease of peanut caused by *Tomato spotted wilt virus* in the southeastern U.S. *Virus Research.* 159: 101–109.

Culbreath, A.K., B.L. Tillman, D.W. Gorbet, C.C. Holbrook, and C. Nischwitz. 2008. Response of new field-resistant peanut cultivars to twin-row pattern or in-furrow applications of phorate for management of spotted wilt. *Plant Dis.* 92: 1307–1312.

Dai, H.H., S.S. Chen, C.Y. Yang, and C. Yu. 2011. Development and comparison of three PCR methods for detecting peanut stunt virus. *Agric. Sci.* 29: 57–61.

Damayanti, T.A. and R.A. Naidu. 2009. Identification of peanut bud necrosis virus and tomato spotted wilt virus in Indonesia for the first time. *Plant Pathol.* 58(4): 782.

Dang, P.M., D.L. Rowland, and W.H. Faircloth. 2009. Comparison of ELISA and RT-PCR assays for the detection of tomato spotted wilt virus in peanut. *Peanut Sci.* 36: 133–137.

Dang, P.M., B.T. Scully, M.C. Lamb, and B.Z. Guo. 2010. Analysis and RT-PCR identification of viral sequences tags from different peanut tissues. *Plant Pathol. J.* (Faisalabad) 9: 14–22.

de Avila, A.C., P. deHaan, R. Kormelink, R. Resundl, R.W. Goldbach, and D. Peters. 1993. Classification of tospoviruses based on phylogeny of nucleoprotein gene sequences. *J. Gen. Virol.* 74: 153–159.

de Breuil, S., M.S. Nievas, F.J. Giolitti, L.M. Giorda, and S.L. Lenardon. 2008. Occurrence, prevalence and distribution of viruses infecting peanut in Argentina. *Plant Dis.* 92: 1237–1240.

Delfosse, P. 2000. Epidemiology and management of the Indian peanut clump virus, PhD dissertation. Université catholique de Louvain, Louvain-la-Neuve, Belgium.

Delfosse, P., F. Nguyen, A.M. Paridaens, C. Bragard, A. Legrève, and C.M. Rush. 2005. Soil treatment with carbofuran mitigates peanut clump virus disease incidence and is toxic to the vector *Polymyxa graminis.* In: *Proceedings of the Sixth Symposium of the International Working Group on Plant Viruses with Fungal Vectors*, Alma Mater Studiorum, Università di Bologna, Bolognà, Italy, September 5–7, 2005, pp. 125–128.

Delfosse, P., A.S. Reddy, K.T. Devi et al. 2002. Dynamics of *Polymyxa graminis* and Indian peanut clump virus (IPCV) infection on various monocotyledonous crops and groundnut during the rainy season. *Plant Pathol.* 51: 546–560.

Demski, J.W. and D.V.R. Reddy. 1988. Peanut stripe virus in the USA. In: *First Coordinator's Meeting on Peanut Stripe Virus*, Malang, Indonesia, June 9–12, 1987 (Patancheru, India: ICRISAT), pp. 10–11.

Demski, J.W., D.V.R. Reddy, G. Sowell Jr., and D. Bays. 1984. Peanut stripe virus—A new seed borne potyvirus from China infecting groundnut (*Arachis hypogaea* L.). *Ann. Appl. Biol.* 105: 495–501.

Demski, J.W., D.V.R. Reddy, S. Wongkaew et al. 1993. Peanut stripe virus: Peanut Collaborative Research Support Program, Griffin, GA, USA and ICRISAT, Patancheru, India. Information Bulletin No. 38. ICRISAT, Patancheru, India.

Demski, J.W. and D. Warwick. 1986. Direct test of peanut seed for the detection of peanut stripe virus. *Peanut Sci.* 13: 38–40.

Deom, C.M. 1999. Engineered resistance. In: *Encyclopedia of Virology*, 2nd edn., eds., A. Granoff and R. Webster. Academic Press, London, U.K.

Deom, C.M., R.A. Naidu, A.J. Chiyembekeza, B.R. Ntare, and P. Subrahmanyam. 2000. Sequence diversity within the three agents of groundnut rosette disease. *Phytopathology* 90: 214–219.

Dieryck, B., P. delfosse, G. Otto et al. 2005. Peanut clump virus and *Polymyxa graminis* interactions with pearl millet (*Pennisetum glaucum* [L.] R. Br.) and sorghum (*Sorghum bicolor* [L.] Moench). *Parasitica* 61: 25–34.

Dieryck, B., G. Otto, D. Doucet, A. Legrève, P. Delfosse, and C. Bragard. 2009. Seed, soil and vegetative transmission contribute to the spread of pecluviruses in Western Africa and the Indian sub-continent. *Virus Res.* 141: 184–189.

Dieryck, B., J. Weyns, D. Doucet, C. Bragard, and A. Legrève. 2011. Acquisition and transmission of peanut clump virus by *Polymyxa graminis* on cereal species. *Phytopathology* 101: 1149–1158.

Dieryck, B., J. Weyns, V. Van Hese, C. Bragard, and A. Legrève. 2008. Peanut clump virus transmission by *Polymyxa graminis* under controlled conditions. *Commun. Agric. Appl. Biol. Sci.* 73: 71–74.

Dietzgen, R.G., N.N. Mitter, C.M. Higgins et al. 2004. Harnessing RNA silencing to protect peanuts from stripe disease. In: *Proceedings of the Fourth International Crop Science Congress*, Brisbane, Queensland, Australia, September 26–October 1, 2004.

do Nascimento, L.C., V. Pensuk, N.P. da Costa et al. 2006. Evaluation of peanut genotypes for resistance to Tomato spotted wilt virus by mechanical and thrips inoculation. *Pesq. Agropec. Bras.* 41: 937–942.

Dunoyer, P., C. Ritzenthaler, O. Hemmer, P. Michler, and C. Fritsch. 2002. Intracellular localization of the peanut clump virus replication complex in tobacco BY-2 protoplasts containing green fluorescent protein-labeled endoplasmic reticulum or Golgi apparatus. *J. Gen. Virol.* 76: 865–874.

Elahinia, S.A., N. Shahraeen, H.R.M. Alipour, M. Nicknejad, and H. Pedramfar. 2008. Identification and determination of some properties of peanut mottle virus using biological and serological methods in Guilan Province. *J. Agric. Sci.* (Guilan) 1(10): 11–21.

German, T.L., D.E. Ullman, and J.W. Moyer. 1992. Tospoviruses: Diagnosis, molecular biology, phylogeny and vector relationships. *Annu. Rev. Phytopathol.* 30: 315–348.

Gillaspie, A.G. Jr., R.N. Pittman, D.L. Pinnow, and B.G. Cassidy. 2000. Sensitive method for testing peanut seed lots for peanut stripe and peanut mottle viruses by immunocapture-reverse transcription-polymerase chain reaction. *Plant Dis.* 84: 559–561.

Golnaraghi, A., R. Pourrahim, N. Shahraeen, and S. Farzadfar. 2002. First report of groundnut bud necrosis virus in Iran. *Plant Dis.* 86: 561.

Gopal, K., R. Jagadeswar, G.P. Babu, and H.D. Upadhyaya. 2004. Sources of resistance to bud necrosis disease in groundnut. *Int. Arachis Newsl.* 24: 36–38.

Gopal, K., V. Muniyappa, and R. Jagadeshwar. 2007. Management of peanut bud necrosis disease in groundnut through manipulation of date of sowing and plant population. *J. Plant Dis. Sci.* 2: 157–161.

Gopal, K., V. Muniyappa, and R. Jagadeshwar. 2010. Management of peanut bud necrosis disease in groundnut through intercropping with cereal and pulse crops. *Arch. Phytopathol. Plant Prot.* 43: 883–891.

Gopal, K., V. Muniyappa, and R. Jagadeeshwar. 2011a. Weed and crop plants as of reservoirs of peanut bud necrosis tospovirus and its occurrence in South India. *Arch. Phytopathol. Plant Prot.* 44: 1213–1224.

Gopal, K., V. Muniyappa, and R. Jagadeeshwar. 2011b. Management of peanut bud necrosis disease in groundnut by botanical pesticides. *Arch. Phytopathol. Plant Prot.* 44: 1233–1237.

Gorbet, D.W. 2007a. Registration of 'ANorden' peanut. *J. Plant Registrations* 1: 123–124.

Gorbet, D.W. 2007b. Registration of 'AP-3' peanut. *J. Plant Registrations* 1: 126–127.

Gururaj, S.P., V. Kenchanagoudar, V.B. Naragund, and M.K. Naik. 2005. Management of Peanut bud necrosis disease through intercropping. *Indian Phytopathol.* 58: 207–211.

Hema, M. and P. Sreenivasulu. 2002. Influence of peanut bud necrosis tospovirus [PBNV] infection on chlorophyll bleaching enzymes of groundnut (*Arachis hypogaea* L.) leaves. *J. Mycol. Plant Pathol.* 32: 245–246.

Hemmer, O., P. Dunoyer, K. Richards, and C. Fritsch. 2003. Mapping of viral RNA sequences required for assembly of peanut clump virus particles. *J. Gen. Virol.* 84: 2585–2594.

Herbert, D.A., S. Malone, S. Aref, and R.L. Brandenburg. 2007. Role of insecticides in reducing thrips injury to plants and incidence of tomato spotted wilt virus in Virginia market-type peanut. *J. Econ. Entomol.* 100: 1241–1247.

Herbert, T.T. 1967. Epidemiology of peanut stunt virus in North Carolina (abstr.). *Phytopathology* 57: 46.

Herselman, L., R. Thwaites, F.M. Kimmins, B. Courtois, P.J.A. van der Merwe, and S.E. Seal. 2004. Identification and mapping of AFLP markers linked to peanut (*Arachis hypogaea* L.) resistance to the aphid vector of groundnut rosette disease. *Theor. Appl. Genet.* 109: 1426–1433.

Higgins, C.M., R.G. Dietzgen, H.M. Akin, I. Sudarsono, K. Chen, and Z. Xu. 1999. Biological and molecular variability of peanut stripe potyvirus. *Curr. Top. Virol.* 1: 1–26.

Higgins, C.M., R.M. Hall, N. Mitter, A. Cruickshank, and R.G. Dietzgen. 2004. Peanut stripe potyvirus resistance in peanut (*Arachis hypogaea* L.) plants carrying viral coat protein gene sequences. *Transgenic Res.* 13: 59–67.

Hapsoro, D., H. Aswidinnoor, Jumanto, R. Suseno, and Sudarsono. 2005. Transformasi tanaman kacang tanah (*Arachis hypogaea* L.) dengan gen cp PStV dengan bantuan Agrobacterium. *J. Agrotropika* 10: 85–93.

Hapsoro, D., H. Aswidinnoor, Jumanto, R. Suseno, and Sudarsono. 2007a. Resistance to peanut stripe virus (PStV) in transgenic peanuts (*Arachis hypogaea* L.) carrying PStV cp gene was stabile up to seven generations of selfing. *Biota* 12: 83–91.

Hapsoro, D., H. Aswidinnoor, Jumanto, R. Suseno, and Sudarsono. 2007b. Transgene identity and number of integration sites and their correlation with resistance to PStV in transgenic peanuts carrying peanut stripe virus (PStV) coat protein gene. *J. Hama dan Penyakit Tumbuhan Tropika* 7: 39–47.

Hapsoro, D., H. Aswidinnoor, Jumanto, R. Suseno, and Sudarsono. 2008. Inheritance of resistance to PStV in transgenic peanuts containing cp PStV gene. *J. HPT Tropika* 8: 31–38.

Hapsoro, D., H. Aswidinnoor, R. Suseno, Jumanto, and Sudarsono. 2010. Pyramiding important disease-resistant character by hybridization of transgenic and non-transgenic peanuts (*Arachis hypogaea* L.). *J. HPT Tropika* 10: 91–99.

Holbrook, C.C. and A.K. Culbreath. 2007. Registration of 'Tifrunner' peanut. *J. Plant Registrations* 1: 124.

Holbrook, C.C. and A.K. Culbreath. 2008. Registration of 'Georganic' peanut. *J. Plant Registrations* 2: 17.

Holbrook, C.C., P. Timper, and A.K. Culbreath. 2008b. Registration of peanut germplasm line TifGP-1 with resistance to the root-knot nematode and tomato spotted wilt virus. *J. Plant Registrations* 2: 57.

Holbrook, C.C., P. Timper, A.K. Culbreath, and C.K. Kvien. 2008a. Registration of 'Tifguard' peanut. *J. Plant Registrations* 2: 92–94.

Hull, R. and A.N. Adams. 1968. Groundnut rosette and its assistor virus. *Ann. Appl. Biol.* 62: 139–145.

Jain, R.K., I. Lahiri, and A. Varma. 2000. Peanut stripe virus prevalence, detection and serological relationship. *Indian Phytopathol.* 53: 14–18.

Jain, R.K., A.N. Pandey, M. Krishnareddy, and B. Mandal. 2005. Immunodiagnosis of groundnut and watermelon bud necrosis viruses using polyclonal antiserum to recombinant nucleocapsid protein of groundnut bud necrosis virus. *J. Virol. Methods* 130: 162–164.

Jain, R.K., K. Umamaheswaran, A.I. Bhat, H.X. Thien, and A.S. Ahlawat. 2002. Necrosis disease on cowpea, mung bean and tomato is caused by groundnut bud necrosis virus. *Indian Phytopathol.* 55: 354.

Kalyani, G., A.S. Reddy, P.L. Kumar et al. 2007. Sources of resistance to tobacco streak virus in wild *Arachis* (*Fabaceae: Papilionoidae*) genotypes. *Plant Dis.* 91: 1585–1590.

Kendre, M.S., N.R. Patange, P.S. Neharkar, and S.M. Telang. 2000. Occurrence of *Thrips palmi*, a vector of bud necrosis disease of groundnut in Marathwada region. *J. Soils Crops* 10: 226–230.

Kesmala, T., S. Jogloy, S. Wongkaew, C. Akkasaeng, and A. Patanothai. 2006. Evaluation of ten peanut genotypes for resistance to peanut bud necrosis virus (PBNV). *Songklanakarin J. Sci. Technol.* 28: 459–467.

Kesmala, T., S. Jogloy, S. Wongkaew, C. Akkasaeng, N. Vorasoot, and A. Patanothai. 2004. Heritability and phenotypic correlation of resistance to peanut bud necrosis virus (PBNV) and agronomic traits in peanut. *Songklanakarin J. Sci. Technol.* 26: 129–138.

Kesmala, T., S. Jogloy, S. Wongkaew, C. Akkasaeng, N. Vorasoot, and A. Patanothai. 2003. Combining ability analysis for peanut bud necrosis virus (PBNV) resistance and agronomic traits of peanuts. *Thai J. Agric. Sci.* 36: 419–428.

Kiss, L., E. Balázs, and K. Salánki. 2009. Characterisation of black locust isolates of peanut stunt virus (PSV) from the Pannon ecoregion show the frequent occurrence of the fourth taxonomic PSV subgroup. *Eur. J. Plant Pathol.* 125: 671–677.

Kolte, S.J. 1984. *Diseases of Annual Edible Oilseed Crops*, Vol. I: Peanut Diseases. CRC Press, Boca Raton, FL.

Kuhn, C.W. 1965. Symptomatology, host range, and effect on yield of a seed transmitted peanut virus. *Phytopathology* 55: 880–884.

Kulkarni, M.S., K.H. Anahosur, and S. Kulkarni. 2003. Managing groundnut bud necrosis virus disease using coconut and sorghum leaf extracts. *Plant Pathol. Newsl.* 21: 2–3.

Kumar, N.R., C.S. Reddy, M.V. Reddy, and K.V.M. Krishnamurthy. 2006. Host range of peanut bud necrosis virus (PBNV) isolate causing leaf curl disease in blackgram. *J. Mycol. Plant Pathol.* 36: 90–91.

Kumar, P.L., R.D.V.J. Prasad Rao, A.S. Reddy, K.J. Madhavi, K. Anitha, and F. Waliyar. 2008. Emergence and spread of tobacco streak virus menace in India and control strategies. *Indian J. Plant Prot.* 36: 1–8.

Kumar, P.L. and F. Waliyar, eds. 2007. Diagnosis and detection of viruses infecting ICRISAT mandate crops. *Methods Manual*. International Crops Research Institute for the Semi-Arid Tropics, Patancheru, India, 133pp.

Kumari, K.V.S.M., R.D.V.J.P. Rao, B. Rajeswari, and B.M. Reddy. 2003. Occurrence of peanut bud necrosis virus on soybean (*Glycine max* (L.) Merr) in Andhra Pradesh. *Indian J. Plant Prot.* 31: 141–142.

Kunkalikar, S., B. Beradpatil, M. Kurulekar, P. Rajagopalan, and R. Anandalakshmi. 2012. Mixed Groundnut bud necrosis virus and Okra yellow vein mosaic virus infection of okra in India. *Plant Health Prog.* doi:10.1094/PHP-2012-1023-01-RS.

Liao, J.Y., C.A. Chang, R.S. Lai, and T.C. Deng. 2004. Identification of peanut stripe virus infecting Uraria crinite. *Plant Prot. Bull.* (Taipei) 46: 379–390.

Liu, Y.Y., S.S. Hou, W.C. Cheng et al. 2010. Cloning and analysis of the cp gene of two Peanut mottle virus isolates from Quingdao. *Acta Phytopathol. Sin.* 40: 647–650.

Lokesh, B., P.R. Rashmi, B.S. Amruta, S. Dhamaiah, M.R.N. Murthy, and N. Savithri. 2010. NSs encoded by groundnut bud necrosis virus is bifunctional enzyme. *PLoS One* 5(3): e9757.

Magbanua, Z.V., H.D. Wilde, J.K. Roberts et al. 2000. Field resistance to tomato spotted wilt virus in transgenic peanut (*Arachis hypogaea* L.) expressing an antisense nucleocapsid gene sequence. *Mol. Breed.* 6: 227–236.

Mandal, B., H.R. Pappu, A.S. Csinos, and A.K. Culbreath. 2006. Response of peanut, pepper, tobacco, and tomato cultivars to two biologically distinct isolates of Tomato spotted wilt virus. *Plant Dis.* 90: 1150–1155.

Mandal, B., H.R. Pappu, and A.K. Culbreath. 2001. Factors affecting mechanical transmission of tomato spotted wilt virus to peanut (*Arachis hypogaea* L.). *Plant Dis.* 85: 1259–1263.

Mandal, B., H.R. Pappu, A.K. Culbreath, C.C. Holbrook, D.W. Gorbet, and J.W. Todd. 2002. Differential response of selected peanut (*Arachis hypogaea*) genotypes to mechanical inoculation by Tomato spotted wilt virus. *Plant Dis.* 86: 939–944.

Manjunatha, L., M.S. Patil, T.R. Kavitha, L.S. Vanitha, and S.R. Vijaya Mahantesha. 2010a. Mechanical transmission of groundnut bud necrosis virus. *Environ. Ecol.* 28(4A): 2456–2458.

Manjunatha, L., M.S. Patil, P.R. Thimmegowda, S.R. Vijaya Mahantesa, and Basamma. 2010b. Effect of antiviral principle on groundnut bud necrosis virus. *J. Plant Dis. Sci.* 5: 12–15.

Mckern, N.M., H.K. Edskes, C.W. Ward, P.M. Strike, O.W. Barnett, and D.D. Shukla. 1991. Coat protein of potyviruses. 7. Amino acid sequences of peanut stripe virus. *Arch. Virol.* 119: 25–35.

Mehta, R., T. Radhakrishnan, A. Kumar et al. 2013. Coat protein- mediated resistance of peanut (*Arachis hypogaea* L.) to peanut stem necrosis disease through *Agrobacterium*-mediated genetic transformation. *Indian J. Virol.* 24: 1–9. doi:10.1007/S 13337-013-0157-9.

Miller, L.E. and J.L. Troutman. 1966. Stunt disease of peanuts in Virginia. *Plant Dis. Rep.* 50: 139–143.

Misari, S.M., J.M. Abraham, J.W. Demski et al. 1988. Aphid transmission of the viruses causing chlorotic rosette and green rosette diseases of peanut in Nigeria. *Plant Dis.* 72: 250–253.

Mozingo, R.W., T.A. Cofflet, P.M. Phipps, D.L. Coker, S. Machado, and S.E. Petrie. 2006. Registration of 'CHAPS' peanut. *Crop Sci.* 46: 2711–2712.

Murakami, M., M. Gallo-Meagher, D.W. Gorbet, and R.L. Meagher. 2006. Utilizing immunoassays to determine systemic tomato spotted wilt virus infection for elucidating field resistance in peanut. *Crop Prot.* 25: 235–243.

Murant, A.F. 1989. Groundnut rosette assistor virus. CMI/AAB descriptions of plant viruses, No. 345.

Murant, A.F. 1990. Dependence of groundnut rosette virus on its satellite RNA as well as on groundnut rosette assistor luteovirus for transmission by *Aphis craccivora. J. Gen. Virol.* 71: 2163–2166.

Murant, A.F. and I.K. Kumar. 1990. Different variants of the satellite RNA of groundnut rosette virus are responsible for the chlorotic and green forms of groundnut rosette disease. *Ann. Appl. Biol.* 117: 85–92.

Murant, A.F., R. Rajeshwari, D.J. Robinson, and J.H. Raschke. 1988. A satellite RNA of groundnut rosette virus that is largely responsible for symptoms of groundnut rosette disease. *J. Gen. Virol.* 69: 1479–1486.

Murant, A.F., D.J. Robinson, and M.E. Taliansky. 1998. Groundnut rosette virus. CMI/AAB descriptions of plant viruses, No. 355.

Nagaraja, R., K.V.K. Murthy, and Nagaraju. 2005a. Serological diagnosis of weeds and thrips harboring on them for the presence of Peanut Bud Necrosis Virus (PBNV). *Environ. Ecol.* 23: 107–110.

Nagaraja, R., R. Venugopal, K.V.K. Murthy, K.S. Jagadish, and Nagaraju. 2005b. Evaluation of groundnut genotypes against Peanut Bud Necrosis Virus (PBNV) and its thrips vector at Bangalore. *Environ. Ecol.* 23: 118–120.

Naidu, R.A., H. Bottenberg, P. Subrahmanyam, F.M. Kimmins, D.J. Robinson, and J.M. Thresh. 1998a. Epidemiology of groundnut rosette virus disease: Current status and future research needs. *Ann. Appl. Biol.* 132: 525–548.

Naidu, R.A., C.-C. Hu, R.E. Pennington, and S.A. Ghabrial. 1995. Differentiation of eastern and western strains of peanut stunt cucumovirus based on satellite RNA support and nucleotide sequence homology. *Phytopathology* 85: 502–507.

Naidu, R.A. and F.M. Kimmins. 2007. The effect of Groundnut rosette assistor virus on the agronomic performance of four groundnut (*Arachis hypogaea* L.) genotypes. *Phytopathology* 155(6): 350–356.

Naidu, R.A., F.M. Kimmins, C.M. Deom, P. Subrahmanyam, A.J. Chiyemekeza, and P.J.A. van der Merwe. 1999. Groundnut rosette: A virus disease affecting groundnut production in Sub-Saharan Africa. *Plant Dis.* 83: 700–709.

Naidu, R.A., J.S. Miller, M.A. Mayo, S.V. Wesley, and A.S. Reddy. 2000. The nucleotide sequence of Indian peanut clump virus RNA 2: Sequence comparisons among pecluviruses. *Arch. Virol.* 145: 1857–1866.

Naidu, R.A, D.J. Robinson and F.M. Kimmins. 1998b. Detection of each of the causal agents of groundnut rosette disease in plants and vector aphids by RT-PCR. *J. Virol. Methods* 76:9–18.

Naidu, R.A., S. Sawyer, and C.M. Deom. 2003. Molecular diversity of RNA-2 genome segments in pecluviruses causing peanut clump disease in West Africa and India. *Arch. Virol.* 148: 83–98.

Nigam, S.N. and K.R. Bock. 1990. Inheritance of resistance to groundnut rosette virus in groundnut (*Arachis hypogaea* L.). *Ann. Appl. Biol.* 117: 553–560.

Nigam, S.N., R.D.V.J. Prasada Rao, P. Bhatnagar-Mathur, and K.K. Sharma. 2012. Genetic management of virus diseases in peanut. *Plant Breed. Rev.* 35: 293–356

Nischwitz, C., A.L. Maas, S.W. Mullis, A.K. Culbrearh, and R.D. Gitaitis. 2007. First report of Peanut mottle virus in forage peanut (*Arachis hypogaea* L.) in North America. *Plant Dis.* 91: 632.

Ntare, B.R. and P.E. Olorunju. 2001. Variation in yield and resistance to groundnut rosette disease in early and medium maturing groundnut genotypes in Nigeria. *African Crop Sci. J.* 9: 451–461.

Ntare, B.R., P.E. Olorunju, and G.L. Hildebrand. 2002. Progress in breeding early maturing peanut cultivars with resistance to groundnut rosette diseases in West Africa. *Peanut Sci.* 29: 17–23.

Obrepalska-Steplowska, A., K. Nowaczyk, M. Budziszewska, A. Czerwoniec, and H. Pospieszny. 2008. The sequence and model structure analysis of three Polish peanut stunt virus strains. *Virus Genes* 36: 221–229.

Olatinwo, R.O., J.O. Paz, S.L. Brown, R.C. Kemerait Jr., A.K. Culbreath, and G. Hoogenboom. 2009. Impact of early spring weather factors on the risk of tomato spotted wilt in peanut. *Plant Dis.* 93: 783–788.

Olatinwo, R.O., J.O. Paz, R.C. Kemerait Jr., A.K. Culbreath and G Hoogenboom. 2010. El Niño-Southern Oscillation (ENSO): impact on tomato spotted wilt intensity in peanut and the implication on yield. *Crop Protection.* 29: 448–453.

Olorunju, P.E., C.W. Kuhn, J.W. Demski, S.M. Misari, and O.A. Ansa. 1991. Disease reactions and yield performance of peanut genotypes grown under groundnut rosette and rosette-free field environments. *Plant Dis.* 75: 1269–1273.

Olorunju, P.E., C.W. Kuhn, J.W. Demski, S.M. Misari, and O.A. Ansa. 1992. Inheritance of resistance in peanut to mixed infections of groundnut rosette virus (GRV) and groundnut rosette assistor virus and a single infection of GRV. *Plant Dis.* 76: 95–100.

Olorunju, P.E., B.R. Ntare, S. Pande, and S.V. Reddy. 2001. Additional sources of resistance to groundnut rosette disease in groundnut germplasm and breeding lines. *Ann. Appl. Biol.* 159: 259–268.

Otto, G., B., Dieryck, P. de Hertog et al. 2005. Epidemiological significance of seed transmission of peanut clump virus to pearl millet. In: *Proceedings of the Sixth Symposium of the International Working Group on Plant Viruses with Fungal Vectors*, Alma Mater Studiorum Università di Bologna, Bolognà, Italy, September 5–7, 2005, pp. 142–145.

Paguio, O.R. and C.W. Kuhn. 1973. Strains of peanut mottle virus. *Phytopathology* 63: 976–980.

Pensuk, V., N. Daengpluang, S. Wongkaew, S. Jogloy, and A. Patanothai. 2002b. Evaluation of screening procedures to identify peanut resistance to Peanut bud necrosis virus (PBNV). *Peanut Sci.* 29: 47–51.

Pensuk, V., S. Jogloy, and A. Patanothai. 2010. Effects of temperature and relative humidity on the effectiveness of peanut bud necrosis virus inoculation on peanut. *Plant Pathol. J.* (Faisalabad) 9: 188–193.

Pensuk, V., S. Jogloy, S. Wongkaew, and A. Patanothai. 2002c. Effectiveness of artificial inoculation methods for screening peanut (*Arachis hypogaea* L.) genotypes for resistance to Peanut bud necrosis virus (PBNV). *Thai J. Agric. Sci.* 35: 379–389.

Pensuk, V., S. Jogloy, S. Wongkaew, and A. Patanothai. 2004. Generation means analysis of resistance to peanut bud necrosis caused by peanut bud necrosis tospovirus in peanut. *Plant Breed.* 123(1): 90–92.

Pensuk, V., S. Wongkaew, S. Jogloy, and A. Patanothai. 2002a. Combining ability for resistance in groundnut (*Arachis hypogaea*) to Peanut bud necrosis tospovirus (PBNV). *Ann. Appl. Biol.* 141: 143–146.

Persley, D.M., L. McMichael, and D. Spence. 2001. Detection of Peanut stripe virus in post-entry quarantine in Queensland. *Austr. Plant Pathol.* 30: 377.

Pietersen, G. and H.M. Garnett. 1992. Some properties of a peanut mottle virus (PMoV) isolate from soybeans in South Africa. *Phytophylactica* 24: 211–215.

Poledate, A.S., P. Laohasiriwong, N. Jaisil et al. 2007. Gene effects for parameters of peanut bud necrosis virus (PBNV) resistance in peanut. *Pak. J. Biol. Sci.* 10: 1501–1506.

Pranav, C., P.U. Krishnaraj, and M.S. Kuruvinashetty. 2008. Identification of peanut bud necrosis virus in sunflower. *J. Plant Dis. Sci.* 3: 56–59.

Prasad Rao, R.D.V.J.P., B.S. Babu, M. Sreekant, and V.M. Kumar. 2003. ELISA and infectivity assay based survey for the detection of peanut bud necrosis virus in mungbean and urdbean in Andhra Pradesh. *Indian J. Plant Prot.* 31: 26–28.

Prasad Rao, R.D.V.J., G.P. Ribeiro, R. Pittman, D.V.R. Reddy, and J.W. Demski. 1993. Reaction of Arachis germplasm to peanut stripe, peanut mottle and tomato spotted wilt viruses. *Peanut Sci.* 20: 115–118.

Prasad Rao, R.D.V.J. and A.S. Reddy. 2005. Virus diseases of groundnut. In: *Diseases of Oilseed Crops*, eds., G.S. Saharan, N. Mehta, and M.S. Sangwan. Indus Publishing Co., New Delhi, India, pp. 153–176.

Prasad Rao, R.D.V.J., A.S. Reddy, F. Waliyar, P. Sreenivasulu, and P. Lavakumar. 2004. Peanut stem necrosis. In: *Serological and Nucleic Acid-Based Methods for the Detection of Plant Viruses*, eds., P. Lavakumar, A.T. Jones, and F. Waliyar, Training Course, April 12–20, 2004, Methods Manual. Virology Unit, ICRISAT, Patancheru, India, pp. 26–29.

Pundhir, V.S., M. Akram, A. Mohammad, and H. Rajshekhara. 2012. Occurrence of stem necrosis disease in potato caused by groundnut bud necrosis virus. *Potato J.* 39: 81–83.

Puttaraju, H.R., H.S. Prakash, and H.S. Shetty. 2001. Detection of Peanut mottle potyvirus in leaf and seed of peanut and its effect on yield. *Indian Phytopathol.* 54: 479–480.

Puttha, R., S. Jogloy, S. Wongkaew, J. Sanitchon, T. Kesmala, and A. Patanothai. 2008. Heritability, phenotypic and genotypic correlation of peanut bud necrosis virus resistance and agronomic traits in peanut. *Asian J. Plant Sci.* 7: 276–283.

Raja, P. and R.K. Jain. 2006. Molecular diagnosis of Groundnut bud necrosis virus causing bud blight of tomato. *Indian Phytopathol.* 59: 359–362.

Rajeshwari, R., A.F. Murant, and P.R. Massalski. 1987. Use of monoclonal antibody to potato leaf roll virus for detecting groundnut rosette assistor virus by ELISA. *Ann. Appl. Biol.* 111: 353–358.

Ramana, C.V., P.V. Rao, R.D.V.J.P. Rao, S.S. Kumar, I.P. Reddy, and Y.N. Reddy. 2011. Genetic analysis for peanut bud necrosis virus (PBNV) resistance in tomato (*Lycopersicon esculentum* Mill.). *Acta Hort.* 914: 459–463.

Ratna, A.S., A.S. Rao, A.S. Reddy et al. 1991. Studies on transmission of peanut clump virus disease by *Polymyxa graminis*. *Ann. Appl. Biol.* 118: 71–78.

Ravi, K.S., A. Buttgereitt, A.S. Kitkaru, S. Deshmukh, D.E. Lesemann, and S. Winder. 2001. Sunflower necrosis disease from India is caused by an ilar virus related to tobacco streak virus. *Plant Pathol.* 5: 800.

Reddy, A.S., P.L. Kumar, and F. Waliyar. 2005. Rate of transmission of Indian peanut clump virus to groundnut by mechanical inoculation. *J. SAT Agric. Res.* 1: 1–3.

Reddy, A.S., R.D.V.J. Prasad Rao, K. Thirumala-Devi et al. 2002. Occurrence of Tobacco streak virus on peanut (*Arachis hypogaea* L.) in India. *Plant Dis.* 86: 173–178.

Reddy, A.S., L.J. Reddy, N. Mallikarjuna et al. 2000. Identification of resistance to Peanut bud necrosis virus (PBNV) in wild *Arachis* germplasm. *Ann. Appl. Biol.* 137: 135–139.

Reddy, A.S., K. Subramanyam, P.L. Kumar, and F. Waliyar. 2007. Assessment of Tobacco streak virus (TSV) transmission through seed in Groundnut and Sunflower. *J. Mycol. Plant Pathol.* 37: 136–137.

Reddy, B.V.B., Y. Sivaprasad, and D.V.R.S. Gopal. 2011. First report of Groundnut bud necrosis virus on *Calotropis gigantea*. *J. Plant Pathol.* 93(Suppl. 4): S4.84.

Reddy, D.V.R. 1991. Groundnut viruses and virus diseases: Distribution, identification and control. *Rev. Plant Pathol.* 70: 665–678.

Reddy, D.V.R., A.A.M. Buiel, T. Satayanarayana et al. 1995. Peanut bud necrosis disease. An overview. In: *Recent Studies on Peanut Bud Necrosis Disease*, eds., A.A.M. Buiel, J.E. Parlevliet, and J.M. Lenne. International Crops Research Institute for the Semi-Arid Tropics, Patancheru, India, pp. 3–7.

Reddy, D.V.R., A.F. Murant, G.H. Duncan, O.A. Ansa, J.W. Demski, and C.W. Kuhn. 1985a. Viruses associated with chlorotic rosette and green rosette diseases of groundnut in Nigeria. *Ann. Appl. Biol.* 107: 57–64.

Reddy, D.V.R., A.F. Murant, J.H. Raschke, M.A. Mayo, and O.A. Ansa. 1985b. Properties and partial purification of infective material from plants containing groundnut rosette virus. *Ann. Appl. Biol.* 107: 65–78.

Reddy, D.V.R., R.A. Naidu, R.D.V.J. Prasad Rao et al. 1994. Current research on groundnut viruses at ICRISAT Asia Center. In: *Working Together on Groundnut Virus Diseases*, eds., D.V.R. Reddy, D. McDonald, and J.P. Moss. International Crops Research Institute for the Semi-Arid Tropics, Hyderabad, India, pp. 53–54.

Reddy, D.V.R., B.L. Nolt, H.A. Hobbs et al. 1988. Clump virus in India: Isolates, host range, transmission and management. In: *Viruses with Fungal Vectors*, eds., J.I. Cooper and M.J.C. Asherr. Association of Applied Biology, Wellesbourne, U.K., pp. 239–246.

Reddy, D.V.R., A.S. Ratna, M.R. Sudarshana, F. Poul, and L. Kiran Kumar. 1992. Serological relationship and purification of bud necrosis virus, a tospovirus occurring in peanut. *Ann. Appl. Biol.* 120: 279–286.

Reddy, D.V.R., J.A. Wightman, R.J. Beshear et al. 1991. Bud necrosis diseases of groundnut caused by tomato spotted wilt virus. Information Bulletin No. 3. International Crops Research Institute for the Semi-Arid Tropics, Patancheru, India.

Riley, D.G., S.V. Joseph, R. Srinivasan, and S. Diffie. 2011. Thrips vectors of tospoviruses. *J. Integr. Pest Manag.* 1(2): 2011. doi:10.1603/IPM10020.

Riniker, S.D., R.L. Brandenburg, G.G. Kennedy, T.G. Isleib, and D.L. Jordan. 2008. Variation among Virginia market-type peanut genotypes in susceptibility to Tomato spotted wilt virus vectored by thrips (*Thysanoptera: Thripidae*). *Peanut Sci.* 35: 92–100.

Robinson, D.J., E.V. Ryabov, S.K. Raj, I.M. Roberts, and M.E. Taliansky. 1999. Satellite RNA is essential for encapsidation of groundnut rosette umbravirus RNA by groundnut rosette assistor luteovirus coat protein. *Virology* 254: 104–114.

Rowland, D., J. Dorner, R. Sorensen, J.P. Jr. Beasley, and J. Todd. 2005. Tomato spotted wilt virus in peanut tissue types and physiological effects related to disease incidence and severity. *Plant Pathol.* 54: 431–440.

Saleh, N., N.M. Horn, D.V.R. Reddy, and K.J. Middleton. 1989. Peanut stripe virus in Indonesia. *Neth. J. Plant Pathol.* 95: 123–127.

Saritha, R.K. and R.K. Jain. 2007. Nucleotide sequence of the S and M RNA segments of a Groundnut bud necrosis virus isolate from Vigna radiata in India. *Arch. Virol.* 152: 1195–1200.

Satyanarayana, T., S.E. Mitchell, D.V.R. Reddy et al. 1996a. On peanut bud necrosis tospovirus SRNA: Complete nucleotide sequences, genome organization, and homology to other tospoviruses. *Arch. Virol.* 141: 85–98.

Satyanarayana, T., S.E. Mitchell, D.V.R. Reddy et al. 1996b. The complete nucleotide sequence and genome organization of RNA segment of peanut bud necrosis tospovirus and comparison with other tospoviruses. *J. Gen. Virol.* 77: 2347–2352.

Schwach, F., G. Adam, and C. Heinze. 2004. Expression of modified nucleocapsid-protein of Tomato spotted wilt virus (TSWV) confers resistance against TSWV and Groundnut ringspot virus (GRSV) by blocking systemic spread. *Mol. Plant Pathol.* 5: 309–316.

Scott, K.P., M.J. Farmer, D.J. Robinson, L. Torrance, and A.F. Murant. 1996. Comparison of the coat protein of groundnut rosette assistor virus with those of other luteoviruses. *Ann. Appl. Biol.* 128: 77–83.

Sibiya, J., W. Manyangarirwa, and S.E. Albrechtsen. 2002. First record of peanut mottle virus from bambara groundnut seed in Zimbabwe. *African Plant Prot.* 8: 1–2.

Singh, M.K., V. Chandel, V. Hallan, R. Ram, and A.A. Zaidi. 2009. Occurrence of peanut stripe virus on patchouli and raising of virus-free patchouli plants by meristem tip culture. *J. Plant Dis. Prot.* 116: 2–6.

Sivaprasad, Y., B.V.B. Reddy, C.V.M.N. Kumar, K.R. Reddy, and D.V.R.S. Gopal. 2011. First report of Groundnut bud necrosis virus infecting Taro (*Colocasia esculenta*). *Austr. Plant Dis. Notes* 6: 30–32.

Spiegel, S., I. Sobolev, A. Dombrovsky et al. 2008. Characterization of a Peanut mottle virus isolate infecting peanut in Israel. *Phytoparasitica* 36(2): 168–174.

Sreekanth, M., M. Sriramulu, R.D.V.J.P. Rao, B.S. Babu, and T.R. Babu. 2002. Effect of sowing date on *Thrips palmi* Karny population and peanut bud necrosis virus incidence in greengram (*Vigna radiata* L. Wilczek). *Indian J. Plant Prot.* 30: 16–21.

Sreekanth, M., M. Sriramulu, R.D.V.J.P. Rao, B.S. Babu, and T.R. Babu. 2004a. Evaluation of certain new insecticides against *Thrips palmi* (Karny), the vector of peanut bud necrosis virus (PBNV) on mungbean (*Vigna radiata* L. Wilczek). *Int. Pest Control* 46: 315–317.

Sreekanth, M., M. Sriramulu, R.D.V.J.P. Rao, B.S. Babu, and T.R. Babu. 2004b. Effect of intercropping on *Thrips palmi* (Karny) population and peanut bud necrosis virus (PBNV) incidence in mungbean (*Vigna radiata* L. Wilczek). *Indian J. Plant Prot.* 32: 45–48.

Sreekanth, M., M. Sriramulu, R.D.V.J.P. Rao, B.S. Babu, and T.R. Babu. 2006b. Effect of different temperatures on the developmental biology of Thrips palmi, the vector of Peanut bud necrosis virus on green gram. *Indian J. Plant Prot.* 34: 136–137.

Sreekanth, M.M., M. Sreeramulu, R.D.V.J.P. Rao, B.S. Babu, and T.R. Babu. 2006a. Virus-vector relationships of Peanut bud necrosis virus and Thrips palmi in greengram. *Indian J. Plant Prot.* 34: 66–71.

Subrahmanyam, P., G.L. Hildebrand, R.A. Naidu, L.J. Reddy, and A.K. Singh. 1998. Sources of resistance to groundnut rosette disease in global groundnut germplasm. *Ann. Appl. Biol.* 132: 473–485.

Subrahmanyam, P., R.A. Naidu, L.J. Reddy, P.L. Kumar, and M. Ferguson. 2001. Resistance to groundnut rosette disease in wild *Arachis* species. *Ann. Appl. Biol.* 139: 45–50.

Subrahmanyam, P., P.J.A. van der Merwe, A.J. Chiyembekeza, and S. Chandra. 2002. Integrated management of groundnut rosette disease. *African Crop Sci. J.* 10: 99–110.

Subrahmanyam, P., S. Wongkaew, D.V.R. Reddy et al. 1992. Field diagnosis of groundnut diseases. Information Bulletin No. 36. International Crops Research Institute for the Semi-Arid Tropics, Patancheru, India, 84pp.

Sun, M.K.C. and T.T. Herbert. 1972. Purification and properties of a severe strain of peanut mottle virus. *Phytopathology* 62: 832.

Suzuki, M., M. Yoshida, T. Yoshinuma, and T. Hib. 2003. Interaction of replicase components between Cucumber mosaic virus and Peanut stunt virus. *J. Gen. Virol.* 84: 1931–1939.

Taliansky, M.E. and D.J. Robinson. 1997. Trans-acting untranslated elements of groundnut rosette virus satellite RNA are involved in symptom production. *J. Gen. Virol.* 78: 1277–1285.

Taliansky, M.E. and D.J. Robinson. 2003. Molecular biology of umbraviruses: Phantom warriors. *J. Gen. Virol.* 84: 1951–1960.

Taliansky, M., I.M. Roberts, N. Kalinina et al. 2003. An umbraviral protein, involved in long-distance RNA movement, binds viral RNA and forms unique, protective ribonucleoprotein complexes. *J. Virol.* 77: 3031–3040.

Taliansky, M.E., D.J. Robinson, and A.F. Murant. 1996. Complete nucleotide sequence and organization of the RNA genome of groundnut rosette umbravirus. *J. Gen. Virol.* 77: 2335–2345.

Taliansky, M.E., D.J. Robinson, and A.F. Murant. 2000. Groundnut rosette disease virus complex: Biology and molecular biology. *Adv. Virus Res.* 55: 357–400.

Taliansky, M.E., E.V. Ryabov, and D.J. Robinson. 1998. Two distinct mechanisms of transgenic resistance mediated by groundnut rosette virus satellite RNA sequences. *Mol. Plant-Microbe Interact.* 11: 367–374.

Thiara, S.K., S.S. Cheema, and S.S. Kang. 2004. Pattern of bud necrosis disease development in groundnut crop in relation to different dates of sowing. *Plant Dis. Res.* (Ludhiana) 19(2): 125–129.

Thien, H.X., A.I. Bhat, and R.K. Jain. 2003. Mungbean necrosis disease caused by a strain of Groundnut bud necrosis virus. *Indian Phytopathol.* 56: 54–60.

Thouvenel, J.-C., M. Dollet, and C. Fauquet. 1976. Some properties of peanut clump, a newly discovered virus. *Ann. Appl. Biol.* 84: 311–320.

Thouvenel, J.-C. and C. Fauquet. 1981a. Further properties of peanut clump virus and studies on its natural transmission. *Ann. Appl. Biol.* 97: 99–107.

Thouvenel, J.-C. and C. Fauquet. 1981b. Peanut clump virus. CMI/AAB descriptions of plant viruses, No. 235. Commonwealth Agricultural Bureaux/Association of Applied Biologists, Farnham Royal, U.K.

Tillman, B.L., D.W. Gorbet, and P.C. Andersen. 2007. Influence of planting date on yield and spotted wilt of runner market type peanut. *Peanut Sci.* 34: 79–84.

Tonsomros, Y., S. Jogloy, S. Wongkaew, C. Akkasaeng, T. Kesmala, and A. Patanothai. 2006. Heritability, phenotypic and genotypic correlations of Peanut bud necrosis virus (PBNV) reaction parameters in peanut. *Songklanakarin J. Sci. Technol.* 28: 469–477.

van der Merwe, P.J.A. and P. Subrahmanyam. 1997. Screening of rosette-resistant short duration groundnut breeding lines for yield and other characteristics. *Int. Arachis Newsl.* 17: 14–15.

van der Merwe, P.J.A., P. SubrahmanKayam, F.M. Kimmins, and J. Willekens. 2001. Mechanism of resistance to groundnut rosette. *Int. Arachis Newsl.* 21: 43–46.

Venkat, H., P. Gangatirkar, A.A. Ka rande, M. Krishnareddy, and H.S. Savithri. 2008. Monoclonal antibodies to the recombinant nucleocapsid protein of a groundnut bud necrosis virus infecting tomato in Karnataka and their use in profiling the epitopes of Indian tospovirus isolates. *Curr. Sci.* 95: 952–957.

Venkataravana, P., N. Jankiraman, D. Nuthan, and V.L. Madhu Prasad. 2008. Genetic enhancement of multiple resistance to foliar diseases in groundnut. *Legume Res.* 31(2): 155–156.

Venkatesan, S., J.A.J. Raja, S. Maruthasalam et al. 2009. Transgenic resistance by N gene of a Peanut bud necrosis virus isolate of characteristic phylogeny. *Virus Genes* 38(3): 445–454.

Vijayalakshmi, K., J.A. Wightman, D.V.R. Reddy, and D.D.R. Reddy. 2000. Effect of different temperatures on the biology of *Thrips palmi* Karny, the vector of peanut bud necrosis virus. *J. Entomol. Res.* 24: 83–85.

Waliyar, F., P.L. Kuma, and B.R. Ntare et al. 2007. A century of research on groundnut rosette disease and its management. Information Bulletin No. 75. International Crops Research Institute for the Semi-Arid Tropics, Patancheru, India, 40pp. ISBN: 978-92-9066-501-4.

Wang, H.L., Y.Y. Chang, and C.A. Chang. 2005. Molecular sequencing and analysis of the viral genome of peanut stripe virus Ts strain. *Taiwan Plant Pathol. Bull.* 14: 211–220.

Wangai, A.W., S.S. Pappu, H.R. Pappu, C.M. Deom, and R.A. Naidu. 2001. Distribution and characteristics of groundnut rosette disease in Kenya. *Plant Dis.* 85: 470–474.

Waterhouse, P.M., M.B. Wang, and T. Lough. 2001. Gene silencing as an adaptive defence against viruses. *Nature* 411: 834–842.

Wiatrak, P.J., D.L. Wright, J.J. Marois, S. Grzes, W. Koziara, and J.A. Pudelko. 2000. Conservation tillage and thimet effects on tomato spotted wilt virus in three peanut cultivars. *Proc. Soil Crop Sci. Soc.* (Florida) 59: 109–111.

Wongkaew, S. and M. Dollet. 1990. Comparison of peanut stripe virus isolates using symptomatology on particular hosts and serology. *Oleagineux* 45: 267–278.

Xu, Z. and Z. Zhang.1986. Effect of infection by three major peanut viruses on the growth and yields of peanut. *Sci. Agric. Sin.* 4: 51–56.

Xu, Z., K. Chen, Z. Zhang, and J. Chen. 1991. Seed transmission of peanut stripe virus in peanut. *Plant Dis.* 75: 723–726.

Xu, Z., Z. Zhang, K. Chen, and J. Chen. 1992. Characteristics of strains of peanut stunt virus by host reactions and pathogenicity to peanut. *Oil Crops of China* 4: 25–29.

Xu, Z.Y., L.Y. Yan, K.R. Chen, and M. Prins. 2004. Nucleotide sequence analyses of RNA 3 of peanut stunt virus Mi strain. *J. Agric. Biotechnol.* 12: 436–441.

Yan, L.Y., Z.Y. Xu, K.R. Chen, R. Goldbach, and M. Prins. 2007. Quick construction of inverted repeat vector of peanut stripe virus cp gene by gateway system. *J. Agric. Biotechnol.* 15: 356–357.

Yan, L.Y., Z.Y. Xu, Z.Y. Liao, and S. Bo. 2012. Transgenic tobacco plants resistant to two peanut viruses via RNA mediated virus resistance. *J. Acta Phytopathol. Sin.* 42(1): 37–44.

Yang, H., P. Ozias-Akins, A.K. Culbreath et al. 2004. Field evaluation of Tomato spotted wilt virus resistance in transgenic peanut (*Arachis hypogaea*). *Plant Dis.* 88: 259–264.

Yayock, J.Y., H.W. Rossel, and C. Harkness. 1976. A review of the 1975 groundnut rosette epidemic in Nigeria. In: *Samaru Conference*, Paper 9. Institute of Agricultural Research, Ahmadu Bello University, Zaria, Nigeria.

Zimmermann, G. 1907. Über eine Krankheit der Erdnüsse (*Arachis hypogaea*). *Der Pflanzer* 3: 129–133.

4 Other Diseases of Peanut

BACTERIAL WILT OF PEANUT

SYMPTOMS

Affected young plants show conspicuous and sudden rapid wilting, and death of roots becomes evident. The whole plant or only a branch of the plant may be wilted. Leaflets of affected plants curl up at the ends and become slightly flaccid, while leaves retain their green color with little fading of normal green color (Mehan et al. 1994). Veins become greener than the lamina, giving leaflets a stripped appearance. A unilateral effect, as can be observed under natural conditions, is noted and only one branch may be affected under artificial conditions also. Wilt symptoms can be observed 3 weeks after sowing, the peak of disease occurrence being 40–50 days after sowing. Plants may wilt entirely over days under conducive conditions, such as after high temperatures. There can also be latent infection. This disease can be distinguished from other wilt diseases by placing infected tissue, freshly sectioned, into water to observe masses of bacterial ooze streaming out. The rapid wilting also distinguishes this from fungal wilts.

Bacterial oozing may be seen on root, stem, and lower branches and the oozing becomes evident as streaks of brown or black discoloration. The affected tissues then become black and show necrosis. When young plants are infected, the pods may remain small, or pods of such plants become wrinkled and may show rotting. When external symptoms are not evident, the infection can be detected in cross sections of stems and roots. Brown pigment formation in host tissue in the cut xylem and pith regions is considered a diagnostic criterion for the identification of the disease.

GEOGRAPHICAL DISTRIBUTION AND LOSSES

Bacterial wilt (BW) disease (*Ralstonia solanacearum*) is widely distributed in tropical, subtropical, and some warm temperate region in the world and poses a great threat to peanut production in China and Southeast Asia including Indonesia and Vietnam (Shan et al. 1998, Doan et al. 2006). More than 10% of the area under peanut is affected in southern China. In Shandong Province in China, where large-seeded cultivars dominate, BW is a very serious problem in Rizhao and Linyi. The disease is reported to affect about 35,000 ha crop area in these two areas (Zhang et al. 2008). Yield reduction generally ranges from 10% to 20%; however, in heavily infested field, over 50% yield losses are not uncommon. The disease is more severe in lands not used for paddy rice in Southeast Asia. In extreme cases, the disease may even cause total crop losses (Mehan et al. 1994, Yu et al. 2011). Although the disease had been reported from several African countries in the 1930s and 1940s, it is not considered economically important there with the exception of Uganda, East Africa (Mehan et al. 1994, Elphinstone 2005, Mace et al. 2007). The worldwide distribution of *R. solanacearum* has been summarized by Commomwealth Agricultural Bureaux International and the European and Mediterranean Plant Protection Organization in the updated series of distribution of maps of plant diseases (Schell 2000).

Pathogen: *Ralstonia solanacearum* (Smith) Yabuuchi et al.

R. solanacearum is the causal pathogen of BW, formerly known as *Pseudomonas solanacearum* E.F. Smith, with similarities in most aspects except that it does not produce a fluorescent pigment like *Pseudomonas* (Fegan and Prior 2005). *R. solanacearum* is a Gram-negative, non-spore-forming, rod-shaped, strictly aerobic bacterium that is 0.5–0.7 × 1.5–2.0 μm in size. For most strains, the optimal growth temperature is between 28°C and 32°C, and it does not grow above 41°C. On solid agar media, individual bacterial colonies are usually visible after 36–48 h growth at 28°C, and colonies of the normal or virulent type are white or cream colored, irregularly shaped, highly fluidal, and opaque. Occasionally, colonies of the mutant or nonvirulent type appear; these are uniformly round, smaller, and butyrous, or dry. A tetrazolium chloride (TZC) medium can differentiate the two colony types. On this medium, virulent colonies appear white with pink enters and nonvirulent colonies are a uniform dark red (Champoiseau et al. 2009).

Variability and Pathotypes in *R. solanacearum* Species Complex

R. solanacearum is classified into five races based on the hosts affected and five biovars based on the ability to use or oxidize several hexose alcohols. Race 1 strains (biovars 1, 3, and 4) are pathogenic to a broad range of hosts, including tomato (*Solanum lycopersicum*), tobacco (*Nicotiana tabacum*), and peanut (*Arachis hypogaea*) and occur in Asia, Australia, and Americas; race 2 strains (biovars 1 and 3) have a more limited host range than race 1 and infect banana (*Musa acuminata*), plantain (*Musa paradisiaca*), heliconia (*Heliconia* spp.), and other plants in the Musaceae family and occur in the Caribbean, Brazil, and the Philippines; race 3 strains (biovar 2) occur in cool upland areas in the tropics and cause severe wilt in potato (*Solanaum tuberosum*), tomato, and geranium (*Geranium* spp.) and occur worldwide, but it is not generally reported in North America covering the United States and Canada and is, therefore, the focus of sanitation and plant quarantine of management practices to prevent the introduction of or spread of the pathogen; race 4 strains (biovars 3 and 4) infect ginger and occur in Asia; and race 5 strains infect mulberry (*Morus alba*) and occur in China (Denny and Hayward 2001). Pathogen diversity and the relationship among races, biovars, and phylotypes have been described (Chen et al. 2000, Alvarez 2005, Fegan and Prior 2005), but this classification system has proved to be unsatisfactory. There are no laboratory tests to define the *race* of an isolate because host ranges of strains are broad and often overlap. The biovars do not correspond to phylogenetically coherent groups, with the exception of biovar 2A, which corresponds to R3bv2.

A phylogenetically meaningful classification scheme has now been developed based on DNA sequence analysis. This scheme divides the species complex into four phylotypes that broadly reflect the ancestral relationships and geographical origins of the strains. Phylotype I strains originated in Asia, phylotype II strains originated in the Americas, phylotype III strains in Africa, and phylotype IV strains in Indonesia. Phylotypes are further subdivided into sequevars based on the sequence of the endoglucanase gene. Multilocus sequence typing and other analyses have confirmed that this system of classification reflects the phylogeny of the group (Prior and Fegan 2005). The 5.8-megabase (Mb) genome of *R. solanacearum* has been completely sequenced. The genome encodes many proteins potentially associated with their role in pathogenicity (Salanoubat et al. 2002).

Diagnosis

Diagnosis can be determined from a section of stem pruned from near the base of a suspect plant. Immediately after pruning the stem, suspend it in a glass of clean water for several minutes. Milky threads will begin to leak from the stem and the water will quickly become white if BW is present. A diagnostic schedule involving direct seed plating and grow-out of peanut seeds for 4 weeks and the leaf bits and twig pieces on tetrazolium chloride agar (TZCA) can be useful for the detection of BW infection in imported peanut germplasm seeds (Anitha et al. 2004). Serological methods are generally quick and reliable but suffer from problems with specificity or sensitivity or both. Additionally, they do not distinguish live cells from dead cells. A number of *R. solanacearum*–specific nucleic acid–based methods that use the polymerase chain reaction (PCR) amplification can detect both living and dead cells, which are more specific and simple than serological approaches.

Epidemiology and Disease Cycle

The bacterium survives in the soil and can maintain infectious populations over several years. Alternative weed hosts may also play a role in survival and over seasoning. The pathogen infects roots through wounds and colonizes the vascular tissue causing plugging of the xylem and leaf wilting. There is often an association between nematode infection and BW, where the nematodes create wounds in the root tissue to allow an entry point for the bacterium to infect the plant. The bacteria get access to the wounds partially by flagellar-mediated swimming motility and chemotaxis attraction toward root exudates (Yao and Allen 2006). Unlike many phytopathogenic bacteria, *R. solanacearum* potentially requires only one entry site to establish a systemic infection that results in the development of BW. After invading a susceptible host, *R. solanacearum* multiplies and moves systemically within the plant before BW symptoms occur. When the pathogen gets into the xylem, tyloses may form to block the axial migration of bacteria within the plant that may lead to vascular dysfunction. *R. solanacearum* possesses genes for all six protein secretion pathways that have been characterized in Gram-negative bacteria; the best known is Type III secretion system (T3SS or TTS), which secretes infection-promoting effect or proteins (T3 Es) into the host cells. Despite being just one of several secretion systems, the T3SS is necessary for *R. solanacearum* to cause disease. Molecular determinants involved in pathogenicity, virulence, and host range specificity have been described using representative strains of the main phylogenetic groups of *R. solanacearum* (Genin and Denny 2012).

Temperature is a major determinant in the distribution of this pathogen, which is widespread in tropical, subtropical, and warm temperate regions where the mean soil temperature is greater than 15°C (Hayward 1991). Wet soil increases the incidence of disease and water movement contributes to the dissemination of inoculum with water movement. The incidence and rate of wilting therefore increase with high temperatures and soil moisture. Continuous cropping of susceptible plants will also favor infection. Young plants are more rapidly diseased than older ones. BW is more prevalent in slightly acid to acid soils. For example, the disease is more prevalent in the acid soil area south of the Sichuan Basin and yellow alluvial soil, sandy alluvial soil, but no peanut BW occurs in purple soil in China (Chui et al. 2004). The pathogen is disseminated by contaminated farming equipment, in soil on tires and footwear, drainage water carrying inoculum through the soil, infested seed, and plants raised in infected soil, spreading the pathogen to new areas (Ziang et al. 2007).

Disease Management

Host Plant Resistance

Compared to many other crops, a relatively broad genetic diversity of resistance to BW has been found in the cultivated peanut, and the development of BW-resistant peanut cultivars has been more successful. Planting resistant cultivars is deemed as the sole economically viable means for effective control (Ding et al. 2011, Tang and Zhou 2000, Zhou et al. 2003, Yu et al. 2011). Some of the BW-resistant peanut genotypes mostly reported from China are R15, R16, R87, R106, K81 (Wang et al. 2009), Xiaohongmao (Liao et al. 2003), Ju and Zhonghua 6 (Liu et al. 2011), and Yuanza 9102 (Bang et al. 2011), and moderately resistant ones are Jihua 1012, Quancha 10, Quancha 646, Yucyou 193, and 38F5-45-21-CS1 (Yuan et al. 2002). It is significant that some of the highly resistant landraces have been identified in dragon-type peanuts mostly related to *A. hypogaea* subsp. *hirsuta* collected from South China where BW is generally serious. The dragon varieties have had been traditionally cultivated in many regions of China for at least 600 years before the varieties of three other types (Virginia, Spanish, and Valencia) were introduced in the late nineteenth century. Interestingly, all the BW-resistant landraces had been from South China and no BW-resistant germplasm has been collected from the BW-free north regions of China. The evolution of resistance, therefore, seems to be associated with regional disease pressure, and the BW in South China must have been a major factor influencing the natural selection of the dragon lines in the region. The resistance to BW in tested dragon lines is dominant, and the degree of dominance is higher in the dragon lines compared to the Spanish- or Valencia-resistant genotypes, and both additive and dominant genes are involved in resistance of dragon to BW, though a significant cytoplasmic effect is associated with the resistance in dragon types (Liao et al. 1986, Shan et al. 1998). As the dragon lines possess some other desirable traits such as drought tolerance, good flavor, and high oleic and linoleic acid ratio, they appear to be more promising in improving resistance to BW in peanuts (Muitia et al. 2006). However, latent infections (infection without visible symptoms) by *R. solanacearum* have been found in some resistant cultivars/dragon-type landraces that affect root proliferations and reduce symbiotic nitrogen fixation and tolerance to drought and yield, which explain the low productivity of the crop in infested areas, and this becomes a challenging task to the breeders (Liao et al. 1998, Huang et al. 2011, Jiang et al. 2013).

Most of the resistant germplasm lines identified are small-seeded genotypes with low yield potential; transferring BW resistance to high-yielding adapted peanut cultivars has therefore become an urgent task (Yu et al. 2011). A large-seeded peanut cultivar Rihua 1 is resistant to BW (Zhang et al. 2008, Ding et al. 2012).

Understanding the mechanism underlying BW resistance at the molecular level should hasten the breeding process (Liao 2001, Chen et al. 2008a). The DNA polymorphism among the promising peanut genotypes resistant to BW has been assessed by simple sequence repeats (SSR) and amplified fragment length polymorphism (AFLP) analysis. There is enough polymorphism in the peanut genotypes with BW resistance based on SSR and AFLP analysis (Jiang et al. 2006, 2007, Mace et al. 2007). Thus far, in peanut, there have been several reports regarding the identification of DNA markers related to BW resistance (Yu et al. 2011). Although several DNA markers related to BW resistance have been identified, the map distances are too large to be used in peanut-breeding programs (Yu et al. 2011). Peng et al. (2011) identified 119 transcription-derived fragments (TDFs) after root wounding inoculation with *R. solanacearum*, from Yuanza 9102 (a Spanish-type peanut cultivar with BW resistance) and Zhonghua 12 (a susceptible Spanish-type peanut cultivar) using cDNA-AFLP, and further studied their expression patterns. A total of 98 TDFs have been cloned and sequenced.

Chemical Control

The use of chemicals, antibiotics, and soil fumigation has shown little effect on the control of BW disease.

Cultural Control

In regions where the disease is endemic, cultural methods appear to be effective under some conditions for reducing bacterial population of *R. solanacearum* and subsequent disease control. Crop rotation of at least 2–5 years involving different nonhost crops particularly paddy rice, maize, and sugarcane may be used for significant disease reduction (Machmud and Hayward 1993, Nawangsih et al. 2012). Intercropping can be better for small farmers as cultivation of beans/maize can reduce disease incidence. Controlling weeds that have the potential to serve as inoculum reservoirs in conjunction with crop rotation can also be effective. Control of root-knot nematode population and cultural practices that minimize root damage can also reduce BW severity. Modification of soil pH by using a combination of organic amendment, fertilizers, and soil solarization is also effective in disease control (Machmud and Hayward 1993, Nawangsih et al. 2012).

Biological Control

Pseudomonas fluorescens strains B16 and VK18 and that of *Bacillus subtilis* strain B11 have been identified as promising biocontrol agents for BW control, the B16 strain yielding more effective results in increased yield of peanut by 0.7–0.94 tons/ha. B16 has a positive effect on the growth and yield of peanut and can replace 20% mineral nitrogen, phosphorus, potassium (NPK) fertilizer without significant changes in crop yield under Vietnam conditions (Doan et al. 2006). Similar results have been obtained with other strains of plant growth–promoting bacteria in the control of BW in Indonesia (Nawangsih et al. 2012). An endophytic *Bacillus amyloliquefaciens* strain BZ6-1, isolated from the stem of healthy peanut plants from *R. solanacearum*–infested fields, has been found to suppress *R. solanacearum* greatly, and the field trials demonstrate the control efficiency of strain BZ6-1 against BW by 62.3% (Wang et al. 2011).

ROOT KNOT OF PEANUT

Symptoms

Under field conditions, areas of root-knot nematode-affected peanut plants are usually round to oblong in shape, and rows of infected plants may never overlap as those of healthy plants. It is not uncommon for plants to wilt and eventually die in areas where nematode populations are high. Foliar symptoms of the root-knot disease may be expressed anytime during the growing season. These symptoms of nematode damage on peanut plants include stunting, yellowing, wilting, and even plant death. Generally, however, root-knot nematode damage symptoms are most evident in a peanut crop beginning about 100 days after planting and during or after periods of hot weather. Stunting of the plants results as the nematode larvae feed in the vascular system of the peanut, causing the formation of giant cells that disrupt the vascular system. Affected plants, when uprooted, show the presence of galls on pods and roots. The feeding roots are deformed. The galls on the roots usually are similar in size and shape to the nodules formed by nitrogen-fixing bacteria. On pods and pegs, the galls are corky and variable in shape.

Geographical Distribution and Losses

Root-knot nematodes (*Meloidogyne* spp.) are among the most serious plant pests in the world. Several species of root-knot nematodes are pathogenic on peanut and cause considerable yield loss annually. Of these, *Meloidogyne arenaria* (Neal) Chitwood race 1, *M. hapla* Chitwood, and *M. javanica* (Treub) Chitwood race 3 are the major pathogenic species of peanut (Abdel-Momen and Starr 1997, Minton and Baujard 1998; Koenning et al. 1999). These three species are known to occur in many peanut-producing regions, including North, Central, and South America, Africa, Asia, Europe, and Australia. *M. arenaria* and *M. javanica* are common in warm peanut-growing regions, whereas *M. hapla* occurs mainly in cool regions. In the United States, *M. arenaria* and *M. hapla* exist throughout the peanut-producing areas. *M. arenaria* is the predominant species parasitizing peanut in the southern regions, especially in Alabama, Florida, Georgia, Texas, and South Carolina in the United States where up to 40% of the fields are infested and yield losses in heavily infested fields can exceed 30%. All three species may cause significant losses in the yield and quality of peanut (Abdel-Momen and Starr 1997). Individual peanut fields heavily infested with the root-knot nematode have sustained yield losses greater than 75%. *M. hapla* is the most prevalent species in more northerly states, including North Carolina, Virginia, and Oklahoma (Koenning and Barker 1992). Populations of *M. javanica* parasitic on peanut are common in Egypt (Tomaszewski et al. 1994) and India (Sharma et al. 1995), but they are rare in the United States, having been described from only a few fields in Florida, Georgia, and Texas (Lima et al. 2002). Only one root-knot nematode, *M. arenaria* race 1, is a major nematode pest of peanut, and unlike most plants, peanut is a poor host or a nonhost to other commonly found root-knot nematodes in Florida (*M. incognita* and *M. javanica*).

PATHOGEN: (*M. arenaria*) LIFE CYCLE

The egg of a root-knot nematode develops into a vermiform first-stage juvenile that undergoes one molt into a second-stage juvenile. The second-stage juvenile hatches from the egg, moves freely in the soil, penetrates the root just behind the root cap, migrates intercellularly in the root, and establishes a feeding site within the developing vascular cylinder. As it feeds on the nematode-induced giant cell system, the second-stage juvenile loses its mobility and begins to increase in girth. After it has imbibed a sufficient quantity of sustenance, the flask-shaped second-stage juvenile molts three times without feeding and matures into a saccate adult female. Females of *M. arenaria* reproduce by mitotic parthenogenesis; as soon as they are mature adults, they begin producing eggs. The second-stage larvae that cannot find a suitable host plant can survive in the soil up to 18 months unless affected by adverse conditions or attacked by predacious fungi.

Male second-stage juveniles undergo metamorphosis during the third molt into elongate vermiform fourth-stage juveniles. The fourth-stage juvenile male remains enclosed in the cuticle of the second and third stages where it molts again to form an adult vermiform male. The male escapes from the cuticles and the root system. It moves freely in the soil, not feeding, only mating with mature adult females. As populations of *M. arenaria* reproduce by mitotic parthenogenesis, males serve no reproductive function.

The length of one generation of *M. arenaria* is greatly affected by temperature. At very high temperatures (>29°C), the life cycle takes approximately 3 weeks, but at very cool temperatures, it can be extended to 2–3 months.

DIAGNOSIS

Morphology of perineal patterns, shape and measurements of the stylet of the female, shape and measurements of the head and stylet of the male, and measurements of the second-stage juveniles are useful characters for species identification. Additional host range tests may be necessary to confirm the identification of the species and determination of the host race. Hosts of *M. arenaria* in the North Carolina differential host range test include tobacco (*Nicotiana tabacum* cv. NC95), pepper (*Capsicum annuum* cv. California Wonder), tomato (*Lycopersicon esculentum* cv. Rutgers), and watermelon (*Citrullus vulgaris* [*C. lanatus*] cv. Charleston Gray). Host race 1 populations infect and reproduce on peanut, whereas host race 2 populations do not.

A DNA probe that is specific for *M. arenaria* has been developed and may be useful for the diagnosis of this species. Cytological and biochemical characterization provide additional characters for the identification of *M. arenaria*.

DISEASE MANAGEMENT

Host Plant Resistance

Resistance to *M. arenaria* and *M. javanica* is highly correlated, indicating that, in many peanut genotypes, the same gene or genes may confer resistance to both species, or the resistance genes for each species are closely linked. But resistance to *M. hapla* is not correlated with resistance to *M. arenaria* or *M. javanica*. The mechanisms of resistance to *M. hapla* may be different from that of *M. arenaria* (Dong et al. 2008). The resistance genes in peanut may be related to differential recognition by the plant of the three *Meloidogyne* spp. Resistance to all three *Meloidogyne* spp. exists within cultivated peanut (*A. hypogaea*), either with or without introgressed genes from wild species (Simpson 1991). Breeding cultivars with resistance to root-knot nematode, however, has been slower because no meaningful resistance has been found in the peanut germplasm collection of *A. hypogaea*. Genes conferring resistance to peanut root-knot nematode have not been found in cultivated peanut, but a number of other *Arachis* spp. have been identified that are highly resistant or immune to the peanut root-knot nematode. Successful crosses to transfer a high level of nematode resistance into *A. hypogaea* have been reported by several researchers (Starr et al. 1995, Garcia et al. 1996, Church et al. 2000, Muitia et al. 2006). The resistance had been obtained from a wild species *Arachis cardenasii*. Two peanut germplasm lines "GP-NC WS 5 and GP-NC WS 6" have been derived from *A. hypogaea* × *A. cardenasii* interspecific cross (Stalker et al. 2002). The first of the two root-knot nematode-resistant peanut cultivars developed by the Texas Agricultural Experiment Station are (1) "COAN" released in 1999 (Simpson and Starr 2001) and (2) "NemaTAM" released in 2002 (Simpson et al. 2003). The resistance in COAN is controlled by a single dominant gene and is expressed as a reduction in nematode reproduction. Although nematodes invade the roots of COAN, most emigrate from the roots, but the few that remain in the roots develop to reproductive adults (Bendezu and Starr 2003). NemaTAM has greater yield potential than COAN and possesses the same level of resistance to the peanut root knot. Both COAN and NemaTAM have been proven resistant to the peanut root-knot nematode in southeastern United States. Neither peanut variety, however, could be successfully grown in southeastern United States because they are highly susceptible to tomato spotted wilt virus (TSWV). Later, some promising peanut germplasm lines with resistance to both the root knot and TSWV could be successfully developed (Holbrook et al. 2003), and in 2008, the USDA released a cultivar, Tifguard, which has resistance to both TSWV and root-knot nematode (Holbrook et al. 2008b). Similarly, another runner-type peanut germplasm line TifGP-1 (PI 648354) is reported to be resistant to both the root-knot nematode and TSWV (Holbrook et al. 2008a). The root-knot nematode resistance present in Tifguard is derived from the single dominant gene in COAN. The University of Georgia and University of Florida field trials have found excellent root-knot nematode resistance with the Tifguard cultivar and good final peanut yields in root-knot-infested fields. Interaction between root-knot nematode *M. arenaria* and *Cylindrocladium* black rot (CBR) fungal pathogen *Cylindrocladium parasiticum* in runner peanut genotypes such as "C724-19-15," "C724-19-25," and "Georgia-O2C" that possess different levels of resistance to nematode and *C. parasiticum* reveals that *C. parasiticum* greatly increases mortality on "C724-19-25" and "Georgia-O2C" but not on "C724-19-15" in the presence of *M. arenaria*, indicating that root-knot resistance in peanut can be broken down due to fungal infection caused by *C. parasiticum* (Dong et al. 2009). The peanut germplasm line "TifGP-2," a nematode-susceptible sister line of nematode-resistant *Tifguard*, that is, peanut closely related sister lines with and without nematode resistance, can be valuable research tools to obtain better understanding of the interactions of nematodes with other pathogens of peanut (Holbrook et al. 2008c, 2012).

Chemical Control

Nematicides have often been used for limiting the damage that nematodes cause on plants. Nematicides are usually used as a soil treatment before planting. However, a few nematicides can be applied after planting. These chemicals are relatively expensive and they require costly equipment and trained personnel to apply them. Peanut crops that are good hosts of *M. arenaria* can be protected with soil application of nematicides such as aldicarb at 31 g ai/100 m row (Timper et al. 2001). The effectiveness is dependent on adequate amount of soil moisture. If an optimum amount of water is available, the optimum effect is achieved; if too much or too little water is present, very little control is achieved.

Cultural Control

Meloidogyne species are obligate parasites and populations decline rapidly in the absence of a host. Rotation of susceptible host crop plants with those that are immune or poor hosts is a useful way to reduce the effect that *M. arenaria* has on plant growth. Unfortunately, the nonhost, when it does occur, is usually less profitable than the susceptible crop. *M. arenaria* has a very large host range, and nonhosts or cultivars that have been reported resistant should be used with caution because of the innate variability that occurs in the root-knot nematodes. Switchgrass (*Panicum virgatum*), for example, do not support the population of root-knot nematode but support the population of nonparasitic nematodes (Kokalis-Burelle et al. 2002). Populations of *M. arenaria* are lower in peanut in the cotton–cotton–peanut than in peanut–peanut–peanut, corn–corn–peanut, or bahiagrass–bahiagrass–peanut cropping systems (Timper et al. 2001). Other agronomic and economic factors are also important in the selection of a rotation crop. An adequate weed control program is absolutely necessary for a crop rotation scheme to be effective because many weed species serve as suitable hosts. While good crop rotation should be continued to reduce all peanut diseases, the advent of resistant peanut varieties will help reduce the need for costly nematicides in peanut production. Soil organic amendment with crab shell chitin has been found not only to reduce *M. arenaria* population but also enhances soil microbial activity and promotes plant growth by 67.7% resulting in reduction in root-knot index on peanut, the number of galls per plant being negatively correlated with the accumulation of total phenols and activities of chitinase and peroxidase subsequently increasing the yield of peanut over 56% (Kalaiarasan et al. 2008b).

Biological Control

Numerous attempts have been made to control root-knot nematodes with parasitic and predacious organisms or various organic amendments, with varying degrees of success. Naturally occurring organisms, such as *Pasteuria penetrans*, which are obligate parasites of *Meloidogyne*, may prove to be effective for biological control (Timper et al. 2001). Peanut seed treatment with *P. fluorescens* at 10 g/kg of seed inhibits *M. arenaria* development in the peanut roots due to reduced and poor development of giant cells (Kalaiarasan et al. 2008a). Possibility of use of facultative Gram-negative symbiotic bacteria belonging to the *Xenorhabdus* species isolated from the gut of the nematode *Steinernema riobrave* is indicated in the control of root-knot nematode of peanut (Vyas et al. 2008). Metabolites of *Xenorhabdus* species so obtained from *S. riobrave* contain the proteins of high molecular weight, 76–90 kDa apart from regular proteins, and appear to play a role in the suppression of root-knot development in peanut.

PEANUT WITCHES' BROOM

Axillary buds of affected plants proliferate to produce numerous small stuff leaves, and the internodes are reduced. The pod stalks (gynophores) grow upward, showing a loss of positive geotropism resulting in the loss of pod formation. The peanut witches' broom (PnWB), first discovered in a geographically isolated area, the Penghu Islands, in 1975 in Taiwan and now reported or suspected to be prevalent in most Asian countries, is a plant disease associated with plant pathogenic phytoplasmas.

Phytoplasmas are a group of phytopathogenic bacteria that are transmitted by sap-feeding leafhopper insect vectors. Phylogenetically, phytoplasmas are related to the animal pathogenic mycoplasmas. Both groups are unique among bacteria in their lack of cell wall and are assigned to the class Mollicutes. However, unlike mycoplasmas that can be cultured and are amenable to genetic manipulations, the *in vitro* cultivation of phytoplasmas has remained unsuccessful despite decades of efforts. The inability to culture phytoplasmas outside of their host has resulted in the designation of the *Candidatus* (*Ca.*) status to their taxonomic assignment and also greatly hampered the research progress in characterizing these pathogens. The PnWB is caused by *Ca. phytoplasma asteris*–PnWB group 16SrII, and it is the first representative of the 16SrII group (Chung et al. 2013).

For the detection and identification of phytoplasmas, polyclonal and monoclonal antibodies, DNA probes, and PCR primers have been developed for various phytoplasmas (Chen and Lin 1997). Besides detection, extrachromosomal DNA, insertion sequence, and various genes of PnWB phytoplasma have been cloned (Chuang and Lin 2000, Wei and Lin 2004, Chi and Lin 2005, Chu et al. 2006, Chen et al. 2008b). With advancement in genomic science, genome sequence has been adopted as a powerful tool to characterize the gene contents of the uncultivated bacteria. Four open reading frames (ORFs) have been identified in the order of hrcA, grpE, dnaK, and dnaJ through the PCR-based technique. Chromosomal arrangement of these genes in PnWB phytoplasma is identical to those of aster yellows witches' broom phytoplasma, onion yellows phytoplasma, and other bacteria phylogenetically related to phytoplasma (Chu et al. 2007a,b). It is also indicative that three rpoD homologous sigma factor genes may exist in PnWB phytoplasma (Chen et al. 2008b). Whole-genome shotgun sequencing of PnWB phytoplasma has also been done (Chung et al. 2013).

RecA protein, the product of recA gene, a key protein involved in DNA recombination and DNA repair of eubacteria, has been cloned and analyzed from phytoplasma-associated PnWB. Gene organization, the nucleotide sequence, and a sequence in the conserved regions of the ORF are similar to those of the other recA genes of eubacteria. Therefore, this gene from phytoplasma-associated PnWB is identified as a putative recA gene (Chu et al. 2006).

Phytoplasmas related to the *Ca. phytoplasma asteris*–PnWB group have been found to cause symptoms of phytoplasma diseases like leaf roll, rosetting, shoot proliferation, and phyllody in Japanese plum trees (Zirak et al. 2009), sweet cherry (Zirak et al. 2010) in Iran, and sesame in Turkey (Cengiz et al. 2013).

REFERENCES

Abdel-Momen, S.M. and J.L. Starr. 1997. Damage functions for three *Meloidogyne* species on *Arachis hypogaea* in Texas. *J. Nematol.* 29: 478–482.

Alvarez, A.M. 2005. Diversity and diagnosis of *Ralstonia solanacearum*. In: *Bacterial Wilt Disease and the Ralstonia solanacearum species Complex*, eds., C. Allen, P. Prior, and A.C. Hayward. APS Press, St. Paul, MN, pp. 437–448.

Anitha, K., S.K. Chakrabarty, G.A. Girish, R.D.V.J.P. Rao, and K.S. Varaprasad. 2004. Detection of bacterial wilt infection in imported groundnut germplasm. *Indian J. Plant Prot.* 32: 147–148.

Bang, W.F., L.V. Jian-Wei, and X.B. Lin. 2011. Differential expression of genes related to bacterial resistance in peanut (*Arachis hypogaea*). *J. Hereditas* 33: 389–396.

Bendezu, I.F. and J.L. Starr. 2003. Mechanism of resistance to *Meloidogyne arenaria* in the peanut cultivar COAN. *J. Nematol.* 35: 115–118.

Cengiz, I., Y. Engin, C. Mursel, U. Rustem, and U. Bulent. 2013. Molecular diagnosis of peanut witches' broom group (16 SrII) phytoplasma infecting sesame in Turkey. *Curr. Opin. Biotechnol.* 24: 39–40.

Champoiseau, P.G., J.B. Jones, and C. Allen. 2009. *Ralstonia solanacearum* race 3 biovar 2 causes tropical losses and temperate anxieties. Online. *Plant Health Progr.* doi:10.1094/PHP-2009-0313-01-RV.

Chen, B.Y., H.F. Jiang, X. Ren, B.S. Liao, and J.Q. Huang. 2008a. Identification and molecular traits of *Arachis* species with resistance to bacterial wilt. *Acta Agric. Boreali Sin.* 23: 170–175.

Chen, M.F. and C.P. Lin. 1997. DNA probes and PCR primers for the detection of a phytoplasma associated with peanut witches' broom. *Eur. J. Plant Pathol.* 103: 137–145.

Chen, S.K., P.W. Chu, K.C. Ho, W.Y. Chen, and C.P. Lin. 2008b. Molecular cloning of sigma factor gene of phytoplasma associated with peanut witches' broom. *Plant Pathol. Bull.* 17: 279–287.

Chen, X.M., F.P. Hu, and Y.R. Wu. 2000. Pathotypes and biotypes of *Ralstonia solanacearum* isolated from bacterial wilt of peanut in Fujian Province. *J. Fujian Agric. Univ.* 29: 470–473.

Chi, K.L. and C.P. Lin. 2005. Cloning and analysis of polC gene of phytoplasma associated with peanut witches' broom. *Plant Pathol. Bull.* 14: 51–58.

Chu, P.W., S.K. Chen, W.Y. Chen, and C.P. Lin. 2007a. Sequence analysis of genes coding for the molecular chaperones GrpE, DnaK, and DnJ from phytoplasma associated with peanut witches' broom. *Plant Pathol. Bull.* 16: 215–224.

Chu, Y., C.C. Holbrook, P. Timper, and P. Ozias-Akins. 2007b. Development of PCR-based molecular marker to select for nematode resistance in peanut. *Crop Sci.* 47: 841.

Chu, Y.R., W.Y. Chen, and C.P. Lin. 2006. Cloning and sequence analyses of rec A gene of phytoplasma associated with peanut witches' broom. *Plant Pathol. Bull.* 15: 211–218.

Chuang, J.G. and C.P. Lin. 2000. Cloning of gyrB and gyrA genes of phytoplasma associated with peanut witches' broom. *Plant Pathol. Bull.* 9: 157–166.

Chung, W.C., L.L. Chen, W.S. Lo, C.P. Lin, and C.H. Kuo. 2013. Comparative analysis of the peanut witches' broom phytoplasma genome reveals horizontal transfer of potential mobile units and effectors. *PLoS One* 8(4): e62770. doi:10.1371/journal.pone0062770.

Church, G.T., C.E. Simpson, M.D. Burow, A.H. Paterson, and J.L. Starr. 2000. Use of RFLP markers for identification of individuals homozygous for resistance to *Meloidogyne arenaria* in peanut. *Nematology* 2: 575–580.

Chui, F.H., M.F. Lai, and X.P. Zen. 2004. Distribution and control and disease-resistant breeding on blight in peanut in Sichuan and Chongqing. *Southwest China J. Agric. Sci.* 17: 741–745.

Denny, T.P. and A.C. Hayward. 2001. Gram-negative bacteria: Ralstonia. In: *Laboratory Guide for Identification of Plant Pathogenic Bacteria*, 3rd edn., eds., N.W. Schaad, J.B. Jones, and W. Chun. APS Press, St. Paul, MN, pp. 51–74.

Ding, Y.F., C.T. Wang, Y.Y. Tang et al. 2012. Isolation and analysis of differentially expressed genes from peanut in response to challenge with *Ralstonia solanacearum. Electron. J. Biotechnol.* 15(5): Article 1.

Ding, Y.F., C.T. Wang., D.L. Zhao et al. 2011. Simple sequence repeats to identify true hybrids in bacterial wilt susceptible × resistant crosses in groundnut. *Electron. J. Plant Breed.* 3: 367–371.

Doan, T.T., T.H. Nguyen, W. Zeller, and C. Ullrich. 2006. Status of research on biological control of tomato and groundnut bacterial wilt in Vietnam. *Mitteilungen aus der Biologischen Bundesanstalt für Land- und Forstwirtschaft* 408: 105–111.

Dong, W.B., T.B. Brenneman, C.C. Holbrook, P. Timper, and A.K. Culbreath. 2009. The interaction between *Meloidogyne arenaria* and *Cylindrocladium parasiticum* in runner peanut. *Plant Pathol.* 58: 71–79.

Dong, W.B., C.C. Holbrook, P. Timper, T.B. Brenneman, Y. Chu, and P. Ozias-Akins. 2008. Resistance in peanut cultivars and breeding lines to three root-knot nematode species. *Plant Dis.* 92: 631–638.

Elphinstone, J.G. 2005. The current bacterial wilt situation: A global overview. In: *Bacterial Disease and the Ralstonia solanacearum Species Complex*, eds., C. Allen, P. Prior, and A.C. Hayward. APS Press, St. Paul, MN, pp. 9–28.

Fegan, M. and P. Prior. 2005. How complex is the *Ralstonia solanacearum* species complex? In: *Bacterial Wilt Disease and the Ralstonia solanacearum Species Complex*, eds., C. Allen, P. Prior, and A.C. Hayward. APS Press, St. Paul, MN, pp. 449–461.

Garcia, G.M., H.T. Stalker, E. Shroeder, and G. Kochert. 1996. Identification of RAPD, SCAR, and RFLP markers tightly linked to nematode resistance genes introgressed from *Arachis cardenasii* into *Arachis hypogaea. Genome* 39: 836–845.

Genin, S. and P. Denny. 2012. Pathogenomics of the *Ralstonia solanacearum* species complex. *Annu. Rev. Phytopathol.* 50: 67–89.

Hayward, A.C. 1991. Biology and epidemiology of bacterial wilt caused by *Pseudomonas solanacearum. Annu. Rev. Phyopathol.* 29: 65–87.

Holbrook, C.C., W.B. Dong, P. Timper, A.K. Culbreath, and C.K. Kvien. 2012. Registration of peanut germplasm line TifGP-2, a nematode-susceptible sister line of Tifguard. *J. Plant Registrations* 6: 2008–2011.

Holbrook, C.C., P. Timper, and A.K. Culbreath. 2003. Resistance to Tomato spotted wilt virus and root-knot nematode in peanut interspecific breeding lines. *Crop Sci.* 43: 1109–1113.

Holbrook, C.C., P. Timper, and A.K. Culbreath. 2008a. Registration of peanut germplasm line TifGP-1 with resistance to the root-knot nematode and tomato spotted wilt virus. *J. Plant Registrations* 2: 57.

Holbrook, C.C., P. Timper, A.K. Culbreath, and C.K. Kvien. 2008b. Registration of 'Tifguard' peanut. *J. Plant Registrations* 2: 92–94.

Holbrook, C.C., P. Timper, W.B. Dong, C.K. Kvien, and A.K. Culbreath. 2008c. Development of near-isogenic peanut lines with and without resistance to the peanut root-knot nematode. *Crop Sci.* 48: 194–198.

Huang, J.Q., I.Y. Yan, X.W. Ye, Y. Lei, and B.S. Liao. 2011. Development of *Ralstonia solanacearum* quantification method and its application in peanut with bacterial wilt disease. *Sci. Agric. Sin.* 44: 58–66.

Jiang, H., X. Ren, Y. Chen et al. 2013. Phenotypic evaluation of the Chinese mini-mini core collection of peanut (*Arachis hypogaea* L.) and assessment for resistance to bacterial wilt disease caused by *Ralstonia solanacearum. Plant Genet. Resour.* 11: 77–83.

Jiang, H.F., B.S. Liao, X.P. Ren et al. 2006. Genetic diversity assessment in peanut genotypes with bacterial wilt resistance. *Acta Agron. Sin.* 32: 1156–1165.

Jiang, H.F., B.S. Liao, X.P. Ren et al. 2007. Comparative assessment of genetic diversity of peanut (*Arachis hypogaea* L.) genotypes with various levels of resistance to bacterial wilt through SSR and AFLP analyses. *J. Genet. Genomics* 34: 544–554.

Kalaiarasan, P., P.L. Lakshmanan, and R. Samiyappan. 2008b. Organic crab shell chitin-mediated suppression of root-knot nematode, *Meloidogyne arenaria* in groundnut. *Indian J. Nematol.* 38: 196–202.

Kalaiarasan, P., P.L. Lakshmanan, R. Samiyappan, and M.J. Sudheer. 2008a. Influence of *Pseudomonas fluorescens* on histopathological changes caused by *Meloidogyne arenaria* in groundnut (*Arachis hypogaea* L.). *Indian J. Nematol.* 38: 18–20.

Koenning, S.R. and K.R. Barker. 1992. Relative damage and functions and reproductive potentials of Meloidogyne arenaria and M.hapla on peanut. *J. Nematol.* 24:187–192.

Koenning, S.R., C. Overstreet, J.W. Noling, P.A. Donald, J.O. Becker, and B.A. Fortnum. 1999. Survey of crop losses in response to phytoparasitic nematodes in the United States. *Suppl. J. Nematol.* 31: 587–618.

Kokalis-Burelle, N., W.F. Mahaffee, R. Rodriguez-Kábana, J.W. Klopper, and K.L. Bowen. 2002. Effects of switchgrass (*Panicum virgatum*) rotations with peanut (*Arachis hypogandea* L.) on nematode populations and soil microflora. *J. Nematol.* 34: 98–105.

Lima, R. D, J.A. Brito, D.W. Dickson, W.T. Crow, C.A. Zamora, and M.L. Mendes. 2002. Enzymatic characterization of *Meloidogyne* spp. associated with ornamentals and agronomic crops in Florida, USA (abstr.). *Nematology* 4: 173.

Liao, B.S. 2001. Molecular markers for bacterial wilt resistance in plants and its potential for breeding. *Chin. J. Oil Crops Sci.* 23: 66–68.

Liao, B.S., Y. Lei, S.Y. Wang, D. Li, H.F. Jiang, and X.P. Ren. 2003. Aflatoxin resistance in bacterial wilt resistant groundnut germplasm. *Int. Arachis Newsl.* 23: 24.

Liao, B.S., W.R. Li, and D.R. Sun. 1986. A study on inheritance of resistance to bacterial wilt in groundnut. *Oil Crops China* 3: 1–8.

Liao, B.S., Z.H. Shan, N.X. Duan et al. 1998. Relationship between latent infection and groundnut wilt resistance. In: *Bacterial Wilt Disease: Molecular and Ecological Aspects*, eds., P.H. Prior, C. Allen, and J. Elphinstone. Springer, New York, pp. 294–299.

Liu, J.W., H.F. Jiang, J.Q. Huang, W.F. Peng, Y.N. Chen, and Z.Q. Li. 2011. Analysis of gene expression profiles in response to *Ralstonia solanacearum* SSH in Peanut. *Acta Botanica Boreali–Occidentalia Sinica* 31: 1517–1523.

Mace, E.S.,W. Yuejin, L. Bos hou, H. Upadhyaya, S. Chandra, and J.H. Crouch. 2007. Simple sequence repeat (SSR)-based diversity analysis of groundnut (*Arachis hypogaea* L.) germplasm resistant to bacterial wilt. *Plant Genet. Resour.: Charact. Util.* 5: 27–36.

Machmud, M. and A.C. Hayward. 1993. Present status of ground bacterial wilt research in Indonesia. In: *Groundnut Bacterial Wilt Proceedings of the Second Working Group Meeting*, ed., V.K. Mehan, Tainan, Taiwan, November 1992.

Mehan, V.K., B.S. Liao, Y.J. Tan, A. Robinson-Smith, D. McDonald, and A.C. Hayward. 1994. Bacterial Wilt of groundnut. Information Bulletin No. 35. ICRISAT, Patancheru, India, 28pp.

Minton, N.A. and P. Baujard. 1990. Nematode parasites of peanut. In: *Plant Parasitic Nematodes in Subtropical Agriculture*, eds., M. Luc, R.A. Sikora, and J. Bridge. CABI, Wallingford, U.K., pp. 285–320.

Muitia, A., Y. López, J.L. Starr, A.M. Schubert, and M.D. Burow. 2006. Introduction of resistance to root-knot nematode (*Meloidogyne arenaria*) into High-Oleic Peanut. *Peanut Sci.* 33: 97–103.

Nawangsih, A.A., R. Aditya, B. Tjahjono, H. Negishi, and K. Suyama. 2012. Bioefficacy and characterization of plant growth-promoting bacteria to control the bacterial wilt disease of peanut in Indonesia. *J. ISSAAS* 18: 185–192.

Peng, W.-F., J.W. Lv, X.P. Ren et al. 2011. Differential expression of genes related to bacterial wilt resistance in peanut (*Arachis hypogaea* L.). *Hereditas* (Beijing) 33: 359–396.

Prior, P. and M. Fegan. 2005. Recent developments in the phylogeny and classification of *Ralstonia solanacearum*. *Acta Hort.* 695: 127–136.

Salanoubat, M. S. Genin, F. Artiguenave et al. 2002. Genome sequence of the plant pathogen *Ralstonia solanacearum*. *Nature* 415: 497–502.

Schell, M.A. 2000. Control of virulence and pathogenicity genes of *Ralstonia solanacearum* by an elaborate survey network. *Annu. Rev. Phytopathol.* 38: 263–292.

Shan, J.H., N.X. Duan, H.F. Jiang, Y.J. Tan, D. Li, and B.S. Liao. 1998. Inheritance of resistance to bacterial wilt in Chinese Dragon groundnuts. In: *Bacterial Wilt Disease: Molecular, Ecological Aspects*, eds., P.H. Prior, C. Allen, and J. Elphinstone. Springer, New York, pp. 303–305.

Sharma, S.B., D.H. Smith, and D. McDonald. 1995. Host races of *Meloidogyne javanica*, with preliminary evidence that the "groundnut race" is widely distributed in India. *Int. Arachis Newsl.* 15: 43–45.

Simpson, C.E. 1991. Pathways for introgression of pest resistance into *Arachis hypogaea* L. *Peanut Sci.* 18: 22–26.

Simpson, C.E. and J.L. Starr. 2001. Registration of COAN Peanut. *Crop Sci.* 41: 918.

Simpson, C.E., J.L. Starr, G.T. Church, M.D. Burrow, and A.H. Paterson. 2003. Registration of Nema Tam Peanut. *Crop Sci.* 43: 1561.

Stalker, H.T., M.K. Beute, B.B. Shew, and K.R. Barker. 2002. Registration of two root-knot nematode-resistant peanut germplasm lines. *Crop Sci.* 42: 312–313.

Starr, J.L., C.E. Simpson and T.A. Lee. 1995. Resistance to Meloidogyne arenaria in advanced generation breeding lines of peanut. *Peanut Sci.* 22: 59–61.

Tang, R.H. and H.Q. Zhou. 2004. Resistance to bacterial wilt in some interspecific derivatives of hybrids between cultivated peanut and wild species. *Oil Crops China* 22: 61–65.

Timper, P., N.A. Minton, A.W. Johnson et al. 2001. Influence of cropping systems on stem rot (*Scleotium rolfsii*), *Meloidogyne arenaria* and the nematode antagonist *Pasteuria penetrans* in peanut. *Plant Dis.* 85: 767–772.

Tomaszewski, E.K., M.A.M. Khalili, A.A. El-Deeb, T.O. Powers, and J.L. Starr. 1994. *Meloidogyne javanica* parasitic on peanut. *J. Nematol.* 26: 436–441.

Vyas, R.V., B. Patel, A. Maghodia, and D.J. Patel. 2008. Significance of metabolites of native *Xenorhabdus*, a bacterial symbiont of *Steinernema*, for suppression of collar rot and root knot diseases of groundnut. *Indian J. Biotechnol.* 7: 371–377.

Wang, C.T., X.Z. Wang, Y.Y. Tan et al. 2009. Field screening of groundnut genotypes for resistance to bacterial wilt in Shandong province in *China. J. SAT Agric. Res.* 7: 1–5.

Wang, X.B., Y.M. Luo, W.L. Wu, and Z.G. Li. 2011. Identification, antimicrobial activity and field control efficacy of an endophytic bacteria strain against peanut bacterial wilt. *Chin. J. Biol. Control.* 27: 88–92.

Wei, H.C. and C.P. Lin. 2004. Cloning and insertion sequence of phytoplasma associated with witches' broom using random sequencing. *Plant Pathol. Bull.* 13: 143–154.

Yao, J. and C. Allen. 2006. Chemotaxis is required for virulence and competitive fitness in the bacterial wilt pathogen *Ralstonia solanacearum. J. Bacteriol.* 188: 3697–3708.

Yu, S.L., C.T. Wang, Q.L. Yang et al. 2011. *Peanut Genetics and Breeding in China.* Shanghai Science and Technology Press, Shanghai, China, 565pp.

Yuan, Z.S., F.P. Hu, Y.C. Hong, and X.Q. Cai. 2002. Test on resistance of peanut varieties or lines to *Ralstonia solanacearum. J. Fujian Agric. Forest. Univ.* 31: 174–176.

Zhang, D.W., N. Zhang, Q. Wang et al. 2008. Breeding and cultivation of Rihua 1, a new peanut variety resistant to bacterial wilt disease. *Shandong Agric. Sci.* 9: 103–104.

Zhou, G.Y., X.Q. Liang, Y.C. Li, and X.X. Li. 2003. Evaluation of peanut cultivars for resistance to bacterial wilt and analysis of family tree. *J. Peanut Sci.* 32: 25–28.

Ziang, H., B. Liao, X. Ren et al. 2007. Comparative assesment of genetic diversity of peanut (Arachis hypogaea L.) Genotypes with various levels of resistance to bacterial wilt through SSR and AFLP analyses. *J. Genet. Genomics* 34: 544–554.

Zirak, L., M. Bahar, and A. Ahoonmanesh. 2009. Molecular characterization of phytoplasmas related to peanut witches' broom and stolbur groups infecting plump in Iran. *J. Plant Pathol.* 91: 713–716.

Zirak, L., M. Bahar, and A. Ahoonmanesh. 2010. Characterization of phytoplasmas related '*Candidatus Phytoplasma asteris*' and peanut WB group associated with sweet cherry diseases in Iran. *J. Phytopathol.* 158: 63–65.

Section III

Rapeseed–Mustard

Rapeseed–mustard is an important group of edible oilseed crops constituting oilseed Brassicas and crucifers, namely, *Brassica juncea*, *Brassica rapa* subsp. *trilocularis* (yellow sarson), *Brassica rapa* subsp. *dichotoma* (toria and brown sarson), *Brassica napus* (oilseed rape), *Brassica carinata* (Karan rai), *Brassica nigra* (Banarasi rai), and *Eruca vesicaria* subsp. *sativa* Miller (*taramira*). These cultivated *Brassica* has two types of genomes, that is, the diploid (elementary) and amphidiploid genomes. The elementary species include *Brassica rapa* (AA, $2n$ = 20,468-516 Mbp), *B. nigra* (BB, $2n$ = 16,468-760 Mbp), and *Brassica oleracea* (CC, $2n$ = 18,599-618 Mbp), and the DNA sequence variation reveals 51.2 Mb of the *Brassica* A and C genomes based on 10 diverse rapeseed–mustard genotypes (Ahmad et al. 2002, Clarke et al. 2013).

Globally, on 36.4 m ha, 72.5 mt of rapeseed–mustard seed is produced (FAO 2014). However, a wide gap exists between the potential yield and that realized at the farmers' field, which is largely because of the number of biotic and abiotic stresses to which the rapeseed–mustard crop is exposed. Diseases are major hurdles toward achieving higher production in rapeseed–mustard. The intensive cultivation of the crop with more inputs has further compounded the problem, and now the incidence of diseases and insect pests have become more frequent and widespread. Severe outbreak of diseases deteriorates the quantity and quality of seed and oil content drastically in different oilseed *Brassica* crops. The expression of full inherent genetic potential of a genotype is governed by inputs that go into the production system. This can be very well illustrated with examples that involve disease management of rapeseed–mustard. The loss in oilseed crops due to biotic stresses is about 19.9%, out of which diseases cause severe yield reduction at different growth stages. Various plant pathogens are reported to distress the crops. Among them, 18 are considered to be economically important in different parts of the world. To overcome such losses, it is essential to know the causal agents, their behavior, and the means to attack the vulnerable phase of the pathogen.

5 Rapeseed–Mustard Diseases

Alternaria BLIGHT

SYMPTOMS

Symptoms of the disease are characterized by the formation of spots on leaves, stem, and siliquae. The pathogen has been reported to affect seed germination and quality and quantity of oil (Meena et al. 2010a). On seedlings, symptoms include dark stem lesions immediately after germination that can result in damping-off or stunted seedlings. Generally, disease appeared at 40–45 days after sowing (DAS), and most critical stages have been reported at 75 and 45 days of plant growth (Meena et al. 2004). Spots produced by *Alternaria brassicae* appear to be usually gray in color compared with black sooty velvety spots produced by *Alternaria brassicicola*. Spots produced by *Alternaria raphani* show distinct yellow halos around them. However, the symptoms may vary with host and environment. Symptoms are first visible on lower leaves with appearance of black points, which later enlarge to develop into prominent, round, and concentric spots of various sizes. With progress of the disease, symptoms appear on middle and upper leaves with smaller-sized spots (Figure 5.1), when defoliation of lower leaves occur. Later, round black conspicuous spots appear on siliquae (Figure 5.2) and stem. These spots may coalesce leading to complete blackening of siliquae or weakening of the stem with the formation of elongated streaks. Rotting of the seed may be seen just beneath the black spot on siliqua of yellow or brown sarson. Spots on mustard siliqua are brownish black with a distinct gray center. When older plants become infected, symptoms often occur on the older leaves, since they are closer to the soil and are more readily infected as a consequence of rain splash or wind-blown rain. Fruit-bearing branches and pods show dark or blackened spots that result in yield loss due to premature pod ripening and shedding of the seeds. The infection of *Alternaria* blight on leaves and silique reduces the photosynthetic area drastically. The phase of infection on silique adversely affects the normal seed development, weight, color, oil content, and the quality of seed. *Alternaria* blight–infected leaves of Indian mustard showed significant decrease in oil, triglyceride, 18:2, and 22:1 fatty acid content and also in the level of different lipid classes (phospholipids, glycolipids, and sterols) (Atwal et al. 2005). Two *Alternaria* toxins, namely, alternariol and alternariol monomethyl ether, were found in high concentration in the seeds infected with *Alternaria* species (Gwiazdowski and Wickiel 2009, Jajor et al. 2011). Several biochemical constituents are found to impart resistance to rapeseed–mustard against *A. brassicae*. Total sugars, reducing sugars, flavonol, and chlorophyll content were present in high amount in healthy leaves, while total phenol, *o*-dihydroxy phenol, carotenoids, and protein content rose with increase in infection of *A. brassicae* (Neeraj and Verma 2010, Mathpal et al. 2011, Gupta et al. 2012, Prakash et al. 2012). However, total sugar, total phenol, and *ortho*-dihydroxyphenol were higher in chlorotic areas than necrotic areas of infected leaves. Flavonol and chlorophyll content was observed lower in different infected parts of leaves than in healthy ones and was prominently lower in necrotic areas than chlorotic zones (Atwal et al. 2004).

Activities of some oxidative enzymes, namely, peroxidase (PO) and polyphenol oxidase, increased in *B. juncea* leaves after infection (Chawla et al. 2001). Total phenol content and specific activities of phenylalanine ammonia lyase (PAL) and tyrosine ammonia lyase (TAL) were higher in *Alternaria* blight–infected leaves and siliquae walls compared to healthy leaves, which suggests their possible involvement in plant protection against the disease (Gupta and Kaushik 2002). Transpiration in oilseed rape reduced after infection with *Alternaria* species (Baranowski et al. 2009).

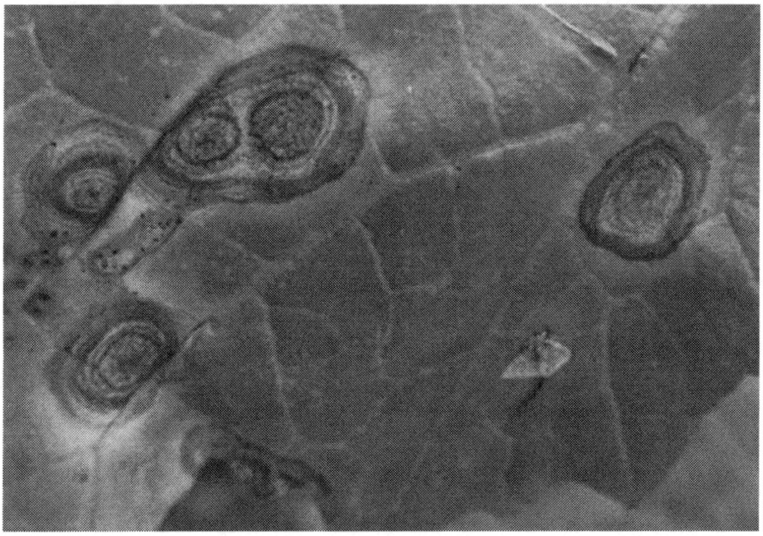

FIGURE 5.1 Alternaria blight on leaves.

FIGURE 5.2 Alternaria blight on pods.

Geographical Distribution and Losses

Rapeseed–mustard crops are ravaged by *Alternaria* blight or black spot; the most common widespread and destructive disease is caused mostly by *A. brassicae* (Berk.) Sacc. infecting all aboveground parts of the plant, reported from all the continents of the world and considered an important constraint in husbandry of oilseed Brassicas. The disease has been reported from Argentina, Australia, Austria, Bangladesh, Brazil, Bhutan, Bulgaria, Canada, Chile, China, Cyprus,

Czechoslovakia, Denmark, Ethiopia, Finland, France, Germany, Ghana, Hong Kong, Hungary, India, Ireland, Italy, Jamaica, Japan, Kenya, Libya, Malawi, Malaya, Mauritius, Morocco, Mozambique, Myanmar, Nepal, the Netherlands, New Zealand, Nicaragua, Nigeria, Norway, Pakistan, Papua New Guinea, Philippines, Poland, Romania, Russia, Saudi Arabia, Scotland, Sierra Leone, Singapore, Spain, South Africa, Sri Lanka, Sudan, Sweden, Switzerland, Taiwan, Tanzania, Trinidad, Turkey, Uganda, the United Kingdom, the United States, Zambia, and Zimbabwe (Saharan et al. 2005). Pathogens of the disease, *A. brassicicola* and *A. raphani*, are also encountered but rarely. Though total destruction of the crop due to the disease is rare and usually yield losses at harvest are 5%–15%, they can reach up to 47% (Kolte et al. 1987) accompanied by reduction in seed quality, namely, seed size and viability. Severity of *Alternaria* blight on oilseed Brassicas differs among seasons and regions as also between individual crops within a region.

Pathogen

It is easy to recognize *Alternaria* sp. by the morphology of their large conidia. They are catenate, formed in chains or solitary, typically ovoid to obclavate, often beaked, pale brown to brown, multi-celled, and muriform (Simmons 2007). Formation of chlamydospores is reported in *A. brassicae* and *A. raphani*, while microsclerotia are found to be produced by the former. Although the use of species-group designation does not resolve definitive species boundaries within *Alternaria*, advantages of its use are that it organizes at the subgeneric level the morphologically diverse assemblage of *Alternaria* species and permits the generalized discussion of morphologically similar species without becoming overrestricted due to nomenclatural uncertainty. Cultural, morphological, pathogenic, and molecular variation in isolates of *A. brassicae* has been indicated by several workers (Gupta et al. 2004a, Mehta et al. 2005, Patni et al. 2005a, Khan et al. 2007a,b, Singh et al. 2009, Goyal et al. 2011a, 2013a, Khulbe et al. 2011, Kumawat et al. 2011, Sharma et al. 2013a,b). Some extracellular enzymes, namely, cellulase and pectinases (polygalacturonase [PG] and pectin methyl esterase), are produced by *A. brassicae* under different cultural conditions (Atwal and Sangha 2004a). However, their exact role in pathogenesis is not known. *Alternaria longipes* and *Alternaria napiforme* are also reported on rapeseed–mustard from India (Kolte 1985). *A. brassicicola* produce some detoxifying enzymes, namely, brassinin hydrolases, which are dimeric protein of 120 kDa and catalyze the detoxification of brassinin, a phytoalexin produced in crucifers after fungal infection (Pedras et al. 2009a). The most selective phytotoxic metabolite, namely, brassicicolin A, and the major phytoalexin, namely, spirobrassinin, were produced by *A. brassicicola* in liquid cultures and in infected leaves of *B. juncea*, respectively (Pedras et al. 2009b).

Epidemiology and Disease Cycle

Efficient, economical, and environment-friendly control of *Alternaria* blight may be obtained through the knowledge of its timing of attack in relation to weather factors, which may enable prediction of its occurrence so as to allow growers to take timely action in an efficient manner for crop management. Weather is an exceptionally important factor in the severity of *Alternaria* blight of oilseed Brassicas. Preliminary work indicates effects of temperatures, relative humidity (RH), and sunshine hours on the occurrence of the blight on the oilseed Brassicas (Chattopadhyay 2008). These reports indicate relationships between different weather factors and *Alternaria* blight occurrence through empirical models. Severity of *Alternaria* blight on leaves and pods was higher in later sown crops (Chattopadhyay et al. 2005). Gupta et al. (2003) found positive correlation between sowing date of rapeseed–mustard crop and *Alternaria* blight disease severity. A delayed sowing results in coincidence of the vulnerable growth stage of plants with warm (maximum temperature: 18°C–26°C; minimum temperature: 8°C–12°C) and humid (mean RH > 70%) weather. Initiation of *Alternaria* blight disease on leaves of mustard occurred during 36–139 DAS, highest being at 45 and 75 DAS. Initiation of the disease on pod occurred between 67 and 142 DAS, highest being at 99 DAS. Severity of *Alternaria* blight disease on leaves was favored by a maximum temperature of 18°C–27°C in the preceding week, minimum temperature of 8°C–12°C, mean temperature >10°C, >92% morning RH, >40% afternoon RH, and mean RH of >70%. Disease severity on pods was positively influenced by 20°C–30°C maximum temperature, >14°C mean temperature, >90% morning RH, >70% mean RH, >9 h sunshine, and >10 h of leaf wetness (Chattopadhyay et al. 2005). The regional and crop-specific models devised thereby could predict the crop age at first appearance of *Alternaria* blight on the leaves and pods, the peak blight severity on leaves and pods, and the crop age at peak blight severity on leaves and pods at least 1 week ahead of first appearance of the disease on the crop, thus allowing growers to take necessary action (Chattopadhyay et al. 2005). Sporulation of *A. brassicae* has been reported to be favored by darkness (Humpherson-Jones and Phelps 1989).

Survival of the pathogen on diseased seed or affected plant debris in tropical or subtropical India (Mehta et al. 2002) and in Nepal (Shrestha et al. 2003) has been ruled out, although under temperate conditions, the pathogen is known to survive noncrop season on seeds and infected crop debris (Humpherson-Jones and Maude 1982). *A. brassicae* was observed predominantly in the seed coat and rarely in embryos of *Brassica campestris* var. toria and *B. juncea* (Shrestha et al. 2000). In Indian subcontinent, oilseeds *Brassica* are sown from late August to November depending on the crop, prevailing temperature, and availability of soil moisture for seed germination. Crop harvest occurs from February to May. Off-season crops are grown in nontraditional areas from May to September and this, coupled with harboring of the fungal pathogen by vegetable *Brassica* crops and alternative hosts (*Anagallis arvensis*, *Convolvulus arvensis*), could be a reason for carryover of the *A. brassicae* from one crop season to another (Mehta et al. 2002). Thus, airborne spores of *A. brassicae* form the primary source of inoculum of this polycyclic disease (Kolte 1985). During the crop season, the pathogen(s) could have several cycles, whereby it is known as a polycyclic disease. Germ tube from germinated spores of *A. brassicae*, *A. brassicicola*, *A. raphani*, and *Alternaria alternata* generally penetrate the undamaged tissues of many brassicaceous hosts directly (Kolte 1985, Goyal et al. 2013a), although indirect penetration through stomata has been reported in *A. brassicae* (Kolte 1985, Goyal et al. 2013b). Black spot lesions develop within 48 h after inoculation. According to Tewari (1986), *A. brassicae* in rapeseed becomes subcuticular after direct penetration. This is followed by colonization of epidermal and the mesophyll cells. In the leaves of rapeseed, the pathogen heavily colonizes the necrotic center and is not present in the chlorotic area indicating that a diffusible metabolite may be directly or indirectly responsible for leaf chlorosis. The plasma membrane is the first target of the diffusible metabolites. Subsequently, the chloroplasts are directly or indirectly affected leading to leaf chlorosis. Recently, Goyal et al. (2013b) found that conidia of *A. brassicae* germinated on the upper epidermis of *B. juncea* leaf, by producing one or several germ tubes and penetrating the host directly without the formation of appressorium. The mycelia ramified, colonized mesophyll and palisade tissue caused necrosis of the cells by producing toxins or metabolites that resulted in the formation of necrotic spots and reduction in photosynthetic area of different plant parts. This infection also decreased the amount of all the cell constituents like lignin, lipids, suberin, and protein, except phenolic compound in all the tissues of *Alternaria*-infected *B. juncea* leaves as compared to healthy leaves.

Disease Management

Host Plant Resistance

Several sources of tolerance against *Alternaria* blight have been reported (AICRP-RM 1986–2014; Chattopadhyay and Séguin-Swartz 2005). A short stature *B. juncea* cv. Divya was found tolerant to *Alternaria* blight (Kolte et al. 2000). Among the different species of oilseed, *Brassica*, *B. juncea*, and *B. rapa* are more susceptible than *B. carinata* and *B. napus*. Lines found tolerant to the disease in *B. juncea* include PAB-9511, PAB-9534, JMM-915, EC-399296, EC-399301, EC-399299, EC-399313, PHR-1, PHR-2, Divya, PR-8988, PR-9024, and RN-490; those in *B. carinata* are HC-1, PBC-9221 (Kiran), NRCDR-515, and DLSC-1; the ones of *B. napus* happen to be PBN-9501,

PBN-9502, PBN-2001, and PBN-2002 (Kolte 1985, AICRP-RM 1986–2014, Patni et al. 2005b, Kolte et al. 2006, Kumar and Kolte 2006). Sources of resistance to *A. brassicae* have been spotted in wild Brassicas, namely, *Brassica alba* (Kolte 1985, Chattopadhyay and Séguin-Swartz 2005), *Camelina sativa*, *Capsella bursa-pastoris*, *Eruca sativa*, *Neslia paniculata* (Chattopadhyay and Séguin-Swartz 2005), *Brassica desnottesii*, *Coincya pseuderucastrum*, *Diplotaxis berthautii*, *Diplotaxis catholica*, *Diplotaxis cretacea*, *Diplotaxis erucoides*, and *Erucastrum gallicum* (Sharma et al. 2002). Three cultivars, namely, Yajima kabu, Saishin, and Shimofusa kabu of *B. rapa*, were identified as resistant against *A. brassicicola* on pods, and these are useful not only for breeding programs for *B. rapa* but also for *B. napus*, a derivative from *B. rapa* and *B. oleracea*, and has little genetic variation due to the limited size of the descendent population (Doullah et al. 2009). A breeding program using number and size of lesions to find the differential response of different genotypes against *Alternaria* blight has been proposed to genetically enhance the level of resistance by Yadav et al. (2008). Resistance to *Alternaria* blight of rapeseed–mustard is found to be associated with factors like phenolic compounds, namely, polyphenol oxidase, PO, catalase in leaves, higher sugars and N, lower in resistant species (Chattopadhyay 2008) or discouragement to conidial retention on plant surface like high deposits of epicuticular wax that forms a physical barrier as a hydrophobic coating to reduce the deposition of waterborne inoculum, and reduce rate of conidia germination and germ-tube formation. *B. napus* (Tower, HNS-3), *B. carinata* (HC-2), and *B. alba* have more wax on plant/leaf surface compared to *B. rapa* (BSH-1, YSPB-24) and *B. juncea* (RH-30) (Chattopadhyay 2008). Two phytoalexins, namely, camalexin and brassinin, and two isothiocyanates (ITCs), namely, allyl- and benzyl-ITCs, were reported to be have antifungal activity at different developmental stages of *Alternaria* blight pathogens, namely, *A. brassicae* and *A. brassicicola* of crucifers (Sellam et al. 2007). Wild crucifers are found to elicit phytoalexins on challenge inoculation (Conn et al. 1988). Activities of some compounds related to camalexin ($C_{11}H_8N_2S$) and 6-methoxycamalexin ($C_{12}H_{10}N_2SO$) are found to be most toxic to *A. brassicae*. Phytotoxin destruxin B elicits phytoalexin response in *B. alba*. Parada et al. (2007) reported that destruxin B is not a host-selective toxin and does not induce accessibility of host plants to *A. brassicae*. Resistance to *A. brassicae* is found to be layered and multicomponent with sensitivity to host-specific toxin destruxin B, quantitative and qualitative elicitation of phytoalexins, hypersensitive reaction, and Ca sequestration determining the fate of host–pathogen interaction (Chattopadhyay and Séguin-Swartz 2005). The resistant *Brassica* varieties also produce some metabolites, namely, sesquiterpenes, deoxyuvidin B, albrassitriol, isoalbrassitriol, and brassicadiol (Saharan et al. 2003).

Molecular Breeding

Since resistance to *Alternaria* blight is governed by additive or polygenes, breeding for resistance to these diseases could involve pyramiding of minor genes, introgression of genes from material found resistant, reciprocal recurrent selection or diallel selective mating (Krishnia et al. 2000), wide hybridization (*B. alba*), molecular breeding (viz., from *C. sativa* by somatic hybridization; transgenic expressing *Trichoderma harzianum* endochitinase gene: Mora and Earle 2001), pollen culture, and sensitivity test to destruxin B. While in studies on the mechanism of tolerance to *Alternaria* blight some have indicated the effect of additive genes or polygene or cluster gene (Krishnia et al. 2000) with resistance being controlled by nuclear genes of partial dominance, there has also been indication of components of resistance being significantly correlated to each other regarding slow blighting (Kumar and Kolte 2001), and dominance (h) having a predominant role in genetic control of time of appearance; additive × dominance predominant for other disease progression factors, namely, area under the disease progress curve (Lakshmi and Gulati 2002, Chattopadhyay and Séguin-Swartz 2005, Meena et al. 2011a). The chitinase enzyme when overexpressed degrades the cell wall of invading fungal pathogens and plays an important role in plant defense response. Indian mustard, which has been transformed with chitinase gene tagged with an overexpressing 35S cauliflower mosaic virus (CaMV) promoter, showed delay in the onset of disease as well as reduction in number and size of lesions (Mondal et al. 2003). Transgenics of Indian mustard with barley antifungal genes class II chitinase and type I ribosome-inactivating protein, which coexpressed in plants, showed some resistance against *A. brassicae* infection through delayed onset of the disease and restricted number, size, and expansion of lesions as compared to wild plants (Chhikara et al. 2012). The pathogenesis-related (PR) proteins are toxic to invading fungal pathogens but are present in plant in trace amount. Thus, overexpression of these proteins may increase resistance to pathogenic fungi in several crops. Indian mustard plants transformed with class I basic glucanase gene showed restricted number, size, and spread of lesion caused by *A. brassicae*. This gene produces a PR protein glucanase that hydrolyzes a major cell-wall component, glucan, of pathogenic fungi and acts as a plant defense barrier (Mondal et al. 2007). Various technologies, namely, embryo rescue, somatic hybridization, somaclonal variations, genetic transformation, molecular markers, and signal transduction, have been used for incorporation of resistance against this pathogen in oilseed Brassicas by Aneja and Agnihotri (2013).

Rapid advances in techniques of tissue culture, protoplast fusion, embryo rescue, and genetic engineering make transfer of disease resistance traits across wide crossability barriers possible. A cDNA encoding hevein (chitin-binding lectin from *Hevea brasiliensis*) was transferred into *B. juncea* cv. RLM-198. Southern analysis of the putative transgenics showed integration of the transgene. Northern and Western analyses proved that the integrated transgene is expressed in the transgenics. In whole plant bioassay under glasshouse conditions, transgenics were found to possess parameters that are associated with resistance such as longer incubation and latent period, smaller necrotic lesion size, lower disease intensity, and delayed senescence (Kanrar et al. 2002). Insight has been gained into genes being expressed during *Alternaria* infection of *Brassica* (Cramer and Lawrence 2004). The authors used suppression subtractive hybridization between RNA isolated from the spores of *A. brassicicola* incubated in water and on the leaf surface of an ecotype of *Arabidiopsis thaliana* followed by cloning and sequencing of cDNA clones that were differentially expressed. One gene (P3F2), only expressed during infection, was identified, although its function remains to be determined. A similar approach with other pathogens could lead to advances in the understanding of pathogenicity.

Induced Host Resistance

Systemic acquired resistance (SAR) is induced by inoculation with avirulent race of *A. brassicae* (Vishwanath et al. 1999, Vishwanath and Vineeta 2007). Pretreatment of *Brassica* plants with beta-aminobutyric acid (BABA) induced resistance in plants and is thought to be mediated through an enhanced expression of PR protein genes, independent of salicylic acid (SA) and jasmonic acid accumulation (Kamble and Bhargava 2007). Some elicitors like benzothiadiazole (BTH) alone and in combination with SA may play a significant role in eliciting the defense-related enzymes, namely, PO, PAL, and superoxide dismutase (SOD) and phenolics, which may help in the reduction of disease severity by empowering the plant to restrict the invasion of *A. brassicae* in *B. juncea* (Sharma et al. 2008a, Sharma and Sohal 2010). Spray of SA on leaves increased the total sugar content but decreased the starch content in the leaves, which was linked with induction of disease resistance by maintaining a healthy flora of saprophytic microbes that are active against pathogens (Atwal and Sangha 2004b).

Mutation breeding may be one of the feasible techniques for breeding pathogen-resistant cultivars in the absence of a useful donor for resistance to the pathogen in the available germplasm of crops. Some *A. brassicicola*–resistant *B. napus* plants were regenerated from selected and unselected calli after mutation with gamma rays (physical) and ethyl methanesulfonate (chemical) mutagens (Sharma et al. 2012a).

Cultural Control

Hot-water treatment of seeds reduced the growth of *Alternaria* (Humpherson-Jones and Maude 1982). However, spores of these fungi can survive on the leaf tissue for 8–12 weeks and that on stem tissue till 23 weeks. Hence, fields that are replanted soon after harvest often coincided with a large amount of inoculum, which is likely to affect the crop's emergence and early growth stages (Humpherson-Jones 1992). Thus, rotation with noncruciferous crops and eradication of cruciferous weed hosts can help control these pathogens while fungicide spray in fields needs to be done at the same time. Early sowing (Meena et al. 2002) of well-stored clean certified seeds after deep ploughing at 45 cm row spacing (Kumar and Kumar 2006), clean cultivation, timely weeding and maintenance of optimum plant population, avoidance of irrigation at flowering, and pod formation stages may help manage the disease. Sowing of seeds should not be done by broadcasting method, because it increases disease severity, which could be decreased on leaves and pods by applying 40 kg K/ha along with the recommended dose of nitrogen (N) or sulfur (40 kg/ha) or along with recommended NPK (Kumar and Kumar 2006). However, a higher dose of N makes the crop susceptible to disease. Soil application of K as basal has been found to check *Alternaria* blight disease in mustard (Khatun et al. 2011). It has been reported that sulfur, zinc, and boron decrease the development of *Alternaria* blight and increase seed yield of mustard crop (Khatun et al. 2010). Application of some micronutrient, namely, B at 1 g/L, Mo at 1 g/L, S at 2 g/L, and Zn at 2 g/L, in various combinations reduced *Alternaria* blight disease and increased yield of rapeseed–mustard (Mondal 2008). Inorganic fertilizers, namely, phosphorus (P) and potassium (K), also decrease the disease, while N increases it (Singh 2004a). These are significant at a time when growers report increasing prevalence of *Alternaria* blight disease and decline in its control.

Biological Control

The GR isolate of *Trichoderma viride* was at par with mancozeb in checking blight severity on mustard leaves and pods (Meena et al. 2004). Conidial suspension of *T. viride* was more effective in comparison to culture filtrate in reduction of disease intensity on leaves and pods (Reshu and Khan 2012). *Bacillus subtilis* strain UK-9 isolated from reclaimed soil caused morphological alternations in vegetative cells and spores by disruption, lysis of cell wall of the pathogen, which resulted in reduction in disease severity, and spore germination on leaves (Sharma and Sharma 2008). Seed treatment with bioagents resulted in increase in lipid (phospholipids, glycolipids, and sterol) and protein content in seeds from treated plants. However, seed treatment and foliar spray with bioagents on leaves of Indian mustard enhanced the content of dry matter, total phenol, *ortho*-dihydroxyphenols, starch, total soluble sugars, reducing sugars, total lipids, and different membrane lipids in the leaves but the total protein content decreased after treatment with biocontrol agents at 30 and 60 DAS, which could be associated with defense mechanisms and enhanced growth of the plants (Sharma et al. 2010a,b).

Effect of Plant Extracts

Extracts of several plants have been evaluated against *A. brassicae* (Patni and Kolte 2006, Bhatiya and Awasthi 2007, Meena and Sharma 2012a, Sasode et al. 2012). The level of efficacy of *Azadirachta indica* extract increases as the number of sprays increases (Mohiddin et al. 2008). Spray of garlic bulb and neem leaf extract at flowering stage suppressed disease incidence (DI) and increased yield of mustard crop (Ferdous et al. 2002). Application of 1% (w/v) aqueous bulb extract of *Allium sativum* at 45 and 75 DAS in checking the disease severity on leaves and pods was at par ($P < 0.05$) with mancozeb as also in highest seed yield (Meena et al. 2004, 2008, 2011b, 2013, Yadav 2009). Two foliar sprays of *Eucalyptus globosus* at 2% (w/v) at 75 and 90 DAS could be done for eco-friendly management of black spot disease of rapeseed–mustard (Chandra et al. 2009). Foliar spray of extract of *Calotropis procera* leaves, *A. indica* kernel, and *A. sativum* bulbs may induce resistance against *A. brassicae* by increasing soluble phenol, sugar content, and soluble proteins, namely, PO, polyphenol oxidase, and PAL content in mustard leaves (Surendra et al. 2012). Among several essential oils evaluated, that of *Mentha piperita* provided complete inhibition of fungal growth at 2000 μg/mL, followed by oil of *Cyperus scariosus* (Dhaliwal et al. 2003).

Chemical Control

Spray with iprodione (Rovral) was effective in checking silique infection due to *A. brassicae*. Both reduction in disease and increase in seed yield and test weight were observed by the application of iprodione (Chattopadhyay and Bhunia 2003, Alam et al. 2010), and its residues in the edible parts of plants were lower than the maximum residue level that indicated the safety of this fungicide at the recommended rate (Mukherjee et al. 2003). Higher number (3–4) of iprodione sprays exerted significant reducing effect on the number of spots per siliquae (Hossain and Rahman 2006). Nowadays, there is a need to adopt new molecules of fungicides for the control of such pathogen keeping in mind their fungicidal resistance. Mycelial growth, conidial germination, and germ-tube elongation revealed the existence of *A. brassicicola* isolates highly resistant (EC_{50} > 100 mg/L) to both dicarboximides (e.g., iprodione and procymidone) and phenylpyrroles (e.g., fludioxonil) (Vasilescu et al. 2004). Application of fungicides on seeds reduced the content of two *Alternaria* toxins, namely, alternariol and alternariol methyl ester (Gwiazdowski and Wickiel 2011). Benlate (Anwar and Khan 2001), Contaf (Singh and Maheshwari 2003), and mancozeb (Meena et al. 2004) were effective in reducing disease severity on leaves and increasing seed yield in mustard. Two consecutive foliar sprays of mancozeb 75 WP (0.2%) followed by one spray of metalaxyl + mancozeb (ridomil MZ 72: 0.25%) resulted in high-seed yield and 1000-seed weight (Singh and Singh 2006). The highest net profit as well as the highest cost–benefit has been obtained with carbendazim/zineb (1:3.2) combination followed by carbendazim/captan (1:1.3) combination (Khan et al. 2007c). Seed treatment with carbendazim and foliar spray of metalaxyl + mancozeb (ridomil MZ 72 WP) was found most effective in reducing disease severity and in increasing seed yield (Prasad et al. 2009b). Exposure of *Alternaria* blight–affected leaves to high concentration (214.5 $\mu g/m^3$) of SO_2 resulted in suppression of the disease (Khan and Khan 2010).

WHITE RUST

The most widely recognized fungal species, *Albugo candida* (Pers.) Roussel, had been thought to be the exclusive white rust pathogen of the Brassicaceae, infecting as many as 63 genera and 241 plant species (Choi et al. 2009). According to the USDA-ARS Systematic Botany and Mycology Laboratory, *A. candida* was recorded on more than 300 hosts (Farr et al. 2004). Only recently was it realized that a high degree of genetic diversity is present within *Albugo* on Brassicaceae (Voglmayr and Riethmüller 2006, Choi et al. 2008) and that several of the observed lineages might constitute distinct species (Choi et al. 2011). The host specificity of *A. candida* has been recorded from Australia (Kaur et al. 2008a), Canada, Germany (Kolte 1985), India (Saharan et al. 2005), Japan, Romania, and the United States (Kolte 1985). Following the recent lectotypification of *A. candida*, the taxonomic status of which had previously been unclear (Choi et al. 2007), two specialized *Albugo* species parasitic to Brassicaceae have been described within *Albugo* (Choi et al. 2007, 2008). It was also demonstrated that *A. candida* has a broad host range extending over more than a dozen genera of the Brassicaceae and into the Cleomaceae, as the type of *Albugo chardonii* W. Weston (Kolte 1985) was found to be nested within *A. candida* (Choi et al. 2007). Capers (*Capparis spinosa*) are affected by white blister rust attributed to *Albugo capparidis* or, applying a broad species concept, to *A. candida* (Choi et al. 2009).

Symptoms

Disease appearing on leaves is characterized by the appearance of white or creamy yellow–raised pustules up to 2 mm in diameter, which later coalesce to form patches. The pustules are found scattered on the lower surface of the leaves. The part of upper surface corresponding to the lower surface is tan yellow, which enable recognition of the affected leaves. After the complete development of the pustule (Figure 5.3), it ruptures and releases a chalky dust of spores (sporangia). With aging of white rust pustules, affected leaves become senescent when necrosis around or in the pustule can

FIGURE 5.3 White rust on leaves.

FIGURE 5.4 Staghead (hypertrophied inflorescence) caused by *Albugo*.

be seen. Such rust pustules are also observed on the surface of well-developed siliquae. These are noted in local infection. Unlike other crucifers (Mundkur 1959), thickening or hypertrophy of the affected leaves is usually not seen in rapeseed–mustard. However, in systemic infection or infection through stem or flower, hypertrophy and hyperplasia are observed, which result in the formation of stagheads (Figure 5.4). Affected flowers become malformed, petals become green like sepals, and stamens may be transformed to leaf-like club-shaped sterile or carpelloid structures, which are found to persist on the flower rather than falling early as in normal plants. Ovules and pollen grains are usually atrophied leading to complete sterility. Association of symptoms of downy mildew with that of white rust is frequented. Whole plant infection at very early plant growth stage due to systemic infection is stunted, thickened with no branching, and beared white rust pustules on surface. Thickening of stem may be due to the modification of cortex into large thin-walled cells with fewer intercellular spaces. In floral parts also, there is an increase in size and number of cells of parenchymatous tissue with few intercellular spaces, lesser differentiation of tissue and organs, and increased accumulation of nutrients. Multiplication and spore production by the pathogen result in consumption of the accumulated nutrients leading to collapse and death of cells, drying of affected plants (Kolte 1985).

Geographical Distribution and Losses

White rust caused by *A. candida* (Pers. ex. Fr.) Kuntz. can result in yield loss up to 47% on oilseeds *Brassica* (Kolte 1985) and 89.9% in *B. juncea* (Varshney et al. 2004) with each percent of disease severity and staghead formation causing reduction in seed yield of about 82 and 22 kg/ha, respectively (Meena et al. 2002). The highest avoidable yield loss up to 28.2% with highest disease intensity, namely, 70.8%, has been reported in late sown *B. juncea* var. Varuna (Singh and Bhajan 2005). The disease has been reported from Australia, Austria, Brazil, Canada, China, Egypt, Fiji, Finland, Germany, India, Iran, Japan, Korea, the Netherlands, New Zealand, Pakistan, Palestine, Poland, Portugal, Romania, Saudi Arabia, Sweden, Spain, Turkey, Venezuela, the United Kingdom, and the United States (Meena et al. 2014).

Pathogen

The obligate parasite *A. candida* or *Cystopus candidus* causes white rust of oilseeds *Brassica* and several other crucifers (Choi et al. 2009). Mycelium is aseptate, intercellular with nuclei-free globular haustoria (Coffey 1975). Masses of mycelia beneath host epidermis form a palisade of cylindrical-shaped sporangiophores, which are thick walled at base and free laterally. The sporangiophores give rise to chains of spherical hyaline, smooth, and 12–18 μ diameter sporangia in a basipetal succession, which germinate to give rise to concave biflagellate zoospores or at times to germ tubes (Walker 1957). Temperature most favorable for the germination of sporangia is 10°C (Khunti et al. 2004). Once zoospores are released from the sporangium, they exhibit chemotactic, electrotactic, and autotactic or autoaggregation responses to target new hosts for infection (Walker and West 2007). Thereafter, zoospores come to rest, retract their flagella, and encyst and germinate by the formation of a germ tube. If germination occurs on a susceptible host, then the germ tube penetrates through stomata to form an intercellular mycelium (Walker 1957). Oogonia and antheridia are formed from the mycelium in intercellular spaces, particularly in systemically affected plants (Webster 1980). Oogonia are globose, terminal or intercalary, each containing up to 100 nuclei; its contents defined into a peripheral zone of periplasm and a single central oosphere. Antheridia are clavate, each with 6–12 nuclei on the sides of an oogonium (Heald 1926, Walker 1957). The heterothallic fungus produces restings spores or oospores, which are highly differentiated with five-layered cell wall and at maturity are tuberculate, 40–55 μ in diameter. Germination of oospores has been described by de Bary (1887) and Vanterpool (1959). Germination by sessile vesicles was the most common. Treatment of oospores with 200 ppm $KMnO_4$ for 10 min induced increased germination (Verma and Bhowmik 1988). Oospores do not appear to require any dormancy period. Meena and Sharma (2012b) used animal gut enzymes (1% β-glucuronidase and arylsulfatase, Sigma make) for the germination of the oospores from hypertrophied plant tissue. Physiological specialization of the pathogen has also been reported. Zoospores produced from germinating oospores constitute the primary source of inoculum for the infection of rapeseed–mustard (Verma and Petrie 1980), particularly when mixed with seeds. There appears to be no dormancy in the case of oospores, as they have been found to germinate just 2 weeks after their collection from affected tissues. The most likely primary infection site is the emerging cotyledon. The production of large masses of zoosporangia on cotyledons seems to require establishment of a large mycelial base inside the host tissue, and in the *Albugo–Brassica* system, such a base apparently develops with a minimum disturbance of the host's synthetic abilities (Harding et al. 1968). The sporangia become visible after the epidermis is ruptured, as a white powdery mass, which can readily be dispersed by wind to cause secondary infection. If sporangia alight on a suitable host leaf or stem surface, they are capable of germinating within a few hours in films of water to form biflagellate zoospores, about eight per sporangium. After swimming for a time, a zoospore encysts and then forms a germ tube, which penetrates the host epidermis. A further crop of sporangia may be formed within 10 days. The establishment and maintenance of a compatible relationship between *A. candida* and its hosts seem to hinge on the successful formation of the first haustorium. In the susceptible host, such as *B. juncea*, the first haustorium forms within 16–18 h after inoculation. Haustoria are small and capitate with spherical heads averaging 4 μ in diameter and are connected to hyphae by slender stalks about 2 μ in length. A haustorium usually originates near the tip of a young hypha, which then continues its growth leaving the haustorium as a side branch. After the formation of the first haustorium in the susceptible host–parasite combination, hyphal growth rate increases rapidly. An encapsulation similar to that observed in *Raphanus sativus* is seen only infrequently around haustoria in a virulent *Albugo*-susceptible *B. juncea* system. In a susceptible host, the hyphae appear to grow around palisade mesophyll cells as a downward spiral, penetrating individual cells with a variable number of haustoria. Verma et al. (1975) found the presence of as many as 14 haustoria in a single cell in "green island" tissue of artificially infected *B. juncea* cotyledons. In the susceptible host, most of the intercellular spaces appear to get occupied by mycelium within 3 days after inoculation. Some biochemical compounds, namely, total phenols, total sugars, and reducing and nonreducing sugars, are generally negatively correlated with DI, but with low disease levels, these are not always consisent (Singh 2005, Mishra et al. 2009). Phenol content reduced more in infected inflorescences than leaves and cotyledons, while free amino acids increased more in infected leaves than inflorescences and cotyledons. But total mineral contents did not show significant variation in both infected and healthy plant parts. However, total chlorophyll content decreased in cotyledons and leaves and increased in the malformed inflorescences (Singh 2005). Studies on quantitative changes in amino acid content of white rust–induced hypertrophies of the mustard plant indicated possible breakdown of protein due to pathogen to release tryptophane and subsequently increasing IAA content of such tissues. However, decrease in IAA, free proline, total proteins, and phenolic compounds in the infected host tissue are also reported. One protein, peptidyl-prolyl *cis-trans* isomerase (PPIase) isoform CYP20-3, was detected only in the susceptible variety and increased in abundance in response to the pathogen. PPIases play an important role in pathogenesis by suppressing the host cell's immune response (Kaur et al. 2011a). Cellulase, endo-polymethyl galacturonase (PMG), and endo-PGs were produced in *B. juncea* leaves infected with *A. candida*. Swelling and disruption of subcellular particles rich in lysomal acid hydrolases were produced by acid phosphatase activity centered primarily in the infected tissues of *B. juncea*. Acid phosphatase activity in antheridia, oogonia, and oospores of *A. candida* indicates that this enzyme plays a role in the synthesis of fungal organs. Increase in PO (Jain et al. 2009), invertase, alpha-amylase, IAA oxidase (Singh et al. 2011a), PAL, TAL, and lipoxygenase (LOX) (Jain et al. 2002) enzyme activity has been reported in infected leaves. However, nitrate reductase enzyme activity increased in infected cotyledons, leaves, and inflorescences (Singh 2005). Erucic acid content is positively correlated with disease infection, which indicates that low-erucic lines are resistant, medium-erucic lines are tolerant, and high-erucic lines are highly susceptible to the disease (Malik et al. 2004). Beta-1,3-glucanase was found to be induced more in the resistant cultivars of *B. juncea* after inoculation with *A. candida* (Kapoor et al. 2003). Infection of the resistant lines with *A. candida* showed significant increase in the glucosinolates as compared to susceptible lines (Pruthi et al. 2001). Much work remains to be done on mutual interaction of *Albugo* and associated microorganisms. Some of the *Alternaria* species are known to produce toxins and these could conceivably have an adverse effect on the survival of *A. candida*. Pathogenic and genetic diversity among different *A. candida* isolates collected from *B. juncea*, *B. rapa*, *B. oleracea*, *B. tournefortii*, *Raphanus raphanistrum*, *R. sativa*, *E. vesicaria* subsp. *sativa*, *C. bursa-pastoris*, and *Sisymbrium irio*, from different locations in western Australia, has been reported (Kaur et al. 2011b).

Under Indian conditions, the *A. candida* isolate obtained from *B. rapa* was distinct in pathogenicity from the one obtained from *B. juncea* (Kolte et al. 1991). Verma et al. (1999) identified two new races of *A. candida* in India, namely, race 12 from *B. juncea* and race 13 from *B. rapa* var. *toria*, using 14 crucifer host differentials. Twenty distinct pathotypes of *A. candida*, 17 from *B. juncea* (AC 18–AC 34), 2 from *B. rapa* var. *brown sarson* (AC 35–AC 36), and one from *B. nigra* (AC 37), have been identified (Jat 1999). Four distinct and new pathotypes of *A. candida*, namely, AC I4 from RL 1359, AC 15 and AC 16 from Kranti, and AC 17 from RH 30 cultivars of *B. juncea*, have been identified on the basis of their differential interactions on 11 host differentials by Gupta and Saharan (2002). Out of these, pathotypes like AC 23, AC 24, and AC 17 infect only one, two, and three host differentials, respectively, indicating a limited virulence potential, while pathotypes of wider virulence, namely, AC 29, AC 27, AC 30, AC 18, and AC 21 infected 21, 18, 16, 12, and 10 host differentials, respectively (Jat 1999, Gupta and Saharan 2002).

Epidemiology and Disease Cycle

Berkenkamp (1980) reported an increase in DI due to soil application of trifluralin herbicide. The optimum temperature for disease development ranged 12°C–18°C. Only 3 h of wetness was required for disease development at 12°C–22°C. First appearance of white rust disease (*A. candida*) on leaves and pods (staghead formation) of mustard occurred 36–131 DAS, highest being at 50 and 70 DAS and 60 and 123 DAS, respectively. Severity of white rust disease on leaves was favored by >40% afternoon RH, >97% morning RH, and 16°C–24°C maximum daily temperature. Staghead formation was significantly and positively influenced by 20°C–29°C maximum daily temperature, further aided by >12°C minimum daily temperature, and >97% morning RH. It was possible to predict the highest severity of white rust disease of the crop season in the initial weeks after sowing with the models developed by stepwise regression (Chattopadhyay et al. 2011). By using hourly weather data, a simple weather-based forewarning model to evaluate DI has been developed (Kumar and Chakravarty 2008). Oospores formed in infected plants overwintered in plant debris and soil, which function as the source of primary inoculum of the pathogen (Butler and Jones 1961, Verma et al. 1975). Oospores have also been observed in naturally infected senesced leaves of *B. juncea* and *B. rapa* var. toria. Oospores can remain viable for over 20 years under dry storage conditions (Verma and Petrie 1975). Possibility of survival and spread of the pathogen by means of oospores, sporangia, and mycelia carried externally on seeds have been reported (Petrie 1975, Meena et al. 2014). The oospores germinate by releasing zoospores, which infect lower leaves during crop season. Secondary infection of the pathogen occurs through sporangia produced on leaves.

Disease Management

Host Plant Resistance

Though a few sources of host resistance ([*B. juncea*: PWR-9541, JMMWR 941-1-2, PAB-9534, PAB-9511, PHR-1, PHR-2, EC-129126, EC-399299, EC-399301, EC-399300, EC-399296, BIO YSR] [*B. rapa*: PT-303, Tobin] [*B. carinata*: HC-1, 2, 3, 4, 5, NRCDR-515, PBC-9921, BC-2, DLSC-1] [*B. napus*: TOWER, EC-33897, EC-339000, DGS 1, GS-7055, HNS-4, GSL-441, PBN-2001, PBN-2002] [*E. sativa*: RTM-1471] [*B. alba*: Exotic-1, Exotic-2]) have been identified (Kolte 1985, AICRP-RM 1986–2014, Chattopadhyay and Séguin-Swartz 2005, Kumar and Kalha 2005), their success is limited to a few pockets keeping in view the volatile race pattern of the pathogen. Three cultivars of Indian mustard PBR 181, EC-399301, and EC-399299 and two cultivars of *B. campestris* were found resistant against white rust (Yadav and Sharma 2004). Meena et al. (2005) also found 38 genotypes free from white rust infection, out of 90 genotypes of 5 oileferous Brassicas, that is, *B. juncea*, *B. napus*, *B. carinata*, *B. campestris,* and *E. sativa* in India. In the Terai region of Uttar Pradesh (India), 5 genotypes, namely, WRR-98-01, NDRS-2004, NDRS-2013, NDRS-2005, and NDRS-2007, were found disease free, and 18 genotypes, namely, RC-781, RIK-75-5, RIK-78-4, CSR-721, PI-43, YRT-3, NDRE-7, NNRE-10, NDYR-29 × NDRE-04, NDRE-190 × NDRE-4, CSCN-5, CSCN-3, CSCN-10, CSCN-12, NSRS-2006, NDRS-2009, NDRS-12, and NDRS-2014, were found resistant (Sinah and Mall 2007). Yellow-seeded mustard (*B. juncea*) variety *T4*, YRT-3184 and rapeseed (*B. rapa* var. yellow sarson), type 6 (Chattopadhyay and Séguin-Swartz 2005) have been reported to be resistant to white rust infection. Three Australian genotypes JM 018, JM 06021, and JM 06026; JR 049 and JN 033; and some Chinese genotypes, namely, RK 2, Ringot, RH 13, Amora III, Quianxianjiecai, Yilihuang, Hatianyoucai, Jinshahuang, Manushuang, *B. juncea* 1, 2, and 3, were resistant to white rust (Singh et al. 2010). Some lines of *B. juncea* var. Cutlass, namely, RESJ-1052, RESJ-1004, RESJ-1005, RESJ-1033, and RESJ-1051, have been found to be resistant to all the Indian isolates as well as 2 V (Canadian isolate). These resistant sources with combined resistance to different white rust isolates could be putative donors for further oilseed *Brassica* crop improvement programs (Awasthi et al. 2012).

Transfer of white rust resistance in rapeseed–mustard from *B. carinata* to *B. juncea* could be partially successful by growing disease-free plants under high disease pressure followed by their repeated backcrossing with *B. juncea* cultivar (Singh et al. 1988). Resistance to white rust in rapeseed–mustard is dominant, governed by one or two genes with either dominant-recessive epistasis or complete dominance at both gene pair but either gene when dominant is epistatic to the other. These genes could be located on the same locus or different loci (Kumar et al. 2002). Resistance to the disease at true leaf infection and susceptibility at the cotyledonary leaf stage of the same genotype EC-399301 of *B. juncea* appears to be governed by two independent genes. Hence, screening for white rust resistance at the cotyledonary leaf stage needs to be carefully considered (Mishra et al. 2009). Interspecific crosses between *B. juncea* and *B. napus* suggested that resistance in WW-1507 and ISN-114 to *A. candida* was controlled by a single dominant gene (Jat 1999). In their study of three interspecific crosses between *B. juncea* and *B. napus,* Subudhi and Raut (1994) revealed digenic control with epistatic interaction for white rust resistance trait and a close association of parental species and different grades of leaf waxiness. Sachan et al. (1995), in their study using diallel crosses between two white rust–resistant Canadian *B. juncea* cvs. Domo and Cutlass and two susceptible *B. juncea* Indian cvs. Kranti and Varuna, reported that F1 hybrids, except susceptible × susceptible, were resistant; segregation pattern for resistance in F2 and test crosses was under the control of a single dominant gene in Domo and Cutlass, and that a recessive gene for susceptibility was present in Kranti and Varuna. Sridhar and Raut (1998) reported a monogenic inheritance showing complete dominance in four crosses and lack of dominance in seven crosses attempted between *B. juncea* and resistance sources derived from different species. According to Jat (1999), the resistance was dominant in all the crosses except susceptible × susceptible, where it was recessive. Under controlled conditions, inoculation with three different races of *A. candida* on F2 population of crosses from resistant × resistant revealed that the resistant genes may be located on the same locus or on different loci. Partial resistance in *B. napus* to *A. candida* was controlled by a single recessive gene designated as *wpr* with a variable expression (Bansal et al. 2005). White rust resistance in *B. juncea* (Somers et al. 2002, Manjunath et al. 2007, Singh et al. 2012, Yadava et al. 2012) and avirulence in race AC 2 of *A. candida* to *B. rapa* cv. Torch (Adhikari et al. 2003) is governed by single dominant gene. The resistance of *B. napus* var. *Regent* is conditioned by independent dominant genes at three loci designated as AC 7-1, AC 7-2, and AC 7-3. Two loci also controlled resistance in *B. napus* to *A. candida* race AC 2 collected from *B. juncea*. The Chinese *B. napus* accession 2282-9, susceptible to AC 7, has one locus controlling resistance to an isolate of *A. candida* collected from *B. carinata*. These studies indicated that only one allele for resistance was sufficient to condition an incompatible reaction in this pathosystem. In addition, a single locus controlling resistance to AC 2 in *B. napus* and *B. rapa* was mapped using restriction fragment length polymorphism (RFLP) marker (Chattopadhyay 2008). A dominant allele at a single locus or two tightly linked loci were reported to confer resistance to both races AC 2 and AC 7 of *A. candida* (Kole et al. 2002). According to Borhan et al. (2008), a dominant white rust–resistant gene, *WRR* 4, encodes a toll-interleukin receptor–nucleotide-binding site–leucine-rich repeat protein that confers broad-spectrum resistance in *A. thaliana* to four races (AC 2, AC 4, AC 7, and AC 9) of *A. candida.* Four Chinese (CBJ-001, CBJ-002, CBJ-003, and CBJ-004) and two Australian (JR049 and JM 06011) genotypes were consistently highly resistant to an *A. candida* pathotype prevailing in Australia throughout the different plant growth stages (Li et al. 2007a, 2008a, 2009a).

Molecular Breeding

White rust susceptible cultivars of *B. juncea* and *B. napus* transformed with WRR4 gene from *A. thaliana* showed resistance to the corresponding *A. candida* races for each host species, which indicates that this gene could be a novel source of white rust resistance in oilseed Brassicas (Borhan et al. 2010). However, there is a need to guard against the danger of breakdown of resistance due to mixed infection with *Hyaloperonospora parasitica* (Singh et al. 2002a,b). Resistant genes have been mapped and identified on the chromosomes of *B. juncea,* namely, *ACr* (Cheung et al. 1998), *AC*-21 (Prabhu et al. 1998), *AC*-2 (Varshney et al. 2004), *ACB*1-*A*4.1, and *ACB*1-*a*5.1 (Massand et al. 2010); *B. rapa*, namely, *ACA*1 (Kole et al. 1996); *B. napus*, namely, *ACA*1 (Ferreira et al. 1994) and *AC 2V1* (Somers et al. 2002); and *A. thaliana*, namely, *RAC*-1, *RAC*-2, *RAC*-3, and *RAC*-4 (Borhan et al. 2001, 2008), effective against one or more than one race of *A. candida.* A single gene (*Acr*) responsible for conferring resistance to *A. candida* was mapped on a densely populated *B. juncea.* Two closely linked RFLP markers identified (X42 and X83) were 2.3 and 4 cM from the *Acr* locus, respectively (Cheung et al. 1998). Kole et al. (2002) have worked out the linkage mapping of genes controlling resistance to white rust in *B. napus.* A tightly linked marker for white rust resistance was developed using amplified FLP (AFLP) in conjunction with bulk segregant analysis (Varshney et al. 2004). A polymerase chain reaction (PCR)-based cleaved amplified polymorphic sequence (CAPS) marker for closely linked random amplified polymorphic DNA (RAPD) marker $OPB06_{1000}$ was developed. Data obtained on 94 recombinant inbred lines revealed that the CAPS marker for $OPBO6_{1000}$ and AFLP marker E-AAC/M-CAA_{350} flank the *Ac2*(t) gene at 3.8 and 6.7 cM, respectively. Validation of the CAPS marker in two different F_2 populations of crosses Varuna × BEC-144, and Varuna × BEC-286 established its utility in marker-assisted selection for white rust resistance. The use of both flanking markers in marker-assisted selection was estimated to only allow 25% misclassification, thus providing greater selection efficiency than traditional approaches (Varshney et al. 2004).

Induced Host Resistance

Tirmali and Kolte (2011, 2012) found nonconventional chemicals (plant defense activators) effective in reducing disease index on leaves and staghead incidence. A plant defense activator, BTH, was found effective in protection from staghead development against the challenge inoculation with *A. candida* (Kaur and Kolte 2001, Kumar 2009).

Cultural Control

Crop rotation with nonhost crops helps in managing this pathogen. Roguing and burning of disease-affected plants, particularly stagheads, help in minimizing inoculum buildup in soil. Overirrigation of crop should be avoided that helps in reducing DI. Clean, healthy, and certified seed should be used to avoid seed-borne white rust disease. Soil application of K as basal at 40 kg/ha resulted in significantly ($P < 0.05$) lesser white rust on leaves and number of stagheads than control. Early sowing of seeds may help decrease DI and staghead formation and increase seed yield (Yadav et al. 2002, Meena et al. 2004, Thapak and Dantre 2004, Biswas et al. 2007). Suitable date of sowing based on location and other epidemiological considerations needs to be decided to enable the escape of the disease in different locations.

Biological Control and Effect of Plant Extracts

In recent years, an increasing consciousness about environmental pollution due to pesticides and development of fungicide-resistant strains in plant pathogens have challenged plant pathologists to search for eco-friendly tools in disease management. Aqueous bulb extract of *A. sativum* 1% (w/v), an isolate of *T. viride*, as seed treatment and in combination as respective foliar sprays was statistically at par with that of mancozeb, combination of metalaxyl 35 ES 6 mL/kg seed treatment + 0.2 g/L spray of combination of metalaxyl + mancozeb in checking the rust severity on leaves and number of stagheads per plant (Meena et al. 2003). Inhibition of oospore development in *A. candida* by a natural bioagent *Psuedomonas syringe* under field condition has been reported (Tewari et al. 2000). Spray with the extract of *Eucalyptus* spp. (Kumar 2009) leaves can effectively manage the disease.

Chemical Control

Metalaxyl (Khunti et al. 2001, Biswas et al. 2007), metalaxyl + mancozeb (ridomil MZ) (Pandya et al. 2000, Godika et al. 2001, Yadav 2003), aluminium tris (Girish et al. 2007), and combination of metalaxyl 35 ES 6 mL/kg seed treatment + 0.2 g/L spray of combination of metalaxyl + mancozeb at 50 and 65 DAS (Kolte 1985, Meena et al. 2003) are reported to be able to manage the disease.

DOWNY MILDEW

Symptoms

Symptoms of the disease appear on all aboveground parts but usually on leaves and inflorescence. Usually a few days after sowing, small angular translucent light green lesions first appear on cotyledonary or the first true leaves during seedling stage and at times could be even restricted to these leaves with subsequently emerging ones not showing any symptom. Such lesions later enlarge and develop into grayish white, irregular necrotic patches on the leaves bearing downy growth of the pathogen (conidia and conidiophores) on its undersurface. In severe attack, the affected leaves dry up and shrivel. The extent to which the necrosis occurs depends upon the type of crop species. Leaf symptoms (Figure 5.5) at the seedling stage, as mentioned earlier, are more conspicuous on

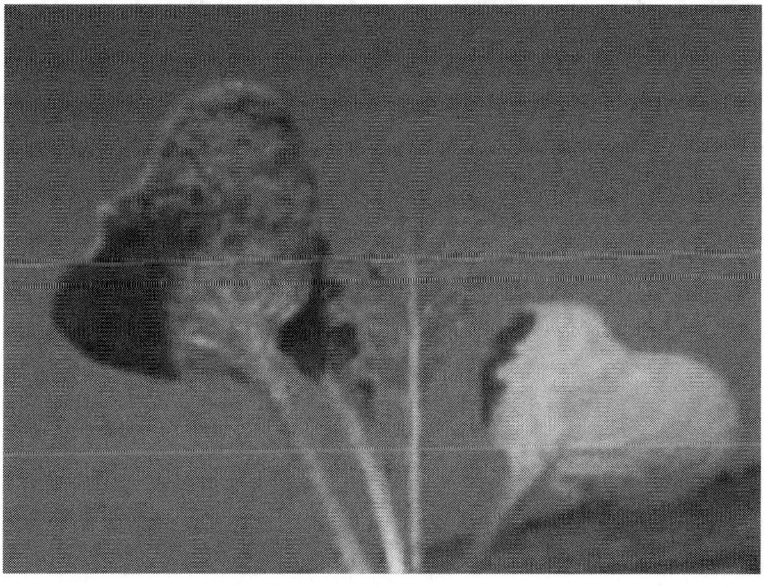

FIGURE 5.5 Downy mildew affected leaves at seedling stage.

B. juncea compared to *B. rapa*. Very late in the season, downy growth may be seen on siliquae as well (Figure 5.6). Thickening of the peduncle/inflorescence (Figure 5.7) due to the disease suggest hypertrophy of affected cells, pith of the stem being more affected than the cortex (Vasudeva 1958). Formation of oospores in the inflorescence takes place as it dries up. The disease is also found to be associated with white rust symptoms on leaves and inflorescence. Systemic infection results in

FIGURE 5.6 Downy mildew affected pods.

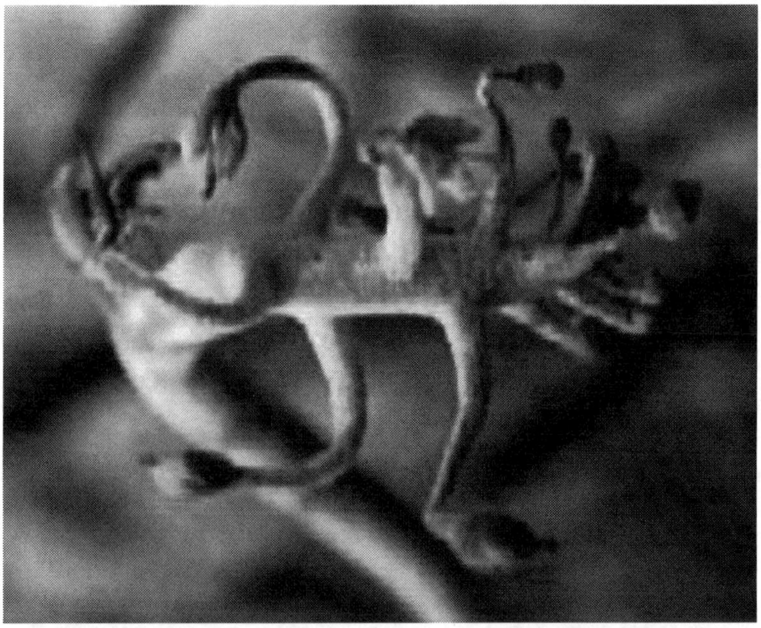

FIGURE 5.7 Staghead (hypertrophied inflorescence) caused by *Hyaloperonospora*.

thickened stunted growth of plant bearing profuse sporulation (Kolte 1985). In all the infected plant parts of oilseeds *Brassica*, total sugar, total phenol, total protein, and chlorophyll and nitrate reductase activity decreased, while chlorophyll content only in inflorescence and total free amino acids in the infected plant parts increased. However, total mineral content did not differ in infected and healthy plants (Singh 2000, 2004).

Geographical Distribution and Losses

The disease is found to appear more frequently in varying proportions, wherever rapeseed–mustard cultivation has been intensified. The disease has been reported from Argentina, Australia, Austria, Brazil, Bulgaria, Canada, Chile, China, Cuba, Cyprus, Czechoslovakia, Denmark, Ethiopia, Fiji, Finland, France, Germany, Greece, Hong Kong, Hungary, India, Iran, Iraq, Ireland, Israel, Italy, Jamaica, Japan, Kenya, Korea, Malaysia, Mexico, Nepal, the Netherlands, New Zealand, Norway, Panama, Pakistan, Philippines, Poland, Portugal, Romania, Russia, South Africa, Switzerland, Taiwan, Turkey, Uganda, the United Kingdom, the United States, Vietnam, and Yugoslavia (Kolte 1985, Saharan et al. 2005). Reports of its occurrence either alone (Porter 1926) or in association with white rust on leaves or inflorescence have been made, which could result in losses up to 58% (Kolte 1985). Seedling death could be even up to 75% when infection occurs at cotyledonary stage and congenial weather conditions are prevalent.

Pathogen

The causal pathogenic fungus *H. parasitica* (Pers.) Constant is an obligate parasite affecting all crucifers though variation exists in conidial size and other fungal structures among strains infecting different species of cruciferae. Mycelium is hyaline, coenocytic, remains intercellular in host, produces large, lobed intracellular haustoria, often branched, which nearly fill the entire cell. Erect conidiophores singly or in groups of determinate growth emerge vertically through the epidermis on the undersurface of the leaves through the stomata. Conidiophores are hyaline with a flattened base, stout main axis, twisted at a point crossing the stomata, and measure 100–300 μm. At the tip, conidiophores are dichotomously branched six to eight times, sterigmata slender, and acutely pointed. Conidia are hyaline, broadly elliptic to globose, 24–27 μm × 12–22 μm. A single conidium is borne at the tip of each branch, and the same is deciduous. Detachment of conidia is possibly caused by hygroscopic twisting of conidiophore related to changes in humidity. Conidiophore wall is uniformly thick. Spherical oogonia and tendril-like antheridia are developed on separate hyphae in hypertrophied tissue to produce oospores that enable the survival of the pathogen for long times withstanding harsh conditions. On germination of the oospore or conidia, the germ tube penetrates the host tissue directly or through the stomata. This pathogen, in the absence of fungal reproductive structures, during early interaction of seedlings, in infected young leaves packed in sealed plastic bags, and seed stocks can be diagnosed by a fast and reliable molecular identification technique, that is, multiplex PCR amplification of full internal transcribed spacer (ITS) and ITS2 regions of *H. parasitica* (Casimiro et al. 2004).

Epidemiology and Disease Cycle

The disease is favored by cool (8°C–16°C) and moist weather with low-light intensity and high (152 mm) rainfall (Kolte 1985). The oospores that survive in infected crop residue, soil, and seed serve as primary source of inoculum. These oospores germinate to infect cotyledonary and primary leaves. Systemic infection could result in staghead formation. Secondary infection of the plants occurs through airborne conidia or waterborne zoospores produced from germinating sporangia.

Disease Management

Host Plant Resistance

EC-129126 (*B. juncea*), PBN-9501, PBN-2002, and GSL-1 (*B. napus*) are reported to be resistant to the disease (AICRP-RM 1986–2014, Nashaat et al. 2004, Chattopadhyay and Séguin-Swartz 2005). However, there is danger of breakdown of resistance due to mixed infection with white rust. Seven genotypes, namely, *Sinapis alba*, *B. carinata* (HC-1), *B. juncea* (DIR-1507 and DIR-1522), and *B. napus* (GS-7027, Midas, and Tower) exhibited stable resistance to the disease (Dang et al. 2000). Genotypes RC 17, RC 346, RC 89, RC 110, and RC 280 were also resistant to the disease (Singh and Singh 2005). Australian genotypes JM06014 and JM06015 and Indian genotypes JM 3 and Kranti were resistant to downy mildew (Singh et al. 2010). Two Australian spring-type oilseed rape genotypes Pioneer 45Y77 and Pioneer 46Y78 were resistant to *H. parasitica* (Ge et al. 2008). Walters et al. (2005) observed three oxylipins, namely, TriHOE1, TriHOE2, and 13-HOT, in incompatible interaction between *B. napus* and downy mildew pathogen while these were not observed in compatible interaction, which indicates their involvement in signaling and/or as antimicrobial compounds in rapeseed-resisting infection by the pathogen. The resistance of some genotypes of rapeseed–mustard to Indian isolate of *H. parasitica* seems to be conditioned by a single dominant gene (Nashaat et al. 2004).

Cultural Control and Effect of Plant Extracts

Selective picking of affected hypertrophied racemes immediately after formation followed by their destruction; rotation with noncruciferous crops could also be helpful. Suitable planting time need to be worked out as per location. Seed treatment and successive foliar spray with garlic bulb extract was found an eco-friendly alternative to manage this disease (Kolte 1985, Bhatt et al. 2009).

Chemical Control

Seed treatment with metalaxyl 35 SD at 6 g/kg and spray of metalaxyl at 0.01% ai were effective in managing the disease (Chattopadhyay 2008). Gopal (2003) found that the seed treatment with metalaxyl along with its two sprays could reduce the DI significantly and increase the seed yield.

Sclerotinia ROT

Symptoms

Based on the symptoms, the disease has been named white blight, white rot, stem blight, stalk break, stem canker, or rape canker. Usually, under natural conditions, the stem of the plant is seen affected more frequently, though all aboveground parts are subject to attack by the disease. Symptoms on the stem become visible as elongated water-soaked lesions that later are covered by a cottony mycelial growth of the fungus (Figure 5.8). Infected plants are at times overlooked until the fungus grows completely throughout the stem to rot it. When the stem is completely girdled by such lesions, the plant wilts and dries. Foliage may show little sign of attack while at times may even start on leaves, which wilts and droops downward and then moves on to the stem. Sometimes, the infection is restricted to a smaller area of pith, which results in slow stunting of the plant and premature ripening rather than their sudden collapse. Such plants under field conditions can be easily identified because of premature ripening. The affected stem tends to shred; numerous grayish-white to black, spherical sclerotia appear either on the surface or in the pith of the affected stem. When the crop is at seed maturity, the plants tend to lodge, touching the siliquae with the soil level. Such plants, though remain free from stem or aerial infection throughout, show rotting of the siliquae with profuse fungal growth, along with sclerotial bodies just above the soil level. Appearance of the disease at an early stage of crop growth results in the death of whole plant (Kolte 1985). Sharma and Sharma (2001) reported significant reduction in seed germination percentage, radicle growth,

FIGURE 5.8 Sclerotinia rot-affected stem.

plumule growth, plant height, secondary branches, and number of silique on primary and secondary branches in Indian mustard cultivars in Haryana, India. Younger plants (up to of 40 days) are highly susceptible as compared to older ones (Ghasolia and Shivpuri 2009). After infection, Alizadeh et al. (2006) observed an increase in glucosinolate and erucic acid content, while the test weight and oleic acid decrease in rapeseed oil.

Geographical Distribution and Losses

Rot of mustard caused by *Sclerotinia sclerotiorum* (Lib.) de Bary has become important in recent times with high (up to 66%) DI and severe yield losses (up to 39.9%) leading to discouragement of growers of the crop (Chattopadhyay et al. 2003), although reports of even 100% yield loss due to the disease are available (Saharan et al. 2005). *Sclerotinia* rot is also a serious threat to oilseed rape production with substantial yield losses worldwide including Australia, Europe, India and North America (McCartney and Lacey 1999, Hind et al. 2003, Koch et al. 2007, Malvarez et al. 2007, Singh et al. 2008). There may be a great variation in losses in yield in the same area from year to year. Yield losses vary with the percentage of plants infected and the stage of growth of the crop at the time of infection. Plants infected at the early flowering stage produce little or no seeds and those infected at the late flowering stage set seed and may suffer little yield reduction. For predicting yield (Y), a linear equation (R^2: 0.89) was fitted on DI ($Y = 310.25 - 2.04^{**}\text{DI}$) using *Sclerotinia* susceptible cv. Rohini (Chattopadhyay et al. 2003). The disease affects broad-leaved crop species and is most common in temperate regions of the world. The first record of its occurrence on rapeseed and mustard appears to have been made from India (Shaw and Ajrekar 1915). Since then, frequent occurrences of the disease in severe form have been reported from Argentina (Gaetan and Madia 2005), Brazil, Canada (loss up to 28%), China, Denmark, Finland (Kolte 1985),

Florida (Young et al. 2012), France, Germany (Kolte 1985), Greece (Tziros et al. 2008), India (Kolte 1985), Italy (Corato and de Baviello 2000), Sweden (Kolte 1985), Texas (Isakeit et al. 2010), and the United Kingdom (Kolte 1985). The disease has been found to be causing severe losses in Rajasthan, Bihar, Uttar Pradesh, Uttarakhand, and Haryana states of India. In one of the surveys conducted at the Directorate of Rapeseed–Mustard Research, Indian Council of Agricultural Research, India (Jha and Sharma 2003), *Sclerotinia* rot has been rated as the most important of eight problems being faced by the farmers of Bharatpur district (Rajasthan, India) in mustard culture. Due to this disease, Shukla (2005a) reported 50.88% yield loss in mustard crop. The pathogen is reported to have a wide host range, known to infect about 408 plant species (Boland and Hall 1994) with no proven source of resistance against the disease reported till date in any of the hosts.

Pathogen

The pathogen is *S. sclerotiorum* (Lib.) de Bary (Syn. *Sclerotinia libertiana* Fuckel; *Whetzelinia sclerotiorum* [Lib.] Korf and Dumont). Mycelium is thin, 9–18 μ in diameter with lateral branches of smaller diameter than the main hyphae. The vegetative hyphae are multinucleate ($n = 8$). Mycelial growth rate on solid agar media is fast and forms a moderate to abundant amount of aerial mycelium. The sclerotia are black, round or semispherical in shape measuring 3–10 μm. These are formed terminally and produced in one or two concentric rings on agar culture media. Sclerotia can be easily detached from the medium. The fungus does not have an obvious true conidial stage, though formation of microconidia in culture media has been reported. The mature sclerotium consists of an outer-pigmented rind and a medulla of prosenchymatous tissues partly embedded in a gelatinous matrix. Several other aspects on the morphology of sclerotium development, physiological and biochemical aspects of sclerotia formation, maturation, and structure of sclerotium have been reviewed by Willets and Wong (1980).

The sclerotial germination is mycelogenic (by mycelium) or carpogenic (by the formation of apothecia). On germination, the sclerotia form stalked apothecia. One to several apothecia may grow from a single sclerotium. The hymenium is made of palisades of asci and paraphyses. The asci measure 119–162.4 μ × 6.4–10.9 μ in size. These are inoperculate, cylindrical, narrow, rounded at the apex with eight ascospores in each ascus. Ascospores are uniform in size ($n = 8$). They measure 10.2–14.0 μ × 6.4–7.7 μ in size. Each ascospore is hyaline, ellipsoid, and has smooth walls. The spores are bi- or triguttulate. The paraphyses are about 100 μm long, 1–2 μm in diameter, slightly swollen at their tips, multinucleate, sparsely septate, and occasionally branched at the bases. The effects of some factors, such as age of sclerotium, temperature, light, and moisture, on apothecial production and ontogeny of apothecia have been reviewed by Willets and Wong (1980). The fungus grows over a range of 0°C–35°C, optimum being 20°C–25°C. Initiation and development of apothecia occur over a range of 10°C–20°C, while 20°C ± 1°C was found better for the same on the sterilized moist sand substrate (Goswami et al. 2012). It attacks field, forage, vegetable and ornamental crops, trees and shrubs, and numerous herbaceous weeds. There is little or no evidence of physiological specialization (Kolte 1985), though variation in pathogenicity has been reported (Goyal et al. 2013c).

Ascospores discharged from the apothecia at the base of the plants in soil constitute important primary source of infection. These ascospores could be stored at −80°C in 30%–40% glycerol for up to 12 months and used as a reliable source of inoculum for pathogenicity test (Olivier and Seguin-Swartz 2006). Mycelium in soil or those arising from sclerotia is a less important initial source of infection because of the low competitive saprophytic ability of the fungus (Kolte 1985). The ascospore can germinate in the presence of a thin film of water, in less than 24 h at 5°C–30°C, optimum being 5°C–10°C. The ascospore gives rise to infection hypha, and initial penetration of the tissue takes place directly by mechanical pressure through the cuticle, or the infection hypha may also penetrate already wounded or injured tissue. After the entrance of the fungus into the host, the mycelia cause enzymatic dissolution of the cell wall in advance, and cells die some distance ahead of the invading hyphae. Pectolytic enzymes are responsible for tissue maceration indirectly damaging the cell membrane, which results in subsequent death of cells (Kolte 1985). Production of cell wall–degrading PMG and cell (C_x) enzymes by *S. sclerotiorum*–infecting *Brassica* plants has been reported (Kolte 1985) that results in the colonization of plant tissues. Virulence of different isolates appears to be associated with the activity of PMG and C_x enzymes. These PGs can activate defense reactions in hosts, having no relation with the enzyme activity (Wang et al. 2008a). Li et al. (2004) reported that an sspg1d gene (endo-PG) is highly expressed in the pathogen under pathogenic conditions. Activities of PO and SOD enzymes were higher in resistant cultivars than susceptible cultivars of rape (Qi et al. 2004). A selective phytotoxin, sclerin, has been found to be produced by *S. sclerotiorum* in three susceptible cruciferous species, namely, *B. napus*, *B. juncea*, and *S. alba* (Pedras and Ahiahonu 2004), but not in a resistant species, namely, *E. gallicum*. However, three phytoalexins, namely, indole-3-acetonitrile, arvelexin, and 1-methoxyspirobrassinin, were found to be elicited in *E. gallicum* in response to the pathogen infection. The role of protease activity in infection of plants of *B. juncea* is also reported (Kolte 1985). A gene (ssv263) encoding a hypothetical, novel protein with unknown function (Liang et al. 2013), and an arabinofuranosidase/beta-xylosidase precursor (Yajima et al. 2009) from *S. sclerotiorum* has been identified as a possible virulence factor in the pathogen. A phytoalexin-detoxifying gene, namely, brassinin glucosyltransferase 1, that detoxifies brassinin phytoalexin in the host has been found to be induced in *S. sclerotiorum* in response to the infection of the host (Sexton et al. 2009). It appears that invasion of tissues of *B. juncea* is also related to the infection process, mediated by production of a toxin that is identified as oxalic acid (Kolte 1985); the oxalic acid is formed in culture filtrate as well as in infected *B. juncea* plants, which is reported to be thermostable, translocatable, and treatment of the host plant with culture filtrate results in disease. Oxalic acid may play a significant role in activating the glucosinolate–myrosinase system during infection (Rahmanpour et al. 2010). Resistance at the cellular level in *B. napus* against the pathogen is a result of retardation of pathogen development on the plant surface and within plant tissues. In resistant lines, formation of appresoria and infection cushions is suppressed that caused extrusion of protoplast from hyphal cells and produces a hypersensitive reaction. In susceptible lines, calcium oxalate crystals are found throughout the leaf tissues, while they are mainly confined to the upper epidermis of the resistant lines, and starch deposits are also more prevelant in susceptible lines (Garg et al. 2010a). Modulation of 32 proteins involved in photosynthesis and metabolic pathways, protein folding and modifications, hormone signaling, and antioxidant defense has been observed in *B. napus* in response to *Sclerotinia* infection (Liang et al. 2008).

According to Huang et al. (2008), on the surface of leaves and stems, infection cushions of different sizes develop that are often flattened and increased in diameter. These infection cushions and network of mycelia are covered by mucilage produced by the pathogen. After removing the infection cushions, numerous penetration pegs enter the cuticle of leaves and stems through the pores. Small changes are observed in cuticle. After penetration, the hyphae grow inter- and intracellularly between the cuticle and epidermal cells and also colonize xylem and phloem. Pathogen may secrete cell wall–degrading enzymes, namely, cellulases, pectinases, and xylanases, which degrade cellulose, pectin, and xylan in the host cell walls during infection and spread in the host tissues. Increase in proline content and PO activity during decrease in malonaldehyde content, free amino acid contents, and polyphenol oxidase activity and conductivity have positive correlation with rape resistance to the pathogen (Zhao et al. 2006).

Cultural, morphological, pathogenic characteristics and carpogenic germination of *S. sclerotiorum* have been studied by some workers (Goswami et al. 2012). Potato dextrose agar medium was the best supporting mycelia growth of the fungus and produced maximum number of sclerotia (Nguyen et al. 2006). Differences in the morphology of *S. sclerotiorum* isolates have previously been observed by Li et al. (2003a), where isolates producing tan sclerotia were identified. Sexton et al. (2006) demonstrated genotypic diversity utilizing microsatellite markers among *S. sclerotiorum* isolates of oilseed rape crops from Southeast Australia. Very few reports exist to date describing dark-pigmented isolates of *S. sclerotiorum*, such as those from Canada and southwestern region of the United States (Lazarovits et al. 2000, Sanogo and Puppala 2007). Recently, the molecular biology approaches enable to evaluate similarity and differences between strains within plant pathogens. Akram et al. (2008) worked on variability among isolates of *S. sclerotiorum*. Irzykowski et al. (2004) observed genetic diversity in natural populations of *S. sclerotiorum* from China by using RAPD molecular marker in addition to comparison of sequences of ITS1-5.8s-ITS2 region. Sharma et al. (2013b) used RAPD analyses of 15 geographical isolates, while Chen et al. (2010a) did sequence-related amplified polymorphism analyses of 76 isolates and observed high polymorphism among them. Genetic diversity based on morphological characteristics (Barari et al. 2011) was assessed by using rep-PCR genomic fingerprinting (Karimi et al. 2011) among geographically different isolates of *S. sclerotiorum*. Pathogenic diversity and genetic structure of the pathogen have been assessed through comparison in virulence and mycelial compatibility between isolates (Karimi et al. 2012). In Turkey, genetic and morphological diversity has been demonstrated for the first time within a population of *S. sclerotiorum*–infecting oilseed rape (Mert-Turk et al. 2007).

Morphogenic and pathogenic diversity among 38 isolates of *S. sclerotiorum* from different locations of Rajasthan, India has been reported (Ghasolia and Shivpuri 2007). Nie et al. (2010) and Ling et al. (2011) observed pathogenic diversity among 495 isolates of *S. sclerotiorum* from different regions of Shaanxi Province and 24 isolates of *S. sclerotiorum* from different regions of Anhui region of China. Garg et al. (2010b) studied the pathogenicity of morphologically different isolates of *S. sclerotiorum* from different regions of western Australia. High degree of pathogenic and genetic diversity has been observed among 17 isolates of *S. sclerotiorum* from India and the United Kingdom (Goyal et al. 2013c). Hence, similar holistic study should be conducted with higher number of *S. sclerotiorum* isolates from different geographical regions, which could provide a better picture of divergence among the pathogen and could be helpful in the generation of resistant material against the stem rot in oilseed Brassicas.

Inoculum of *S. sclerotiorum* can be detected in field-based air samples (using a Burkard spore trap) and from petals by PCR assay of nuclear ribosomal ITS sequences (Freeman et al. 2002). The presence of *S. sclerotiorum* on plants may be detected by using immunological detection method, namely, dimeric single-chain fragment variable (scFv) antibody with affinity for the pathogen (Yajima et al. 2008) and polyclonal antibody-based immunoassay (Bom and Boland 2000). Petal infection by *S. sclerotiorum* can be rapidly detected by real-time PCR (RT-PCR) (Yin et al. 2009) and nested PCR (Qin et al. 2011) techniques. Although detection of ascospores of *S. sclerotiorum* can be done by using passive trap, volumetric trap, and PCR techniques (Rogers et al. 2008, 2009) can also be used for the quantification of the ascospores (Penaud et al. 2012).

Epidemiology and Disease Cycle

Sclerotinia rot was positively correlated with increase in soil moisture and RH (R^2: 0.87 and 0.99, respectively), both at sowing and during the flowering period (50–60 DAS). Gupta et al. (2004b) found delay in sowing of rapeseed–mustard crop and reduced DI. High seeding rate and plant density increase the potential for lodging, which may be responsible for plant-to-plant spread of this disease (Jurke and Fernando 2008). Combination of cool weather and high soil moisture during the critical stage of 60–70 days age of crop favored higher incidence on Indian mustard (Sharma et al. 2009). Sharma et al. (2010c) observed petal infection with ascospores during full bloom stage and found rainfall as an important factor in carpogenic infection of *S. sclerotiorum* in *B. juncea*. Detection of healthy Indian mustard crop and its early differentiation from *Sclerotinia* rot–affected *B. juncea* plants was possible using remote-sensing technique, which could help in multistage disease tracking and forecasting (Dutta et al. 2006, Bhattacharya and Chattopadhyay 2013). Singh et al. (2000) developed a stepwise multiple linear regression model for *Sclerotinia* rot of Indian mustard. Under epidemiological study of *Sclerotinia* rot, based on DI and 10 independent weather variables, a multiple linear regression model has been described. The equation of the fitted model is percent

Sclerotinia rot incidence = −11.2351 + 0.9529*BSSH + 4.93924*Eva + 3.83308*pH + 0.60885*RF (mm) − 0.406458*RH 720 + 0.524095*RH 1420 + 0.17386*Soil moisture (%) − 0.30461*T_{max} − 0.677744*T_{min} − 2.19556*WS (DRMR 2010). The ScleroPro system is easy to handle and fully computerized, and based on the weather and field-site-specific data, this program has been available for growers and advisors since 2006 (Koch et al. 2007).

The pathogen primarily survives from one crop period to another in the soil through sclerotia. Such sclerotial bodies get mixed with the soil through affected plant debris after the crop is harvested, or when seeds contaminated with the sclerotial bodies are sown in the soil. Samples have been found to contain up to 432 sclerotia/kg seed, and a certain level of sclerotia in soil is reported to be maintained by the formation of secondary sclerotia (Kolte 1985). There are reports that the fungus can also survive through either mycelium or ascospores in dead or live plants, on testa of seed (Kolte 1985). Some wild plants that act as primary source of inoculum are hogweed (*Heracelum sphondylium* L.), cow parsley (*Anthriscus sylvestris* [L.] Holfm.), *Chenopodium* spp., and *Asphondilia* spp., which are also found to be infected and carry over the pathogen.

The disease is also known to be airborne, while seed treatments were assessed for protection against seed- and soil borne nature of the pathogen. Since aerial infection, apart from that taking place in the soil, is dependent entirely on continued production and dissemination of ascospores, epidemics are common in areas of continuously cool moist weather concurrent with the susceptible stage of crop, particularly the flowering period. Fields sown with rapeseed for 2 years favor more germination of sclerotia than fields sown with the crop for 1 year. Pollen and petals of rapeseed are known to stimulate ascospore germination. Rapeseed crops do not appear to restrict the movement of airborne ascospore. Ascospores are carried into the air current as high as 147.0 cm above the soil level. Spores could be trapped at a horizontal distance of 150 m from source indicating that the ascospores being airborne are carried to sufficient distance and cause spread of the disease from field to field. There did not appear to be a correlation between total rainfall and ascospore incidence. It is observed that ascospores on pollen grains of rapeseed adhere tightly. Honeybee-carried pollen and pollen in honeycombs have also been reported to carry ascospores. However, in view of the readily available wind-borne inoculum, the relative importance of transfer of spores by the honeybees is of less significance (Kolte 1985).

Spray of herbicide barban on rapeseed crop, used for managing *Orobanche*, increased its susceptibility to infection by *S. sclerotiorum*, possibly through altering the physiology of the plant as the herbicide has no inhibitory effect on the *B. juncea* plants. The disease has been noted to be high in plots, where *Orobanche* incidence is low and *vice versa*. Susceptibility of the plants to *Sclerotinia* rot is more when ammonium sulfate was applied, while thiourea spray showed less intensity of the disease (Kolte 1985).

Disease Management

Host Plant Resistance

Because of the wide host range and lack of tissue specificity, breeding resistant varieties appears to be less successful. However, differences in general growth habit and morphological characters of plants might be important characteristics for tolerance of the disease. For example, the *Omi nature* variety of *B. napus*, of medium height, early maturity, with a stiff stem and many branches, was resistant to the disease. The *Isuzu* variety of *B. napus* had a high degree of resistance (Kolte 1985). Another character, that is, stem diameter of the plants, may be a useful parameter for tolerance of the pathogen (Li et al. 2006a). It has also been reported that the high-glucosinolate lines are more susceptible to *S. sclerotiorum* as compared to the low-glucosinolate ones (Song and Guan 2008). Several genotypes of rapeseed–mustard have been screened against *Sclerotinia* rot caused by *S. sclerotiorum* by using different methods under natural and artificial conditions (Ghasolia and Shivpuri 2005a, Chand and Rai 2009, Prasad et al. 2009a, Sharma et al. 2012b). Responses of some genotypes (e.g., cv. Charlton) were observed relatively consistent irrespective of the isolates of the pathogen, whereas highly variable responses were observed in some other genotypes (e.g., Zhongyouang No. 4, Purler) against the same isolates. Genotypes with higher levels of resistance need to be included in oilseed *Brassica* breeding programs to enhance the level of field resistance in cultivated *B. napus* and *B. juncea.* Although complete resistance has not been identified in canola, partial field resistance to *Scleritinia* rot in Chinese cultivars Zhongyou 821 (Li et al. 1999) and Zhongshuang No. 9 (Wang et al. 2003) has been identified. Four cultivars of *B. napus*, namely, BOH 2600, Bermuda, Capio, and Mohican, were found resistant to *S. sclerotiorum* after a 3-year study (Starzycka et al. 2004). For the three consecutive cropping seasons, eight genotypes, namely, Hyola-401, PBN-9501, PWR-9541, Kiran, RH-9401, RH-492, RW-8410, and PAB-9511, were found resistant (percent DI [PDI] < 1%) to moderately resistant reaction (PDI = 1%–10%) to *S. sclerotiorum* (Ghasolia and Shivpuri 2005a). Genotype Ringot I of *B. juncea* was reported resistant to the rot (Goyal et al. 2011b). Other resistant *B. napus* genotypes that have been previously reported are 06-6-3792 (China), ZY004 (China), RT 108 (Australia) with mean stem lesion lengths < 3.0 cm (Li et al. 2007b, 2008b), and ZY006 with mean stem lesion lengths < 0.45 cm (Li et al. 2008b). In addition, the levels of resistance reported previously in *B. juncea* were, in particular, far lower, for example, *B. juncea* JM 06018 and JM 06006 with mean stem lesion lengths of 4.8 cm (Li et al. 2008b), as compared with *B. napus* genotypes. However, the situation has begun to improve because of the screening of the sources of *Sclerotinia* resistance from Chinese native cultivars (Li et al. 2009b). Garg et al. (2010c) reported high levels of resistance against *S. sclerotiorum* in introgression lines derived from *Erucastrum cardaminoides*, *Diplotaxis tenuisiliqua*, and *Eriospermum abyssinicum*. The novel sources of resistance identified in this study are a highly valuable resource that can be used in oilseed *Brassica* breeding programs to enhance resistance in *B. napus* and *B. juncea* cultivars against *Sclerotinia* rot. Results indicate that more than one *S. sclerotiorum* isolate should be included in any screening program to identify host resistance. It has also been reported that concentration and culture time of mycelial suspension of the pathogen and time for maintaining high RH after inoculation play a major role in disease development in the inoculated plants (Zang et al. 2010). Unique genotypes, which show relatively consistent resistant reactions (e.g., cv. Charlton) across different isolates, are the best for commercial exploitation to breed for resistance in oilseed *Brassica* against *Sclerotinia* rot (Garg et al. 2010c).

Molecular Breeding

PG inhibitor genes (Bnpgipl and Bnpgip2) (Li et al. 2003b, Hegedus et al. 2008) and EIN3 gene (Xu et al. 2009a) in *B. napus* may play an important role in resistance to *S. sclerotiorum.* Early induction of germin-like genes, namely, BnGLP3 and BnGLP12, that participates in an oxidative burst could play a vital role in defense of *B. napus* against the pathogen (Rietz et al. 2012). This oxidative burst can be detected in vivo in infected oilseed rape by using a modified platinum electrode on which Pt microparticles were dispersed and coated with a poly(*o*-phenylenediamine) film (Xu et al. 2009b). Sequential activation of salicylic and jasmonic acid signaling has been found to be associated with defense in oilseed rape against the pathogen (Wang et al. 2012). The SA levels in *Sclerotinia* rot–infected oilseed rape can be detected by using copper nanoparticles–modified gold electrode (Wang et al. 2010a). The LOX2 gene (Ren et al. 2010a) and PDF1.2 gene (Ji et al. 2009) in *B. napus* may be involved in jasmonate-mediated defense against *S. sclerotiorum.* Transformed lines of canola showed improved resistance to *S. sclerotiorum* with A9Ss gene from *Pseudomonas alcaligenes* strain A9 (Guo et al. 2006), oxalate oxidase gene (Zou et al. 2007, Dong et al. 2008), hrf2 gene (harpinXooc protein) from *Xanthomonas oryzae* pv. *oryzicola* (Ma et al. 2008), Ovd gene from *Orychophragmus violaceus* (Wu et al. 2009), pathogen-specific scFv antibody (Yajima et al. 2010), and pgipl gene (PG-inhibiting proteins [PGIPs]) from bean cv. Daneshjoo (Abedi et al. 2011). Introduction of glucose oxidase gene into *B. napus* has also been reported to increase resistance to the pathogen (He et al. 2007).

Cultural Control

Management is difficult, inconsistent, and uneconomical due to the presence of wide host range and long-term survival of the resting structures. Since the disease is carried over through sclerotia with crop debris and refuse of the plants stimulate sclerotial formation, it is advisable to collect and burn all the infected stubbles to kill the sclerotia (Vasudeva 1958). For sowing, sclerotia-free clean seeds should be used. In view of the airborne infection through ascospores and a wide host range of about 408 species, use of crop rotation appears to be a less successful method for managing the disease. However, deep summer ploughing and crop rotation with nonsusceptible hosts (rice, maize), use of only recommended dose of nitrogenous fertilizer, irrigation and keeping plant population within limits of recommendation, and flooding of soil, if possible, appear to minimize the sclerotial population in the soil, which subsequently might prove useful in control of the disease resulting from soil borne inoculum (Yuan et al. 2009). Avoidance of overcrowding of plants in a row to minimize plant-to-plant contact through root and stem to aid reduction of disease spread by mycelial means appears effective. Keeping a check on broad-leaf weeds like *Chenopodium* spp. is important in checking the disease. Occurrence of *Sclerotinia* blight can be reduced or avoided by late sowing of the rape (Fei et al. 2002). Late sowing might be helpful under the conditions of Canada by shortening the overlap between phenological susceptibility and exposure to maximum ascospore load (Kolte 1985). In Rajasthan (India), some conditions like sandy soil with late sowing and less irrigation were helpful in lowering DI (Ghasolia et al. 2004), while in Haryana (India), conditions like late sowing with presowing and three supplemental irrigations at branching, flowering, and pod formation were helpful in reducing DI (Sharma et al. 2001). Soil application of compost inhibited carpogenic germination of *S. sclerotiorum* and reduced *Sclerotinia* infection (Couper et al. 2001). Extracts of five organic amendments, namely, sunflower cake, safflower cake, mustard cake, neem cake, and farmyard manure, significantly reduced mycelial growth of the *S. sclerotiorum* (Tripathi et al. 2010). The enzymatic hydrolysis of glucosinolates in the *Brassica* releases ITCs, which could be potentially useful in curbing the pathogen (Kurt et al. 2011). Combination effect of micronutrients, namely, B at 1 g/L, Mo at 1 g/L, S at 2 g/L, and Zn at 2 g/L, in reduction of *Sclerotinia* rot incidence and increase in yield in rapeseed–mustard have been reported (Mondal 2008). Role of N in incidence of *Sclerotinia* rot of oilseeds *Brassica* is confusing (Gupta et al. 2004c, Shukla 2005b).

Biological Control

Chattopadhyay et al. (2002) reported a few of the biological treatments (seed treatment with *T. viride* and *A. sativum* aqueous bulb extract) to be effective against the disease coupled with effects of growth promotion and better plant stand and yields, which surpassed the efficacy of carbendazim on *Sclerotinia*-infested farmers' fields (Meena et al. 2006). Integration of the seed treatment with foliar sprays reaped better reduction of the disease (Chattopadhyay et al. 2004, 2007, Yadav 2009). *Trichoderma atroviride* showed coil formation and penetration of pathogen hyphae (Matroudi et al. 2009). Soil application of *T. harzianum* at 15 g/kg soil simultaneously or 7 days prior to the pathogen resulted in low disease intensity (Mehta et al. 2012). Wu and Wang (2000) reported that W-1 strain of *Caseobacter* spp. can manage the pathogen. Carpogenic germination of sclerotia of the pathogen could be reduced by using a bioagent *Gliocladium virens* (Ghasolia and Shivpuri 2005b). Antifungal activity of 11-3-1 strain of *Streptomyces longisporoflavus* against *S. sclerotiorum* has been observed (Han et al. 2012). The *Pseudomonas fluorescens* P13 isolated from oilseed rape field soil produced hydrogen cyanide (Li et al. 2011), and *Pseudomonas chlororaphis* PA-23 induced canola plants to produce more hydrolytic enzymes, namely, chitinase and beta-1,3-glucanase (Fernando et al. 2007), in response to the infection of *S. sclerotiorum*, thus was effective against the pathogen. However the control of pathogen by *Psuedomonas* strain DF41 is dependent upon lipopeptide production and the presence of a functional Gac system in the bioagent (Berry et al. 2010). Chitinase activity of different genotypes of *B. napus* is significantly correlated with their *Sclerotinia* rot scores and suggested that chitinase can be used in breeding program for improving disease resistance in rape.

Seed treatment with a bacterial strain, namely, *Mesorhizobium loti* MP6, isolated from root nodules of *Mimosa pudica* resulted in enhanced seed germination, early vegetative growth, and seed yield with drastic decline in incidence of *Sclerotinia* rot (Chandra et al. 2007). Y1 (Yan et al. 2005), NJ-18 (Yang et al. 2009a), YS45 (Zhang et al. 2009), Tu-100 (Hu et al. 2005, 2011), and EDR2 (Gao et al. 2013) strains of *B. subtilis* and BS6 strain of *Bacillus amyloliquefaciens* (Fernando et al. 2007) have been promising against this disease in oilseed rape. A new antifungal protein produced by *Bacillus licheniformis* W10 could be used as biofungicide to curb this disease (Sun et al. 2007). A bioagent, namely, *Coniothyrium minitans*, that destroys the hyphae (Jiang et al. 2000) and the sclerotia (Cael et al. 2001, Penaud and Michi 2009) of *S. sclerotiorum* has been used to control the disease. This bioagent degrades oxalic acid to nullify the pH effect thereof. Further, this may stimulate the production of beta-1,3-glucanase by the bioagent and may improve the mycoparasitism of the agent on *S. sclerotiorum* to result in protection of the plants from infection by the pathogen (Ren et al. 2007). Water-assisted application of *C. minitans* at the time of transplanting oilseed rape seedlings has been found effective in suppressing carpogenic germination of the pathogen (Yang et al. 2009b). Treatment of soil with this bioagent was found effective in reducing ascospore production by the pathogen (Huang and Erickson 2004). Drenching of deep soil with spores of *C. minitans* could be done before or after crop planting for increasing long-term efficacy on a regular basis on infected plots (Luth et al. 2012). Li et al. (2006b) reported aerial application of this bioagent as an effective method to curb the mycelial growth of the pathogen on petals. Tautomycin produced by *Streptomyces spiroverticillatus* and other related compounds, namely, 2,3-dimethylmaleic anhydride, diphenylmaleic anhydride, and dimethyl maleate, have significant potential against the pathogen (Chen et al. 2011). Sawdust soil–based bioformulation of *P. fluorescens* PS1 caused morphological alternation by hyphal perforation that enable to curb the disease (Aeron et al. 2011).

Chemical Control

In addition to contamination of seed, viable sclerotia present a potential quarantine hazard in export of seed. Viable sclerotia in infested seed of oilseed *Brassica* could be eradicated by fumigation with methyl bromide (Kolte 1985). In order to check the secondary spread of the disease, the possibility of control of the disease through foliar sprays of chemicals has been investigated. Since the pathogen is soil borne, application of chemicals to soil for managing the disease is not only of limited value but also hazardous to environment. Certain chemicals such as quintozene, fentin acetate, and calcium cyanamide have been found effective to inhibit the apothecial development of the fungus. The efficacy of calcium cyanamide in controlling the disease by 40%–90% has been confirmed under field conditions in Germany (Kolte 1985). Ridomil MZ (mancozeb + metalaxyl) as a seed dresser effected highest germination with no postemergence mortality by *S. sclerotiorum* (Pathak and Godika 2002a). Seed treatment at sowing and foliar spray at first budding/flowering with 0.2% of benomyl proved best on farmers' field (Chaudhary et al. 2010). Use of carbendazim at 0.25% as foliar spray could be effective in controlling this disease (Kolte 2005). Application of fungicides at full flowering phase (Jajor et al. 2010) by using venturi nozzle technology (Kutcher and Wolf 2006) was effective in reducing infection by the pathogen. Foliar spray of zinc pyrithione curbed the pathogen (Wang and Yang 2007).

Since no single method can effectively manage *S. sclerotiorum*, the best approach to control the pathogen is by integration of various eco-friendly measures. In recent years, an increasing consciousness about environmental pollution due to pesticides, and development of fungicide-resistant strains in *S. sclerotiorum* (Penaud et al. 2003) has challenged plant pathologists to search for eco-friendly tools for *Sclerotinia* rot management. Boscalid (trade name *Cantus* in China) is a new broad-spectrum fungicide belonging to carboxamides class. It inhibits the enzyme succinate ubiquinone reductase (Complex II), also known as succinate dehydrogenase, in the mitochondrial electron transport chain (Wang et al. 2009, Zhang et al. 2009, Gu et al. 2012). Use of such methyl benzimidazole fungicides in oilseed Brassicas to manage *Sclerotinia* rot has been reported to result in widespread fungicide-resistant strains of *S. sclerotiorum* (Penaud et al. 2003). A new fungicide, namely, prochloraz-manganese chloride, is found to be effective in delaying both myceliogenic and carpogenic germination of *S. sclerotiorum*; thus, it has both protective and therapeutic effects on the disease (Ren et al. 2010b).

POWDERY MILDEW

Symptoms

Powdery mildew appears in the form of dirty-white, circular, floury patches on both sides of lower leaves (Figure 5.9) of the infected plants. Under favorable environmental conditions (relatively higher temperature), the floury patches increase in size and coalesce to cover the entire stem and leaves. Severely affected plants remain poor in growth and produce less siliquae. Green siliquae also show white patches in the initial stage of infection. Later, such siliquae become completely covered with a white mass of mycelia and conidia. Severely diseased siliquae remain small in size and produce small shrivelled fewer seeds at the base with twisted sterile tips. As the season advances, under favorable conditions, cleistothecia may be formed on both sides of affected leaves, stems, and siliquae, which become visible in the form of black scattered and/or concentrated bodies (Kolte 1985).

Geographical Distribution and Losses

Occurrence of powdery mildew on oilseeds *Brassica* is reported from France, Germany, India, Japan, Argentina, Australia, Sweden, Turkey, the United Kingdom, and the United States (Kolte 1985, Gaetan and Madia 2004, Kaur et al. 2008b). It is generally believed that the disease does not cause much damage to oilseed *Brassica* crops except in occasional severe outbreaks, when all the leaves and siliquae get covered with the powdery growth of the fungus at early phonological stage. In certain states of India such as Gujarat, Haryana, Madhya Pradesh, Rajasthan, and Uttar Pradesh, the disease has been found to occur quite severely, possibly as an effect of climate change (Kumar et al. 2013) resulting in considerable loss in yield. Kohire et al. (2008a) observed 40% yield loss in

FIGURE 5.9 Powdery mildew.

Indian mustard. Considering the differences in disease intensity from year to year, it appears that the loss is proportional to the disease intensity, which varies considerably depending on the stage at which it occurs.

Pathogen

The pathogen is *Erysiphe cruciferarum* Opiz ex. Junell. The mycelium is ectophytic. Penetration is confined to the epidermal cells, in which haustoria form, the remainder of the fungus being extramatrical. Conidiophores arise from the superficial hyphae on the host surface. Conidiophore is septate, and conidia are borne singly. Conidia are ellipsoid to cylindrical. Ripe conidia fall off quickly and are disseminated by wind. Conidial size ranges 8.3–20.8 μm × 20.8–45.8 μm with an average range of 12.58–14.9 μm × 31.0–36.9 μm. Conidia germinate at optimum of 20°C–25°C by the formation of two types of straight germ tubes, one is short with slightly lobed appressoria and another is long with unlobed appressoria. The length of germ tube ranges 20–30 μm. Cleistothecia are globose to subglobose with numerous hypha-like brownish septate appendages. They are pinkish brown when young and turn brown to dark brown on reaching maturity. Cleistothecia measure 83.2–137.3 μm in diameter (av. 104.4–119.1 μm) on different species and varieties of *Brassica*. The number of asci varies 3–8 per cleistothecium, each ascus producing 2–6 ascospores. Asci are subglobose to broadly ovate, not stalked, light brown to yellowish in color, and measure 25.0–37.4 μm × 41.6–66.6 μm with an average range of 31.7–34.5 μm × 52.3–62.0 μm on different species and varieties. Ascospores are ovoid and measure 19–22 μm × 11–13 μm.

Epidemiology and Disease Cycle

The disease is favored by relatively dry weather conditions (Kohire et al. 2008b). Initiation of powdery mildew disease in mustard occurred during 50–120 DAS. Severity of the disease was favored by >5 days of ≥9.1 h of sunshine, >2 days of morning RH of <90%, afternoon RH 24%–50%, minimum temperature >5°C, and a maximum temperature of 24°C–30°C. Regression analysis showed maximum temperature and afternoon RH of the week preceding the date of observation was positively and negatively linked, respectively, to the disease severity (R^2: 0.9) (Desai et al. 2004). It is possible to forecast the occurrence of the disease using weather-based models (Laxmi and Kumar 2011). Cleistothecial formation appears to be favored by alternating low and moderate temperature, low nutrition of the host, low RH, dry soil, and aging of the host (Kolte 1985). Late sowing and frequent crop irrigation increase the incidence and severity of disease (Kohire et al. 2008c). *B. rapa*, *B. nigra*, *B. juncea*, *C. bursa-pastoris*, *Coronopus didymus*, and *R. sativus* have been found susceptible to *E. cruciferarum* (Kolte 1985). The pathogenic fungus is likely to carry over from season to season through cleistothecia or as mycelium in volunteer host plants. These cleistothecia release ascospores that cause infection on lower leaves during favorable condition. The secondary spread of the disease occurs through the conidia produced in infected leaves.

Disease Management

Host Plant Resistance

Limited sources of resistance against the powdery mildew has been reported in *B. alba*, *B. alboglabra*, *B. rapa* var. brown sarson, *B. chinensis*, *B. japonica*, and *E. sativa* (Kolte 1985). Five genotypes, namely, RN-490, RN-505, PBC-9221, PBN-9501, and PBN-9502, were resistant against powdery mildew (Pathak and Godika 2002b). Two Australian (JM06009, JM06012) and two Indian (JM3, Kranti) genotypes were resistant to powdery mildew (Singh et al. 2010).

Molecular Breeding

Transgenic plants of *B. napus* expressing the bacterial catalase katE in the chloroplast could inhibit the growth of *E. cruciferarum*. These plants revealed constitutive expression of the catalase enzymes, PO, polyphenoloxidase, and high levels of free polyamines like putrescine, spermidine, and spermine (El-Awady et al. 2008).

Cultural Control

Choice of suitable planting dates according to the locale appears to offer a promising method of managing the disease. Irrigation scheduling only at the 50% branching stage could manage the disease (Hingole and Mayee 2003).

Biological Control

Some trends of efficacy of seed treatment and foliar sprays by *T. viride* and aqueous bulb extract of *A. sativum* against the disease have been reported by Meena et al. (2003, 2013), which may need confirmation.

Chemical Control

Should the disease become serious, it no doubt could be managed by dusting plants with sulfur. Karathane when sprayed thrice at 10-day intervals also gives good control of the disease (Kolte 1985). Foliar spray of carbendazim at 0.25% could be effective in controlling this disease (Kolte 2005). Dinocap or tridemorph (0.1%) could be sprayed thrice on leaves to reduce the disease severity significantly, which may also increase the yield of mustard (Shete et al. 2008). Spray of 0.04% tridemorph followed by 0.05% hexaconazole, 0.05% tebuconazole, and 0.20% wettable sulfur on leaves was found effective against this disease (Patel and Patel 2008).

BLACKLEG OR STEM CANKER

Symptoms

Severe infection of the pathogen can cause seedling death, but stem cankering may occur at any plant growth stage on any plant part. During infection, the pathogen grows systemically down toward the tap root of the plant. Blackleg disease causes two distinct types of symptoms, namely, leaf lesions and stem canker. Stem cankering is the major reason of yield loss associated with blackleg. Root rot symptom on oilseed rape has also been reported in Australia as an extension of the stem canker disease caused by *Leptosphaeria maculans*. It appeared before flowering and increased in severity during flowering and at maturity. Infection of *B. napus* roots by *L. maculans* can occur via invasion of cotyledons or leaves by airborne ascospores and directly by the entry of hyphae at sites of lateral root emergence in the soil. The pathogen grew within stem and hypocotyl tissue during the vegetative stages of plant growth and proliferated into the roots within xylem vessels at the onset of flowering. Hyphae grew in all tissues in the stem and hypocotyl but were restricted mainly to xylem tissue in the root (Sprague et al. 2007, 2009).

Symptoms appear first as water-soaked lesions on cotyledons, hypocotyls, and leaves of the host. These lesions turn white to gray color, round to irregular in shape, and become dotted with numerous pinhead-sized black asexual fruiting bodies called pycnidia. When in a mature state and under moist conditions, the pycnidia exude spores in pink ooze on the host. This disease can be distinguished from *A. brassicae* infection by the presence of pycnidia, which are not formed by the *A. brassicae* on *Brassica* crops. Black lesions are generally also seen on the leaves and deep brown lesions with a dark margin that can be seen on the base of stem (Marcroft and Bluett 2008). In severe epidemic conditions, the pathogenic fungus girdles the stem at the crown, leading to lodging and death of the plant. Typical lesions of blackleg can also occur on pods. Pod infection may lead to premature pod shatter and seed infection. The seed beneath pod lesions may be sunken or shrivelled and pale gray in color. Li et al. (2008b) reported cytological changes, namely, condensation of cytoplasm, shrinkage in cell size, nuclear DNA fragmentation, shrinkage and condensation of the cytoplasm, chromatin fragmentation, and lobing of the nucleus due to hypersensitive reaction in cotyledon and stem tissues of *B. napus*, respectively, after infection by an avirulent strain of *L. maculans*.

Li et al. (2008c) reported that *L. maculans* may elicit apoptosis as a dependent component of pathogenesis in susceptible *B. napus*, and that the pathogen may use apoptotic cells as a source of nutrition for reproduction and further growth. Some proteins, namely, SOD, nitrate reductase, and carbonic anhydrase, were identified as being unique in the resistant plants, and upon pathogen challenge, some other proteins like photosynthetic enzymes (fructose bisphosphate aldolase, triose phosphate isomerase, sedoheptulose bisphosphatase), dehydroascorbate reductase, peroxiredoxin, malate dehydrogenase, glutamine synthetase, *N*-glyceraldehyde-2-phosphotransferase, and PPIase were observed to be increased in the resistant plants that were generated by an interspecific cross between the highly susceptible *B. napus* and the highly resistant *B. carinata* plants (Subramanian et al. 2005).

Geographical Distribution and Losses

Blackleg or stem canker is one of the major diseases of *Brassica* crops such as turnip rape (*B. rapa* L.), cabbage (*Brassica oleracea* L.), rapeseed (*B. napus* L.), and Indian mustard (*B. juncea* L.) grown in temperate regions of the world. For the first time it was reported on stems of red cabbage (Tode 1791). This disease is found in all continents and its world-wide importance and spread has been reported (Fitt et al. 2006). It has been reported as a serious disease also in Argentina (Gaetan 2005a), Australia (Lamey 1995), Brazil (Fernando et al. 2003), Canada (Lamey 1995), China (West et al. 2000), France (Lamey 1995), Greece (Vagelas et al. 2009), Latvia (Bankina et al. 2008), Lithuania (Brazauskiene et al. 2011, 2012), Poland (Kaczmarek et al. 2009a, Dawidziuk and Jedryczka 2011), the United Kingdom (Stonard et al. 2010a), and the United States (Mendoza et al. 2011). Yield losses up to 20% due to blackleg disease were recorded in Canada (Petrie 1978). Pedras et al. (1995) indicated crop losses due to blackleg in Canada alone exceed $30 million annually. It is the most important global disease of *B. napus* crops and causes annual yield losses of more than $900 million in Europe, North America, and Australia (West et al. 2001, Howlett 2004, Fitt et al. 2006). Both spring and winter types are affected by blackleg disease, particularly in Australia, Europe, and North America. Under epiphytotic conditions, this disease can cause yield losses of up to 90% (Kolte 1985, Sosnowski et al. 2004, Marcroft and Bluett 2008).

Pathogen

Blackleg or stem canker is caused by the heterothallic ascomycete fungus *L. maculans* (Desm.) Ces. et de Not. (anamorph: *Phoma lingam* Tode ex. Fr.). Another species, namely, *Leptosphaeria biglobosa* Shoem et Brun, has also been identified as causal agent of blackleg of canola in Australia (Wouw et al. 2008, Zhou et al. 2010), Canada (El-Hadrami et al. 2010), Lithuania (Brazauskiene et al. 2011, 2012), Poland (Kaczmarek et al. 2009, Dawidziuk and Jedryczka 2011), the United Kingdom (Stonard et al. 2010b), and the United States (Dilmaghani et al. 2009). Both species differ in their biochemical and molecular characteristics as well as in pathogenicity (Kaczmarek et al. 2009a), but are often found together in infected tissues of the same host (Dawidziuk et al. 2010). *L. maculans* can infect a wide variety of cruciferous crops, including cabbage, oilseed rape, and cruciferous weeds. Up to 28 crucifer species have been reported as hosts (Petrie 1969). *L. maculans* reproduces both sexually by forming pseudothecia and asexually by forming pycnidia on host species. The development of pseudothecia on stubble and the subsequent discharge of ascospores are greatly influenced by the genotype of the crop species (Marcroft et al. 2003). Pseudothecia of the pathogen are black, immersed, globose with protruding ostioles, ranging 300–500 μm in diameter, and are normally found on woody plant tissues. Asci are cylindrical to clavate, sessile or short stipitate measuring 80–125 × 15–22 μm; the ascus wall is bitunicate. Ascospores are hyaline and spindle shaped when young but yellow tan and five septate at maturity measuring 30–70 μm × 4–9 μm (Boerema 1976, Sawatsky 1989).

Two types of pycnidia of *P. lingam* (anamorph phase), designated Group I and Group II, have been found on *Brassica* spp. As a parasite, the pathogen produces Group I pycnidia with a pseudosclerenchymatous wall structure, which are initially closed developing a papillate opening (sometimes with a neck), while as a saprophyte, the pathogen produces Group II pycnidia with a pseudoparenchymatous wall structure. Group II pycnidia have dark walls, are often irregular in shape and may or may not have a papilla. Hyaline, oval, single-celled pycnidiospores measuring 1–2 × 2.5–5 μm are exuded from the pycnidia (Boerema 1976). Under high-humidity conditions, ascospores and pycnidiospores adhere to cotyledons or young leaves and germinate to produce hyphae that penetrate through stomata and wounds (Chen and Howlett 1996, West et al. 2001, Hua et al. 2004) and grow into substomatal cavities without forming appressoria (Hammond et al. 1985). After entering into substomatal cavities, the fungus grows between the epidermis and palisade layer and then into intercellular spaces in the mesophyll of lamina. The pathogen then reaches the vascular strands of the petiole and grows systemically within the plant without forming symptoms. The pathogen moves down the petiole and into the stem where it eventually invades and kills the cells of the stem cortex more commonly at the crown and cause the stem canker symptom (Hammond et al. 1985, Sprague et al. 2007, Travadon et al. 2009).

L. maculans exists in two forms, namely, avirulent and virulent. The avirulent form usually infects plants near maturity and causes only superficial disease symptoms, which results in shallow stem lesions, and rarely forming extended cankers that girdle the stem, while the virulent form attacks the crop earlier and causes severe stem canker and economic yield losses. It is especially virulent on *B. napus*. If basal infection begins early, stem cankers appear from flowering onward. As the season progresses, cankers penetrate, deepen, and may girdle stem bases, often completely severing the plant.

A pathogenicity gene that encodes isocitrate lyase has been identified in *L. maculans*. Isocitrate lyase is a component of the glyoxylate cycle and is essential for the successful colonization of *B. napus* (Idnurm and Howlett 2002). Sexton et al. (2000) have cloned a gene encoding endo-PG, pg1, and two genes encoding cellulases, cel1 and cel2, in *L. maculans*. The sp1 and sp2 genes are also expressed in *L. maculans* during the infection of *B. napus* plants, and later the gene secretes a serine protease with protease activity (Wilson and Howlett 2005). The Lmpma 1 gene of *L. maculans* encodes a plasma membrane H+-ATPase isoform, which is essential for pathogenicity toward oilseed rape (Remy et al. 2008). Remy et al. (2009) reported that the Lmepi gene encodes a highly conserved UDP-glucose-4-epimerase enzyme of the Leloir pathway, which is involved in galactose metabolism, and indicated a link between this primary metabolism and pathogenicity in *L. maculans* toward oilseed rape.

A non-host-selective phytotoxin, Sirodesmin PL, which causes blackleg disease of canola, is produced by *L. maculans* (Elliott et al. 2007). The selective phytotoxin maculansin A has been isolated from *L. maculans*, which was more toxic to resistant (*B. juncea* cv. Cutlass) than the susceptible plants (*B. napus*). However, it did not elicit phytoalexin production either in resistant or in susceptible plants (Pedras and Yu 2008).

The pathogen induced production of chitinase in cotyledons of the host in a time-dependent manner. This enzyme started to accumulate before symptom appearance. The proteins, namely, antioxidant enzymes, photosynthetic, and metabolic enzymes, and those involved in protein processing and signaling were found to be significantly affected by the pathogen in the host. The enzymes specifically involved in the detoxification of free radicals increased in response to the pathogen in the tolerant *B. carinata*, whereas no such increase was observed in the susceptible *B. napus* (Sharma et al. 2008b). *Brassica* plants produce some phytoalexins, namely, brassinin and camalexin, in response of *L. maculans*. Camalexin was found a substantially stronger inhibitor of the pathogen than brassinin (Pedras et al. 2007).

For the first time, variability for virulence in *L. maculans* was reported by Cunningham (1927). Pathogenic variability among *L. maculans* isolates of oilseed rape has been reported in southern Australia (Sosnowski et al. 2001) and western Canada (Kutcher et al. 2007). Australian populations of *L. maculans* have a high level of genetic variability as compared to European and North American isolates (Kutcher et al. 1993), along with a high diversity of avirulence genes (Balesdent et al. 2005). However, a low degree of genetic differentiation between isolates of *L. maculans* from seven sites in both eastern and western Australia has been observed by using AFLP marker (Barrins et al. 2004). Genetic diversity among different Australian, European, and North American isolates of *L. maculans* has also been studied by using AFLP marker (Purwantara et al. 2000). Molecular analyses of populations of *L. maculans* have shown high gene flow within and between populations. Pathogenic and genetic variation in *L. maculans* isolates on rapeseed–mustard has been reported in Australia, and microsatellite marker has been developed to study genetic variation in the Australian *L. maculans* population (Hayden et al. 2003). Isolates of *L. maculans* are usually classified either on the basis of their aggressiveness or by pathogenicity groups (Koch et al. 1991). Chen and Fernando (2006a) reported five pathogenicity groups (PG1, PG2, PG3, PG4, and PGT) of *L. maculans* on the basis of a series of inoculations on canola cultivars (*Westar*, *Glacier*, and *Quinta*) in western Canada and the United States. The PG3 and PG2 of *L. maculans* have also been reported on winter rape in Hungary (Szlavik et al. 2006) and Iran (Mirabadi et al. 2009), respectively. Nine races of *L. maculans* have been described and designated as AvrLm 1–9 (Mitrovic and Trkulja 2010).

Kenyon et al. (2004) developed a method for the detection of systemic growth of *L. maculans* in oilseed rape using quantitative RT-PCR. Sosnowski et al. (2006) used a quantitative PCR assay for the detection of *L. maculans* in soil. PCR-based molecular diagnostic techniques enabled detection, identification, and accurate quantification of airborne inoculum at the species level. Species-specific primers targeted at the ITS region of *L. maculans* and *L. biglobosa* were used to detect the quantity of the pathogen by traditional end point and quantitative RT-PCR methods, the latter being comparatively more sensitive, especially in years with low ascospore numbers (Kaczmarek et al. 2009b). The pathogen *L. maculans* on *B. napus* seeds was detected by using PCR-based techniques (Chen et al. 2010b, Yi et al. 2010). The ratio between airborne propagules of species, namely, *L. maculans* and *L. biglobosa*, of oilseed rape in Poland was evaluated by using a molecular approach based on species-specific primers and quantitative RT-PCR (Kaczmarek et al. 2009a).

Epidemiology and Disease Cycle

Infection of *B. napus* by *L. maculans* and subsequent development of leaf and stem lesions is influenced by cultivar resistance and weather conditions. Agronomic practices such as cultivar choice and fungicide use may also indirectly influence phoma stem canker epidemics at the regional level (Stonard et al. 2010b). Temperature and rainfall affect not only the development of pathogen but also the resistant response of the host (Fitt et al. 2008a). Elliott et al. (2011) for the first time isolated the *L. macunlans* and *L. biglobosa canadensis* isolates from *B. juncea* stuble in Australia. Dry climates lengthen the persistence of infected debris and may synchronize the release of airborne ascospores with seedling emergence; disease spread within plants is most rapid in regions with high temperatures from flowering to harvest (West et al. 2001). Kruse and Verreet (2005) reported that precipitation is of particular importance for *L. maculans* ascospore release during September in Germany. This effect was reduced in October and November while the influence of temperature increased. A very close correlation could be established between *L. maculans* ascospore release and leaf infection in autumn (September–November: $r = 0.82^{**}$). The correlation between the autumn infection of the leaves and root collar was highly significant (October–November: $r = 0.83^{**}$). Huang et al. (2006) reported that Rlm6-mediated resistance to *L. maculans* in *B. napus* leaves is affected by ambient temperature. High humidity and moderate temperatures during vegetative growth promote disease development (Ghanbarnia et al. 2009). Under high RH condition, *L. biglobosa* can cause increase in disease, which may coincide with reduced accumulation of lignin at early stages of infection (El-Hadrami et al. 2010). According to Brazauskiene et al. (2007), the abundance of ascospores in the air depended on the weather factors, especially the amount and frequency of rainfall. During the daily period, the abundance of ascospore spread was influenced by the ambient RH. Magyar et al. (2006) found that high RH, rainfall, melting snow, and moderate wind act as the most important factors in the dissemination of the ascospores.

Dawidziuk et al. (2012) reported that higher winter temperatures may increase the ability of pseudothecia to release ascospores and the discharge of ascospores of the pathogen into the air, and cause early plant infections. This in turn will increase the number of infected plants, the DI at harvest, and reduce the yield of oilseed rape. Differences in climate, especially temperature, and cultural conditions may affect the proportions of *L. maculans* and *L. biglobosa* in stem base lesions of oilseed rape in the United Kingdom (Stonard et al. 2008). Lo-Pelzer et al. (2009) reported that it is possible to forecast the quantity of available primary inoculum for a given disease severity.

A regional preharvest forecast for stem canker incidence and a crop-specific risk assessment method that predict the onset of *Phoma* leaf spotting using postharvest weather data and thermal time relationships for canker development and canker severity, have been developed (Gladders et al. 2004). Evans et al. (2006) developed an empirical model to predict the date when incidence (percentage plants affected) of phoma leaf spot can be expected to reach 10% on oilseed rape and to guide timing of fungicide applications against this disease to prevent pathogen spread from leaf to stem and the subsequent development of damaging stem cankers. Two weather-based models (Improved Blackleg Sporacle and SporacleEzy) were developed to predict the first seasonal release of ascospores of *L. maculans* or *L. biglobosa* from oilseed rape debris under many climates and thus could contribute to the development of the strategies for the control of the disease (Salam et al. 2003, 2007). A forecasting system for a autumn application of fungicide against the important rape pathogen *L. maculans* has been developed (Bremer 2007). Ghanbarnia et al. (2009) developed a nonlinear model to evaluate the combined effect of total rainfall and average maximum temperature per week on the mean blackleg disease severity of canola. However, Fitt et al. (2008b) developed a model to describe the spread of *L. maculans* across Alberta Province, Canada and was used to estimate the potential spread of *L. maculans* across the oilseed rape growing areas of Yangtze River, China and its associated costs.

The System for Forecasting Disease Epidemics (SPEC) is a joint initiative of the Institute of Plant Genetics PAS and DuPont Poland for stem canker forecasting in the world. It monitors the concentration of ascospores of the *L. maculans–L. biglobosa* species complex as a tool for disease prevention against stem canker (blackleg), and it is addressed to oilseed rape farmers, associated farm service personnel, breeders, commercial company representatives as well as to students and researchers with an interest in plant pathology and plant protection (Jedryczka et al. 2004, 2006, 2008).

L. maculans has a very complicated life cycle. It survives as a saprophyte by forming mycelium, pycnidia, and pseudothecia on crop residues, mainly on stubble (Hall 1992) subsisting from one season to the next. The inoculum production of *L. maculans* decreases with the increasing burial duration in field soil over 10 months, before ceasing, which may be due to associated microbiota (Naseri et al. 2008). This pathogen has both a teleomorph (ascospores) and an anamorph (pycnidiospores) phases on host species and can complete several disease cycles during a single growing season. In Australia and Europe, the main sources of infection of seedlings are infected seed, and for mature plants are wind-dispersed ascospores that are produced within pseudothecia on crop residues during summer. Maturation of pseudothecia is greatly affected by wetness (Liu et al. 2007). Soil borne ascospores and pycnidiospores of *L. maculans* were also able to cause seedling death, even after the spores had remained in a plant growth medium for up to 21 days before sowing (Li et al. 2007c). Ascospores can travel up to 8 km in Australia (Bokor et al. 1975) and 1.5 km in the United Kingdom (Gladders and Musa 1980) and enter into the host through stomata to infect the plant. Soon after the infection, they produce gray whitish lesions and black pycnidia on the leaves. During the growing season, these pycnidia produce conidia or pycnidiospores that are dispersed by rain splash. These spores cause a secondary infection, which is usually less severe than primary infection with ascospores. However, in western Canada and Poland, asexual pycnidiospores are the primary source of inoculum (Ghanbarnia et al. 2011). The pathogen overwinters as pseudothecia and mycelium in the stubble. In spring, the pseudothecia release their ascospores and the cycle repeats itself.

Krause et al. (2006) reported that the severity of *P. lingam* stem infection increased significantly with increasing number of oviposition punctures of *Ceutorhynchus napi*, which is one of the most destructive insect pests of winter oilseed rape in Central Europe. Females of *C. napi* deposit their eggs into the top of elongating stems that cause punctures in the stem and thought to predispose the stems to early secondary infections by *P. lingam* (*L. maculans*).

Disease Management

Host Plant Resistance

Two types of genetic resistance to *L. maculans* are usually identified in *Brassica*, that is, qualitative resistance (monogenic/race-specific/vertical resistance) that is expressed at the seedling stage and the quantitative one (polygenic/race-nonspecific/horizontal resistance) that is expressed in the adult plants. Qualitative resistance controlled by single major dominant gene has been reported in several spring and winter cultivars of *B. napus*, namely, Cresor, Maluka, Dunkeld, Maluka, Skipton, and Major (Stringam et al. 1992, Dion et al. 1995, Ferreira et al. 1995, Mayerhofer et al. 1997, Rimmer et al. 1999, Raman et al. 2012). Eighteen major genes for resistance to *L. maculans*, *Rlm1* to *Rlm11*, *RlmS*, *LepR1* to *LepR4*, *BLMR1*, and *BLMR2*, have been identified in *Brassica* species; *B. rapa*, *B. napus*, *B. juncea*, and *B. nigra* (Rimmer and van den Berg 1992, Balesdent et al. 2002, 2013, Yu et al. 2005, Delourme et al. 2006, Rimmer 2006, Van de Wouw et al. 2008, Yu et al. 2008a, Long et al. 2011, Raman et al. 2012). Six of them, *Rlm1*, *Rlm2*, *Rlm3*, *Rlm4*, *Rlm7*, and *Rlm9*, were identified in *B. napus*, all of them except *Rlm2* were clustered genetically on chromosome A07 (Delourme et al. 2004). *Rlm2* was mapped on chromosome A10 (Delourme et al. 2006). The *Rlm5* and *Rlm6* were identified in *B. juncea*, *Rlm8* and *Rlm11* in *B. rapa*, and *Rlm10* was identified in *B. nigra*. Four resistance genes, *LepR1*, *LepR2*, *LepR3*, and *LepR4*, were introgressed into *B. napus* from *B. rapa* subsp. *sylvestris*. Recently, two genes *BLMR1* and *BLMR2* were identified in Surpass 400, which is an Australian cultivar developed from an interspecific cross between wild *B. rapa* subsp. *sylvestris* and *Brassica oleracea* subsp. *alboglabra* (Buzza and Easton 2002, Long et al. 2011).

Christianson et al. (2006) reported that resistance to *L. maculans* in *B. juncea* populations is controlled by two independent genes, one of them being dominant and positioned on linkage group J13 and a recessive gene positioned on linkage group J18 based on segregation for resistance in the F2 population. In *B. rapa*, it is governed by three specific genes, namely, Rlm1, Rlm2, and Rlm7 (Leflon et al. 2007). Saal et al. (2004) identified a *B. juncea*–derived recessive gene termed rjlm2 that conferred resistance to *L. maculans* in oilseed rape. Gladders et al. (2006) reported that *B. napus* lines with Rlm6 resistance gene gave very effective control of leaf spot and stem canker caused by *L. maculans* in Europe, while Stachowiak et al. (2006) found both Rlm6 and Rlm7 resistant genes effective for the same. However, populations of *L. maculans* in Europe are known to have a high frequency of virulence to overcome resistance genes Rlm1–4 and Rlm9, and therefore, quantitative resistance makes an important contribution to stem canker control. Li et al. (2003c) reported the breakdown of a *B. rapa* subsp. *sylvestris* single dominant blackleg resistance gene in rape field of western Australia. Sprague et al. (2006) also reported that the *B. napus* cultivars derived from *B. rapa* ssp. *sylvestris* with single major gene resistance showed higher disease severity than cultivars with polygenic resistance in South Australia. Jedryczka et al. (2009) reported that Rlm6 and Rlm7 resistance genes for genetic protection of rapeseed against the present population of *L. maculans* in Poland.

Till date, nine resistance genes (Rlm1–9) have been identified in *Brassica* species (Gout et al. 2006). The corresponding nine avirulence genes designated as AvrLm1–9 have been identified in *L. maculans* (Balesdent et al. 2006), mapped at four independent loci, thereby revealing two clusters of three-and four-linked avirulence genes (Gout et al. 2006). The avirulence gene, AvrLepR1, of *L. maculans* corresponds to a resistance gene LepR1 of *B. napus*, and this plant gene control dominant, race-specific resistance to this pathogen (Ghanbarnia et al. 2012). Tollenaere et al. (2012) identified and characterized candidate Rlm4 blackleg resistance genes in *B. napus* by using next-generation sequencing technique. This major qualitative resistant locus (Rlm4) was mapped on chromosome A7 by using simple sequence repeat marker (Raman et al. 2012). *B. napus* cv. Surpass 400 was reported to have a single dominant resistant gene to *L. maculans* (Li and Cowling 2003), while Wouw et al. (2009) found at least two resistance genes, one of which is Rlm1 in *B. napus* cv. Surpass 400, with *sylvestris-derived* resistance. Two blackleg resistance genes, namely, LepR1 and LepR2, were mapped on N2 and N10 linkage groups of DHP95 and DHP96 lines of *B. napus*, respectively. The LepR1 generally conferred a higher level of cotyledon resistance than LepR2, because LepR1 prevented hyphal penetration, while LepR2 reduced hyphal growth and inhibited sporulation (Yu et al. 2005). The resistant gene LepR3 was found in *B. napus* cv. *Surpass 400* (Yu et al. 2008a). This gene provides race-specific resistance to the fungal pathogen *L. maculans*. LepR3 is the first functional *B. napus* disease resistance gene to be cloned and encodes a receptor-like protein. It has also been demonstrated that avirulence toward LepR3 is conferred by AvrLm1 avirulence gene, which is responsible for both the Rlm1- and LepR3-dependent resistance responses in *B. napus* (Larkan et al. 2013, Rouxel and Balesdent 2013).

The pathogen *L. maculans*, carrying AvrLm1 avirulence gene, when inoculated on *B. napus* plants carrying Rlm1 resistance gene, increased the biosynthesis of SA and ethylene (ET) and induced expression of the SA-associated genes ICS1, WRKY70, and PR-1, and ET-associated genes ASC2a, HEL, and CHI (Sasek et al. 2012a). Huang et al. (2009) found that quantitative resistance to *L. maculans* operates during colonization of *B. napus* stems by the pathogen.

Sinapis arvensis contains high resistance against various aggressive isolates of the blackleg fungus; so this species is valuable for the transfer of blackleg resistance to oilseed rape (*B. napus*) (Snowdon et al. 2000). *Brassica* species containing the B genome (i.e., winter *B. napus*, *B. nigra*, *B. juncea,* and *B. carinata*) are resistant to blackleg disease. Promising recombinant katanning early maturing (KEM) breeding lines derived from *B. napus* × *B. juncea* crosses were crossed with the spring-type *B. napus* cv. Dunkeld, which has useful polygenic resistance to blackleg. KEM recombinant lines showing regular meiotic behavior and a high level of blackleg resistance were screened using isolates of *L. maculans* having different AvrLm genes, which indicated *B. juncea* resistance gene Rlm6 had been introgressed into a *B. napus* spring-type cv. Dunkeld carrying polygenic resistance. The combination of both resistances would enhance the overall efficacy of resistance against *L. maculans* (Chevre et al. 2008).

Several *B. napus* and *B. juncea* germplasm from Australia, China, and India have been evaluated against Australian populations of *L. maculans*. *B. napus* genotypes from Australia were found more resistant than the Chinese and Indian genotypes (Li et al. 2008d). Two cultivars, namely, Aviso and Twister, of *B. napus* were found resistant to *L. maculans* in all seasons of the United Kingdom (Stonard et al. 2007). Light et al. (2011) reported that winter *B. napus* and *B. nigra* lines have outstanding potential for improving blackleg disease resistance under Australian conditions. Two blackleg-resistant lines, 16S and 61446, have been developed through interspecific hybridization between *B. napus* and *B. rapa* subsp. *sylvestris* and backcrossing to *B. napus* (Yu et al. 2013). In these lines, resistance to *L. maculans* is controlled by a single recessive gene (at LepR4 locus), and resistance alleles are allelic. Line 16S that carry LepR4a was found highly resistant, while line 61446 that carry LepR4b was found moderately resistant to stem canker under field conditions.

Molecular Breeding

Ananga et al. (2006) demonstrated that RAPD primers could be effectively used to identify DNA markers that are associated with blackleg disease resistance, which might also exist in the A and C genomes. Derivation of double haploid lines with superior levels of resistance to *L. maculans* compared with parental populations, and their multiyear, multisite (in locations with high pathogen diversity) evaluations could be an efficient practice to develop lines with high resistance to blackleg disease (Delourme et al. 2008). Dusabenyagasani and Fernando (2008) developed a sequence characterized amplified region marker available for marker-assisted selection in breeding canola for resistance against blackleg caused by PG3 of *L. maculans*.

Transgenic *B. napus* plants expressing pea DRR206 constitutively are resistant to the PG2 of *L. maculans* (Wang and Fristensky 2001). Kazan et al. (2002) reported that transgenic canola expressing MiAMP1 from the seeds of *Macadamia integrifolia* may be useful for the management of blackleg disease. A transgenic oilseed rape, *B. napus* cv. Hanna, with increased blackleg resistance has been developed by transferring Lm1 gene from *B. nigra* (Wretblad et al. 2003).

Induced Host Resistance

Resistance in canola can be induced by either pre- or coinoculation with the weakly aggressive isolates of the *L. biglobosa* and *L. maculans* (Chen and Fernando 2006b, Li et al. 2006c, El-Hadrami and Daayf 2009). After inoculation, the accumulated hydroxycinnamates act as precursors for the synthesis of lignin and phenylamide phytoalexins that could explain the restricted development of further inoculated highly aggressive isolates of the pathogen. Pretreatment of *B. napus* leaves with ascospores of *L. biglobosa* or chemical defense activators, namely, acibenzolar-*S*-methyl or menadione sodium bisulfite (MSB), delayed the appearance of *L. maculans Phoma* leaf spot lesions on the plants (Liu et al. 2006). MSB induced resistance locally and systemically (Borges et al. 2003). Treatment of *B. napus* plants with the SAR-inducing chemical benzo-(1,2,3)-thiadiazole-7-carbothioic acid *S*-methyl ester (BTH) significantly enhanced resistance against *L. maculans* (Potlakayala et al. 2007). Abscisic acid (Kaliff et al. 2007) and BABA (Sasek et al. 2012b) can induce callose- and SA-independent resistance, respectively, in *B. napus* against *L. maculans*. The gacS gene of *P. chlororaphis* was found to be responsible for antifungal and biocontrol activity against *L. maculans* of canola. Some low level of induced systemic resistance was observed in *P. chlororaphis* biocontrol of blackleg of canola (Ramarathnam et al. 2011).

Cultural Control

Various management practices such as crop rotation, careful stubble and residue management, time of sowing, use of certified seed, hot-water treatment of seeds, and control of volunteer cruciferous weeds have been recommended. A rotation including barley, field peas, and wheat for 3 years following oilseed rape helped eliminate potential sources of pathogen inoculum of *L. maculans* under all tillage systems (Turkington et al. 2000). The appropriate combination of rotation and tillage may lower airborne inoculum and reduce infection of rape by *L. maculans* (Guo et al. 2005, 2008), while Marcroft et al. (2003, 2004) reported that canola crops should be sown at distances greater than 100 m and preferably 500 m from last season's canola stubble, rather than extending rotation length between crops. Low seed rate and row spacing can increase the percentage of infestation of rape stem by *P. lingam* (Pusz 2007). Infested residue should be buried deep, and a shallow tillage or direct seeding method should be used in the spring to avoid bringing infected canola residue back to the surface. Since the primary infection of the plants occur by the airborne ascospores, canola should not be seeded within 1 km of infested land for 3–4 years (McGee and Emmett 1977). Soil borne ascospores and pycnidiospores can be managed by allowing the sand to dry between infestations and sowing by adding a 20 mm layer of uninfested sand over the top of the infested sand, respectively (Li et al. 2007c). The control of volunteer oilseed rape and susceptible cruciferous weeds should be done to prevent the establishment of the pathogen in fields. Wild mustard (*S. arvensis*) is highly susceptible to the virulent isolate of *L. maculans* (Petrie 1979).

Damage from blackleg could be minimized by sowing canola crops as early as possible before the onset of maturation of pseudothecia thus avoiding major ascospore showers at the seedling stage of maximum susceptibility and by doing fungicide protection in case of a late break season (Khangura and Barbetti 2004). Sowing of *B. napus* cultivars with different complements of resistance genes in subsequent years, that is, rotation of resistance genes minimizes disease pressure by manipulating fungal populations (Marcroft et al. 2012). Sprague et al. (2010) reported that defoliation of plants before stem elongation tended to develop less disease than defoliation during the reproductive phase of plant growth. In the future, this management strategy could be applicable in canola crops defoliated by grazing animals.

Biological Control

A bacterial isolate *Paenibacillus polymyxa* can inhibit the growth of *L. maculans* by producing antifungal peptides (Beatty and Jensen 2002). *Cyathus striatus* can reduce the production of the initial inoculum (pseudothecia) of the pathogen on rape stubble (Maksymiak and Hall 2000, 2002). A mixture of biological agents, namely, designated strain 17-1 (an associative endorhizosphere bacterial strain stimulating plant growth and protecting plants from pathogens) and 38-22, gave the highest increase in disease resistance and the best yield of spring rape (Farniev et al. 2009). Seed treatment with a commercial biofungicide of *Serratia plymuthica* reduced *L. maculans* activity by 50% in Germany (Marquardt and Ehlers 2010). Seed treatment with *S. plymuthica* and *P. chlororaphis* bioagents reduced mean disease by 71.6% and 54.0%, respectively, in canola (Abuamsha et al. 2011). More frequent treatments with commercial product, namely, Trifender WP of *Trichoderma asperellum* bioagent, during vegetation of oilseed rape could be effective against the blackleg disease (Kowalska and Remlein-Starosta 2011).

Chemical Control

Canola plants after 3–5 leaf growth stage are known to be less susceptible to blackleg than seedlings, so the protection of seedlings at that growth stage is an important method to manage this disease. Seed treatment with some fungicides, namely, thiram, fenpropimorph, benomyl, thiabendazole, and iprodione, has been found effective in managing the disease. However, these treatments do not protect plants grown in infested fields.

Ballinger et al. (1988) found that flutriafol applied as a fertilizer dressing on superphosphate granules significantly reduced the levels of stem canker in areas where the disease was prevalent. However, in western Canada, this fungicide had only limited efficacy (Xi et al. 1989). Dressing of canola seeds with fluquinconazole fungicides before sowing was found effective against *L. maculans* in situations of high disease severity, and grain yield increased when cultivars had lower blackleg resistance (Marcroft and Potter 2008). Chemical treatment of canola residues is a significant method to reduce the disease pressure on seedling. A number of chemical fungicides, such as fluquinconazole, flutriafol, and glufosinate ammonium (glufosinate), were able to delay pseudothecial development and decreased the subsequent ascospore discharge by more than 95% (Wherrett et al. 2003). However, impact (flutriafol) at 0.5 and 1 g/L, roundup (glyphosate) at 40 g/L, and copper sulfate inhibited the development of pseudothecia of *L. maculans* on canola residues and subsequently reduced ascospores production by 99% (Khangura 2004). Application of fungicides tended to be more beneficial at higher N rates and on upper slope positions since incidence was greatest under these conditions (Kutcher and Malhi 2004, Kutcher et al. 2005).

Pretreatment of host leaves with acibenzolar-*s*-methyl decreased the incidence of *Phoma* leaf lesions on seedling leaves (Liu et al. 2007). Azoxystrobin (Amistar 250 SC at 0.7 dm^3/ha) was found effective in decreasing blackleg infection of rape (Ratajkiewicz et al. 2009). Cytokinin, especially 6-benzyl amino purine, is able to significantly reduce disease symptoms and mycelial growth within plant tissues (Sharma et al. 2010c). CaraxReg is an innovative combination of 210 g/L mepiquat chloride and 30 g/L metconazole and is approved for the control of blackleg on oilseed rape (Gerber et al. 2010). Eckert et al. (2010) studied the effect of flusilazole, tebuconazole, and methyl benzimidazole carbamate fungicides (benomyl and carbendazim) on the germination of ascospores, conidia, and germ-tube growth of *L. maculans* and *L. biglobosa*. Triazole-based fungicides, namely, metconazole, protioconazole, tebuconazole, and flusilazole, were most efficient, and mixture of protioconazole and tebuconazole or flusilazole and carbendazim were very active against the *L. maculans* on oilseed rape (Jedryczka and Kaczmzrek 2011). The most effective timings for the application of flusilazole + carbendazim were when leaves 7–11 were present on most plants and at least 10% of plants were affected by phoma leaf spot. Two half-dose applications of fungicide reduced *Phoma* stem canker and increased yield more than a single full dose application when *Phoma* leaf spot epidemics were early (Steed et al. 2007). Kaczmarek et al. (2009c) found that the application of fungicides at the time following the maximum ascospores concentration significantly reduced DI and caused the highest increase of yield. Early treatment of carbendazim + flusilazole fungicides was found more effective than the late treatment (Hood et al. 2007). Only one spray of flusilazole fungicide may result in the highest reduction of infected plants when it was done on the day of the highest ascospore release till no longer than 3 weeks afterward (Kaczmarek et al. 2011). Application of flusilazole fungicide on oilseed rape increased the glucobrassicin, protein content, and yield while decreased the total alkenyl glucosinolate content in seeds (Brachaczek et al. 2011).

DAMPING-OFF AND SEEDLING BLIGHT

Symptoms

A necrotic lesion 1–2 cm long may be seen at the base of the stem, with girdling sometimes taking place near the soil level. The taproot may be discolored and sometimes wire-stem symptoms may be seen. Salmon-colored spore masses of *Fusarium* are often observed on affected tissues. Sometimes, the symptoms are confined to roots consisting of light-brown lesions on the taproot and at the bases of larger lateral roots. Girdling of the main root may take place, which may lead to loss of the entire root system. Damping-off and seedling blight are mostly encountered due to the use of infested seed. Primary lesions consisting of small, circular necrotic spots, along with secondary lesions with large irregular borders, appear on leaves. Under high-moisture conditions, whitish hyphae appeared on the stems (Yang et al. 2004).

Geographical Distribution and Losses

Several species of fungi are involved in causing seed rot and seedling blight around the world. Among them, *Rhizopus stolonifer* is reported to be a more important cause. Postemergence mortality is not frequent, with *Pythium aphanidermatum*, *Pythium butleri*, *Rhizoctonia solani*, *Sclerotium rolfsii*, *Macrophomina phaseolina*, and *Fusarium* spp. being the pathogens involved in India, causing 6%–15% incidence (Kolte 1985, Khan and Kolte 2002). Bottom rot of *B. campestris* L. caused by *R. solani* has also been reported in Japan (Eimori et al. 2005). They mostly survive on crop debris and soil as different resting structures to infect the following crop.

Disease Management

Molecular Breeding

Transgenic *B. napus* plants expressing pea DRR206 were found resistant against biotrophic root pathogen *R. solani* (Wang and Fristensky 2001). Development of transgenics by the transfer from bean (*Phaseolus vulgaris*) cv. Goli of pgip2 gene, which encodes PGIPs, can be useful in the future (Akhgari et al. 2012).

Cultural Control

Drainage from the crop field should be ensured at the time of sowing the crop in order to avoid water stagnation. Clean cultivation and removal of crop debris before sowing are important to manage the problem.

Biological Control

Significant combined effect of *B. napus* green manuring as well as of *Trichoderma* seed treatment against different soil pathogenic fungi (*Pythium* and *Rhizoctonia*) could be useful (Galletti et al. 2006). The mutant strain of *T. viride* 1433, namely, Tvm6, can be used to control *P. aphanidermatum* pathogen of mustard (Khare et al. 2010). Root colonization by *P. fluorescens* can prevent the establishment of *R. solani* on the root system (Tehrani et al. 2007, Zanjani et al. 2011).

Chemical Control

Seed treatment with thiophanate methyl 70 WP at 2 g/kg ensured better plant stand with protection against *S. rolfsii*, *R. solani*, and *Fusarium oxysporum* (Khan and Kolte 2002). Seed treatment with Metalaxyl 35 SD 6 g/kg + carbendazim 1 g ai/kg or with any other suitable seed protectant fungicide may be helpful in increasing the stand of the crop. Application of glyphosate herbicide 10 days before seeding increased seedling emergence and seed yield of canola in field infested with *R. solani* (Rashid et al. 2013).

CLUBROOT

Symptoms

At the initial stages the affected plants show normal healthy growth, but as the disease develops, the plants become stunted showing pale green or yellowish leaves. The plant is then killed within a short time. When the plants are pulled, overgrowth (hypertrophy/hyperplasia) of the main and lateral roots (Figure 5.10) becomes visible in the form of small or spindle or spherical-shaped knobs, called clubs. Depending on the type of root of a species, the shape of the club varies. When many

FIGURE 5.10 Club root.

infections occur close together, the root system is transformed into various-shaped malformations. The swollen roots contain large numbers of resting spores and plasmodia. The older, more particularly the larger, clubbed roots disintegrate before the end of the season.

Geographical Distribution and Losses

Incidence and severity is greater in regions with severe winters than in regions with spring-type climates. It occurs more frequently in soils, which are acidic and poorly drained. More damage due to the disease results on vegetable crops such as cabbage (*B. oleracea* L.) and turnip (*B. rapa* var. *rapifera*) than on oilseed rape (*B. rapa* var. *oleracea*) and mustard (*B. juncea*). Woronin (1878) was the first to study the disease in a systematic manner, life cycle of the fungus, and its relation to host tissue in detail. Walker (1952) described the disease in detail on cabbage. On oilseeds *Brassica*, the disease is reported to occur in East Germany, Malaya, New Zealand, Poland, Sweden, the United Kingdom, and the United States (Kolte 1985). The disease has been reported from the hills of Darjeeling (Chattopadhyay and Sengupta 1952) and Nilgiri (Rajappan et al. 1999) in India on vegetable Brassicas. On *B. rapa* var. yellow sarson (Laha et al. 1985) and var. toria (Das et al. 1987), the disease has been reported from West Bengal and Orissa, respectively, with losses in yield being up to 50% (Chattopadhyay 1991). For the first time, this disease has also been reported on *E. sativa* in Brazil (Lima et al. 2004). This disease has also been reported on canola in Australia (Khangura and Wright 2012) and on rapeseed in Luxembourg (Desoignies et al. 2009). Internationally, this disease causes up to 50% yield loss and is considered a serious disease of rapeseed in France, Canada, Czechoslovakia, Sweden, the United Kingdom, and Germany (Donald and Porter 2003). In southern districts of New Zealand, losses due to clubroot on rape are reported high, and this factor has been the major cause of decline in crop acreage in that country (Lobb 1951). It has been reported to cause 10.2% yield loss on rape in China (Wang et al. 2008b) and 70%–90% in Canada (Pageau et al. 2006).

Pathogen

The pathogen is *Plasmodiophora brassicae* Woronin, which is an obligately biotrophic fungus. Biology of the pathogen has been reviewed (Kolte 1985). There is no evidence that pathotypes of *Plasmodiophora* exist with a single genus or group of related genera within the host family, Cruciferae. Hence, the taxonomic concept *formae speciales* has not been applied to *P. brasssicae*. There is also much morphological variation to justify the taxonomic division of the species on the basis of morphology. Genetic and pathogenic variability in the field isolates of *P. brassicae* has been reported by some workers (Xue et al. 2008, Strehlow et al. 2010). The fungus has a plasmodial vegetative stage characterized by a naked, amoeboid, multinucleate protoplast without a definite cell wall. The plasmodium is produced only in the cells of the host plant and remains intracellular, with two distinct phases. The first, the primary one, usually results from infection by primary zoospores derived from the resting spores, and the secondary one results from infection by secondary zoospores derived from a zoosporangium.

The resting spore is hyaline, spherical, and measures up to 4 μ in diameter. It germinates by giving rise to single biflagellate primary zoospores (the first motile stage) having one long and one short flagella. The zoospore swims by means of its flagella, the long flagellum trailing and short flagellum pointing forward. This zoospore penetrates the host root hairs, and there it develops into a primary plasmodium in the affected cell. The plasmodium formed in this manner later cleaves into multinucleate portions surrounded by separate membranes, and each portion develops into zoosporangia. The zoosporangia come out of the host tissue through pores formed in the host cell wall. About 4–8 biflagellate secondary zoospores are formed upon germination of a single zoosporangium. Each secondary zoospore, except for their small size, is indistinguishable from the primary zoospore. The exact role of the secondary zoospores is not known, but it is likely that the secondary zoospores pair and unite to produce a zygote to cause fresh infection of the roots, producing new plasmodium called secondary or zoosporangial plasmodium, which in turn forms resting spores (Kolte 1985).

Though there are no *formae speciales* in *P. brassicae*, the fungus shows a lot of variation in pathogenicity. Physiologic specialization in *P. brassicae* was first demonstrated by Honig (1931). Information on the variation of the fungus has been reviewed, and a uniform set of differential hosts is described and proposed for research as an international approach for the identification of physiologic races of *P. brassicae*. Such a set of host genotypes is referred to as the European Clubroot Differential (ECD) set. The set consists of 15 different host varieties: 5 each of *B. rapa*, *B. napus*, and *B. oleracea*. Using the ECD set, 34 physiologic races have been identified in Europe (Kolte 1985).

The resting spores in soil serve as primary source of inoculum. Infection of the host takes place when uninucleate primary biflagellate zoospores are released on the germination of the resting spores. Germination of resting spores was most favored at 24°C, 2.6 pH, and 5 days of dark period (Wang et al. 2002). The zoospores may collide several times with a root hair before becoming attached; later, it appears to be attached at a point opposite to the origin of the flagella through adhesorium. The zoospores then encyst and penetrate the root hair or epidermal cells. The process of penetration of such cells appears to be direct but it has not been ascertained whether enzymes or toxins are involved in pathogenesis. After the entrance of the pathogen through root hairs, the formation of plasmodium and the subsequent development of zoosporangia take place in the infected tissue as described earlier. Then the zoospores derived from the zoosporangia are believed to reinfect the root and initiate formation of secondary plasmodia. Primary zoospores can directly cause secondary infection when the host is already in primary infection (Feng et al. 2013). Whether the secondary plasmodia penetrate the cell wall or if they are transformed passively from cell to cell during cell division is not certain. The plasmodium has no specialized feeding structure such as haustoria. It remains immersed in the host cytoplasm surrounded by a thin plasmodial envelope. There is also no evidence for phagocytic inclusion of the host cell organelles with the plasmodium. The plasmodium enlarges, and repeated nuclear division takes place, and the cells containing these become hypertrophied, although the host nucleus remains active. Hypertrophy of the host cells is apparently brought about by increased DNA synthesis and restriction of the cell division process. Presence of plasmodia in the host (*B. napus*) cell is associated with increased nuclei, at least in callus culture. Galling of susceptible *B. rapa* roots is the result of *P. brassicae* infection, enabling the enzyme glucosinolase to act on glucobrassicin, the indole glucosinolate. It is that the formation of the auxins, 3-indole acetonitrile and/or 3-indoleacetic acid, the characteristic extensive proliferation of tissue takes place. Since crucifers commonly contain indole glucosinolates, it has been suggested that this explains their susceptibility to galling. It appears that there is a close correlation between increase in the oxidative process and gall growth. As the galls develop on roots of the rape plant, the activity of glucose-6-phosphogluconate dehydrogenase, aldolase, triose phosphate isomerase, isocitrate dehydrogenase, and malate dehydrogenase increase is to reach a peak at 28–33 DAS. Then there is accumulation of glucose-6-phosphate, pyruvate, ketoglutarate, and malate in the affected cells. At sporulation, the activity of the aforementioned enzymes and concentration of the metabolites is decreased. During pathogenesis, these phenomena parallel the vegetative growth of the fungus. The metabolic regulation of phytoanticipins and phytoalexins has been found to be correlated with the infection period in the infected roots of oilseed canola (Pedras et al. 2008). Infectious pathogen spores can be detected by one-step PCR protocol, a quantitative/semiquantitative PCR-based technique in the soil and on the seed or tubers harvested from disease infested fields (Cao et al. 2007, Perek et al. 2010, Yin et al. 2010, Rennie et al. 2011).

Epidemiology and Disease Cycle

Development of the disease is favored by high soil moisture and cool weather; however, the disease can occur at any soil temperature between 9°C and 30°C. While development of clubroot was not observed at or below 17°C, it was slower above 26°C than at 23°C–26°C temperature (Gossen et al. 2012).

The fungus survives in the form of resting spores in soil. After the death of the galls, the resting spores are released in the soil; the pathogen thus becomes soil borne and is dispersed in soil as resting spores through farm implements, footwear, floodwater, etc. There is no evidence that the fungus lives as a saprophyte, yet soils are known to remain infested for 10 years or longer without the presence of a host. The pathogen can also survive on cruciferous weeds, namely, *C. bursa-pastoris.* Some of the noncruciferous hosts are also affected by *P. brassicae.* They are *Agrostis* spp., *Dactylis* sp., *Holcus* sp., *Lolium* spp., *Papaver* sp., and *Rumex* sp. Whether these noncruciferous plants play any part in maintaining the continuity of the disease in the absence of a cruciferous host is not known. However, it has been reported that secondary zoospores produced on *Lolium* spp. can infect canola (Feng et al. 2012). Wallenhammar (2010) studied the presence of clubroot in soil samples from 190 fields using a bioassay based on baiting the soils with *B. rapa* subsp. *pekinensis* (Chinese cabbage) *Granaat.* Clubroot incidence was significantly decreased after *Brassica* crops ceased to be grown. The half-life of spore inoculum was determined to 3.6 years for a field with 100% infestation. The level of infestation declined to below the detection level after a period of 17.3 years. Observations on yield loss from *P. brassicae* infections in spring oilseed rape (*B. napus* L.) are reported. In field tests of partly resistant cultivars of spring oilseed turnip (*B. rapa* L.), multiplication of clubroot was moderate.

Repeated cropping of susceptible host results in greater gall mass, reduced plant height, and increased numbers of resting spores in the soil mix compared to resistant host (Hwang et al. 2013). Increase in inoculum density, inoculation of young seedling could increase the disease severity and decrease the plant height and seed yield (Hwang et al. 2011a). Clubroot infection decreases abundance of adenosine kinase, which is involved in cytokinin homeostasis and also reduces host lignin biosynthesis. Enzymes level of ROS metabolism also declined sharply at 12 h after infection but increased at 24–72 h. These observations exhibit major changes in crop metabolism shortly after infection, which may result in the susceptibility of the host (Cao et al. 2008).

Disease Management

Host Plant Resistance

The control of the disease is difficult because of the longevity of resting spores in the soil. Among different methods, use of resistant varieties appears important to manage the disease. Certain kinds of *Brassica* spp. seem to have a natural resistance to the disease. Some genotypes of *B. juncea*, *B. rapa* var. toria, and *B. rapa* var. yellow sarson were found resistant in field condition (Sharma et al. 2012c). Deora et al. (2012) found 45H29 cultivar of canola resistant to disease. Hasan et al. (2012) observed pathotype-specific resistance in diploid species, namely, *B. rapa* (AA), *B. nigra* (BB), and *B. oleracea* (CC), and in the amphidiploid *B. napus* (AACC). Among *B. rapa* genotypes, turnip was most resistant, followed by winter- and spring-type oilseed rape. Contrastingly, rutabaga group of *B. napus* was observed homogeneous for resistance to Canadian *P. brassicae* pathotypes. The European winter canola (*B. napus*) cultivar Mendel has been used for the development of open-pollinated as well as hybrid canola cultivars (Rahman et al. 2011). It appears that resistance in *B. rapa* lines of a known genotype is associated with hypersensitive cortical cell death following invasion of *P. brassicae* from infected root hairs. Black mustard (*B. nigra* L.) is commonly reported as a resistant host due to volatile mustard oil, which remains to be proven (Kolte 1985). However, a positive correlation between clubroot susceptibility and clubroot-induced accumulation of several amino acids was found (Wagner et al. 2012). Sowing of a resistant variety could reduce the inoculum potential, while the cropping of susceptible variety increased the same (Hwang et al. 2011b, 2012a).

There are several physiologic races of *P. brassicae*, which vary in their ability to infect *Brassica* spp., and this complicates the problem of breeding-resistant varieties. Resistant varieties bred for clubroot resistance in one country may be completely susceptible to strains of the pathogens derived from another. Differential hosts used in ECD set are resistant to some races and susceptible to others. Development of resistant varieties through interspecific hybridization appears to be logical.

Resistance to *P. brassicae* Race 3 was successfully transferred from the turnip rape (*B. rapa*) variety Wasslander to rape (*B. napus*) variety Nevin by production of the fertile species, *Brassica napocampestris*, followed by two generations of back crossing of Nevin. Some cultures of *B. napus* (GSL-1, WBBN-1, WBBN-2, PCRS-80, WW-1507, ISN-700, MNS-3), *B. carinata* (HC-1, HC-4, HC-5, 9221, PC-3, PCC-2, PPSC-1, PC-5), and *B. nigra* (ACCBN-479) are reported resistant to the disease (Chattopadhyay et al. 2001).

Molecular Breeding

Suwabe et al. (2003) found that clubroot resistance in *B. rapa* is under oligogenic control, and at least two loci, that is, Crrl and Crr2, are necessary for resistance. A resistance gene Crrla has been identified in *B. rapa* L. (Hatakeyama et al. 2013). Yu et al. (2008b) reported that the resistant character in the resynthesized *B. napus* line HW243 is controlled by a single dominant gene for resistance to the disease. But durability of resistance seems unlikely to be profitable due to the development of newer pathotypes of *P. brassicae* keeping in view the faster rate of sexual reproduction in the pathogen. Wu et al. (2012) estimated the expression of stage-specific genes, namely, Pb-YPT, Pb-Brip9, and Pb-PSA, during infection of the pathogen in *B. rapa* by RT-PCR.

Development of transgenic lines by introducing thaumatin-like protein Hv-TLP8 from barley into oilseed rape via *Agrobacterium*-mediated transformation has been reported, which exhibits enhanced resistance to the pathogen (Reiss et al. 2009).

Cultural Control

Development of clubroot is significantly affected by cultivars, sowing date, soil moisture, and infection date (Wang et al. 2002). In Germany, pot experiments conducted (under field conditions) indicated that clubroot in yellow mustard can be reduced from 100% to 66% by mixing 50% compost into naturally infested soil. Early sowing can reduce infection compared to the late sowing of the crop (Hwang et al. 2012b). In view of the long viability of resting spores in soil, short-term crop rotation is not feasible, and traces of the pathogen could be detectable after more than 19 years of host plant absence, making its eradication very difficult (Rastas et al. 2012). Since *P. brassicae* also infects cruciferous weeds such as *C. bursa-pastoris*, it may be important to control the weeds in order to check the incidence of the disease. Use of 10–30 mg/kg boron and calcium nitrate in soil of pH 6.5 or 7.3 was effective in reducing clubroot severity (Ruaro et al. 2009).

Growing the crop in fields known to be infested with the clubroot pathogen should be avoided. General measures aimed at mitigating the incidence of the disease through improved drainage and application of lime brings about control of the disease. Spores of *P. brassicae* do not germinate or germinate very poorly in alkaline soils. On this basis, amendment of infested soil with lime is suggested. The amendment is done so as to raise the pH of the soil to 7.2. Treating the soil with lime 1 kg/m^2 area has been reported to control clubroot in mustard (AICRP-RM 2000). Application of worm cast as base fertilizer could effectively control the clubroot of rape, with an efficacy of 56.3%–61.4%, decrease the soil acidity, and increase the soil organic content (Wang et al. 2010b) Plant tolerance and resistance can be effectively increased against clubroot disease of mustard by using nutrients, namely, B, Mo, and Ca. These nutrients result in increase in yield and reduction in average weight of clubs per plant (Sen 2005, Deora et al. 2011). Nitrogenous fertilizers, namely, calcium ammonium nitrate and calcium nitrate, could be used to control the disease and increase shoot dry weight and seed yield (Bhattacharya and Mandal 2006).

Biological Control

Mixing or pouring of some antagonists, namely, *Serratia* spp. and *Trichoderma* spp., by using mushroom compost as a carrier for the antagonists was found effective in reduction of infection by up to 40% (Preiss et al. 2010). Lahlali et al. (2013) found that the biofungicide serenade (*B. subtilis*) suppresses the disease on canola via antibiosis and induced host resistance.

Chemical Control

Although certain chemicals like azoxystrobin, benomyl, fluazinam, flusulfamide, methyl thiophanate, quintozene, limestone, and other soil fumigants are known to be effective against *P. brassicae*, such disease management methods are not feasible and economical because of the high cost of chemicals and their application.

Fusarium WILT

Symptoms

The leaves of the affected plants show drooping, vein clearing, and chlorosis, followed by wilting, drying, resulting in the death of the plant. The symptoms progress from the base upward. The expression of the disease symptoms varies with the age of the plants. In the early stage of development, affected plants do not show all the typical symptoms. Plants affected in preflowering and early flowering stages show defoliation, and stems of such plants externally develop longitudinal ridges and furrows, which are generally not observed in the later stages. Diseased plants often show stunting, which is more pronounced when the plants are attacked in preflowering stages. Such plants have small pods with no seeds. Unilateral development of the disease is also observed in some of the cases when only one side of the plant shows symptoms of the disease. Roots of the diseased plants show no external abnormality or decay of the tissue until the plants are completely dried. Vascular tissues of stem and root show the presence of the mycelium and/or microconidia of the pathogen. Such tissues show browning of their walls and their plugging with a dark gummy substance, which is one of the characteristic symptoms of vascular wilts. At later stages of the disease, epidermis of roots sloughs off. The diseased plants eventually collapse and die (Kolte 1985).

Geographical Distribution

Mustard is affected by *Fusarium* wilt caused by *F. oxysporum* f. sp. *conglutinans* (Wr.) Snyder and Hansen. The first authentic report of *F. oxysporum* f. sp. *conglutinans* as the cause of the disease in *B. juncea* was made from India, followed by another on *B. nigra* (Kolte 1985). It has also been reported on canola in Argentina (Gaetan 2005b).

Pathogen

The causal fungal pathogen is *F. oxysporum* f. sp. *conglutinans*. Two types of cultures were isolated as Group A and B isolates and found that both were pathogenic to *B. rapa* var. toria, *B. rapa* var. yellow sarson, *Brassica oleracea* var. *botrytis*, *B. oleracea* var. *capitata*, *E. sativa*, *Matthiola incana*, *B. nigra*, *S. alba*, *Symphytum officinale*, and *R. sativus*. Susceptibility of *B. carinata*, *Crambe abyssinica*, and *C. hispanica* has been reported from the United States (Kolte 1985). Devi et al. (2009) reported *Fusarium moniliforme* as causal agent of *Fusarium* wilt of rapeseed from four districts of Manipur, India.

Disease Management

Extracts of plants *Vitex trifolia* and *Artemisia nilagirica* were found to have significant fungicidal properties (Devi et al. 2009). Seed treatment with carbendazim at 0.1% ai or a suitable biofungicide could be effective in managing the disease.

OTHER FUNGAL DISEASES

Rotting of seed is reported to be caused by *Nematospora sinecauda* (Oram et al. 2003), while white leaf spot caused by *Pseudocercosporella capsellae* has also been reported, when grayish white to brownish lesions on leaf (often with a distinct brown margin) and some grayish stem lesions occur (Eshraghi et al. 2005).

Bacterial Stalk Rot

Symptoms

Symptoms of the disease are characterized by the appearance of water-soaked lesions at the collar region of plants, which is usually accompanied by a white frothing. The tender branches are also affected as the lesions advance further to cover larger areas. The leaves show signs of water stress and wither. The affected stem and branches, particularly the pith tissues, become soft, pulpy, and produce dirty white ooze with a foul smell. The infected collar region becomes sunken and turns buff white to pale brown. Badly affected plants topple at the basal region within a few days.

Geographical Distribution and Losses

The first report about the occurrence of stalk rot caused by *Erwinia carotovora* (Jones) Holland appears to have been made in *B. juncea* in Rajasthan (India) by Bhowmik and Trivedi (1980). On an average, about 40%–60% of plants may be affected by the disease. Presence of the disease in fodder varieties of *Brassica* spp. is also observed. Vigorously growing succulent plants, due to an extra dose of N, as well as those growing in poorly drained soil are more severely affected. Root rot caused by *Erwinia carotovora* pv. *carotovora* (Jones) Bergy is an emerging threat for rapeseed–mustard production system, recently reported from the farmers' field in some pockets of India (AICRP-RM 2006–2008, Meena et al. 2010b).

Pathogen

The bacterium is Gram negative, rod shaped with blunt ends, capsulated, and motile with peritrichous flagella. It forms grayish, circular, translucent, shining, smooth colonies on nutrient agar with a raised centre and wavy margin. The bacterium can infect *B. oleracea* var. *botrytis*, *Daucus carota*, *Lycopersicon esculentum*, and *Nicotiana tabacum*.

Epidemiology and Disease Cycle

The disease is favored by warm and humid weather. It usually appears after first irrigation in mustard. The pathogen survives on diseased plant debris in soil.

Disease Management

Bacterial stalk rot can be managed to some extent by using cultural practices, namely, crop rotation with nonhost crops, deep ploughing in summer months, and roguing and burning of diseased debris. These practices help in minimizing inoculum buildup in soil. Early sowing of crops, removal of weeds, and avoidance of overirrigation are effective in reducing the DI. This disease can be reduced by spraying streptocycline 100 ppm and copper oxychloride at 0.2%.

Bacterial Rot

Symptoms

Symptoms appear when the plants are 2 months old. In the initial stages, dark streaks of varying length are observed either near the base of the stems or 8–10 cm above the ground level. These streaks gradually enlarge and girdle the stem. Finally, the diseased stem becomes very soft and hollow due to severe internal rotting, and this often results in total collapse of the plant. Sometimes, cracking of the stem is observed before the toppling of the plant. Occasionally, symptoms appear on leaves. Lower leaves show the symptoms first, which include midrib cracking and browning of the veins; when extensive, it brings about withering of the leaves. Profuse exudation of yellowish fluid from affected stems and leaves may also occur. Blackened veins and V-shaped necrotic lesions on the leaf margins are surrounded by yellow halos. The advanced phases of the disease include lesion enlargement, foliar chlorosis, and death of leaves. The disease develops from the lower leaves to the apex, resulting in complete leaf necrosis and defoliation. The affected plants, on stripping, show a dark brown crust full of bacterial ooze. The black rot does not cause any disagreeable odor.

Geographical Distribution and Losses

Patel et al. (1949) first observed the black rot symptoms in *B. juncea* in India under natural conditions. In 1970, 100% incidence on cauliflower was reported, and during 1969, National Seed Corporation of India suffered heavy losses of about 0.5 million rupees in 10 ha of cauliflower seed crop due to combined infection of stump rot and black rot. The disease is now reported to occur in a severe form (60% incidence) in the Indian State of Haryana (Kolte 1985). Occurrence of the disease has also been reported in Brazil, Canada, Germany, Serbia (Popovic et al. 2013), Mozambique (Bila et al. 2013), Sweden, and the United States. In fact, monoculture is presently the dominant form of crop management worldwide, which plays a major role in disease progression (Zhu et al. 2000).

Pathogen

The pathogen is *Xanthomonas campestris* pv. *campestris* (Pammel) Dowson. The bacterium is a short rod with rounded ends, occurring singly, rarely in pairs. In culture on potato dextrose agar, it measures 1.5 μ (1.2–2.1 μ) × 0.7 μ (0.5–1.0 μ). It is motile with a single polar flagellum, Gram negative, not acid fast, aerobic, and capsulated without spore formation. On nutrient dextrose agar, colony is dark yellow, circular, nonfluidic, convex, and opaque. The thermal death point is 50°C–58°C. Genetic and pathogenic variability among the isolates of *X. campestris* pv. *campestris* has been identified by several workers (Gaetan and Lopez 2005, Miguel-Wruck et al. 2010, Singh et al. 2011b, Raghavendra et al. 2013). A DNA probe has been developed for rapid identification of strain of this pathogen in plant tissues (Shih et al. 2000). This pathogen can be specifically and rapidly detected by several methods, namely, multiplex polymerase chain reaction (Berg et al. 2005, Leu et al. 2010), multiplex RT-PCR assay (Berg et al. 2006), Bio-PCR (Singh and Dhar 2011), classical biochemical assays, enzyme-linked immunosorbent assay (ELISA) with monoclonal antibodies, Biolog identification system, and PCR with specific primers and pathogenicity tests (Bila et al. 2013).

Epidemiology and Disease Cycle

The host range includes *B. alba*, *B. rapa* var. brown sarson, *B. rapa* var. yellow sarson, *B. carinata*, *B. chinensis*, *B. hirta*, *B. napus*, *B. nigra*, *B. oleracea*, *B. rapa*, *B. tourneforti*, and *R. sativus*. The pathogen does not infect *E. sativa* and *C. sativa* (Kolte 1985). Details of the mode of penetration and the infection process have not been studied using rapeseed–mustard plants. However, it is believed that the pathogen overwinters in diseased plant refuse or in seed and penetrates the host through either stomata or hydathodes and establishes the infection in a similar manner as in other crucifers (Berg et al. 2005).

Disease Management

Host Plant Resistance

Race-specific resistance to the pathogen has been found in Brassicas with B and D genomes (Ignatov et al. 2001). Griffiths and Nickels (2001) reported that a single dominant gene may control the resistance. In progeny of *B. carinata*, the resistance is conferred by a major dominant gene Rb that can be used for breeding purposes (Ignatov et al. 2001). Some new alien addition lines resistant to black rot have been generated by somatic hybridization between cauliflower and black mustard (*B. nigra*) (Wang et al. 2011). Some degree of resistance to races 1 and 4 of *X. campestris* pv. *campestris* in different *B. napus* crops, mainly in underexplored pabularia group, has been identified (Lema et al. 2011). Resistance was identified in five accessions of *B. carinata* (PI 193460, PI 193959, PI 194254, PI 280230, PI 633077) and four accessions of *B. nigra* (PI 197401, A 25399, A 25401, PI 458981) determined by repeated symptomless responses after inoculation. Five accessions of *B. rapa* (PI 633154, A9285, PI 340208, PI 597831, PI 173847) represent promising new sources of resistance to the pathogen. Incomplete resistance was identified in an accession of *E. sativa* (PI 633207), *Lepidium* spp. (PI 633265), *S. arvensis* (PI 296079), and two accessions of *B. napus* (PI 469733 and PI 469828). These identified accessions represent germplasm that can be used in breeding for resistance to *Xcc* in the future (Griffiths et al. 2009).

Biological Control

Foliar spray or the combined seed soaking and soil drenching with *Pseudomonas aeruginosa* (KA19 strain) and *Bacillus thuringiensis* (SE strain) are also found effective in reducing black rot lesions compared to untreated control (Mishra and Arora 2012).

Chemical Control

Captafol spray (0.2% ai) at 20-day intervals is reported to give good control of the disease; aureomycin (chlorotetracycline) 200 μg/mL was most effective in reducing the infection from 85% to about 15% resulting in an increase in yield by 60%. Among fungicides, carboxin was most effective, reducing the infection by 79% with a corresponding increase in yield by 49%. Spray application of copper oxychloride is also reported to give a considerable degree of control of the disease (Kolte 1985).

OTHER BACTERIAL DISEASES

Bacterial leaf spots are also reported to be caused by *Pseudomonas viridiflava*, which result in white and corky brown spots on leaves and sometimes water-soaked spots on the lower leaf surface (Myung et al. 2010). Brownish-black color leaf spots are also reported to be triggered by *Pseudomonas syringae* pv. *maculicola* (Peters et al. 2004) and *Pseudomonas cannabina* pv. *alisalensis* (Bull and Rubio 2011).

Mosaics

Symptoms

Symptoms on *B. juncea* appear as vein clearing, green vein banding, mottling, and severe puckering of the leaves. The affected plants remain stunted and do not produce flowers, or very few flowers are produced on such plants. When siliquae are formed, they remain poorly filled and show shrivelling, which results in decrease in yield and oil content (Jasnic and Bagi 2007). Symptoms of vein clearing, stunting, and pod malformation have been observed (Sahandi et al. 2004). According to Sahandi et al. (2004), the symptoms appear as systemic conspicuous vein clearing, vein banding, yellowing, and distortion of young leaves. During the later stages of infection, numerous raised or nonraised dark green islands of irregular outline appear in the chlorotic area between the veins, giving rise to a mottled appearance. Curvature of the midrib and distortion of the leaf blade on affected leaves can also be a prominent symptom. Plants infected early are usually stunted and killed, but those infected late show reduced growth only slightly (Kolte 1985). The number of primary branches, seeds per pod, and percentage oil per seed is reduced; the glucosinolate concentration in the oil is also significantly increased in infected plants (Stevens et al. 2008).

Geographical Distribution and Losses

Occurence of the virus disease has been reported on different cruciferous plants in Iran (Shahraeen et al. 2002, Tabarestani et al. 2010), New Zealand (Kolte 1985), Taiwan (Chen et al. 2000), on winter oilseed rape in Austria (Graichen et al. 2000), and on *B. nigra* (Thurston et al. 2001) and *B. rapa* ssp. *sylvestris* (Pallett et al. 2002) in the United Kingdom. Biswas and Chowdhury (2005) reported the disease in *B. juncea* and *R. sativus* in the Himalayan regions of West Bengal, India, and indicated that it may be a strain of turnip yellow virus (TuYV).

Yield could decrease by more than 70% (Jasnic and Bagi 2007). Over 30% of the crop has been reported to be destroyed by the disease in China resulting in 37%–90% loss in yield (Kolte 1985). In Australia, yield loss due to TuYV on oilseed rape may reach up to 46% (Stevens et al. 2008). In Iran, canola field infection with turnip mosaic virus (TuMV), CaMV, and beet western yellows virus (BWYV) was 1.7%–8.3% (Tabarestani et al. 2010).

Pathogen

Some of the more common crucifer mosaic diseases are caused by viruses included in turnip virus I group. On rapeseed–mustard, the mosaic diseases caused by this virus group are described under different names, namely, (1) rape mosaic in China and Canada, (2) mustard mosaic in the United States and Trinidad, (3) Chinese sarson mosaic in India, (4) *B. nigra* virus in the United States, and (5) turnip mosaic in China, Germany, Hungary, Soviet Union, and the United Kingdom (Kolte 1985).

Occurence of six viruses, namely, BWYV, CaMV, turnip crinkle virus, TuMV, turnip rosette virus, and turnip yellow mosaic virus (TYMV), were detected on *B. nigra* (Thurston et al. 2001) and *B. rapa* ssp. *sylvestris* (Pallett et al. 2002) in the United Kingdom. TuMV, CaMV, and BWYV are considered as the most important viruses of canola in Iran (Tabarestani et al. 2010) and of rape (including TYMV) in Europe (Mamula 2008). Cai et al. (2009) found two new strains, youcai mosaic virus–Br and oilseed rape mosaic virus (ORMV)-Wh, which are related to the ORMV cluster of tobamoviruses and distantly to tobacco mosaic virus as a pathogen of oilseed rape in China. The pathogen can be detected by tissue blot immunoassay (Coutts and Jones 2000), ELISA (Thurston et al. 2001, Pallett et al. 2002), double-antibody sandwich-ELISA (Chen et al. 2000, Graichen et al. 2000, Sahandi et al. 2004), antibody sandwich-ELISA and PCR (Farzadfar et al. 2005), and RT-PCR technique (Tabarestani et al. 2010). Polyclonal antiserum has been produced against CaMV isolate that can also be used in indirect ELISA system for virus survey and identification (Sahandi et al. 2004). Graichen et al. (2000) detected no TuYV infection in virus-like symptoms having plants and infection in symptomless plants, which highlights the necessity of serological testing of plant samples for the determination of virus infection in oilseed rape.

Epidemiology and Disease Cycle

Disease intensity varies among years and depends on weather conditions during season, source of inoculum, and vector population (Jasnic and Bagi 2007). Coutts and Jones (2000) found wild radish (*R. raphanistrum*) as a substantial virus reservoir in canola field in Southwest Australia. If TuMV and CaMV viruses are transferred from *S. arvensis* to oilseed rape with aphid, then wild mustard could be a reservoir of these virus infections for oilseed rape under natural condition (Dikova 2008).

TuYV can be transmitted by *Myzus persicae* (act as main vector), *Brachycorynella asparagi*, *Cavariella aegopodii*, *Macrosiphoniella sanborni*, *Macrosiphum albifrons*, *Myzus nicotianae*, *Nasonovia ribisnigri*, *Pentatrichopus fragaefolii*, *Rhopalosiphum maidis*, *Acyrthosiphon pisum* (green race), *Aphis gossypii*, *Aulacorthum circumflexum*, *Aulacorthum solani*, *Brevicoryne brassicae*, *Rhopalosiphum padi*, and *Sitobion avenae* (Schliephake et al. 2000). Green peach aphid (*M. persicae*), turnip or mustard aphid (*Lipaphis erysimi*), cabbage aphid (*B. brassicae*) and cowpea aphid (*Aphis craccivora*) may serve as vector of BWYV, TuMV, and CaMV on rape and Indian mustard in New South Wales, Australia (Hertel et al. 2004). Stevens et al. (2008) found peach-potato aphid, *M. persicae*, as vector of TuYV and reported 72% of winged *M. persicae* carry this virus on oilseed rape in Australia. He also reported that milder autumn and winter conditions favor the development of the aphid vectors and encourage virus spread. Green peach aphid requires minimum 0.5–1.0 h for acquisition and inoculation process for BWYV transmission on mustard. This transmission is influenced by temperature, and the highest transmission rate could be obtained at 20°C–25°C. After virus acquisition, aphid could retain BWYV for at least 2 weeks. The virus could not pass onto the progeny of its vector. BWYV replicated well in mustard at 15°C–25°C (Chen 2003).

Maling et al. (2010) modified a previously developed hybrid mechanistic/statistical model, which was used to predict vector activity and epidemics of vector-borne viruses, to simulate virus epidemics in the BWYV-*B. napus* pathosystem in a Mediterranean-type environment.

Disease Management

Coutts et al. (2010) found that *B. napus*, which has some resistance to BWYV, can be used in conjunction with imidacloprid seed dressings as component of an integrated pest management strategy to manage BWYV in *B. napus* crops. The genetically engineered cross protection of *Brassica* crops with weak strain Bari-1 Gene VI of CaMV and its genetic regularity have been studied by Gong et al. (2001). Lehmann et al. (2003) were able to induce coat protein–mediated resistance to TuMV in *B. napus*. Control measures include the elimination of inoculum source, isolation from areas, which contain inoculum, aphid control, and growing virus resistant or tolerant oilseed rape genotypes (Jasnic and Bagi 2007). Seed dressing with imidacloprid in sufficient amount (525 g ai/100 kg of seed) before sowing is a good prospect for the control of BWYV and *M. persicae* in *B. napus* crops (Jones et al. 2007). Seed treatment of oilseed rape with thiamethoxam is an excellent alternative to insecticide spray for controlling TuYV transmission by *M. persicae* (Dewar et al. 2011).

Phyllody and Aster Yellows

Symptoms

The characteristic symptom is the transformation of floral parts into leafy structures. The corolla becomes green and sepaloid. The stamens turn green and become indehiscent. The gynoecium is borne on a distinct gynophore and produces no ovules in the ovary. In addition, there are some leafy structures attached to the false septum. The affected plants may show varying degrees of severity of the disease, and the affected part of the raceme does not form siliquae. Some plants may show only terminal portion of the branches affected with the disease, whereas in others, the whole branches show the symptoms.

Geographical Distribution and Losses

Under natural conditions, the phyllody has been reported to occur on oilseed rape (*B. napus*) in Greece (Maliogka et al. 2009), Italy (Rampin et al. 2010), Poland (Zwolinska et al. 2011), Iran (Salehi et al. 2011), Canada on *B. rapa* (Olivier et al. 2006), India on toria (*B. rapa* var. toria), and yellow sarson (*B. rapa* var. yellow sarson) in the states of Punjab, Haryana, New Delhi, and Uttar Pradesh (Kolte 1985, Azadvar and Baranwal 2010) in India. In Canada, aster yellows (AY) has also been reported on *B. napus* and *B. rapa* (Olivier et al. 2010). Yield loss may go up to 90% (Kolte 1985).

Pathogen

The phyllody disease is reported to be caused by the *Candidatus Phytoplasma asteris* (*Phytoplasma asteris*) phytoplasma of subgroup 16Srl-A, 16Srl-B, and 16SrlX-C (Olivier et al. 2006, Maliogka et al. 2009, Azadvar and Baranwal 2010, Rampin et al. 2010, Zwolinska et al. 2011). The AY disease is reported to be caused by the phytoplasma of 16Srl-A and 16Srl-B subgroup (Olivier et al. 2010). Phytoplasma are nonhelical, mycoplasma-like bacteria that lack cell walls. They almost exclusively inhabit the phloem sieve-tube elements of the infected plant and is transmitted from plant to plant by phloem-feeding homopteran insects mainly plant hoppers (*Laodelphax striatellus*) (Azadvar and Baranwal 2010) or leafhoppers (*Circulifer haematoceps*) (Salehi et al. 2011), and less frequently psyllids and Cicadellidae (Jajor 2007). Seed transmission of phytoplasma in winter oilseed rape has also been reported (Calari et al. 2011). This pathogen can be detected by several methods, namely, nested PCR (Olivier et al. 2006, Azadvar et al. 2011, Calari et al. 2011), PCR, and RFLP (Wang and Hiruki 2001, Zwolinska et al. 2011). Sequencing and phylogenetic analysis of 16S rRNA, a part of 23S rRNA, partial sec A genes, rp gene and 16S–23S intergenic spacer region, and RFLP pattern of 1.25 kb 16S rDNA sequences of the pathogen has also been done (Azadvar and Baranwal 2010).

Epidemiology

Early planting of toria in late August or at its normal planting time in September has been shown to favor the development of the disease in toria under Indian conditions. As high as 24% incidence of the disease, depending on the variety, can be seen in plants sown in August (Kolte 1985). The disease development is favored by prolonged dry and warm weather. The pathogen survives on alternate hosts like sesame, which serves as primary source of infection. The disease is transmitted through leafhopper and seeds. The disease spreads by repeated cycles of secondary infection through the process of transmission.

Disease Management

Weed management in and around fields that serve as hosts for the pathogen and increasing planting population may reduce the incidence of phyllody. Roguing and destruction of the infected plants will reduce further spread of disease. The population of insect vectors should be controlled by using appropriate insecticides when they are at their peak. Two sprays of dimethoate or metasystox at 0.1% at an interval of 15 days starting from the initiation of symptoms should be done to manage the insect vectors.

REFERENCES

Abedi, A., M. Motalebi, M. Zamani, and K. Piri. 2011. Transformation of canola (R line Hyola 308) by pgip1 gene from bean cv. Daneshjoo to improve resistance to *Sclerotina sclerotiorum*. *Iran. J. Field Crop Sci.* 42: 53–61.

Abuamsha, R., M. Salman, and R.U. Ehlers. 2011. Effect of seed priming with *Serratia plymuthica* and *Pseudomonas chlororaphis* to control *Leptosphaeria maculans* in different oilseed rape cultivars. *Eur. J. Plant Pathol.* 130: 287–295.

Adhikari, T.B., J.Q. Liu, S. Mathur, C.R.X. Wu, and S.R. Rimmer. 2003. Genetic and molecular analyses in crosses of race 2 and race 7 of *Albugo candida*. *Phytopathology* 93: 959–965.

Aeron, A., R.C. Dubey, D.K. Maheshwari, P. Pandey, V.K. Bajpai, and S.C. Kang. 2011. Multifarious activity of bioformulated *Pseudomonas fluorescens* PS1 and biocontrol of *Sclerotinia sclerotiorum* in Indian rapeseed (*Brassica campestris* L.). *Eur. J. Plant Pathol.* 131: 81–93.

Ahmad, H., S. Hasnain, and A. Khan. 2002. Evolution of genomes and genomic relationship among rapeseed-mustard. *Biotechnology* 1: 78–87.

AICRP-RM (All India Coordinated Res. Project Rapeseed-Mustard). 1986–2014. Annual reports. Directorate of Rapeseed-Mustard Research (ICAR), Bharatpur, India.

Akhgari, A.B., M. Motallebi, and M.R. Zamani. 2012. Bean polygalacturonase-inhibiting protein expressed in transgenic *Brassica napus* inhibits polygalacturonase from its fungal pathogen *Rhizoctonia solani*. *Plant Prot. Sci.* 48: 1–9.

Akram, A., S.M. Iqbal, N. Ahmed, U. Iqbal, and A. Ghafoor. 2008. Morphological variability and mycelia compatibility among the isolates of *Sclerotinia sclerotiorum* associated with stem rot of chickpea. *Pak. J. Bot.* 40: 2663–2668.

Alam, K.H., M.M. Haque, F.M. Aminuzzaman, A.N.F. Ahmmed, and M.R. Islam. 2010. Effect of different fungicides and plant extracts on the incidence and severity of gray blight of mustard. *Int. J. Sustain. Agric. Technol.* 6(8): 6–9.

Alizadeh, N., A. Babai-Ahary, Y. Assadi, M. Valizadeh, and B. Passebaneslam. 2006. The effects of sclerotinia stem rot of oilseed rape on the production and quality of extracted oil. *J. Sci. Technol. Agric. Nat. Res.* 10: 485–495.

Ananga, A.O., E. Cebert, K. Soliman, R. Kantety, R.P. Pacumbaba, and K. Konan. 2006. RAPD markers associated with resistance to blackleg disease in *Brassica* species. *Afr. J. Biotechnol.* 5: 2041–2048.

Aneja, J.K. and A. Agnihotri. 2013. Alternaria blight of oilseed brassicas: Epidemiology and disease control strategies with special reference to use of biotechnological approaches for attaining host resistance. *J. Oilseed Brassica* 4: 1–10.

Anwar, M.J. and S.M. Khan. 2001. Efficacy of some fungicides for the control of leaf blight on rapeseed & mustard. *Pak. J. Phytopathol.* 13: 167–169.

Atwal, A.K., Ramandeep, S.K. Munshi, and A.P.S. Mann. 2004. Biochemical changes in relation to Alternaria leaf blight in Indian mustard. *Plant Dis. Res.* (*Ludhiana*) 19: 57–59.

Atwal, A.K. and M.K. Sangha. 2004a. Production of extracellular enzymes by *A. brassicae* under different cultural conditions. *Cruciferae Newsl.* 25: 73–74.

Atwal, A.K. and M.K. Sangha. 2004b. Effect of salicylic acid spray on sugar metabolites in *B. juncea* leaves infected with *A. brassicae*. *Cruciferae Newsl.* 25: 75–76.

Atwal, A.K., R. Toor, and R.K. Raheja. 2005. Lipid compositional changes due to Alternaria blight in the leaves of Indian mustard, *Brassica juncea* L. *J. Oilseeds Res.* 22: 108–110.

Awasthi, R.P., N.I. Nashaat, S.J. Kolte, A.K. Tewari, P.D. Meena, and R. Bhatt. 2012. Screening of putative resistant sources against Indian and exotic isolates of *Albugo candida* inciting white rust in rapeseed-mustard. *J. Oilseed Brassica* 3: 27–37.

Azadvar, M. and V.K. Baranwal. 2010. Molecular characterization and phylogeny of a phytoplasma associated with phyllody disease of toria (*Brassica rapa L.* subsp. *dichotoma* (Roxb.)) in India. *Indian J. Virol.* 21: 133–139.

Azadvar, M., V.K. Baranwal, and D.K. Yadava. 2011. Transmission and detection of toria [*Brassica rapa* L. subsp. *dichotoma* (roxb.)] phyllody phytoplasma and identification of a potential vector. *J. Gen. Plant Pathol.* 77: 194–200.

Balesdent, M.H., A. Attard, M.L. Kuhn, and T. Rouxel. 2002. New avirulence genes in the phytopathogenicfungus *Leptosphaeria maculans*. *Phytopathology* 92: 1122–1133.

Balesdent, M.H., M.J. Barbetti, H. Li, K. Sivasithamparam, L. Gout, and T. Rouxel. 2005. Analysis of *Leptosphaeria maculans* race structure in a worldwide collection of isolates. *Phytopathology* 95: 1061–1071.

Balesdent, M.H., I. Fudal, B. Ollivier, P. Bally, J. Grandaubert, F. Eber, A.M.M. Leflon, and T. Rouxel. 2013. The dispensable chromosome of *Leptosphaeria maculans* shelters an effector gene conferring avirulence towards *Brassica rapa*. *New Phytol.* 198: 887–898.

Balesdent, M.H., K. Louvard, X. Pinochet, and T. Rouxel. 2006. A large-scale survey of races of *Leptosphaeria maculans* occurring on oilseed rape in France. [Special Issue: Sustainable strategies for managing *Brassica napus* (oilseed rape) resistance to *Leptosphaeria maculans*, phoma stem canker]. *Eur. J. Plant Pathol.* 114: 53–65.

Ballinger, D.J., F.A. Salisbury, J.F. Kollmorgen, T.D. Potter, and D.R. Coventry. 1988. Evaluation of rates of fiutriafol for control of blackleg of rapeseed. *Aust. J. Agric. Anim. Husb.* 28: 517–519.

Bankina, B., Z. Gaile, O. Balodis, and R. Vitola. 2008. Phoma blackleg (stem canker) of oilseed rape in Latvia. *Agronomijas Vestis* 10: 93–97.

Bansal, V.K., J.P. Tewari, G.R. Stringam, and M.R. Thiagarajah. 2005. Histological and inheritance studies of partial resistance in the *Brassica napus–Albugo candida* host–pathogen interaction. *Plant Breed.* 124: 27–32.

Baranowski, P., W. Mazurek, M. Jedryczka, and D. Babula-Skowronska. 2009. Temperature changes of oilseed rape (*Brassica napus*) leaves infected by fungi of *Alternaria* sp. *Rosliny Oleiste* 30: 21–33.

Barari, H., V. Alavi, E. Yasari, and S.M. Badalyan. 2011. Study of genetic variations based on the morphological characteristics, within the population of *Sclerotinia sclerotiorum* from the major oilseed planting areas in Iran. *Int. J. Biol.* 3(2): 61–66.

Barrins, J.M., P.K. Ades, P.A. Salisbury, and B.J. Howlett. 2004. Genetic diversity of Australian isolates of *Leptosphaeria maculans*, the fungus that causes blackleg of canola (*Brassica napus*). *Aust. Plant Pathol.* 33: 529–536.

Beatty, P.H. and S.E. Jensen. 2002. *Paenibacillus polymyxa* produces fusaricidin-type antifungal antibiotics active against *Leptosphaeria maculans*, the causative agent of blackleg disease of canola. *Can. J. Microbiol.* 48: 159–169.

Berg, T., L. Tesoriero, and D.L. Hailstones. 2005. PCR-based detection of *Xanthomonas campestris* pathovars in Brassica seed. *Plant Pathol.* 54: 416–427.

Berg, T., L. Tesoriero, and D.L. Hailstones. 2006. A multiplex real-time PCR assay for detection of *Xanthomonas campestris* from brassicas. *Lett. Appl. Microbiol.* 42: 624–630.

Berkenkamp, B. 1980. Effects of fungicides and herbicides on staghead of rape. Canadian. *J. Plant Sci.* 60: 1039–1040.

Berry, C., W.G.D. Fernando, P.C. Loewen, and T. R. De Kievit. 2010. Lipopeptides are essential for *Pseudomonas* sp. DF41 biocontrol of *Sclerotinia sclerotiorum. Biol. Control* 55: 211–218.

Bhatiya, B.S. and R.P. Awasthi. 2007. In vitro evaluation of some antifungal plant extracts against *Alternaria brassicae* causing Alternaria blight of rapeseed-mustard. *J. Plant Dis. Sci.* 2: 126–131.

Bhatt, R., R.P. Awasthi, and A.K. Tewari. 2009. Management of downy mildew and white rust diseases of mustard. *Pantnagar J. Res.* 7: 54–59.

Bhattacharya, B.K. and C. Chattopadhyay. 2013. A multi-stage tracking for mustard rot disease combining surface meteorology and satellite remote sensing. *Comput. Electron. Agric.* 90: 35–44.

Bhattacharyya, T.K. and N.C. Mandal. 2006. Management of clubroot (*Plasmodiophora brassicae*) of rapeseed and mustard by nitrogenous fertilizers. *Ann. Plant Prot. Sci.* 14: 260–261.

Bhowmik, T.P. and B.M. Trivedi. 1980. A new bacterial stalk rot of *Brassica. Curr. Sci.* 49: 674–675.

Bila, J., C.N. Mortensen, M. Andresen, J.G. Vicente, and E.G. Wulff. 2013. *Xanthomonas campestris* pv. *campestris* race 1 is the main causal agent of black rot of Brassicas in Southern Mozambique. *Afr. J. Biotechnol.* 12: 602–610.

Biswas, C., R. Singh, and R.B. Tewari. 2007. Management of white rust (*A. candida*) of mustard (*Brassica juncea*) by altering sowing date and fungicides. *Indian J. Agric. Sci.* 77: 626–628.

Biswas, K.K. and A. Chowdhury. 2005. Rai mosaic disease in rai (*Brassica juncea*) and radish (*Raphanus sativus*) in the Himalayan hilly regions of West Bengal. *J. Mycopathol. Res.* 43: 263–266.

Boerema, G.H. 1976. The *Phoma* species studied in culture by Dr. R.W.G. Dennis. *Trans. Br. Mycol. Soc.* 67: 289–219.

Bokor, A., M.J. Barbetti, and P.M. Wood. 1975. Blackleg of rapeseed. *J. Dept. Agric. West. Aust.* 16: 7–10.

Boland, G.J. and R. Hall. 1994. Index of plant hosts of *Sclerotinia sclerotiorum. Can. J. Plant Pathol.* 16: 94–108.

Bom, M. and G.J. Boland. 2000. Evaluation of polyclonal-antibody-based immunoassays for detection of *Sclerotinia sclerotiorum* on canola petals, and prediction of stem rot. *Can. J. Microbiol.* 46: 723–729.

Borges, A.A., H.J. Cools, and J.A. Lucas. 2003. Menadione sodium bisulphite: A novel plant defence activator which enhances local and systemic resistance to infection by *Leptosphaeria maculans* in oilseed rape. *Plant Pathol.* 52: 429–436.

Borhan, M.H., E. Brose, J.L. Beynon, and E.B. Holub. 2001. White rust (*Albugo candida*) resistance loci on three Arabidopsis chromosomes are closely linked to downy mildew (*Peronospora parasitica*) resistance loci. *Mol. Plant Pathol.* 2: 87–95.

Borhan, M.H., N. Gunn, A. Cooper, S. Gulden, M. Tor, S.R. Rimmer, and E.B. Holub. 2008. WRR4 encodes a TIR-NB-LRR protein that confers broad-spectrum white rust resistance in *Arabidopsis thaliana* to four physiological races of *Albugo candida. Mol. Plant Microbe Interact.* 21: 757–768.

Borhan, M.H., E.B. Holub, C. Kindrachuk, M. Omidi, G. Bozorgmanesh-Frad, and S.R. Rimmer. 2010. WRR4, a broad-spectrum TIR-NB-LRR gene from *Arabidopsis thaliana* that confers white rust resistance in transgenic oilseed brassica crops. *Mol. Plant Pathol.* 11: 283–291.

Brachaczek, A., J. Kaczmarek, K. Michalski, and M. Jedryczka. 2011. The influence of the fungicide containing flusilazole on yield and seed quality of winter oilseed rape. *Rosliny Oleiste* 32: 167–180.

Brazauskiene, I., E. Petraitiene, A. Piliponyte, and G. Brazauskas. 2011. Diversity of *Leptosphaeria maculans/L. biglobosa* species complex and epidemiology of phoma stem canker on oilseed rape in Lithuania. *J. Plant Pathol.* 93: 577–585.

Brazauskiene, I., E. Petraitiene, A. Piliponyte, and G. Brazauskas. 2012. The peculiarities of phoma stem canker (*Leptosphaeria maculans/L. biglobosa* complex) infections in winter and spring oilseed rape (*Brassica napus* L.). *Zemdirbyste* (*Agriculture*) 99: 379–386.

Brazauskiene, I., E. Petraitiene, and E. Povilioniene. 2007. Investigation of indicators of Phoma stem canker (*Leptosphaeria maculans*) epidemiology and disease incidence in winter rape. *Zemdirbyste* (*Agriculture*) 94: 176–188.

Bremer, H. 2007. Development of a forecasting system to control stem canker *Phoma lingam* (Teleomorph: *Leptosphaeria maculans*) on winter oilseed rape in autumn. *Gesunde Pflanzen.* 59: 161–169.

Bull, C.T. and I. Rubio. 2011. First report of bacterial blight of crucifers caused by *Pseudomonas cannabina* pv. *alisalensis* in Australia. *Plant Dis.* 95: 1027.

Butler, E.J. and S.G. Jones. 1961. *Plant Pathology*. McMillan, London, U.K.

Buzza, G. and A. Easton. 2002. A new source of blackleg resistance from *Brassica sylvestris*. In: *GCIRC Technical Meeting*, Poznan, Poland. Poland Bulletin No. 18.

Cael, N., A. Penaud, and Y. Decroos. 2001. New elements in the fight against Sclerotinia on oilseed rape: A new biological fungicide studied in France. *Phytoma* 539: 12–16.

Cai, L., K.R. Chen, X.J. Zhang, L.Y. Yan, M.S. Hou, and Z.Y. Xu. 2009. Biological and molecular characterization of a crucifer Tobamovirus infecting oilseed rape. *Biochem. Genet.* 47: 451–461.

Calari, A., S. Paltrinieri, N. Contaldo, D. Sakalieva, N. Mori, B. Duduk, and A. Bertaccini. 2011. Molecular evidence of phytoplasmas in winter oilseed rape, tomato and corn seedlings. *Bull. Insectol.* 64(Suppl.): S157–S158.

Cao, T.S., S. Srivastava, M.H. Rahman, N.N.V. Kav, N. Hotte, M.K. Deyholos, and S.E. Strelkov. 2008. Proteome-level changes in the roots of *Brassica napus* as a result of *Plasmodiophora brassicae* infection. *Plant Sci.* 174: 97–115.

Cao, T.S., J. Tewari, and S.E. Strelkov. 2007. Molecular detection of *Plasmodiophora brassicae*, causal agent of clubroot of crucifers, in plant and soil. *Plant Dis.* 91: 80–87.

Casimiro, S., M. Moura, L. Ze-Ze, R. Tenreiro, and A.A. Monteiro. 2004. Internal transcribed spacer 2 amplicon as a molecular marker for identification of *Peronospora parasitica* (crucifer downy mildew). *J. Appl. Microbiol.* 96: 579–587.

Chand, P. and D. Rai. 2009. Sources of resistance against sclerotinia rot in rapeseed-mustard. *Int. J. Plant Prot.* 2: 101–102.

Chandra, B., R.P. Awasthi, and A.K. Tiwari. 2009. Eco-friendly disease management of Alternaria blight (*Alternaria brassicae*) of rapeseed and mustard. *Environ. Ecol.* 27: 906–910.

Chandra, S., K. Choure, R.C. Dubey, and D.K. Maheshwari. 2007. Rhizosphere competent Mesorhizobiumloti MP6 induces root hair curling, inhibits *Sclerotinia sclerotiorum* and enhances growth of Indian mustard (*Brassica campestris*). *Braz. J. Microbiol.* 38: 124–130.

Chattopadhyay, A.K. 1991. Studies on the control of clubroot disease of rapeseed-mustard in West Bengal. *Indian Phytopathol.* 44: 397–398.

Chattopadhyay, A.K. and C.K. Bhunia. 2003. Management of Alternaria leaf blight of rapeseed-mustard by chemicals. *J. Mycopathol. Res.* 41: 181–183.

Chattopadhyay, A.K., A.K. Moitra, and C.K. Bhunia. 2001. Evaluation of *Brassica* species for resistance to *Plasmodiophora brassicae* causing club root of rapeseed mustard. *Indian Phytopathol.* 54: 131–132.

Chattopadhyay, C. 2008. Management of diseases of rapeseed-mustard with special reference to Indian conditions. In: *Sustainable Production of Oilseeds: Rapeseed-Mustard Technology*, eds., A. Kumar, J.S. Chauhan, and C. Chattopadhyay. Agrotech Publishing Academy, Udaipur, India, pp. 364–388.

Chattopadhyay, C., R. Agrawal, A. Kumar, L.M. Bhar, P.D. Meena, R.L. Meena, S.A. Khan et al. 2005. Epidemiology and forecasting of Alternaria blight of oilseed *Brassica* in India—A case study. *Z. Pflanzenk. Pflanzen.* 112: 351–365.

Chattopadhyay, C., R. Agrawal, A. Kumar, R.L. Meena, K. Faujdar, N.V.K. Chakravarthy, A. Kumar, P. Goyal, P.D. Meena, and C. Shekhar. 2011. Epidemiology and development of forecasting models for White rust of *Brassica juncea* in India. *Arch. Phytopathol. Plant Prot.* 44: 751–763.

Chattopadhyay, C., P.D. Meena, and S. Kumar. 2002. Management of Sclerotinia rot of Indian mustard using ecofriendly strategies. *J. Mycol. Plant Pathol.* 32: 194–200.

Chattopadhyay, C., P.D. Meena, and R.L. Meena. 2004. Integrated management of Sclerotinia rot of Indian Mustard. *Indian J. Plant Prot.* 32: 88–92.

Chattopadhyay, C., P.D. Meena, R.K. Sastry and R.L. Meena. 2003. Relationship among pathological and agronomic attributes for soilborne diseases of three oilseed crops. *Indian J. Plant Prot.* 31: 127–128.

Chattopadhyay, C. and G. Séguin-Swartz. 2005. Breeding for disease resistance in oilseed crops in India. *Annu. Rev. Plant Pathol.* 3: 101–142.

Chattopadhyay, C., R. Vijay Kumar, and P.D. Meena. 2007. Biomanagement of Sclerotinia rot of *Brassica juncea* in India—A case study. *Phytomorphology* 57: 71–83.

Chattopadhyay, S.B. and S.K. Sengupta. 1952. Addition to fungi of Bengal. *Bull. Bot. Soc. Bengal* 6: 57–61.

Chaudhary, P., U. Rawat, and R. Govila. 2010. Control of *Sclerotinia sclerotiorum* (Lib.) de Bary in Indian mustard with fungicides. *J. Plant Dev. Sci.* 2: 45–49.

Chawla, H.K.L., V. Gupta, and G.S. Saharan. 2001. Changes in activities of oxidative enzymes in *Brassica juncea* leaves during interaction with *Alternaria brassicae*. *Cruciferae Newsl.* 23: 55–56.

Chen, B.Y., Q. Hu, C. Dixelius, G.Q. Li, and X.M. Wu. 2010a. Genetic diversity in *Sclerotinia sclerotiorum* assessed with SRAP markers. *Biodivers. Sci.* 18: 509–515.

Chen, C.Y. and B.J. Howlett. 1996. Rapid necrosis of guard cells is associated with the arrest of fungal growth in leaves of Indian mustard (*Brassica juncea*) inoculated with avirulent isolates of *Leptosphaeria maculans*. *Physiol. Mol. Plant Pathol.* 48: 73–81.

Chen, G.Y., C.P. Wu, B. Li, H. Su, S.Z. Zhen, and Y.L. An. 2010b. Detection of *Leptosphaeria maculans* from imported Canola seeds. *J. Plant Dis. Prot.* 117: 173–176.

Chen, T.H. 2003. Occurrence, identification, aphid transmission and ecology of Beet western yellows virus in Taiwan. *Plant Pathol. Bull.* 12: 43–56.

Chen, T.H., S.F. Wu, T.J. Chen, and H.L. Chen. 2000. Occurrence and serodiagnosis of virus diseases of crucifers in Taiwan. *Plant Pathol. Bull.* 9: 39–46.

Chen, X.L., X.H. Zhu, Y.C. Ding, and Y.C. Shen. 2011. Antifungal activity of tautomycin and related compounds against *Sclerotinia sclerotiorum*. *J. Antibiot.* 64: 563–569.

Chen, Y. and W.G.D. Fernando. 2006a. Prevalence of pathogenicity groups of *Leptosphaeria maculans* in western Canada and North Dakota, USA. *Can. J. Plant Pathol.* 28: 533–539.

Chen, Y. and W.G.D. Fernando. 2006b. Induced resistance to blackleg (*Leptosphaeria maculans*) disease of canola (*Brassica napus*) caused by a weakly virulent isolate of *Leptosphaeria biglobosa*. *Plant Dis.* 90: 1059–1064.

Cheung, W.Y., R.K. Gugel, and B.S. Landry. 1998. Identification of RFLP markers linked to the white rust resistance gene (*Acr*) in mustard (*Brassica juncea* (L.) Czern. and Coss.). *Genome* 41: 626–628.

Chevre, A.M., H. Brun, F. Eber, J.C. Letanneur, P. Vallee, M. Ermel, I. Glais, H. Li, K. Sivasithamparam, and M.J. Barbetti. 2008. Stabilization of resistance to *Leptosphaeria maculans* in *Brassica napus-B. juncea* recombinant lines and its introgression into spring-type *Brassica napus*. *Plant Dis.* 92: 1208–1214.

Chhikara, S., D. Chaudhury, O.P. Dhankher, and P.K. Jaiswal. 2012. Combined expression of a barley class II chitinase and type I ribosome inactivating protein in transgenic *Brassica juncea* provides protection against *Alternaria brassicae*. *Plant Cell Tissue Organ Cult.* 108: 83–89.

Choi, Y., H. Shin, S. Ploch, and M. Thines. 2011. Three new phylogenetic lineages are the closest relatives of the widespread species *Albugo candida*. *Fungal Biol.* 115: 598–607.

Choi, Y.J., H.D. Shin, S.B. Hong, and M. Thines. 2007. Morphological and molecular discrimination among *Albugo candida* materials infecting *Capsella bursa-pastoris* world-wide. *Fungal Divers.* 27: 11–34.

Choi, Y.J., H.D. Shin, S. Ploch, and M. Thines. 2008. Evidence for uncharted biodiversity in the *Albugo candida* complex, with the description of a new species. *Mycol. Res.* 112: 1327–1334.

Choi, Y.J., H.D. Shin, and M. Thines. 2009. The host range of *Albugo candida* extends from Brassicaceae through Cleomaceae to Capparaceae. *Mycol. Prog.* 8: 329–335.

Christianson, J.A., S.R. Rimmer, A.G. Good, and D.J. Lydiate. 2006. Mapping genes for resistance to *Leptosphaeria maculans* in *Brassica juncea*. *Genome* 49: 30–41.

Clarke, W.E., I.A. Parkin, H.A. Gajardo, D.J. Gerhardt, E. Higgins, C. Sidebottom, A.G. Sharpe et al. 2013. Genomic DNA enrichment using sequence capture microarrays: A novel approach to discover sequence nucleotide polymorphisms (SNP) in *Brassica napus* L. *PLoS One* 8(12): e81992. doi:10.1371/journal.pone.0081992.

Coffey, M.D. 1975. Ultra structural features of the haustorial apparatus of the white blister fungus, *Albugo candida*. *Can. J. Bot.* 53: 1285–1299.

Commonwealth Mycological Institute. 1978. Distribution maps of plant diseases. Map No. 73. Edition 4, Surrey, England.

Conn, K.L., J.P. Tewari, and J.S. Dahiya. 1988. Resistance to *Alternaria brassicae* and phytoalexin-elicitation in rapeseed and other crucifers. *Plant Sci.* 55: 21–25.

Corato, U. and G. de Baviello. 2000. Occurrence of *Sclerotinia sclerotiorum* on *Brassica carinata* in Southern Italy (Basilicata). *Inf. Fitopatol.* 50: 61–63.

Couper, G., A. Litterick, and C. Leifert. 2001. Control of *Sclerotinia* within carrot crops in NE Scotland: The effects of irrigation and compost application on sclerotia germination. http://www.abdn.ac.uk/organic/pdfs/sclerotia.pdf. Accessed on 26 September, 2014.

Coutts, B.A. and R.A.C. Jones. 2000. Viruses infecting canola (*Bassica napus*) in south-west Australia: Incidence, distribution, spread, and infection reservoir in wild radish (*Raphanus raphinistrum*). *Aust. J. Agric. Res.* 51: 925–936.

Coutts, B.A., C.G. Webster, and R.A.C. Jones. 2010. Control of Beet western yellows virus in *Brassica napus* crops: Infection resistance in Australian genotypes and effectiveness of imidacloprid seed dressing. *Crop Pasture Sci.* 61: 321–330.

Cramer, R.A. and C.B. Lawrence. 2004. Identification of *Alternaria brassicicola* genes expressed in planta during pathogenesis of *Arabidopsis thaliana*. *Fungal Genet. Biol.* 41: 115–128.

Cunningham, G.H. 1927. Dry-rot of swedes and turnips: Its causes and control. Wellington, New Zealand. *New Zeal. Dept. Agric. Bull.* 133: 1–51.

Dang, J.K., M.S. Sangwan, N. Mehta, and C.D. Kaushik. 2000. Multiple disease resistance against four fungal foliar diseases of rapeseed-mustard. *Indian Phytopathol.* 53: 455–458.

Das, S.N., S.K. Mishra, and P.K. Swain. 1987. Reaction of some toria varieties to *Plasmodiophora brassicae*. *Indian Phytopathol.* 40: 120.

Dawidziuk, A. and M. Jedryczka. 2011. Mathematical modeling of life cycles of pathogenic microorganisms as a tool to protect oilseed rape against fungal diseases. *Rosliny Oleiste* 32: 181–194.

Dawidziuk, A., J. Kaczmarek, and M. Jedryczka. 2012. The effect of winter weather conditions on the ability of pseudothecia of *Leptosphaeria maculans* and *L. biglobosa* to release ascospores. *Eur. J. Plant Pathol.* 134: 329–343.

Dawidziuk, A., I. Kasprzyk, J. Kaczmarek, and M. Jedryczka. 2010. Pseudothecial maturation and ascospore release of *Leptosphaeria maculans* and *L. biglobosa* in south-east Poland. *Acta Agrobot.* 63: 107–120.

De Bary, A. 1887. *Comparative Morphology and Biology of the Fungi, Mycetezoa and Bacteria* (English translation). Clarendon Press, Oxford, U.K., pp. 135–136.

Delourme, R., H. Brun, M. Ermel, M.O. Lucas, P. Vallee, C. Domin, G. Walton, H. Li, K. Sivasithamparam, and M.J. Barbetti. 2008. Expression of resistance to *Leptosphaeria maculans* in *Brassica napus* double haploid lines in France and Australia is influenced by location. *Ann. Appl. Biol.* 153: 259–269.

Delourme, R., A.M. Chevre, H. Brun, T. Rouxel, M.H. Balesdent, J.S. Dias, P. Salisbury, M. Renard, and S.R. Rimmer. 2006. Major gene and polygenic resistance to *Leptosphaeria maculans* in oilseed rape (*Brassica napus*). *Eur. J. Plant Pathol.* 114: 41–52.

Delourme, R., M.L. Pilet-Nayel, M. Archipiano, R. Horvais, X. Tanguy, T. Rouxel, H. Brun, M. Renard, and M.H. Balesdent. 2004. A cluster of major specific resistance genes to *Leptosphaeria maculans* in *Brassica napus*. *Phytopathology* 94: 578–583.

Deora, A., B.D. Gossen, and M.R. McDonald. 2012. Infection and development of *Plasmodiophora brassicae* in resistant and susceptible canola cultivars. *Can. J. Plant Pathol.* 34: 239–247.

Deora, A., B.D. Gossen, F. Walley, and M.R. McDonald. 2011. Boron reduces development of clubroot in canola. *Can. J. Plant Pathol.* 33: 475–484.

Desai, A.G., C. Chattopadhyay, R. Agrawal, A. Kumar, R.L. Meena, P.D. Meena, K.C. Sharma, M.S. Rao, Y.G. Prasad, and Y.S. Ramakrishna. 2004. *Brassica juncea* powdery mildew epidemiology and weather based forecasting models for India—A case study. *Z. Pflanzenk. Pflanzen.* 111: 429–438.

Desoignies, N., M. Eickermann, P. Delfosse, F. Kremer, N. Godart, L. Hoffmann, and A. Legreve. 2009. First report of *Plasmodiophora brassicae* on rapeseed in the Grand Duchy of Luxembourg. *Plant Dis.* 93: 1220.

Devi, N.A., R.K.T. Devi, and N.I. Singh. 2009. Fungicidal properties of some plant extracts against Fusarium wilt of rapeseed. *J. Interacad.* 13: 404–407.

Dewar, A.M., M.F. Tait, and M. Stevens. 2011. Efficacy of thiamethoxam seed treatment against aphids and turnip yellows virus in oilseed rape. *Aspects Appl. Biol.* 106: 195–202.

Dhaliwal, H.J.S., T.S. Thind, and C. Mohan. 2003. Fungitoxic potential of ten essential oils against *Alternaria brassicae* (Berk.) Sacc. *Pesticide Res. J.* 15: 142–144.

Dikova, B.A. 2008. *Sinapis arvensis* L. as a source of viruses—Cauliflower mosaic virus (CaMV) and Turnip mosaic virus (TuMV) infecting oilseed rape. *Acta Phytopathol. Entomol. Hung.* 43: 93–99.

Dilmaghani, A., M.H. Balesdent, J.P. Didier, C. Wu, J. Davey, M.J. Barbetti, H. Li et al. 2009. The *Leptosphaeria maculans-Leptosphaeria biglobosa* species complex in the American continent. *Plant Pathol.* 58: 1044–1058.

Dion, Y., R.K. Gugel, G.F.W. Rakow, G. Séguin-Swartz, and B.S. Landry. 1995. RFLP mapping of resistance to the blackleg disease [causal agent, *Leptosphaeria maculans* (Desm.) Ces. Et De Not.] in canola (*Brassica napus* L.). *Theor. Appl. Genet.* 91: 1190–1194.

Donald, C. and I. Porter. 2003. Clubroot (*Plasmodiophora brassicae*) an imminent threat to the Australian canola industry. In: *Thirteenth Biennial Australian Research Assembly on Brassicas. Proceedings of Conference*, Tamworth, New South Wales, Australia, September 8–12, 2003, pp. 114–118.

Dong, X.B., R.Q. Ji, X.L. Guo, S.J. Foster, H. Chen, C.H. Dong, Y.Y. Liu, Q.O. Hu, and S.Y. Liu. 2008. Expressing a gene encoding wheat oxalate oxidase enhances resistance to *Sclerotinia sclerotiorum* in oilseed rape (*Brassica napus*). *Planta* 228: 331–340.

Doullah, M.A.U., M.B. Meah, G.M. Mohsin, A. Hassan, T. Ikeda, H. Hori, and K. Okazaki. 2009. Evaluation of resistance in *Brassica rapa* to dark pod spot (*Alternaria brassicicola*) using the in vitro detached pod assay. *SABRAO J. Breed. Genet.* 41: 101–113.

DRMR (Directorate of Rapeseed-Mustard Research). 2010. Annual Report 2009–10. Sewar, Bharatpur, India.

Dusabenyagasani, M. and W.G.D. Fernando. 2008. Development of a SCAR marker to track canola resistance against blackleg caused by *Leptosphaeria maculans* pathogenicity group 3. *Plant Dis.* 92: 903–908.

Dutta, S., B.K. Bhattacharya, D.R. Rajak, C. Chattopadhyay, N.K. Patel, and J.S. Parihar. 2006. Disease detection in mustard crop using EO-1 hyperion satellite data. *J. Indian Soc. Remote Sensing* (*Photonirvachak*) 34: 325–329.

Eckert, M.R., S. Rossall, A. Selley, and B.D.L. Fitt. 2010. Effects of fungicides on in vitro spore germination and mycelial growth of the phytopathogens *Leptosphaeria maculans* and *L. biglobosa* (phoma stem canker of oilseed rape). *Pest Manag. Sci.* 66: 396–405.

Eimori, K., R. Shimada, and J. Takeuchi. 2005. First occurrence of bottom rot of *Brassica campestris* L. (rapifera group) and damping-off of *Raphanus sativus* L. (daikon group) caused by *Rhizoctonia solani* in Japan. *Annu. Rep. Kanto-Tosan Plant Prot. Soc.* 52: 35–38.

El-Awady, M., E.A.M. Reda, W. Haggag, S.Y. Sawsan, and A.M. El-Sharkawy. 2008. Transgenic canola plants over-expressing bacterial catalase exhibit enhanced resistance to *Peronospora parasitica* and *Erysiphe polygoni*. *Arab. J. Biotechnol.* 11: 71–84.

El-Hadrami, A. and F. Daayf. 2009. Priming canola resistance to blackleg with weakly aggressive isolates leads to an activation of hydroxycinnamates. *Can. J. Plant Pathol.* 31: 393–406.

El-Hadrami, A., W.G.D. Fernando, and F. Daayf. 2010. Variations in relative humidity modulate *Leptosphaeria* spp. pathogenicity and interfere with canola mechanisms of defence. *Eur. J. Plant Pathol.* 126: 187–202.

Elliott, C.E., D.M. Gardiner, G. Thomas, A. Cozijnsen, A. Wouw, and B.J. van de Howlett. 2007. Production of the toxin sirodesmin PL by *Leptosphaeria maculans* during infection of *Brassica napus*. *Mol. Plant Pathol.* 8: 791–802.

Elliott, V.L., S.J. Marcroft, R.M. Norton, and P.A. Salisbury. 2011. Reaction of *Brassica juncea* to Australian isolates of *Leptosphaeria maculans* and *Leptosphaeria biglobosa* 'canadensis'. *Can. J. Plant Pathol.* 33: 38–48.

Eshraghi, L., M.P. You, and M.J. Barbetti. 2005. First report of white leaf spot caused by *Pseudocercosporella capsellae* on *Brassica juncea* in Australia. *Plant Dis.* 89: 1131.

Evans, N., A. Baierl, P. Gladders, B. Hall, and B.D.L. Fitt. 2006. Prediction of the date of onset of phoma leaf spot epidemics on oilseed rape in the UK. *Bull. OILB/SROP* 29: 287–292.

Farniev, A.T., I.V. Alikova, and B.Z. Kulova. 2009. The role of biological preparations in increase of disease resistance and yield of spring rape. *Kormoproizvodstvo* 9: 11–15.

Farr, D.F., A.Y. Rossman, M.E. Palm, and E.B. McCray. 2004. *Online Fungal Databases*. Systematic Botany & Mycology Laboratory, ARS, USDA, Beltsville, MD. http://nt.ars-grin.gov/fungaldatabases.

Farzadfar, S., R. Pourrahim, A.R. Golnaraghi, and A. Ahoonmanesh. 2005. Occurrence of Cauliflower mosaic virus in different cruciferous plants in Iran. *Plant Pathol.* 54: 810.

Fei, W.X., Q.S. Li, X.J. Wu, S.M. Hou, F.X. Chen, W.X. Wang, and B.C. Hu. 2002. A study on control of sclerotinia blight [*Sclerotinia sclerotiorum*] of rape with agronomic measures. *Chin. J. Oil Crop Sci.* 24: 47–49.

Feng, J., S.F. Hwang, and S.E. Strelkov. 2013. Studies into primary and secondary infection processes by *Plasmodiophora brassicae* on canola. *Plant Pathol.* 62: 177–183.

Feng, J., Q. Xiao, S.F. Hwang, S.E. Strelkov, and B.D. Gossen. 2012. Infection of canola by secondary zoospores of *Plasmodiophora brassicae* produced on a nonhost. *Eur. J. Plant Pathol.* 132: 309–315.

Ferdous, S.M., M.B. Meah, and M.M. Hossain. 2002. Comparative effect of plant extracts and fungicide on the incidence of Alternaria blight of mustard. *Bangladesh J. Train. Dev.* 15: 207–210.

Fernando, W.G.D., S. Nakkeeran, Y. Zhang, and S. Savchuk. 2007. Biological control of *Sclerotinia sclerotiorum* (Lib.) de Bary by *Pseudomonas* and *Bacillus* species on canola petals. *Crop Prot.* 26:100–107.

Fernando, W.G.D., P.S. Parks, G. Tomm, L.V. Viau, and C. Jurke. 2003. First report of blackleg disease caused by *Leptosphaeria maculans* on canola in Brazil. *Plant Dis.* 87: 314.

Ferreira, M.E., P.H. Williams, and T.C. Osborn. 1994. RFLP mapping of *Brassica napus* using double haploid lines. *Theor. Appl. Genet.* 89: 615–621.

Ferreira, M.E., P.H. Williams, and T.C. Osborn. 1995. Mapping of locus controlling resistance to *Albugo candida* in *Brassica napus* using molecular markers. *Phytopathology* 85: 218–220.

Fitt, B.D.L., H. Brun, M.J. Barbetti, and S.R. Rimmer. 2006. World-wide importance of phoma stem canker (*Leptosphaeria maculans and L. biglobosa*) on oilseed rape (*Brassica napus*). *Eur. J. Plant Pathol.* 114: 3–15.

Fitt, B.D.L., N. Evans, P. Gladders, Y.J. Huang, and J.S. West. 2008a. Phoma stem canker and light leaf spot on oilseed rape in a changing climate. *Aspects Appl. Biol.* 88: 143–145.

Fitt, B.D.L., B.C. Hu, Z.Q. Li, S.Y. Liu, R.M. Lange, P.D. Kharbanda, M.H. Butterworth, and R.P. White. 2008b. Strategies to prevent spread of *Leptosphaeria maculans* (phoma stem canker) onto oilseed rape crops in China; costs and benefits. *Plant Pathol.* 57: 652–664.

Freeman, J., E. Ward, C. Calderon, and A. McCartney. 2002. A polymerase chain reaction (PCR) assay for the detection of inoculum of *Sclerotinia sclerotiorum*. *Eur. J. Plant Pathol.* 108: 877–886.

Food and Agriculture Organization (FAO). 2014. Food and agricultural commodities production. http://www.faostat.fao.org/site/339/default.aspx. Retrieved September 12, 2014.

Gaetan, S.A. 2005b. Occurrence of Fusarium wilt on canola caused by *Fusarium oxysporum* f. sp. conglutinans in Argentina. *Plant Dis.* 89: 432.

Gaetan, S. and N. Lopez. 2005. First outbreak of bacterial leaf spot caused by *Xanthomonas campestris* on canola in Argentina. *Plant Dis.* 89: 683.

Gaetan, S. and M. Madia. 2004. First report of canola powdery mildew caused by *Erysiphe polygoni* in Argentina. *Plant Dis.* 88: 1163.

Gaetan, S. and M. Madia. 2005. Occurrence of stem rot on canola caused by *Sclerotinia sclerotiorum* in Argentina. *Plant Dis.* 89: 530.

Gaetan, S.A. 2005a. First outbreak of blackleg caused by *Phoma lingam* in commercial canola fields in Argentina. *Plant Dis.* 89: 435.

Galletti, S., P.L. Burzi, E. Sala, S. Marinello, and C. Cerato. 2006. Combining Brassicaceae green manure with *Trichoderma* seed treatment against damping-off in sugarbeet. *Bull-OILB/SROP* 29: 71–75.

Gao, X.N., Y.F. Chen, Q.M. Han, H.Q. Qin, Z.S. Kang, and L.L. Huang. 2013. Identification and biocontrol efficacy of endophytic bacterium EDR2 against *Sclerotinia sclerotiorum*. *J. Northwest A&F Univ.—Nat. Sci. Edn.* 41: 175–181.

Garg, H., C. Atri, P.S. Sandhu, B. Kaur, M. Renton, S.K. Banga, H. Singh, C. Singh, M.J. Barbetti, and S.S. Banga. 2010c. High level of resistance to *Sclerotinia sclerotiorum* in introgression lines derived from hybridization between wild crucifers and the crop *Brassica* species *B. napus* and *B. juncea*. *Field Crops Res.* 117: 51–58.

Garg, H., L.M. Kohn, M. Andrew, H. Li, K. Sivasithamparam, and M.J. Barbetti. 2010b. Pathogenicity of morphologically different isolates of *Sclerotinia sclerotiorum* with *Brassica napus* and *B. juncea* genotypes. *Eur. J. Plant Pathol.* 126: 305–315.

Garg, H., H. Li, K. Sivasithamparam, J. Kuo, and M.J. Barbetti. 2010a. The infection processes of *Sclerotinia sclerotiorum* in cotyledon tissue of a resistant and a susceptible genotype of *Brassica napus*. *Ann. Bot. Lond.* 106: 897–908.

Ge, X.T., H. Li, S. Han, K. Sivasithamparam, and M.J. Barbetti. 2008. Evaluation of Australian *Brassica napus* genotypes for resistance to the downy mildew pathogen, *Hyaloperonospora parasitica*. *Aust. J. Agric. Res.* 59: 1030–1034.

Gerber, M., M. Muller, W. Rademacher, A. Mittnacht, H. Platen, M. Strey, and A. Buckenauer. 2010. CARAXReg.—A new fungicide with morphoregulatory activity in oilseed rape. *Julius-Kuhn-Arch.* 428: 147–148.

Ghanbarnia, K., W.G.D. Fernando, and G. Crow. 2009. Developing rainfall- and temperature-based models to describe infection of canola under field conditions caused by pycnidiospores of *Leptosphaeria maculans*. *Phytopathology* 99: 879–886.

Ghanbarnia, K., W.G.D. Fernando, and G. Crow. 2011. Comparison of disease severity and incidence at different growth stages of naturally infected canola plants under field conditions by pycnidiospores of *Phoma lingam* as a main source of inoculum. *Can. J. Plant Pathol.* 33: 355–363.

Ghanbarnia, K., D.J. Lydiate, S.R. Rimmer, G.Y. Li, H.R. Kutcher, N.J. Larkan, P.B.E. McVetty, and W.G.D. Fernando. 2012. Genetic mapping of the *Leptosphaeria maculans* avirulence gene corresponding to the Lepr1 resistance gene of *Brassica napus*. *Theor. Appl. Genet.* 124: 505–513.

Ghasolia, R.P. and A. Shivpuri. 2005a. Screening of rapeseed-mustard genotypes for resistance against Sclerotinia rot. *Indian Phytopathol.* 58: 242.

Ghasolia, R.P. and A. Shivpuri. 2005b. Studies on carpogenic germination of sclerotia of *Sclerotinia sclerotiorum* (Lib.) de Bary, causing Sclerotinia rot of Indian mustard. *Indian Phytopathol.* 58: 224–227.

Ghasolia, R.P. and A. Shivpuri. 2007. Morphological and pathogenic variability in rapeseed and mustard isolates of *Sclerotinia sclerotiorum*. *Indian Phytopathol.* 60: 76–81.

Ghasolia, R.P. and A. Shivpuri. 2009. Inoculum-disease relationships in Sclerotinia rot of Indian mustard (*Brassica juncea*). *Indian Phytopathol.* 62: 199–203.

Ghasolia, R.P., A. Shivpuri, and A.K. Bhargava. 2004. Sclerotinia rot of Indian mustard in Rajasthan. *Indian Phytopathol.* 57: 76–79.

Girish, G., M.B. Gowda, M. Saifulla, M. Mahesh, and Satheesh 2007. Management of white rust and Alternaria leaf blight of mustard using fungicides. *Environ. Ecol.* 25S(Special 3A): 830–833.

Gladders, P., H. Brun, X. Pinochet, M. Jedryczka, I. Happstadius, and N. Evans. 2006. Studies on the contribution of cultivar resistance to the management of stem canker (*Leptosphaeria maculans*) in Europe. *Bull. OILB/SROP* 29: 293–300.

Gladders, P., C. Dyer, B.D.L. Fitt, N. Evans, F. Bosch, J.M. van den Steed, A. Baierl et al. 2004. Development of a decision support system for phoma and light leaf spot in winter oilseed rape ('password' project). *HGCA Project Rep.* 357: 213pp.

Gladders, P. and T.M. Musa. 1980. Observations on the epidemiology of *Leptosphaeria maculans* stem canker in winter oilseed rape. *Plant Pathol.* 29: 28–37.

Godika, S., J.P. Jain, and A.K. Pathak. 2001. Evaluation of fungitoxicants against Alternaria blight and white rust diseases of Indian mustard (*Brassica juncea*). *Indian J. Agric. Sci.* 71: 497–498.

Gong, Z.H., Y.K. He, X.J. Song, G.H. Zhang, and G.H. Zhang. 2001. Diagnosis and genetic analysis of resistance to cauliflower mosaic virus in Brassica crops which was transformed with CaMV gene VI. *Acta Hortic. Sin.* 28: 466–468.

Gopal, K. 2003. Fungicidal control of downy mildew of mustard. *Ann. Agric. Res.* 24: 533–536.

Gossen, B.D., K.K.C. Adhikari, and M.R. McDonald. 2012. Effects of temperature on infection and subsequent development of clubroot under controlled conditions. *Plant Pathol.* 61: 593–599.

Goswami, K., A.K. Tewari, and R.P. Awasthi. 2012. Cultural, morphological and pathogenic characteristics and carpogenic germination of *Sclerotinia sclerotiorum*, the cause of Sclerotinia rot of rapeseed-mustard. *Pantnagar J. Res.* 10: 40–45.

Gout, L., I. Fudal, M.L. Kuhn, F. Blaise, M. Eckert, L. Cattolico, M.H. Balesdent, and T. Rouxel. 2006. Lost in the middle of nowhere: The AvrLm1 avirulence gene of the Dothideomycete *Leptosphaeria maculans*. *Mol. Microbiol.* 60: 67–80.

Goyal, P., M. Chahar, M. Barbetti, S.Y. Liu, and C. Chattopadhyay. 2011b. Resistance to Sclerotinia rot caused by *Sclerotinia sclerotiorum* in *Brassica juncea* and *B. napus* germplasm. *Indian J. Plant Prot.* 39: 60–64.

Goyal, P., M. Chahar, A.P. Mathur, A. Kumar, and C. Chattopadhyay. 2011a. Morphological and cultural variation in different oilseed *Brassica* isolates of *Alternaria brassicae* from different geographical regions of India. *Indian J. Agric. Sci.* 81: 1052–1059.

Goyal, P., C. Chattopadhyay, A.P. Mathur, A. Kumar, P.D. Meena, S. Datta, and M.A. Iquebal. 2013a. Pathogenic and molecular variability among Brassica isolates of *Alternaria brassicae* from India. *Ann. Plant Prot. Sci.* 21: 349–359.

Goyal, P., A. Kumar, M. Chahar, M.A. Iquebal, S. Datta, and C. Chattopadhyay. 2013c. Pathogenic and genetic variability among Brassica isolates of *Sclerotinia sclerotiorum* from India and UK. *Ann. Plant Prot. Sci.* 21: 377–386.

Goyal, P., A.P. Mathur, and C. Chattopadhyay. 2013b. Histopathology in *Brassica juncea—Alternaria brassicae* interaction and localization of histochemicals in Alternaria blight-infected *B. juncea* leaves. *Ann. Plant Prot. Sci.* 21: 322–328.

Graichen, K., F. Rabenstein, and E. Kurtz. 2000. The occurrence of turnip yellows virus in winter oilseed rape in Austria. *Pflanzenschutzberichte* 59: 35–46.

Griffiths, P.D., L.F. Marek, and L.D. Robertson. 2009. Identification of crucifer accessions from the NC-7 and NE-9 plant introduction collections that are resistant to black rot (*Xanthomonas campestris* pv. *campestris*) races 1 and 4. *HortScience* 44: 284–288.

Griffiths, P.D. and J.L. Nickels. 2001. Association of a molecular polymorphism with black rot resistance derived from Ethiopian mustard. *Cruciferae Newsl.* 23: 57–58.

Gu, B.C., G.M. Zhu, X.G. Yue, T. Xiao, and Y.L. Pan. 2012. Action mode and control effect of Boscalid on rape sclerotinia stem rot. *Acta Agric. Jiangxi* 24: 114–117.

Guo, X., W.G.D. Fernando, and M. Entz. 2008. Dynamics of infection by *Leptosphaeria maculans* on canola (*Brassica napus*) as influenced by crop rotation and tillage. *Arch. Phytopathol. Plant Prot.* 41: 57–66.

Guo, X.L., C.H. Dong, X.J. Hu, Y.Y. Liu, and S.Y. Liu. 2006. Transformation of oilseed rape with a novel gene from a bacterial strain to improve resistance to *Sclerotinia sclerotiorum*. *Acta Hortic.* 706: 265–268.

Guo, X.W., W.G.D. Fernando, and M.H. Entz. 2005. Effects of crop rotation and tillage on the blackleg disease of canola. *Can. J. Plant Pathol.* 27: 53–57.

Gupta, K. and G.S. Saharan. 2002. Identification of pathotypes of *Albugo candida* with stable characteristic symptoms on Indian mustard. *J. Mycol. Plant Pathol.* 32: 46–51.

Gupta, K., G.S. Saharan, N. Mehta, and M.S. Sangwan. 2004a. Identification of pathotypes of *Alternaria brassicae* from Indian mustard [*Brassica juncea* (L.) Czern. and Coss]. *J. Mycol. Plant Pathol.* 34: 15–19.

Gupta, M., B. Summuna, S. Gupta, and S.A. Mallick. 2012. Assessing the role of biochemical constituents in resistance to Alternaria blight in rapeseed-mustard. *J. Mycol. Plant Pathol.* 42: 463–468.

Gupta, R., R.P. Awasthi, and S.J. Kolte. 2003. Influence of sowing dates and weather factors on development of Alternaria blight on rapeseed-mustard. *Indian Phytopathol.* 56: 398–402.

Gupta, R., R.P. Awasthi, and S.J. Kolte. 2004b. Influence of sowing dates on the incidence of Sclerotinia stem rot of rapeseed-mustard. *Ann. Plant Prot. Sci.* 12: 223–224.

Gupta, R., R.P. Awasthi, and S.J. Kolte. 2004c. Effect of nitrogen and sulphur on incidence of Sclerotinia rot of mustard. *Indian Phytopathol.* 57: 193–194.

Gupta, S.K. and C.D. Kaushik. 2002. Metabolic changes in mustard leaf and siliqua wall due to the infection of Alternaria blight (*Alternaria brassicae*). *Cruciferae Newsl.* 24: 85–86.

Gwiazdowski, R. and G. Wickiel. 2009. Occurrence of Alternaria mycotoxins in seeds of different rape cultivars. *Prog. Plant Prot.* 49: 934–937.

Gwiazdowski, R. and G. Wickiel. 2011. The influence of fungicide protection on the decrease of *Alternaria* toxins amount in the seeds of winter oilseed rape. *Prog. Plant Prot.* 51: 1409–1414.

Hall R. 1992. Epidemiology of blackleg of oilseed rape. *Can. J. Plant Pathol.* 14: 46–55.

Hammond, K.E., B.G. Lewis, and T.M. Musa. 1985. A systemic pathway in the infection of oilseeed rape plants by *Leptosphaeria maculans*. *Plant Pathol.* 34: 557–565.

Han, L.R., H.J. Zhang, B.W. Gao, J. He, and X. Zhang. 2012. Antifungal activity against rapeseed Sclerotinia stem rot and identification of actinomycete strain 11-3-1. *Acta Phytophyl. Sin.* 39: 97–102.

Harding, H., P.H. Williams, and S.S. McNabola. 1968. Chlorophyll changes, photosynthesis and ultra structure of chloroplasts in *Albugo candida* induced "green islands" on detached *Brassica juncea* cotyledons. *Can. J. Bot.* 46: 1229–1234.

Hasan, M.J., S.E. Strelkov, R.J. Howard, and H. Rahman. 2012. Screening of *Brassica* germplasm for resistance to *Plasmodiophora brassicae* pathotypes prevalent in Canada for broadening diversity in clubroot resistance. *Can. J. Plant Sci.* 92: 501–515.

Hatakeyama, K., K. Suwabe, R.N. Tomita, T. Kato, T. Nunome, H. Fukuoka, and S. Matsumoto. 2013. Identification and characterization of Crr1a, a gene for resistance to clubroot disease (*Plasmodiophora brassicae* Woronin) in *Brassica rapa* L. *PLoS One* 8: e54745.

Hayden, H.L., A.J. Cozijnsen, S.J. Marcroft, and B.J. Howlett. 2003. Pathogenic and genetic variation in Leptosphaeria maculans in Australia. In: *Thirteenth Biennial Australian Research Assembly Brassicas. Proceedings of Conference*, Tamworth, New South Wales, Australia, September 8–12, 2003, pp. 140–143.

He, X.L., G.H. Liu, B.L. Zhang, J.M. She, Q.X. Ruan, Q.H. Xue, and W.C. Ni. 2007. *Agrobacterium tumefaciens* mediated introduction of glucose oxidase gene into *Brassica napus* L. *Jiangsu J. Agric. Sci.* 23: 546–551.

Heald, F.D. (ed.) 1926. Diseases due to downy mildew and allies. In: *Manual of Plant Diseasesed.* McGraw Hill Book Company, Inc., New York, Chapter 16, pp. 390–426.

Hegedus, D.D., R.G. Li, L. Buchwaldt, I. Parkin, S. Whitwill, C. Coutu, D. Bekkaoui, and S.R. Rimmer. 2008. *Brassica napus* possesses an expanded set of polygalacturonase inhibitor protein genes that are differentially regulated in response to *Sclerotinia sclerotiorum* infection, wounding and defense hormone treatment. *Planta* 228: 241–253.

Hertel, K., M. Schwinghamer, and R. Bambach. 2004. Virus diseases in canola and mustard. Agnote—NSW Agriculture DPI 495, 1st edn., Dubbo, New South Wales, Australia, 6pp.

Hind, T.L., G.J. Ash, and G.M. Murray. 2003. Prevalence of Sclerotinia stem rot of canola in New South Wales. *Aust. J. Exp. Agric.* 43: 163–168.

Hingole, D.G. and C.D. Mayee. 2003. Influence of irrigation scheduling and land layout on white rust and powdery mildew diseases and yield of mustard. *J. Soils Crops* 13: 185–186.

Honig, F. 1931. Der Kohlkropferreger (*Plasmmodiophora brassicae* Wor.). *Gartenbauwenissenchaft* 5: 116–225.

Hood, J.R., N. Evans, S. Rossall, M. Ashworth, and B.D.L. Fitt. 2007. Effect of spray timing of flusilazole on the incidence and severity of phoma leaf spot (*Leptosphaeria maculans* and *L. biglobosa*) on winter oilseed rape. *Aspects Appl. Biol.* 83: 29–33.

Hossain, M.D. and M.Z. Rahman. 2006. Effect of number of Rovral spray on the yield, incidence and seed infection of Alternaria in mustard. *Int. J. Sustain. Agric. Technol.* 2: 38–41.

Howlett, B.J. 2004. Current knowledge of the interaction between *Brassica napus* and *Leptosphaeria maculans*. *Can. J. Plant Pathol.* 26: 245–252.

Hu, X.J., D.P. Roberts, M.L. Jiang, and Y.B. Zhang. 2005. Decreased incidence of disease caused by *Sclerotinia sclerotiorum* and improved plant vigor of oilseed rape with *Bacillus subtilis* Tu-100. *Appl. Microbiol. Biotechnol.* 68: 802–807.

Hu, X.J., D.P. Roberts, J.E. Maul, S.E. Emche, X. Liao, X.L. Guo, Y.Y. Liu, L.F. McKenna, J.S. Buyer, and S.Y. Liu. 2011. Formulations of the endophytic bacterium *Bacillus subtilis* Tu-100 suppress *Sclerotinia sclerotiorum* on oilseed rape and improve plant vigor in field trials conducted at separate locations. *Can. J. Microbiol.* 57: 539–546.

Hua, L., K. Sivasithamparam, M.J. Barbetti, and J. Kuo. 2004. Germination and invasion by ascospores and pycnidiospores of *Leptosphaeria maculans* on spring-type *Brassica napus* canola varieties with varying susceptibility to blackleg. *J. Gen. Plant Pathol.* 70: 261.

Huang, H.C. and R.S. Erickson. 2004. Effect of soil treatment of fungal agents on control of apothecia of *Sclerotinia sclerotiorum* in canola and safflower fields. *Plant Pathol. Bull.* 13: 1–6.

Huang, L., H. Buchenauer, Q. Han, X. Zhang, and Z. Kang, Z. 2008. Ultrastructural and cytochemical studies on the infection process of *Sclerotinia sclerotiorum* in oilseed rape. *J. Plant Dis. Prot.* 115: 9–16.

Huang, Y.J., R.H. Liu, Z.Q. Li, N. Evans, A.M. Chevre, R. Delourme, M. Renard, and B.D.L. Fitt. 2006. Effects of temperature on Rlm6-mediated resistance to *Leptosphaeria maculans* in *Brassica napus*. *Bull. OILB/SROP* 29: 307–316.

Huang, Y.J., E.J. Pirie, N. Evans, R. Delourme, G.J. King, and B.D.L. Fitt. 2009. Quantitative resistance to symptomless growth of *Leptosphaeria maculans* (phoma stem canker) in *Brassica napus* (oilseed rape). *Plant Pathol.* 58: 314–323.

Humpherson-Jones, F.M. 1992. The development of weather-related disease forecasts for vegetable crops in the UK: Problems and prospects. *Eur. Plant Prot. Organ. Bull.* 21: 425–429.

Humpherson-Jones, F.M. and R.B. Maude. 1982. Studies on the epidemiology of *Alternaria brassicicola* in *Brassica oleracea* seed production crops. *Ann. Appl. Biol.* 100: 61–71.

Humpherson-Jones, F.M. and K. Phelps. 1989. Climatic factors influencing spore production in *Alternaria brassicae* and *Alternaria brassicicola*. *Ann. Appl. Biol.* 114: 449–458.

Hwang, S.F., H.U. Ahmed, S.E. Strelkov, B.D. Gossen, G.D. Turnbull, G. Peng, and R.J. Howard. 2011a. Seedling age and inoculum density affect clubroot severity and seed yield in canola. *Can. J. Plant Sci.* 91: 183–190.

Hwang, S.F., H.U. Ahmed, Q. Zhou, A. Rashid, S.E. Strelkov, B.D. Gossen, G. Peng, and G.D. Turnbull. 2013. Effect of susceptible and resistant canola plants on *Plasmodiophora brassicae* resting spore populations in the soil. *Plant Pathol.* 62: 404–412.

Hwang, S.F., H.U. Ahmed, Q. Zhou, S.E. Strelkov, B.D. Gossen, G. Peng, and G.D. Turnbull. 2011b. Influence of cultivar resistance and inoculum density on root hair infection of canola (*Brassica napus*) by *Plasmodiophora brassicae*. *Plant Pathol.* 60: 820–829.

Hwang, S.F., H.U. Ahmed, Q. Zhou, S.E. Strelkov, B.D. Gossen, G. Peng, and G.D. Turnbull. 2012a. Assessment of the impact of resistant and susceptible canola on *Plasmodiophora brassicae* inoculum potential. *Plant Pathol.* 61: 945–952.

Hwang, S.F., T. Cao, Q. Xiao, H.U. Ahmed, V.P. Manolii, G.D. Turnbull, B.D. Gossen, G. Peng, and S.E. Strelkov. 2012b. Effects of fungicide, seeding date and seedling age on clubroot severity, seedling emergence and yield of canola. (Special Issue: Plant adaptation to environmental change.) *Can. J. Plant Sci.* 92: 1175–1186.

Idnurm, A. and B.J. Howlett. 2002. Isocitrate lyase is essential for pathogenicity of the fungus *Leptosphaeria maculans* to canola (*Brassica napus*). *Eukaryot. Cell* 1: 719–724.

Ignatov, A., Y. Kuginuki, K. Hida, and J.D. Taylor. 2001. Resistance of brassicas with B and D genome to *Xanthomonas campestris* pv. *campestris*, casual agent of black rot. *J. Russ. Phytopathol. Soc.* 2: 59–61.

Irzykowski, W., J.M. Sun, Q.S. Li, T.C. Gao, S.M. Hou, A. Aguedo and M. Jedryczka. 2004. DNA polymorphism in *Sclerotinia sclerotiorum* isolates from oilseed rape in China. *Bull OILB/SROP* 27: 67–74.

Isakeit, T., J.E. Woodward, C. Niu, and R.J. Wright. 2010. First report of Sclerotinia stem rot of canola caused by *Sclerotinia sclerotiorum* in Texas. *Plant Dis.* 94: 792.

Jain, S., V. Jain, and A.S. Rathi. 2009. Changes in antioxidant pathway enzymes in Indian mustard leaves due to white rust infection. *Indian Phytopathol.* 62: 499–504.

Jain, S., P. Pruthi, V. Jain, and H.K.L. Chawla. 2002. Changes in phenylalanine ammonia lyase, tyrosine ammonia lyase and lipoxygenase activities in the leaves of *Brassica juncea* L. infected with *Albugo candida*. *Physiol. Mol. Biol. Plants* 8: 261–266.

Jajor, E. 2007. Phytoplasma disease of rape. *Ochrona Roslin* 52(7/8): 48–49.

Jajor, E., M. Korbas, J. Horoszkiewicz-Janka, and M. Wojtowicz. 2010. Influence of weather conditions and date of fungicidal protection on the occurrence of *Sclerotinia sclerotiorum* on oilseed rape. *Prog. Plant Prot.* 50: 1334–1339.

Jajor, E., G. Wickiel, and J. Horoszkiewicz-Janka. 2011. Fungi of the *Alternaria* genus and their toxic metabolites in seeds of winter oilseed rape. *Prog. Plant Prot.* 51: 1633–1638.

Jasnic, S. and F. Bagi. 2007. Viral diseases of oilseed rape. *Biljni Lekar* (*Plant Doctor*) 35: 458–461.

Jat, R.R. 1999. Pathogenic variability and inheritance of resistance to *Albugo candida* in oilseed *Brassica*. Ph.D. thesis, CCSHAU, Hisar, India, 129pp.

Jedryczka, M. and J. Kaczmarek. 2011. Efficiency of selected fungicides in protection against stem canker of Brassicas in optimal and late spray time. *Prog. Plant Prot.* 51: 1639–1643.

Jedryczka, M., J. Kaczmarek, and J. Czernichowski. 2006. Development of a decision support system for control of stem canker of oilseed rape in Poland. *Bull. OILB/SROP* 29: 267–276.

Jedryczka, M., J. Kaczmarek, A. Dawidziuk and A. Brachaczek. 2008. System for forecasting disease epidemics—Aerobiological methods in Polish agriculture. *Aspects Appl. Biol.* 89: 65–70.

Jedryczka, M., R. Matysiak, R. Bandurowski, and D. Rybacki. 2004. SPEC—The decision support system against stem canker of oilseed rape in Poland. *Rosliny Oleiste* 25: 637–644.

Jedryczka, M., A. Stachowiak, J. Olechnowicz, Z. Karolewski, and A. Podlesna. 2009. The comparison of composition of avirulence genes and races in collections of phytopathogenic fungus *Leptosphaeria maculans* in Poland. *Rosliny Oleiste* 30: 197–205.

Jha, S.K. and A.K. Sharma. 2003. Problem identification in rapeseed-mustard. *Sarson News* 7: 7.

Ji, R.Q., X.B. Dong, H. Feng, M. Barbetti, R.C. Gao, Y.Y. Liu, and S.Y. Liu. 2009. Expression of PDF1.2 in oxalate oxidase transgenic oilseed rape. *Plant Physiol. Commun.* 45: 479–482.

Jiang, D.H., G.Q. Li, Y.P. Fu, X.H. Yi, and D.B. Wang. 2000. Biocontrol of reinfection of oilseed rape stem rot caused by *Sclerotinia sclerotiorum* by *Coniothyrium minitans* and its survival on leaf of oilseed rape (*Brassica napus*). *Acta Phytopathol. Sin.* 30: 60–65.

Jones, R.A.C., B.A. Coutts, and J. Hawkes. 2007. Yield-limiting potential of Beet western yellows virus in *Brassica napus*. *Aust. J. Agric. Res.* 58: 788–801.

Jurke, C.J. and W.G.D. Fernando. 2008. Effects of seeding rate and plant density on sclerotinia stem rot incidence in canola. *Arch. Phytopathol. Plant Prot.* 41: 142–155.

Kaczmarek, J., A. Brachaczek, and M. Jedryczka. 2011. The influence of the fungicide containing flusilazole on the effectiveness of the protection of oilseed rape against stem canker of brassicas. *Rosliny Oleiste* 32: 153–166.

Kaczmarek, J., M. Jedryczka, B.D.L. Fitt, J.A. Lucas, and A.O. Latunde-Dada. 2009a. Molecular detection of the primary inoculum of the pathogenic fungi *Leptosphaeria maculans* and *L. biglobosa* in Lower Silesia air samples. *Rosliny Oleiste* 30: 9–20.

Kaczmarek, J., M. Jedryczka, B.D.L. Fitt, J.A. Lucas, and A.O. Latunde-Dada. 2009b. Analyses of air samples for ascospores of *Leptosphaeria maculans* and *L. biglobosa* by light microscopy and molecular techniques. *J. Appl. Genet.* 50: 411–419.

Kaczmarek, J., A. Maczynska, A. Brachaczek, and M. Jedryczka. 2009c. Optimization of time of fungicide sprays against stem canker of oilseed rape in 2007/2008. *Prog. Plant Prot.* 49: 1749–1752.

Kaliff, M., J. Staal, M. Myrenas, and C. Dixelius. 2007. ABA is required for *Leptosphaeria maculans* resistance *via* ABI1- and ABI4-dependent signaling. *Mol. Plant Microbe Interact.* 20: 335–345.

Kamble, A. and S. Bhargava. 2007. beta-Aminobutyric acid-induced resistance in *Brassica juncea* against the necrotrophic pathogen *Alternaria brassicae*. *J. Phytopathol.* 155: 152–158.

Kanrar, S., J.C. Venkateswari, P.B. Kirti, and V.L. Chopra. 2002. Transgenic expression of hevein, the rubber tree lectin, in Indian mustard confers protection against *Alternaria brassicae*. *Plant Sci.* 162: 441–448.

Kapoor, A., H.R. Singal, and S. Jain. 2003. Induction, purification and characterization of beta-1,3-glucanase from *Brassica juncea* L. infected with *Albugo candida*. *J. Plant Biochem. Biotechnol.* 12: 157–158.

Karimi, E., N. Safaie, and M. Shams-Bakhsh. 2011. Assessment of genetic diversity among *Sclerotinia sclerotiorum* populations in canola fields by rep-PCR. *Trakia J. Sci.* 9: 62–68.

Karimi, E., N. Safaie, and M. Shams-Bakhsh. 2012. Mycelial compatibility groupings and pathogenic diversity of *Sclerotinia sclerotiorum* (Lib.) de Bary populations on canola in Golestan Province of Iran. *J. Agric. Sci. Technol.* 14: 421–434.

Kaur, A. and S.J. Kolte. 2001. Protection of mustard plants against the staghead phase of white rust by foliar treatment with benzothiadiazole, an activator of plant defense system. *J. Mycol. Plant Pathol.* 31: 133–138.

Kaur, P., R. Jost, K. Sivasithamparam, and M.J. Barbetti. 2011a. Proteome analysis of the *Albugo candida–Brassica juncea* pathosystem reveals that the timing of the expression of defence related genes is a crucial determinant of pathogenesis. *J. Exp. Bot.* 62: 1285–1298.

Kaur, P., C.X. Li, M.J. Barbetti, M.P. You, H. Li, and K. Sivasithamparam. 2008b. First report of powdery mildew caused by *Erysiphe cruciferarum* on *Brassica juncea* in Australia. *Plant Dis.* 92: 650.

Kaur, P., K. Sivasithamparam, and M.J. Barbetti. 2011b. Host range and phylogenetic relationships of *Albugo candida* from cruciferous hosts in Western Australia, with special reference to *Brassica juncea*. *Plant Dis.* 95: 712–718.

Kaur, P., K. Sivasithamparam, and M.J. Barbetti. 2008a. Pathogenic behaviour of strains of *Albugo candida* from *Brassica juncea* (Indian mustard) and *Raphanus raphanistrum* (wild radish) in Western Australia. *Aust. J. Plant Pathol.* 37: 353–356.

Kazan, K., A. Rusu, J.P. Marcus, K.C. Goulter, and J.M. Manners. 2002. Enhanced quantitative resistance to *Leptosphaeria maculans* conferred by expression of a novel antimicrobial peptide in canola (*Brassica napus* L.). *Mol. Breed.* 10: 63–70.

Kenyon, D., J. Thomas, and C. Handy. 2004. Feasibility of using quantitative PCR for assessing resistance to stem canker in oilseed rape cultivars. *Bull. OILB/SROP* 27: 109–117.

Khan, M.M., R.U. Khan, and F.A. Mohiddin. 2007a. Variation in occurrence and morphology of *Alternaria brassicae* (Berk.) Sacc. Causing blight in rapeseed and mustard. *Ann. Plant Prot. Sci.* 15: 414–417.

Khan, M.M., R.U. Khan, and F.A. Mohiddin. 2007b. In vitro effect of temperature and culture media on the growth of *Alternaria brassicae* infecting rapeseed-mustard. *Ann. Plant Prot. Sci.* 15: 526–527.

Khan, M.M., R.U. Khan, and F.A. Mohiddin. 2007c. Studies on the cost-effective management of Alternaria blight of rapeseed-mustard (*Brassica* spp.). *Phytopathol. Mediterr.* 46: 201–206.

Khan, M.R. and M.M. Khan. 2010. Effect of intermittent exposures of SO_2 on the leaf blight caused by *Alternaria brassicicola* on Indian mustard. *Agric. Ecosyst. Environ.* 139: 728–735.

Khan, R.U. and S.J. Kolte. 2002. Some seedling diseases of rapeseed-mustard and their control. *Indian Phytopathol.* 55: 102–103.

Khangura, R. 2004. Influence of chemicals on inoculum production by *Leptosphaeria maculans* on canola (*Brassica napus* L.) residues. *J. Food Agric. Environ.* 2: 188–192.

Khangura, R.K. and M.J. Barbetti. 2004. Time of sowing and fungicides affect blackleg (*Leptosphaeria maculans*) severity and yield in canola. *Aust. J. Exp. Agric.* 44: 1205–1213.

Khangura, R.K. and D.G. Wright. 2012. First report of club root caused by *Plasmodiophora brassicae* on canola in Australia. *Plant Dis.* 96: 1075–1076.

Khare, A., B.K. Singh, and R.S. Upadhyay. 2010. Biological control of *Pythium aphanidermatum* causing damping-off of mustard by mutants of *Trichoderma viride* 1433. *J. Agric. Technol.* 6: 231–243.

Khatun, F., M.S. Alam, M.A. Hossain, and S. Alam. 2010. Effect of sulphur, zinc and boron on the severity of Alternaria leaf blight of mustard. *Bangladesh J. Plant Pathol.* 26: 23–29.

Khatun, F., M.S. Alam, M.A. Hossain, S. Alam, and P.K. Malaker. 2011. Effect of NPK on the incidence of Alternaria leaf blight of mustard. *Bangladesh J. Agric. Res.* 36: 407–413.

Khulbe, A., R.P. Awasthi, and A.K. Tewari. 2011. Morphological and cultural diversity in isolates of *Alternaria brassicae* infecting rapeseed and mustard. *Pantnagar J. Res.* 9: 206–209.

Khunti, J.P., R.R. Khandar, and M.F. Bhoraniya. 2001. Fungicidal control of white rust (*Albugo cruciferarum* S.F. Gray) of mustard {*Brassica juncea* (L.)} Czern and Coss. *Agric. Sci. Dig.* 21: 103–105.

Khunti, J.P., R.R. Khandar, and M.F. Bhoraniya. 2004. Effect of media and temperature on germination of sporangia of *Albugo cruciferarum* causing white rust of mustard. *Adv. Plant Sci.* 17: 93–96.

Koch, E., K. Song, T.C. Osborn, and P.H. Williams. 1991. Relationship between pathogenicity based on restriction fragment length polymorphism in *Leptosphaeria maculans*. *Mol. Plant Microbe Interact.* 4: 341–349.

Koch, S., S. Dunker, B. Kleinhenz, M. Röhrig, and A. Tiedemann. 2007. A crop loss-related forecasting model for Sclerotinia stem rot in winter oilseed rape. *Phytopathology* 97: 1186–1194.

Kohire, O.D., R. Ahmed, S.S. Chavan, and V.C. Khilare. 2008a. Yield losses amongst four varieties of mustard due to powdery mildew in Maharashtra. *J. Phytol. Res.* 21: 331–332.

Kohire, O.D., R. Ahmed, S.S. Chavan and V.C. Khilare. 2008c. Influence of sowing dates, irrigation and spraying on the epidemiology of powdery mildew of mustard in Maharashtra. *J. Phytol. Res.* 21: 323–324.

Kohire, O.D., V.O.K. Patil, R. Ahmed, S.S. Chavan, and V.C. Khilare. 2008b. Epidemiology of powdery mildew of mustard in Marathwada. *J. Plant Dis. Sci.* 3: 235–236.

Kole, C., R. Teutonico, A. Mengistu, P.H. Williams, and T.C. Osborn. 1996. Molecular mapping of a locus controlling resistance to *Albugo candida* in *Brassica rapa*. *Phytopathology* 86: 367–369.

Kole, C., P.H. Williams, S.R. Rimmer, and T.C. Osborn. 2002. Linkage mapping of genes controlling resistance to white rust (*Albugo candida*) in *Brassica rapa* (syn. *campestris*) and comparative mapping to *Brassica napus* and *Arabdiopsis thaliana*. *Genome* 45: 22–27.

Kolte, S.J. 1985. *Diseases of Annual Edible Oilseed Crops*, Vol. II: Rapeseed-Mustard and Sesame Diseases. CRC Press Inc., Boca Raton, FL, 135pp.

Kolte, S.J. 2005. Tackling fungal diseases of oilseed brassicas in India. *Brassica* 7(1/2): 7–13.

Kolte, S.J., R.P. Awasthi, and Vishwanath. 1987. Assessment of yield losses due to Alternaria blight in rapeseed and mustard. *Indian Phytopathol.* 40: 209–211.

Kolte, S.J., R.P. Awasthi, and Vishwanath. 2000. Divya mustard: A useful source to create Alternaria black spot tolerant dwarf varieties of oilseed brassicas. *Plant Var. Seeds* 13: 107–111.

Kolte, S.J., D.K. Bordoloi, and R.P. Awasthi. 1991. The search for resistance to major diseases of rapeseed and mustard in India. In: *GCIRC Eighth International Rapeseed Congress*, Saskatoon, Saskatchewan, Canada, pp. 219–225.

Kolte, S.J., N.I. Nashaat, A. Kumar, R.P. Awasthi, and J.S. Chauhan. 2006. Indo-UK collaboration on oilseeds: Towards improving the genetic base of rapeseed-mustard in India. *Indian J. Plant Genet. Res.* 19: 346–351.

Kowalska, J. and D. Remlein-Starosta. 2011. Research of nonchemical methods of winter oilseed rape protection in Poland. *J. Res. Appl. Agric. Eng.* 56: 220–223.

Krause, U., B. Koopmann, and B. Ulber. 2006. Impact of rape stem weevil, *Ceutorhynchus napi*, on the early stem infection of oilseed rape by *Phoma lingam*. *Bull. OILB/SROP* 29: 323–328.

Krishnia, S.K., G.S. Saharan, and D. Singh. 2000. Genetic variation for multiple disease resistance in the families of interspecific cross of *Brassica juncea* × *Brassica carinata*. *Cruciferae Newsl.* 22: 51–53.

Kruse, T. and J.A. Verreet. 2005. Epidemiological studies on winter oilseed rape (*Brassica napus* L. var. napus) infected by *Phoma lingam* (teleomorph *Leptosphaeria maculans*) and the effects of different fungicide applications with FolicurReg. (tebuconazole). *Z. Pflanzenk. Pflanzen.* 112: 17–41.

Kumar, A. 2009. Eco-friendly options for the management of white rust in Indian mustard under mid hill conditions of North-Western India. *Indian Phytopathol.* 62: 44–48.

Kumar, A., V. Kumar, B.K. Bhattacharya, N. Singh, and C. Chattopadhyay. 2013. Integrated disease management: Need for climate-resilient technologies. *J. Mycol. Plant Pathol.* 43: 28–36.

Kumar, B. and S.J. Kolte. 2001. Progression of Alternaria blight of mustard in relation to components of resistance. *Indian Phytopathol.* 54: 329–331.

Kumar, B. and S.J. Kolte. 2006. Development of Alternaria blight in genotypes of Indian mustard (*Brassica juncea* (L.) Czern. & Coss.) under field conditions. *Indian Phytopathol.* 59: 314–317.

Kumar, G. and N.V.K. Chakravarty. 2008. A simple weather based forewarning model for white rust in Brassica. *J. Agrometeorol.* 10: 75–80.

Kumar, N. and A. Kumar. 2006. Effect of cultural practices on Alternaria blight in *Brassica juncea* and *B. napus*. *Indian J. Agric. Sci.* 76: 389–390.

Kumar, S. and C.S. Kalha. 2005. Evaluation of rapeseed-mustard germplasm against white rust and Alternaria blight. *Ann. Biol.* 21: 73–77.

Kumar, S., G.S. Saharan, and D. Singh. 2002. Inheritance of resistance in inter and intraspecific crosses of *Brassica juncea* and *Brassica carinata* to *Albugo candida* and *Erysiphe cruciferarum*. *J. Mycol. Plant Pathol.* 32: 59–63.

Kumawat, R.C., A.C. Mathur, and G.L. Kumawat. 2011. Physiological studies of Alternaria blight of Indian mustard caused by *Alternaria brassicicola* L. *Green Farming* 2: 574–576.

Kurt, S., U. Gunes, and E.M. Soylu. 2011. In vitro and in vivo antifungal activity of synthetic pure isothiocyanates against *Sclerotinia sclerotiorum*. *Pest Manag. Sci.* 67: 869–875.

Kutcher, H.R., M. Keri, D.L. McLaren, and S.R. Rimmer. 2007. Pathogenic variability of *Leptosphaeria maculans* in western Canada. *Can. J. Plant Pathol.* 29: 388–393.

Kutcher, H.R. and S.S. Malhi. 2004. Effect of topography, N fertilization and fungicide application on diseases, yield and seed quality of canola in North-Central Saskatchewan. In: *Proceedings of the Seventh International Conference on Precision Agriculture Precision Resources Management*, Hyatt Regency, Minneapolis, MN, July 25–28, 2004, pp. 1000–1007.

Kutcher, H.R., S.S. Malhi, and K.S. Gill. 2005. Topography and management of nitrogen and fungicide affects diseases and productivity of canola. *Agronomy J.* 97: 533–541.

Kutcher, H.R., C.G.J. van den Berg, and S.R. Rimmer. 1993. Variation in pathogenicity of *Leptosphaeria maculans* on *Brassica* spp. based on cotyledon and stem reactions. *Can. J. Plant Pathol.* 15: 253–258.

Kutcher, H.R. and T.M. Wolf. 2006. Low-drift fungicide application technology for sclerotinia stem rot control in canola. *Crop Prot.* 25: 640–646.

Laha, J.N., I. Naskar, and B.D. Sharma. 1985. A new record of clubroot disease of mustard. *Curr. Sci.* 54: 1247.

Lahlali, R., G. Peng, B.D. Gossen, L. McGregor, F.Q. Yu, R.K. Hynes, S.F. Hwang, M.R. McDonald, and S.M. Boyetchko. 2013. Evidence that the biofungicide serenade (*Bacillus subtilis*) suppresses clubroot on canola via antibiosis and induced host resistance. *Phytopathology* 103. 245–254.

Lakshmi, K. and S.C. Gulati. 2002. Inheritance of components of horizontal resistance to *Alternaria brassicae* (Berk.) Sacc. in Indian mustard, *Brassica juncea* (L.) Czern and Coss. *J. Oilseeds Res.* 19: 17–21.

Lamey, H.A. 1995. Survey of blackleg and sclerotinia stem rot of canola in North Dakota in 1991 and 1993. Special report. *Am. Phytol. Soc.* 79: 322–324.

Larkan, N.J., D.J. Lydiate, I.A.P. Parkin, M.N. Nelson, D.J. Epp, W.A. Cowling, S.R. Rimmer, and M.H. Borhan. 2013. The *Brassica napus* blackleg resistance gene LepR3 encodes a receptor-like protein triggered by the *Leptosphaeria maculans* effector AVRLM1. *New Phytol.* 197: 595–605.

Laxmi, R.R. and A. Kumar. 2011. Forecasting of powdery mildew in mustard (*Brassica juncea*) crop using artificial neural network approach. *Indian J. Agric. Sci.* 81: 855–860.

Lazarovits, G., A.N. Starratt, and H.C. Huang. 2000. The effect of tricyclazole and culture medium on production of the melanin precursor 1,8-dihydroxynaphthalene by *Sclerotinia sclerotiorum* isolate SS7. *Pesticide Biochem. Physiol.* 67: 54–62.

Leflon, M., H. Brun, F. Eber, R. Delourme, M.O. Lucas, P. Vallee, M. Ermel, M.H. Balesdent, and A.M. Chevre. 2007. Detection, introgression and localization of genes conferring specific resistance to *Leptosphaeria maculans* from *Brassica rapa* into *B. napus*. *Theor. Appl. Genet.* 115: 897–906.

Lehmann, P., C.E. Jenner, E. Kozubek, A.J. Greenland, and J.A. Walsh. 2003. Coat protein-mediated resistance to *Turnip mosaic virus* in oilseed rape (*Brassica napus*). *Mol. Breed.* 11: 83–94.

Lema, M., P. Soengas, P. Velasco, M. Francisco, and M.E. Cartea. 2011. Identification of sources of resistance to *Xanthomonas campestris* pv. *campestris* in *Brassica napus* crops. *Plant Dis.* 95: 292–297.

Leu, Y.S., W.L. Deng, W.S. Yang, Y.F. Wu, A.S. Cheng, S.T. Hsu, and K.C. Tzeng. 2010. Multiplex polymerase chain reaction for simultaneous detection of *Xanthomonas campestris* pv. *campestris* and *X. campestris* pv. *raphani*. *Plant Pathol. Bull.* 19: 137–147.

Li, C.X., S.J. Barker, D.G. Gilchrist, J.E. Lincoln, and W.A. Cowling. 2008c. *Leptosphaeria maculans* elicits apoptosis coincident with leaf lesion formation and hyphal advance in *Brassica napus*. *Mol. Plant Microbe Interact.* 21: 1143–1153.

Li, C.X. and W.A. Cowling. 2003. Identification of a single dominant allele for resistance to blackleg in *Brassica napus* 'Surpass 400'. *Plant Breed.* 122: 485–488.

Li, C.X., H. Li, A.B. Siddique, K. Sivasithamparam, P. Salisbury, S.S. Banga, S. Banga et al. 2007b. The importance of the type and time of inoculation and assessment in the determination of resistance in *Brassica napus* and *B. juncea* to *Sclerotinia sclerotiorum*. *Aust. J. Agric. Res.* 58: 1198–1203.

Li, C.X., H. Li, K. Sivasithamparam, T.D. Fu, Y.C. Li, S.Y. Liu, and M.J. Barbetti. 2006a. Expression of field resistance under Western Australian conditions to *Sclerotinia sclerotiorum* in Chinese and Australian *Brassica napus* and *Brassica juncea* germplasm and its relation with stem diameter. *Aust. J. Agric. Res.* 57: 1131–1135.

Li, C.X., S.Y. Liu, K. Sivasithamparam, and M.J. Barbetti. 2009b. New sources of resistance to Sclerotinia stem rot caused by *Sclerotinia sclerotiorum* in Chinese and Australian *Brassica napus* and *B. juncea* germplasm screened under Western Australian conditions. *Aust. Plant Pathol.* 38: 149–152.

Li, C.X., K. Sivasithamparam, and M.J. Barbetti. 2009a. Complete resistance to leaf and staghead disease in Australian *Brassica juncea* germplasm exposed to infection by *Albugo candida* (white rust). *Aust. Plant Pathol.* 38: 63–66.

Li, C.X., K. Sivasithamparam, G. Walton, P. Fels, and M.J. Barbetti. 2008a. Both incidence and severity of white rust disease reflect host resistance in *Brassica juncea* germplasm from Australia, China and India. *Field Crops Res.* 106: 1–8.

Li, C.X., K. Sivasithamparam, G. Walton, P. Salisbury, W. Burton, S.S. Banga, S. Banga et al. 2007a. Expression and relationships of resistance to white rust (*Albugo candida*) at cotyledonary, seedling, and flowering stages in *Brassica juncea* germplasm from Australia, China and India. *Aust. J. Agric. Res.* 58: 259–264.

Li, C.X., N. Wratten, P.A. Salisbury, W.A. Burton, T.D. Potter, G. Walton, H. Li et al. 2008d. Response of *Brassica napus* and *B. juncea* germplasm from Australia, China and India to Australian populations of *Leptosphaeria maculans*. *Aust. Plant Pathol.* 37: 162–170.

Li, G.Q., H.C. Huang, A. Laroche, and S.N. Acharaya. 2003a. Occurrence and characterization of hypovirulence in the tan sclerotial isolates of S10 of *Sclerotinia sclerotiorum*. *Mycol. Res.* 107: 1350–1360.

Li, G.Q., H.C. Huang, H.J. Miao, R.S. Erickson, D.H. Jiang, and Y.N. Xiao. 2006b. Biological control of sclerotinia diseases of rapeseed by aerial applications of the mycoparasite *Coniothyrium minitans*. *Eur. J. Plant Pathol.* 114: 345–355.

Li, H., M.J. Barbetti, and K. Sivasithamparam. 2006c. Concomitant inoculation of an avirulent strain of *Leptosphaeria maculans* prevents break-down of a single dominant gene-based resistance in *Brassica napus* cv. Surpass 400 by a virulent strain. *Field Crops Res.* 95: 206–211.

Li, H., H.B. Li, Y. Bai, J. Wang, M. Nie, B. Li, and M. Xiao. 2011. The use of *Pseudomonas fluorescens* P13 to control sclerotinia stem rot (*Sclerotinia sclerotiorum*) of oilseed rape. *J. Microbiol.* 49: 884–889.

Li, H., K. Sivasithamparam, and M.J. Barbetti. 2003c. Breakdown of a *Brassica rapa* subsp. *silvestris* single dominant blackleg resistance gene in *B. napus* rapeseed by *Leptosphaeria maculans* field isolates in Australia. *Plant Dis.* 87: 752.

Li, H., K. Sivasithamparam, and M.J. Barbetti. 2007c. Soilborne ascospores and pycnidiospores of *Leptosphaeria maculans* can contribute significantly to blackleg disease epidemiology in oilseed rape (*Brassica napus*) in Western Australia. *Aust. Plant Pathol.* 36: 439–444.

Li, H., K. Sivasithamparam, M.J. Barbetti, S.J. Wylie, and J. Kuo. 2008b. Cytological responses in the hypersensitive reaction in cotyledon and stem tissues of *Brassica napus* after infection by *Leptosphaeria maculans*. *J. Gen. Plant Pathol.* 74: 120–124.

Li, R.G., R. Rimmer, L. Buchwaldt, A.G. Sharpe, G. Seguin-Swartz, and D.D. Hegedus. 2004. Interaction of *Sclerotinia sclerotiorum* with *Brassica napus*: Cloning and characterization of endo- and exo-polygalacturonases expressed during saprophytic and parasitic modes. *Fungal Genet. Biol.* 41: 754–765.

Li, R.G., R. Rimmer, M. Yu, A.G. Sharpe, G. Seguin-Swartz, D. Lydiate, and D.D. Hegedus. 2003b. Two *Brassica napus* polygalacturonase inhibitory protein genes are expressed at different levels in response to biotic and abiotic stresses. *Planta* 217: 299–308.

Li, Y.C., J. Chen, R. Bennett, G. Kiddle, R. Wallsgrove, Y.J. Huang, and Y.H. He. 1999. Breeding, inheritance and biochemical studies on *Brassica napus* cv. Zhougyou 821: Tolerance to *Sclerotinia sclerotiorum* (stem rot). In: *Proceedings of the 10th International Rapeseed Congress*, Canberra, Australia, eds., N. Wratten and P.A. Salisbury, 61pp.

Liang, Y., S. Srivastava, M.H. Rahman, S.E. Strelkov, and N.N.V. Kav. 2008. Proteome changes in leaves of *Brassica napus* L. as a result of *Sclerotinia sclerotiorum* challenge. *J. Agric. Food Chem.* 56: 1963–1976.

Liang, Y., W. Yajima, M.R. Davis, N.N.V. Kav, and S.E. Strelkov. 2013. Disruption of a gene encoding a hypothetical secreted protein from *Sclerotinia sclerotiorum* reduces its virulence on canola (*Brassica napus*). *Can. J. Plant Pathol.* 35: 46–55.

Light, K.A., N.N. Gororo, and P.A. Salisbury. 2011. Usefulness of winter canola (*Brassica napus*) race-specific resistance genes against blackleg (causal agent *Leptosphaeria maculans*) in southern Australian growing conditions. *Crop Pasture Sci.* 62: 162–168.

Lima, M.L.P., A.C. CafeFilho, N.L. Nogueira, M.L. Rossi, and L.R. Schuta. 2004. First report of clubroot of *Eruca sativa* caused by *Plasmodiophora brassicae* in Brazil. *Plant Dis.* 88: 573.

Ling, P., N.X. Gao, Z.M. Gao, L.Y. Wang, and H.F. Yu. 2011. Comparison of the biological characteristics of different pathogenic strains in *Sclerotinia sclerotiorum*. *J. Anhui Agric. Univ.* 38: 916–919.

Liu, S.Y., Z. Liu, B.D.L. Fitt, N. Evans, S.J. Foster, Y.J. Huang, A.O. Latunde-Dada, and J.A. Lucas. 2006. Resistance to *Leptosphaeria maculans* (phoma stem canker) in *Brassica napus* (oilseed rape) induced by *L. biglobosa* and chemical defence activators in field and controlled environments. *Plant Pathol.* 55: 401–412.

Liu, Z., A.O. Latunde-Dada, B.D.L. Fitt, and A.M. Hall. 2007. Effects of pre-treatment with acibenzolar-*s*-methyl or *Leptosphaeria biglobosa* on development of phoma stem canker (*L. maculans*) epidemics on winter oilseed rape. *Aspects Appl. Biol.* 83: 71–74.

Lobb, W.R. 1951. Resistant type of rape for areas with club root. *N. Z. J. Agric.* 82: 65–66.

Long, Y., Z. Wang, Z. Sun, D.W. Fernando, P.B. McVetty, and G. Li. 2011. Identification of two blackleg resistance genes and fine mapping of one of these two genes in a *Brassica napus* canola cultivar 'Surpass 400'. *Theor. Appl. Genet.* 122: 1223–1231.

Lo-Pelzer, E., J.N. Aubertot, O. David, M.H. Jeuffroy, and L. Bousset. 2009. Relationship between severity of blackleg (*Leptosphaeria maculans/L. biglobosa* species complex) and subsequent primary inoculum production on oilseed rape stubble. *Plant Pathol.* 58: 61–70.

Luth, P., J.P. Boulon, and U. Eiben. 2012. New ways of applying the fungal antagonist *Coniothyrium minitans* strain CON/M/91-08 to protect oilseed rape against sclerotinia stem rot. In: *10e Conférence Internationale sur les Maladies des Plantes*, Tours, France, December 3–5, 2012, pp. 692–702.

Ma, L.L., R. Huo, X.W. Gao, D. He, M. Shao, and Q. Wang. 2008. Transgenic rape with hrf2 gene encoding harpin Xooc resistant to *Sclerotinia sclerotinorium*. *Sci. Agric. Sin.* 41: 1655–1660.

Magyar, D., T. Barasits, G. Fischl, and W.G.D. Fernando. 2006. First report of the natural occurrence of the teleomorph of *Leptosphaeria maculans* on oilseed rape and airborne dispersal of ascospores in Hungary. *J. Phytopathol.* 154: 428–431.

Maksymiak, M.S. and A.M. Hall. 2000. Biological control of *Leptosphaeria maculans* (anamorph *Phoma lingam*) causal agent of blackleg/canker on oil seed rape by *Cyanthus striatus*, a bird's nest fungus. In: *The BCPC Conference: Pests and Diseases*, Vol. 1. *Proceedings of the International Conference*, Brighton Hilton Metropole Hotel, Brighton, U.K., November 13–16, 2000, pp. 507–510.

Maksymiak, M.S. and A.M. Hall. 2002. Biocontrol of canker on oilseed rape by reduction and inhibition of initial inoculum. In: *The BCPC Conference: Pests and Diseases*, Vols. 1 and 2. *Proceedings of the International Conference*, Brighton Hilton Metropole Hotel, Brighton, U.K., November 18–21, 2002, pp. 769–772.

Malik, R.S., R. Malik, and M. Kumar. 2004. White rust (*Albugo candida*) infection in relation to erucic acid content in mustard [*Brassica juncea* (L.) Czern & Coss]. *Indian J. Genet. Plant Breed.* 64: 331–332.

Maling, T., A.J. Diggle, D.J. Thackray, K.H.M. Siddique, and R.A.C. Jones. 2010. An epidemiological model for externally acquired vector-borne viruses applied to Beet western yellows virus in *Brassica napus* crops in a Mediterranean-type environment. *Crop Pasture Sci.* 61: 132–144.

Maliogka, V.I., J.T. Tsialtas, A. Papantoniou, K. Efthimiou, and N.I. Katis. 2009. First report of a phytoplasma associated with an oilseed rape disease in Greece. *Plant Pathol.* 58: 792.

Malvarez, M., I. Carbone, N.J. Grunwald, K.V. Subbarao, M. Schafer, and L.M. Kohn. 2007. New populations of *Sclerotinia sclerotiorum* from lettuce in California and peas and lentils in Washington. *Phytopathology* 97: 470–483.

Mamula, D. 2008. Virus diseases of oilseed rape. *Glasilo Biljne Zastite* 8: 321–323.
Manjunath, H., D.S. Phogat, and D. Singh. 2007. Inheritance of white rust (*Albugo candida*) resistance in Indian mustard. *Nat. J. Plant Improv.* 9: 96–98.
Marcroft, S. and C. Bluett. 2008. Blackleg of canola, Ag1352. Agriculture Notes, Department of Primary Industries, State of Victoria, Australia, ISSN 1329-8062.
Marcroft, S., S. Sprague, P. Salisbury, and B. Howlett. 2003. Survival and dissemination of *Leptosphaeria maculans* in south-eastern Australia. In: *Thirteenth Biennial Australian Research Assembly Brassicas. Proceedings of Conference*, Tamworth, New South Wales, Australia, September 8–12, 2003, pp. 131–133.
Marcroft, S.J. and T.D. Potter. 2008. The fungicide fluquinconazole applied as a seed dressing to canola reduces *Leptosphaeria maculans* (blackleg) severity in south-eastern Australia. *Aust. Plant Pathol.* 37: 396–401.
Marcroft, S.J., S.J. Sprague, S.J. Pymer, P.A. Salisbury, and B.J. Howlett. 2004. Crop isolation, not extended rotation length, reduces blackleg (*Leptosphaeria maculans*) severity of canola (*Brassica napus*) in south-eastern Australia. *Aust. J. Exp. Agric.* 44: 601–606.
Marcroft, S.J., S.J. Sprague, P.A. Salisbury, and B.J. Howlett. 2003. Production of blackleg inoculum from stubble of different *Brassica* species. In: *Thirteenth Biennial Australian Research Assembly on Brassicas. Proceedings of Conference*, Tamworth, New South Wales, Australia, September 8–12, 2003, pp. 124–127.
Marcroft, S.J., A.P. Wouw, P.A. van de Salisbury, T.D. Potter, and B.J. Howlett. 2012. Effect of rotation of canola (*Brassica napus*) cultivars with different complements of blackleg resistance genes on disease severity. *Plant Pathol.* 61: 934–944.
Marquardt, D. and R.U. Ehlers. 2010. Reduction of winter oilseed rape pathogens by seed treatment with the rhizobacterium *Serratia plymuthica. Julius-Kuhn-Archiv* 428: 100.
Massand, P.P., S.K. Yadava, P. Sharma, A. Kaur, A. Kumar, N. Arumugam, Y.S. Sodhi et al. 2010. Molecular mapping reveals two independent loci conferring resistance to *Albugo candida* in the east European germplasm of oilseed mustard *Brassica juncea. Theor. Appl. Genet.* 121: 137–145.
Mathpal, P., H. Punetha, A.K. Tewari, and S. Agrawal. 2011. Biochemical defense mechanism in rapeseed-mustard genotypes against Alternaria blight disease. *J. Oilseed Brassica* 2: 87–94.
Matroudi, S., M.R. Zamani, and M. Motallebi. 2009. Antagonistic effects of three species of *Trichoderma* sp. on *Sclerotinia sclerotiorum*, the causal agent of canola stem rot. *Egypt. J. Biol.* 11: 37–44.
Mayerhofer, R., V.K. Bansal, M.R. Thiagarajah, G.R. Stringam, and A.G. Good. 1997. Molecular mapping of resistance to *Leptosphaeria maculans* in Australian cultivars of *Brassica napus. Genome* 40: 294–301.
McCartney, H.A. and M.E. Lacey. 1999. Timing and infection of sunflowers by *Sclerotinia sclerotiorum* and disease development. *Aspects Appl. Biol.* 56: 151–156.
McGee, D.E. and R.W. Emmett. 1977. Blackleg (*Leptosphaeria maculans*) (Desm.) Ces. et de Not. of rapeseed in Victoria: Crop losses and factors, which affect disease severity. *Aust. J. Agric. Res.* 28: 47–51.
Meena, P.D., R.P. Awasthi, C. Chattopadhyay, S.J. Kolte, and A. Kumar. 2010a. Alternaria blight: A chronic disease in rapeseed-mustard. *J. Oilseed Brassica* 1: 1–11.
Meena, P.D., C. Chattopadhyay, A. Kumar, R.P. Awasthi, R. Singh, S. Kaur, L. Thomas, P. Goyal, and P. Chand. 2011b. Comparative study on the effect of chemicals on Alternaria blight in Indian mustard—A multi-location study in India. *J. Environ. Biol.* 32: 375–379.
Meena, P.D., C. Chattopadhyay, S.S. Meena, and A. Kumar. 2011a. Area under disease progress curve and apparent infection rate of Alternaria blight disease of Indian mustard (*Brassica juncea*) at different plant age. *Arch. Phytopathol. Plant Prot.* 44: 684–693.
Meena, P.D., C. Chattopadhyay, F. Singh, B. Singh, and A. Gupta. 2002. Yield loss in Indian mustard due to white rust and effect of some cultural practices on Alternaria blight and white rust severity. *Brassica* 4: 18–24.
Meena, P.D., R.B. Gour, J.C. Gupta, H.K. Singh, R.P. Awasthi, R.S. Netam, S. Godika et al. 2013. Non-chemical agents provide tenable, eco-friendly alternatives for the management of the major diseases devastating Indian mustard (*Brassica juncea*) in India. *Crop Prot.* 53: 169–174.
Meena, P.D., R.L. Meena, and C. Chattopadhyay. 2008. Eco-friendly options for management of Alternaria blight in Indian mustard (*Brassica juncea*). *Indian Phytopathol.* 61: 65–69.
Meena, P.D., R.L. Meena, C. Chattopadhyay, and A. Kumar. 2004. Identification of critical stage for disease development and biocontrol of Alternaria blight of Indian mustard (*Brassica juncea*). *J. Phytopathol.* 152: 204–209.
Meena, P.D., K. Mondal, A.K. Sharma, C. Chattopadhyay, and A. Kumar. 2010b. Bacterial rot: A new threat for rapeseed-mustard production system in India. *J. Oilseed Brassica* 1: 39–41.
Meena, P.D. and P. Sharma. 2012a. Antifungal activity of plant extracts against *Alternaria brassicae* causing blight of *Brassica* spp. *Ann. Plant Prot. Sci.* 20: 256–257.
Meena, P.D. and P. Sharma. 2012b. Methodology for production and germination of oospores of *Albugo candida* infecting oilseed *Brassica. Vegetos* 25: 115–119.

Meena, P.D., A.K. Sharma, S.K. Jha, and C. Chattopadhyay. 2006. Impact of fungal diseases on yield of Indian mustard—Farmers' perception and on-farm success of garlic clove extract in management of Sclerotinia rot. *Indian J. Plant Prot.* 34: 229–232.

Meena, P.D., P.R. Verma, G.S. Saharan, and M.H. Borhan. 2014. Historical perspectives of white rust caused by *Albugo candida* in oilseed Brassica. *J. Oilseed Brassica* 5: 1–41.

Meena, R.L., K.L. Jain, and P. Rawal. 2005. Evaluation of rapeseed-mustard genotypes for resistance to white rust (*Albugo candida*). *Resist. Pest Manag. Newsl.* 14: 29–30.

Meena, R.L., P.D. Meena, and C. Chattopadhyay. 2003. Potential for biocontrol of white rust of Indian mustard (*Brassica juncea*). *Indian J. Plant Prot.* 31: 120–123.

Mehta, N., N.T. Hieu, and M.S. Sangwan. 2012. Efficacy of various antagonistic isolates and species of Trichoderma against *Sclerotinia sclerotiorum* causing white stem rot of mustard. *J. Mycol Plant Pathol.* 42: 244–250.

Mehta, N., A.K. Khurana, and M.S. Sangwan. 2005. Characterization of *Alternaria brassicae* isolates from Haryana on the basis of nutritional behaviour. In: *Integrated Plant Disease Management: Challenging Problems in Horticultural and Forest Pathology*, eds., R.C. Sharma and J.N. Sharma, Scientific Publishers, Jodhpur, India, pp. 295–303.

Mehta, N., M.S. Sangwan, M.P. Srivastava, and R. Kumar. 2002. Survival of *Alternaria brassicae* causing Alternaria blight in rapeseed-mustard. *J. Mycol. Plant Pathol.* 32: 64–67.

Mendoza, L.E., R. Del, A. Nepal, J.M. Bjerke, M. Boyles, and T. Peeper. 2011. Identification of *Leptosphaeria maculans* pathogenicity group 4 causing blackleg on winter canola in Oklahoma. *Plant Dis.* 95: 614.

Mert-Turk, F., M. Ipek, D. Mermer, and P. Nicholson. 2007. Microsatellite and morphological markers reveal genetic variation within a population of *Sclerotinia sclerotiorum* from oilseed rape in the Canakkale Province of Turkey. *J. Phytopathol.* 155: 182–187.

Miguel-Wruck, D.S., J.R. de Oliveira, and L.A. dos Santos Dias. 2010. Host specificity in interaction *Xanthomonas campestris* pv. *campestris*—brassicas. *Summa Phytopathol.* 36: 129–133.

Mirabadi, A.Z., K. Rahnama, and A. Esmaailifar. 2009. First report of pathogenicity group 2 of *Leptosphaeria maculans* causing blackleg of oilseed rape in Iran. *Plant Pathol.* 58: 1175.

Mishra, K.K., S.J. Kolte, N.I. Nashaat, and R.P. Awasthi. 2009. Pathological and biochemical changes in *Brassica juncea* (mustard) infected with *Albugo candida* (white rust). *Plant Pathol.* 58: 80–86.

Mishra, S. and N.K. Arora. 2012. Evaluation of rhizospheric *Pseudomonas* and *Bacillus* as biocontrol tool for *Xanthomonas campestris* pv. *campestris*. *World J. Microbiol. Biotechnol.* 28: 693–702.

Mitrovic, P. and V. Trkulja. 2010. *Leptosphaeria maculans* and *Leptosphaeria biglobosa*—Causal agents of phoma stem cancer and dry root rot of oilseed rape. *Glasnik Zastite Bilja* 33: 34–45.

Mohiddin, F.A., M.R. Khan, and M.M. Khan. 2008. Comparative efficacy of various spray schedules of some plant extracts for the management of Alternaria blight of Indian mustard (*Brassica juncea* (L.) Czern & Coss). *Trends Biosci.* 1: 50–51.

Mondal, G. 2008. Effect of plant nutrients on incidence of Alternaria blight and sclerotinia rot diseases of rapeseed and mustard. *J. Interacad.* 12: 441–445.

Mondal, K.K., R.C. Bhattacharya, K.R. Koundal, and S.C. Chatterjee. 2007. Transgenic Indian mustard (*Brassica juncea*) expressing tomato glucanase leads to arrested growth of *Alternaria brassicae*. *Plant Cell Rep.* 26: 247–252.

Mondal, K.K., S.C. Chatterjee, N. Viswakarma, R.C. Bhattacharya, and A. Grover. 2003. Chitinase-mediated inhibitory activity of *Brassica* transgenic on growth of *Alternaria brassicae*. *Curr. Microbiol.* 47: 171–173.

Mora, A.A. and E.D. Earle. 2001. Resistance to *Alternaria brassicicola* in transgenic broccoli expressing a *Trichoderma harzianum* endochitinase gene. *Mol. Breed.* 8: 1–9.

Mukherjee, I., M. Gopal, and S.C. Chatterjee. 2003. Persistence and effectiveness of iprodione against Alternaria blight in mustard. *Bull. Environ. Contam. Toxicol.* 70: 586–591.

Mundkur, B.B. 1959. *Fungi and Plant Diseases*. MacMillan, London, U.K., 75pp.

Myung, I.S., Y.K. Lee, S.W. Lee, W.G. Kim, H.S. Shim, and D.S. Ra. 2010. A new disease, bacterial leaf spot of rape, caused by atypical *Pseudomonas viridiflava* in South Korea. *Plant Dis.* 94: 1164–1165.

Naseri, B., J.A. Davidson, and E.S. Scott. 2008. Survival of *Leptosphaeria maculans* and associated mycobiota on oilseed rape stubble buried in soil. *Plant Pathol.* 57: 280–289.

Nashaat, N.I., A. Heran, R.P. Awasthi, and S.J. Kolte. 2004. Differential response and genes for resistance to *Peronospora parasitica* (downy mildew) in *Brassica juncea* (mustard). *Plant Breed.* 123: 512–515.

Neeraj and S. Verma. 2010. Quantitative changes in sugar and phenolic contents of *Brassica* leaves induced by *Alternaria brassicae* infection. *Int. J. Plant Prot.* 3: 28–30.

Nguyen, D.C. and N.P. Dohroo. 2006. Morphological, cultural and physiological studies on *Sclerotinia sclerotiorum* causing stalk rot of cauliflower. *Omonrice* 14: 71–77.

Nie, F.J., Y.X. Zuo, L.L. Huang, H.Q. Qin, X.N. Gao, and Q.M. Han. 2010. Pathogenicity of *Sclerotinia sclerotiorum* isolated from rapeseed in Shaanxi Province. *Acta Phytophyl. Sin.* 37: 499–504.

Olivier, C.Y., B. Galka, and G. Seguin-Swartz. 2010. Detection of aster yellows phytoplasma DNA in seed and seedlings of canola (*Brassica napus* and *B. rapa*) and AY strain identification. *Can. J. Plant Pathol.* 32: 298–305.

Olivier, C.Y. and G. Seguin-Swartz. 2006. Development of a storage method for ascospores of *Sclerotinia sclerotiorum*. *Can. J. Plant Pathol.* 28: 489–493.

Olivier, C.Y., G. Seguin-Swartz, D. Hegedus and T. Barasubiye. 2006. First report of "*Candidatus phytoplasma asteris*"-related strains in *Brassica rapa* in Saskatchewan, Canada. *Plant Dis.* 90: 832.

Oram, R.N., J.P. Edlington, D.M. Halsall, and P.E. Veness. 2003. Pseudomonas blight and Nematospora seed rot of Indian mustard. In: *Thirteenth Biennial Australian Research Assembly Brassicas. Proceedings of Conference*, Tamworth, New South Wales, Australia, September 8–12, 2003, pp. 146–150.

Pageau, D., J. Lajeunesse, and J. Lafond. 2006. Impact of clubroot [*Plasmodiophora brassicae*] on the yield and quality of canola. *Can. J. Plant Pathol.* 28: 137–143.

Pallett, D.W., M.I. Thurston, M. Cortina-Borja, M.L. Edwards, M. Alexander, E. Mitchell, A.F. Raybould, and J.I. Cooper. 2002. The incidence of viruses in wild *Brassica rapa* ssp. *sylvestris* in southern England. *Ann. Appl. Biol.* 141: 163–170.

Pandya, R.K., M.L. Tripathi, and R. Singh. 2000. Efficacy of fungicides in the management of white rust and Alternaria blight of mustard. *Crop Res.* (*Hisar*) 20: 137–139.

Parada, R.Y., K. Oka, D. Yamagishi, M. Kodama, and H. Otani. 2007. Destruxin B produced by *Alternaria brassicae* does not induce accessibility of host plants to fungal invasion. *Physiol. Mol. Plant Pathol.* 71: 48–54.

Patel, J.S. and S.J. Patel. 2008. Influence of foliar sprays of fungicides, phytoextracts and bioagent on powdery mildew and yield of mustard. *Karnataka J. Agric. Sci.* 21: 462–463.

Patel, M.K., S.G. Abhyankar, and Y.S. Kulkarni. 1949. Black rot of cabbage. *Indian Phytopathol.* 2: 58.

Pathak, A.K. and S. Godika. 2002a. Efficacy of seed dressing fungicides in the management of stem rot disease of mustard caused by *Sclerotinia sclerotiorum* (Lib.) de Bary. *Cruciferae Newsl.* 24: 93–94.

Pathak, A.K. and S. Godika. 2002b. Multiple disease resistance in different genotypes of rapeseed-mustard. *Cruciferae Newsl.* 24: 95–96.

Patni, C.S. and S.J. Kolte. 2006. Effect of some botanicals in management of Alternaria blight of rapeseed-mustard. *Ann. Plant Prot. Sci.* 14: 151–156.

Patni, C.S., S.J. Kolte, and R.P. Awasthi. 2005a. Cultural variability of *Alternaria brassicae*, causing Alternaria blight of mustard. *Ann. Plant Physiol.* 19: 231–242.

Patni, C.S., S.J. Kolte, and R.P. Awasthi. 2005b. Screening of Indian mustard (*Brassica juncea* (Linn) Czern and Coss) genotypes to *Alternaria brassicae* (Berk) sac. isolates based on infection rate reducing resistance. *J. Interacad.* 9: 498–507.

Pedras, M.S.C. and P.W.K. Ahiahonu. 2004. Phytotoxin production and phytoalexin elicitation by the phytopathogenic fungus *Sclerotinia sclerotiorum*. *J. Chem. Ecol.* 30: 2163–2179.

Pedras, M.S.C., P.B. Chumala, W. Jin, M.S. Islam, and D.W. Hauck. 2009a. The phytopathogenic fungus *Alternaria brassicicola*: phytotoxin production and phytoalexin elicitation. *Phytochemistry* 70: 394–402.

Pedras, M.S.C., R.S. Gadagi, M. Jha, and V.K. Sarma-Mamillapalle. 2007. Detoxification of the phytoalexin brassinin by isolates of *Leptosphaeria maculans* pathogenic on brown mustard involves an inducible hydrolase. *Phytochemistry* 68: 1572–1578.

Pedras, M.S.C., Z. Minic, and V.K. Sarma-Mamillapalle. 2009b. Substrate specificity and inhibition of brassinin hydrolases, detoxifying enzymes from the plant pathogens *Leptosphaeria maculans* and *Alternaria brassicicola*. *FEBS J.* 276: 7412–7428.

Pedras, M.S.C., J.L. Taylor, and V.M. Morales. 1995. Phomalignin A and other yellow pigments in *Phoma lignin* and *P. wasabiae*. *Phytochemistry* 38: 1215–1222.

Pedras, M.S.C. and Y. Yu. 2008. Structure and biological activity of maculansin A, a phytotoxin from the phytopathogenic fungus *Leptosphaeria maculans*. *Phytochemistry* 69: 2966–2971.

Pedras, M.S.C., Q.A. Zheng, and S. Strelkov. 2008. Metabolic changes in roots of the oilseed canola infected with the biotroph *Plasmodiophora brassicae*: Phytoalexins and phytoanticipins. *J. Agric. Food Chem.* 56: 9949–9961.

Penaud, A., J. Carpezat, M. Leflon, and X. Pinochet. 2012. Trapping and quantification of ascospores of *Sclerotinia sclerotiorum* for sclerotinia risk analysis. [French]. In: *10e Conference Internationale sur les Maladies des Plantes*, Tours, France, December 3–5, 2012, pp. 87–94.

Penaud, A., B. Huguet, V. Wilson, and P. Leroux. 2003. Fungicides resistance of *Sclerotinia sclerotiorum* in French oilseed rape crops. In: *Abstracts, 11th International Rapeseed Congress*, Copenhagen, Denmark, July 6–10, 2003, p. 156.

Penaud, A. and H. Michi. 2009. *Coniothyrium minitans*, a biological agent for integrated protection. *OCL—Oleagineux, Corps Gras, Lipides* 16: 158–163.

Perek, A., K. Pieczul, E. Jajor, and M. Korbas. 2010. Application of PCR reaction for identification of *Plasmodiophora brassicae* Wor. (clubroot) and pathotype P1. *Prog. Plant Prot.* 50: 705–708.

Peters, B.J., G.J. Ash, E.J. Cother, D.L. Hailstones, D.H. Noble, and N.A.R. Urwin. 2004. *Pseudomonas syringae* pv. *maculicola* in Australia: Pathogenic, phenotypic and genetic diversity. *Plant Pathol.* 53: 73–79.

Petrie, G.A. 1969. Variability in *Leptosphaeria maculans* (Desm.) Ces. and De Not., the cause of blackleg of rape. Ph.D. Thesis, Univ. Saskatchewan, Saskatoon, Canada.

Petrie, G.A. 1975. Prevalence of oospores of *Albugo cruciferarum* in *Brassica* seed samples from western Canada, 1967–73. *Can. Plant Dis. Surv.* 55: 19–24.

Petrie, G.A. 1978. Occurrence of a highly virulent strain of blackleg on rape in Saskatchewan (1975–77). *Can. Plant. Dis. Surv.* 58: 21–25.

Petrie, G.A. 1979. Blackleg of rape. *Can. Agric.* 24: 22–25.

Popovic, T., J. Balaz, M. Starovic, N. Trkulja, Z. Ivanovic, M. Ignjatov, and D. Josic. 2013. First report of *Xanthomonas campestris* pv. *campestris* as the causal agent of black rot on oilseed rape (*Brassica napus*) in Serbia. *Plant Dis.* 97: 418.

Porter, R.H. 1926. A preliminary report of surveys for plant diseases in East China. *Plant Dis. Rep.* 46(Suppl.): 153–166.

Potlakayala, S.D., D.W. Reed, P.S. Covello, and P.R. Fobert. 2007. Systemic acquired resistance in canola is linked with pathogenesis-related gene expression and requires salicylic acid. *Phytopathology* 97: 794–802.

Prabhu, K.V., D.J. Somers, G. Rakow, and R.K. Gugel. 1998. Molecular markers linked to white rust resistance in mustard *Brassica juncea*. *Theor. Appl. Genet.* 97: 865–870.

Prakash, A.T., H. Punetha, and R. Mall. 2012. Investigation on metabolites changes in *Brassica juncea* (Indian mustard) during progressive infection of *Alternaria brassicae*. *Pantnagar J. Res.* 10: 211–216.

Prasad, R., S. Kumar, and D.R. Chandra. 2009a. Sclerotial germination and inoculation methods for screening of rapeseed-mustard genotypes against stem rot (*Sclerotinia sclerotiorum*). *Ann. Plant Prot. Sci.* 17: 136–138.

Prasad, R., K.K. Maurya, and S.B.L. Srivastava. 2009b. Eco-friendly management of Alternaria blight of Indian mustard, *Brassica juncea* L. *J. Oilseeds Res.* 26: 65.

Preiss, U., H. Mather-Kaub, and G. Albert. 2010. Reduction of the adverse effect of *Plasmodiophora brassicae* Wor. through the use of microorganisms and mushroom compost. *Julius-Kuhn-Archiv* 428: 439–440.

Pruthi, V., H.K.L. Chawla, and G.S. Saharan. 2001. *Albugo candida* induced changes in phenolics and glucosinolates in leaves of resistant and susceptible cultivars of *Brassica juncea*. *Cruciferae Newsl.* 23: 61–62.

Purwantara, A., J.M. Barrins, A.J. Cozijnsen, P.K. Ades, and B.J. Howle. 2000. Genetic diversity of isolates of the *Leptosphaeria maculans* species complex from Australia, Europe and North America using amplified fragment length polymorphism analysis. *Mycolog. Res.* 104: 772–781.

Pusz, W. 2007. The influence of plant density on infestation of winter oilseed rape by pathogenic fungi. *Prog. Plant Prot.* 47: 287–290.

Qi, S.W., C.Y. Guan, and C.L. Liu. 2004. Relationship between some enzyme activity and resistance to *Sclerotinia sclerotiorum* of rapeseed cultivars. *Acta Agron. Sin.* 30: 270–273.

Qin, L., Y. Fu, J. Xie, J. Cheng, D. Jiang, G. Li, and J. Huang. 2011. A nested-PCR method for rapid detection of *Sclerotinia sclerotiorum* on petals of oilseed rape (*Brassica napus*). *Plant Pathol.* 60: 271–277.

Raghavendra, B.T., D. Singh, D.K. Yadava, K.K. Mondal, and P. Sharma. 2013. Virulence analysis and genetic diversity of *Xanthomonas campestris* pv. *campestris* causing black rot of crucifers. *Arch. Phytopathol. Plant Prot.* 46: 227–242.

Rahman, H., A. Shakir, and M.J. Hasan. 2011. Breeding for clubroot resistant spring canola (*Brassica napus* L.) for the Canadian prairies: Can the European winter canola cv. Mendel be used as a source of resistance? *Can. J. Plant Sci.* 91: 447–458.

Rahmanpour, S., D. Backhouse, and H.M. Nonhebel. 2010. Reaction of glucosinolate-myrosinase defence system in *Brassica* plants to pathogenicity factor of *Sclerotinia sclerotiorum*. *Eur. J. Plant Pathol.* 128: 429–433.

Rajappan, K., B. Ramaraj, and S. Natarajan. 1999. Knol-khol—A new host for club-root disease in the Nilgiris. *Indian Phytopathol.* 52: 328.

Raman, R., B. Taylor, S. Marcroft, J. Stiller, P. Eckermann, N. Coombes, A. Rehman et al. 2012. Molecular mapping of qualitative and quantitative loci for resistance to *Leptosphaeria maculans* causing blackleg disease in canola (*Brassica napus* L.). *Theor. Appl. Genet.* 125: 405–418.

Ramarathnam, R., W.G.D. Fernando, and T. de Kievit. 2011. The role of antibiosis and induced systemic resistance, mediated by strains of *Pseudomonas chlororaphis*, *Bacillus cereus* and *B. amyloliquefaciens*, in controlling blackleg disease of canola. *BioControl* 56: 225–235.

Rampin, E., N. Mori, L. Marini, F. Zanetti, G. Mosca, V. Girolami, N. Contaldo, and A. Bertaccini. 2010. Phytoplasma infection on rape in Italy. *Informatore Agrario* 66: 59–60.

Rashid, A., S.F. Hwang, H.U. Ahmed, G.D. Turnbull, S.E. Strelkov, and B.D. Gossen. 2013. Effects of soil-borne *Rhizoctonia solani* on canola seedlings after application of glyphosate herbicide. *Can. J. Plant Sci.* 93: 97–107.

Rastas, M., S. Latvala, and A. Hannukkala. 2012. Occurrence of *Plasmodiophora brassicae* in Finnish turnip rape and oilseed rape fields. *Agric. Food Sci.* 21: 141–158.

Ratajkiewicz, H., R. Kierzek, Z. Karolewski, and M. Wachowiak. 2009. The effect of adjuvants, spray volume and nozzle type on azoxystrobin efficacy against *Leptosphaeria maculans* and *L. biglobosa* on winter oilseed rape. *J. Plant Prot. Res.* 49: 440–445.

Reiss, E., J. Schubert, P. Scholze, R. Kramer, and K. Sonntag. 2009. The barley thaumatin-like protein Hv-TLP8 enhances resistance of oilseed rape plants to *Plasmodiophora brassicae. Plant Breed.* 128: 210–212.

Remy, E., M. Meyer, F. Blaise, M. Chabirand, N. Wolff, M.H. Balesdent, and T. Rouxel. 2008. The Lmpma1 gene of *Leptosphaeria maculans* encodes a plasma membrane H+-ATPase isoform essential for pathogenicity towards oilseed rape. *Fungal Genet. Biol.* 45: 1122–1134.

Remy, E., M. Meyer, F. Blaise, U.K. Simon, D. Kuhn, M.H. Balesdent, and T. Rouxel. 2009. A key enzyme of the leloir pathway is involved in pathogenicity of *Leptosphaeria maculans* toward oilseed rape. *Mol. Plant Microbe Interact.* 22: 725–736.

Ren, L., K.R. Chen, C.Y. Wang, L.X. Luo, J.G. Jia, J. Wang, and X.P. Fang. 2010b. Potential of prochloraz-manganese chloride in controlling Sclerotinia stem rot of oilseed rape. *Sci. Agric. Sin.* 43: 4183–4191.

Ren, L., G.Q. Li, Y.C. Han, D.H. Jiang, and H.C. Huang. 2007. Degradation of oxalic acid by *Coniothyrium minitans* and its effects on production and activity of beta-1,3-glucanase of this mycoparasite. *Biol. Control* 43: 1–11.

Ren, Q.H., J.Y. Huang, H. Mao, B.M. Tian, and S.Y. Liu. 2010a. *Brassica napus* LOX2 gene expression induced by methyl jasmonate (MeJA), benzothiadiazole (BTH), and *Sclerotinia sclerotiorum. J. Henan Agric. Sci.* 7: 22–25.

Rennie, D.C., V.P. Manolii, T. Cao, S.F. Hwang, R.J. Howard, and S.E. Strelkov. 2011. Direct evidence of surface infestation of seeds and tubers by *Plasmodiophora brassicae* and quantification of spore loads. *Plant Pathol.* 60: 811–819.

Reshu and M.M. Khan. 2012. Role of different microbial-origin bioactive antifungal compounds against *Alternaria* spp. causing leaf blight of mustard. *Plant Pathol. J.* (Faisalabad) 11: 1–9.

Rietz, S., F.E.M. Bernsdorff, and D.G. Cai. 2012. Members of the germin-like protein family in *Brassica napus* are candidates for the initiation of an oxidative burst that impedes pathogenesis of *Sclerotinia sclerotiorum. J. Exp. Bot.* 63: 5507–5519.

Rimmer, S.R. 2006. Resistance genes to *Leptosphaeria maculans* in *Brassica napus. Can. J. Plant Pathol. Rev. Can. Phytopathol.* 28: S288–S297.

Rimmer, S.R. and C.G.J. van den Berg. 1992. Resistance to oilseed *Brassica* spp. to blackleg caused by *Leptosphaeria maculans. Can. J. Plant Pathol.* 14: 56–66.

Rimmer, S.R., M.H. Borhan, B. Zhu, and D. Somers. 1999. Mapping resistance genes in *Brassica napus* to *Leptosphaeria maculans*. In: *Proceedings of the 10th International Rapeseed Congress*, Canberra, Australia. http://www.regional.org.au/au/gcirc/3/47.htm.

Rogers, S.L., S.D. Atkins, and J.S. West. 2009. Detection and quantification of airborne inoculum of *Sclerotinia sclerotiorum* using quantitative PCR. *Plant Pathol.* 58: 324–331.

Rogers, S.L., S.D. Atkins, A.O. Latunde-Dada, J.F. Stonard, and J.S. West. 2008. A quantitative-PCR method to detect and quantify airborne ascospores of *Sclerotinia sclerotiorum. Aspects Appl. Biol.* 89: 77–79.

Rouxel, T. and M.H. Balesdent. 2013. From model to crop plant-pathogen interactions: Cloning of the first resistance gene to *Leptosphaeria maculans* in *Brassica napus. New Phytol.* 197: 356–358.

Ruaro, L., V. da C. Lima Neto, and P.J. Ribeiro Jr. 2009. Influence of boron, nitrogen sources and soil pH on the control of club root of crucifers caused by *Plasmodiophora brassicae. Trop. Plant Pathol.* 34: 231–238.

Saal, B., H. Brun, I. Glais, and D. Struss. 2004. Identification of a *Brassica juncea*-derived recessive gene conferring resistance to *Leptosphaeria maculans* in oilseed rape. *Plant Breed.* 123: 505–511.

Sachan, J.N., S.J. Kolte, and B. Singh. 1995. Genetics of resistance to white rust (*Albugo candida* race-2) in mustard (*Brassica juncea*). In: *GCIRC Ninth Inernational Rapeseed Congress*, Cambridge, U.K., pp. 1295–1297.

Sahandi, A., N. Shahraeen, S. Ghorbani, and R. Pourrahim. 2004. Identification and studies of some biological and serological properties of *Cauliflower Mosaic Virus* (CaMV) isolated from oilseed rape from West Azarbaejan Province. *J. Agric. Sci.* 1: 101–111.

Saharan, G.S., N. Mehta, and M.S. Sangwan. 2003. Nature and mechanism of resistance to Alternaria blight in rapeseed-mustard system. *Ann. Rev. Plant Pathol.* 2: 85–128.

Saharan, G.S., N. Mehta, and M.S. Sangwan, eds. 2005. *Diseases of Oilseed Crops*. Indus Publ. Co., New Delhi, India, 643pp.

Salam, M.U., B.D.L. Fitt, J.N. Aubertot, A.J. Diggle, Y.J. Huang, M.J. Barbetti, P. Gladders et al. 2007. Two weather-based models for predicting the onset of seasonal release of ascospores of *Leptosphaeria maculans* or *L. biglobosa*. *Plant Pathol.* 56: 412–423.

Salam, M.U., R.K. Khangura, A.J. Diggle, and M.J. Barbetti. 2003. Blackleg sporacle: A model for predicting onset of pseudothecia maturity and seasonal ascospore showers in relation to blackleg of canola. *Phytopathology* 93: 1073–1081.

Salehi, M., K. Izadpanah, and M. Siampour. 2011. Occurrence, molecular characterization and vector transmission of a phytoplasma associated with rapeseed phyllody in Iran. *J Phytopathol.* 159: 100–105.

Sanogo, S. and N. Puppala. 2007. Characterization of darkly pigmented mycelial isolates of *Sclerotinia sclerotiorum* on Valencia peanut in New Mexico. *Plant Dis.* 91: 1077–1082.

Sasek, V., M. Novakova, P.I. Dobrev, O. Valentova, and L. Burketova. 2012b. Beta-aminobutyric acid protects *Brassica napus* plants from infection by *Leptosphaeria maculans*. Resistance induction or a direct antifungal effect? *Eur. J. Plant Pathol.* 133: 279–289.

Sasek, V., M. Novakova, B. Jindrichova, K. Boka, O. Valentova, and L. Burketova. 2012a. Recognition of avirulence gene AvrLm1 from hemibiotrophic ascomycete *Leptosphaeria maculans* triggers salicylic acid and ethylene signaling in *Brassica napus*. *Mol. Plant Microbe Interact.* 25: 1238–1250.

Sasode, R.S., S. Prakash, A. Gupta, R.K. Pandya, and A. Yadav. 2012. In vitro study of some plant extracts against *Alternaria brassicae* and *Alternaria brassicicola*. *J. Phytol.* 4: 44–46.

Sawatsky, W.M. 1989. Evaluation of screening techniques for resistance to *Leptosphaeria maculans* and genetic studies of resistance to the disease in *Brassica napus*. M.Sc. thesis, University of Manitoba, Winnipeg, Manitoba, Canada.

Schliephake, E., K. Graichen, and F. Rabenstein. 2000. Investigations on the vector transmission of the Beet mild yellowing virus (BMYV) and the Turnip yellows virus (TuYV). *Z. Pflanzenk. Pflanzen.* 107: 81–87.

Sellam, A., B. Iacomi-Vasilescu, P. Hudhomme, and P. Simoneau. 2007. In vitro antifungal activity of brassinin, camalexin and two isothiocyanates against the crucifer pathogens *Alternaria brassicicola* and *Alternaria brassicae*. *Plant Pathol.* 56: 296–301.

Sen, P. 2005. Antagonistic effect of Ca, B and Mo on club-root disease of rape-mustard. *Indian Agriculturist* 49: 13–16.

Sexton, A.C., Z. Minic, A.J. Cozijnsen, M.S.C. Pedras, and B.J. Howlett. 2009. Cloning, purification and characterisation of brassinin glucosyltransferase, a phytoalexin-detoxifying enzyme from the plant pathogen *Sclerotinia sclerotiorum*. *Fungal Genet. Biol.* 46: 201–209.

Sexton, A.C., M. Paulsen, J. Woestemeyer, and B.J. Howlett. 2000. Cloning, characterization and chromosomal location of three genes encoding host-cell-wall-degrading enzymes in *Leptosphaeria maculans*, a fungal pathogen of *Brassica* spp. *Gene* 248: 89–97.

Sexton, A.C., A.R. Whitten, and B.J. Howlett. 2006. Population structure of *Sclerotinia sclerotiorum* in an Australian canola field at flowering and stem-infection stages of the disease cycle. *Genome* 49: 1408–1415.

Shahraeen, N., S. Farzadfar, R. Kamran, and B. Naseri. 2002. Occurrence of viral diseases of oil seed rape plant (canola) in Iran. *Appl. Entomol. Phytopathol.* 69: 189–191.

Sharma, G., V.D. Kumar, A. Haque, S.R. Bhat, S. Prakash, and V.L. Chopra. 2002. *Brassica* coenospecies: A rich reservoir for genetic resistance to leaf spot caused by *Alternaria brassicae*. *Euphytica* 125: 411–419.

Sharma, M., S.K. Gupta, and A. Pratap. 2012a. Induced mutagenesis and in vitro cell selection for resistance against Alternaria blight in *Brassica napus* L. *Indian J. Agric. Res.* 46: 369–373.

Sharma, M., B.S. Sohal, and P.S. Sandhu. 2008a. Induction of defense enzymes and reduction in Alternaria blight by foliar spray of BTH in *Brassica juncea* (L.) Coss. *Crop Improv.* 35: 167–170.

Sharma, N., N. Hotte, M.H. Rahman, M. Mohammadi, M.K. Deyholos, and N.N.V. Kav. 2008b. Towards identifying *Brassica* proteins involved in mediating resistance to *Leptosphaeria maculans*: A proteomics-based approach. *Proteomics* 8: 3516–3535.

Sharma, N., M.H. Rahman, Y. Liang, and N.N.V. Kav. 2010c. Cytokinin inhibits the growth of *Leptosphaeria maculans* and *Alternaria brassicae*. *Can. J. Plant Pathol.* 32: 306–314.

Sharma, N. and S. Sharma. 2008. Control of foliar diseases of mustard by *Bacillus* from reclaimed soil. *Microbiol. Res.* 163: 408–413.

Sharma, P., J.S. Chauhan, and A. Kumar. 2012b. Evaluation of Indian and exotic Brassica germplasm for tolerance to stem rot caused by *Sclerotinia sclerotiorum*. *J. Mycol. Plant Pathol.* 42: 297–302.

Sharma, P., S. Deep, S. Sharma, and D.S. Bhati. 2013a. Genetic variation of *Alternaria brassicae* (Berk.) Sacc., causal agent of dark leaf spot of cauliflower and mustard in India. *J. Gen. Plant Pathol.* 79: 41–45.

Sharma, P., P.D. Meena, A. Kumar, C. Chattopadhyay, and P. Goyal. 2009. Soil and weather parameters influencing Sclerotinia rot of *Brassica juncea*. In: *Souvenir and Abstracts, Fifth International Conference of IPS on Plant Pathology in the Globalized Era*, IARI, New Delhi, India, November 10–13, 2009, p. 97.

Sharma, P., P.D. Meena, S. Kumar, and J.S. Chauhan. 2013b. Genetic diversity and morphological variability of *Sclerotinia sclerotiorum* isolates of oilseed Brassica in India. *Afr. J. Microbiol. Res.* 7: 1827–1833.

Sharma, P., P.K. Rai, P.D. Meena, S. Kumar, and S.A. Siddiqui. 2010c. Relation of petal infestation to incidence of *Sclerotinia sclerotiorum* in *Brassica juncea*. In: *Abstracts, National Conference on Recent Advances Integrated Disease Management Enhancing Food Production*, S.K.R.A.U., Bikaner, India, October 27–28, 2010, p. 76.

Sharma, P., S.A. Siddiqui, P.K. Rai, P.D. Meena, J. Kumar, and J.S. Chauhan. 2012c. Evaluation of Brassica germplasm for field resistance against clubroot (*Plasmodiophora brassicae* Woron). *Arch. Phytopathol. Plant Prot.* 45: 356–359.

Sharma, S. and G.R. Sharma. 2001. Influence of white rot (*Sclerotinia sclerotiorum*) on growth and yield parameters of Indian mustard (*Brassica juncea*) varieties. *Indian J. Agric. Sci.* 71: 273–274.

Sharma, S. and B.S. Sohal. 2010. Foliar spray of benzothiadiazole and salicylic acid on *Brassica juncea* var. RLM619 to combat Alternaria blight in field trials. *Crop Improv.* 37: 87–92.

Sharma, S., J. Singh, G.D. Munshi, and S.K. Munshi. 2010a. Biochemical changes associated with application of biocontrol agents on Indian mustard leaves from plants infected with Alternaria blight. *Arch. Phytopathol. Plant Prot.* 43: 315–323.

Sharma, S., J. Singh, G.D. Munshi, and S.K. Munshi. 2010b. Effects of biocontrol agents on lipid and protein composition of Indian mustard seeds from plants infected with *Alternaria* species. *Arch. Phytopathol. Plant Prot.* 43: 589–596.

Sharma, S., J.L. Yadav, and G.R. Sharma. 2001. Effect of various agronomic practices on the incidence of white rot of Indian mustard caused by *Sclerotinia sclerotiorum*. *J. Mycol. Plant Pathol.* 31: 83–84.

Sharma, S.R. and S.J. Kolte. 1994. Effect of soil-applied NPK fertilizers on severity of black spot disease (*Alternaria brassicae*) and yield of oilseed rape. *Plant Soil* 167: 313–320.

Shaw, F.J.F. and S.L. Ajrekar. 1915. The genus *Rhizoctonia* in India. *Mem. Dept. Agric. India Bot. Scr.* 7: 177.

Shete, M.H., G.N. Dake, A.P. Gaikwad, and N.B. Pawar. 2008. Chemical management of powdery mildew of mustard. *J. Plant Dis. Sci.* 3: 46–48.

Shih, H.D., Y.C. Lin, H.C. Huang, K.C. Tzeng, and S.T. Hsu. 2000. A DNA probe for identification of *Xanthomonas campestris* pv. *campestris*, the causal organism of black rot of crucifers in Taiwan. *Bot. Bull. Acad. Sin.* 41: 113–120.

Shrestha, S.K., S.B. Mathur, and L. Munk. 2000. *Alternaria brassicae* in seeds of rapeseed and mustard, its location in seeds, transmission from seeds to seedlings and control. *Seed Sci. Technol.* 28: 75–84.

Shrestha, S.K., L. Munk, and S.B. Mathur. 2003. Survival of *Alternaria brassicae* in seeds and crop debris of rapeseed and mustard in Nepal. *Seed Sci. Technol.* 31: 103–109.

Shukla, A.K. 2005a. Estimation of yield losses to Indian mustard (*Brassica juncea*) due to Sclerotinia stem rot. *J. Phytol. Res.* 18: 267–268.

Shukla, A.K. 2005b. Sclerotinia rot—Its prevalence in Indian mustard at different levels of nitrogen. *Indian Phytopathol.* 58: 493–494.

Simmons, E.G. 2007. *Alternaria: An Identification Manual*. CBS Biodiversity Series No. 6, Utrecht, the Netherlands.

Sinah, P. K. and A.K. Mall. 2007. Screening of rapeseed and mustard genotypes against white rust (*Albugo cruciferarum* S.F. Gray). *Vegetos* 20: 59–62.

Singh, D. and S. Dhar 2011. Bio-PCR based diagnosis of *Xanthomonas campestris* pathovars in black rot infected leaves of crucifers. *Indian Phytopathol.* 64: 7–11.

Singh, D., S. Dhar, and D.K. Yadava. 2011b. Genetic and pathogenic variability of Indian strains of *Xanthomonas campestris* pv. *campestris* causing black rot disease in crucifers. *Curr. Microbiol.* 63: 551–560.

Singh, D. and V.K. Maheshwari. 2003. Effect of Alternaria leaf spot disease on seed yield of mustard and its management. *Seed Res.* 31: 80–83.

Singh, D., V.K. Maheshwari, and A. Gupta. 2002a. Field evaluation of fungicides against white rust of *Brassica juncea*. *Agric. Sci. Dig.* 22: 267–269.

Singh, D., R. Singh, H. Singh, R.C. Yadav, N. Yadav, M. Barbetti, and P. Salisbury. 2009. Cultural and morphological variability in *Alternaria brassicae* isolates of Indian mustard, *Brassica juncea* L. Czern & Coss. *J. Oilseeds Res.* 26: 134–137.

Singh, D., H. Singh, and T.P. Yadava. 1988. Performance of white rust (*Albugo candida*) resistant genotypes developed from interspecific crosses of *B. juncea* × *B. carinata*. *Cruciferae Newsl.* 13: 110–111.

Singh, D.K., K. Kumar, and P. Singh 2012. Heterosis and heritability analysis for different crosses in *Brassica juncea* with inheritance of white rust resistance. *J. Oilseed Brassica* 3: 18–26.

Singh, D.P. 2004a. The effect of inorganic fertilizers and weather parameters on Alternaria blight of rapeseed-mustard under screen house conditions. *Ann. Agric. Bio Res.* 9: 229–234.

Singh, H.V. 2000. Biochemical basis of resistance in *Brassica* species against downy mildew and white rust of mustard. *Plant Dis. Res.* 15: 75–77.

Singh, H.V. 2004. Biochemical transformation in *Brassica* spp. due to *Peronospora parasitica* infection. *Ann. Plant Prot. Sci.* 12: 301–304.

Singh, H.V. 2005. Biochemical changes in *Brassica juncea* cv. Varuna due to *Albugo candida* infection. *Plant Dis. Res.* (*Ludhiana*) 20: 167–168.

Singh, R., B.S. Dahiya, and N.N. Tripathi. 2000. Weather factors associated with Sclerotinia stem-rot of Indian mustard and development of a linear model for its prediction. *Cruciferae Newsl.* 22: 63–64.

Singh, R., D. Singh, H. Li, K. Sivasithamparam, N.R. Yadav, P. Salisbury, and M.J. Barbetti. 2008. Management of Sclerotinia rot of oilseed Brassica—A focus on India. *Brassica* 10: 1–27.

Singh, R., D. Singh, P. Salisbury, and M.J. Barbetti. 2010. Field evaluation of indigenous and exotic *Brassica juncea* genotypes against Alternaria blight, white rust, downy mildew and powdery mildew diseases in India. *Indian J. Agric. Sci.* 80: 155–159.

Singh, R.B. and R. Bhajan. 2005. Occurrence, avoidable yield loss and management of white rust, *Albugo candida*, in late sown mustard, *Brassica juncea* L. Czern Coss. *J. Oilseeds Res.* 22: 111–113.

Singh, R.B. and R.N. Singh. 2005. Status and management of foliar diseases of timely sown mustard in mid-eastern India. *Plant Dis. Res.* (Ludhiana) 20: 18–24.

Singh, R.B. and R.N. Singh. 2006. Spray schedule for the management of Alternaria blight and white rust of Indian mustard (*Brassica juncea*) under different dates of sowing. *Indian J. Agric. Sci.* 76: 575–579.

Singh, U.S., N.I. Nashaat, K.J. Doughty, and R.P. Awasthi. 2002b. Altered phenotypic response to *Peronospora parasitica* in *Brassica juncea* seedlings following prior inoculation with an avirulent isolate of *Albugo candida*. *Eur. J. Plant Pathol.* 108: 555–564.

Singh, Y., D.V. Rao, and A. Batra. 2011a. Enzyme activity changes in *Brassica juncea* (L.) Czern. & Coss. in response to *Albugo candida* Kuntz. (Pers.). *J. Chem. Pharm. Res.* 3: 18–24.

Snowdon, R.J., H. Winter, A. Diestel, and M.D. Sacristan. 2000. Development and characterisation of *Brassica napus-Sinapis arvensis* addition lines exhibiting resistance to *Leptosphaeria maculans*. *Theor. Appl. Genet.* 101: 1008–1014.

Somers, D.J., G. Rakow, and S.R. Rimmer. 2002. *Brassica napus* DNA markers linked to white rust resistance in *Brassica juncea*. *Theor. Appl. Genet.* 104: 1121–1124.

Song, Z.R. and C.Y. Guan. 2008. Relationship between glucosinolate characteristics and resistance to *Sclerotinia sclerotiorum* in *Brassica napus* L. *J. Hunan Agric. Univ.* 34: 462–465.

Sosnowski, M.R., E.S. Scott, and M.D. Ramsey. 2001. Pathogenic variation of South Australian isolates of *Leptosphaeria maculans* and interactions with cultivars of canola (*Brassica napus*). *Aust. Plant Pathol.* 30: 45–51.

Sosnowski, M.R., E.S. Scott, and M.D. Ramsey. 2004. Infection of Australian canola cultivars (*Brassica napus*) by *Leptosphaeria maculans* is influenced by cultivar and environmental conditions. *Aust. Plant Pathol.* 33: 401–411.

Sosnowski, M.R., E.S. Scott, and M.D. Ramsey. 2006. Survival of *Leptosphaeria maculans* in soil on residues of *Brassica napus* in South Australia. *Plant Pathol.* 55: 200–206.

Sprague, S.J., M.H. Balesdent, H. Brun, H.L. Hayden, S.J. Marcroft, X. Pinochet, T. Rouxel, and B.J. Howlett. 2006. Major gene resistance in *Brassica napus* (oilseed rape) is overcome by changes in virulence of populations of *Leptosphaeria maculans* in France and Australia. (Special issue: Sustainable strategies for managing *Brassica napus* (oilseed rape) resistance to *Leptosphaeria maculans* (phoma stem canker). *Eur. J. Plant Pathol.* 114: 33–40.

Sprague, S.J., B.J. Howlett, and J.A. Kirkegaard. 2009. Epidemiology of root rot caused by *Leptosphaeria maculans* in *Brassica napus* crops. *Eur. J. Plant Pathol.* 125: 189–202.

Sprague, S.J., J.A. Kirkegaard, S.J. Marcroft, and J.M. Graham. 2010. Defoliation of *Brassica napus* increases severity of blackleg caused by *Leptosphaeria maculans* implications for dual-purpose cropping. *Ann. Appl. Biol.* 157: 71–80.

Sprague, S.J., M. Watt, J.A. Kirkegaard, and B.J. Howlett. 2007. Pathways of infection of *Brassica napus* roots by *Leptosphaeria maculans*. *New Phytol.* 176: 211–222.

Sridhar, K. and R.N. Raut. 1998. Differential expression of white rust resistance in Indian mustard (*Brassica juncea*). *Indian J. Genet. Plant Breed.* 58: 319–322.

Stachowiak, A., J. Olechnowicz, M. Jedryczka, T. Rouxel, M.H. Balesdent, I. Happstadius, P. Gladders, A. Latunde-Dada, and N. Evans. 2006. Frequency of avirulence alleles in field populations of *Leptosphaeria maculans* in Europe. (Special issue: Sustainable strategies for managing *Brassica napus* (oilseed rape) resistance to *Leptosphaeria maculans* (phoma stem canker).) *Eur. J. Plant Pathol.* 114: 67–75.

Starzycka, E., M. Starzycki, H. Cichy, A. Cicha, G. Budzianowski, and H. Szachnowska. 2004. Resistance of some winter rapeseed (*Brassica napus* L.) cultivars to *Sclerotinia sclerotiorum* (Lib.) de Bary infection. *Rosliny Oleiste* 25: 645–654.

Steed, J.M., A. Baierl, and B.D.L. Fitt. 2007. Relating plant and pathogen development to optimise fungicide control of phoma stem canker (*Leptosphaeria maculans*) on winter oilseed rape (*Brassica napus*). *Eur. J. Plant Pathol.* 118: 359–373.

Stevens, M., G. McGrann, and B. Clark. 2008. Turnip yellows virus (syn Beet western yellows virus): An emerging threat to European oilseed rape production? *HGCA Res. Rev.* 69: 36pp.

Stonard, J.F., K. Downes, E. Pirie, B.D.L. Fitt, N. Evans, and A.O. Latunde-Dada. 2007. Resistance of current and historical oilseed rape cultivars to phoma stem canker (*Leptosphaeria maculans*) and light leaf spot (*Pyrenopeziza brassicae*) in 2003/04, 2004/05 and 2005/06 UK growing seasons. *Aspects Appl. Biol.* 83: 35–38.

Stonard, J.F., A.O. Latunde-Dada, Y.J. Huang, J.S. West, N. Evans, and B.D.L. Fitt. 2010b. Geographic variation in severity of phoma stem canker and *Leptosphaeria maculans/L. biglobosa* populations on UK winter oilseed rape (*Brassica napus*). *Eur. J. Plant Pathol.* 126: 97–109.

Stonard, J.F., B. Marchant, A.O. Latunde-Dada, Z. Liu, N. Evans, P. Gladders, M. Eckert, and B.D.L. Fitt. 2008. Effect of climate on the proportions of *Leptosphaeria maculans* and *L. biglobosa* in phoma stem cankers on oilseed rape in England. *Aspects Appl. Biol.* 88: 161–164.

Stonard, J.F., B.P. Marchant, A.O. Latunde-Dada, Z. Liu, N. Evans, P. Gladders, M.R. Eckert, and B.D.L. Fitt. 2010a. Geostatistical analysis of the distribution of *Leptosphaeria* species causing phoma stem canker on winter oilseed rape (*Brassica napus*) in England. *Plant Pathol.* 59: 200–210.

Strehlow, B., U. Preiss, R. Horn, and C. Struck. 2010. Genetic variability of the causal agent of clubroot, *Plasmodiophora brassicae*, in different regions of Germany. *Julius-Kuhn-Archiv* 428: 396–397.

Stringam, G.R., V.K. Bansal, M.R. Thiagarajah, and J.P. Tewari. 1992. Genetic analysis of blackleg (*Leptosphaeria maculans*) resistance in *Brassica napus* L. using the doubled haploid method. In: *XIII Eucarpia Congress*, Angers, France, July 6–11, 1992.

Subramanian, B., V.K. Bansal, and N.N.V. Kav. 2005. Proteome-level investigation of *Brassica carinata*-derived resistance to *Leptosphaeria maculans*. *J. Agric. Food Chem.* 53: 313–324.

Subudhi, P.K. and R.N. Raut. 1994. White rust resistance and its association with parental species type and leaf waxiness in *Brassica juncea* × *Brassica napus* crosses under the action of EDTA and gamma-ray. *Euphytica* 74: 1–7.

Sun, Q.L., X.J. Chen, Y.H. Tong, Z.L. Ji, H.D. Li, and J.Y. Xu. 2007. Inhibition of antifungal protein produced by *Bacillus licheniformis* W10 to *Sclerotinia sclerotiorum* and control of rape stem rot by the protein. *J. Yangzhou Univ. Agric. Life Sci. Edn.* 28: 82–86.

Surendra, S., S.L. Godara, S. Gangopadhayay, and K.S. Jadon. 2012. Induced resistance against *Alternaria brassicae* blight of mustard through plant extracts. *Arch. Phytopathol. Plant Prot.* 45: 1705–1714.

Suwabe, K., H. Tsukazaki, H. Iketani, K. Hatakeyama, M. Fujimura, T. Nunome, H. Fukuoka, S. Matsumoto, and M. Hirai. 2003. Identification of two loci for resistance to clubroot (*Plasmodiophora brassicae* Woronin) in *Brassica rapa* L. *Theor. Appl. Genet.* 107: 997–1002.

Szlavik, S., T. Barasits, and W.G.D. Fernando. 2006. First report of pathogenicity group-3 of *Leptosphaeria maculans* on winter rape in Hungary. *Plant Dis.* 90: 684.

Tabarestani, A.Z., M. Shamsbakhsh, and N. Safaei. 2010. Distribution of three important aphid borne canola viruses in Golestan province. *Iran. J. Plant Prot. Sci.* 41: 251–259.

Tehrani, A.S., M. Ahmadzadeh, M. Farzaneh, and S.A. Sarani. 2007. Effect of powder formulations of two strains of *Pseudomonas fluorescens* for control of rapeseed damping-off caused by *Rhizoctonia solani* in greenhouse and field conditions. *Iran. J. Agric. Sci.* 37: 1121–1130.

Tewari, J.P. 1986. Subcuticular growth of *Alternaria brassicae* in rapeseed. *Can. J. Bot.* 64: 1227–1231.

Tewari, J.P., I. Tewari and S.C. Chatterjee. 2000. Inhibition of oospore development in *A. candida* by *Pseudomonas syringae. Bulletin OILB / srop* 23 (6): 51–53.

Thapak, S.K. and R.K. Dantre. 2004. Stag head infection due to white rust disease in mustard under Bastar situation. *J. Interacad.* 8: 356–358.

Thurston, M.I., D.W. Pallett, M. Cortina-Borja, M.L. Edwards, A.F. Raybould, and J.I. Cooper. 2001. The incidence of viruses in wild *Brassica nigra* in Dorset (UK). *Ann. Appl. Biol.* 139: 277–284.

Tirmali, A.M. and S.J. Kolte. 2011. Protection of mustard plants against staghead phase of white rust with biotic and abiotic inducers and glucosinolate content in seeds. *J. Plant Dis. Sci.* 6: 175–177.

Tirmali, A.M. and S.J. Kolte. 2012. Induction of host resistance in mustard with non-conventional chemicals against white rust (*Albugo candida*). *J. Plant Dis. Sci.* 7: 27–31.

Tode, H.I. 1791. *Fungi Mecklenburgenses Selecti.* Fasciculus. II (51, Plate XVI, Fig. 126).

Tollenaere, R., A. Hayward, J. Dalton-Morgan, E. Campbell, J.R.M. Lee, M.T. Lorenc, S. Manoli et al. 2012. Identification and characterization of candidate Rlm4 blackleg resistance genes in *Brassica napus* using next-generation sequencing. (Special issue: Next generation sequencing technologies.). *Plant Biotechnol. J.* 10: 709–715.

Travadon, R., B. Marquer, A. Ribule, I. Sache, J.P. Masson, and H. Brun. 2009. Systemic growth of *Leptosphaeria maculans* from cotyledons to hypocotyls in oilseed rape: Influence of number of infection sites, competitive growth and host polygenic resistance. *Plant Pathol.* 58: 461–469.

Tripathi, S.C., A.K. Tripathi, and C. Chaturvedi. 2010. Effect of different organic amendments on the growth of *Sclerotina sclerotiorum* (Lib.) de Bary *in vitro. Vegetos* 23: 191–193.

Turkington, T.K., G.W. Clayton, H. Klein-Gebbinck, and D.L. Woods. 2000. Residue decomposition and blackleg of canola: Influence of tillage practices. *Can. J. Plant Pathol.* 22: 150–154.

Tziros, G.T., G.A. Bardas, J.T. Tsialtas, and G.S. Karaoglanidis. 2008. First report of oilseed rape stem rot caused by *Sclerotinia sclerotiorum* in Greece. *Plant Dis.* 92: 1473.

Vagelas, I., K. Tsinidis, S. Nasiakou, G. Papageorgiou, I. Ramniotis, and A. Papachatzis. 2009. Phoma stem canker (blackleg) a threat to oilseed rape (*Brassica napus* L.) in Greece. *Analele Universitatii din Craiova—Biologie, Horticultura, Tehnologia Prelucrarii Produselor Agricole, Ingineria Mediului* 14: 357–360.

Van de Wouw, A.P., S.J. Marcroft, M.J. Barbetti, H. Li, P.A. Salisbury, L. Gout, T. Rouxel, B.J. Howlett, and M.H. Balesdent. 2008. Dual control of avirulence in *Leptosphaeria maculans* towards a *Brassica napus* cultivar with '*Sylvestris*-Derived' resistance suggests involvement of two resistance genes. *Plant Pathol.* 58: 305–313.

Vanterpool, T.C. 1959. Oospore germination in *Albugo candida. Can. J. Bot.* 39: 30–31.

Varshney, A., T. Mohapatra, and R.P. Sharma. 2004. Development and validation of CAPS and AFLP markers for white rust resistance gene in *Brassica juncea. Theor. Appl. Genet.* 109: 153–159.

Vasilescu, B., H. Avenot, N. Bataillé-Simoneau, E. Laurent, M. Guénard, and P. Simoneau. 2004. In vitro fungicide sensitivity of *Alternaria* species pathogenic to crucifers and identification of *Alternaria brassicicola* field isolates highly resistant to both dicarboximides and phenylpyrroles. *Crop Prot.* 23: 481–488.

Vasudeva, R.S. 1958. Diseases of rape and mustard. In: *Rape and Mustard*, ed., D.P. Singh. Indian Central Oilseed Committee, Hyderabad, India, pp. 77–86.

Verma, P.R. and Petrie, G.A. 1975. Germination of oospoes of *Albugo candida. Can. J. Bot.* 53: 836–842.

Verma, P.R. and G. Petrie. 1980. Effect of seed infestation and flower bud inoculation on systemic infection of turnip rape by *Albugo candida. Can. J. Plant Sci.* 60: 267–271.

Verma, P.R., Harding, H., Petrie, G.A. and Williams, P.H. 1975. Infection and temporal development of mycelium of *Albugo candida* in cotyledons of four *Brassica* spp. *Can. J. Bot.* 53: 1016–1020.

Verma, P.R., G.S. Saharan, A.M. Bartaria, and A. Shivpuri. 1999. Biological races of *Albugo candida* on *Brassica juncea* and *Brassica rapa* var. Toria in India. *J. Mycol. Plant Pathol.* 29: 75–82.

Verma, U. and T.P. Bhowmik. 1988. Oospores of *A. candida* (Pers. ex. Lev.) Kuntze: Its germination and role as the primary source of inoculum for the white rust disease of rapeseed and mustard. *Int. J. Trop. Plant Dis.* 6: 265–269.

Vishwanath, S.J. Kolte, M.P. Singh, and R.P. Awasthi. 1999. Induction of resistance in mustard (*Brassica juncea*) against Alternaria black spot with an avirulent *Alternaria brassicae* isolate-D. *Eur. J. Plant Pathol.* 105: 217–220.

Vishwanath and Vineeta. 2007. Systemic resistance induction against Alternaria leaf blight through avirulent strain of *Alternaria brassicea* in mustard. *Pantnagar J. Res.* 5: 66–69.

Voglmayr, H. and A. Riethmüller. 2006. Phylogenetic relationships of *Albugo* species (white blister rusts) based on LSU rDNA sequence and oospore data. *Mycol. Res.* 110: 75–85.

Wagner, G., S. Charton, C. Lariagon, A. Laperche, R. Lugan, J. Hopkins, P. Frendo et al. 2012. Metabotyping: A new approach to investigate rapeseed (*Brassica napus* L.) genetic diversity in the metabolic response to clubroot infection. *Mol. Plant Microbe Interact.* 25: 1478–1491.

Walker, C.A. and P.V. West. 2007. Zoospore development in the oomycetes. *Fungal Biol. Rev.* 21: 10–18.

Walker, J.C. 1952. *Diseases of Vegetable Crops*. McGraw-Hill, New York, 323pp.

Walker, J.C. 1957. *Plant Pathology*. McGraw-Hill Book Co., Inc., New York, pp. 214–219.

Wallenhammar, A.C. 2010. Monitoring and control of *Plasmodiophora brassicae* in spring oilseed *Brassica* crops. *Acta Hortic.* 867: 181–190.

Walters, D.R., T. Cowley, H. Weber, and N.I. Nashaat. 2005. Changes in oxylipins in compatible and incompatible interactions between oilseed rape and the downy mildew pathogen *Peronospora parasitica*. *Physiol Mol. Plant Pathol.* 67: 268–273.

Wang, G.X., H. Yan, X.Y. Zeng, X.G. Sheng, Y. Tang, S. Han, M. Zong, K. Lu, and F. Liu. 2011. New alien addition lines resistance to black rot generated by somatic hybridization between cauliflower and black mustard. *Acta Hortic. Sin.* 38: 1901–1910.

Wang, H.Z., G.H. Liu, Y.B. Zheng, X.F. Wang, and Q. Yang. 2003. Breeding of *Brassica napus* cultivar Zhongshuang No. 9 with resistance to *Sclerotinia sclerotiorum* and dynamics of its important defence enzyme activity. In: *Proceedings of the 11th International Rapeseed Congress*, Copenhagen, Denmark, 43pp.

Wang, J., Y. Huang, X.L. Hu, Y.Z. Niu, X.L. Li, and Y. Liang. 2008b. Study on symptom, yield loss of clubroot and modality of *Plasmodiophora brassicae* in rape. *Chin. J. Oil Crop Sci.* 30: 112–115.

Wang, J.X., H.X. Ma, Y. Chen, X.F. Zhu, W.Y. Yu, Z.H. Tang, C.J. Chen, and M.G. Zhou. 2009. Sensitivity of *Sclerotinia sclerotiorum* from oilseed crops to boscalid in Jiangsu Province of China. *Crop Prot.* 28: 882–886.

Wang, K. and C. Hiruki. 2001. Molecular characterization and classification of phytoplasmas associated with canola yellows and a new phytoplasma strain associated with dandelions. *Plant Dis.* 85: 76–79.

Wang, S., S.J. Zheng, L. Shen, H.M. Liao, and L.M. Luo. 2010b. Evaluation of control effect on rape clubroot of wormcast. *Southwest China J. Agric. Sci.* 23: 1910–1913.

Wang, X.Y., H.J. Peng, M.Q. Gao, H.B. Han, J.H. Zhang, and C.G. Xiao. 2002. Preliminary identification of the pathogen of root tumour of stem mustard and factors influencing the incidence of the disease. *Southwest China J. Agric. Sci.* 15: 75–78.

Wang, X.Y., E.D. Yang, C.K. Qi, S. Chen, J.F. Zhang, and Q. Yang. 2008a. Construction of cDNA expression library of oilseed rape and screening and identification of interaction partner of PG, a virulence factor from *Sclerotinia sclerotiorum*. *Acta Agron. Sin.* 34: 192–197.

Wang, Y. and B. Fristensky. 2001. Transgenic canola lines expressing pea defence gene DRR206 have resistance to aggressive blackleg isolates and to *Rhizoctonia solani*. *Mol. Breed.* 8: 263–271.

Wang, Z., F. Wei, S.Y. Liu, Q. Xu, J.Y. Huang, X.Y. Dong, J.H. Yu, Q. Yang, Y.D. Zhao, and H. Chen. 2010a. Electrocatalytic oxidation of phytohormone salicylic acid at copper nanoparticles-modified gold electrode and its detection in oilseed rape infected with fungal pathogen *Sclerotinia sclerotiorum*. *Talanta* 80: 1277–1281.

Wang, Z., X.L. Tan, Z.Y. Zhang, S.L. Gu, G.Y. Li, and H.F. Shi. 2012. Defense to *Sclerotinia sclerotiorum* in oilseed rape is associated with the sequential activations of salicylic acid signaling and jasmonic acid signaling. *Plant Sci.* 184: 75–82.

Wang, Z.Y. and H.F. Yang. 2007. Toxicity of ZPT to sclerotinia blight of rape and field control efficacy. *Acta Agric. Shanghai* 23: 121–122.

Webster, J. 1980. *Introduction to Fungi*, 2nd edn. Cambridge University Press, Cambridge, U.K., 189pp.

West, J.S., N. Evans, S. Liu, B. Hu, and L. Peng. 2000. *Leptosphaeria maculans* causing stem canker of oilseed rape in China. *Plant Pathol.* 49: 800.

West, J.S., P.D. Kharbanda, M.J. Barbetti, and B.D.L. Fitt. 2001. Epidemiology and management of *Leptosphaeria maculans* (phoma stem canker) on oilseed rape in Australia, Canada and Europe. *Plant Pathol.* 50: 10–27.

Wherrett, A.D., K. Sivasithamparam, and M.J. Barbetti. 2003. Sporulation behaviour of *Leptosphaeria maculans* on canola (*Brassica napus*) residue and potential to manipulate timing and quantity of ascospore discharge with chemicals. In: *Thirteenth Biennial Australian Research Assembly Brassicas. Proceedings of Conference*, Tamworth, New South Wales, Australia, September 8–12, 2003, pp. 129–130.

Willets, H.J. and J.A.L. Wong. 1980. The biology of *Sclerotinia sclerotiorum*, *S. trifoliorum*, and *S. minor* with emphasis on specific nomenclature. *Bot. Rev.* 46: 101–165.

Wilson, L.M. and B.J. Howlett. 2005. *Leptosphaeria maculans*, a fungal pathogen of *Brassica napus*, secretes a subtilisin-like serine protease. *Eur. J. Plant Pathol.* 112: 23–29.

Woronin, M. 1878. *Plasmodiophora brassicae*. Urheberder der Kohlpflanzen-Hemie. *Jahrb. Wiss Bot.* 11: 548–574.

Wouw, A.P., S.J. van de Marcroft, M.J. Barbetti, H. Li, P.A. Salisbury, L. Gout, T. Rouxel, B.J. Howlett, and M.H. Balesdent. 2009. Dual control of avirulence in *Leptosphaeria maculans* towards a *Brassica napus* cultivar with 'sylvestris-derived' resistance suggests involvement of two resistance genes. *Plant Pathol.* 58: 305–313.

Wouw, A.P., V.L. van de Thomas, A.J. Cozijnsen, S.J. Marcroft, P.A. Salisbury, and B.J. Howlett. 2008. Identification of *Leptosphaeria biglobosa* 'canadensis' on *Brassica juncea* stubble from northern New South Wales, Australia. *Aust. Plant Dis. Notes* 3: 124–128.

Wretblad, S., S. Bohman, and C. Dixelius. 2003. Overexpression of a *Brassica nigra* cDNA gives enhanced resistance to *Leptosphaeria maculans* in *B. napus*. *Mol. Plant Microbe Interact.* 16: 477–484.

Wu, J., L.T. Wu, Z.B. Liu, L. Qian, M.H. Wang, L.R. Zhou, Y. Yang, and X.F. Li. 2009. A plant defensin gene from *Orychophragmus violaceus* can improve *Brassica napus*' resistance to *Sclerotinia sclerotiorum*. *Afr. J. Biotechnol.* 8: 6101–6109.

Wu, J.S. and J.S. Wang. 2000. Control effect of W-1, a bacterial strain with detoxification activity, against *Sclerotonia sclerotonium* and its primary identification. *J. Southwest Agric. Univ.* 22: 487–489.

Wu, L.Y., J. Siemens, S.K. Li, J. Ludwig-Muller, Y.J. Gong, L. Zhong. and J.M. He. 2012. Estimating *Plasmodiophora brassicae* gene expression in lines of *B. rapa* by RT-PCR. *Sci. Hortic.* 133: 1–5.

Xi, K.I., R.A.A. Morrall, P.R. Verma, and N.D. Wescott. 1989. Efficacy of seed treatment and feftilizer coating with flutrifol against blackleg of canola. *Can. J. Plant Pathol.* 11: 199.

Xu, L.M., J.Y. Huang, X.Q. Liu, R. Qin, and S.Y. Liu. 2009a. Cloning of *Brassica napus* EIN3 gene and its expression induced by *Sclerotinia sclerotiorum*. *Agric. Sci. Technol. Hunan* 10: 33–36.

Xu, Q., S.Y. Liu, Q.J. Zou, X.L. Guo, X.Y. Dong, P.W. Li, D.Y. Songm, H. Chen, and Y.D. Zhao. 2009b. Microsensor in vivo monitoring of oxidative burst in oilseed rape (*Brassica napus* L.) leaves infected by *Sclerotinia sclerotiorum*. *Anal. Chim. Acta* 632: 21–25.

Xue, S., T. Cao, R.J. Howard, S.F. Hwang, and S.E. Strelkov. 2008. Isolation and variation in virulence of single-spore isolates of *Plasmodiophora brassicae* from Canada. *Plant Dis.* 92: 456–462.

Yadav, M.S. 2003. Efficacy of fungitoxicants in the management of Alternaria blight and white rust of mustard. *J. Mycol. Plant Pathol.* 33: 307–309.

Yadav, M.S. 2009. Biopesticidal effect of botanicals on the management of mustard diseases. *Indian Phytopathol.* 62: 488–492.

Yadav, M.S., S.S. Dhillon, and J.S. Dhiman. 2002. Effect of date of sowing, varieties and chemical treatments on the development of Alternaria blight and white rust of mustard. *J. Res.* 39: 528–532.

Yadav, R. and P. Sharma. 2004. Genetic diversity for white rust (*Albugo candida*) resistance in rapeseed-mustard. *Indian J. Agric. Sci.* 74: 281–283.

Yadav, R., S.S. Meena, and S.C. Chatterjee. 2008. Quantitative variability for lesion size and lesion number as components of resistance against *Alternaria brassicae* in Rapeseed-Mustard. *Indian Phytopathol.* 61: 83–86.

Yadava, D.K., M. Vignesh, V. Sujata, N. Singh, R. Singh, B. Dass, M.S. Yadav, T. Mohapatra, and K.V. Prabhu. 2012. Understanding the genetic relationship among resistant sources of white rust, a major fungal disease of *Brassica juncea*. *Indian J. Genet. Plant Breed.* 72: 89–91.

Yajima, W., Y. Liang, and N.N.V. Kav. 2009. Gene disruption of an arabinofuranosidase/beta-xylosidase precursor decreases *Sclerotinia sclerotiorum* virulence on canola tissue. *Mol. Plant Microbe Interact.* 22: 783–789.

Yajima, W., M.H. Rahman, D. Das, M.R. Suresh, and N.N.V. Kav. 2008. Detection of *Sclerotinia sclerotiorum* using a monomeric and dimeric single-chain fragment variable (scFv) antibody. *J. Agric. Food Chem.* 56: 9455–9463.

Yajima, W., S.S. Verma, S. Shah, M.H. Rahman, Y. Liang, and N.N.V. Kav. 2010. Expression of anti-sclerotinia scFv in transgenic *Brassica napus* enhances tolerance against stem rot. *New Biotechnol.* 27: 816–821.

Yan, L.Y., L.C. Zhou, Y.J. Tan, Z.H. Shan, M.Z. Shen, and C. Leonid. 2005. Isolation and identification of biocontrol bacteria from rhizosphere of rapeseed. *Chin. J. Oil Crop Sci.* 27: 55–57, 61.

Yang, D.J., B. Wang, J.X. Wang, Y. Chen, and M.G. Zhou. 2009a. Activity and efficacy of *Bacillus subtilis* strain NJ-18 against rice sheath blight and Sclerotinia stem rot of rape. *Biol. Control* 51: 61–65.

Yang, G.H., X.Q. Chen, H.R. Chen, S. Naito, A. Ogoshi. and J.F. Zhao. 2004. First report of foliar blight in *Brassica rapa* subsp. chinensis caused by *Rhizoctonia solani* AG-4. *Plant Pathol.* 53: 260.

Yang, L., G.Q. Li, D.H. Jiang, and H.C. Huang. 2009b. Water-assisted dissemination of conidia of the mycoparasite *Coniothyrium minitans* in soil. *Biocontrol Sci. Technol.* 19: 779–796.

Yi, J.P., G.L. Zhou, L.P. Yin, Z.B. Chen, and J.Z. Zheng. 2010. Detection of *Leptosphaeria maculans* (*Phoma lingam*) in oilseed rape samples imported from Australia. *Acta Phytopathol. Sin.* 40: 628–631.

Yin, Q., J. Song, Y. Liu, H.Y. Liu, and X.Q. Huang. 2010. Establishment of PCR rapid detection of *Plasmodiophora brassicae* resting spores in soils. *Southwest China J. Agric. Sci.* 23: 390–392.

Yin, Y.N., L.S. Ding, X. Liu, J.H. Yang, and Z.H. Ma. 2009. Detection of *Sclerotinia sclerotiorum* in planta by a real-time PCR assay. *J. Phytopathol.* 157: 465–469.

Young, H.M., P. Srivastava, M.L. Paret, H. Dankers, D.L. Wright, J.J. Marois, and N.S. Dufault. 2012. First report of Sclerotinia stem rot caused by *Sclerotinia sclerotiorum* on *Brassica carinata* in Florida. *Plant Dis.* 96: 1581.

Yu, F., D.J. Lydiate, and S.R. Rimmer. 2005. Identification of two novel genes for blackleg resistance in *Brassica napus*. *Theor. Appl. Genet.* 110: 969–979.

Yu, F., D.J. Lydiate, and S.R. Rimmer. 2008a. Identification and mapping of a third blackleg resistance locus in *Brassica napus* derived from *B. rapa* subsp. *sylvestris*. *Genome* 51: 64–72.

Yu, F.Q., R.K. Gugel, H.R. Kutcher, G. Peng, and S.R. Rimmer. 2013. Identification and mapping of a novel blackleg resistance locus LepR4 in the progenies from *Brassica napus* × *B. rapa* subsp. sylvestris. *Theor. Appl. Genet.* 126: 307–315.

Yu, Q.Q., L.S. Tian, Y.Z. Niu, and S.X. Guo. 2008b. Preliminary study on inheritance of clubroot resistance in a resythesized *Brassica napus* line. *Southwest China J. Agric. Sci.* 21: 1313–1315.

Yuan, W.H., N. Liu, X.W. Wang, and H. Zheng. 2009. Effects of cultural measure on resistance and tolerance to *Scleratinia sclerotiorum* in double-low rapeseed. *Acta Agric. Univ. Jiangxiensis* 31: 855–857.

Zang, X.P., Y.P. Xu, and X.Z. Cai. 2010. Establishment of an inoculation technique system for *Sclerotinia sclerotiorum* based on mycelial suspensions. *J. Zhejiang Univ.* (*Agric. Life Sci.*) 36: 381–386.

Zanjani, M.H.M., M. Ahmadzadeh, A.S. Tehrani, K. Behboudi, and R.S. Riseh. 2011. Effects of *Rhizoctonia solani* on root colonization of canola by *Pseudomonas fluorescens* strain UTPF86. *Iran. J. Plant Prot. Sci.* 42: 163–170.

Zhang, Y., C. Bai, G.H. Ran, Z.Y. Zhang, Y.H. Chen, and G. Wu. 2009. Characterization of endophytic bacterial strain YS45 from the citrus xylem and its biocontrol activity against Sclerotinia stem rot of rapeseed. *Acta Phytopathol. Sin.* 39: 638–645.

Zhao, X.H., C.L. Chen, C.X. Jiao, L. Gan, and J.W. Lu. 2006. Physiological and biochemical responses to *Sclerotinia sclerotiorum* among different rape varieties. *J. Huazhong Agric. Univ.* 25: 488–492.

Zhou, G.L., L.L. Shang, C. Yu, L.P. Yin, D.S. Xu, and J.P. Yi. 2010. Detection of *Leptosphaeria maculans* and *L. biglobosa* in oilseed rape samples imported from Australia. *Acta Phytophyl. Sin.* 37: 289–294.

Zhu, Y., C. Hairu, F. Jinghua, W. Yunyue, L. Yan, C. Jianbing, X.F. Jin et al. 2000. Genetic diversity and disease control in rice. *Nature* 406: 718–722.

Zou, Q.J., S.Y. Liu, X.Y. Dong, Y.H. Bi, Y.C. Cao, Q. Xu, Y.D. Zhao, and H. Chen. 2007. In vivo measurements of changes in pH triggered by oxalic acid in leaf tissue of transgenic oilseed rape. *Phytochem. Anal.* 18: 341–346.

Zwolinska, A., K. Krawczyk, T. Klejdysz, and H. Pospieszny. 2011. First report of '*Candidatus phytoplasma asteris*' associated with oilseed rape phyllody in Poland. *Plant Dis.* 95: 1475.

Section IV

Sunflower

The cultivated sunflower (*Helianthus annuus* L. var *macrocarpus* (DC) Ck II) belongs to the Compositae (an Asteraceae) family. It is a sparingly branched annual herb about 1.0–3.0 m in height. It is insensitive to photoperiods. The basic chromosome number of *H. annuus* is 20 pairs ($2n = 40$). Sunflower has a large genome (3600 Mbp) with abundant repetitive sequences (Baack et al. 2005, Kane et al. 2011).

Sunflowers are referred to as *composites* because what looks like a single large sunflower head is actually an inflorescence composed of a composite of many tiny, usually 1000–2000, individual flowers joined to a common base called the receptacle. The flowers around the circumference are lingulate ray florets with neither stamens nor pistil. The fertile disk florets are located within the head. Each disk floret is a perfect flower. The flowering behavior facilitates cross-pollination, and insects, particularly the bees, represent the essential vector of sunflower pollen. The degree of cross-pollination may be to the extent of 100%. The sunflower seed is a specific type of elongated rhomboid indehiscent achene, which may be white, black, or striped gray and black. The oil content is more than 40%. The oil is characterized by a high concentration of linoleic acid and a moderate level of oleic acid. While the sunflower seed could still be harvested for edible oil, the woody stalk could be used as a biofuel. By producing food and biomass, such a crop would be both economically and politically viable.

The center of origin of the sunflower is believed to be North America from where it has spread to Europe and Asia. Now it is grown in all continents except the Antarctica. Europe and America account for nearly 70% and 80% of the total production, respectively (Harter et al. 2004, Damodaran and Hegde 2007). Sunflower cultivation in Asian countries is comparatively recent. Asia accounts for nearly 20%–22% of the global sunflower and contributes to about 18% of the production. The productivity of sunflower in Asia is about 1.0 ton/ha, which is lower than the world average. India is the largest grower of sunflower in the Asian continent. This is a short-duration crop that is adaptable to a wide range of agroclimatic situations, having high yield potential, suitable for cultivation in all seasons due to its day neutral nature and can fit well in various intercropping and sequence cropping systems. However, the average yield of this crop in India is lowest; it is less than half the world average and static hovering around 0.5–0.6 ton/ha.

The crop performs well and yields more oil in temperate zones. It grows well in a well-drained soil, ranging in texture from sandy to clay. The emergence of new diseases and large climatic variations, particularly recurrence of drought stress during critical growth stages, has affected stability and yield on a regular basis. With continuous cropping in the same field area, the crops suffer from diseases resulting in large losses in yield. Sunflower diseases are described in the following chapter.

6 Sunflower Diseases

RUST

Symptoms

Symptoms of sunflower rust appear on all the aboveground plant parts but are more prevalent on leaves. Small, orange to yellowish spots appear in compact circular groups followed by brownish, circular to elongated, and pulverulent uredinia scattered over the upper and lower surfaces of the leaf. Uredial pustules usually appear first on the lower leaves. They are small, circular, 0.5–1.00 mm in diameter, powdery, orange to black in color, and usually surrounded by chlorotic areas (Figure 6.1). The uredia may coalesce to occupy large areas on the affected plant parts. Usually late in the crop season, as the plant approaches maturity or is subjected to physiological stress, teliospores appear in the uredia and develop into telia on the affected senescent tissues, and the black rust stage appears.

In the case of highly resistant varieties, no uredia are produced, and only small chlorotic or necrotic flecks develop at the point of infection.

Geographical Distribution and Losses

The first report of sunflower rust described by Lewis von Schweinitz from the southeastern United States dates from 1822 (Sackston 1981). The disease now occurs in virtually all the sunflower-growing areas of the world and is more common in temperate and subtropical regions (Kolte 1985). It is, however, considered an important disease of sunflower in Argentina (Gutierrez et al. 2012), Australia (Sendall et al. 2006), Canada (Rashid 2004, Gulya and Markell 2009), Cuba (Perez et al. 2002), India (Mayee 1995, Amaresh and Nargund 2002a), Israel (Shtienberg and Johar 1992), Pakistan (Mukhtar 2009), Russia, South Africa (Los et al. 1995), Turkey (Tan 1994, 2010), and the United States (Harveson 2010, Friskop et al. 2011). It has also been recorded in almost all European and adjacent Mediterranean countries engaged in sunflower production (Sackston 1978).

Severe infection can decrease head size, seed size, oil content, and yield. On an average, in North American conditions, yield losses ranging from 25% to 50% have been reported in areas of intensive sunflower cultivation. Loss estimates are based on field observations or yield comparisons of resistant and susceptible varieties in performance trials (Zimmer et al. 1973). Still further, the severity of sunflower rust in the states of Manitoba in Canada in 2003 (Rashid 2004) and in Nebraska, North Dakota, South Dakota, and Minnesota in the United States has increased steadily from 17% to 77% through the 1990s and 2000s, and dramatic yield reductions have been recorded in localized *hot spots* (Gulya et al.1990a, Gulya and Markell 2009, Harveson 2010, Friskop et al. 2011).

Quantitative assessment of the effect of rust on yield of sunflower in Australia has shown reduction up to 76% (Middleton and Obst 1972, Brown et al. 1974). Siddiqui and Brown (1977) from Australia reported that oil yield losses in sunflower are generally influenced by the growth stages of plant when infection occurs and by the degree of intensity of infection. The effect of rust on growth parameters varies according to the moisture stress to which the plant is subjected. In Kenya, the disease caused 60% yield losses in severe cases; consequently, sunflower acreage dropped dramatically from 24,280 ha in 1949 to 1,420 ha in 1952 (Singh 1974).

The quality of seed is also adversely affected by reduction in test weight and oil content and by increased hull to kernel ratio (Middleton and Obst 1972).

FIGURE 6.1 Sunflower rust. Note the minute uredopustules on the leaf. (Courtesy of Dr. Chander Rao and Dr. Varaprasad, DOR, Hyderabad, India.)

PATHOGEN: *Puccinia helianthi* SCHW.

Classification
Kingdom: Fungi
Phylum: Basidiomycota
Class: Urediniomycetes
Subclass: Incertae sedis
Order: Uredinales
Family: Pucciniaceae
Genus: *Puccinia*
Species: *helianthi*
Binomial name: *Puccinia helianthi* Schwein.

P. helianthi is a macrocylic heterothallic, autoceious fungus. Production of all the stages of spore forms on sunflower has been reported from important sunflower-producing countries (Kolte 1985, Sendall et al. 2006) and India (Mathar et al. 1975). The morphological characteristics of different fruiting structures of the fungus have been described in detail by Baily (1923) and Sendall et al. (2006). The life cycle of *P. helianthi* is represented by five (a–e) spore types:

a. *Urediniospores*: These are unicellular, dikaryotic, repeating spores produced in uredosori in 5–7 days after infection. They are brown and vary from subglobose to obovate in shape measuring 25–32 μm × 10–25 μm in size. The wall of these spores is cinnamon brown, 1–2 μm thick, and finely echinulate, usually with two equatorial germ pores. The spores are often slightly thickened at the apex and base. The urediniospores best germinate at 18°C–20°C by giving rise to germ tubes from equatorial germ pores. Germination of fresh spores is little affected by light intensity (2,200–4,300 lux) during spore production, but increasing light intensity is unfavorable for germination of urediniospores (Sood and Sackston 1972).
b. *Teliospores*: These are diploid resting spores produced in teliosori. They are bicelled, smooth, oblong, elliptical, and slightly constricted at the septum and measure 40–60 μm × 18–30 μm in size. The wall of the spore is smooth, chestnut brown, 1.5–3 μm thick at the sides, and 8–12 μm thick above with an apical pore. The spores are pedicellate; the pedicel may be colorless or pale luteous, fragile, and 60–150 μm. The teliospores produced at

FIGURE 6.2 Aecial cups of *P. helianthi* under Argentina conditions. (Courtesy of Dr. Tom Gulya, USDA-ARS, Northern Crops Research Lab, Fargo, ND.)

lower temperature (10°C) begin to germinate about 15 days after their formation, but those formed at higher temperature do not germinate (Hennessy and Sackston 1970).

c. *Basidiospores* (sporidia): These are monokaryotic, nonrepeating spores produced on promycelium.
d. *Pycniospores*: These are haploid gametes; are small, oval, and hyaline; appear shining and viscous in mass; and represent the sexual spores (spermatia) produced in flask-shaped (1 mm in diam.) pycnia (sexual stage) formed on young seedlings on the cotyledons or on true leaves, primarily on the first leaf in about 8–14 days following infection resulting from inoculation with sporidia (Bailey 1923). Insects transfer nectar containing haploid pycniospores of one mating type (+) to the receptive hyphae (–) of the opposite mating type, thereby affecting cross-fertilization.
e. *Aeciospores*: These are unicellular, dikaryotic, nonrepeating spores produced in aecia. Aecia develop on the abaxial surface of the leaf under the fertilized pycnia and discharge dikaryotic aeciospores (Figure 6.2). Aeciospore infection results in the formation of uredinia that produce prodigious numbers of dikaryotic urediniospores. The urediniospores are disseminated by wind and become airborne to cause infection to nearby neighboring or distant sunflower plants.

Variability, Host Specificity, and Pathotypes

Variability and differences in pathogenicity among isolates of *P. helianthi* have been clearly demonstrated as reviewed by Kolte (1985) and Pandey et al. (2005). Though *P. helianthi* is common on cultivated sunflower (*H. annuus*), it also attacks a number of other *Helianthus* species such as *H. decapetalus*, *H. petiolaris*, *H. subcanescens*, and *H. tuberosus* (Parmelee 1977, Shopov 1980). The occurrence of distinct physiologic races of *P. helianthi* on the cultivated sunflower was established first by Sackston (1962) using well-defined 0–4 infection types on a set of host differentials and provided a basis for the explanation of pathogenic differences observed by earlier workers on a variety of *Helianthus* species. A modified Sackston's (1962) numerical rating system described by Yang et al. (1989) can be currently used for rust evaluation as follows: 0 = immune, no uredia or hypersensitive flecks; 1 = high resistance, presence of hypersensitive flecks or lesions, or pustules smaller than 0.2 mm in diameter with or without chlorotic halo; 2 = resistant, pustules smaller than 0.4 mm; 3 = susceptible, pustules 0.4–0.6 mm in diameter; and 4 = highly susceptible, pustules larger than 0.6 mm in diameter. Reactions 0, 1, and 2 are classed as resistant, while reactions 3 and 4 are rated as susceptible. Rust reaction can be thus rated visually on the basis of both pustule size (infection type) and leaf area covered by pustules (severity). Recent advances allow every isolate to be characterized by its virulence and avirulence toward all known resistance genes in the host. Differentiation of such pathotypes is by reference to their interaction with sunflower cultivars carrying specific genes/genetic factors R1, R2, R3, R4, R5, and R6 conditioning the response (Putt and Sackston 1963, Limpert et al. 1994, Qi et al. 2011). This high-resolution technique obviates the need for detailed race description. New races or, in modern terms, new virulence combinations (*pathotypes*) appear frequently in response to the selection pressure extended by cultivars resistant at the time of introduction. Surveys of virulence, that is, of pathotypes (races) in the late 1990s and in the 2000s, have been carried out in Argentina (Huguet et al. 2008, Moreno et al. 2011), Australia (Kong et al. 1999, Sendall et al. 2006), Canada (Rashid 2004, Gulya and Markell 2009), South Africa (Los et al. 1995, Anonymous 2010), Turkey (Tan 1994, 2010), and the United States (Qi et al. 2011).

In North America

In North America (NA) particularly in Canada and the United States, four NA races (1, 2, 3, and 4) of *P. helianthi* were identified by Sackston (1962) using three standard Canadian sunflower rust differential lines, following a maximum of 2^3 races to be differentiated. Later, the sequential numbering system of race identification was changed to a coded triplet system to produce a virulence formula using a set of nine differentials, which allows theoretical 2^9 races to be identified, assuming no duplication of genes between lines (Gulya et al. 1990a, Gulya and Markell 2009). An international ad hoc committee approved the use of these lines and triplet code for rust race nomenclature (Gulya and Masirevic 1988). To compare the older race classification system and the triplet coding system, the previous NA race 1 corresponds to race 100 of the coded triplet system, NA race 2 to race 500, NA race 3 to race 300, and NA race 4 to race 700. The differentials used in this system include inbred lines S37-388; Canadian lines MC 90, MC 29, and P-386; and lines HA-R1 through HA-R5. S37-388 is universally susceptible to all races, and others have different reaction patterns and are all derived from diverse pedigrees (Gulya and Masirevic 1996, Rashid 2006).

In Australia

Kong et al. (1999) have given a chronological record of the appearance of major pathotypes in Australia from 1978 to 1997, and 23 pathotypes have been recognized, mostly from commercial crops. Frequent shifts in virulence have occurred since the first change was determined in 1983. These have resulted in successive *boom and bust* cycles where commercial sunflower hybrids with resistance to the prevalent pathotypes became susceptible during the rapid shifts in virulence. Almost all pathotypes identified since 1986 trace to a common progenitor, Aus 4. Results of virulence surveys (avirulence and virulence) data accumulated over 25 years revealed that diverse pathotypes of *P. helianthi* evolve in wild sunflower populations providing a continuum of genetically heterogeneous hosts on which *P. helianthi* can potentially complete its sexual cycle. This results in sexual recombination in the causal fungus in seasons that favor completion of the sexual cycle and subsequent selection of recombinant pathotypes and that mutation too contributes steadily to the development of new virulence genes in the population of *P. helianthi* (Gulya 2006, Sendall et al. 2006). Many new pathotypes have been identified due to an increase in the number and availability of differential hosts. At present, 21 differential hosts are routinely used for pathotype identification, whereas only 4 were available in 1983. Many of these differential hosts have not been characterized genetically, but based on their rust reaction, many are presumed to contain more than a single gene for rust resistance.

In Other Countries: Argentina, Turkey, and India

The occurrence of different races of *P. helianthi* in Argentina was first recognized in 1957. Races 100, 300, and 500 were discovered during the 1960 decade and the race 700 during the 1970s. There has been a significant change in the composition of the races of *P. helianthi* in Argentina since 1985. Interestingly, presently (in the 2000s), no isolate of *P. helianthi* belongs to the group of races 100, 300, or 500. In fact all the collected isolates belong to the group of race 700, the first predominant race being 700 followed by race 740. Other variants of race 700, namely, 701, 704, 720, 744, and 760, have also been reported (Huguet et al. 2008, Moreno et al. 2011). This indicates that the deployment of several rust resistance genes (viz., R1 and R2) in sunflower commercial hybrids during the last 30 years determined a selection pressure over rust populations and an associated drift in the frequency of virulence genes. In Turkey, Tan (2010) accomplished race identification of *P. helianthi* under field conditions where seedlings of 23 differential genotypes were naturally infected in the main sunflower production area and concluded the prevalence of rust race 1 (= newly designated as race 100) and race 3 (= race 300) in Turkey, and races 2 (= race 500) and 4 (= race 700) being nonprevalent in that country. Pathogenic variability in *P. helianthi* in India has been studied, but the reports about variability for this pathosystem appear to be scarce (Patil et al. 1998, 2002).

Epidemiology and Disease Cycle

Sunflower rust can occur at any time during the growing season, but disease onset is dependent on the environment and inoculum source. When the disease occurs early, it is usually the result of primary infection originating from primary inoculum sources such as sporidia from germination of surviving teliospores on previous sunflower crop or wild sunflowers, or from aeciospores on volunteer seedlings or from urediniospores formed on volunteer seedlings in high-altitude areas and carried through air currents. The secondary infection occurs usually through repeatedly produced urediniospores in a crop season, and late-season epidemics are generally a result of urediniospores blown in from distant fields (Kolte 1985).

When teliospores act as primary sources, they germinate early in the spring by producing a promycelium from each cell bearing four haploid sporidia also referred to as basidia. The sporidia are of (+) and (–) mating groups, which results through meiosis, while the teliospores undergo the germination process (Kolte 1985, Pandey et al. 2005). Under favorable conditions, when a sporidium comes in contact with the surface of the cotyledon, leaf petiole, or hypocotyl of a sunflower, it produces a germ tube that penetrates directly and establishes the infection, resulting in the development of flask-shaped pycnium producing pycniospores. The (+)- and (–)-type sporidial infections result in the development of their respective types of pycnia and pycniospores where receptive hyphae (–), the female, and spermogonia (+), the male, cross-fertilize, and mating between these two opposite types occurs through insects or rainwater to produce a dikaryotic thallus, which subsequently forms aecia with binucleate aeciospores in about 8–10 days. Aeciospores become airborne and infect sunflower foliage usually near where they are produced. They germinate at 6°C–25°C with an optimum temperature of 16°C for 1 h. But the establishment of infection requires 10 h. Aeciospores, like urediniospores, usually germinate by producing a single germ tube from one of the germ pores, within 4 h after inoculation if the free moisture is present. The germ tube forms an irregularly shaped appressorium over stomata 6–8 h after inoculation. The infection peg is then formed from the lower surface of the appressorium and penetrates the substomatal vesicle, from which two or more infection hyphae arise. When the infection hypha contacts a cell, a septum is formed and a haustorial mother cell is produced, from which knob-shaped to elongated numerous haustoria are formed in the host cell establishing a nutritional relationship with it (Sood and Sackston 1972). The invading hypha grows rapidly in susceptible varieties and culminates in the aggregation of hyphae under the epidermis, resulting in the formation of uredosori containing dikaryotic urediniospores, the economically important stage of the disease cycle. Urediniospores can be disseminated to long distances by wind and infect most of plant tissues. The process of infection through urediniospores is similar to the one described for aeciospores. In favorable conditions of free moisture (dew) and warm temperatures (12°C–29°C), the uredial stage repeats its cycle every 10–14 days. Because infection is favored by free moisture, infection may be most severe in leaf depressions, on leaf veins, where moisture persists. When temperature falls beyond the favorable range for infection and disease development, the repeating cycle (uredinia) slows and stops. Late-season cold temperatures or host maturity will initiate the changes from the uredinial stage into the overwintering telial stage. Once the telia occur, the disease cycle for that growing season ceases. In the spring, teliospores germinate and produce sporidia, which are visible by microscopic observation only. Sporidia will infect leaves, leading to the formation of pycnium, and the cycle repeats.

Factors Affecting Infection and Disease Development

Kolte (1985) reviewed factors affecting sunflower rust infection and disease development. A day temperature range of 25°C–30.5°C with relative humidity of 86%–92% promotes greater rust intensity, and the relative humidity is positively correlated with the severity of rust. Water-congested sunflower plants are more susceptible to *P. helianthi.* A day temperature of 25°C and a night temperature of 18°C have been found more conducive for the development of the disease under Canadian conditions. Temperature also affects the incubation period. The incubation period following infection through the uredospores is reported as 5, 8, and 7 h at temperatures of 18°C, 14°C, and 22°C, respectively.

Light intensity in the range of 1200–2000 fc influences the maximum production of pustules. Darkness at the time of inoculation and throughout the early stages of infection tends to diminish the intensity of symptoms.

The severity of rust is reported to be less on 15-day-old plants and increases with age, the maximum being on 75-day-old plants. Susceptibility during senescence is directly related to reduced protein synthesis and not to changes in protein content of leaves.

Excess nitrates in the soil or in solution encourage rust infection and defoliation by the rust, but boron and other micronutrients applied to soil reduce its incidence.

Disease Management

Host Resistance

Among the various strategies for rust control, the deployment of diverse resources of resistance in commercial hybrids and varieties remains the most effective approach. Cultivated sunflower originated from the genus *Helianthus* that consists of 51 wild species, 14 annual and 37 perennial, and all are native to the Americas. A large amount of genetic variation in terms of host resistance exists in the wild species, providing genetic diversity for improvement. The origin of most rust resistance genes present in the cultivated sunflower can be traced to wild species mainly *H. annuus*, *H. argophyllus*, and *H. petiolaris* (Hennessy and Sackston 1970, Zimmer and Rehder 1976, Jan et al. 1991, Quresh et al. 1991, 1993, Quresh and Jan 1993, Gulya et al. 2000). One line (PS 1089) derived from *H. argophyllus* × cultivated sunflower and two lines (PS 2011 and PS 2032) derived from *H. petiolaris* × cultivar crosses are reported to be immune to the prevalent races in India (Sujatha et al. 2003). Thus, several sources of rust resistance are known, and the R1, R2, R3, R4, R5, and R6 genes have been characterized and used widely to develop rust-resistant commercial hybrids and varieties (Seiler 1992, Rashid 2006, Sendall et al. 2006, Gulya and Markell 2009, Lawson et al. 2010, Qi et al. 2011, 2012). Genome localization of sunflower rust resistance genes has been documented (Bulos et al. 2012, 2013).

The rust resistance genes R1 and R2 were the first to be discovered in sunflowers and originated from the wild sunflower. Gene R1 present in the inbred lines MC 69 and MC 90 conferred resistance to rust races 100 and 500 (old races 1 and 2). A sequence characterized amplified region (SCAR) marker SCTO6 (950) was found that cosegregates with rust resistance gene R1 and mapped to linkage group (LG) 8 (Lawson et al. 1998, Yu et al. 2003). However, R1 gene is no longer effective against current virulent races (Qi et al. 2011). In contrast, the gene R2 present in inbred line MC 29, an old Canadian line, showed resistance to 90% of 300 rust isolates tested in the United States in the years 2007 and 2008 including race 336, the predominant race in North America. However, MC 29 is moderately susceptible to race 777, the most virulent race currently known in North America (Qi et al. 2011). R2 has been used in Australian sunflower breeding program and provides resistance to all known Australian races (Sendall et al. 2006, Lawson et al. 2010).

Rust resistance gene R3 identified in the line PhRR3 conferred resistance to two Australian rust races (Goulter 1990). Selecting from Argentinean open-pollinated varieties, five multirace resistant lines, HA-R1 to HA-R5, were released in 1985 (Gulya 1985). The R4 locus is located on LG 13 in *H. annuus* (Sendall et al. 2006). The rust resistance gene R4 present in the germplasm line HA-R3 was derived from an Argentinean interspecific pool with Russian open-pollinated varieties crossed with *H. annuus*, *H. argophyllus*, and *H. petiolaris* (Gulya 1985, de Romano and Vazquez 2003). These lines HA-R1, HA-R4, and HA-R5 were also reported to carry alleles of the R4 locus, whereas the line HA-R2 had a different gene, R5 (Miller et al. 1988). HA-R2 was a selection from the Argentinean open-pollinated cultivar Impira INTA. This cultivar was developed from the interspecific cross between *H. argophyllus* and *H. annuus* cultivar Saratov selection Pergamino (de Romano and Vazquez 2003). Therefore, it is believed that the R5 gene in HA-R2 originated from *H. argophyllus*. Gene R5 conferred resistance to 86% of 300 rust isolates tested in the United States in the years 2007 and 2008, including the predominant race 336, but conferred susceptibility to race 777. Sunflower rust-resistant lines previously released by the USDA were evaluated for their reaction to current virulent races. This evaluation identified nine germplasm lines, HA-R6, HA-R8, RHA 397, RHA 464, PH3, PH4, PH5, TX16 R, and RFANN-1742, that were resistant to both races 336 and 777 (Miller and Gulya 2001, Jan et al. 2004, Jan and Gulya 2006, Hulke et al. 2010, Qi et al. 2011). Similarly germplasm lines in sunflower have been screened for rust resistance in India (Velazhahan et al. 1991).

It is thus revealed from the previous text that incorporating effective resistance genes into sunflower inbred lines and commercial hybrids should mitigate the threat posed by current virulent rust races not only in the North American situation but also in rest of the sunflower-growing countries in the world. The potential to reduce losses due to rust stimulates genetic efforts to develop molecular markers linked to effective rust resistance genes in order to facilitate marker-assisted selection (MAS) (Tang et al. 2002, 2003, Knapp 2003). Molecular markers have been identified for a number of sunflower rust-resistant R genes. These markers have been used to detect resistance genes in breeding lines and wild sunflower. For example, two SCAR markers, SCT06950, which is associated with the rust resistance gene R1, and SCX20 (600), which is linked to the Radv gene, mapped to linkage groups (LGs) 8 and 13, respectively (Yu et al. 2003, Qi et al. 2011). Lawson et al. (2010) reported mapping of the R2 gene to sunflower LG 9. Qi et al. (2011) mapped the R4 gene to a large nucleotide-binding site and leucine-rich repeat (NBS-LRR) cluster on LG13. Molecular mapping of the gene R5 has not been reported in the literature. However, this gene is reported to be associated with two simple sequence repeats (SSRs) as reported by Sendall et al. (2006). A germplasm line HA R2 carrying the rust resistance gene R6 was released as a multirace rust-resistant line in 1985 but has not been widely used in commercial hybrid production. R6 remains effective against the prevalent rust races of sunflower in North America. Molecular marker analysis demonstrated by Qi et al. (2012) revealed that the LG2 markers showed association with rust resistance. Genotyping of the 94 F2 individuals (progenies derived from the crosses HA 89 with HA-R2) with 23 polymorphic SSR markers from LG2 confirmed the R6 location on LG2, flanked by two SSR markers, ORS1197-2 and ORS653a, at 3.3 and 1.8 cM of genetic distance, respectively. The markers for R6 developed by Qi et al. (2012) will provide a useful tool for speeding up deployment of the R6 gene in commercial sunflower hybrid production.

The future of the development of sunflower inbred lines or hybrids with high levels of durable resistance will depend on the ability to select genotypes that have combinations of effective resistance genes. Knowledge of virulence evolution of the pathogen population and available DNA markers closely linked to host R genes is a prerequisite for successful gene pyramids. MAS is the choice. To improve the efficiency of MAS, it is important that the recombination frequency between the target gene and the marker be as low as possible. Developing a molecular marker that is located within the rust resistance gene (gene-specific marker) will eventually solve the problem, and the durable genetic resistance through gene pyramiding will be effective for the management of rust.

Chemical Control

Though fungicides may be considered as a last alternative in controlling the rust disease, lack of genetic resistance to some races of causal fungus, *P. helianthi*, necessitates the use of effective fungicides to reduce the impact of rust disease on sunflower yield and quality of seed. Earlier fungicides like dithiocarbamates (maneb, zineb, mancozeb), elemental sulfur, Bordeaux mixture, benodanil, and oxycarboxin were established to be effective for the control of the disease (Kolte 1985). Now newer fungicides, tebuconazole 39 (Folicur) at 0.125 kg/ha, pyraclostrobin 25 (Headline) at 0.15 kg/ha, prothioconazole 48 (Proline) at 0.2 kg/ha, boscalid 25 (Lance) at 0.25 kg/ha, and propiconazole 12.5 + trifloxystrobin 12.5 (Stratego) at 0.18 kg/ha, are reported to be highly effective in managing the sunflower rust (Gulya 1991, Shtienberg 1995, Markell 2008, NDSU 2009). The action threshold for fungicide application is 3% uredopustule coverage on upper leaves (Shtienberg 1995). This, therefore, means fungicidal management of rust should be considered when rust is found on the upper leaves and the plant is in the range of vegetative stages up to the R6 growth stage, or when rust pustules cover 5% of the lower leaves at or before flowering. Recommendations from fungicide research trials indicate that when rust has infected the upper four leaves at 1% or less, then fungicides like Headline and Quadris can be used. If the infection of the upper four leaves is 3% or greater, then Folicur may be used. All effective fungicides (Proline, Folicur, Headline, and Stratego) can reduce the rust incidence and severity as expressed in area under the disease progress curve (AUDPC). Triazoles (Proline and Folicur) tend to have lower AUDPC values than strobilurins (boscalid). The AUDPC due to the triazole group of fungicide spray is reduced by 50%, consequently increasing the crop yield by 10%–20%. Confection sunflower, with their higher value and greater susceptibility, would more likely pay back the cost of fungicide application. The effectiveness of early or late applications or both may vary between years depending on the earliness of the rust infection and disease development. An infection on the upper leaves at the growth stage of R6 or later will not likely have a negative yield effect (Shtienberg 1995, NDSU 2009), and hence, fungicide spray may not be necessary.

In the United States, no fungicides have the federal label or use against sunflower rust. A *specific exemption* under any emergency may be granted by the Environmental Protection Agency (EPA) in some years for use of a specific fungicide for that single year (NDSU 2009). Interestingly, the Colorado State in that country supported the request for the recommendation of use of Folicur 3.6 F when rust epidemics threaten the crop.

Cultural Control

Cultural management practices include plowing under or early-season management of volunteer sunflower plants carrying infection and all crop remnants by removal and destruction by fire. Growing of the sunflower crop 500 m away from the site of the previous years' plot is useful in minimizing the incidence and severity of the sunflower rust (Mitov 1957). Sunflowers should not be planted 2 years in a row in the same field. If possible, avoid planting next to a field that had sunflower last year. If rust occurs on volunteer plants in the vicinity of a planted field, they should be destroyed as soon as possible to prevent the spores from blowing into the planted field. Depending on the occurrence of disease in a particular locality, the choice of planting dates may be used to advantage to avoid the disease (Kolte 1985). Early planting and short-season hybrids will generally have less rust. Avoid dense plant stand and high-nitrogen fertilization (Perez et al. 2002).

Controlling wild sunflowers is a very important step in the management of sunflower rust. All 51 species of *Helianthus* found in North America are hosts to the rust pathogen. All spore stages readily occur on wild sunflowers, which increase the sunflower rust problem in two ways. First, when the early spore stages appear on wild and volunteer sunflowers, the onset of uredinia is earlier. This allows more infection cycles to take place, which creates a greater yield loss potential. Second, sexual recombination occurs when the pathogen completes its sexual cycle. This may result in new races that overcome available resistance. Therefore, removal of wild sunflower populations around fields is desirable and strongly recommended.

DOWNY MILDEW

SYMPTOMS

Disease symptoms of various kinds depending on the age of tissue, level of inoculum, environmental conditions (moisture and temperature), and cultivar reaction become evident as seedling damping-off, systemic symptoms, local foliar lesions, and basal root or stem galls.

Damping-Off

When susceptible sunflower plants are subject to subterranean infection by the downy mildew fungus, damping-off in the seedling stage occurs, particularly under cool (12°C–13°C) and very wet soil conditions. Seedlings are killed before or soon after emergence, resulting in reduced plant stands under field conditions. Affected plants dry and become windblown.

Systemic Symptoms

Sunflower plants carrying systemic infection are severely stunted. Close correlation has been found between fungal growth and height of the seedlings following inoculation, depending on the susceptibility and direction of spread of the fungus. In a susceptible variety, the pathogen tends to colonize the whole plant. Leaves of affected plants bear abnormally thick, downward-curled leaves that show prominent yellow and green epiphyllous mottling (Figure 6.3). A hypophyllous downy growth of

FIGURE 6.3 Downy mildew of sunflower. (Courtesy of Dr. Chander Rao and Dr. Varaprasad, DOR, Hyderabad, India.)

the fungus, consisting of the conidiophores and conidia, develops and covers large areas that are concurrent with the epiphyllous yellow spot (Cvjetkovic 2008). The stem becomes brittle. The systemically infected sunflower plants show loss of phototropic and negative geotropic responses. Such plants also show pronounced reduction of the development of secondary rootlets. Flower heads of the affected plants remain sterile and produce no seeds, or only occasionally the seeds are produced on such heads. When the older plants are infected, the symptom expression may be delayed until flowering without visible chlorotic symptoms on leaves.

Local Foliar Lesions

Small, angular greenish-yellow spots appear on leaves as a result of secondary infection through zoospores liberated from wind-borne zoosporangia. The spots may enlarge and coalesce to infect a larger part of the leaf. Plants are susceptible to such infection for a longer period than to systemic infection. The fungal growth becomes visible at the lower surface of the diseased area and persists for some time under humid conditions. Such local foliar symptoms usually do not result in systemic symptoms and are, therefore, considered to be of less economic importance.

Basal Root or Stem Galls

Development of basal gall symptoms occurs independently of the infection that results in systemic symptoms. The root infection may result in the formation of galls at the base of the plants on primary roots. Such roots are discolored, scurfy, and hypertrophied; the number of fibrous secondary roots is reduced, and the plants become susceptible to drought. The percentage of plants with basal gall symptoms seldom exceeds 3% in a particular field. Such plants are less vigorous and subject to lodging. Lodging of plants with basal gall symptoms results in fracturing directly through the galled area, thus causing total loss of a particular plant.

Geographical Distribution and Losses

Downy mildew of sunflower generally is found in more temperate regions where emerging seedlings are exposed to low temperature and abundant precipitation. Like sunflower, its major host, the causal fungus *Plasmopara halstedii* (Farl.) Berl. and de Toni, is considered to be of North American, Western Hemisphere, origin. The disease was first described in the northeastern United States in 1882, and in 1892, it was found on *H. tuberosus* in Russia (Kolte 1985). As the sunflower expanded to other countries, the disease has followed it closely, especially after World War II. The fungus has been distributed by seed trade rapidly and is reported to occur in various sunflower-growing countries all over the world except in Australia and New Zealand, though downy mildew on *Arctotheca* and *Arctotis* in Australia and New Zealand has been attributed to *P. halstedii* (Constantinescu and Thines 2010). It takes first place on its economic importance for sunflower production in the United States, Canada, and European countries. Epidemic outbreak of the disease in 2007 and 2008 caused 85% losses in sunflower yield in Turkey (Göre 2009). The overall loss directly attributed to downy mildew was estimated to be a half million dollars in the Red River Valley in 1970. When long periods of precipitation and cool weather follow the planting, losses to the extent of 80% have been reported from major sunflower-producing areas of eastern Europe. In France, where the sunflower is grown continuously for 2 years, about 70%–80% of plants have been reported to show incidence of the disease. In certain fields, about 90% incidence of the disease has been reported from Yugoslavia. Systemic infections are most destructive, occasionally causing 50%–95% yield reduction (Rahim 2001). Yield losses may be due to the total loss of seedlings, resulting from damping-off of the seedlings induced by the disease. Yield losses may also become more noticeable and serious when large field areas such as low spots are affected. Yield losses are generally additive, the combination of plant mortality, lighter, and fewer seeds produced by surviving plants, and lower oil content.

Pathogen: *Plasmopara halstedii* (Farl.) Berl. and de Toni

Classification
Domain: Eukaryota
Kingdom: Chromista
Phylum: Oomycota
Class: Oomycetes
Order: Peronosporales
Family: Peronosporaceae
Genus: *Plasmopara*
Species: *halstedii*

The pathogen *P. halstedii* (synonym *P. helianthi*) is an obligate parasite conveniently used for a group of closely related pathogens that cause downy mildew on sunflowers and many other genera and species of the subfamilies Asteroidae and Cichorioideae of the family Compositae.

The sporangiophores are slender, monopodially branched at nearly right angles with three sterigmata at the very end bearing ovoid to ellipsoid zoosporangia singly at the tips of branches. It is interesting that entirely new types of sporangia are formed on sunflower roots differing from those produced on leaves. The sporangiophores emerge through stomata on leaves. The size of sporangia is variable as is the number of biflagellate zoospores released by one single sporangium. The zoosporangia germinate by the formation of biflagellate zoospores or by germ tubes. The sporangia germinate in 2% sucrose in tap or distilled water. The temperature range for the germination of zoosporangia is 5°C–28°C, with an optimum temperature range being 16°C–18°C. The zoosporangia formed at low temperature (8°C) may show low germination (1%–6%), whereas those formed at high temperature (27°C) are reported to show high germination (86%–95%). The vegetative thallus is composed of intercellular hyphae that produce globular haustoria that penetrate into the host cells allowing the obligate biotrophic fungus to absorb nutrients.

Sexual reproduction is by means of oogamy resulting in the formation of thick-walled oospores in the intercellular spaces of roots, stems, and seeds that act as surviving structures. The oospores are brown with a slightly paler wall and measure about 27–32 μm in diameter.

Physiological Races

The fungus completes the sexual cycle annually, affording maximum opportunity for the recombination of virulence genes and the development of new races, which is evident from reports of work done by several workers from different parts of the world. The identification and nomenclature of these races are based on the reaction of a set of differential lines (Sackston et al. 1990, Tourvieille de Labrouhe et al. 2000). Physiological races of *P. halstedii* were first reported by Zimmer (1974) and distribution of races appeared to be geographically separated. For example, in 1991, a total eight races of the fungus were reported. Races 1, 4, and 6 were confined in Europe and races 2, 3, and 4 in Asia. Race 5 was confined to greenhouse; race 7 was reported in Argentina, and race 8 was reported in North Dakota in the United States (Gulya et al. 1991). Currently worldwide, 36 races have been identified controlled by 15 dominant resistance genes, and such a set of 15 differential host lines is RHA 265, RHA 274, RHA 464, DM 2, PM 17, 803, HAR-4, HAR-5, HA 335, HA 337, RHA 340, HA 419, HA 428, HA 458, and TX 16; among them about 6–7 are the world's dominant races, the four races (DM 700, DM 710, DM 730, DM 770) being the prominent ones (Gulya 2007a, Gulya et al. 2011, Viranyi and Spring 2011). In the United States, 11 races have been identified (2000–2008), but no isolate of *P. halstedii* from that country could overcome the PI (6) gene (HA 335) since it was released in 1988 until 2009 when the first *hot* race (DM 734) attacking the PI (6) gene was identified, and in 2010, it was further detected to be prevalent in North Central Dakota and Minnesota in the United States. Four more *hot races* (DM 314, DM 704, DM 714, DM 774) that are able to overcome the PI (6) gene (HA 335, HA 336) and PI 7 gene (HA 337, HA 338, HA 339) have been identified to be prevalent mostly in North Dakota and also in Minnesota in the United States.

In France, race 100 was first identified in 1965 and was well controlled by two resistance-specific genes PI (1) and PI (2). Zimmer and Fick (1974) found that the gene PI (1) provides resistance against race 100 and PI (2) against races 100 and 300. These two genes controlled the downy mildew population in Europe until 1998 when new races emerged (710 and 703) in France (Tourviellie de Labrouhe et al. 2000, Delmotte et al. 2008, Jocic' et al. 2012). Later research showed that these races were introduced from the United States via infected seeds (Roeckel-Drevet et al. 2003). Since then, a monitoring network that includes breeders and extension partners has been conducted by the French Ministry of Agriculture allowing to follow the evolution of the pathogen. Thus, 15 races more could be identified, especially race DM 304 (the first race in France to overcome the PI (6) gene) since 2000 and in 8 years (2000–2008). Six more *hot races* (DM 307, DM 314, DM 334, DM 704, DM 707, DM 714) have been identified in France (Sakr et al. 2009, Sakr 2010, Tourvieille de Labrouhe et al. 2010, Ahmed et al. 2012).

In Bulgaria, during 1988–2000, over a period of 12 years, only two downy mildew races were known. Now there are five races 300, 330, 700, 721, and 731; race 700 is dominant in the largest area (Shindrova 2013). In Romania, for approximately 35 years, there existed only two races, but in the last decade, five races of the downy mildew pathogen have been reported (Teodorescu et al. 2013). Races 100, 300, 310, 330, 710, 703, 730, and 770 have been identified in Spain. Race 703 is of high virulence in the northeast, while Race 310 seems to occur over the south, the main sunflower-growing region of the country (Molinero-Ruiz et al. 2003). In Hungary, five races (100, 700, 730, 710, 330) are prevalent (Kinga et al. 2011). In Russia, seven races of the pathogen could be identified in the Krasnodar region of the Russian Federation, and it is determined that against a background of dominant race 330, races 710 and 730 are also economically significant. A conclusion has been made about the necessity of separate testing on resistance of sunflower to these races and extraction of a material with complex resistance to them (Antonova et al. 2010). In Italy, HA 335 containing the efficient genes for resistance to *P. halstedii* never shows any symptoms under varied favorable climatic conditions (Raranciuc and Pacureanu-Joita 2006). In Serbia, race 100 was the race until 1990. In 1991, the presence of race 730 was confirmed in that country (Lac'ok 2008). The most predominant single race in sunflower-growing Karnataka, Andhra Pradesh, and Maharashtra states of India appears to be race 100 (Kulkarni et al. 2009).

In the last decade, advanced tools of biotechnology have enabled discernment of intraspecific groups of *Plasmopara* on the molecular level and led to the shift from a morphological to a phylogenetic species concept (Spring and Thines 2004, Viranyi and Spring 2011). With molecular markers based on the partial sequence of the nuclear internal transcribed spacer (ITS) regions, Spring et al. (2006) and Thines et al. (2005) detected polymorphism between profiles of races 100, 310, and 330, as well as between groups of populations representing races 700, 701, 703, 710, and 730. Giresse et al. (2007) found high genetic variability between isolates from France and Russia using single nucleotide polymorphism (SNP) markers, whereas Sakr (2010) utilized expressed sequence tags (EST)-derived markers to determine the genetic relationship between races. Evidence for asexual genetic recombination in *P. halstedii* is also reported (Spring and Zipper 2006).

Disease Cycle and Epidemiology

Pathogen survives through oospores in the residue of the preceding sunflower crop in soil or through oospores on seeds from the systemically infected plants. Some oospores have been reported to remain dormant up to 14 years. Overwintering oospores in plant residues in soil or seed germinate mostly under wet conditions the following spring. Primary infection is effected during seed germination in the soil and the emergence of sunflower seedlings. It may be caused by fungus mycelium or oospores present on infected seeds, or by oospores present in infected soil into which healthy seeds are sown. Starting from a single oospore that germinates and gives rise to a single sporangium, zoospore differentiation and release follow. In the presence of free water, the zoospore swarms rapidly and, if a host tissue (root, root hair, stem, or less commonly leaf) is available, settles on an infection site where encystment and subsequent germination take place. Penetration of the host is direct through the epidermis. Once established, the fungus grows intercellularly, and in a compatible host/pathogen combination, it starts with systemic colonization toward the plant apex. Systemic mycelium may be present in all plant tissues except meristems. When conditions are favorable, asexual sporulation takes place by means of sporangiophores arising primarily through stomata or other openings on the invaded tissue. Oospores are also produced in infected plant parts, primarily in roots and stem.

The number of diseased plants depend on the amount of inoculum on seeds and in soil. No matter if primary infection starts from seeds or soil, the course of disease development in infected plants is identical. The fungus develops in unison with the development of young plants. It penetrates the root, stem, and cotyledons and reaches the meristematic tissue at the top of young plants. The fungus develops inside the infected plants intercellularly, in all plant parts, invading the young tissues and depriving the infected plants of assimilates and water. This is why infected plants lag behind healthy ones in growth and development. This way of fungus expansion inside the plant tissues is called a systemic infection. It begins with the infection of the germ and ends with the infection of the head and seeds. The fungus penetrates all parts of the seed (husk, endosperm, and germ), which then produces a new infected seedling. In that way, conditions are created for the occurrence of the disease in the subsequent sunflower-growing season.

For the development of the downy mildew of sunflower, rain is the critical factor during the first fortnight of growth, because only then are the seedlings susceptible to systemic infection. The period of maximum susceptibility to systemic infection is as short as 5 days at 22°C–25°C under greenhouse conditions. Under field conditions where mean air temperature during emergence is 13.2°C, plants remain susceptible for at least 15 days, provided enough rain during this period becomes available to provide soil water for only a few hours (Raranciuc and Pacureanu-Joita 2006). The percentage of infected plants is increased with depth of sowing.

The age of the sunflower seedlings is also an important factor in the development of systemic symptoms of downy mildew. Susceptibility of the seedlings decreases as the age advances; 3-day-old seedlings are the most susceptible to systemic infection. Therefore, any environmental factor that favors rapid seedling development shortens the interval of maximum susceptibility. Although seedling development is directly proportional to soil temperature, the range of soil temperature that normally prevails during the spring planting season is not a factor-limiting infection by downy mildew, particularly in the Red River Valley area of the United States and Canada (Kolte 1985).

Spread of the disease in relation to soil type has been studied. The heavy clay soils and flat topography of the Red River Valley area result in poor drainage, which favors downy mildew. The spread of the disease under field conditions preferentially follows the line of slope. Besides, tillage and running water are likely to be important factors in the spread of the disease. Sunflower plants suffering from boron deficiency become more susceptible to downy mildew.

Disease Management

Host Plant Resistance

Host plant resistance using race-specific genes designated as Pl, of which 22 have been described, is the most effective (Gulya 2007). Genes that confer resistance to downy mildew are dominant and often form clusters (As-Sadi et al. 2012, Vincourt et al. 2012). A number of Pl genes have been reported (Pl (1) to PI (15), Plv, Plw, Plx-z, Mw, Mx, Plarg, Pl HA-R4), and the position of 11 genes has been determined on the SSR genetic map (Mulpuri et al. 2009, Jocić et al. 2012, Liu et al. 2012). Sunflower and *P. halstedii* have a typical *gene-for-gene relationship* for each virulence gene presented in the pathogen exists a corresponding resistance gene in the host plant. If the plant has an effective resistance gene that will counteract the virulence gene in the pathogen, the infection will be stopped near the penetration site expressing a hypersensitive reaction (HR) that is manifested as an extensive cell death in the infected tissue. The constant evolution of new physiological races, due to pathogenic variability and selection pressure resulting from the use of resistant hybrids and seed treatment fungicides, continuously challenges breeders to identify and introduce new resistance genes or gene clusters. Wild sunflower species have been a plentiful source of genes for downy mildew resistance. Downy mildew can be controlled by single, race-specific major dominant genes. Multirace resistant germplasms from wild sunflower species have been developed. The multitude of genes from the wild species for downy mildew resistance is supported by the number of germplasm releases that incorporate protection against ever-evolving pathotypes of downy mildew that infect cultivated sunflower.

It is, however, not advisable to use only one resistance gene in developing new cultivars. Rather, several different resistance genes should be employed, either by growing different hybrids carrying the different resistance genes or by pyramiding such genes. This strategy may extend the life cycle of each gene by keeping the selection pressure less effective against all known races minimizing the development of a new race. For most of these resistance genes, sequence-specific markers have been developed, which facilitate their detection and make the selection process faster and more reliable. A considerable number of sunflower hybrids that are genetically resistant to downy mildew have been released for commercial cultivation from time to time (Shirshikar 2008). Some of these hybrids are Sungene-85, MSFH-47, Pro-009, Prosun-09, SH-416, DRSF-108, PCSH-243, PRO-011, SCH-35 or Maruti, NSH-23, Sunbred-2073, NSSH-303, K-678, and MISF-93. Most commercial hybrids marketed as *downy mildew resistant* become susceptible to the new races; a few hybrids, however, can be bred that show resistance to major prevailing races in a specified sunflower-growing region (Raranciuc and Pacureanu-Joita 2006, Seiler 2010).

By combining the parial resistance provided by minor genes with specific resistance genes, durable resistance could be achieved (Labrouhe et al. 2008, Tourvieille de Labrouhe et al. 2008, Vear et al. 2008) or by introducing genes from different clusters with different origins in a single genotype (Jocić et al. 2010). Defeated hypostatic genes may be resistant to such new races. Hence, the combination of these defeated genes with novel genes, to which the pathogen has not been exposed, will extend the useful life of the defeated genes and will provide more durable resistance (Lawson et al. 1998). A great number of researchers have contributed to better understanding of the mechanisms involved in downy mildew resistance. The developments in biotechnology resulted in molecular markers for detecting PI genes and provided means for MAS. Candidate resistance genes have been proposed. For example, a marker derived from a bacterial artificial chromosome (BAC) clone has been found to be very tightly linked to the gene conferring resistance to race 300, and the corresponding BAC clone has been sequenced and annotated. It contains several putative genes including three toll-interleukin receptor–nucleotide-binding site–leucine-rich repeat (TIR–NBS–LRR) genes. However, only one TIR–NBS–LRR appeared to be expressed and thus constitutes a candidate gene for resistance to *P. halstedii* race 300 (Franchel et al. 2012). Resistance to *P. halstedii* can be of two types: Type I resistance can restrict the growth of the pathogen as in the case of" PI ARG gene "controlled by TIR–NBS–LRR genes and Type II resistance cannot restrict the growth of the pathogen, allowing the pathogen to invade, and subsequently an HR occurs as in the case of the "PI 14" gene controlled by coiled-coil CC–NBS–LRR genes (Radwan et al. 2011). MAS could be used for detecting not only major but also minor genes and would bring researchers a step closer to achieving sustainable resistance to downy mildew (Jocić et al. 2012).

Chemical Control

A wide range of commercial fungicides are available in the market with different modes of action for the management of downy mildew (Gisi 2002, Gisi and Sierotzki 2008). Coating of seeds with metalaxyl derivatives is most frequently used, as it provides protection at the time of primary infection, that is, at early stages of development of sunflower especially if a new pathogenic race occurs. Seed treatment with the aforementioned chemical at the rate of 3–6 g/kg seed is reported to give complete check of downy mildew of sunflower. In certain situations, the plants may remain completely protected throughout the growing period by following the aforementioned treatment. Strobilurins (especially trifloxystrobin) exhibit high activity against *P. halstedii* and is a promising group of fungicides for controlling sunflower downy by seed treatment and foliar spray (Sudisha et al. 2010), though other fungicides are not considered effective for the control of the foliar portion of this disease and are not generally recommended. Concerning issues related to the use of chemical management include emergence of pathogen races resistant to fungicides as reported from several countries, notably from France, Germany, Turkey, and Hungary (Viranyi and Spring 2011); negative environmental effects of fungicides; and the economic feasibility of the disease management measures.

Seed treatment, combined with the use of a downy mildew–resistant hybrid or cultivar, offers the best promise for the management of the disease.

Induced Host Resistance

Besides the traditional management strategies, alternative or supplementary methods are reported to be effective in providing protection against sunflower downy mildew. One such possible solution is the use of systemic acquired resistance (SAR), that is, activation of the defense system of the plants. Commercially available immunoactivator Bion 50 W (benzo(1,2,3)-thiadiazole-7-carbothioic acid *S*-methyl ester) at 320 mg/L has been found to reduce the infection of sunflower by *P. halstedii* (Tosi et al. 1999, Korosi et al. 2009, 2011). Seed treatment with beta-aminobutyric acid (BABA) at the concentration of 50 mM also induces resistance to *P. halstedii* in sunflower (Nandeshkumar et al. 2009). Chitosan-induced resistance is also found to be effective against downy mildew in sunflower (Nandeshkumar et al. 2008). Induction of resistance by culture filtrate of *Trichoderma harzianum* against the disease is also reported (Nagaraju et al. 2012). This method of using specific chemical compounds for triggering plant defense mechanisms proves to be effective in diminishing the severity of infection of downy mildew in genotypes without genetic resistance (Gisi 2002, Ba'n et al. 2004).

Cultural Control

The emergence of pathogen strains resistant to chemicals and the occurrence of new races able to overcome specific resistances have led to include cultural practices for a more sustainable management of downy mildew. The choice of planting sites and optimum sowing time should be such that seedlings emerge rapidly and such that it reduces chances of free soil water during the period of susceptibility. For example, sunflower hybrid planting seed is almost exclusively produced in California. Due to the lack of summer rains and furrow irrigation, California-produced seed is relatively disease free, and thus, it regularly meets phytosanitary restrictions imposed by many countries (Gulya et al. 2012). Fields should be selected such that these are at least 500 m away from a field on which sunflower had been grown the previous year (Jocic' et al. 2012). Seed meant for sowing should be clean, and the seed should be obtained from a disease-free area; crop rotation is possible but not feasible, since the pathogen persists in soil and plant residues for 5 or more years, but proper crop rotation, that is, maintaining intervals of 4–5 years between two sunflower crops in the same field with other nonsusceptible crops, appears to be a quite desirable practice. The incidence of downy mildew increased from 42% in the second year to 100% in the fourth without crop rotation in Spain. While in plots where wheat and sunflower are grown alternately, the incidence has been reported to go only up to 15%–16%. Since wild and volunteer sunflowers and weed are hosts for the pathogen, eliminating these plants will help reduce overall inoculum built-up in the fields. Sowing should be performed at optimum time with avoidance of late planting (Covarelli and Tosi 2006, Jocic' et al. 2010).

Biological Control Hypovirulence in *P. halstedii*

Plasmopara halstedii virus (PhV) is an isometric virus found in the oomycete *P. halstedii* (Gulya et al. 1990a, 1992, Mayhew et al. 1992). The fully sequenced virus genome consists of two ss(+) RNA strands encoding for the virus polymerase and the coat protein (CP), respectively. Most of the field isolates of *P. halstedii* from different countries show morphologically and biochemically indistinguishable virions (Heller-Dohmen et al. 2008). The virions are isometric and measure 37 nm in diameter with one polypeptide of 36 kDa capsid protein and two segments of ssRNA (3.00 and 1.6 kb) that have been found to harbor PhV. The complete nucleotide sequence of PhV has been established and it shows similarities to the *Sclerophthora macrospora* virus (SmV) and viruses within the *Tombusviridae* family as well as *Nodaviridae* (Heller-Dohmen et al. 2011). The presence of PhV leads to hypovirulence effects by weakening the aggressiveness of *P. halstedii* (Grasse et al. 2013). The PhV thus offers a great promise for obtaining a biological control of downy mildew disease of sunflower though practical utility of such an effect is yet to be investigated.

Regulatory Control

P. halstedii is listed as a plant quarantine pathogen (Ioos et al. 2012). Plants grown from infected seed, although they show no visible systemic symptoms, have been reported to produce infected seed and disseminate the pathogen. In addition, as seen earlier, there exists a physiological specialization within the *P. halstedii*, suggesting the importance of prevention of chance introduction of the more prevalent and widely virulent North American race of *P. halstedii* into areas where it does not occur. In Australia and perhaps in South Africa and in India, strict quarantine regulations have precluded its introduction.

ALTERNARIASTER BLIGHT

Symptoms

Symptoms of the disease are characterized by the development of dark brown to black, circular-to-oval spots, varying from 0.2 to 5.0 mm in diameter. The spots are surrounded by a necrotic chlorotic zone with a gray-white necrotic center marked with concentric rings (Figure 6.4). Initially the spots are small and they gradually increase in size, making their first appearance on the lower leaves. As the plant grows, the spots subsequently are developed on middle and upper leaves. At the later stages, elongated spots are formed on petioles, stem, and ray florets. Under high humidity conditions, the spots enlarge in size and coalesce resulting in blighting of leaves and sometimes rotting of flower heads.

Geographical Distribution and Losses

Alternariaster blight (formerly termed as *Alternaria* blight) of sunflower was first described in Uganda in 1943 and has since been recognized as a potentially destructive disease in most of the

FIGURE 6.4 *Alternaria* leaf spot of sunflower. (Courtesy of Dr. Tom Gulya, USDA-ARS, Northern Crops Research Lab, Fargo, ND.)

sunflower-growing areas of the world. It is reported from Argentina, Australia, Brazil, Bulgaria, India, Japan, Romania, Tanzania, Yugoslavia, South Africa, and the United States (Amabile et al. 2002, Calvet et al. 2005, Berglund 2007, Singh and Ferrin 2012).

In subtropical sunflower-growing areas, *Alternariaster* blight is considered as a major disease and can cause yield losses from 15% to 90% (Berglund 2007). The disease has been reported to reduce the seed and oil yields by 27%–80% and 17%–33%, respectively, in India. A negative correlation between increase in disease intensity (25%–96%) and yield components and oil content has been established (Kolte 1985, Chattopadhyay 1999). The most affected components due to the disease are the number of seeds per head, followed by the seed yield per plant. The disease also affects the quality of the sunflower seeds by adversely affecting the seed germination and vigor of the seedlings (Amaresh and Nargund 2004, Wagan et al. 2006). The loss in seed germination varies from 23% to 32% (Ahamad et al. 2000, Pandey and Saharan 2005). The nature of yield reduction is determined to some extent by the stage of plant growth when the disease epidemic develops. For example, the relationship between severity and yield in the R3 (second phase of inflorescence elongation) growth stage has proved that plants with disease severity higher than 10% show yield lower than 500 kg/ha regardless of the sowing dates. This value can therefore be used as a damage threshold for the disease (Leite et al. 2006).

Pathogen: *Alternariaster helianthi* (Hansf.) Simmons (= *Alternaria helianthi* (Hansf.) Tubaki and Nishihara)

Classification
Kingdom: Fungi
Division: Ascomycota
Class: Dothideomycetes
Order: Pleosporales
Family: Leptosphaeriaceae
Genus: *Alternariaster*
Species: *helianthi*

The pathogen has been first described as a member of the genus *Alternariaster* by Simmons (2007) and has been renamed *Alternariaster helianthi* (Hansford) Simmons (formerly *Alternaria helianthi* and *Helminthosporium helianthi*) as type and has hitherto been monotypic based on the absence of conspicuous internal pigmented, circumhilar ring found commonly in conidia and conidiophores of the *Alternaria* fungus. The phylogenetic analysis made by Alves et al. (2013) confirms the segregation of *Alternariaster* from *Alternaria* by showing that *Alternariaster* is a well-delimited taxon belonging to the Leptosphaeriaceae instead of the Pleosporaceae to which *Alternaria* belongs (Schoch et al. 2009). The mycelium is a septate, rarely branched, brown, and 2.5–5.0 μm in width. The conidiophores are hypophyllous, solitary or in small groups, straight to slightly sinuous, 100–225 × 7.5–10 μm, simple 3–6 septate, pale to chestnut brown, smooth, conidiogenous cells tretic, integrated, and terminal to intercalary and sympodial (Alves et al. 2013). The conidia are dry, solitary, and cylindrical to subcylindrical, occasionally with cells of different sizes, 60–115 × 11–29 μm, with rounded apex and base, transversally 5–9 septate (1–2 longitudinal or oblique septa), often deeply constricted at septa, eguttulate, subhyaline to pale brown, smooth, and thickened and darkened hilum (Figure 6.5). Germ tubes are oriented particularly to the main axis of the conidium and also polar (Alves et al. 2013). The conidia are not produced in chains, but 2–3 conidia in short chains may be observed in the culture as well as on the diseased host bits, on incubation in moist chambers. Genetic variability in isolates of *A. helianthi* has been assessed by random amplified polymorphic DNA (RAPD) analyses, which reveal the presence of six genetically distinct groups in India. The isolates Ah-1, Ah-7, and Ah-14 are reported to be genetically distinct (Prasad et al. 2009). In general, potato-dextrose agar (PDA) has been used for isolation by several workers, and it appears that the fungus produces very scanty mycelial growth and moderate to

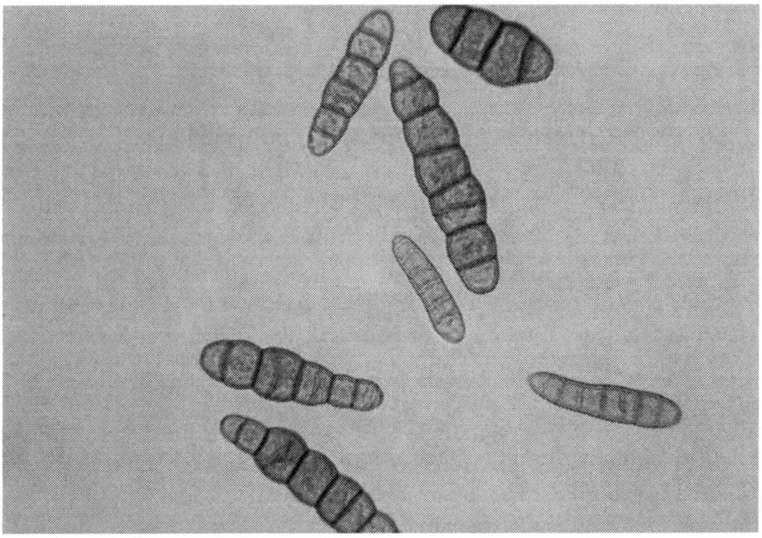

FIGURE 6.5 *A. helianthi* spores. (Courtesy of Dr. Tom Gulya, USDA-ARS, Northern Crops Research Lab, Fargo, ND.)

abundant sporulation on PDA. Comparatively good mycelial growth could be obtained on sunflower leaf extract agar medium (SLEAM) and Richards agar medium. Autoclaved carrot discs have been proved excellent for sporulation and luxuriant growth of *A. helianthi*. SLEAM with 2% sucrose and sterilized carrot disc supports maximum sporulation (Sujatha et al. 1997). On potato–carrot agar, the colony is raised centrally, with aerial mycelium felted, while having a wide periphery of flat sparse olivaceous buff to greenish glaucous mycelium with irregular margins (Alves et al. 2013). The estimated minimum temperatures for mycelial growth rate and for conidium germination are 5.5°C and 7.9°C, respectively, while the maximum temperatures are 32.9°C and 40.0°C, respectively (Leite and Amorim 2002). The optimum temperature for growth of the fungus in culture is 26°C, and it sporulates at the temperature range of 5°C and 35°C–40°C, with an optimum sporulation temperature of 20°C. Sporulation and mycelial development of the fungus occur best at pH 5.3–5.9 (Kolte 1985).

Epidemiology and Disease Cycle

The pathogen is seed borne and can, therefore, be introduced into new areas from infected seed (Salustiano et al. 2006, Micheli et al. 2007). Udayashankar et al. (2012) have developed the species-specific PCR-based diagnostic technique that provides a quick, simple, powerful, reliable alternative to conventional method in the detection and identification of *A. helianthi*. Locally, however, infested stubbles and crop debris left on the top of the soil from one growing season to the next is the most important source of inoculum from which primary infections are established. On such plants, the fungus overwinters as mycelium. Since sunflower can be grown throughout the year in all crop seasons, volunteer plants of infected sunflower or overwintering sunflower may also be an important primary source of inoculum. The exact process of penetration and infection at the host tissue level is through the cuticle and cell wall, and junctions between epidermal cells are the most frequent sites of appressoria formation (Romero and Subero 2003). The pathogen produces a specific toxin in culture and produces typical symptoms of the disease when inoculated on leaves (Kalamesh et al. 2012). The toxin inhibits seed germination as well as root and shoot growth under in vitro conditions (Madhavi et al. 2005a).

Relative lesion density and severity are influenced by temperature and leaf wetness duration. The disease appears to be more severe at a temperature of 25°C. The minimum temperature for disease development, estimated by generalized beta function, is 13.0°C, and the maximum is 35.8°C. Relative lesion density increases with increasing periods of leaf wetness, as described by a logistic model (Leite and Amorim 2002). Positive significant correlation with relative humidity and a negative significant correlation with air temperatures and sunshine hours have been observed for disease development (Das et al. 1998, Amaresh and Nargund 2004). Thus, hot weather and frequent rain during milk and wax stages of sunflower plant development favor *Alternariaster* blight infection.

Disease Management

Host Plant Resistance

Attempts to identify the sources of resistance to the disease have been made by several workers (Madhavi et al. 2005b, Murthy et al. 2005, Gopalkrishnan et al. 2010). Wild species such as *H. tuberosus*, *H. occidentalis*, *H. resinosus*, and *H. argophyllus* are highly resistant to the disease and can be used in breeding for disease resistance (Madhavi et al. 2005b, Sujatha and Prabakaran, 2006, Prasad et al. 2009). The pathogen has been found to be restricted to epidermal cells in resistant wild sunflower as well as increase accumulation of phenols (Madhavi et al. 2005a). Sources of resistance to the disease have also been located in several germplasm accessions and hybrids. These are HPM-15R, HPM-116, and HPM-140 (Amaresh and Nargund 2000); LC-985, Performer, Select, Lc1029, and LC1093 (Raranciuc and Pacureanu 2002); 135, 1171, P-1019, 347, 446, 1039, 1210, and 1483 (Mesta et al. 2005); PEH-K04 hybrids 43, 50, 60, 77, 80, 81, 84, 92, and 98 (Nagaraju et al. 2005); RHA 587 and ARG × RHA 587 (Reddy et al. 2006); EC 68414 (Dawar and Jain 2010); sunflower hybrid parental lines CMS7-1A, DRS 9, DRS 63, and DRS 34; and four hybrids CMS7-1A × DRS 22, CMS7-1A × DRS 9, DCMS 15 × DRS 9, and DCMS 15 × DRS 63 (Sujatha et al. 2008). Higher peroxidase activities are recorded in sunflower genotypes with high threshold levels of resistance and lesser in susceptible genotype indicating strong evidence for the important role of peroxidase enzymes in the central defense system against necrotrophic pathogen *A. helianthi*, which could be used as a reliable biomarker for assessing resistance (Anjana et al. 2007, 2008). A number of sunflower genotypes are reported to possess partial resistance to the disease. The gametophytic selection combined with the conventional sporophytic selection can be considered as an effective tool in population improvement program to achieve a high level of resistance in a relatively short time (Chikkodi and Ravikumar 2000, Shobana Rani and Ravikumar 2006). Selection for resistant pollen on the stigmatic surface results in a corresponding increase in progeny resistance and successive pollen selection to further improve disease resistance of progeny. Repeated cycles of selection are required to achieve a useful level of resistance in sunflower, since resistance to *Alternariaster* in sunflower is polygenetically controlled (Chikkodi and Ravikumar 2000). Resistance to *Alternariaster* blight can be inducted or improved in the progenies derived through mutagenic treatment when seeds of sunflower genotypes are treated with 20 and 30 Kr of gamma rays (Oliveria et al. 2004, Patil and Ravikumar 2010, Shobharani and Ravikumar 2010). SAR in sunflower against *Alternariaster* blight can be inducted due to foliar application of salicylic acid at the concentration of 20 mM and Bion (acibenzolar at 0.05–5.0 mM). A lag period of 3–7 days is required for the induction of SAR (Ratnam et al. 2004a,b).

Chemical Control

Chemical management with protective (nonsystemic) fungicides such as iprodione, chlorothalonil, and mancozeb each at 0.2% spray as well as with therapeutic (systemic) fungicides such as hexaconazole, carbendazim, and propiconazole each at 0.1% spray has been found effective against the disease. But systemic fungicides are more effective than the nonsystemic ones (Amaresh and Nargund 2000, 2002, Amaresh et al. 2000, 2004, Singh 2000). Seed treatment with a mixture of carbendazim + iprodione in a 1:1 ratio at 0.3% followed by foliar spray; systemic fungicide hexaconazole (0.1%) gives best management of the disease with high yield (Rao et al. 2007, 2009). Combination of carbendazim + mancozeb in the ratio of 1:1 at 0.2% spray can be used most effectively in the management of the disease avoiding the possibility of fungicide-resistant strains of the pathogen (Singh 2002, Mathivanan and Prabavathy 2007).

Cultural Control

Sanitation measures like selecting pathogen-free healthy seed and destruction of crop residues from previously affected crop help to reduce the initial inoculum intensity that can delay the onset of the disease by 11 days (Leite et al. 2005, Jurkovic et al. 2008). Occurrence and severity of the disease depend on the season and planting dates. This can be selectively used in disease management. For example, late August to mid-September planting of sunflower in most sunflower-growing states in India remains free from most major diseases with only traces of *Alternariaster* infection (Singh 2002, Amaresh et al. 2003, Mesta et al. 2009, Gadhave et al. 2011). Such a planting date is recommended for raising disease-free seed crop of sunflower.

Biological Control

Efficacy of *Pseudomonas fluorescens* as seed dresser can be enhanced by biopriming the sunflower seed for the effective and eco-friendly management of *Alternariaster* blight of sunflower (Rao et al. 2009). Antagonistic fungi *Gliocladium virens* (Anitha and Murugesan 2001) and *Trichoderma virens* (Mathivanam et al. 2000) are reported to be effective in managing the infection of *A. helianthi* in sunflower. Antibiosis is indicated as the mechanism of antagonistic effect of *G. virens* on the pathogen. Prior infection of sunflower plants with sunflower mosaic virus reduces the severity of *Alternariaster* blight of sunflower (Bhardwaj and Mohan 2005).

Sclerotinia WILT AND STEM ROT

SYMPTOMS

Symptoms of the disease appear in three different phases on the sunflower—(a) basal stalk rot and wilt, (b) midstalk rot, and (c) head rot—and they are rather considered as three distinct diseases caused by the same pathogen.

Basal Stalk Rot and Wilt

Basal stalk rot and wilt are triggered through root infection from the fungus present in the soil and can appear in sunflower seedlings, but usually they appear during anthesis and seed development stages when the plants attain a height of about 5–6 ft (Figure 6.6). At first, wilted plants are scattered in the field, but later they are commonly found in series within rows. This disease usually appears in patches within the field. The incipient *Sclerotinia* wilt in sunflower is characterized by rotting through the taproot or through the hypocotyl axis (Darvishzadeh et al. 2012). Water-soaked lesions occur on the taproot at the soil line and on some fibrous root. If moisture conditions remain conducive, lesions on the stem below the soil level get covered with dense white growth of the fungus, which can be seen with loosely attached black sclerotial bodies that are irregular in size and shape. Similar bodies are found in stem and root piths, the occurrence of which is a reliable diagnostic feature of the disease.

Midstalk Rot

Plants infected later in the season may not wilt, and the only exterior symptoms may be a small brown lesion at the stem base or at any part of the stem, often in the upper half, which often results in breakage of the stem at the point of infection. Stems of severely diseased plants shred into vascular strands, becoming straw colored as they dry (Figure 6.7). Such stems are weak, and the plants lodge easily. Symptoms of stem rot in the upper half of the stem are usually due to airborne ascospores. Infection may start initially in leaf axils before progressing down the petiole to the stalk. The rotted part of the stem may or may not show the presence of sclerotia.

FIGURE 6.6 Basal stalk rot of sunflower caused by *S. sclerotiorum*. Note the fungal growth and sclerotia on the affected stem. (Courtesy of Dr. Tom Gulya, USDA-ARS, Northern Crops Research Lab, Fargo, ND.)

FIGURE 6.7 Midstalk rot of sunflower caused by *S. sclerotiorum*. (Courtesy of Dr. Tom Gulya, USDA-ARS, Northern Crops Research Lab, Fargo, ND.)

FIGURE 6.8 Head rot of sunflower caused by *S. sclerotiorum*. (Courtesy of Dr. Tom Gulya, USDA-ARS, Northern Crops Research Lab, Fargo, ND.)

Head Rot

Sometimes, the plants may remain healthy until the flower heads are produced. The flower heads, once formed, during the long period of their formation from budding to seed maturity stage, may be attacked by the fungus. The symptoms may become visible in any part of the receptacle. The affected flower portions may show the presence of a conspicuous white mycelial growth of the fungus, making evident the spread of the rot throughout the flower head. The head may ultimately be shredded resembling a broom-like appearance, and most of the tissue of the flower head is converted into a continuous mat of sclerotial tissue (Figure 6.8). Severely affected heads show incomplete filling of the head with seed. The head rot may be partial or complete. Seeds formed on partially affected heads may show the presence of sclerotia on their surfaces.

Geographical Distribution and Losses

Sclerotinia disease (Basal stalk rot/wilt/head rot) is one of the most damaging diseases of oilseed sunflower distributed all over the world in temperate regions and under cool tropical conditions, often at intermediate altitude. It has become economically important in all sunflower-growing areas in North American countries (Canada, the United States, and Mexico), East Europe, and other countries like Argentina (Fusari et al. 2012), Croatia (Ćosić and Postic 2008), Serbia (Maširević and Jasnic 2006b), Turkey (Tozlu and Demirci 2011), Egypt (El-Deeb et al. 2000), Iran (Bolton et al. 2006, Davar et al. 2010), Tanzania and South Africa (Anonymous 2010), and South Asia and China (Pandey and Saharan 2005). Yield loss depends on the sunflower development stage in which the disease occurs. If infection occurs in the early sunflower development stage, the yield loss will be approximately equal to the disease occurrence percentage. Disease occurrence percentage and yield losses caused by *Sclerotinia* wilt can even reach 100% because it causes whole plant devastation (Lamey et al. 2000, Saharan and Mehta 2008). Sunflower plants infected at the beginning of the flowering stage can lose up to 98% of their potential yield, while plants infected 8 weeks after flowering can lose only 12% of their potential yield (Maširević and Gulya 1992). In the United States, annual losses on all crops caused by *Sclerotinia sclerotiorum* exceed $200 million, while in 1999, *Sclerotinia* head rot epidemic on sunflower caused crop loss valued at $100 million (Bolton et al. 2006). In Serbia, *Sclerotinia* wilt is the most common form of *Sclerotinia* disease and appears in sunflower crop more frequently than the other two forms. Its average frequency in Serbia is about 15%–20%, but in some years, the frequency can reach even around 50% (Tančić et al. 2011). Seed quality, as measured by test weight, oil, and protein content, is also adversely affected by the disease in partially infected plants at the near maturity stage of the crop with increase in shell percentage resulting in reduction in economic value (Eva and Andrej 2000, Zandoki and Turoczi 2004). The presence of sclerotia in seed can reduce the grade and market value of the crop. No toxins are produced by *Sclerotinia* in sunflower seed, but heavy contamination with sclerotia is considered unacceptable for human or animal consumption.

Pathogen(s): *Sclerotinia sclerotiorum* (Lib.) de Bary, *Sclerotinia trifoliorum* Fuckel, *Sclerotinia minor* Jagger

Classification
Kingdom: Fungi
Phylum: Ascomycotina
Class: Leotiomycetes
Subclass: Leotiomycetidae
Order: Helotiales
Family: Sclerotiniaceae
Genus: *Sclerotinia*
Species, *S. mino*r Jagger, *S. sclerotiorum* (Lib.) de Bary

S. sclerotiorum was first recognized as a sunflower pathogen in 1861 in the Unites States (Kolte 1985). *S. minor* is another species reported from South America, Australia, Canada, and California (United States) causing root rot and wilt on sunflower, but is much less commonly found than *S. sclerotiorum. S. trifoliorum* has also been reported to be associated with the disease in Chile and Russia (formerly Soviet Union). They produce a fluffy white mycelium on and in infected plant parts. This mycelium aggregates itself into sclerotia, which are the structures that allow *Sclerotinia* species to survive in soil in the absence of a plant host. *S. minor* has uniformly round sclerotia measuring 0.5–2 mm, while those of *S. sclerotiorum* produce larger and irregular sclerotia, some measuring 1–5 cm. Sclerotia produced by *S. sclerotiorum* in heads are very similar in size and shape to sunflower seeds. Sclerotia exhibit either myceliogenic (eruptive) or carpogenic germination, the former giving rise to white vegetative hyphae that extend from sclerotia that have been stimulated to germinate by host plant exudates and the latter to apothecia as described in detail by several researchers (Bolton et al. 2006, Saharan and Mehta 2008). *S. minor* sclerotia rarely form apothecia, germinating instead by the direct emergence of hyphae (*myceliogenic* germination); *S. sclerotiorum usually* germinates carpogenically; and only occasionally, it germinates myceliogenically. At soil depths of up to 2 cm, apothecia can extend from the sclerotia of *S. sclerotiorum/S. trifoliorum* to reach the soil surface. A single sclerotium can produce as many as eight apothecia. Apothecia are tan to light brown, flesh-colored discs averaged about 2–8 mm in diameter and may be difficult to see. The asci are cylindrical hyaline and are produced in tightly packed masses at the upper surface of the apothecium. The asci measure 66–136 μm in length and 6–10 μm in width. The ascospores, only visible with a microscope, are monostichous, ellipsoid, one celled, thin walled, clear, or nonpigmented numbering eight per ascus and measuring 7.4–11 × 3.7–4.6 μm in size. The paraphyses are filiform. The ascospore morphology may be somewhat differing between *S. sclerotiorum* and *S. trifoliorum.*

Variability in the Pathogen

Isolates of *S. sclerotiorum* differ significantly in aggressiveness (Ekins et al. 2005, 2007, Zandoki et al. 2006), variation in oxalic acid production (Durman et al. 2005), and mycelial compatibility groups (MCGs) (Durman et al. 2003, 2005, Zandoki et al. 2006). Aggressiveness is positively correlated to colony radial growth, percent large sclerotia, and dry weight per sclerotium (Durman et al. 2003). However, there appears to be no correlation between genetic diversity among isolates and virulence differentiation (Li et al. 2005). The population structure of *S. sclerotiorum* on sunflower in Australia shows that the sclerotia, all eight ascospores within an ascus, are of only one genotype as revealed through multicopy restriction fragment length polymorphisms (RFLPs), MCGs, and RAPDs. Single and multicopy RFLP analyses have shown that majority of sunflower plants are infected by only one genotype (Ekins et al. 2011). Interestingly, isolates of *S. sclerotiorum* from the United Kingdom are reported to form a population that is significantly different from other populations (Li et al. 2009). Two very distinct sclerotia-producing strains of *S. sclerotiorum*, one as a normal strain (normal black sclerotia with white medulla) and the other as an aberrant strain (tan sclerotia with brown medulla), are known to be prevalent in Russia (formerly the Soviet Union) and Canada (Huang and Yeung 2002). The tan sclerotia produced by the aberrant strain have been found to have no dormancy and more than 85% of the sclerotia germinate myceliogenically on moist sand at 16°C–20°C with or without chilling treatment. Serotonin (5-hydroxytryptamine) is present in large amount in normal black sclerotia but absent or present in small quantity in abnormal sclerotia. Abnormal sclerotia instead contain a large amount of 5-hydroxyindole acetic acid (Kolte 1985, Huang and Yeung 2002).

Epidemiology and Disease Cycle

The pathogen is a facultative parasite and attacks over 400 plant species of 75 botanical families ensuring all time possibility of alternative sources of primary infection (Lazar et al. 2011). Sclerotia are the most important means of perennation. The survival time in soil is very variable, but 5–6 years is thought to be an upper limit. Survival of mycelium in seeds may also occur, but epidemiologically, it is of little consequence. Under most conditions, myceliogenic germination is of limited importance because only limited saprophytic spread occurs in natural nonsterile field soils. However, where sclerotia and susceptible plants are in close proximity, devastating stem base infections or root rot may result. Most sclerotial germination occurs at optimum of 24% soil moisture when the sclerotia are embedded at lower depths of soil up to 5 cm where the average temperature of soil (5–10 cm in depth) during the growing season of sunflower in rain-fed condition remains to be 30°C (Irany et al. 2001). A prerequisite for carpogenic germination is a period of chilling to break dormancy followed by rising temperatures and a high humidity. In temperate latitudes, apothecia typically mature during spring and early summer, although there are many reports relating to other seasons of the year. Conditions suitable for carpogenic germination of *S. minor* probably occur in southern regions in Australia, and carpogenic germination is probably a rare event in northern regions, and if it does occur, it probably does not coincide with first flower bud and anthesis stages in sunflower crops (Ekins et al. 2011). The apothecial stripes elongate in response to light and the ascospores are wind dispersed. Ascospores landing on potential hosts such as sunflower need water for germination, a requirement of 16–24 h being typical. Germination is possible throughout the range of 0°C–25°C, with an optimum at 15°C–20°C, and the pathogen is unable to cause infection at 30°C–35°C (Raj and Saharan 2001, Vuong et al. 2004). Continuous wetness on leaves within the canopy or on flowers on sunflower for a period of 42–72 h is needed for ascospore infection of the capitulum, and symptoms appear about 5 weeks later. This threshold can be used to define regions at risk. Head rot due to *S. sclerotiorum*, however, is best developed at 80% relative humidity for shorter periods of 16–24 h (Raj and Saharan 2000a). Disease appearance significantly vary depending on the quantity of rainfall, high crop density, sowing dates, temperature over the vegetation, and selection of sunflower hybrid for sowing (Alexandrov and Angelova 2004, Simic et al. 2008). Also, it seems that an exogenous nutrient base is required for infection. Wounded, dead, or senescent tissues are readily colonized and serve as a food base from which infection of healthy tissues can take place. Ascospores are thought to be discharged along with mucilage that can cement the spores to host tissue, more particularly the senescent petals and other flower parts that provide a major avenue of infection from the site where the flower parts lodge, the sunflowers being most susceptible during the first flower bud stage coincidence of flowering, and ascospore release becomes a major factor of epidemiological significance in the occurrence of airborne infection causing stem or head rot (Raj and Saharan 2000b). Germinated ascospores produce appressoria that can vary from simple lobed forms to complex multibranched cushion-like structures. Entry is usually by direct penetration through the cuticle assisted by extensive endopolygalacturonase pectolytic and cellulolytic enzymes during the early phase of colonization causing dissolution of the host cell structure resulting in the development of stem or head rot symptoms (Cotton et al. 2002). *S. sclerotiorum* secretes several acid proteases, and one of the genes, acpl, encoding an acid protease, has been cloned and sequenced. The acpl gene is expressed in plant infection, which is low at the beginning of infection but increases suddenly at the stage of necrosis spreading, suggesting thereby that glucose and nitrogen starvation together with acidification can be considered as key factors controlling *S. sclerotiorum* gene expression during pathogenesis (Poussereau et al. 2001a). Similarly, another gene asps encoding aspartyl protease is expressed in the beginning of infection of *S. sclerotiorum* in sunflower (Poussereau et al. 2001b).

The toxic metabolite, oxalic acid, produced by the fungus also plays an important role in the development of wilt symptoms. A positive correlation has been found between oxalic acid and shikimate dehydrogenase activity during the infection process caused by *S. sclerotiorum* in sunflower (Enferadi et al. 2011). Oxalic acid has been shown to move systemically in the plant and accumulate to critical level, and this elicits the wilt syndrome. Metabolic profiles of sunflower genotypes with contrasting response to *S. sclerotiorum* infection have been studied (Peluffo et al. 2010). There is induction of glycerol synthesis in *S. sclerotiorum* that exerts a positive effect on osmotic protection of fungal cells that favors fungal growth in plant tissues (Jobic et al. 2007). Monoculture with high level of N fertilization and irrigation exacerbate the disease (Gergely et al. 2002). Sclerotia from such affected sunflower plants are returned to the soil as the host decomposes, or they may be distributed by cultural operations, harvesting, etc. In most regions, the absence of a conidial stage and the environmental requirements for apothecium formation restrict *S. sclerotiorum* to a single annual infection cycle, and the disease is referred to as a single-cycle disease.

Disease Management

Host Plant Resistance

In Cultivated and Wild Helianthus *Species Germplasm*

There have been great efforts in searching for tolerance to midstalk rot in both cultivated sunflower and wild sunflower species through artificial inoculation methods (Castano et al. 2001, Becelaere and van Miller 2004, Vasić et al. 2004, Giussani et al. 2008). Screening parental inbred lines for resistance to *S. sclerotiorum* is an important step in developing sunflower hybrids with improved resistance to the disease (Hahn 2002, Huang 2002). A number of sunflower lines and hybrids with various levels of tolerance have been reported (Ronicke et al. 2004, Binsfeld et al. 2005, Castaño and Giussani 2006, Reimonte and Castano 2008), but complete resistance has not yet been observed. The level of tolerance is not yet considered adequate for the control of the disease, which is polygenic and under additive control, so that breeding programs have to combine favorable genes from different sources (Castaño et al. 2001, Becelaere and van Miller 2004, Davar et al. 2010). This confirms the need to consider different isolates in the stem rot resistance breeding programs (Darvishzadeh 2012), and selection for resistance to the disease could start at the inbred line development stage.

Sunflower is an unusual host in that it is prone to both head rot and stalk rot, and since resistance to each phase is independent, this doubles the breeding efforts (Vear et al. 2007). The most resistant breeding lines and commercial hybrids exhibit as low as 10%–15% head rot or stalk rot compared to 90%–100% on susceptible material. Near immunity to stalk rot is observed in most perennial *Helianthus* species and less so in annual species (Gulya 2007b, Silva et al. 2007).

The genotypes that show broad partial resistance to the disease are restorer lines RHA 439 and RHA 440 and maintainer line HA 441 (Miller and Gulya 2006); inbred lines SWS-B-04 (Ronicke et al. 2005a,b), R-28 from *H. argophyllus* (Baldini et al. 2002, 2004, Verzea et al. 2004), HA 302 (Rodriguez et al. 2004), 765, KS 7 (Wang et al. 2010), and 3146 (Wang et al. 2010); maintainer lines HA 451 and HA 452 (Miller et al. 2006); restorer lines RHA 453, RHA 4555 (Miller et al. 2006), and TUB-5-3234 (Micic et al. 2005a); inbred line NDBLOS sel (Micic et al. 2004, 2005b); and two hybrids Pioneer 6480 and Pioneer 6479 (Mosa et al. 2000). Four sunflower hybrids have been developed possessing resistance to ascospore penetration and mycelia extension in the capitulum tissue and could, therefore, be recommended for cultivation in the province of Buenos Aires in Argentina without increasing the risk of *S. sclerotiorum* attack (Godoy et al. 2005).

Molecular Breeding and Transgenic Sunflower for Resistance to Sclerotinia *Diseases in Sunflower*

The SSR markers associated with partial resistance to different isolates could be used in pyramiding polygenes in sunflower disease breeding programs (Micic et al. 2005a,b, Darvishzadeh 2012). Utilization of molecular markers to aid breeders in selecting genotypes with desirable traits through MAS has proved to be very effective. For example, in numerous studies, DNA markers associated with different traits have been reported. Baldini et al. (2002, 2004) used single-marker regression and identified several amplified fragment length polymorphism (AFLP) and SSR markers associated with basal stem resistance to *S. sclerotiorum* in sunflower. Markers of introgressed zones of *H. argophyllus*, *H. debilis*, *H. praecox*, and *H. petiolaris* in the resistant lines are assumed to be good candidates to identify the segments carrying stalk rot–resistant quantitative trait loci (QTLs). The possibility of detecting *H. petiolaris* accessions with a high level of resistance to *S. sclerotiorum* than others is indicated (Caceres et al. 2006). Independent QTLs, other than that for stalk rot resistance, have been identified for head rot resistance (Ronicke et al. 2005a, Yue et al. 2008). However, the prospects of MAS for resistance to *S. sclerotiorum* are limited due to the complex genetic architecture of the trait. The MAS can be superior to classical phenotypic selection only with low marker costs and fast selection cycles (Micic et al. 2004). Attempts have been made to establish resistance against *S. sclerotiorum* by genetic engineering (Scleonge et al. 2000, Schnabl et al. 2002, Hu et al. 2003, Lu et al. 2003, Sawahel and Hagran 2006). These studies are based on a gene controlling the production of an enzyme oxalate oxidase (OXOX). Oxalate is a phytotoxin secreted by *S. sclerotiorum* (Vasic et al. 2002). It weakens the plant tissue and crops with natural resistance to *S. sclerotiorum* such as wheat, barley, maize, or rice, producing OXOX, which breaks down and detoxifies the phytotoxin produced by *S. sclerotiorum*. Contrary to such crops, sunflower has a very low OXOX activity. An OXOX gene from wheat has been isolated and inserted into sunflower plants via *Agrobacterium*-mediated transformation. The *Sclerotinia*-induced lesions in transgenic sunflower are found to be significantly smaller than those in the control leaves (Hu et al. 2003). Compared with the original line, this gene increased resistance, but in general, the level of resistance is not better than in lines obtained by conventional breeding. Therefore, it should be possible to combine the transgenic lines with natural resistance to provide a level of resistance higher than in the available commercial hybrids (Bazzalo et al. 2000). Transgenic sunflower plants constituently expressing OXOX gene exhibit enhanced resistance against the oxalic acid (OA) generating fungus *S. sclerotiorum* (Hu et al. 2003). It is, however, apprehended that OXOX transgene will more likely diffuse naturally after its escape from the host plants (Burke and Rieseberg 2003). Since OA plays a vital role in the establishment of pathogenicity, attempts made to degrade OA will enhance resistance against *S. sclerotiorum* by increasing the production of H_2O_2 mediated through oxidative burst. Such genetically modified cultivars may become a major means of *Sclerotinia* stalk rot management in the future (Link and Johnson 2012). Accumulation of phenolic compounds, their deposition on cell walls and lignifications, is a well-characterized mechanism of disease resistance against *S. sclerotiorum* (Prats et al. 2003, Rodríguez et al. 2004). Conceivably, resistant plants also have higher associated levels of phenylalanine ammonia lyase (PAL), which facilitates the biosynthesis of important phenolic derivatives such as lignin, and shikimic acid and the related enzymatic activity of shikimic dehydrogenase (SKDH), which are useful in identifying a biochemical paradigm that provides a clear correlation to disease-resistant genotypes (Enferadi et al. 2011). Accumulation of scopoletin, one of the coumarins as phytoalexins, may well confer head rot resistance with minimal plant damage and might be one of the basis for resistance to *S. sclerotiorum* (Prats et al. 2006, 2007).

Chemical Control

Foliar infection from airborne ascospores and lack of genetic resistance to *Sclerotinia* head rot need to identify foliar fungicide applications to reduce the impact on sunflower yield and quality. Systemic (azoxystrobin, benomyl, topsin, boscalid, and penthiopyrad) and protectant (iprodione, procymidone, vinclozolin, and fluazinam) fungicides have been demonstrated to be successful and economical if properly timed to manage *Sclerotinia* diseases of sunflower particularly the head rot disease (Link and Johnson 2012). Results suggest that plant coverage rather than systemic movement of the chemical is important for good management. Fungicides, applied as protectants before infection, especially during the bloom period, are effective in inhibiting infection by ascospores in fields with a history of infestation with *S. sclerotiorum* (Rashid 2011). The number of fungicide applications required for disease management depends on the length of the crop season duration of the cultivar or hybrid and the period of time that *weak* tissues (flower petals) are available for colonization by ascospores. If only one application is made, the early application is more effective than the late application. Better results are obtained from a two-application system, one at flowering and another 15 days later (Dietz 2011). In order to be effective, it is necessary that fungicides penetrate deep into the canopy to adequately cover the flowers and the places on the plant where the senescing petals might adhere or become lodged. Among the previously mentioned fungicides, penthiopyrad (a new Group 7 active ingredient) has preventive, residual, and postinfection activity. The strength of this group of fungicide is coupled with the activity that is both translaminar and locally systemic. Penthiopyrad goes through the plant tissue to attack fungal pathogen. It penetrates internally from the upper sprayed leaf surface to the lower unsprayed surface and provides an extended period of control of *Sclerotinia* infection.

Cultural Control

Well-drained sunny field sites away from the previous year's infested plot should be preferred for sowing. Certified seeds should be used to ensure the purity of the seeds without any contamination of sclerotia. The type of tillage operations may affect disease incidence. There is evidence that minimum or reduced tillage that maintains sclerotia on or near the soil surface may promote microbial degradation of sclerotia, whereas deep burial of sclerotia promotes their survival. The number of apothecia, however, may be reduced by tillage practices that bury the sclerotia deep in the soil, such as with a moldboard plow. If sclerotia are buried by deep tillage, use shallow tillage in subsequent years to avoid bringing the sclerotia back near the soil surface. Tillage operations also redistribute sclerotia throughout the soil and can actually increase disease incidence by creating a more uniform distribution of sclerotia within a field (Nelson and Lamey 2000). Crop rotation with nonhost crops such as wheat, barley, beets, and flax reduces the number of sclerotia in the soil by loss of viability over time. In addition, sclerotia may germinate in the absence of a host crop, but without subsequent host infection, new sclerotia are not returned to the soil and numbers are gradually reduced. Crop rotation is most effective when initiated before the fungus becomes a serious problem in a field. If numbers of sclerotia in a field are low, rotations of 3–5 years with a nonhost crop may be sufficient (Rashid 2003). Once the pathogen is well established in a field, and the soil is highly infested with sclerotia, crop rotation may be of less value because of the long survival time of these propagules. When a crop is irrigated, the goal is to manage irrigation events to reduce the frequencies of 12–24 h periods of leaf wetness, especially during the bloom period, when flower petals can become colonized by the ascospores of *S. sclerotiorum*. To reduce disease due to *S. minor*, hyphal germination of sclerotia can be reduced by allowing the soil surface to dry thoroughly between irrigation events. Each irrigation event must therefore provide sufficient water to allow for a prolonged dry period. Cropping practices that reduce the intensity and duration of a disease-favorable microclimate within the canopy can lessen the disease's severity. Factors that may influence the microclimate include row spacing and orientation, nitrogen fertilizers, and cultivar selection. Studies on row spacing in sunflower crops consistently show that *Sclerotinia* wilt/basal rot incidence is lower in crops with wide row widths than those planted in narrow rows. Consequently, the management goal is to space rows at the distance that will maintain plant densities for maximum yield while providing for adequate room to facilitate air movement to reduce high-moisture microclimates within the canopy. Because infection by ascospores of *S. sclerotiorum* and *S. trifoliorum* requires an extended period of free moisture, orienting rows parallel to the direction of the prevailing winds also may be of some value in quickly drying the canopy after a rain or irrigation event. In addition, to avoid dense crop canopies, applied nitrogen should not exceed the optimal rate for a particular crop. Lastly, when choices are available, cultivars that mature early and have a more upright, as opposed to a vining (prostrate), growth habit can provide *avoidance* or *escape* resistance, generally resulting in less disease (Rashid 2003, Turkington et al. 2011, Link and Johnson 2012). Deep burial of sclerotia prevents them from producing apothecia. One must avoid bringing these buried sclerotia to the surface in following seasons. Once they return to the soil surface and are still viable, they can again cause disease. Selected nonhost crops in rotation with maize will reduce inoculum.

Biological Control

Sclerotia of *S. sclerotiorum* are subject to attack by soil microorganisms such as *Coniothyrium minitans*, *Talaromyces flavus* (teleomorph of *Penicillium vermiculatum*), *Sporidesmium sclerotivorum*, *Trichoderma viride* (Ashofteh et al. 2009, Link and Johnson 2012, Tozlu and Demirci 2011), *T. harzianum* (Singh et al. 2004), *Bacillus* sp. (Yu et al. 2006), *P. fluorescens* (Behboudi et al. 2005), and certain isolates of *Actinomycetes* (Baniasadi et al. 2009). Among these antagonists, only *Coniothyrium minitans* and *Trichoderma* spp. have been practically used for biological control of the sunflower wilt caused by *S. sclerotiorum*. In the fields effectively, it appears that secretion of β-1, 3-glucanase from *C. minitans* degrades and lyses sclerotial tissues. *C. minitans* will produce hundreds of pycnidia on the surface of a colonized sclerotium giving it the aspect of a spiny, irregular surface. Usually, few hyphal threads will grow out of an infected sclerotium. This mycoparasite will spread as conidia in the soil. *C. minitans* has a good saprophytic ability and can grow on plant residues or be easily cultured on artificial media. *C. minitans* has been released as a commercial product for suppression of the wilt phase of the disease. In practice, dried spores of this antagonist are sprayed either onto pathogen-infested crop debris at the end of a season or onto the soil surface before planting, and the disease control is economical. Use of micronutrient zinc solely or in combination with molybdenum improves the biocontrol activity of *P. fluorescens* strain UPPF 61 (Ashofteh et al. 2009, Heidari-Tajabadi et al. 2011). Another biocontrol agent (Agate-25K) based on *Pseudomonas chlororaphis* is reported to be effective in the control of *Sclerotinia* disease of sunflower in Russia (Vinokurova 2000). The head rot phase of the disease has been successfully controlled by field testing of honeybee (*Apis mellifera*)–dispersed *Trichoderma* formulation (a mixture of six isolates of *Trichoderma* including *T. koningii*, *T. aureoviride*, and *T. longibrachiatum*) containing *Trichoderma* conidia and viable hyphal fragments, industrial talc, and milled corn kernels in Argentina (Escande et al. 2002). An isolate of *Epicoccum purpurascens* (*E. nigrum*) well adapted to the fluctuating conditions typical of natural environments could contribute to achieving an acceptable level of control of head rot (Pieckenstain et al. 2001). Interestingly under Argentina conditions, the microorganisms, particularly the fungal flora that colonize florets of *Sclerotinia*-tolerant sunflower varieties, play a part in an indirect mechanism that protects flowers from ascospore germination and pathogen growth (Rodriguez et al. 2001). Spontaneously occurring hypovirulence in the tan sclerotial isolate S10 of *S. sclerotiorum* from sunflower in Manitoba, Canada, has been characterized, and the preliminary in vitro transmission test indicated that the hypovirulence in the hypovirulent isolate is transmissible, but double-stranded ribonucleic acids (dsRNAs) have not been detected in hypovirulent and virulent isolates derived from S10. The existence of dsRNA-free hypovirulence in S10 progenies suggests that another hypovirulence mechanism may exist in *S. sclerotiorum* (Li et al. 2003). There is, however, a great potential of making use of this typical phenomenon of hypovirulence in the biological control of *Sclerotinia* diseases of sunflower.

Antifungal protein, trypsin inhibitor (serin proteinases), is a potent antifungal compound associated with sunflower seeds, can completely inhibit the germination of *S. sclerotiorum* ascospores at a concentration of 14 μm/mL indicating the possibility of its use in disease management (Mendieta et al. 2004).

CHARCOAL ROT

Symptoms

The most obvious and common symptom of the disease, under field conditions, is the sudden wilting of plants, which usually appears after pollination, though such plants may have become infected very early in the season (Figure 6.9). Symptoms first observed in plants approaching physiological maturity consist of silvery gray lesions girdling the stem at the soil line, reduced head diameter, and premature plant death (Gulya et al. 2010, Mahmoud and Budak 2011). Pith in the lower stem is completely absent or compressed into horizontal layers. Black spherical microsclerotia are observed in the pith area of the lower stem, underneath the epidermis, and on the exterior of the taproot (Figure 6.10). The pathogen generally affects the fibrovascular system of the roots and basal internodes and impedes the transport of nutrients and water to the upper parts of plants. Progressive wilting, premature aging, loss of vigor, and reduced yield are characteristic features of *M. phaseolina* infection. The internal stem shows a shredded appearance. Later, the vascular bundles become covered with small black flecks or microsclerotia of the fungus.

FIGURE 6.9 Charcoal rot–affected sunflower plants under field conditions. (Courtesy of Dr. Chander Rao and Dr. Varaprasad, DOR, Hyderabad, India.)

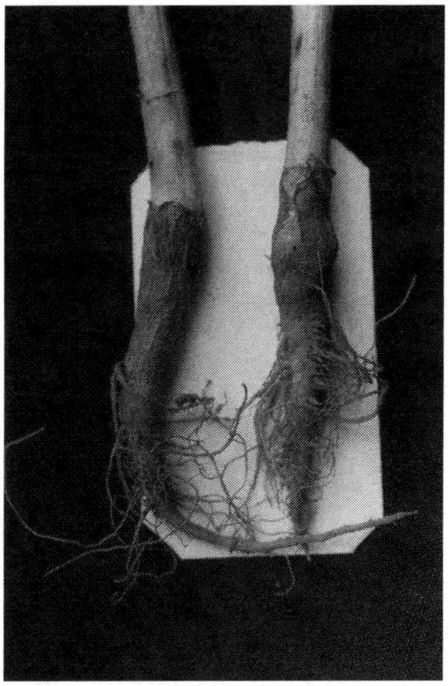

FIGURE 6.10 Charcoal rot–affected root of sunflower (left) in comparison to healthy root (right). (Courtesy of Dr. Chander Rao and Dr. Varaprasad, DOR, Hyderabad, India.)

Geographical Distribution and Losses

The charcoal rot of sunflower is widely distributed throughout tropical, subtropical, and warm temperate regions. It is widespread throughout Latin America, Eastern and Southern Africa, Egypt, West Asia, Middle East including Iran and Turkey (Habib et al. 2007, Mahmoud and Budak 2011, Ijaj et al. 2012), and South Asia, more particularly in Pakistan (Khan 2007). With change in climate, the diseases are also reported to occur in the otherwise relatively cooler regions of the United States (Gulya et al. 2002, Ullah et al. 2011, Weems et al. 2011) and Europe (Sarova et al. 2003, Bokor 2007, Veverka et al. 2008, Csondes et al. 2012). Crop loss estimates are available to the extent of 64% in the Krasnodar region of Russia, 46% in India (Kolte 1985), and 90% in Pakistan (Khan 2007). Under favorable conditions, total failure of the crop in specific areas has been recorded (Khan 2007, Ijaz et al. 2013). The overall yield losses in all varieties at flowering, ripening, and sowing stages are reported to be in the range of 7%–45%, 6%–41%, and 5%–37%, respectively, in Pakistan (Wagan et al. 2004). It is thus evident that continuous increasing trend of charcoal rot is alarming for farmers and authorities in sunflower business not only in Pakistan (Khan et al. 2003) but also in neighboring Iran (Rafiei et al. 2013). Significant decrease in yield is reported with increasing population density of the pathogen. Decrease in seed yield is reported to be 41%, 62%, and 79% at low, moderate, and high pathogen densities, respectively, in Pakistan (Khan et al. 2005a). *M. phaseolina* grows well on sunflower seeds and has been shown to increase the content of oil and free fatty acids in the seeds, as well as discoloration of the oil.

Pathogen: *Macrophomina phaseolina* (Tassi) Goid

Classification
Kingdom: Fungi
Phylum: Ascomycota
Class, Dothideomycetes
Subclass: Incertae sedis
Order: Botryosphaeriales

Family: Botryosphaeriaceae
Genus: *Macrophomina*
Species: *phaseolina* (Tassi) Goid

The details of morphology and culture characteristics of the pathogen are described in Chapter 2 under peanut diseases. High levels of pathogenic variability and genetic diversity have been observed between *M. phaseolina* isolates from different geographical origins or even different hosts after characterization with different markers (RAPDs, RFLPs, and AFLPs) (Almeida et al. 2003, Tancic et al. 2012), although isolates from the same species and same location had related pathotypes. Chlorate-sensitive and chlorate-resistant types of isolates of *M. phaseolina* are known (Mohmmad et al. 2001, Aboshosha et al. 2007), and variation among isolates in pathogenicity is evident (Khan et al. 2005b, Csondes et al. 2010).

Genetic diversity of *M. phaseolina* from Hungary indicates the coexistence of different haplotypes in such country. There appears to be a geographical dominance of a given haplotype and closer genetic relationship might exist between spatially distinct haplotypes (Csondes et al. 2012). Significant pathogenic and genetic variability has been observed within the Iranian isolates obtained from sunflower (Rayatpanah et al. 2012a,b).

Epidemiology and Disease Cycle

In addition to its survival through sclerotia in soil or in the form of sclerotia carried in crop residues, it is also reported to be seed borne in sunflowers (Csondes 2011). However, the pathogen is less aggressive in the preemergence phase, but more serious in the postemergence phase and later stages of growth (Arafa et al. 2000).

M. phaseolina forms appressoria on the epidermis of sunflower. They may aid in both mechanical and chemical penetration, which is direct. Penetration of the adult stem is by mass action of hyphae, which is intra- and intercellular. Initially, infection is restricted to the root epidermal cells and cortical cells, cell configuration including organelles become distorted, and intercellular spaces are occupied by the hyphae, which appear amorphous, with intracellular invasion occurring later (Naz and Ashraf 2006). The incubation period appears to be 6–10 days in plants inoculated with sclerotial suspension and 3–5 days with pycnidiospores.

The role played by various pectolytic and some cellulolytic enzymes in the infection of sunflower plants by the fungus has been well studied. The possibility of production of a non-host-specific toxin by *M. phaseolina* is indicated. The development of necrotic spots on leaves due to *M. phaseolina* is attributed to this toxin. It is reported that the pathogen does not grow beyond the necrotic regions on the inoculated leaves, and the virulence of the different isolates has not been found to be correlated with toxin production in culture. Sunflower plants are most susceptible to charcoal rot at reproductive stage (Suriachandraselvan and Seetharaman 2003). Infection of sunflower seed by *M. phaseolina* takes place when anthesis in the outer quarter of the inflorescence radius is complete. The infection progresses during the seed development stages before the seed reaches maturity. Maximum seed infection occurs when the seed in the outer quarter remains soft. The incidence of the disease increases with increasing salinity level of irrigation water (Fayadh et al. 2011). Moisture stress and higher temperature and periods of drought also favor the development of the disease (Alexandrov and Koteva 2001). At lower temperature (20°C–25°C), seedling mortality due to the disease varies in the range of 8%–67%, whereas at 30°C–35°C, the mortality rate varies from 75% to 100% (Kolte 1985). Disease incidence increases with increase in plant density and sclerotial population in soil (Perez et al. 2002).

Disease Management

Host Plant Resistance

Little is known about the relative resistance of most sunflower varieties. It is demonstrated that sunflower varieties respond differently to artificial as well as to natural infections in the field. This indicates the possibility of control of charcoal rot by breeding for resistance. A number of sunflower genotypes,

SF-87, PTH-1, and SMT (Hafeez and Ahmad 2001), A-43, G-100, G-133, G-17, G-33, G-29, G-10, and G-78 (Khan et al. 2010), CMS 19× R 43, B line 1052/1, and CMS 350/1× R 43 (Dalili et al. 2009), and Giza 102 (Aboshosha et al. 2008, El-Hai et al. 2009), have been found to be resistant in various degrees to charcoal disease, and sunflower genotypes RF81-74*AF80-460/2/1, RF81-1/2*AF81-112, RF81-1/2*AF80-452/2/2, RF81-06/1*AF80-448/1/2, HYSUN33, and AZARGOL show significantly low incidence of charcoal rot (0.33%) under sick soil conditions in Iran (Rafiei et al. 2013). Protein analysis, peroxidase activity, and peroxidase isozyme pattern derived from the sunflower cultivar Giza 102 can be used as genetic markers for host resistance studies in sunflower to *M. phaseolina* (Aboshosha et al. 2008). Sunflower plants regenerated from tolerant callus from hypocotyl explants from a tissue culture medium exhibit more resistance against *M. phaseolina* (Ramadan et al. 2011).

Chemical Control

Seed treatment with carbendazim, thophanate methyl, and thiabendazole each at 2.5 g/kg of seed has been reported to be effective to manage sunflower crops from the seed-borne infection and increasing seed germination (Bhutta et al. 2001). Similarly, fenpropimorph (Corbel) is effective in managing the *M. phaseolina* infection as seed treatment (Piven' et al. 2002, 2004). Other measures of chemical control as used for charcoal rot of peanut may be useful for sunflower crop also, and phosphorus as calcium superphosphate has been reported to be effective in decreasing the incidence of the disease. Maximum control of *M. phaseolina* infection has been obtained when sunflower seeds are coated with Na alginate in combination with Ca carbonate and gum arabic followed by Na alginate in combination with Ca carbonate and carboxymethyl cellulose (Muhammad and Shanaz 2012). Spermine (SP) as seed soaking and/or foliar application of K and/or Zn is helpful in reducing the harmful effects of charcoal of sunflower (El-Metwally and Sakr 2010). Seed soaking or foliar spray of antioxidants (citric acid and salicylic acid at 10 mM) and micronutrients (manganese and zinc at 2 g/L) significantly reduces the incidence of charcoal rot (El-Hai et al. 2009).

Cultural Control

The use of clean seed, the application of organic matter, balanced NPK fertilizer application, long rotations with nonhost crops, avoidance of excessively dense plant populations, and sanitation including the burial of debris by hand or by plough have been suggested as cultural practices against charcoal rot (Aleksandrov 2000, Bistrichanov et al. 2000, Alexandrov and Koteva 2001). Soil amended with nursery fertilizers (urea, diammonium phosphate (DAP), and frutan at 0.1%) in combination with seed treatment with gamma rays (60 cobalt) emitting gamma rays for 2 min results in reduced charcoal rot infection (Naheed et al. 2011). There is a possibility of exploiting the allelopathic effect of *Chenopodium* species (*C. album*, *C. murale*, *C. ambro*) against *M. phaseolina* infecting sunflower (Muhammad and Javaid 2007).

Biological Control

Seed coating with antagonist *Trichoderma reesei* and cotton cake or with *T. harzianum* and mustard cake has proved to be effective in protecting sunflower plants from charcoal rot (Muhammad et al. 2010, Muhammad and Zaki 2010, Ullah et al. 2010). Application of *T. viride* (4 g/kg of seed), + 10 kg/ha soil + FYM (12 tons/ha), and neem cake has also been found effective in reducing charcoal rot incidence (Mani and Hepziba 2003, Mani et al. 2005, Sudha and Prabhu 2008, Suthinraj et al. 2008). The highest disease suppression (61%) has been reported with rice straw composted with cow manure and inoculation with *T. harzianum*. Amendment of compost with *T. harzianum* accelerates composting and can improve disease suppression effect (Morsy and El-Korany 2007). A urea, *Rhizobium*, and *T. harzianum* combination gives better biocontrol effect against the pathogen (Siddiqui et al. 2000). Combined use of *Pseudomonas aeruginosa* with sea weeds significantly decreases the infection caused by *M. phaseolina* (Shahnaz et al. 2007). VAM fungus *Scutellospora auriglobosa* is consistently associated with sunflower variety Helico 250 under Pakistan conditions, and this VAM fungus is found to increase the growth of sunflower with reduction in the incidence of charcoal rot (Jalaluddin et al. 2008).

Effect of Plant Extracts

Efficacy of some plant extracts (*Eucalyptus camaldulensis*, *Azadrichta indica*, *Allium sativum*, and *Datura alba*) and plant products against *M. phaseolina* has been experimentally demonstrated (Arshad et al. 2008, Ullah et al. 2007).

STEM NECROSIS DISEASE

Symptoms

The disease is observed at all growth stages starting from seedlings to mature plant. The characteristic field symptoms of the disease include mosaic on leaves that leads to extensive necrosis of leaf lamina, petiole, stem, and floral calyx and complete death of seedlings eventually (Figure 6.11). Early infection either kills the plant or causes severe stunting with malformed head filled with chaffy seeds (Ravi et al. 2001). Necrosis at the bud formation stage makes the capitulum bend and twist resulting into complete failure of seed setting and maturation (Figure 6.12).

Geographical Distribution and Losses

Sunflower necrosis disease (SND) is becoming a potential threat to sunflower cultivation in the Indian subcontinent. The disease was first recorded in parts of Karnataka state in 1997. Since then, the disease has become increasingly important in Andhra Pradesh, Karnataka, Maharashtra, and Tamil Nadu, the four major sunflower-growing states of India, and is a limiting factor in sunflower production; up to 80% of the plants of some open-pollinated varieties and hybrids were affected during the 1999 survey in sunflower-growing areas, and yield losses ranging from 30% to 100% have been reported (Shirshikar 2010). Early-infected plants remain stunted and develop malformed heads with poor or no seed setting, resulting in complete loss of the crop (Papaiah Sardaru et al. 2013). There has been a continuous threat to sunflower production in India due to tobacco streak virus (TSV) epidemics and reduction of over 40% in the yield since 1997, amounting to annual loss of Rs. 76 crores (Jain et al. 2003). The disease is also reported to occur in Australia and the Netherlands (Sharman et al. 2008) and in Iran (Hosseini et al. 2010, 2012).

FIGURE 6.11 Stem necrosis of sunflower caused by the sunflower stem necrosis virus. (Courtesy of Dr. R.K. Jain, IARI, New Delhi, India.)

Pathogen

The disease is caused by a strain related to TSV infecting sunflower: TSV-SF (genus, *Ilarvirus*; family, Bromoviridae). In electron microscopy, *Ilarvirus*-like particles can be detected in crude sap of SND-affected sunflower and *Chenopodium quinoa* plants inoculated with leaf extracts prepared from SND-affected sunflower plants. In addition to several other herbaceous virus indicator plants, groundnut, cowpea, and cotton, which are significant crops in India, become infected. Back transmission to healthy sunflower seedlings with leaf extracts of systemically infected indicator plants results in identical symptoms of SND, hence confirming the ilar-like virus as the causative agent of SND (Ravi et al. 2001, Prasada Rao et al. 2009). Thus, all the experimental data unequivocally prove that the virus causing SND in many sunflower varieties and in many different growing regions in India is a strain of TSV. The association of a tospovirus, antigenically related to groundnut bud necrosis (GBNV) and watermelon silver mottle (WSMV) viruses, with the disease has been reported earlier (Jain at al. 2000, Venkata Subbiah et al. 2000). TSV first described by Johnson (1936) is the type species of the genus *Ilarvirus*, of the family Bromoviridae that includes viruses having tripartite quasi isometric particles of size 27–35 nm. The virus has three nucleoprotein particles designated as RNA-l (3.4 kb), RNA-2 (3.1 kb), and RNA-3 (2.2 kb). RNAs 1–3 are genomic and encodes proteins la (119 kDa), 2a (91 kDa), and 3a (32 kDa), respectively, whereas RNA-4a (0.9 kb) and RNA-4 (1.0 kb) are subgenomic expressed from RNA-2 and RNA-3. RNA 4a encodes 2b (22 kDa) and CPs (28 kDa), respectively. The TSV genome is infectious only in the presence of its CP or RNA-4. None of the SND causing TSV-SF full genomes could be sequenced, but many researchers have sequenced and reported full-length RNA3 that hosts the movement protein and CP gene (Bag et al. 2008).

Transmission

Mechanical/Sap Transmission

The virus can be transmitted by mechanical or sap inoculation from sunflower to sunflower. Sap extracted in 0.05 M phosphate buffer with 0.075 thioglycerol as inhibitor is more efficient in transmitting the virus (Lokesh et al. 2008b, Pankaja et al. 2011). In general, *Ilarvirus* has a wide host range as they are efficiently sap transmissible to many of the host plants belonging to Amaranthaceae, Chenopodiaceae, and Fabaceae. A rapid and efficient sap inoculation method for tobacco streak virus (TSV-SF) has been developed for screening a large number of sunflower genotypes for resistance to the disease (Sundaresha et al. 2012).

Vector Transmission

The major mode of transmission of TSV-SF is by infected pollen, which can spread by wind or carried by thrips, which transport infected pollen on their bodies (Chander Rao and Shanta Laxmi Prasad 2009). The virus–vector specificity relationship is yet to be established for this virus. Pollen and thrips collected from TSV-infected *Parthenium* weed released together show 58.3% and 70% disease incidence at vegetative and flowering stages of the sunflower crop. *Thrips palmi* successfully transmits the virus to sunflower test plants on acquisition access period (AAP) of 2–3 days and inoculation access period (IAP) of 3–5 days (Lokesh et al. 2008b). A single thrip has been found enough to acquire and transmit the virus from an infected to healthy sunflower plant, and it is revealed that the vector *T. palmi* could acquire the virus with an AAP of 3 days from the cotyledonary leaves of an infected sunflower plant, with a resultant 16.67% transmission. Similarly, an IAP of 6 days is necessary for successful transmission of the virus with 13.33% transmission (Pankaja et al. 2010b, 2011).

Seed Transmission

Certain strains of TSV are known to be transmitted in the seed of a range of host species (Prasada Rao et al. 2009). However, transmission of the TSV isolate occurring in India is not spread through the seed (Papaiah Sardaru et al. 2013, Prasada Rao et al. 2009, Bhat et al., 2002a

Pankaja et al. 2010a). Even in the absence of seed transmission, primary inocula of the TSV are provided by secondary hosts and weed hosts prevalent in and around the sunflower fields by the thrip vector.

Diagnosis

Polyclonal antiserum against TSV-SF has been developed for the rapid diagnosis of TSV using the direct antigen coating-enzyme linked immunosobent assay (DAC-ELISA) method (Ramiah et al. 2001a,b). The serological electroblot immunoassay diagnosis method for CP of the sunflower necrosis virus (Bhat et al. 2002a) and an efficient reverse transcription-polymerase chain reaction (Bhat et al. 2002b, Srinivasan and Mathivanan 2011a) have been developed. Recently, Sarovar et al. (2010a) have reported a high-efficiency immunocapture reverse transcription-polymerase chain reaction (IC-RT-PCR) for RNA3 of TSV-SF, and they have also developed a serological and probe-based blotting technique for the detection of TSV-infected sunflower plants (Pankaja et al. 2010a, Sarovar et al. 2010b). The serological relationship has been confirmed by Western blot analysis and immunoelectron microscopy (IEM) decoration assays using sunflower necrosis virus (SNV) and TSV antisera in reciprocal tests. In RT-PCR, using oligonucleotide primers deduced from conserved sequences within TSV RNA 3 and flanking the entire CP region, an approximately 1000 bp dsDNA fragment could be amplified from SNV-infected sunflowers. A sequence analysis of cloned sunflower necrosis virus (SNV) PCR fragments revealed nucleotide identities of approximately 90% with TSV RNA 3 and a CP amino acid homology between SNV and TSV of more than 90%.

Epidemiology, Host Range, and Disease Cycle

The virus survives throughout the year on several weeds, namely, *Parthenium hysterophorus*, *Tridax procumbens*, *Phyllanthus* sp., *Euphorbia geniculata*, and *Digera arvensis*. A total of 12 weeds, namely, *D. arvensis*, *A. aspera*, *Lagasca mollis*, *P. hysterophorus*, *A. hispidum*, *A. conyzoides*, *C. bengalensis*, *E. geniculata*, *Phyllanthus niruri*, *Malvastrum coromandelianum*, *Abutilon indicum*, and *Physalis minima*, have been found to be infected with the natural infection of TSV. Of these, *Parthenium* is the most widely distributed and is a symptomless carrier of TSV and produces several flushes of flowers during its life cycle ensuring continuous supply of TSV-infected pollen. It hosts the virus as well as thrips and produces copious pollen throughout the season and acts as a primary source of inocula initiating and sustaining the TSV infection during a crop season. Besides, thrips colonizing flowers of these plants can become externally contaminated with pollen and movement of these thrips to new hosts results in introduction of the virus into fields. Windblown pollen of *Parthenium* contaminates the leaves and thrips arriving independently may well contribute to infection. Epidemiological studies on SND indicated the positive correlation between the thrips population and the weather parameters, namely, maximum and minimum temperature and sunshine and dry spells, whereas negative correlation was observed with rainfall and relative humidity. Disease incidence is positively correlated with thrips population and minimum temperature, relative humidity, and rainfall. However, negative correlation with maximum temperature can be observed (Upendhar et al. 2006, 2009). The disease incidence is higher in kharif and summer seasons, whereas it is low in rabi season. The sunflower cultivars sown during July and August show high necrosis incidence compared to postrainy season, that is, September onward (Shishikar 2003).

Disease Management

Host Plant Resistance

The most economical and convenient way to manage TSV is to grow resistant varieties. So far, complete resistant varieties/hybrids are not available in sunflower. The sap inoculation technique has been optimized for large-scale screening of sunflower genotypes against SND.

Systematic studies have been undertaken for the identification of reliable sources of resistance to SND in wild sunflowers (Sujatha 2006). Babu et al. (2007) screened 30 hybrids along with their parents against SND under natural conditions, using a 0–4 scale. Fourteen hybrids (CMS 378A × RHA 265, CMS 378A × DSI 218, CMS 378A × RHA 344, CMS 234A × RHA 265, CMS 234A × RHA 271, CMS 234A × RHA 344, CMS 234A × RHA 345, CMS 234A × RHA 346, CMS 7-1A × RHA 345, DCMS 41 × RHA 274, DCMS 41 × SF 216, DCMS 42 × RHS 273, DCMS 42 × RHA 859, and DCMS 43 × DSI 218) and two parents (CMS 378 A and CMS 234A) recorded resistant reaction. In general, hybrids indicated better tolerance than the populations and inbreds. Among the 96 genotypes screened, only 8 (RHA 284, RHA 5D-1, RHA 265, RHA 859, RHA 297, RHA 365, CR-1, and R-214-NBR) have not been found to be infected by the disease (Ajith Prasad 2004).

Transgenic Approach

Pradeep et al. (2012) amplified, cloned, and sequenced the CP gene of TSV from sunflower (*H. annuus* L.). In their study, a 421 bp fragment of the TSV CP gene could be amplified and gene constructs encoding the hairpin RNA (hpRNA) of the TSV CP sequence has been subcloned into the binary vector pART27. This gene construct was then mobilized into the *Agrobacterium tumefaciens* strain LBA4404 via triparental mating using pRK2013 as a helper. Sunflower (cv. Co 4) and tobacco (cv. Petit Havana) plants were transformed with *A. tumefaciens* strain LBA4404 harboring the hpRNA cassette, and in vitro selection was performed with kanamycin. The integration of the transgene into the genome of the transgenic lines was confirmed by PCR analysis. Infectivity assays with TSV by mechanical sap inoculation demonstrated that both the sunflower and tobacco transgenic lines exhibited resistance to TSV infection and accumulated lower levels of TSV compared with nontransformed controls (Papaiah Sardaru et al. 2013).

Cultural Control

Removal of virus sources especially weeds that germinated with early rains, in fallow lands, on road sides, and on field bunds helps in reducing secondary inoculum thereby reducing the TSV incidence. Moreover, sunflower and groundnut should not be grown side by side or at least synchronization of the flowering period of sunflower with groundnut crop should be avoided as sunflower crop provides infective pollen inoculum with TSV. Similarly, removal of early-infected sunflower will not reduce disease incidence as early-infected sunflower does not produce flowers. TSV-susceptible crops like marigold and chrysanthemum should not be grown adjacent to sunflower fields. Natural barriers such as tall grasses in the field protected the adjacent crops from the disease. The tall grasses might obstruct not only wind-borne-infected pollen from outside weeds but also wind-borne thrips. Sowing 7–11 rows of fast-growing cereals (pearl millet, sorghum, or maize) as border crop around fields that obstruct the movement of thrips from landing on crop plants were found to reduce disease incidence in sunflower (Chander Rao et al. 2002, Basappa et al. 2005, Lokesh et al. 2008c). Mesta et al. (2004) reported that the use of border crop-like sorghum reduced the incidence of SND from 18% to 37%. Bare patches in the field attract thrips landing. Optimum plant population discourages thrips landing on the sunflower crop indicating that maintenance of optimum plant population is one of the options for the management of TSV infection (Papaiah Sardaru et al. 2013). The date of sowing of crops mainly depends on rainfall pattern and distribution (Lokesh et al. 2008d). Shirshikar (2003) opined that the incidence of SND could be minimized if sunflower is sown in the postrainy season, that is, from September onward (Shirshikar 2003, Upendhar et al. 2006, 2009). Intercropping with red gram or castor is helpful in reducing disease intensity compared to monocropping of sunflower (Sreekanth et al. 2004).

Chemical Control

Seed treatment with imidacloprid at 5 g/kg seed and imidacloprid (0.5%) spray reduces disease incidence with higher yield compared with other treatments (Lokesh et al. 2008c). Management trial for SND at All India Coordinated Research Project (AICRP) on oilseeds revealed that seed treatment with either imidacloprid at 5 g/kg seed or thiomethoxam at 4 g/kg seed followed

FIGURE 6.12 Stem necrosis of sunflower: Necrotic streaks on stem. (Courtesy of Dr. Varaprasad and Chander Rao, DOR, Hyderabad, India.)

by two sprays at 30 and 45 days found to reduce necrosis disease and increase seed yield significantly over untreated control (Shirshikar et al. 2009, Shirshikar 2010).

Antiviral Compounds

The use of various antiviral materials such as *Prosopis*, goat milk, and *Bougainvillea* in combinations has been used to induce resistance in sunflower against TSV-SF (Lavanya et al. 2009). Among them, *Bougainvillea spectabilis* with goat milk, *Prosopis chilensis* with goat milk, *B.spectabilis* alone, and *P. chilensis* alone are found highly effective in inducing resistance in sunflower against SND. The combinations of treatments that involve plant products with goat milk are reported to be more effective than the individual ones. Significantly enhanced PR proteins like β-1,3-glucanase and oxidative enzymes like peroxidase, polyphenol oxidase, and PAL have been observed in sunflower using previously mentioned antiviral materials.

Biological Control

Plant growth–promoting microbial consortia (PGPMC)–mediated biological management of SND under field conditions has been experimented (Srinivasan et al. 2009, Srinivasan and Mathivanan 2011b). Powder and liquid formulations of two PGPMCs (PGPMC-1, consisting of *Bacillus licheniformis* strain ML2501 + *Bacillus* sp. strain MML2551 + *Pseudomonas aeruginosa* strain MML2212 + *Streptomyces fradiae* strain MML1042; PGPMC-2, consisting of *B. licheniformis* MML2501 + *Bacillus* sp. MML2551 + *P. aeruginosa* MML2212) when evaluated along with farmers' practice (imidacloprid + mancozeb) in farmers' fields, significant disease reduction, increase of seed germination, plant height, and yield parameters have been recorded with an additional seed yield of 840 kg/ha, an additional income of Rs. 10,920/ha with a benefit–cost ratio of 6:1.

OTHER SUNFLOWER DISEASES

Some other diseases of potential importance are briefly given in Table 6.1.

TABLE 6.1
Other Sunflower Diseases of Potential Importance

Disease	Pathogen/Causal Agent	Geographical Distribution	Significant Disease Description	References
Black stem	*Phoma macdonaldii* Boerem.	Argentina, Bulgaria, China	Attacks roots and collars of the plants, resulting in early death. Penetration of the fungus into the root occurs through natural fissures or through the epidermis and appears to be similar in both resistant and susceptible lines, but the colonization rate of the stele is reduced in partially resistant line. *P. macdonaldii* and *Phomopsis helianthi* (*Diaporthe helianthi*) coexist in Argentina in sunflower, and frequently on the same plant, causing overlaying spots on the petiole insertion, with mixed colors.	Al-Fadil et al. (2009), Bistrichanov et al. (2000), Falico de Alcaraz et al. (2000), Frei (2010), Wu et al. (2012)
Head rot	*Rhizopus stolonifer* (Ehren. Fr) Lind.	Russia, Turkey, United States	Mechanical or physical damage on the back of the sunflower head results in head rot disease and significantly reduces the seed yield.	Yldrm et al. (2010), Kolte (1985), Pandey and Saharan (2005)
Head rot	*Rhizopus oryzae* Went & Prinsen Gernling	Australia, Egypt, United States	This is the first report of *R. oryzae* causing head rot on sunflower in New Mexico. Heads were brown to dark brown with discoloration extending down the sepals and peduncles into the stems. The basal parts of the heads were shredded and had grayish, fluffy mycelial mats visible in the lumen, and kernels were mostly seedless.	Sanogo et al. (2010)
Stem canker or gray stem spot	*Diaporthe helianthi* Munt. (anamorph *Phomopsis helianthi* Munt.; Cverk et al.)	Bulgaria, Croatia, Serbia, Russia	Dangerous quarantine disease causes drastic reductions in yield and oil content in the sunflower (*Helianthus annuus* L.) crop. The use of disease-tolerant genotypes and fungicides is the basis of current disease control. However, there is considerable genetic variability of the pathogen that could lead to the occurrence of new strains, which could be more aggressive or more resistant to chemical control.	Bistrichanov et al. (2000), Vasyutin et al. (2003), Vrandecic and Jurkovic (2008), Debaeke and Estragnat (2003, 2009), Says-Lesage et al. (2002)
Gray spot	*Septoria helianthi* Ellis & Kellerman	Brazil, Bulgaria, Pakistan, Serbia	Dissemination of *S. helianthi* conidia is the occurrence of precipitation on the leaves, causing splashing leading conidia from the lower leaves to the upper leaves of the plants. There is a relationship of the severity of occurrence of septoria leaf spot with the occurrence of rainfall and supplementary irrigations in the sunflower crop.	Bistrichanov et al. (2000), Masirevic and Jasnic (2006), Hamid and Jalaluddin (2007), Loose et al. (2012)
Stem black spot	*Phoma oleracea* var. *helianthi-tuberosi* Sacc. (= *Leptosphaeria indquistii* Frezzi)	Argentina	At R6 and R7, nitrogen fertilization significantly increases disease incidence (31% and 12%, respectively) and stem spot number (43% and 22%, respectively). However, the interaction between genotype and nitrogen fertilization appears to be not significant.	Velazquez and Formento (2000)

(*Continued*)

TABLE 6.1 (*Continued*)
Other Sunflower Diseases of Potential Importance

Disease	Pathogen/Causal Agent	Geographical Distribution	Significant Disease Description	References
White blister rust	*Pustula obtusata* (syn. *Pustula tragopogonis*, *Albugo tragopogonis*, *Pustula helianthicola*)	Europe (southwest of France, Belgium) and other parts of the world including Australia, North and South America, South Africa, and China with losses as much as 70%–80%	Spraying the crop with of 64% Sandofan WP (oxadixyl) or metalaxyl fungicide is effective in controlling the disease.	Rost and Thines (2012), Chen et al. (2008), Crepel et al. (2006)
Gray mold	*Botrytis cinerea* Pers. Ex. Fr.	Bulgaria, Japan, Romania, Russia	Sumilex 50 WP (procymidone) gives better protection against *B. cinerea*, and systemic acquired resistance (SAR) is mediated in sunflower by abiotic inducers such as salicylic acid (SA), benzo-(1,2,3)-thiadiazole-7-carbothioic S-methyl ester (BTH), 2,6-dichloroisonicotinic acid (INA), or EDTA for the control of *Botrytis* infection.	Bistrichanov et al. (2000), Tomioka and Sato (2011), Eva (2004), Dmitriev et al. (2003)
Sclerotium wilt	*Sclerotium rolfsii* Sacc.	Most tropical countries and warmer regions in temperate zones	For root and stem rots, biocontrol is effective; *Bradyrhizobium* sp., *Rhizobium* sp., *Trichoderma harzianum*, *T. pseudokoningii*, *T. polysporum*, *T. virens*, and fluorescent *Pseudomonas* are the most potentially useful biocontrol agents.	Bhatia et al. (2005), Cilliers et al. (2002), Fouzia and Shahzad (2011)
Verticillium wilt	*Verticillium albo-atrum* Reinke & Berth, *V. dahliae* Kleb., *V. sulphurellum*	Argentina, Serbia, Canada, United States, China	The phenotypic changes induced by *V. dahliae*'s necrosis and ethylene-inducing proteins (VdNEP) indicate that this protein acts both as a defense elicitor and a pathogenicity factor in sunflower. Rotations especially where sunflower follows potato should be avoided in disease management.	Alkher et al. (2009), Creus et al. (2007), Fu et al. (2012), Gulya et al. (2012), Pandey and Saharan (2005)
Powdery mildew	*Erysiphe cichoracearum* (*Golovinomyces cichoracearum* var. *cichoracearum*; *Podosphaera xanthii* and *Leveillula taurica* are also the causes of powdery mildew in Taiwan)	Brazil, India, Serbia, Taiwan	In general, the disease is more severe at the full flowering and preflowering stages during winter. The highest percent disease control is obtained by two sprays of fungicides difenoconazole (88.64%), penconazole (87.59%), and propiconazole (85.91%).	Akhileshwari et al. (2012), Chen et al. (2008), Almeida et al. (2008)

(Continued)

TABLE 6.1 (*Continued*)
Other Sunflower Diseases of Potential Importance

Disease	Pathogen/Causal Agent	Geographical Distribution	Significant Disease Description	References
Root and collar rot	*Rhizoctonia solani* Kuhn AG-IV (*Thanatephorus cucumeris*)	India	Symptoms include leaf yellowing and wilting, root rot, and death of plants; mycelium can be observed on necrotic crowns. Reported to be seed borne also in sunflower. *Trichoderma harzianum*–mediated biocontrol may be related to alleviating *R. solani*–induced oxidative stress in sunflower.	Lakshmidevi et al. (2010), Raj et al. (2008), Singh et al. (2011), Srinivasan and Visalakchi (2010)
Crown gall	*Agrobacterium tumefaciens* (Smith & Town) Conn.	Almost all sunflower-growing regions	Though the disease appears to be important, it is less commonly observed under natural conditions. It can be easily produced on sunflower by artificial inoculations.	Binboga-Meral (2007)
Bacterial stalk and head rot	*Pectobacterium atrosepticum*	Turkey	Dark and water-soaked necrotic areas develop on stems and heads, and bacteria appear as droplets and ooze from the diseased tissues.	Bastas et al. (2009)
Bacterial stalk rot	*Erwinia chrysanthemi* and *E. carotovora* subsp. *carotovora*	Taiwan	The disease mainly appears on the stem and cause water-soaked symptoms. In some severe cases, the stem pith disintegrates and shows hollow stem symptoms. "Moon bright" and "Sunbright" cultivars show better resistance when inoculated with Ech. Tetracycline has been proved to be the most effective in controlling the disease.	Hseu et al. (2004)
Angular leaf spot/bacterial blight	*Pseudomonas syringae* pv. *helianthi* (Kawamura) Young, Dye, & Wilkie	Brazil, Mexico, Serbia	The bacterium is disseminated and transmitted through the seeds from infected plants. Diseased plants are obtained from either naturally infected seed or artificially inoculated seed, the latter being more effective. Bacteria also inhibit seed germination by 50%. It is concluded that *P. syringae* pv. *helianthi* infects sunflower seed from infected plants and that this is one mechanism of the disease transmission.	Balaz and Popovic (2006), Borba Filho et al. (2007), Maselli et al. (2000, 2002)
Apical chlorosis	*Pseudomonas syringae* pv. *tagetis* (Hellmers) Young, Dye & Wilkie	Mexico, United States	Tagetitoxin-affected leaves are so severely chlorotic that they become white. Tagetitoxin-affected leaves have 99% less chlorophyll per unit fresh mass than healthy plant leaves. Electron microscopy of sections of chlorotic (toxin-affected) sunflower leaf cells reveals that leaf palisade and spongy parenchyma cells possess few chloroplasts and that those present contain disorganized thylakoids and grana and no apparent starch grains. Rates of photosynthetic CO_2 assimilation per unit leaf area in toxin-affected leaves are significantly lower than in healthy leaves.	Kong et al. (2004), Robinson et al. (2004)

(*Continued*)

TABLE 6.1 (*Continued*)
Other Sunflower Diseases of Potential Importance

Disease	Pathogen/Causal Agent	Geographical Distribution	Significant Disease Description	References
Sunflower mosaic	Sunflower mosaic virus (SuMV), a distinct species within the family Potyviridae	India, United States	Most of the seeds from the infected plants are grayish black, whereas those from healthy plants are black. The virus is transmitted by *Myzus persicae* and *Capitphorus elaegni* (*Capitophorus elaeagni*) and also is seed borne in at least one sunflower cultivar. Phylogenetic analysis of the coat protein amino acid sequence revealed that SuMV is most closely related to tobacco etch virus (TEV). There is a significant reduction in *Alternaria* infection when the virus is inoculated 20 days prior to fungal infection.	Gulya et al. (2002), Bharadwaj and Mohan (2005)
Chlorotic mottle	Sunflower chlorotic mosaic virus (SuCMoV)	Argentina	It causes systemic chlorotic mottling symptoms; growth reductions and severe yield losses. Chlorotic symptom development induced by SuCMoV infection is accompanied by changes in different redox-related metabolites and transcripts. Oxidative damage is expressed after symptom development in this host–pathogen combination.	Rodriguez et al. (2012), Lenardon (2008), Arias et al. (2005)
Root knot	*Meloidogyne incognita* (Kofoid & White) Chitwood	Egypt, India, Pakistan, United States	Sunflower plants show yellowing, stunting, and death of plants in the field resulting in 16.44% yield losses. The combined infection with *Meloidogyne incognita* plus any of the fungi such as *R. solani*, *M. phaseolina*, and *F. solani* on sunflower results in significant reduction in the number of root galls and nematode egg masses. Rugby 10 G (cadusafos) and Furadan (carbofuran) are the most effective nematicides useful in the root knot disease management. Soil amendments with certain organic plant materials and biological control agents can be useful in controlling *Meloidogyne incognita* infection.	Rehman et al. (2006), Mokbel et al. (2007), Prasad et al. (2001), Mohammad et al. (2001)
Reniform nematode	*Rotylenchulus reniformis* Linford & Oliveira	India, Pakistan	The minimum damaging threshold of *Rotylenchulus reniformis* on sunflower is 1000 nematodes/plant. Some botanicals (furfural, sugarcane by-product, and extracts of *Calotropis procera* leaf and root) have been found to be effective in minimizing the damage.	Prasad et al. (2001), Singh and Prasad (2010), Ismail and Mohamed (2007)

REFERENCES

Aboshosha, S.S., S.I.A. Alla, A.E. El-Korany, and E. El-Argawy. 2007. Characterization of *Macrophomina phaseolina* isolates affecting sunflower growth in El-Behera Governorate, Egypt. *Int. J. Agric. Biol.* 9: 807–815.

Aboshosha, S.S., S.I.A. Alla, A.E. El-Korany, and E. El-Argawy. 2008. Protein analysis and peroxidase isozymes as molecular markers for resistance and susceptibility of sunflower to *Macrophomina phaseolina. Int. J. Agric. Biol.* 10: 28–34.

Ahamad, S., M. Srivastava, and S.S. Chauhan. 2000. Effect of seed-borne *Alternaria helianthi* on seed germination, seedling infection and management causing blight of sunflower. *Ann. Plant Prot. Sci.* 8: 262–264.

Ahmed, S., D.T. de Labrouhe, and F. Delmotte. 2012. Emerging virulence arising from hybridisation facilitated by multiple introductions of the sunflower downy mildew pathogen *Plasmopara halstedii. Fungal Genet. Biol.* 49: 847.

AjithPrasad, H.N. 2004. Transmission and serological diagnosis of sunflower necrosis virus from various sources and screening for resistance. M.Sc. (Agri.) thesis, University of Agricultural Science, Bangalore, India.

Akhileshwari, S.V., Y.S. Amaresh, M.K. Naik, V. Kantharaju, I. Shankergoud, and M.V. Ravi. 2012. Field evaluation of fungicides against powdery mildew of sunflower. *Karnataka J. Agric. Sci.* 25: 278–280.

Aleksandrov, V. 2000. Influence of sunflower sowing date and sowing density on attack by *Sclerotium bataticola. Rasteniev"dni Nauki* 37: 948–951.

Alexandrov, V. and M. Angelova. 2004. Influence of meteorological conditions and agronomical practices upon attack of sunflower by *Sclerotinia* white rot (*Sclerotinia sclerotiorum* Lib.). *Bulgarian J. Agric. Sci.* 10: 185–189.

Alexandrov, V. and V. Koteva. 2001. Attack on sunflower by charcoal rot (*Sclerotium bataticola* T.) under the influence of climate and mineral fertilization. *Bulgarian J. Agric. Sci.* 7: 271–274.

Al-Fadil, T.A., A. Jauneau, Y. Martinez, M. Rickauer, and G. Dechamp-Guillaume. 2009. Characterisation of sunflower root colonisation by *Phoma macdonaldii. Eur. J. Plant Pathol.* 124: 93–103.

Alkher, H., A. El-Hadrami, K.Y. Rashid, L.R. Adam, and F. Daayf. 2009. Cross-pathogenicity of *Verticillium dahliae* between potato and sunflower. *Eur. J. Plant Pathol.* 124: 505–519.

Almeida, A.M.R., R.V. Abdelnoor, C.A.A. Aris et al. 2003. Genotypic diversity among isolates of *Macrophomina phaseolina* revealed by RAPD. *Fitopatol. Bras.* 28: 279–282.

Almeida, A.M.R., E. Binneck, F.E. Piuga et al. 2008. Characterization of powdery mildews strains from soybean, bean, sunflower, and weeds in Brazil using rDNA-ITS sequences. *Trop. Plant Pathol.* 33: 20–26.

Alves, J.L., J.H.C. Woudenberg, L.L. Durate, P.W. Crous, and R.W. Barreto. 2013. Reappraisal of the genus *Alternariaster* (*Dothideomycetes*). *Persoonia* 31: 77–85.

Amabile, R.F., C.M. Vasconcelos, and A.C. Gomes. 2002. Severity of alternaria in sunflower crop in the Cerrado region of Federal District, Brazil. *Pesq. Agropec. Bras.* 37: 251–257.

Amaresh, Y.S. and V.B. Nargund. 2000a. Management of *alternaria* blight and rust of sunflower. *Karnataka J. Agric. Sci.* 13: 465–467.

Amaresh, Y.S. and V.B. Nargund. 2000b. Screening of sunflower genotypes against *alternaria* leaf blight and rust. *Karnataka J. Agric. Sci.* 13: 468–469.

Amaresh, Y.S. and V.B. Nargund. 2002. Development of disease prediction models in sunflower rust epidemics. *Indian J. Plant Prot.* 30(2): 105–108.

Amaresh, Y.S. and V.B. Nargund. 2004. Assessment of yield losses due to *Alternaria* leaf blight and rust of sunflower. *J. Mycol. Plant Pathol.* 34: 75–79.

Amaresh, Y.S., V.B. Naragund, K.H. Anahosur, S. Kulkarni, and B.V. Patil. 2003. Influence of sowing dates and weather factors on development of *Alternaria* leaf blight of sunflower. *Karnataka J. Agric. Sci.* 16: 68–71.

Amaresh, Y.S., V.B. Nargund, and B.V. Patil. 2004. Integrated management of Alternaria leaf blight of sunflower, *Helianthus annuus* caused by *Alternaria helianthi. J. Oilseeds Res.* 21: 204–205.

Amaresh, Y.S., V.B. Nargund, and B. Somasekhar. 2000. Use of botanicals and fungitoxicants against *Alternaria helianthi*, the casual agent of sunflower, leaf blight. *Plant Dis. Res.* 15: 140–145.

Anitha, R. and K. Murugesan. 2001. Mechanism of action of *Gliocladium virens* on *Alternaria helianthi. Indian Phytopathol.* 54: 449–452.

Anjana, G., K.R. Kini, H.S. Shetty, and H.S. Prakash. 2007. Differential expression of sunflower peroxidase isoforms and transcripts during necrotrophic interaction with *Alternaria helianthi. Russ. J. Plant Physiol.* 54: 513–517.

Anjana, G., K.R. Kini, H.S. Shetty, and H.S. Prakash. 2008. Changes in peroxidase activity in sunflower during infection by necrotrophic pathogen *Alternaria helianthi. Arch. Phytopathol. Plant Prot.* 41: 586–596.

Anonymous. 2010. Sunflower production guideline. Department of Agriculture, Forestry and Fisheries, Pretoria, Republic of South Africa.

Antonova, T.S., M.V. Iwebor, V.T. Rozhkova, N.M. Araslanova, and V.A. Gavrilova. 2010. Results of the evaluation of sunflower's samples from the VIR collection on resistance to races of downy mildew agent wide spread in Krasnodar region of Russian Federation. In: *Proceedings of the International Symposium on "Sunflower Breeding on Disease Resistance"*, Krasnodar, Russia, June 23–24, 2010, pp. 132–136.

Arafa, M.K.M., M.H.A. Hassan, and M.A. Abdel-Sater. 2000. Sunflower seed discoloration and its relation to seed quality, mycotoxin production, and emergence damping-off. *Assiut J. Agric. Sci.* 31: 231–247.

Arias, M.C., C. Luna, M. Rodriguez, S. Lenardon, and E. Taleisnik. 2005. Sunflower chlorotic mottle virus in compatible interactions with sunflower: ROS generation and antioxidant response. *Eur. Plant Pathol.* 113: 223–232.

Arshad, J., A. Muhammad, and S.S. Hashmi. 2008. Antifungal activity of *Datura alba* against *Macrophomlna phaseollna*. *Pak. J. Phytopathol.* 20: 175–179.

Ashofteh, F., M. Ahmadzadeh, and V.F. Mamaghani. 2009. Effect of mineral components of the medium used to grow biocontrol strain UTPF61 of *Pseudomonas fluorescens* on its antagonistic activity against sclerotinia wilt of sunflower and its survival during and after the formulation process. *J. Plant Pathol.* 91: 607–613.

As-sadi, F., S. Carrere, Q. Gascuel et al. 2011. Transcriptomic analysis of the interaction between *Helianthus annuus* and its obligate parasite *Plasmopara halstedii* shows single nucleotide polymorphisms in CRN sequences. *BMC Genomics* 12: 498.

Baack, E.J., K.D. Whitney, and L.H. Rieseberg. 2005. Hybridization and genome size evolution: Timing and magnitude of nuclear DNA content increases in *Helianthus* homoploid hybrid species. *New Phytol.* 167: 623–630. doi:10.1111/j.1469-8137.2005.01433.x. PMID:15998412.

Babu, S.S., A.N. Reddy, K.R. Reddy et al. 2007. Screening of sunflower genotypes against sunflower necrosis disease under field conditions. *Crop Res.* 33: 223–225.

Bag, S., R.S. Singh, and R.K. Jain. 2008. Further analysis of coat protein gene sequences of Tobacco streak virus isolates from diverse locations and hosts in India. *Indian Phytopathol.* 61: 118–123.

Bailey, D.L. 1923. Sunflower rust. *Minn. Agric. Exp. Stn. Bull.* 16: 31pp.

Balaz, J. and T. Popovic. 2006. Bacterial diseases of sunflower. *Biljni Lekar* (*Plant Doctor*) 34: 347–351.

Baldini, M., M. Turi, G.P. Vischi, Vannozzi, and A.M. Olivieri. 2002. Evaluation of genetic variability for *Sclerotinia sclerotiorum* (Lib.) de Bary resistance in sunflower and utilization of associated molecular markers. *Helia* 25: 177–189.

Baldini, M., M. Vischi, M. Turi et al. 2004. Evaluation of genetic variability for *Sclerotinia sclerotiorum* Lib. de Bary resistance in a F2 population from a cross between susceptible and resistant sunflower. *Helia* 27(40): 159–170.

Ba'n, R., F. Vira'nyi, and H. Komja'ti. 2004. Bebzothiadiazole-induced resistance to *Plasmopara halstedii* (Farl.) Berl. et de Toni in sunflower. *Advances in Downy Mildew Research*, Vol. 2, eds., P. Spencer-Phillips and M. Jeger. Kulwer Academic Publishers, London, UK, pp. 265–273.

Baniasadi, F., G.H.S. Bonjar, A. Baghizadeh et al. 2009. Biological control of *Sclerotinia sclerotiorum*, causal agent of sunflower head and stem rot disease, by use of soil borne Actinomycetes isolates. *Am. J. Agric. Biol. Sci.* 4: 146–151.

Basappa, H. and M. Santha Lakshmi Prasad. 2005. Insect pests and diseases of sunflower and their management, ed., D.M. Hedge. Directorate of Oilseeds Research, Hyderabad, India, 80pp.

Bastas, K.K., H. Hekimhan, S. Maden, and M. Tor. 2009. First report of bacterial stalk and head rot disease caused by *Pectobacterium atrosepticum* on sunflower in Turkey. *Plant Dis.* 93: 1352.

Bazzalo, M.E., I. Bridges, T. Galella et al. 2000. Sclerotinia head rot resistance conferred by wheat oxalate Oxidase gene in transgenic sunflower. In: *The Proceedings of the 15th International Sunflower Conference*, Toulouse, France, June 2000, pp. 60–65.

Becelaere, G. and J.F. van Miller. 2004 Combining ability for resistance to Sclerotinia head rot in sunflower. *Crop Sci.* 44: 1542–1545.

Behboudi, K., A. Sharifi-Tehrani, G.A. Hedjaroude, J. Zad, M. Mohammadi, and H. Rahimian. 2005. Effects of fluorescent pseudomonads on *Sclerotinia sclerotiorum*, the causal agent of sunflower root rot. *Iranian J. Agric. Sci.* 36: 791–803.

Berglund, D.R. 2007. Extension publication A-1331. North Dakota Agricultural Experiment Station, North Dakota State University, Fargo, ND.

Bhardwaj, M. and J. Mohan. 2005. Interaction of sunflower mosaic virus and *Alternaria helianthi* in sunflower. *Plant Arch.* 5: 309–310.

Bhat, A.I., R.K. Jain, N. Anil Kumar, M. Ramiah, and A. Varma. 2002b. Serological and coat protein sequence studies suggest that necrosis disease on sunflower in is caused by a strain of *Tobacco streak virus*. *Arch. Virol.* 147: 651–658.

Bhat, A.I., R.K. Jain, and M. Ramiah. 2002a. Detection of *Tobacco streak virus* from sunflower and other crops by reverse transcription polymerase chain reaction. *Indian Phytopathol.* 55: 216–218.

Bhatia, S., R.C. Dubey, and D.K. Maheshwari. 2005. Enhancement of plant growth and suppression of collar rot of sunflower caused by *Sclerotium rolfsii* through fluorescent *Pseudomonas*. *Indian Phytopathol.* 58: 17–24.

Bhutta, A.R., M.M. Rahber-Bhatti, I. Ahmad, and I. Sultana. 2001. Chemical control of seed-borne fungal pathogens of sunflower. *Helia* 24: 35, 67–72.

Binboga-Meral, U. 2007. Tumor formation in sunflower using A281 strain of *Agrobacterium tumefaciens*. *Tarim Bilimleri Dergisi* 13: 166–168.

Binsfeld, P.C., C. Cerboncini, V. Hahn, I. Grone, and H. Schnabl. 2005. Sclerotinia resistance achieved by asexual gene transfer from wild species to *Helianthus annuus* lines. *Documentos—Embrapa Soja* 261: 9–11.

Bistrichanov, S., V. Alexandrov, and I. Hristov. 2000. Impact of some agrotechnical factors on the diseases' spreading on sunflower in the North Western Bulgaria. *Pochvoznanie, Agrokhimiya i Ekologiya* 35: 3.

Bokor, P. 2007. *Macrophomina phaseolina* causing a charcoal rot of sunflower through Slovakia. *Biologia (Bratislava)* 62: 136–138.

Bolton, D.M., B.P.H.J. Thomma, and B.D. Nelson. 2006. *Sclerotinia sclerotiorum* (Lib.) de Bary: Biology and molecular traits of a cosmopolitan pathogen. *Mol. Plant Pathol.* 7: 1–16.

Borba Filho, A.B., L. Kobayasti, and D. Cassetari Neto. 2007. Evaluation of angular leaf spot and alternaria leaf spot in sunflower on Campo Verde—MT. *Documentos—Embrapa Soja* 292: 145–148.

Brown, J.F., P. Kajornchaiyakul, M.Q. Siddiqui, and S.J. Allen. 1974. Effect of rust on growth and yield of sunflower in Australia. In: *Proceedings of the Sixth International Sunflower Conference*, Bucharest, Romania, p. 639.

Bulos, M., E. Altieri, M.L. Ramos, P. Vergani, and C.A. Sala. 2012. Genome localization of rust resistance genes. In: *Proceedings of the 18th International Sunflower Conference*, Mar del Plata, Argentina, pp. 980–995.

Bulos, M., M.L. Ramos, E. Altieri, and C.A. Sala. 2013. Molecular mapping of a sunflower rust gene from HAR6. *Breed. Sci.* 63: 141–146.

Burke, J.M. and L.H. Rieseberg. 2003. Fitness effects of transgenic disease resistance in sunflowers. *Science (Washington)* 300(5623): 1250.

Caceres, C., F. Castano, R. Rodriguez et al. 2006. Variability of *Sclerotinia* responses in *Helianthus petiolaris*. *Helia* 29(45): 43–48.

Calvet, N.P., M.R.G. Ungaro, and R.F. Oliveira. 2005. Virtual lesion of *Alternaria* blight on sunflower. *Helia* 28: 89–99.

Castaño, F. and A. Giussani, 2006. Variability of daily growth of white rot in sunflowers. *J. Genet. Breed.* 60: 67–70.

Castaño, F., F. Vear, and D.T. de Labrouhe. 2001. The genetics of resistance in sunflower capitula to *Sclerotinia sclerotiorum* measured by mycelium infections combined with ascospore tests. *Euphytica* 122: 373–380.

Chander Rao, S., R.D.V.J. Prasad Rao, H. Singh, and D.M. Hedge. 2002. Information bulletin no. 6—On sunflower necrosis disease and its management. Directorate of Oilseeds Research, Hyderabad, India.

Chander Rao, S. and M. Santha Lakshmi Prasad. 2009. Screening methodology for Alternaria leaf blight and sunflower necrosis disease (SND), Training manual. Directorate of Oil seeds Research, Hyderabad, India, p. 42.

Chattopadhyay, C. 1999. Yield loss attributable to Alternaria blight of sunflower (*Helianthus annuus* L.) in India and some potentially effective control measures. *Int. J. Pest Manag.* 45: 15–21.

Chen, R.S., C. Chu, C.W. Cheng, W.Y. Chen, and J.G. Tsay. 2008 Differentiation of two powdery mildews of sunflower (*Helianthus annuus* L.) by a PCR-mediated method based on ITS sequences. *Eur. J. Plant Pathol.* 121: 1–8.

Chikkodi, S.B. and R.L. Ravikumar. 2000. Influence of pollen selection for *Alternaria helianthi* resistance on the progeny performance against leaf blight in sunflower (*Helianthus annuus* L.). *Sex. Plant Reprod.* 12: 222–226.

Cilliers, A.J., Z.A. Pretorius, and P.S. van Wyk. 2002. Mycelial compatibility groups of *Sclerotium rolfsii* in South Africa. *S. Afr. J. Bot.* 68: 389–392.

Constantinescu, O. and M. Thines. 2010. *Plasmopara halstedii* is absent from Australia and New Zealand. *Pol. Bot. J.* 55: 293–298.

Cosic, J. and J. Postic. 2008. White rot of sunflower (*Sclerotinia sclerotiorum* (Lib.) De Bary). *Glasilo Biljne Zastite* 8: 344–347.

Cotton, P., C. Rascle, and M. Fevre. 2002. Characterization of PG2, an early endoPG produced by *Sclerotinia sclerotiorum*, expressed in yeast. *FEMS Microbiol. Lett.* 213: 239–244.

Covarelli, L. and L. Tosi. 2006. Presence of sunflower downy mildew in an integrated weed control field trial. *J. Phytopathol.* 154: 281–285.

Crepel, C., S. Inghelbrecht, and S.G. Bobev. 2006. First report of white rust caused by *Albugo tragopogonis* on sunflower in Belgium. *Plant Dis.* 90: 379.

Creus, C., M.E. Bazzalo, M. Grondona, F. Andrade, and A.J. Leon. 2007. Disease expression and ecophysiological yield components in sunflower isohybrids with and without *Verticillium dahliae* resistance. *Crop Sci.* 47: 703–710.

Csondes, I. 2011. Mycelial compatibility of Hungarian *Macrophomina phaseolina* isolates. *Acta Agron. Hung.* 59(4): 371–377.

Csondes, I., A. Cseh, J. Taller, and P. Poczai. 2012. Genetic diversity and effect of temperature and pH on the growth of isolates from sunflower fields in Hungary. *Mol. Biol. Rep.* 39: 3259–3269.

Csondes, I., S. Kadlicsko, and R. Gaborjanyi. 2010. Diverse virulence of *Macrophomina phaseolina* isolates of different origin on sunflower and pepper plants. *Novenyvedelem* 46: 453–463.

Cvjetkovic, B. 2008. Sunflower downy mildew (*Plasmopara halstedii* (Farlow) Berl.). [Croatian]. *Glasilo Biljne Zastite* 8: 353–356.

Dalili, S.A., S.V. Alavi, and V. Rameeh. 2009. Response of sunflower new genotypes to charcoal rot disease. *Macrophomina pheseolina*. (Tassi) Goidanch. *Seed Plant* 25: 147–155.

Damodaran, T. and D.M. Hegde. 2007. Oilseeds situation: A statistical compendium. Directorate of Oilseed Research, Hyderabad, India.

Darvishzadeh, A.M., A.K. Masouleh, and Y. Ghosta. 2012. The infection processes of *Sclerotinia sclerotiorum* in basall stem tissue of a susceptible genotype of *Helianthus annuus* L. *Not. Bot. Horti Agrobo.* 40: 143–149.

Darvishzadeh, R. 2012. Association of SSR markers with partial resistance to *Sclerotinia sclerotiorum* isolates in sunflower (*Helianthus annuus* L.). *Aust. J. Crop Sci.* 6: 276–282.

Das, N.D., G.R.M. Sarkar, and N.N. Srivastava. 1998. Studies on the progression of *Alternaria* blight disease caused by *Alternaria helianthi* (Hansf.) Tubaki and Nishihara of sunflower. *Ann. Plant Prot. Sci.* 6: 209–211.

Davar, R., R. Darvishzadeh, A. Majd, Y. Ghosta, and A. Sarrafi. 2010. QTL mapping of partial resistance to basal stem rot in sunflower using recombinant inbred lines. *Phytopathol. Mediterr.* 49: 330–341.

Dawar, V. and V. Jain. 2010. Cell wall degrading enzymes and permeability changes in sunflower (*Helianthus annus*) infected with *Alternaria helianthi*. *Int. J. Agric. Environ. Biotechnol.* 3: 321–325.

Debaeke, P. and A. Estragnat. 2003. A simple model to interpret the effects of sunflower crop management on the occurrence and severity of a major fungal disease: *Phomopsis* stem canker. *Field Crops Res.* 83: 139–155.

Debaeke, P. and A. Estragnat. 2009. Crop canopy indicators for the early prediction of *Phomopsis* stem canker (*Diaporthe helianthi*) in sunflower. *Crop Prot.* 28: 792–801.

Delmotte, F., X. Giresse, S. Richard-Cervera et al. 2008. Single nucleotide polymorphisms reveal multiple introductions into France of *Plasmopara* the plant pathogen causing sunflower downy mildew. *Infect. Genet. Evol.* 8(5): 534–540.

de Romano, A.B. and A.N. Vazquez. 2003. Origin of the Argentine sunflower varieties. *Helia* 26: 127–136.

Dietz, J. 2011. Broad spectrum fungicides in the works. New Fluency Agent, Bayer Crop Science (AgAnnex).

Dmitriev, A., M. Tenaand, and J. Jorrin. 2003. Systemic acquired resistance in sunflower (*Helianthus annuus* L.). *Tsitologiya i Genetika* 37: 9–15.

Durman, S.B., A.B. Menendez, and A.M. Godeas. 2003. Mycelial compatibility groups in Buenos Aires field populations of *Sclerotinia sclerotiorum* (*Sclerotiniaceae*). *Aust. J. Bot.* 51: 421–427.

Durman, S.B., A.B. Menendez, and A.M. Godeas. 2005. Variation in oxalic acid production and mycelial compatibility within field populations of *Sclerotinia sclerotiorum*. *Soil Biol. Biochem.* 37: 2180–2184.

Ekins, M.G., E.A.B. Aitken, and K.C. Goulter. 2005. Identification of *Sclerotinia* species. *Aust. Plant Pathol.* 434: 549–555.

Ekins, M.G., E.A.B. Aitken, and K.C. Goulter. 2007. Aggressiveness among isolates of *Sclerotinia sclerotiorum* from sunflower. *Aust. Plant Pathol.* 36: 580–586.

Ekins, M.G., H.L. Hayden, E.A.B. Aitken, and K.C. Goulter. 2011. Population structure of *Sclerotinia sclerotiorum* on sunflower in Australia. *Aust. Plant Pathol.* 40: 99–108.

El-Deeb, A.A., S.M. Abdallah, A.A. Mosa, and M.M. Ibrahim. 2000. *Sclerotinia* diseases of sunflower in Egypt. *Arab Univ. J. Agric. Sci.* 8: 779–798.

El-Hai, K.M.A., M.A. El-Metwally, S.M. El-Baz, and A.M. Zeid. 2009. The use of antioxidants and microelements for controlling damping-off caused by *Rhizoctonia solani* and charcoal rot caused by *Macrophomina phasoliana* on sunflower. *Plant Pathol. J.* (*Faisalabad*) 8: 79–89.

El-Metwally, M.A. and M.T. Sakr. 2010. A novel strategy for controlling damping-off and charcoal rot diseases of sunflower plants grown under calcareous-saline soil using spermine, potassium and zinc. *Plant Pathol. J.* (*Faisalabad*) 9: 1–13.

Enferadi, S.T., Z. Rabiei, G.P. Vannozzi, and G.A. Akbari. 2011. Shikimate dehydrogenase expression and activity in sunflower genotypes susceptible and resistant to *Sclerotinia sclerotiorum* (Lib.) de Bary. *J. Agric. Sci. Technol.* 13: 943–952.

Escande, A.R., F.S. Laich, and M.V. Pedraza. 2002. Field testing of honeybee-dispersed *Trichoderma* spp. to manage sunflower head rot (*Sclerotinia sclerotiorum*). *Plant Pathol.* 51: 346–351.

Eva, B. 2004. Biological control of pathogens *Sclerotinia sclerotiorum* (Lib.) de Bary and *Botrytis cinerea* (Pers.) from sunflower (*Helianthus annuus* L.) crops. *Cercetari Agronomice in Moldova* 36: 73–80.

Eva, B. and E. Andrei. 2000. The effect of *Sclerotinia sclerotiorum* (Lib.) de Bary attack on the morphophysiological traits in sunflower. *Probleme de Protectia Plantelor* 28: 47–67.

Falico de Alcaraz, L., G. Visintin, M.E. Alcaraz, B. Garcia, and C. Caceres. 2000. Association of *Phomopsis* spp. and *Phoma mcdonaldii* on sunflower stems. *Fitopatologia* 35: 169–175.

Fayadh, M.A., H.J. Al-Tememi, and L.A. Bnein. 2011. Effect of some environmental factors on charcoal rot disease of sunflower caused by *Macrophomina phaseolina* (Tassi) Goid. *Arab J. Plant Prot.* 29: 1–6.

Fouzia, C. and S. Shahzad. 2011. Efficacy and persistence of micobial antagonists against *Sclerotium rolfsii* under field conditions. *Pak. J. Bot.* 43: 2627–2634.

Franchel, J., M.F. Bouzidi, G. Bronner, F. Vear, P. Nicolas, and S. Mouzeyar. 2012. Positional cloning of a candidate gene for resistance to the sunflower downy mildew, *Plasmopara halstedii* race 300. *Theor. Appl. Genet.* 126: 359–367.

Frei, P. 2010. *Phoma macdonaldii* on sunflower; biology and treatment. *Julius-Kuhn-Archiv* 428: 350–351.

Friskop, A., S. Markell, T. Gulya et al. May 2011. Sunflower rust: Plant disease management. NDSU Extension Service, Agricultural Experiment Station, North Dakota State University, Hettinger, ND.

Fu, J.H., Q.Y. Guo, and L. Mu. 2012. Research on biological characteristics and pathogenicity of sunflower *Verticillium* wilt pathogen. *Xinjiang Agric. Sci.* 49: 1440–1448.

Fusari, C.M., J.A. Rienzo, C. di Troglia et al. 2012. Association mapping in sunflower for *Sclerotinia* Head Rot resistance. *BMC Plant Biol.* 12: 93.

Gadhave, R.N., T.R. Mogle, and P.S. Nikam. 2011. Influence of dates of sowing on incidence of *Alternaria* blight of sunflower. *J. Plant Dis. Sci.* 6: 124–127.

Gergely, L., Z.B. Vas, T. Szabo, and A. Zalka. 2002. The susceptibility of sunflower F1-hybrids to *Sclerotinia* stem-base rot in Hungarian variety trials. *Tiszantuli Novenyvedelmi Forum*, Debrecen, Hungary, October 16–17, 2002, pp. 57–60.

Giresse, X., D.T. de Labrouhe, and S. Richard-Cervera. 2007. Twelve polymorphic expressed sequence tags-derived markers for *Plasmopara halstedii*, the causal agent of sunflower downy mildew. *Mol. Ecol. Notes* 7: 1363–1365.

Gisi, U. 2002. Chemical control of downy mildews. In: *Advances in Downy Mildew Research*, eds., P.T.N. Spencer-Phillips, U. Gisi, and A. Lebeda. Kluwer Academic Publishers, Dordrecht, the Netherlands, pp.119–159.

Gisi, U. and H. Sierotzki. 2008. Fungicide modes of action and resistance in downy mildews. In: *The Downy Mildews-Genetics, Molecular Biology and Control*, eds., A. Lebeda, P.T.N. Spencer-Philipps, and B.M. Cooke. Springer, Dordrecht, the Netherlands, pp. 157–167.

Giussani, A., F. Castano, and R. Rodriguez. 2008. Performance of six sunflower hybrids and derived inbred lines during white rot development in capitula. *AgriScientia* 25: 35–39.

Godoy, M., F. Castano, J. Re, and R. Rodriguez. 2005. *Sclerotinia* resistance in sunflower: I. Genotypic variations of hybrids in three environments of Argentina. *Euphytica* 145: 147–154.

Goplalkrishnan, G., N. Manivannan, P. Vindhiyavarman, and K. Thiyagarjan. 2010. Evaluation and identification of *Alternaria* species resistant sunflower genotypes. *Electron. J. Plant Breed.* 1: 177–181.

Göre, M.E. 2009. Epidemic outbreaks of downy mildew caused by *Plasmopara halstedii* on sunflower in Thrace, part of the Marmara region of Turkey. *Plant Pathol.* 58: 396.

Goulter, K.C. 1990. Breeding of a rust differential sunflower line. In: *Proceedings of the Eighth Australian Sunflower Association Workshop*, ed., J.K. Kochman. Australian Sunflower Association, Toowoomba, Queensland, Australia, March 19–22, 1990, pp. 120–124.

Grasse, W., R. Zipper, M. Totska, and O. Spring. 2013. *Plasmopara halstedii* virus causes hypovirulence in *Plasmopara halstedii*, the downy pathogen of the sunflower. *Fungal Genet. Biol.* 57: 42–47.

Gulya, T.J. 1985. Registration of five disease resistant sunflower germplasm. *Crop Sci.* 25: 719–720.

Gulya, T.J. 1991. Efficacy of fungicides on sunflower rust severity and yield losses. In: *Proceedings of the Sunflower Research Workshop*. National Sunflower Association, Brismarck, ND, January 8–9, 1990, pp. 15–16.

Gulya, T.J. 2006. Diversity in the sunflower: *Puccinia helianthi* pathosystem in Australia. *Aust. Plant Pathol.* 35: 657–670.

Gulya, T.J. 2007a. Distribution of *Plasmopara halstedii* races from sunflower around the world. In: *Advances in Downy Mildew Research, Proceedings of the Second International Downy Mildew Symposium*, Olomouc, Czech Republic, July 2–6, 2007, Vol. 3, pp. 121–134.

Gulya, T.J. 2007b. Recent advances and novel concepts in management of Sclerotinia diseases on sunflower and other crops. Abstract. *S. Afr. J. Sci.* 103: 1.

Gulya, T.J., T.P. Freeman, and D.E. Mayhew. 1990b. Virus-like particles in *Plasmopara halstedii* sunflower downy mildew. *Phytopathology* 80: 1032.

Gulya, T.J., T.P. Freeman, and D.E. Mayhew. 1992. Ultrastructure of virus-like particles in *Plasmopara halstedii. Can. J. Bot.* 70: 334–339.

Gulya, T.J., G. Kong, and M. Borothers. 2000. Rust resistance in wild *Helianthus annuus* and variation by geographic origin. In: *Proceedings of the 15th International Sunflower Conference*, Toulouse, France, pp. 138–142.

Gulya, T.J., J. Krupinsky, M. Draper, and L.D. Charlet. 2002a. First report of charcoal rot (*Macrophomina phaseolina*) on sunflower in North and South Dakota. *Plant Dis.* 86: 923.

Gulya, T.J. and S. Masirevic. 1988. International standardization of techniques and nomenclature for sunflower rust. In: *Proceedings 10th Sunflower Research Workshop*, National Sunflower Association, Bismarck, ND, p. 10.

Gulya, T.J. and S.G. Markell. 2009. Sunflower rust status-2008 race frequency cross the Midwest and resistance among commercial hybrids. National Sunflower Association, Rust Status 09pdf. www.sunflowernsa.com.

Gulya, T.J., S. Markell, M. McMullen, B. Harveson, and L. Osborne. 2011. New virulent races of downy mildew: Distribution, status of downy mildew resistant hybrids, and USDA sources of resistance. www.sunflowernsa.com./uploads/resources/575/gulya-virulent races downy mildew pdf. Accessed on September 8, 2014

Gulya, T.J. and S. Masirevic. 1996. Inoculation and evaluation methods for sunflower rust. In: *Proceedings of the 18th Sunflower Research Workshop*, National Sunflower Association, Bismarck, ND, pp. 31–38.

Gulya, T., A. Mengistu, K. Kinzer, N. Balbyshev, and S. Markell. 2010. First report of charcoal rot of sunflower in Minnesota, USA. *Plant Health Prog.* doi:10.1094/PHP-2010-0707-02.

Gulya, T.J., S. Rooney-Latham, J.S. Miller et al. 2012. Sunflower diseases remain rare in California seed production fields compared to North Dakota. *Plant Health Prog.* doi:10.1094/PHP-2012-1214-01-RS.

Gulya, T.J., W.E. Sackston, F. Viranyi, S. Masirevic, and K.Y. Rashid. 1991. New races of the sunflower downy mildew pathogen (*Plasmopara halstedii*) in Europe and North and South America. *J. Phytopathol.* 132: 303–311.

Gulya, T.J., P.J. Shiel, T. Freeman, R.L. Jordan, T. Isakeit, and P.H. Berger. 2002b. Host range and characterization of Sunflower mosaic virus. *Phytopathology* 92: 694–702.

Gulya, T.J., R. Venette, J.R. Venette, and H.A. Lamey. 1990a. Sunflower rust. NDSU Ext. Ser. Bull. P4, Fargo, ND.

Gutierrez, A., M. Cantamutto, and M. Poverene. 2012. Disease tolerance in *Helianthus petiolaris*: A genetic resource for sunflower breeding. *Plant Prod. Sci.* 15: 204–208.

Habib, H., S.S. Mehdi, M.A. Anjum, M.E. Mohyuddin, and M. Zafar. 2007. Correlation and path analysis for seed yield in sunflower (*Helianthus annuus* L.) under charcoal rot (*Macrophomina phaseolina*) stress conditions. *Int. J. Agric. Biol.* 9: 362–364.

Hafeez, A. and S. Ahmad. 2001. Screening of sunflower germplasm for resistance to charcoal rot. *Sarhad J. Agric.* 17: 615–616.

Hahn, V. 2002. Genetic variation for resistance to *Sclerotinia* head rot in sunflower inbred lines. *Field Crops Res.* 77: 153–159.

Hamid, M. and M. Jalaluddin 2007. A new report of *Septoria helianthi* leaf spot of sunflower from Sindh. *Pak. J. Bot.* 39: 659–660.

Harter, A.V., K.A. Gardner, D. Falush, D.L. Lentz, R.A. Bye, and L.H. Rieseberg. 2004. Origin of extant domesticated sunflowers in eastern North America. *Nature (Lond.)* 430(6996): 201–205.

Harveson, R. February 2010. Rust of sunflower in Nebraska. NebGuide. University of Nebraska, Lincoln, NE.

Heidari-Tajabadi, F., M. Ahmadzadeh, A. Moinzadeh, and M. Khezri. 2011. Influence of some culture media on antifungal activity of *Pseudomonas fluorescens* UTPF61 against the *Sclerotinia sclerotiorum*. *Afr. J. Agric. Res.* 30: 6340–6347.

Heller-Dohmen, M., J.C. Göpfert, R. Hammerschmidt, and O. Spring. 2008. Different pathotypes of the sunflower downy mildew pathogen *Plasmopara halstedii* all contain isometric virions. *Mol. Plant Pathol.* 9: 777–786.

Heller-Dohmen, M., J.C. Göpfert, R. Hammerschmidt, and O. Spring. 2011. The nucleotide sequence and genome organization of *Plasmopara halstedii* virus. *Virol. J.* 8: 123. doi:10.1186/1743-422X-8-123.

Hennessy, C.M.R. and W.E. Sackston. 1970. Studies on sunflower rust. V. Culture of *Puccinia helianthi* throughout its complete life cycle on detached leaves of sunflower (*Helianthus annuus*). *Can. J. Bot.* 48: 1811.

Hosseini, S., M.K. Habibi, G. Mosahebi, M. Motamedi, and S. Winter. 2012. First report on the occurrence of Tobacco streak virus in sunflower in Iran. *J. Plant Pathol.* 94: 585–589.

Hosseini, S., S. Winter, G.H. Mosahebi, M.K. Habibi, and N.A.D. Habili. 2010. A comparison of biological and molecular characterization of Sudanese-faba bean, Indian and Iranian-sunflower tobacco streak virus isolates. *Iranian J. Plant Prot. Sci.* 41: 41–49.

Hseu, S.H., P.F. Sung, C.W. Wu, S.C. Shih, and C.Y. Lin. 2004. Bacterial stalk rot of sunflower in Taiwan: Varietal resistance and agrochemical screening. *Plant Prot. Bull. (Taipei)* 46: 367–378.

Hu, X., D.L. Bidney, N. Yalpani et al. 2003. Overexpression of a gene encoding hydrogen peroxide-generating oxalate oxidase evokes defense responses in sunflower. *Plant Physiol.* 133: 170–181.

Huang, H.C. 2002. Screening sunflower for resistance to *sclerotinia wilt*. *Plant Pathol. Bull.* 11: 15–18.

Huang, H.C. and J.M. Yeung. 2002. Biochemical pathway for the formation of abnormal sclerotia of *Sclerotinia sclerotiorum*. *Plant Pathol. Bull.* 11: 1–6.

Huguet, N., J. Perez, and F. Quiroz. 2008. *Puccinia helianthi* Schw., infecciones en hibridos comerciales en Argentina y su evolucion durante dos decadas. In: *Proceedings of the 17th International Sunflower Conference*, Cordoba, Spain, June 8–12, 2008. International Sunflower Association, Paris, France, pp. 215–218.

Hulke, B.S., J.F. Miller, and T.J. Gulya. 2010. Registration of the restorer oilseed sunflower germplasm RHA 464 possessing genes for resisatance to downy mildew and sunflower rust. *J. Plant Registration* 4: 249–254.

Ijaz, S., H.A. Sadaqat, and M.N. Khan. 2013. A review of the impact of charcoal rot (*Macrophomina phaseolina*) on sunflower. *J. Agric. Sci.* 151: 222–227.

Ioos, R., C. Fourrier, V. Wilson, K. Webb, J.L. Schereffer, and D.T. de Labrouhe. 2012. An optimized duplex real-time PCR tool for sensitive detection of the quarantine oomycete *Plasmopara halstedii* in sunflower seeds. *Phytopathology* 102: 908–917.

Irany, H., D. Ershad, and A. Alizadeh. 2001. The effect of soil depth, moisture and temperature on sclerotium germination of *Sclerotinia sclerotiorum* and its pathogenicity. *Iranian J. Plant Pathol.* 37: 185–195.

Ismail, A.E. and M.M. Mohamed. 2007. Effect of different concentrations of furfural as a botanical nematicide and the application methods in controlling *Meloidogyne incognita* and *Rotylenchulus reniformis* infecting sunflower. *Pak. J. Nematol.* 25: 45–52.

Jain, R.K., A.I. Bhat, A.S. Byadagi et al. 2000 Association of a *tospovirus* with sunflower necrosis disease. *Curr. Sci.* 79: 1703–1705.

Jain, R.K., A.I. Bhat, and A. Varma. 2003. Sunflower necrosis disease-an emerging viral problem. Technical Bulletin-1, Unit of Virology, IARI, New Delhi, India, 11pp.

Jalaluddin, M., M. Hamid, and S.E. Muhammad. 2008. Selection and application of a VAM-fungus for promoting growth and resistance to charcoal rot disease of sunflower var. Helico-250. *Pak. J. Bot.* 40: 1313–1318.

Jan, C.C. and T.J. Gulya. 2006. Registration of a sunflower germplasm resistant to rust, downy mildew and virus. *Crop Sci.* 46: 1829.

Jan, C.C., Z. Quresh, and T.J. Gulya. 1991. Genetics of rust race 4 resistance derived from *Helianthus petiolaris* and *Helianthus agrophullus* accessions. In: *Proceedings of the Sunflower Research Workshop*, Fargo, ND, January 10–11, 1991. National Sunflower Association, Brismarck, ND, p. 112.

Jan, C.C., Z. Quresh, and T.J. Gulya. 2004. Registration of seven rust resistance sunflower germplasms. *Crop Sci.* 44: 1887–1888.

Jobic, C., A.M. Boisson, E. Gout et al. 2007. Metabolic processes and carbon nutrient exchanges between host and pathogen sustain the disease development during sunflower infection by *Sclerotinia sclerotiorum*. *Planta* 226: 251–265.

Jocić, S., S. Cvejić, N. Hladni, D. Miladinović, and V. Miklič. 2010. Development of sunflower genotypes resistant to downy mildew. *Helia* 33: 173–180.

Jocić, S., D. Miladinović, I. Imerovski et al. 2012. Towards sustainable downy mildew resistance in sunflower. *Helia* 35: 61–72.

Johnson, J. 1936. Tobacco streak, a virus disease. *Phytopathology* 26: 285–291.

Jurkovic, D., J. Cosic, and R. Latkovic. 2008. Sunflower leaf spot (*Alternaria helianthi* (Hansf.) Tub. and Nish.). *Glasilo Biljne Zastite* 8: 341–343.

Kalamesh, B.K.K., Y.G. Shadakshari, and B.K. Athoni. 2012. Isolation of toxin produced by *Alternaria helianthi* causing blight in sunflower. *Int. J. Plant Prot.* 5: 278–282.

Kane, N.C., N. Gill, M.G. King et al. 2011. Progress towards a reference genome for sunflower. *Botany* 89: 429–437.

Khan, S.N. 2007. *Macrophomina phaseolina* as causal agent for charcoal rot of sunflower. *Mycopath* 5: 111–118.

Khan, S.N., N. Ayub, and I. Ahmed. 2003. A re-evaluation of geographical distribution of charcoal rot on sunflower crop in various agroecological zones of Pakistan. *Mycopath* 1: 63–66.

Khan, S.N., A. Ayub, A. Iftikhar, and A. Shehzad. 2005b. Population dynamics interaction of *Macrophomina phaseolina* for mortality in sunflower. *Mycopath* 3: 33–36.

Khan, S.N., A. Ayub, A. Iftkhar, and Y. Tahira. 2005a. Effect of inoculum density of *Macrophomina phaseolina* on grain production of sunflower. *Mycopath* 3: 61–63.

Khan, H.U., M.A. Sahi and S.T. Anjum. 2010. Identification of resistant sources against charcoal rot disease in sunflower advance lines/varieties. *Pakistan J. Phytopathol* 22:105–107.

Kinga, R., J. Biro, A. Kovacs et al. 2011. Appearance of a new sunflower downy mildew race in the South-East region of the Hungarian Great Plain. *Novenyvedelem* 47: 279–286.

Knapp, S.J. 2003. Towards a saturated molecular genetic linkage map for cultivated sunflower. *Crop Sci.* 43: 367–387.

Kolte, S.J. 1985. *Diseases of Annual Edible Oilseed Crops*, Vol. III: *Sunflower, Safflower and Nigerseed Diseases*. CRC Press, Boca Raton, FL.

Kong, G.A., K.C. Goulter, J.H. Kochman et al. 1999. Evolution of pathotypes of *Puccinia helianthi* on sunflower in Australia. *Aust. Plant Pathol.* 28(4): 320–332.

Kong, H., C.D. Patterson, W. Zhang, Y. Takikawa, A. Suzuki, and J. Lydon. 2004. A PCR protocol for the identification of *Pseudomonas syringae* pv. *tagetis* based on genes required for tagetitoxin production. *Biol. Control* 30: 83–89.

Korosi, K., R. Ban, B. Barna, and F. Viranyi. 2011. Biochemical and molecular changes in downy mildew-infected sunflower triggered by resistance inducers. *J. Phytopathol.* 159: 471–478.

Korosi, K., N. Lazar, and F. Viranyi. 2009. Resistance to downy mildew in sunflower induced by chemical activators. *Acta Phytopathol. Entomol. Hung.* 44: 1–9.

Kulkarni, S., Y.R. Hegde, and R.V. Kota. 2009. Pathogenic and morphological variability of *Plasmopara halstedii*, the causal agent of downy mildew in sunflower. *Helia* 32: 85–90.

Labrouhe, D.T., F. de Serre, P. Walser, S. Roche, and F. Vear. 2008. Quantitative resistance to downy mildew (*Plasmopara halstedii*) in sunflower (*Helianthus annuus*). *Euphytica* 164: 433–444.

Lačok, N. 2008. Zdrav suncokret-garancija za dobijanje ulja visokog kvaliteta. In: *Zbornik radova49. Savetovanja: Proizvodnja i prerada uljarica*, Herceg Novi, Crna Gora, pp. 45–51.

Lakshmidevi, N., J. Sudisha, S. Mahadevamurthy, H.S. Prakash, and H.S. Shetty. 2010. First report of the seed-borne nature of root and collar rot disease caused by *Rhizoctonia solani* in sunflower from India. *Aust. Plant Dis. Notes* 5: 11–13.

Lamey, A., J. Knodel, G. Endres, T. Gregoire, and R. Ashley. 2000. Sunflower disease and midge survey. NDSU, Extension Service, Fargo, ND. http://www.ag.ndsu.nodak.edu. Accessed on September 10, 2014.

Lavanya, N., D. Saravanakumar, I. Rajendran, M. Ramiah, T. Raguchander, and R. Samiyappan. 2009. Management of sunflower necrosis virus through anti-viral substances. *Arch. Phytopathol. Plant Prot.* 42: 265–276.

Lawson, W.R., K.C. Goulter, R.J. Henry, G.A. Kong, and J.K. Kochman. 1998. Marker assisted selection for two rust resistance genes in sunflower. *Mol. Breed.* 4: 227–234.

Lawson, W.R., C.C. Jan, T. Shatte, L. Smith, G.A. Kong, and J.K. Kochman. 2010. DNA markers linked to the R2 rust resistance gene in sunflower (*Helianthus annuus* L.) facilitate anticipatory breeding for this disease variant. *Mol. Breed.* 28: 569–576. doi:10.1007/s11032-010-9506-1.

Lazar, A., E.A. Lupu, and C. Leonte. 2011. Researches on resistance some sunflower hybrids to the artificial infection with *Sclerotinia sclerotiorum*. *Seria Agron.* 54: 67–70.

Leite, R.M.V.B.C. and L. Amorim. 2002. Influence of temperature and leaf wetness on the monocycle of *Alternaria* leaf spot of sunflower. *Fitopatol. Bras.* 27: 193–200.

Leite, R.M.V.B.C., L. Amorim, and A. Bergamin Filho. 2005. Effect of sowing date and initial inoculum of *Alternaria helianthi* in sunflower. *Documentos—Embrapa Soja.* 261: 101–104.

Leite, R.M.V.B.C., L. Amorim, and A. Bergamin Filho. 2006. Relationships of disease and leaf area variables with yield in the *Alternaria helianthi*-sunflower pathosystem. *Plant Pathol.* 55: 73–81.

Lenardon, S.L. 2008. Sunflower chlorotic mottle virus. *Characterization, Diagnosis & Management of Plant Viruses*, Vol. 1: *Industrial Crops* (Plant Pathogens Series-1), Stadium Press, Houston, Texas. pp. 349–359.

Li, G.Q., H.C. Huang, A. Laroche, and S.N. Acharya. 2003. Occurrence and characterization of hypovirulence in the tan sclerotial isolate S10 of *Sclerotinia sclerotiorum*. *Mycol. Res.* 107: 1350–1360.

Li, Y.H., H. Wang, J.C. Li, D.J. Wang, and D.R. Li. 2005. Infection of *Sclerotinia sclerotiorum* to rapeseed, soybean and sunflower and its virulence differentiation. *Acta Phytopathol. Sin.* 35: 486–492.

Li, Z.Q., Y.C. Wang, Y. Chen, J.X. Zhang, and W.G.D. Fernando. 2009. Genetic diversity and differentiation of *Sclerotinia sclerotiorum* populations in sunflower. *Phytoparasitica* 37: 77–85.

Limpert, E., B. Clifoeord, A. Dreiseitt et al. 1994. Systems of designations of pathotypes of plant pathogens. *J. Phytopathol.* 140: 359–362.

Link, V.H. and B. Johnson. 2012. White mold. *Plant Health Instructor* doi:10.1034/PHI-1-2007.0809-01 (updated 2007–2012).

Liu, Z., T.J. Gulya, G.J. Seiler, B.A. Vick, and C.C. Jan. 2012. Molecular mapping of the PI(16) downy mildew resistance gene from HA-R4 to facilitate marker- assisted selection in sunflower. *Theor. Appl. Genet.* 125: 121–131.

Lokesh, B.K., G.N. Maraddi, M.B. Agnal, and S.N. Upperi. 2008b. Present status of different modes of transmission of sunflower necrosis virus (SNV) on sunflower, weeds and crop plants. *Int. J. Plant Prot.* 1: 25–28.

Lokesh, B.K., Nagaraju, K.S. Jagadish, and Y.G. Shadakshari. 2008a. Disease reaction of sunflower genotypes against sunflower necrosis virus disease and its vector *Thrips palmi* under field condition. *Environ. Ecol.* 26: 1552–1556.

Lokesh, B.K., Nagaraju, K.S. Jagadish, and Y.G. Shadakshari. 2008c. Efficacy of new chemical molecules and sorghum as border row crop against sunflower necrosis virus disease and its vector *Thrips palmi*. *Int. J. Agric. Sci.* 4: 687–690.

Lokesh, B.K., Nagaraju, K.S. Jagadish, and Y.G. Shadakshari. 2008d. Effects of different dates of sowing and weather parameters on sunflower necrosis virus disease and its thrips vector. *Int. J. Agric. Sci.* 4: 576–579.

Los, O., Z.A. Pretorius, F.J. Kloppers et al. 1995. Virulence of *Puccinia helianthi* on differential sunflower genotypes in South Africa. *Plant Dis.* 79: 859.

Loose, L.H., A.B. Heldwein, I.C. Maldaner, D.D.P. Lucas, F.D. Hinnah, and M.P. Bortoluzzi. 2012. Alternaria and septoria leaf spot severity on sunflower at different sowing dates in Rio Grande do Sul State, Brazil. *Bragantia* 71: 282–289.

Lu, G. 2003. Engineering *Sclerotinia sclerotiorum* resistance in oilseed crops. *Afr. J. Biotechnol.* 2: 509–516.

Madhavi, K.J., M. Sujatha, D.R.R. Reddy, and S. Chander Rao. 2005a. Culture characteristics and histological changes in leaf tissues of cultivated and wild sunflowers infected with *Alternaria helianthi*. *Helia* 28: 1–12.

Madhavi, K.J., M. Sujatha, D.R.R. Reddy, and Chander Rao. 2005b. Biochemical characterization of resistance against *Alternaria helianthi* in cultivated and wild sunflowers. *Helia* 28: 13–23.

Mahmoud, A. and H. Budak. 2011. First report of charcoal rot caused by *Macrophomina phaseolina* in sunflower in Turkey. *Plant Dis.* 95: 223.

Mani, M.T. and S.J. Hepziba. 2003. Biological control of dry root rot of sunflower. *Plant Dis. Res.* (*Ludhiana*) 18: 124–126.

Mani, M.T., S.J. Hepziba, and K. Siddeswaran. 2005. Effect of organic amendments and biocontrol agents against dry root rot of sunflower caused by *Macrophomina phaseolina*. *Adv. Plant Sci.* 18: 23–26.

Markell, S. 2008. Rust damage and control: Early appearance of sunflower rust is a cause of concern in North Dakota. Folicur is now available with Headline and Quandris. National Sunflower Association. htpp://www.sunflowernsa.com/growers,/default/asp?contentID=367. Accessed on September 13, 2014.

Maselli, A., Y. Guevara, and A. Aponte. 2000. Reaction of sunflower (*Helianthus annuus* L.) genotypes to *Pseudomonas syringae* pv. *helianthi*. *Rev. Mexi. Fitopatol.* 18: 111–114.

Maselli, A., Y. Guevara, and L. Subero. 2002. Detection and transmission of *Pseudomonas syringae* pv. *helianthi* through sunflower seed. *Rev. Mexi. Fitopatol.* 20: 114–117.

Maširević, S. and T.J. Gulya. 1992. *Sclerotinia* and *Phomopsis*—Two devastating sunflower pathogens. *Field Crops Res.* 30: 271–300.

Masirevic, S. and S. Jasnic. 2006a. Leaf and stem spot of sunflower. *Biljni Lekar* (*Plant Doctor*) 34: 326–333.

Maširević, S. and S. Jasnic. 2006b. Root, stem and head rot of sunflower. *Biljni Lekar* (*Plant Doctor*) 34: 336–343.

Mathar, A.S., C.L. Subrahmanyam, and P. Ramasamy. 1975. Occurrence of aecial and pycnial stages of *Puccinia helianthi* Schw. in India. *Curr. Sci. India* 44: 284.

Mathivanan, N. and V.R. Prabavathy. 2007. Effect of carbendazim and mancozeb combination on *Alternaria* leaf blight and seed yield in sunflower (*Helianthus annus* L.). *Arch. Phytopathol. Plant Prot.* 40: 90–96.

Mathivanan, N., K. Srinivasan, and S. Chelliah. 2000. Field evaluation of *Trichoderma viride* Pers. ex. S. F. Gray and *Pseudomonas fluorescens* Migula against foliar diseases of groundnut and sunflower. *J. Biol. Control* 14: 31–34.

Mayee, C.D. 1995. Yield and oil reductions in sunflower due to rust. *J. Maharashtra Univ.* 20: 358–360.

Mayhew, D.E., A.L. Cook, and T.J. Gulya. 1992. Isolation and characterization of a mycovirus from *Plasmopara halstedii* virus. *Can. J. Bot.* 70: 1734–1737.

Mendieta, J.R., A.M. Giudici, and L. de la Canal. 2004. Occurrence of antimicrobial serin-proteinases in sunflower seeds. *J. Phytopathol.* 152: 43–47.

Mesta, R.K., V.I. Benagi, I. Shankergoud, and S.N. Megeri. 2009. Effect of dates of sowing and correlation of weather parameters on the incidence of *alternaria* blight of sunflower. *Karnataka J. Agric. Sci.* 22: 441–443.

Mesta, R.K., K. Pramod, and H. Arunkumar. 2004. Management of sunflower necrosis disease by vector control. *Indian Phytopathol.* 37: 381.

Mesta, R.K., G. Sunkad, P. Katti, and I. Shankergoud. 2005. Evaluation of sunflower germplasm against *Alternaria* leaf blight and rust diseases. *Agric. Sci. Dig.* 25: 221–222.

Micheli, A., O.C. Silva, L.C. da Bavoso et al. 2007. Effect of chemical seed treatment on sunflower. *Documentos—Embrapa Soja* 292: 40–43.

Micic, Z., V. Hahn, E. Bauer et al. 2004. QTL mapping of *Sclerotinia* midstalk-rot resistance in sunflower. *Theor. Appl. Genet.* 109: 1474–1484.

Micic, Z., V. Hahn, E. Bauer et al. 2005b. Identification and validation of QTL for *Sclerotinia* midstalk rot resistance in sunflower by selective genotyping. *Theor. Appl. Genet.* 111: 233–242.

Micic, Z., V. Hahn, E. Bauer, C.C. Schon, and A.E. Melchinger. 2005a. QTL mapping of resistance to *Sclerotinia* midstalk rot in RIL of sunflower population NDBLOSsel × CM625. *Theor. Appl. Genet.* 110: 1490–1498.

Middleton, K.J. and N.R. Obst. 1972. Sunflower rust reduces yield. *Aust. Plant Pathol. Soc. Newsl.* 10: 18.

Miller, J.F. and T.J. Gulya. 2001. Registration of three rust resistant sunflower germplasm lines. *Crop Sci.* 37: 1988–1989.

Miller, J.F. and T.J. Gulya. 2006. Registration of two restorer (RHA 439 and RHA 440) and one maintainer (HA 441) *Sclerotinia* tolerant oilseed sunflower germplasms. *Crop Sci.* 46: 482–483.

Miller, J.F., T.J. Gulya, and B.A. Vick. 2006. Registration of two maintainer (HA 451 and HA 452) and three restorer (RHA 453-RHA 455) *sclerotinia*-tolerant oilseed sunflower germplasms. *Crop Sci.* 46: 2727–2728.

Miller, J.F., R.H. Rodriguez, and T.J. Gulya. 1988. Evaluation of genetic materials for inheritance of resistance to race 4 in sunflower. In: *Proceedings of the 12th International Sunflower Conference*, International Sunflower Association, Paris, France, pp. 361–365.

Mitov, N. 1957. A study of the spread of sunflower rust (*Puccinia helianthi* Schw.), Abstr. *Rev. Appl. Mycol.* 36: 591.

Mohammad, A., S. Ahmad, and I. Ahmad. 2001. Pathogenic and cultural variations in *Macrophomina phaseolina*, the cause of charcoal rot in sunflower. *Sarhad J. Agric.* 17: 253–255.

Mokbel, A.A., I.K.A. Ibrahim, M.R.A. Shehata, and M.A.M. El Saedy. 2007. Interaction between certain root-rot fungi and the root-knot nematode, *Meloidogyne incognita* on sunflower plants. *Egypt. J. Phytopathol.* 35: 1–11.

Molinero-Ruiz, L., J.M. Melero-Vara, and J. Dominguez. 2003. Inheritance of resistance of two races of sunflower downy mildew (*Plasmopara halstedii*) in two *Helianthus annuus* L. lines. *Euphytica* 131: 47–51.

Moreno, P.S., A.B. de Romano, M.C. Romano et al. 2011. A survey of physiological races of *Puccinia helianthi* in Argentina. In: *The Proceedings of 18th International Sunflower Conference*, Mar del Plata, Argentina.

Morsy, S.M. and A.E. El-Korany. 2007. Suppression of damping-off and charcoal-rot of sunflower with composted and non-composted agricultural wastes. *Egypt. J. Phytopathol.* 35: 23–38.

Mosa, A.A., A.A. El-Deeb, and M.M. Ibrahim. 2000. Evaluation of sunflower hybrids for resistance to *Sclerotinia sclerotiorum*. *Ann. Agric. Sci.* (*Cairo*) 45: 689–702.

Muhammad, A. and A. Javaid. 2007. Exploitation of allelopathic potential of *Chenopodium* species to control charcoal rot pathogen of sunflower. *Pak. J. Agric. Res.* 20: 130–131.

Muhammad, A., M.J. Zaki, and D. Shahnaz. 2010. Effect of oilseed cakes alone or in combination with *Trichoderma* species for the control of charcoal rot of sunflower (*Helianthus annus* L.). *Pak. J. Bot.* 42: 4329–4333.

Muhammad, A., M.J. Zaki, and D. Shahnaz. 2012. Development of a Na-alginate-based bioformulation and its use in the management of charcoal rot of sunflower (*Helianthus annuus* L.). *Pak. J. Bot.* 44: 1167–1170.

Muhammad, A.M.W. and M.J. Zaki. 2010. Bioefficacy of microbial antagonists against *Macrophomina phaseolina* on sunflower. *Pak. J. Bot.* 42: 2935–2940.

Mukhtar, I. 2009. Sunflower diseases and insect pests in Pakistan: A review. *Afr. Crop Sci. J.* 17: 109–118.

Mulpuri, S., Z. Liu, J. Feng, T.J. Gulya, and C.C. Jan. 2009. Inheritance and molecular mapping of a downy mildew resistance genePI(13) in cultivated sunflower (*Helianthus annuus* L.). *Theor. Appl. Genet.* 119: 795–803.

Murthy, U.K., I.E. Lyngdoh, T. Gopalakrishna, M.B. Shivanna, and D.T. Prasad. 2005. Assessment of heritability of *Alternaria helianthi* resistance trait in sunflower using molecular markers. *Helia* 28: 33–41.

Nagaraju, A., S.M. Murthy, S.N. Raj, S.C. Nayaka, and A.C.U. Shankar. 2012. Induction of resistance by culture filtrates of *Trichoderma harzianum* in sunflower against downy mildew caused by *Plasmopara halstedi. J. Mycol. Plant Pathol.* 42: 513–519.

Nagaraju, J.K.S., K.T. Puttarangaswamy, G. Girish, and Y.G. Shadakshari. 2005. Evaluation of experimental hybrids of sunflower for multiple disease resistance at Bangalore. *Environ. Ecol.* 23S: 749–750.

Naheed, I., D. Shahnaz, M.J. Zaki, T. Marium, and A. Zeeshan. 2011. Combined use of (60Cobalt) gamma irradiated seeds and nursery fertilizers in the control of root rot fungi of crop plants. *Int. J. Biol. Biotechnol.* 8: 521–527.

Nandeeshkumar, P., B.R. Sarosh, K.R. Kini, H.S. Prakash, and H.S. Shetty. 2009. Elicitation of resistance and defense related proteins by beta-amino butyric acid in sunflower against downy mildew pathogen *Plasmopara halstedii. Arch. Phytopathol. Plant Prot.* 42: 1020–1032.

Nandeeshkumar, P., J. Sudisha, K.K. Ramachandra, H.S. Prakash, S.R. Niranjana, and S.H. Shekar. 2008. Chitosan induced resistance to downy mildew in sunflower caused by *Plasmopara halstedii. PMPP Physiol. Mol. Plant Pathol.* 72: 188–194.

Naz, F. and M. Ashraf. 2006. Histopathological studies of sunflower seedlings infected with *Macrophomina phaseolina. Int. J. Biol. Biotechnol.* 3: 107–112.

Nelson, B. and A. Lamey. 2000. Sclerotinia diseases of sunflower. North Dakota State University of Agriculture and University Extension Department, Fargo, ND, p. 840.

North Dakota State University (NDSU). 2009. Evaluation of fungicides and fungicide timing on control of sunflower rust (*Puccinia helianthi*) at three locations in North Dakota—2009 report. http://www.researchgate.net/publication/43291199_Evaluation_of_fungicides_and_fungicide_timing_on_control_of_sunflower_rust_(Puccinia_helianthi)_at_three_locations_in_North_Dakota_in_2009. Accessed on September 14, 2014.

Oliveira, M.F., A. de Tulmann Neto, R.M.V.B.C. Leite, V.B.R. Castiglioni, and C.A.A. Arias. 2004. Mutation breeding in sunflower for resistance to *Alternaria* leaf spot. *Helia* 27: 41–49.

Pandey, R.N. and G.S. Saharan. 2005. Facultative fungal diseases of sunflower. In: *Diseases of Oilseed Crops*, eds., G.S. Saharan, N. Mehta, and M.S. Sangwan. Indus Publishing Co., New Delhi, India, pp. 373–459.

Pandey, R.N., G.S. Saharan, and G.B. Valand. 2005. Obligate fungal diseases of sunflower. In: *Diseases of Oilseed Crops*, eds., G.S. Saharan, N. Mehta, and M.S. Sangwan. Indus Publishing Co., New Delhi, India, pp. 319–372.

Pankaja, N.S., G.V.H. Babu, and Nagaraju. 2010a. Serological diagnosis of crop and weed plants for the presence of Sunflower Necrosis Virus (SNV) through Direct Antigen Coated-Enzyme Linked Immuno Sorbent Assay (DAC-ELISA). *Int. J. Plant Prot.* 3: 213–218.

Pankaja, N.S., G.V.H. Babu, and Nagaraju. 2010b. Virus vector relationship studies of Sunflower Necrosis Virus (SNV) and its vector *Thrips palmi* (Karny). *Int. J. Plant Prot.* 3: 260–263.

Pankaja, N.S., G.V.H. Babu, and Nagaraju. 2011. Transmission modes of sunflower necrosis virus—Determination and confirmation. *Int. J. Plant Prot.* 4: 38–42.

Papaiah Sardaru, A., M.A. Johnson, B. Viswanath, and G. Narasimha. 2013. Sunflower necrosis disease—A threat to Sunflower cultivation in India. *Rev. Ann. Plant Sci.* 2: 543–555.

Parmelee, J.A. 1977. *Puccinia helianthi. Fungi Canadenses* 95: 1–2. Agriculture Canada, Ottawa, Ontario, Canada.

Patil, L.C. and R.L. Ravikumar. 2010. Evaluation of homozygous mutant lines for *alternaria* leaf blight resistance, seed yield and oil content in sunflower. *Karnataka J. Agric. Sci.* 23: 352–353.

Patil, M.A., P.V. Khalikar, and A.P. Suryawanshi. 2002. Pathogenic variability in *Puccinia helianthi*, the cause of sunflower rust in India. Review. Marathwada Agricultural University, Parbhani, India, p. 45.

Patil, P.V., M.R. Kachapur, and K.H. Anahosur. 1998. Identification of physiological races of *Puccinia helianthi* in India. *Indian Phytopathol.* 51: 376–378.

Perez, A., L. Herrera Isla, and O. Saucedo Castillo. 2002. Effect of the population density on the main diseases of sunflower (*Helianthus annuus* L.) in a grey carbonic soil. *Centro Agricola* 29: 38–41.

Peluffo, L., V. Lia, C. Troglia et al. 2010. Metabolic profiles of sunflower genotypes with contrasting response to *Sclerotinia sclerotiorum* infection. *Phytochemistry* 71: 70–80.

Pieckenstain, F.L., M.E. Bazzalo, A.M.I. Roberts, and R.A. Ugalde. 2001. *Epicoccum purpurascens* for biocontrol of Sclerotinia head rot of sunflower. *Mycol. Res.* 105: 77–84.

Piven', V.T., N.G. Mikhailyuchenko, and S.A. Semerenko. 2002. Corbel as seed treatment for sunflower seed. *Zashchita i Karantin Rastenii* 5: 30–31.

Piven', V.T., I.I. Shulyak, and N.V. Muradasilova. 2004. Protection of sunflower. *Zashchita i Karantin Rastenii* 4: 42–51.

Poussereau, N., S. Creton, G. Billon-Grand, C. Rascle, and M. Fevre. 2001a. Regulation of acp1, encoding a non-aspartyl acid protease expressed during pathogenesis of Sclerotinia sclerotiorum. *Microbiology (Reading)* 147: 717–726.

Poussereau, N., S. Gente, C. Rascle, G. Billon-Grand, and M. Fevre. 2001b. aspS encoding an unusual aspartyl protease from *Sclerotinia sclerotiorum* is expressed during phytopathogenesis. *FEMS Microbiol. Lett.* 194: 27–32.

Pradeep, K., V.K. Satya, M. Selvapriya et al. 2012. Engineering resistance against Tobacco streak virus (TSV) in sunflower and tobacco using RNA interference. *Biol. Plantarum* 56: 735–741.

Prasad, D., G. Chawla, and I. Ahmad 2001. Pathogenicity of *Rotylenchulus reniformis* on sunflower. *Ann. Plant Prot. Sci.* 9: 163–164.

Prasad, M.S.L., M. Sujatha, and S.C. Rao. 2009. Analysis of cultural and genetic diversity in *Alternaria helianthi* and determination of pathogenic variability using wild *Helianthus* species. *J. Phytopathol.* 157: 609–617.

Prasda Rao, R.D.V.J.P., K.J. Madhavi, A.S. Reddy et al. 2009. Non-transmission of Tobacco streak virus isolate occurring in India through the seeds of some crop and weed hosts. *Indian J. Plant Prot.* 37: 92–96.

Prats, E., M.E. Bazzalo, A. Leon, and J.V. Jorrin. 2003. Accumulation of soluble phenolic compounds in sunflower capitula correlates with resistance to *Sclerotinia sclerotiorum. Euphytica* 132: 321–329.

Prats, E., M.E. Bazzalo, A. Leon, and J.V. Jorrin. 2006. Fungitoxic effect of scopolin and related coumarins on *Sclerotinia sclerotiorum*. A way to overcome sunflower head rot. *Euphytica* 147: 451–460.

Prats, E., J.C. Galindo, and M.E. Bazzalo. 2007. Antifungal activity of a new phenolic compound from capitulum of a head rot-resistant sunflower genotype. *J. Chem. Ecol.* 33: 2245–2253.

Putt, E.D. and W.E. Sackston. 1963. Studies on sunflower rust. IV. Two genes, R1 and R2, for resistance in the host. *Can. J. Plant Sci.* 43: 490.

Qi, L.L., J. Gulya, B.S. Hulke, and B.A. Vick. 2012. Chromosome location, DNA markers and rust resistance of sunflower gene R. *Mol. Breed.* 30(2): 745–756. doi:10.1007/s 11032-011-9659-6.

Qi, L.L., T.J. Gulya, G.J. Seller et al. 2011. Identification of resistance to new virulent races of rust in sunflowers and validation of DNA markers in the gene pool. *Phytopathology* 101: 241–249. doi:10.1094/PHYTO-06-10-0162.

Quresh, J. and C.C. Jan. 1993. Allelic relationship among genes for resistance to sunflower rust. *Crop Sci.* 33: 235–238.

Quresh, J., C.C. Jan, and T.J. Gulya. 1993. Resistance to sunflower and its inheritance in wild sunflower species. *Plant Breed.* 110: 297–306.

Quresh, Z., C.C. Jan, and T.J. Gulya. 1991. Genetic relationship between wild and cultivated species for rust resistance in sunflower. In: *Proceedings of the Sunflower Research Workshop*. National Sunflower Association, Bismarck, ND, January 8–9, 1991, pp. 110–111.

Radwan, O., M.F. Bouzidi, and S. Mouzeyar. 2011. Molecular characterization of two types of resistance in sunflower to *Plasmopara halstedii*, the causal agent of downy mildew. *Phytopathology* 101: 970–979.

Rafiei, M., A.D.S. Rayatpanah, A. Andarkhor, and M.S.N. Abadi. 2013. Study on reaction of sunflower lines and hybrids to *Macrophomina phaseolina* (Tassi) Goid. causal agent of Charcoal rot disease. *World Appl. Sci. J.* 21: 129–133.

Rahim, M. 2001. Allelic relationship of resistant genes to sunflower downy mildew race 1 in various sunflower inbred lines. *Online J. Biol. Sci.* 4: 201–205.

Raj, K. and G.S. Saharan. 2000a. Effect of relative humidity on infection and development of sunflower head rot. *Res. Crops* 1: 334–336.

Raj, K. and G.S. Saharan. 2000b. Effect of head age for initiation and development of head rot in sunflower. *Ann. Biol.* 16: 153–156.

Raj, K. and G.S. Saharan. 2001. Effect of temperature on the infection and development of sunflower head rot. *Agric. Sci. Dig.* 21: 189–191.

Raj, T.S., D.J. Christopher, R.S. Rajakumar, and S.U. Rani. 2008. Effect of organic amendments and *Trichoderma viride* on growth and root rot incidence of sunflower. *Ann. Plant Prot. Sci.* 16: 242–244.

Ramadan, N.A., A.M. Abdallah, and B.A. Malaabeeda. 2011. Enhancement of acquired resistance and selection of sunflower (*Helianthus annus* L.) plants resistant to the pathogen *Macrophomina phaseolina* (Tassi) Goid. obtained from hypocotyls stem callus culture. *Arab. J. Plant Prot.* 29: 43–50.

Ramiah, M., A.I. Bhat, R.K. Jain et al. 2001a. Isolation of an isometric virus causing sunflower necrosis disease in India. *Plant Dis.* 85(4): 443.

Ramiah, M., A.I. Bhat, R.K. Jain et al. 2001b. Partial characterization of an isometric virus causing sunflower necrosis disease. *Indian Phytopathol.* 54: 246–250.

Rao, M.S.L., S. Kulkarni, S. Lingaraju, and H.L. Nadaf. 2009. Bio-priming of seeds: A potential tool in the integrated management of *Alternaria* blight of sunflower. *Helia* 32: 107–114.

Rao, M.S.L., S. Kulkarni, S.D. Sagar, and V.R. Kulkarni. 2007. Efficacy of seed dressing fungicides and bioagents on *Alternaria* blight and other seed quality parameters of sunflower. *J. Plant Dis. Sci.* 1: 34–36.

Raranciuc, S. and M.J. Pacureanu. 2002. Evaluation of some sunflower genotypes concerning the reaction to *Alternaria* spp. pathogen. *Romanian Agric. Res.* 17–18: 29–33.

Raranciuc, S. and M. Pacureanu-Joita. 2006. Study concerning the main factors which influence the field infection of sunflower downy mildew. *Analele Institutului National de Cercetare-Dezvoltare Agricola Fundulea.* 73: 191–204.

Rashid, K.Y. 2003. Diseases of sunflower. In: *Diseases of Field Crops in Canada*, eds., K.L. Bailey, B.D. Gossen, R.K. Gugel, and R.A.A. Morrall. The Canadian Pyhtopathology Society, Ottawa, Ontario, Canada, pp. 169–176.

Rashid, K.Y. 2004. Local epidemics of sunflower rust in Manitoba in 2003—New races or favourable environmental conditions? http://www.sunflowernsu.com. Accessed on September 11, 2014.

Rashid, K.Y. 2006. Sunflower rust races in Manitoba. Agriculture and Agri-food Canada. http:/www.umanitoba.ca/afs/agronomists/conf/proceedings/2006/Rashid_sunflower_rust_races.pdf. Accessed on September 13, 2014.

Rashid, K.Y. 2011. Fungicides to control Sclerotinia head rot in sunflower. Agriculture and Agri-Food, Canada, Morden Research Station, Morden, Manitoba, Canada.

Ratnam, M.V.S., P.N. Reddy, S.C. Rao, and R.B. Raju. 2004a. Systemic induced resistance in sunflower to *Alternaria* leaf blight by foliar application of SA and Bion. *J. Oilseeds Res.* 21: 104–107.

Ratnam, S.V., P.N. Reddy, V. Muthusubramanian, and K.C. Rao. 2004b. Induction of systemic resistance in sunflower to *Alternaria* leaf blight. *Indian J. Plant Prot.* 32: 114–117.

Ravi, K.S., A. Buttgereitt, A.S. Kitkaru, S. Deshmukh, D.E. Lesemann, and S. Winter. 2001. Sunflower necrosis disease from India is caused by an ilarvirus related to Tobacco streak virus. *Plant Pathol.* 50(6): 800.

Rayatpanah, S., S.A. Dalili, and E. Yasari. 2012a. Diversity of *Macrophomina phaseolina* (Tassi) Goid based on chlorate phenotypes and pathogenicity. *Int. J. Biol.* 4: 54–63.

Rayatpanah, S., S.G. Nanagulyan, S.V. Alav, M. Razavi, and A. Ghanbari-Malidarreh. 2012b. Pathogenic and genetic diversity among Iranian isolates of *Macrophomina phaseolina. Chilean J. Agric. Res.* 72: 40–44.

Reddy, C.V.C.M., A.V.V. Reddy, B. Sinha, and M.S. Lakshmi. 2006. Screening of sunflower genotypes for resistance against *Alternaria* blight. *Asian J. Plant Sci.* 5: 511–515.

Rehman, A., B. Rubab, and M.H. Ullah. 2006. Evaluation of different chemicals against root knot nematode (*Meloidogyne incognita*) on sunflower. *J. Agric. Soc. Sci.* 2: 185–186.

Reimonte, G. and F. Castano. 2008. Susceptibility of sunflower (*Helianthus annuus*) hybrids to mid stem rot and broken stem caused by *Sclerotinia sclerotiorum. Ciencia e Investigacion Agraria* 35: 21–28.

Robinson, J.M., J. Lydon, C.A. Murphy, R. Rowland, and R.D. Smith. 2004. Effect of *Pseudomonas syringae* pv. *tagetis* infection on sunflower leaf photosynthetic and ascorbic acid relations. *Int. J. Plant Sci.* 165: 263–271.

Rodriguez, M., N. Munoz, S. Lenardon, and R. Lascano. 2012. The chlorotic symptom induced by Sunflower chlorotic mottle virus is associated with changes in redox-related gene expression and metabolites. *Plant Sci.* 196: 107–116.

Rodriguez, M.A., N. Venedikian, M.E. Bazzalo, and A. Godeas. 2004. Histopathology of *Sclerotinia sclerotiorum* attack on flower parts of *Helianthus annuus* heads in tolerant and susceptible varieties. *Mycopathologia* 157: 291–302.

Rodriguez, M.A., N. Venedikian, and A. Godeas. 2001. Fungal populations on sunflower (*Helianthus annuus*) anthosphere and their relation to susceptibility or tolerance to *Sclerotinia sclerotiorum* attack. *Mycopathologia* 150: 143–150.

Roeckel-Drevet, P., J. Tourvieille, T.J. Gulya, G. Charmet, P. Nicolas, and D. Tourvieille de Labrouhe. 2003. Molecular variability of sunflower downy mildew, *Plasmopara halstedii*, from different continents. *Can. J. Microbiol.* 49: 492–502.

Romero, B.T. and L.J. Subero. 2003. Pathogenesis of *Alternaria helianthi* in sunflower leaves at different humidity conditions. II. Penetration. *Fitopatol. Venez.* 16: 11–16.

Ronicke, S., V. Hahn, and W. Friedt. 2005a. Resistance to *Sclerotinia sclerotiorum* of 'high oleic' sunflower inbred lines. *Plant Breed.* 124: 376–381.

Ronicke, S., V. Hahn, and R. Horn. 2004. Interspecific hybrids of sunflower as a source of *Sclerotinia* resistance. *Plant Breed.* 123: 152–157.

Ronicke, S., V. Hahn, A. Vogler, and W. Friedt, W. 2005b. Quantitative trait loci analysis of resistance to *Sclerotinia sclerotiorum* in sunflower. *Phytopathology* 95: 834–839.

Rost, C. and M. Thines. 2012. A new species of *Pustula* (Oomycetes, Albuginales) is the causal agent of sunflower white rust. *Mycol. Prog.* 11: 351–359.

Sackston, W.E. 1962. Studies on sunflower rust. III. Occurrence, distribution, and significance of races *Puccinia helianthi* Schw. *Can. J. Bot.* 40: 1449–1458.

Sackston, W.E. 1978. Sunflower disease mapping in Europe and adjacent Mediterranean countries. *Helia Information Bulletin of the Research Network on Sunflower*, No. 1: 21–31.

Sackston, W.E. 1981. The sunflower crop and disease: Progress, problems and prospects. *Plant Dis.* 65(8): 643.

Sackston, W.E., T.J. Gulia, and J.F. Miller. 1990. Proposed international system for designating races of *Plasmopara halstedii. Plant Dis.* 74: 721–723.

Saharan, G.S. and N. Mehta 2008. *Sclerotinia Diseases of Crop Plants: Biology, Ecology and Disease Management*, Vol. LXII. Springer-Verlag GmbH, Heidelberg, Germany.

Sakr, N. 2010. Studies on pathogenicity in *Plasmopara halstedii* (sunflower downy mildew). *Int. J. Life Sci.* 4: 48–59.

Sakr, N., M. Ducher, J. Tourvieille, P. Walser, F. Vear, and D. Tourvieille de Labrouhe. 2009. A method to measure aggressiveness of *Plasmopara halstedii* (Sunflower DownyMildew). *J. Phytopathol.* 157: 133–136.

Salustiano, M.E., J. Machado, and J.E. da C. Pittis et al. 2006. Comparison of two health methods in the detection of *Alternaria helianthi* in sunflower seeds. *Fitopatol. Bras.* 31: 322.

Sanogo, S., B.F. Etarock, S. Angadi, and L.M. Lauriault. 2010. Head rot of sunflower caused by *Rhizopus oryzae* in New Mexico. *Plant Dis.* 94: 638–639.

Sarova, J., I. Kudlikova, Z. Zalud, and K. Veverka. 2003. *Macrophomina phaseolina* (Tassi) Goid. moving north—Temperature adaptation or change in climate? *Zeitschrift fur Pflanzenkrankheiten und Pflanzenschutz* 110: 444–448.

Sarovar, B. and D.V.R. Sai Gopal. 2010b. Development of a probe-based blotting technique for the detection of Tobacco streak virus. *Acta Virol.* 54: 221–224.

Sarovar, B., Y. Siva Prasad, and D.V.R. Sai Gopal. 2010a. Detection of *Tobacco streak virus* by immunocapture-reverse transcriptasepolymerase chain reaction and molecular variability analysis of a part of RNA3 of sunflower, gherkin, and pumpkin from Andhra Pradesh, India. *Sci. Asia* 36: 194–198.

Sawahel, W. and A. Hagran. 2006. Generation of white mold disease-resistant sunflower plants expressing human lysozyme gene. *Biol. Plantarum* 50: 683–687.

Says-Lesage, V., P. Roeckel-Drevet, A. Viguie, J. Tourvieill, P. Nicolas, and D.T. de Labrouhe. 2002. Molecular variability within *Diaporthe/Phomopsis helianthi* from France. *Phytopathology* 92: 308–313.

Scelonge, C., L. Wang, D. Bidney et al. 2000. Transgenic Sclerotinia resistance in sunflower (*Helianthus annuus* L.). In: *The Proceedings of 15th International Sunflower Conference*, Toulouse, France, pp. 66–71.

Schnabl, H., P.C. Binsfeld, C. Cerboncini et al. 2002. Biotechnological methods applied to produce *Sclerotinia sclerotiorum* resistant sunflower. *Helia* 25: 191–197.

Schoch, C.L., P.W. Crous, J.Z. Groenewald et al. 2009. A class-wide phylogenetic assessment of *Dothidiomycetes*. *Stud. Mycol.* 64: 1–15.

Seiler, G.J. 1992. Utilization of wild sunflower species for the improvement of cultivated sunflower. *Field Crops Res.* 30: 192–230.

Seiler, G.J. 2010. Utilization of wild *Helianthus* species in breeding for disease resistance. In: *Proceedings of the International Symposium on "Sunflower Breeding on Disease Resistance"* Krasnodar, Russia, June 23–24, 2010, pp. 36–50.

Sendall, B.C., G.A. Kong, K.C. Goulter et al. 2006. Diversity in the sunflower: *Puccinia helianthi* pathosystem in Australia. *Aust. Plant Pathol.* 35(6): 657–670.

Shahnaz, D., A. Sadia, T. Marium, and M.J. Zaki. 2007. Use of sea weed and bacteria in the control of root rot of mash bean and sunflower. *Pak. J. Bot.* 39: 1359–1366.

Sharman, M., J.E. Thomas, and D.M. Persley. 2008. First report of Tobacco streak virus in sunflower (*Helianthus annuus*), cotton (*Gossypium hirsutum*), chickpea (*Cicer arietinum*) and mung bean (*Vigna radiata*) in Australia. *Aust. Plant Dis. Notes* 3: 27–29.

Shindrova, P. 2013. Investigation on the race composition of downy mildew (*Plasmopara halstedii* Farl. Berlese et de Tony) in Bulgaria during 2007–2008. *Helia* 33: 19–24.

Shirshikar, S.P. 2003. Influence of different sowing dates on the incidence of sunflower necrosis disease. *Helia* 26(39): 109–116.

Shirshikar, S.P. 2008. Response of newly developed sunflower hybrids and varieties to downy mildew disease. *Helia* 31: 19–26.

Shirshikar, S.P, M.H. Chavan, S.K. Deshpande and K.A. Deshpande 2009. Control of sunflower necrosis disease with new chemicals, *J. Oilseeds Res.* (special issue) 26: 484–486.

Shirshikar, S.P. 2010. Sunflower necrosis disease management with thiomethoxam. *Helia* 33(53): 63–68.

Shobana Rani, T. and R.L. Ravikumar. 2006. Sporophytic and gametophytic recurrent selection for improvement of partial resistance to Alternaria leaf blight in sunflower (*Helianthus annuus* L.). *Euphytica* 147: 421–431.

Shobharani, T. and R.L. Ravikumar. 2010. Effect of genotype and mutagen dose on induced variability for resistance to *Alternaria* leaf blight in sunflower, *Helianthus annuus* L. *J. Oilseeds Res.* 27: 26–30.

Shopov, T. 1980. Reaction of some cultivars, lines, and wild species of sunflower to powdery mildew, rust and broom rape. Abstr. *Rev. Plant Pathol.* 59: 2862.

Shtienberg, D. 1995. Rational suppression of sunflower rust: Development and evaluation of an action threshold. *Plant Dis.* 79: 506–510.

Shtienberg, D. and D. Johar. 1992. Fungicidal disease suppression and yield losses associated with sunflower rust in Israel. *Crop Prot.* 11: 529–534.

Siddiqui, I.A., S. Ehteshamul-Haque, M.J. Zaki, and A. Ghaffar. 2000. Effect of urea on the efficacy of *Bradyrhizobium* sp. and *Trichoderma harzianum* in the control of root infecting fungi in mungbean and sunflower. *Sarhad J. Agric.* 16: 403–406.

Siddiqui, M.Q. and J.F. Brown. 1977. Effects of simulated rust epidemics on the growth and yield of sunflower. *Aust. J. Agric. Res.* 28: 389.

Silva, O.C., A. da Micheli, L.C. Bavoso et al. 2007. Evaluation of the sunflower genotypes to stalk and head rot susceptibility in field conditions. *Documentos—Embrapa Soja* 292: 149–152.

Simic, B., J. Cosic, R. Popovic, and K. Vrandecic. 2008. Influence of climate conditions on grain yield and appearance of white rot (*Sclerotinia sclerotiorum*) in field experiments with sunflower hybrids. *Cereal Res. Commun.* 36(Suppl. 5): 63–66.

Simmons, E.G. 2007. *Alternaria: An Identification Manual. CBS Biodiversity Series* 6. Fungal Biodiversity Center, Utrecht, the Netherlands, pp. 667–668.

Singh, B.N., A. Singh, S.P. Singh, and H.B. Singh. 2011. *Trichoderma harzianum*-mediated reprogramming of oxidative stress response in root apoplast of sunflower enhances defence against *Rhizoctonia solani. Eur. J. Plant Pathol.* 131: 121–123.

Singh, J.P. 1974. Chemical control of sunflower rust in Kenya. In: *Proceedings of the Sixth International Sunflower Conference*, Bucharest, Romania.

Singh, R. and D.M. Ferrin. 2012. First report of stem and foliar blight of sunflower caused by *Alternariastr helianthi* in Louisiana. *Plant Dis.* 96: 761–762.

Singh, R.S., S.S. Mann, J. Kaur, and A.R. Kaur. 2004. Variation in antagonistic potentiality of *Trichoderma harzianum* isolates against *Sclerotinia sclerotiorum* causing head rot of sunflower. *Indian Phytopathol.* 57: 185–188.

Singh, S. and D. Prasad. 2010. Management of *Rotylenchulus reniformis* on sunflower through botanicals. *Ann. Plant Prot. Sci.* 18: 220–222.

Singh, S.N. 2000. Relative efficacy of fungicides against seedling mortality and *Alternaria* blight of sunflower (*Helianthus annuus* L.). *J. Mycol. Plant Pathol.* 30: 119–120.

Singh, S.N. 2002. Effect of sowing dates and fungicidal spray on *Alternaria* blight and yield of sunflower. *Indian Phytopathol.* 55: 104–106.

Sood, P.N. and W.E. Sackston. 1972. Studies on sunflower rust. XI. Effect of temperature and light on germination and infection of sunflowers by *Puccinia helianthi. Can. J. Bot.* 50: 1879.

Spring, O., M. Bachofer, M. Thines, A. Riethmüller, M. Göker, and F. Oberwinkler. 2006. Intraspecific relationship of *Plasmopara halstedii* isolates differing in pathogenicity and geographic origin based on ITS sequence data. *Eur. J. Plant Pathol.* 114: 309–315.

Spring, O. and M. Thines. 2004. On the necessity of new characters for classification and systematics of biotrophic Peronosporomycetes. *Planta* 219: 910–914.

Spring, O. and R. Zipper. 2006. Evidence for asexual genetic recombination in sunflower downy mildew, *Plasmopara halstedii*. *Mycol. Res.* 110: 657–663.

Sreekanth, M., M. Sriramulu, R.D.V.J. Prasada Rao, B.S. Babu, and T.R. Babu. 2004. Effect of intercropping on *Thrips palmi* (Karny) population and peanut bud necrosis virus in mungbean. *Indian J. Plant Prot.* 32(1): 45–48.

Srinivasan, K., M. Krishnaraj, and N. Mathivanan. 2009. Plant growth promotion and the control of sunflower necrosis virus disease by the application of biocontrol agents in sunflower. *Arch. Phytopathol. Plant Prot.* 42: 160–172.

Srinivasan, K. and N. Mathivanan. 2011a. Establishment, purification, maintenance and serological diagnosis of Sunflower necrosis virus in callus. *Phytoparasitica* 39: 509–515.

Srinivasan, K. and N. Mathivanan. 2011b. Plant growth promoting microbial consortia mediated classical biocontrol of sunflower necrosis virus disease. *J. Biopest.* 4: 65–72.

Srinivasan, K. and S. Visalakchi. 2010. First report of *Rhizoctonia solani* causing a disease of sunflower in India. *Plant Dis.* 94: 488.

Sudha, A. and S. Prabhu. 2008. Evaluation of organic soil amendments against *Macrophomina phaseolina* (Tassi). *J. Biopest.* 1: 1.

Sudisha, J., S.R. Niranjana, S.L. Sukanya, R. Girijamba, N. Lakshmi Devi, and H. Shekhar Shetty. 2010. Relative efficacy of strobilurin formulations in the control of downy mildew of sunflower. *J. Pest Sci.* 83: 461–470.

Sundaresha, S., R. Sreevathsa, G.B. Balol, G. Keshavareddy, K.T. Rangaswamy, and M. Udayakumar. 2012. A simple, novel and high efficiency sap inoculation method to screen for tobacco streak virus. *Physiol. Mol. Biol. Plants* 18: 365–369.

Suriachandraselvan, M. and K. Seetharaman. 2003. Factors influencing susceptibility of sunflower to charcoal rot disease caused by *Macrophomina phaseolina*. *J. Mycol. Plant Pathol.* 33: 252–255.

Suthinraj, T., S. Usharani, and D.J. Chriostopher. 2008. Effect of organic amendments and *Trichoderma viride* (Pres. ex Gray) on root rot incidence and yield of sunflower (*Helianthus annus* L.). *Adv. Plant Sci.* 21: 61–63.

Sujatha M. 2006. Wild *Helianthus* species used for broadening the genetic base of cultivated sunflower in India. *Helia* 29(44): 77–86.

Sujatha, M. and A.J. Prabakaran. 2006. Ploidy manipulation and introgression of resistance to *Alternaria helianthi* from wild hexaploid *Helianthus* species to cultivated sunflower (*Helianthus annuus* L.) aided by anther culture. *Euphytica* 152: 201–215.

Sujatha, M., A.J. Prabhakaran, and C. Chattopadhyay. 1997. Reaction of wild sunflower and certain interspecific hybrids to *Alternaria helianthi*. *Helia* 20: 15–24.

Sujatha, M., A.J. Prabakaran, S.N. Sudhakara Babu et al. 2003. Differential reaction of recombinant interspecific inbred lines of sunflower to red rust incited by *Puccinia helianthi*. *Helia* 26(39): 25–36.

Sujatha, M., A.V. Reddy, and A. Sivasankar. 2008. Identification of sources of resistance to *Alternaria* blight in sunflower. *Curr. Biotica* 2: 249–260.

Suthinraj, T., S. Usharani, and D.J. Chriostopher. 2008. Effect of organic amendments and *Trichoderma viride* (Pres. ex Gray) on root rot incidence and yield of sunflower (*Helianthus annus* L.). *Adv. Plant Sci.* 21: 61–63.

Tan, A.S. 1994. Occurrence, distribution and identification of rust (*Puccinia helianthi* Schw.) races of sunflower (*Helianthus annuus* L.) in Turkey. *J. Aegean Agric. Res. Inst. Anadolu* 4(1): 26–37.

Tan, A.S. 2010. Identification of rust (*Puccinia helianthi* Schw.) races in sunflower (*Helianthus annuus* L.) in Turkey. *Helia* 33(53): 181–190. doi:10.2298/HEL.1053/1855.

Tancic, S., B. Dedic, A. Dimitrijevic, S. Terzic, and S. Jocic. 2012. Bio-ecological relations of sunflower pathogens—*Macrophomina phaseolina* and *Fusarium* spp. and sunflower tolerance to these pathogens. *Romanian Agric. Res.* 29: 349–359.

Tančić, S., B. Dedic, S. Jocic et al. 2011. Sclerotinia wilt occurrence on sunflower in Vojvodina, Serbia. *Ratarstvo i Povrtarstvo* 48: 353–358.

Tang, S., V.K. Kishore, and S.J. Knapp. 2003. PCR-multiplexes for a genome-like framework of simple sequence repeat marker loci in cultivated sunflower. *Theor. Appl. Genet.* 107: 6–19.

Tang, S., J. Yu, M.B. Slabaugh, D.K. Shintani, and S.J. Knapp. 2002. Simple sequence repeat map of the sunflower genome. *Theor. Appl. Genet.* 105: 1124–1136.

Teodorescu, A., N. Boaghe, G. Dicu, and M. Popa. 2013. Research regarding new races of *Plasmopara halstedii* in Fundulea area. Scientific Papers. Series A. Agonomy, Vol. LVI, pp. 363–365, ISSN: 2285-5785. http://agronomyjournal.usamv.ro/index.php/scientific-papers/9-articles/382-research-regarding-new-races-of-plasmopara-halstedii-in-fundulea-area. Accessed on September 14, 2014.

Thines, M., H. Komjati, and O. Spring. 2005. Exceptional length of ITS in *Plasmopara halstedii* is due to multiple repetitions in the ITS-2 region. *Eur. J. Plant Pathol.* 112: 395–398.

Tomioka, K. and T. Sato. 2011. Gray mold of yacon and sunflower caused by *Botrytis cinerea*. *J. Gen. Plant Pathol.* 77: 217–219.

Tosi, L., R. Luigetti, and A. Zazzerini. 1999. Benzodithiadiazole induces resistance to *Plasmopara helianthi* in sunlower plants. *J. Phytopathol.* 147: 365–370.

Tourvieille de Labrouhe, D., A. Borda, J. Tourvieille et al. 2010. Impact of major gene resistance management for sunflower on fitness of *Plasmopara halstedii* (downy mildew) populations. *Oleagineux, Corps Gras, Lipides* 17: 56–64.

Tourvieille de Labrouhe, D., E. Pilorge, P. Nicolas, and F. Vear. 2000. *Le mildiou du tournesol*. CETIOM-INRA, Versailles, France, 176pp.

Tourvieille de Labrouhe, D., F. Serre, S. Roche, P. Walser, and F. Vear. 2008. Quantitative resistance to downy mildew (*Plasmopara halstedii*) in sunflower (*Helianthus annuus*). *Euphytica* 164: 433–444.

Tozlu, E. and E. Demirci. 2011. Determination of potential biocontrol organisms against *Sclerotinia sclerotiorum* and *S. minor* on sunflower. *Anadolu Tarm Bilimleri Dergisi* 26: 101–106.

Turkington, T.K., H.R. Kutcher, D. McLaren, and K.Y. Rashid. 2011. Managing Sclerotinia in oilseed and pulse crops. *Prairie Soils Crops J.* 4: 105–113.

Udayashankar, A.C., S. Chandra Nayaka, B. Archana et al. 2012. Specific-PCR-based detection of *Alternaria helianthi*: The cause of blight and leaf spot in sunflower. *Arch. Microbiol.* 194: 923–932.

Ullah, H., M.A. Khan, S.T. Sahi, and A. Sohail. 2010. Identification of resistant sources against charcoal rot disease in sunflower advance lines/varieties. *Pak. J. Phytopathol.* 22: 105–107.

Ullah, M.H., M.A. Khan, S.T. Sahi, and A. Habib. 2007. Evaluation of plant extracts against charcoal rot of sunflower caused by *Macrophomina phaseolina* (Tassi) Goid. *Pak. J. Phytopathol.* 19: 113–117.

Ullah, M.H., M.A. Khan, S.T. Sahi, and A. Habib. 2011. Evaluation of antagonistic fungi against charcoal rot of sunflower caused by *Macrophomina phaseolina* (Tassi) Goid. *Afr. J. Environ. Sci. Technol.* 5: 616–621.

Upendhar, S., T.V.K. Singh, and R.D.V.J. Prasada Rao. 2006. Relationship between thrips population, sunflower necrosis disease incidence and weather parameters. *J. Oilseeds Res.* 23: 267–269.

Upendhar, S., T.V.K. Singh, and R.D.V.J. Prasada Rao. 2009. Interrelationship between thrips and necrosis disease in sunflower, *Helianthus annus* L. *J. Oilseeds Res.* 26: 682–684.

Vasic, D., R. Marinković, F. Miladinović, S. Jocić and D. Škorić. 2004. Gene actions affecting sunflower resistance to *Sclerotinia sclerotiorum* measured by sclerotia infections of roots, stems and capitula. In: *The Proceedings of 16th International Sunflower Conference*, Fargo, North Dakota, pp. 603–608.

Vasic, D., D. Skoric, K. Taski, and L. Stosic. 2002. Use of oxalic acid for screening intact sunflower plants for resistance to *Sclerotinia* in vitro. *Helia* 25: 36: 145–152.

Vasyutin, A.S., B.I. Yudin, and V.V. Chumakova. 2003. Control of *Phomopsis* spread in sunflower. *Zashchita i Karantin Rastenii* 9: 28–30.

Vear, F., F. Serre, I. Jouan-Dufournel et al. 2008. Inheritance of quantitative resistance to downy mildew (*Plasmopara halstedii*) in sunflower (*Helianthus annuus* L.). *Euphytica* 164: 561–570.

Vear, F., F. Serre, S. Roche, P. Walser, and D.T. Labrouhe. 2007. Improvement of *Sclerotinia sclerotiorum* head rot resistance in sunflower by recurrent selection of a restorer population. *Helia* 30(46): 1–12.

Velazhahan, R., P. Narayanasamy, and R. Jeyarajan. 1991. Reaction of sunflower germplasm to rust disease in Tamil Nadu. *Indian Phytopathol.* 44: 239–241.

Velazquez, P.D. and N. Formento. 2000. Effect of nitrogen fertilization on "stem black spot" (*Phoma oleracea* var. *helianthi-tuberosi* Sacc.) in four genotypes of sunflower (*Helianthus annuus* L.). *AgriScientia* 17: 41–47.

Venkata Subbaiah, K., D.V.R. Saigopal, and M. Krishna Reddy. 2000. First report of a *tospovirus* on sunflower (*Helianthus annuus* L.) from India. *Plant Dis.* 84: 1343.

Verzea, M., F. Raducanu, M. Baldini, S. Raranciuc, and I. Moraru. 2004. Evaluation of genetic variability of some sunflower genotypes to the *Sclerotinia sclerotiorum* by different resistance tests and AFLP analyses. *Cercetari de Genetica Vegetala si Animala* 8: 35–44.

Veverka, K., J. Palicova, and I. Krizkova. 2008. The incidence and spreading of *Macrophomina phaseolina* (Tassi) Goidanovich on sunflower in the Czech Republic. *Plant Prot. Sci.* 44: 127–137.

Vincourt, P., F. As-sadi, A. Bordat et al. 2012. Consensus mapping of major resistance genes and independent QTL for quantitative resistance to sunflower downy mildew. *Theor. Appl. Genet.* 125: 909–920.

Vinokurova, T.P. 2000. Prospects of use of Agate-25K on sunflower. *Zashchita i Karantin Rastenii* 3: 30–31.

Viranyi, F. and O. Spring. 2011. Advances in sunflower downy mildew research. *Eur. J. Plant Pathol.* 129: 207–222.

Vrandecic, K. and D. Jurkovic. 2008. Sunflower stem canker—*Diaporthe helianthi* Munt.-Cvet. (anamorph *Phomopsis helianthi*). *Glasilo Biljne Zastite* 8: 348–352.

Vuong, T.D., D.D. Hoffman, B.W. Diers, J.F. Miller, J.R. Steadman, and G.L. Hartman. 2004. Evaluation of soybean, dry bean, and sunflower for resistance to *Sclerotinia sclerotiorum*. *Crop Sci.* 44: 777–783.

Wang, Y.G., Z. Feng, X. Cao et al. 2010. Resistance evaluation of different sunflower cultivars to *Sclerotinia sclerotiorum*. *Chin. J. Oil Crop Sci.* 32: 540.

Wagan, K.H., M.A. Pathan, and M.M. Jiskani. 2006. Effect of *Alternaria helianthi* on disease development, germination and seed quality in sunflower. *Pak. J. Agric. Eng. Vet. Sci.* 22: 26–29.

Wagan, K.H., M.A. Pathan, M.M. Jiskani, and H.B. Leghari. 2004. Inoculation of *Macrophomina phaseolina* at three stages in sunflower plant and its effect on yield components of different sunflower varieties. *Mycopath* 2: 21–24.

Weems, J.D., S.A. Ebelhar, V. Chapara, D.K. Pedersen, G.R. Zhang, and C.A. Bradley. 2011. First report of charcoal rot caused by *Macrophomina phaseolina* on sunflower in Illinois. *Plant Dis.* 95: 1318.

Wu, P.S., H.Z. Du, X.L. Zhang, J.F. Luo, and L. Fang. 2012. Occurrence of *Phoma macdonaldii*, the causal agent of sunflower black stem disease, in sunflower fields in China. *Plant Dis.* 96: 1696.

Yang, S.M., W.M. Dowler, and A. Luciano. 1989. A new gene in sunflower for resistance to *Puccinia helianthi*. *Phytopathology* 79: 474–477.

Yldrm, I., H. Turhan, and B. Ozgen. 2010. The effects of head rot disease (*Rhizopus stolonifer*) on sunflower genotypes at two different growth stages. *Turkish J. Field Crops* 15: 94–98.

Yu, J.K., S. Tang, M.B. Slabaugh et al. 2003. Towards a saturated molecular genetic linkage map for cultivated sunflower. *Crop Sci.* 43: 367–387.

Yu, X.Y., G. Wang, N. Zhang, and Y.F. Zhang. 2006. Preliminary study on an antagonistic bacterium of *Sclerotinia sclerotiorum* of sunflower. *J. Jilin Agric. Univ.* 28: 494–497.

Yue, B., S.A. Radi, B.A. Vick et al. 2008. Identifying quantitative trait loci for resistance to *Sclerotinia* head rot in two USDA sunflower germplasms. *Phytopathology* 98: 926–931.

Zandoki, E., S. Szodi, and G. Turoczi. 2006. Mycelial compatibility, aggressiveness and cultural characteristics of *Sclerotinia sclerotiorum*. *Acta Agronomica Ovariensis* 48: 31–39.

Zandoki, E. and G. Turoczi. 2004. Various aggressivity of *Sclerotinia sclerotiorum* strains on sunflower. *Novenyvedelem* 40: 67–70.

Zimmer, D.E. 1974. Physiological specialization between races of *Plamopara halstedii* in America and Europe. *Phytopathology* 65: 1465.

Zimmer, D.E. and G.M. Fick. 1974. Some diseases of sunflower in the United States—Their occurrence, biology and control. In: *Proceedings of the Sixth International Sunflower Conference*, Bucharest, Romania, pp. 673–680.

Zimmer, D.E. and D. Rehder. 1976. Rust resistance of wild *Helianthus* species of the North Central United States. *Phytopathology* 66: 208.

Zimmer, D.E., M.L. Kinman, and G.N. Fick. 1973. Evaluation of sunflower for resistance to rust and *Verticillium* wilt. *Plant Dis. Rep.* 57: 524.

Section V

Sesame

Sesame (*Sesamum indicum* L.; syn. *S. orientale* L.) variously named as *gingelly* or *til* belongs to the family Pedaliaceae. It is an annual, 1.0–1.5 m tall, herbaceous plant, maturing in 70–140 days. The basic chromosome number is 13 pairs ($2n = 26$). Molecular marker techniques such as amplified fragment length polymorphism, random amplified polymorphic DNA (RAPD), inter-simple sequence repeats (ISSR), and simple sequence repeats have been widely used in genetic diversity studies in sesame (Yadava et al. 2012, Pathak et al. 2014). The genome size of *S. indicum* is estimated to be about approximately in the range of 354–369 Mb (Ashi 2006, Wei et al. 2011).

The flowers are solitary, axillary, short-pediceled, and zygomorphic and are borne on the upper stem or branches. Self-pollination is the rule, but natural cross-pollination due to visiting bees may usually be seen to the extent of 5%. The fruit is a capsule and contains numerous small ovate seeds. Sesame seed contains high oil content 45%–52% (Hegde 2009) with 83%–90% unsaturated fatty acids, 20% proteins, and various minor nutrients such as vitamins and minerals and a large amount of characteristic lignans (methylenedioxyphenyl compounds) such as sesamin, sesamol, sesamolin, and tocopherols. Therefore, sesame seeds with high amounts of nutritional components are consumed as a traditional health food for its specific antihypertensive effect and anticarcinogenic, anti-inflammatory, and antioxidative activities. Sesame is thought to have originated in India, though its origin is sometimes traced to southern and southwestern Africa and also to the East Indies. The crop is mainly grown in the tropics and subtropics. Sesame grows on a variety of soils, but good yield is obtained on light, sandy loam, well-drained soils of moderate fertility. Principal sesame-producing countries are India, China, Korea, Iran, Turkey, Burma, and Pakistan in Asia; Egypt and Sudan in Africa; Greece in Europe; Venezuela, Argentina, and Colombia in South America; Nicaragua and El Salvador in Central America; and Mexico and the United States in North America. The largest producer and exporter of sesame seed in 2011 was Myanmar, secondly India, followed by China, Ethiopia, Nigeria, and Uganda. China is the world's largest consumer, and 70% of the world's sesame crop is grown in Asia, followed by Africa having a gross share of 26% in the world (FAOSTAT 2011).

About 65% of the annual sesame crop is processed into oil, and 35% is used in food. The food segment includes about 42% roasted sesame, 36% washed sesame, 12% ground sesame, and 10% roasted sesame seed. Because protein content and oil content are inversely proportional, seeds with an increased oil content have a decreased protein content. The oil quickly permeates and penetrates the skin, entering the blood stream through the capillaries. While in the blood stream, molecules of sesame seed oil maintain good cholesterol (HDL) and assist the body in removing bad cholesterol (LDL).

Different diseases of major economic importance affecting the crop are described in the following chapter.

7 Sesame Diseases

Phytophthora BLIGHT

SYMPTOMS

The disease can attack plants of all ages after they attain 10 days of age. Symptoms appear on all aerial parts of the affected plants. The first symptom is the appearance of water-soaked brown spots on leaves and stems (Figure 7.1). The spots gradually extend in size. Under favorable weather conditions, the brownish discolored spots spread rapidly both upward and downward and also around the stem. The brownish area later turns deep brown and becomes black with the spread of the infection. The capsules are also affected. In humid weather, the white woolly growth of the fungus can be seen on the surface of affected capsules. Capsules on affected branches are poorly formed. The seeds remain shriveled in the case of severe attack.

GEOGRAPHICAL DISTRIBUTION AND LOSSES

Phytophthora blight of sesame was first reported from India by Butler (1918). Widespread occurrence of the disease has now been reported from Argentina, Dominican Republic, Egypt, India, Iran, Sri Lanka, and Venezuela (Verma et al. 2005). The disease has been reported to be of economic importance in the states of Assam, Gujarat, Madhya Pradesh, and Rajasthan in India and in Sri Lanka (Kolte 1985, Pathirana 1992, Kalita et al. 2000, 2002). Since the disease generally kills the affected plants, it can be observed that the net loss is directly proportional to the incidence of the disease. The mortality of the plants due to the disease may be as high as 72%–79%. The disease is becoming increasingly more important in Assam in recent years where the losses in yield in sesame crop range between 51% and 53% (Kalita et al. 2002).

Besides causing blight, the pathogen is found to be associated with vivipary in immature seeds of sesame contained in greed pods of plants raised from naturally infected seeds. It is an unusual phenomenon that besides increasing the seed infection also renders poor-quality seeds. The host–pathogen interaction results in abnormal seedling emergence, which lacks vigor and further survival (Dubey et al. 2011).

PATHOGEN

The pathogen is *Phytophthora parasitica* (Dastur) var. *sesami* Prasad (*P. nicotianae* B. de Haan var. *parasitica* [Dastur] Waterh).

Mycelium of the fungus in young culture is coenocytic and profusely branched, but septa can be observed in 2-month-old cultures. The hyphae are hyaline and are 2–8 μ thick.

The fungus does not form sporangia on culture media, but abundant sporangia can be observed in nature on woolly mycelium growing on infected capsules. The sporangiophores are branched sympodially and bear ovate-to-spherical sporangia terminally. They have a prominent apical papilla and measure 25–50 × 20–35 μ in size.

The mycelium, when floated in tap water, forms zoosporangia readily in 48 h. The zoospores are formed inside, and they clearly get separated within the sporangium. The zoospores are liberated in water if the mycelium is flooded with water. The antheridium can be observed at the base, and attachment is typically amphigynous. The oospores are spherical, smooth, double walled, and hyaline.

FIGURE 7.1 Phytophthora blight of sesame. The first symptom is the appearance of water-soaked brown spots on leaves. (Courtesy of Dr. Anil Kotasthane, IGKV, Raipur, India.)

The fungus grows well on oatmeal/agar at an optimum temperature of 30°C. Culture of the fungus may show tendency to lose virulence if it is maintained on artificial medium for a long period. The fungus is reported to be thiamine deficient. Its growth becomes good when thiamine is incorporated (200 μg/L) in the medium. The fungus grows best at pH 6.5.

Epidemiology and Disease Cycle

The pathogen can survive in mycelial form up to 50°C temperature, and culture having chlamydospores may survive up to 52°C. Viability of the culture can be kept in a refrigerator for 1 year at 5°C. These studies suggest that the fungus can survive in soil during the summer and winter where temperature never rises beyond 50°C or drops below 5°C. The fungus survives in soil during the unfavorable period in the form of dormant mycelium and/or in the form of chlamydospores. In addition to soil, seed also appears to play an important role in the recurrence and spread of the disease. In seed, the mycelium has been located in the embryo. However, there are reports that the fungus reduces seed viability but it is not seed borne (Maiti et al. 1988). The mycelium in the host tissue is inter- or intracellular, but it does not form haustoria. The sporangiophores emerge in groups by rupturing the epidermis, but sometimes they emerge through stomata (Verma et al. 2005). The zoosporangia are formed abundantly if humid weather prevails for 2–3 days but soon stop formation if a dry spell appears. The secondary infection occurs through zoospores. *P. parasitica* var. *sesami* is restricted in its pathogenicity to sesame plants only. Sehgal and Prasad (1966) have shown variation in virulence among various isolates of *P. parasitica* var. *sesami*. Single-zoospore isolates show great variations, under similar conditions of infection, in virulence, which may range from nonpathogenic to highly pathogenic. A few isolates of *P. parasitica* var. *sesami* can lose virulence, but the loss in virulence is not permanent, since a few cultures can regain the loss of virulence after passage through the host. On repeated host passages, the culture can even become more virulent than the original ones (Sehgal and Prasad 1971).

Heavy rains for at least 2 weeks and high humidity (above 90%) for 3 weeks or more favor the development of the disease. When such favorable conditions persist for a longer time, the infection appears quite fast. It is observed that the initial development of the disease is much earlier when the soil temperature is 28°C, while the initial appearance of the disease is delayed with an increase in the soil temperature up to 37°C. The pathogen is favored by 30°C, can tolerate 35°C, but fails to grow at 37°C. Hence, soil temperature of 28°C–30°C is necessary for disease development (Prasad et al. 1970). It is further reported that incidence of the *Phytophthora* blight of sesame shows a close parallelism to the growth of the fungus. Bright sunshine hours for 2–3 days are not favorable for disease development since zoospore formation is stopped under such conditions. The disease appears to become more severe in heavy soils (Verma 2002). The moderate nitrogenous fertilizer application leads to more incidence of *Phytophthora* blight of sesamum (Verma and Bajpai 2001).

Disease Management

Host Plant Resistance

Out of several strains and varieties of sesame (*Sesamum orientate* L.) and five other species, namely, *S. occidentalis* Heer and Regal, *S. indicum* L., *S. laciniatum* Willd., *S. prostratum* Retz., and *S. radiatum* Schum. and Thonn., tested for resistance to the disease, none is identified to be resistant to the disease (Kolte 1985, Choi et al. 1987). However, under All India Coordinated Research Project on Oilseeds, a number of sesame lines over several years of crop season testing have been found to be tolerant to *Phytophthora* blight. These lines are TC-25, JLSC-8, TKG-21, AT-60, AT-64, B-14, Chopra-1, Durga (TKG-6), JLT-3, JLT-7, Lakhora-1, Phule till-1, RT-46, T-12, and T-13. These lines/strains that have shown tolerance over longer duration can be grown to manage the adverse effect of the disease on yield (Verma et al. 2005). In Venezuela, three lines, 71-184-1, 79-129-2, and 71-145-3 (selected from B_4 of Ajinio Atar 55), are reported to be disease resistant. The National Institute of Crop Science in Korea has developed a new black-seeded variety *Kangheuk*, which is a high-yielding, high-lodging, and *Phytophthora* blight–resistant variety (Shim et al. 2012). Epiphytotic conditions and nonavailability of resistant germplasm had prompted the use of gamma ray–induced (450–600 Gy) mutation breeding for the development of *Phytophthora* blight–resistant sesame variety *ANK-S 2* in Sri Lanka (Pathirana 1992).

Chemical Control

Seed-borne infection can be controlled by treating the seed with thiram (0.3%). Secondary infection and further spread of the disease can be brought under control by three sprayings of Bordeaux mixture (3:3:50), each at an interval of 1 week after the appearance of the disease (Verma et al. 2005). Spray application of dithiocarbamate fungicides such as mancozeb (0.3%) or zineb (0.3%) and Fytolan (copper oxychloride) (0.3%) is reported to be effective in the control of the disease (Kalita et al. 2000, 2002).

Cultural Control

Sanitation and clean cultivation should be followed as additional measures to control the disease. Use of sowing date depending upon the prevailing local conditions and crop fields with light soil with proper drainage should be preferred to avoid heavy losses due to disease. The intercropping of sesame with soybean, castor, maize, sorghum, or pearl millet in the ratio of 1:3 or 3:1 shows a low incidence of the disease with higher yield. Application of farm yard manure (FYM) or neem cake with inorganic fertilizers N_{60}, P_{40}, and K_{20} reduces the disease incidence (Verma et al. 2005). Planting of sesame in 0.2 mm wide ridge in plots mulched with vinyl reduces the spread of the disease by at least 30% and increases the yield by 22% (Choi et al. 1984).

Biological Control

Species of *Pseudomonas*, *Bacillus*, and *Streptomyces*, which are most active at 25°C–27°C at field capacity moisture level, can be suppressive to *Phytophthora* species in soil (Erwin 1983). Antagonistic *Trichoderma* species, namely, *T. viride*, *T. harzianum*, and *Pseudomonas fluorescence*, when used as seed treatment, not only reduce the disease significantly but substantially increase the sesame yield (Verma 2002). Application of phosphorus-solubilizing bacteria (PSB) along with neem cake or 50% NPK + FYM or 100% NPK + PSB reduces the disease (Verma and Bajpai 2001).

CHARCOAL ROT

Symptoms

Sesame plants may be attacked immediately after sowing. The germinating seeds may become brown and rot. In the seedling stage, the roots may become brown and rot, resulting in the death of the plants. If the plants survive, the older plants are affected at the base of the stem indicating the formation of lesion that later spreads to the middle portion of the stem and becomes ashy, causing drooping of leaves and top of the plants (Figure 7.2). Such plants make poor growth and remain stunted. The mycelium of the fungus progresses upward in the stem, and as the stem dries, pycnidia are formed as minute black dots. The stem may break off, and the blackening may extend upward on the stem. The capsules are also affected. Such capsules open prematurely, exposing shriveled and discolored seeds (Figure 7.3). Seeds may show the presence of sclerotia on the surface.

Geographical Distribution and Losses

Reports of occurrence of charcoal rot of sesame have been made from all over the sesame-growing areas in the world (Kolte 1985, Verma et al. 2005). The disease is particularly reported to be quite serious, limiting the production of the crop in Ismailia Governorate Region in Upper Egypt (Abdou et al. 2001, El-Bramawy and Wahid 2007); Southeastern Anatolia Region in Turkey (Sağır et al. 2009); in the Portuguesa state in Venezuela (Cardona and Rodriguez 2006, Martinez-Hilders et al. 2013); in the Chandrapur district of Vidarbha region of Maharashtra, the Gwalior Division of Madhya Pradesh, and in the states of Haryana and Chhattisgarh in India (Kolte 1985,

FIGURE 7.2 Severely affected sesame plant showing charcoal symptoms. (Courtesy of Dr. Anil Kotasthane, IGKV, Raipur, India.)

FIGURE 7.3 Charcoal rot–affected sesame plants. Note the ashy color of the stem with infected discolored capsules. (Courtesy of Dr. Anil Kotasthane, IGKV, Raipur, India.)

Deepthi et al. 2014); and in Pakistan (Akhtar et al. 2011). Seedling mortality due to seed-borne infection aggravates the disease problem by reducing the plant stand per unit area, resulting in low yield. About 5%–100% yield loss due to the disease is reported. An estimated yield loss of 57% at about 40% disease incidence is reported (Maiti et al. 1988). In Venezuela, losses in sesame due to charcoal rot have been evaluated resulting up to 65% of seed weight reduction for affected plants (Martinez-Hilders et al. 2013). The importance of the charcoal rot lies not only in affecting the yield and causing quantitative and qualitative losses (Sağır et al. 2009) but also in increasing soil infestation with the causal fungus. For example, sclerotia of *Macrophomina phaseolina* in Venezuelan soils of sesame production areas have been estimated to be up to 200 per gram of soil (Martinez-Hilders et al. 2013). If the disease appears simultaneously with *Phytophthora* blight or with *Fusarium* wilt, the losses in yield usually are very high.

Pathogen

The pathogen is *M. phaseolina* (Tassi) Goid. The morphological and physiological characteristics of the pathogen have been described under chapters on peanut and sunflower diseases. Molecular methods used for determining the level of genetic diversity and polymorphism among *M. phaseolina* populations affecting sesame include random amplified polymorphic DNA (RAPD), amplified fragment length polymorphism, and inter-simple sequence repeats (ISSR). It is revealed that *M. phaseolina* populations in all the major sesame production regions in China (Wang et al. 2011), Iran (Bakhshi et al. 2010, Mahdizadeh et al. 2012), Mexico (Munoz-Cabanas et al. 2005), and Venezuela (Martinez-Hilders and Laurentin 2012, Martinez-Hilders et al. 2013) are highly genetically diverse based on genomic data. High level of genotypic variability is likely due in part to the exposure of the pathogen to diverse environment and a wide host range within these countries. However, no clear association between geographical origin and host of each isolate has been found, though isolates from the same location show a tendency to belong to their respective closer groups indicating closer genetic relatedness (Bakshi et al. 2010).

Leaf, stem, and root extracts of urd and mung beans have been reported to show an inhibitory effect on sclerotial formation of the sesame isolate in vitro (Kolte and Shinde 1973). Ammonium chloride also has an inhibitory effect on sclerotial formation of the fungus (Kolte 1985). The sesame isolates are chlorate sensitive and grow normally with numerous dark microsclerotia production on the potassium chlorate–containing medium (Rayatpanah et al. 2012). Interestingly, the soybean and sunflower isolates are chlorate sensitive and divided into two classes. Class 1 includes the isolates that grow sparsely with a feathery-like pattern, and Class 2 includes the isolates that grow well with a nonfeathery pattern. Isolates with feathery-like pattern are more virulent on soybean and sunflower (Rayatpanah et al. 2012). Two distinct strains, namely, pycnidia-forming and sclerotia-forming strains, have been reported from Bangladesh. The pycnidial strain is reported to be more pathogenic on sesame than the sclerotial strain (Kolte 1985, Verma et al. 2005).

Epidemiology and Disease Cycle

The fungus survives as free sclerotia in soil or as mycelium and sclerotia carried in crop debris. It spreads by the movement of soil and crop debris (Al-Ahmad and Saidawai 1988) and through the sesame seeds. The sesame seed has been found to carry the fungus on and inside the testa as sclerotia or as stromatic mycelium. The standard blotter method, use of a selective medium and scanning electron microscopy, facilitates the detection of seed-borne infection in sesame (Verma et al. 2005, El-Wakil et al. 2011). There is a positive correlation between microsclerotia on sesame seed with percent plant infection (Gupta and Cheema 1990). It is reported that the germinating seed and seedlings stimulate normal sclerotial germination and attract developing mycelium to the host roots. Entry may occur directly through the cuticle and epidermis, infection cushions and appressoria are also reported to be formed on sesame plants prior to infection, and the pathogen produces cell-wall-degrading pectolytic and cellulolytic enzymes. The most aggressive isolate produces more cell-wall-degrading enzymes than the less aggressive isolates (Gabr et al. 1998).

A high soil temperature (35°C) and low osmotic potential reduce plant vigor and favor growth of the fungus and initiation of infection. Maximum temperature of 31.6°C, minimum temperature of 24°C, and relative humidity of 88% favor severe charcoal rot disease development (Deepthi et al. 2014). The response of the sesame crop to stress conditions has been found to be of significant importance in epidemiology, and irrigation reduces infection by reducing drought stress. Periods of drought between heavy rains favor the development of the disease in Africa. Strains of *M. phaseolina* are known for their wide host range and infect a large number of weeds and rotation crops, which function as a source of inoculum and survival of the pathogen (Simosa and Delgado 1991, Kolte 1997).

Disease Management

Host Plant Resistance

Differences in host resistance to charcoal rot through screening of genotypes and molecular marker techniques in sesame have been noted in trials in several Asian and South American countries (Chattopadhyay and Sastry 2000, Melean 2003, El-Fiki et al. 2004b, El-Bramawy and Shaban 2007, El-Bramawy et al. 2009, Gao et al. 2011, Zhang et al. 2012). The results reveal that there is high heritability for host resistance indicating additive gene nature of the resistance characters and consequently a high gain from selection. High resistance gives the lowest seed yield, and great care is taken during selection and pedigree selection for resistance breeding program to *M. phaseolina* (Mahdy et al. 2005b, El-Bramawy and Wahid 2006, 2007, El-Shakhees and Khalifa 2007). In Egypt, sesame line P5 (NM 59) and the F6-derived lines C6.3, C1.10, and C3.8 are the most valuable sources of resistance to charcoal rot disease. The P5 line is resistant to both charcoal rot and *Fusarium* wilt, whereas the F6-derived line C6.3 is the most resistant and top-yielding one against combined infection of charcoal rot and *Fusarium* wilt diseases (Shabana et al. 2014).

Different sesame germplasm lines and cultivars that have been found tolerant or less susceptible to charcoal rot are ORM 7, ORM 14, and ORM 17 (Dinakaran and Mohammed 2001, Subrahmaniyan et al. 2001, Thiyagu et al. 2007); TLC-246, TL6-279, and TLCCCCC-281 (John et al. 2005); ZZM0565, ZZM0570, Xiangcheng dazibai, Xincai Xuankang, Shangshui farm species, and KKU 3 (Zhao et al. 2012); mutants NS 13 P1, NS 163-1, NS 270 P1, and NS 26004 (Akhtar et al. 2011); UBQ5, UF 4A, and alpha-tubulin (Liu et al. 2012); UCLA-1, EXP-1, and DV-9 (Melean 2003); and Aceteru-M, Adnan (5/91), Taka 2, B 35, and mutation 48 (El-Fiki et al. 2004b).

The sources of resistance appear to differ in the mechanism of resistance. Factors such as morphological traits like single stem (Li et al. 1991); medium branch numbers (El-Bramawy 2008, El-Bramawy et al. 2009); creamy or white seed color (El-Bramawy et al. 2009); antifungal nutritional components such as phytin, trypsin inhibitor, and tannins (El-Bramawy and Embaby 2011); certain biochemical factors as faster rate of activity of polyphenol oxidase enzymes in Chinese sesame cultivar *Yuzhi 11* (Liu et al. 2012); and different isoenzyme band patterns (Zhang et al. 2001) have been linked with resistance to charcoal rot in sesame. The mature plant reaction, through hybridization studies, indicated that susceptibility in the mature plant is dominant over tolerance, and it is controlled by 1, 2, or 3 pairs of genes (Kolte 1985).

Chemical Control

Seed treatment with carbendazim (0.1%–0.3%) gives complete control of seed-borne infection of *M. phaseolina* in sesame when used as seed treatment fungicide (Choudhary et al. 2004, Rajpurohit 2004b, Shah et al. 2005, John et al. 2010). Other seed treatment fungicides are thiophanate methyl (John et al. 2010), Benlate or Rizolex T at 3 g/kg seed (El-Deeb et al. 1985, El-Fiki et al. 2004a), mancozeb (Mudingotto et al. 2002), thiram, captan, and carboxin (Verma et al. 2005). Soil treatment with fungicides is effective but impracticable. The integration of fungicide carbendazim seed treatment (0.1%) with carbendazim-tolerant strain of *T. viride* (Tv-Mut) as induced by mutating the native strain of the fungus by UV irradiation and soil supplemented with 20 kg P and 15 kg K/ha show the highest reduction (91.7%) in sesame stem–root rot incidence caused by *M. phaseolina* (Chattopadhyay and Sastry 2002). Aminobutylic acid and potassium salicylate can effectively control charcoal rot in sesame by induction of host resistance against *M. phaseolina* and increasing plant height, indole acetic acid (IAA) content, and peroxidase (PO) activity (Shalaby et al. 2001). Soaking sesame seeds in indole butyric acid at 100 ppm or salicylic acid at 4 mM produce healthy stand of plants.

Cultural Control

The average charcoal rot incidence can be lowered down by choice of sowing date and levels and time of irrigation depending on the local conditions in a particular geographical area. Early sowing by June 10 in Egypt and following *hills-over-furrows* method of sowing and giving only one irrigation during the whole growing season to a crop fertilized with N at 65 kg, P at 200 kg, and K at 50 kg/feddan (0.42 ha) result in significant reduction in charcoal rot incidence (Shalaby and Bakeer 2000). Similar results are evident under early sowing and frequency of irrigation in Turkey (Sağır et al. 2010). Lowering concentration of Ca, Na, Mg, and Fe and increasing concentration of K, Cu, and Zn in the soil by applying chemical fertilizers and organic manure may reduce very much the charcoal root rot incidence (Narayanaswamy and Gokulakumar 2010). Sesame crop grown as mixed or intercropped with green gram in 1:1 is useful in the management of the charcoal rot and results in higher sesame yield in the arid region (Rajpurohit 2002, Ahuja et al. 2009). It is noteworthy to note that green gram (mung) and black gram plant extracts are inhibitory to the growth of *M. phaseolina* (Kolte and Shinde 1973). Six weeks of soil solarization of infested crop field sites in the summer months result in good sesame seed germination and better disease management under Indian conditions (Chattopadhyay and Sastry 2001).

Biological Control

Effect of antagonistic fungi and bacteria isolated from the rhizosphere of sesame is reported to be efficiently more effective in controlling the root rot and stem rot of sesame caused by *M. phaseolina* (El-Bramawy and El-Sarag 2012). Sesame seed treatment with (a) *T. viride* at 4 g/kg of seed (Rajpurohit 2004a, Hafedh et al. 2005, Rani et al. 2009, Zeidan et al. 2011), (b) *T. harzianum* (Pineda 2001, Cardona and Rodriguez 2002, 2006, El-Fiki et al. 2004b, Nair et al. 2006, Sattar et al. 2006, Moi and Bhattachrya 2008), (c) *P. fluorescens* (Jayshree et al. 2000, Moi and Bhattachrya 2008), and (d) *Bacillus subtilis* (Nair et al. 2006, Elewa et al. 2011) has been found effective in the control of charcoal rot disease. Green manure of *Crotolaria* amended with *Trichoderma* constitutes a viable alternative for the control of charcoal rot of sesame (Cardona 2008).

A combination of seed treatment and soil application of the antagonists through the application of clay granules impregnated with *T. harzianum* or *P. fluorescens* at sowing time appears to be much more effective in the control of the charcoal rot (Pineda 2001, Cardona and Rodriguez 2002). Application of vesicular–arbuscular mycorrhizae (VAM), namely, *Glomus* spp., together with biocontrol agents *T. viride* or *B. subtilis* significantly helps in efficient control of root rot (*M. phaseolina*) and *Fusarium* wilt diseases of sesame than individual application of either VAM or antagonists (Elewa et al. 2011, Zeidan et al. 2011). Soil solarization in combination with fungal antagonists *T. pseudokoningii* and *Emericella nidulans* singly or in mixed inocula reduces charcoal rot incidence in sesame significantly (Ibrahim and Abdel-Azeem 2007). Seed treatment with *Azotobacter chrococum* and seed + soil treatment with *Azospirillum* also reduce the disease by about 30% (Verma et al. 2005, Maheshwari et al. 2012).

Effect of Plant Extracts

Extracts of *Thevetia neriifolia* (Bayounis and Al-Sunaidi 2008a), *Azadirachta indica*, *Datura stramonium*, *Nerium oleander*, *Eucalyptus camaldulensis* (Bayounis and Al-Sunaidi 2008b), and *Helichrysum* flower (Shalaby et al. 2001) show inhibitory effect on the growth of *M. phaseolina*, indicating their potential use in the control of the disease. The extracts of *Eucalyptus* (*Eucalyptus rostrata*, *E. camaldulensis*), peppermint (*Mentha piperita*), and thyme (*Thymus serpyllum*), when used in sand culture or under in vitro conditions in growth media and inoculated with *M. phaseolina*, have been found to show increase in sesame seed germination despite the presence of *M. phaseolina* in the culture, indicating potential usefulness of these extracts (Sidawi et al. 2010).

Fusarium WILT

Symptoms

Plants get infected at any stage of the crop development including the damping-off phase in the seedling stage (Fallahpori et al. 2010, 2013). During later stages of the plant, yellowing of the leaves is the first noticeable symptom of the wilt in the field. Leaves become yellowish, droop, and dessicate. Sometimes such leaves show inward rolling of the edges and eventually dry up. The terminal portion dries up and becomes shrunken and bent over. In a severe infection, the entire plant becomes defoliated and dry. In a less severe infection or when mature plants are infected, only one side of the plant may develop symptoms, resulting in partial wilting, and a half stem rot symptom has been reported (Cho and Choi 1987). A blackish discoloration in the form of streaks appears on infected plants. Discoloration of the vascular system is conspicuous in the roots. Roots in the later stages show rotting, wholly or partially corresponding with that side of the plant showing disease symptoms. Numerous pink pinhead-sized sporodochia (containing macroconidia of the fungus) may be seen scattered over the entire dried stem. The capsules of wilted plants also show numerous sporodochia.

Geographical Distribution and Losses

Fusarium wilt of sesame was reported for the first time from North America in 1950 (Armstrong and Armstrong 1950). Since then, the disease is reported to occur in Egypt, Colombia, Greece, India, Iran, Israel, Japan, Korea, Malawi (formerly Nyasaland), former Soviet Union, the United States, and Venezuela. Similar disease has been reported from Pakistan, Peru, Puerto Rico, and Turkey (Kolte 1985, Verma et al. 2005). The disease can be devastating on susceptible varieties of sesame, but many local varieties have been found to have some degree of resistance to local races of the fungus. Epiphytotic occurrence of the disease was reported in 1961 and 1964 in the United States and in 1959 in Venezuela.

Pathogen

The pathogen is *Fusarium oxysporum* (Schelt.) f. *sesami* Jacz. Isolation of the causal fungus could be obtained more easily from the infected dry sample (dry sample isolation) compared to conventional direct isolation technique from freshly infected sesame plants (Su et al. 2012).

The fungus produces profuse mycelial growth on potato dextrose agar. The mycelium is arid, hyaline, septate, and richly branched, turning light pink when old. The microconidia are formed abundantly. They are hyaline, ovoid to ellipsoid, unicellular, and measure 8.5 × 3.25 μ in size. In the old culture, the macroconidia are formed sparsely. They are 3–5, septate, and measure in the range of 35–49 × 4–5 μ in size.

The macroconidia are produced abundantly in sporodochia as they develop on affected plants. The chlamydospores are globose to subglobose, smooth, or wrinkled and measure 7–16 μ in diameter. Physiological studies on the pathogen have been made. The fungus grows best on Richard's medium. It grows at the temperature range of 10°C–30°C with an optimum temperature of 25°C. Nitrate nitrogen and pH 5.6–8 support the maximum growth of the fungus. Illumination inhibits spore germination (Liu et al. 2010).

Epidemiology and Disease Cycle

The fungus is restricted in its host range to sesame. Morphological differences and similarities have been reported in different isolates of *F. oxysporum* f. *sesami*. Three strains have been reported in Venezuela on the basis of morphological differences, but these strains are reported to show a similar degree of pathogenicity. It is revealed that there is a relationship between vegetative compatibility groups of the pathogen and geographic origin of the isolates collected from the different sesame-growing regions (Basirnia and Banihashemi 2005). The pathogenic variation and molecular characterization of *Fusarium* species isolated from wilted sesame have been studied (Li et al. 2012).

The pathogen is reported to be seed and soil borne, and it may persist for many years in the soil. The amount of seed transmission of the pathogen varies in the range of 1%–14% depending on the severity of systemically infected sesame plants (Basirnia and Banihashemi 2006). It appears that the fungus penetrates the host through root hairs and causes trichomycosis. The most virulent isolates produce more cell-wall-degrading enzymes than the less virulent ones (Gabr et al. 1998). The culture filtrate of *F. oxysporum* f. *sesami* has been reported to have an inhibitory effect on sesame. Shoot and root growth is also inhibited by culture filtrate of the fungus indicating the production of toxic substances by the pathogen. Some elements like vanadium, zinc, boron, molybdenum, and manganese are highly inhibitory to *F. oxysporum* f. *sesami* (Gabr et al. 1998).

High soil temperature to a depth of 5–10 cm and 17%–27% water-holding capacity during dry periods is favorable for the development of the disease. Drought stress in the sesame plants predisposes the plants to infection and development of wilt and influences the host genotype reaction to the disease (El-Shemey et al. 2005, Kavak and Boydak 2011). The *Fusarium* wilt of sesame is reported to be associated with nematode attack in South America and with *M. phaseolina* in Egypt, India, and Uganda (Kolte 1985, 1997). The density of the fungus becomes higher in soil under continuous cropping (Paik et al. 1988).

Disease Management

Host Plant Resistance

Sesame *Fusarium* wilt–resistant accessions/genotypes have been identified from different sesame-growing countries. These are NSKMS 260, 261, and 267 and TMV 3 (Badri et al. 2011, Jyothi et al. 2011); S-5, S-4, H-9 (El-Bramawy and Wahid 2007); S2, H4, mutant 8, UNA 130, H1, S1 (El-Bramawy and Wahid 2009); somaclonal strain GZO 25 (Abd-El Moneem et al. 1996); Camdibi, WS-143, WS-313, Birkan (Silme and Cagrgan 2010); and Sanliurfa-63189 (Kavak and Boydak 2006). A considerable degree of variability in differences in resistance to *Fusarium* wilt among sesame genotypes has been reported, and the breeding methodology in sesame improvement (i.e., the selection, pedigree method, and hybridization) depends upon the nature and magnitude of the gene action in controlling the genetic behavior of the disease resistance trait (El-Bramawy et al. 2001, Zhang et al. 2001, El-Bramawy 2003, Ammar et al. 2004, Mahdy et al. 2005b, Kavak and Boydak 2006, El-Shakhess and Khalifa 2007, El-Hamid and El-Bramawy 2010). Parental line P5 and F6-derived lines C1.6, C1.10, C3.8, C6.3, C6.5, and C9.15 are reported to be the most resistant sesame lines to the *Fusarium* wilt disease (Shabana et al. 2014). Higher resistance of a germplasm line, in general, is reported to be lower yielder (El-Hamid and El-Bramawy 2010). However, high-yielding multiple disease-resistant sesame cultivar *Yuzhi 11* has been developed in China through breeding (Wei et al. 1999).

Satisfactory sources of *Fusarium* wilt resistance in sesame have been developed by mutagenesis and mutation breeding techniques (Uzun and Cagrgan 2001, Soner Slme and Cagrgan 2010). For example, *Birkan* is a high-yielding sesame mutant cultivar derived from 400 Gy gamma radiation in Egypt (Silme and Cagrgan 2010).

Sesame accessions with medium branch number and creamy or white seed color are the only covariate that significantly correlates with the infection caused by *F. oxysporum* f. sp. *sesami*, and these traits can be directly used for direct selection of sesame accessions that are resistant to *Fusarium* wilt and charcoal rot diseases (El-Bramawy et al. 2009). Depending on the genotypes, gene action for resistance to *Fusarium* wilt has been found to be additive with high heritability (El-Bramawy 2006, Bayoumi and El-Bramawy 2007, El-Bramawy and Wahid 2007), nonadditive (El-Bramawy and Shaban 2007, 2008, El-Shakhess and Khalifa 2007), and with epistatic effects (Bakheit et al. 2000).

Besides the genetic factors, some wilt-resistant genotypes possess the high value of the pathogen antinutritional factors such as phytic acid, trypsin inhibitor, and tannins (El-Bramawy and Embaby 2011). Many other *Fusarium* wilt–resistant sesame genotypes show significant differences in yield and yield components, total phenol contents, polyphenol oxidase, and PO enzyme activity indicating the importance of biochemical constituents in the expression of mechanism of resistance/tolerance to *Fusarium* wilt in sesame (Ghallab and Bakeer 2001).

Chemical Control

Seed treatment with benomyl or carboxin at 0.2% or with carbendazim (0.25%) or thiram (0.3%) results in significant control of the disease up to about 45 days after seed germination (Ahmed et al. 1989, Shalaby 1997). Sesame seeds soaked in ascorbic acid and salicylic acid (5 mM) for 24 h and sown and then treated with ascorbic acid and salicylic acid 15 days after sowing give best control of the disease through induced host resistance (Abdou et al. 2001).

Cultural Control

Balanced fertilization and insect pest control ensure good growth of the crop and help in the reduction of the disease. Trace elements such as copper, manganese, and zinc decrease the incidence of wilt of sesame (Abd-El-Moneem 1996). In heavily infested soil, at least 5 years should elapse between two sesame crops. Cultivation of sesame in rotation with onion or wheat is helpful in the reduction of the *Fusarium* wilt in sesame (El-Kasim et al. 1991). Sanitation and clean cultivation and choice of sowing dates depending on the known prevailing local conditions are taken into practical use in disease management (Verma et al. 2005). For example, sowing the sesame crop around June 10 through *hills-over-furrow* and fertilizing the crop with NPK (65, 200, and 50 kg/feddan, respectively, in 0.42 ha) and giving one irrigation during the growing season in Egypt are a very useful cultural practice package for the management of the disease in sesame (Shalaby and Bakeer 2000).

Biological Control

Several microbial antagonists such as *T. viride*, *Gliocladim virens* (Kang and Kim 1989, Wuike et al. 1998, Sangle and Bambawale 2004, El-Bramawy and El-Sarag 2012), *Bacillus polymyxa* (Hyun et al. 1999), *B. subtilis* (El-Sayeed et al. 2011), *Enterobacter cloacae* (Abdel-Salam et al. 2007), *Pseudomonas aeruginosa* (Abdel-Salam et al. 2007), *P. putida* and *P. fluorescens* (Farhan et al. 2010), *Streptomyces bikiniensis*, and *S. echinoruber* (Chung and Hong 1991, Chung and Ser 1992) are inhibitory to the growth of *F. oxysporum* f. sp. *sesami* and show high potential for their use in the management of *Fusarium* wilt of sesame. *Trichoderma* species grown on cow dung slurry and cow dung are the most effective in the control of the wilt disease of sesame (Sangle and Bambawale 2004).

Fusarium wilt of sesame can be controlled with application of plant-growth-promoting rhizobacteria, and this practice offers a potential nonchemical means for disease management. A combination of *Azospirillum brasilense*–based Cerialin and *Bacillus megaterium*–based Phosphoren biofertilizers plus Topsin (100 ppm) has been found to give significant reduction of *Fusarium* wilt incidence, with increased morphological characteristics and plant yield (Ziedan et al. 2012). Similarly, a mixture of *P. putida* 2 plus *P. fluorescens* 3 treatment together (Fusant) as biocide and biofertilizer gives better control of the wilt disease with higher sesame crop yield (Farhan et al. 2010). Fertilizer-adaptive variant tetracycline-resistant strain *TRA2* of *Azotobacter chroococcum*, an isolate of wheat rhizosphere, has been found to show plant-growth-promoting attributes and strong antagonistic effect against sesame wilt and charcoal rot pathogens. Seed bacterization with the strain TRA2 results in significant decrease in *Fusarium* wilt disease incidence and increase in vegetative growth of sesame plants (Maheshwari et al. 2012).

Glomus spp. (VAM) protect the sesame plants by colonizing the root system and consequently reduce colonization of fungal pathogens in sesame rhizosphere by stimulation of bacteria belonging to the *Bacillus* group. These bacteria show high antagonistic potential, and this significantly reduces *Fusarium* wilt incidence in sesame (El-Sayeed Ziedan et al. 2011).

Effect of Plant Extracts

Extracts of leaves of thyme, eucalyptus, and garlic reduce the incidence of *Fusarium* wilt disease of sesame. Extract of peppermint (*M. piperita*) leaves not only reduces the wilt incidence but also increases the yield of sesame plant (Sidawi et al. 2010).

Alternaria LEAF SPOT

Symptoms

Symptoms of the disease appear mainly on leaf blades as small, brown, round-to-irregular spots, varying from 1 to 8 mm in diameter. The spots later become larger and darker with concentric zonations demarcated with brown lines inside the spots on the upper surface (Figure 7.4). On the lower surface, the spots are lighter brown in color. Such spots often coalesce and may involve large portions of the blade, which become dry and are shed. Dark brown, spreading, water-soaked lesions can be seen on the entire length of the stem. The lesions also occur on the midrib and even on veins of leaves. In very severe attacks, plants may be killed within a very short period after symptoms are first noted, while milder attacks cause defoliation. Occasionally, seedlings and young plants are killed exhibiting pre- and postemergence damping-off.

FIGURE 7.4 *Alternaria* leaf spot of sesame. (Courtesy of Dr. B.A. Tunwari, Federal University, Wukari, Nigeria, and H. Nahunnaro, Modibbo Adama University of Technology, Yola, Nigeria.)

Geographical Distribution and Losses

Alternaria leaf spot of sesame was first described by Kvashnina (1928) from the North Caucasus region in the former Soviet Union. Kawamura in Japan studied a similar leaf spot pathogen on sesame and named it as *Macrosporium sesami* Kawamura. Mohanty and Behera (1958) from India reported *Alternaria* blight of sesame and found the causal organism to be closely resembling *M. sesami*. However, it differed from *M. sesami* in that some of the spores were catenulate. On the basis of the catenulation, the fungus was placed in *Alternaria* and renamed as *A. sesami* (Kawamura) Mohanty and Behera. In India and in the United States, it was earlier referred only by the name *Alternaria* sp. The first identification of *A. sesami* in the United States was probably made by Leppik and Sowell in 1958.

The *Alternaria* leaf spot is now reported to occur in most of the tropical and subtropical areas of the world. Epiphytotic occurrence of the disease has been reported from the Stoneville area in Mississippi in 1962, the Tallahassee area in Florida in 1958, and the coastal area of Orissa in 1957 and Maharashtra in India in 1975 (Kolte 1985). It is now reported to be of more economic significance in Egypt (El-Bramawy and Shaban 2007, 2008), India (Naik et al. 2007), Kenya (Ojiambo et al. 2000a,b), Nigeria (Enikuomehin et al. 2011), Pakistan (Marri et al. 2012), and Uganda (Mudingotto et al. 2002).

The amount of damage to the sesame plant is dependent on the stage of growth and environmental conditions. Disease severity is negatively correlated with the seed yield, 1000-seed weight, and seeds/capsule (Ojiambo et al. 2000b). The disease causes 20%–40% loss in sesame crop in the state of Uttar Pradesh in India (Kumar and Mishra 1992). It is, however, reported that about 0.1–5.7 g seeds/100 fruits are lost due to the disease under Karnataka conditions in India (Kolte 1985).

Pathogen

The pathogen is *Alternaria sesami* (Kawamura) Mohanty and Behera. The conidiophores of the fungus are pale brown, cylindrical, simple, erect, 0–3 septate, and not rigid, arise singly and measure 30–54 × 4–7 μ, and produce conidia at the apex. The conidia are produced singly or in chains of two. They are straight or slightly curved, obclavate, and yellowish brown to dark or olivaceous brown in color and measure 30–120 × 9–30 μ (excluding the beak). The conidia have 4–12 transverse septa and 0–6 longitudinal septa at which they are slightly constricted and terminate in a long hyaline beak measuring 24–210 × 2–4 μ. The beak may be simple or branched.

The optimal temperature for the growth of the fungus is in the range of 20°C–30°C, and the optimum pH for growth is 4.5. Maximum growth of the fungus is reported on mannitol followed by lactose as carbon sources, and the ammonium form of nitrogen is superior to the nitrate form.

Epidemiology and Disease Cycle

A. sesami mainly survives through seed up to 11 months, and it can also perpetuate in infected debris for nearly 11 months under field conditions (Agarwal et al. 2006, Naik et al. 2007). From infected capsules, *A. sesami* can penetrate into the seed coat, where it remains viable until germination of seed. The spores of the fungus attached to the seeds or capsule may serve to carry and disseminate the pathogen. The disease becomes most severe on plants established from seeds with 8% infection, and the disease severity increases with increased seed infection level (Ojiambo et al. 2000a, 2003, 2008). Though the infection process appears to be similar to other *Alternaria* species, culture filtrate from *A. sesami* reveals the presence of toxin, the tenuazonic acid (Rao and Vijayalakshmi 2000).

Seed infection is observed to be highest in plants inoculated between 8 and 10 weeks of age and least at 1, 6, and 12 weeks of age (Ojiambo et al. 2008). Excessive rainfall favors the development of the disease. The fungus is restricted to sesame in its pathogenicity. Distinct physiological races have not been identified, although differential virulence among isolates of *A. sesami* has been described from India and the United States.

Disease Management

Host Plant Resistance

Development and use of resistant sesame varieties is the best option. Hairy plants on the whole are reported to be free from attack due to *A. sesami*. The disease-resistant genotypes are *S. occidentalis* cvs. Heer, Regel; *S. radiatum* cvs Schum. and Thonn. and *S. malabaricum* (Shekharappa and Patil 2001b); and S-122 (Marri et al. 2012) and RT 273 (Eswarappa et al. 2011). Single dominant allele and 10 kb RAPD marker have been identified for resistance to *Alternaria* leaf spot of sesame (Eswarappa et al. 2011).

The old sesame lines *SI 948* (Kulithalai), *SI-1561*, *1683*, *1737*, *2177*, and *2381*, and Rio are reported to be resistant to the disease. Sesame varieties *Sirogoma* and *Venezuela 51*, NO 4, E-8, JT-7, JT-63-117, A-6-5, JT-66-276, Anand-9, JT-62-10, VT-43, and Anand-74 are also reported to be moderately resistant to the disease (Kolte 1985). Some other sesame genotypes that are moderately resistant to the *Alternaria* leaf spot are Navile-1, 351888, 899, 908, TC28, Madhavi, Co-1-12, Co-1-16, TC-25, and Tarikere (Basavaraj et al. 2007) and MT-15, DORS-102, DS-14, and DS-10, which show multiple-disease resistance including to *Alternaria* leaf spot disease (Jahagirdar et al. 2003). Biparental mating or diallel selective mating and heterosis breeding have been suggested for the development of *Alternaria*-resistant cultivars (El-Bramawy and Shaban 2007, 2008).

Induced Host Resistance

Resistance-inducing chemicals like salicylic acid at 1% conc. and boiagents *T. viride* and *P. fluorescens* induce systemic resistance in sesame against *A. sesami* and result in higher plant vigor index (Savitha et al. 2011, 2012). Aqueous leaf extract of neem (*A. indica*) provides the control of the *Alternaria* leaf spot disease without adversely affecting spore germination of *A. sesami*, and protection of sesame plants against *A. sesami* by neem extract is due to the stimulation of plant natural defense response as the treated sesame plants exhibit significantly higher level of enzymes like phenylalanine ammonia lyase, PO, and contents of phenolic compounds (Guleria and Kumar 2006). Similarly, it is noteworthy to observe that extract of another plant, *Mikania scandens*, when treated on inoculated sesame plants induces host resistance to the delay of the development of the *Alternaria* leaf spot (Lubaina and Murugan 2013a,b).

Chemical Control

Two sprays of mancozeb at 0.25% (Mudingotto et al. 2002, Rajpurohit 2003) or a combination of mancozeb at 0.25% plus methyldemeton at 1 mL/L (Rajpurohit 2004b) or mancozeb at 0.25% plus streptocycline at 0.025% (Shekharappa and Patil 2001a) have been found to be effective in the management of *Alternaria* leaf spot of sesame with increase in yield of sesame crop.

Cultural Control

Salt density at 2%–5% concentration can be used to sort out the infected seed from the seed lots to maintain healthy nucleus seed after further washing and drying the seed (Enikuomehin 2010). Seeds floated at 2% and 5% salt conc. are characteristically discolored, malformed, infected, and lightweight.

Experimental evidence has been presented in Nigeria that intercropping sesame with maize in a single alternate row (1:1) arrangement can be useful in reducing the severity of *Alternaria* leaf blight of sesame (Enikuomehin et al. 2010, 2011).

WHITE LEAF SPOT OR *Cercospora* LEAF SPOT

Symptoms

Small circular spots are scattered on both leaf surfaces. At first, they are minute, and later they increase in size to become 5 mm in diameter with whitish center (white spot) surrounded by a blackish purple margin (Figure 7.5). The spots may enlarge rapidly, coalesce into irregular blotches that often become about 4 cm in diameter, and are concentrically zoned. Under humid conditions, the disease becomes severe involving premature defoliation. The disease causes defoliation particularly in early maturing varieties. On petioles, the spots are elongated. Capsules show more or less circular, brown-to-black lesions (1–7 mm).

Geographical Distribution and Losses

The white spot of sesame is reported from Australia, Brazil, China, Colombia, the Dominican Republic, India, Nicaragua, Sri Lanka, Suriname, the United States, and Venezuela (Kolte 1985, Shivas et al. 1996, Verma et al. 2005).

FIGURE 7.5 *Cercospora* leaf spot of sesame. (Courtesy of Dr. Anil Kotasthane, IGKV, Raipur, India.)

The disease is endemic in most of the sesame-growing areas of Takum, Donga, Wuakeri, Bali, Kurmi, and Karim-Lamido in Taraba state and major sesame-growing regions of Nigeria, which has assumed more serious occurrence in the forest/Savannah transition zone of southwest Nigeria to which the crop has been recently introduced. It is widely prevalent in other countries of Africa (Uwala 1998, Einkuomehin 2005). The losses due to the disease in Nigeria range from 22% to 53% (Einkuomehin et al. 2002). It is reported that the disease severity in India can be as high as 53%–96% resulting in an average yield loss of 20% (Mohanty 1958, Patil et al. 2001).

Pathogen

The pathogen is *Cercospora sesami* Zimmerman (*Mycosphaerella sesamicola*). Stromata are slight to none. Conidiophores are olivaceous, septate, usually single or in fascicles of up to 10, epiphyllous, nodulose, and thickened toward the tip and measure 40–60 × 4 μ. Conidia are hyaline, cylindric, toothed upward, and commonly 7–10 septate and measure 90–135 × 3–4 μ. The pathogen is reported to sporulate well on carrot leaf decoction agar medium. *C. sesami* perpetuates through infected seed and also through plant residues in soil.

Disease Management

Some of the sesame genotypes, namely, IS 4, 15, 21, 29, 41, 41A, 41B, 128, and 128B, FS 150 (H 60-18) from Morocco, ES 234 from Mexico, and ES 242 (Precoz) from Venezuela (Kolte 1985); 65b-58, 60/2/3-1-8B, 69B-392,73a-96B from Nigeria (Poswal and Misari 1994, Nyanapah et al. 1995); and BIC-7-2, Sidhi 54, Rewa 114, and Seoni Malwa from India (Tripathi et al. 1996), have been reported to be resistant to the disease. Two sesame cultivars, E 8 and NCRTBEN-01 from Nigeria, show better stand establishment with certain degree of tolerance to the disease (Nahunnaro and Tunwari 2012a).

Many synthetic fungicides had shown promise in the management of sesame diseases (Shokalu et al. 2002). However, the high cost of such chemicals forbids their use by ordinary farmers. Seed treatment with systemic fungicides like carbendazim (0.15%) or Bayleton (0.15%) is reported to be effective in the control of the seed-borne inoculum. Sesame crop sprayed with carbendazim at 0.1% or Quintal at 0.2 gives best degree of disease management with increase in seed yield by 31.28% (Hoque et al. 2009, Palakshappa et al. 2012). Two sprays of a mixture of mancozeb at 0.2% plus endosulfan 35 EC at 1 mL/L, first spray being given at flower initiation stage and the second at pod formation stage, result in good control of insect pests and *Cercospora* leaf spot disease (Ali and Singh 2003).

Hot-water treatment of seeds at a temperature of 53°C for 30 min gives good control of the disease. Aqueous leaf extract of plants *Aspilia africana*, *Chromolaena odorata*, *A. indica*, and *Allium sativum*, when sprayed once every week, gives significant reduction in disease severity (Enikuomehin 2005, Nahunnaro and Tunwari 2012b). The plant extracts of garlic, *Ocimum*, and *Chromolaena* are comparable to synthetic fungicide (benlate) in reducing the amount of *Cercospora* leaf spot on sesame (Enikuomehin and Peters 2002, Tunwari and Nahunnaro 2014).

Plant debris should be burned after threshing and before plowing. Early-sown crop in the middle of June to first week of July is less affected due to *Cercospora* leaf spot, and these sowing dates are preferred for sowing sesame in wider row spacing of 20–30 cm in India and Nigeria (Tripathi et al. 1998a, Enikuomehin et al. 2002, Verma et al. 2005). Intercropping-induced microclimatic effects influence foliar disease severity including that of *Cercospora* leaf spot of sesame. Grain yield, weight of 1000 seeds, number of capsules/plant, and weight of seed/plant have been observed to be significantly higher in the 1:1 row arrangement than the sole crop or other row arrangements. The study made by Enikuomehin et al. (2008) demonstrates that intercropping sesame with maize in a single alternate row (1:1) arrangement can be used to reduce white leaf spot severity of sesame.

PHYLLODY DISEASE

Symptoms

Affected sesame plants express symptoms, depending on the stage of crop growth and time of infection. A plant infected in its early growth remains stunted to about two-thirds of a normal plant, and the entire plant may be affected. The entire inflorescence is replaced by a growth consisting of short,

FIGURE 7.6 Phyllody of sesame. Note the transformation of flower parts into green leaflike structures. (Courtesy of Dr. Anil Kotasthane, IGKV, Raipur, India.)

twisted leaves closely arranged on a stem with very short internodes. However, when infection takes place at later stages, normal capsules are formed on the lower portion of the plants, and phylloid flowers are present on the tops of the main branches and on the new shoots that are produced from the lower portions.

The most characteristic symptom of the disease is transformation of flower parts into green leaflike structures followed by abundant vein clearing in different flower parts (Figure 7.6). The calyx becomes polysepalous and shows multicostate venation compared to its gamosepalous nature in healthy flowers. The sepals become leaf like but remain smaller in size. The phylloid flowers become actinomorphic in symmetry, and the corolla becomes polypetalous. The corolla may become deep green, depending upon the stage of infection. The veins of the flowers become thick and quite conspicuous. The stamens retain their normal shape, but they may become green in color. Sometimes, the filaments may, however, become flattened, showing its tendency to become leaf like. The anthers become green and contain abnormal pollen grains. In a normal flower, there are only four stamens, but a phylloid flower bears five stamens. The carpels are transformed into a leaf outgrowth, which forms a pseudosyncarpous ovary by their fusion at the margins. This false ovary becomes very enlarged and crop. In Sudan, red varieties of sesame have been found to be affected to the extent of 100%.

Inside the ovary, instead of ovules, there are small petiole-like outgrowths, which later grow and burst through the wall of the false ovary producing small shoots. These shoots continue to grow and produce more leaves and phylloid flowers. The stalk of the phylloid flowers is generally elongated, whereas the normal flowers have very short pedicels. Increased IAA content appears to be responsible for proliferation of ovules and shoots. Sometimes, these symptoms are found to be accompanied with yellowing, cracking of seed capsule, germination of seeds in capsules, and formation of dark exudates on the foliage (Akhtar et al. 2009, Pathak et al. 2012).

Normal-shaped flowers may be produced on the symptomless areas of the plants, but such flowers are usually dropped before capsule formation, or the capsules are dropped later leaving the stalk completely bared.

Geographical Distribution and Losses

Prevalence of the sesamum phyllody erroneously named *leaf curl* is traced since 1908 in Mirpur Khas area of India (now in Pakistan), as cited by Vasudeva and Sahambi (1955), and a detailed historical account of the occurrence and causal agent of the disease has been reviewed earlier by Kolte (1985). It has been reported from India, Iran, Iraq, Israel, Myanmar, Sudan, Nigeria, Tanzania, Pakistan, Ethiopia, Thailand, Turkey, Uganda, Upper Volta, Venezuela, and Mexico (Kolte 1985, Salehi and Izadpanah 1992, Esmailzadeh-Hosseini et al. 2007, Akhtar et al. 2009). The first evidence of association of mycoplasma-like organism (now known as phytoplasma) with the disease was obtained in Upper Volta by Cousin et al. (1971).

Affected plants remain partially or completely sterile, resulting in total loss in yield. As much as 10%–100% incidence of the disease has been recorded in the sesame crop in India. The yield loss due to phyllody in India is estimated to about 39%–74%. The losses in plant yield, germination, and oil content of sesame seeds may be as high as 93.66%, 37.77%, and 25.92%, respectively. It is estimated that a 1% increase in phyllody incidence decreases the sesame yield by 8.4 kg under Coimbatore conditions in India. Robertson (1928) from Burma reported up to 90% incidence of the disease in the Sagaing and Lower Chin districts. A survey conducted in Thailand during 1969 and 1970 indicated that the phyllody was so severe in northeastern Thailand that farmers decreased the acreage for the sesame. Phyllody is a very serious disease, which can inflict up to 80% yield loss with a disease intensity of 1%–80% (Kumar and Mishra 1992, Salehi and Izadpanah 1992). The average phyllody incidence is reported to be about 20% with yield losses in sesame seed yield due to phyllody ranges to be 7%–28% in Pakistan (Sarwar and Haq 2006, Sarwar and Akhtar 2009).

Pathogen: Certain Strains of 16 Sr Taxonomic Group of Phytoplasma

The pathogen is now investigated to be phytoplasma (formerly referred to as mycoplasma-like organism—wall-less bacteria belonging to the class Mollicutes). Light microscopy of hand-cut sections treated with Dienes stain shows blue areas in the phloem region of phyllody-infected sesame plants (Al-Rawi et al. 2001, Akhtar et al. 2009). The phytoplasma pleomorphic bodies are reported to be present in phloem sieve tubes of affected sesame plants. Electron microscopy has revealed that the big pleomorphic bodies, ranging from 100 nm diameter to 625 nm diameter, are present in the sieve tubes. Generally, the phytoplasmas are round, but some may be 1500 nm long and 200 nm wide. Bodies with beaded structures can also be noticed. The phytoplasmas are bounded by a single unit membrane as is typical for the Mollicutes and show ribosome-like structure and DNA-like strands within. Phytoplasma cells contain one circular double-stranded DNA chromosome with a low G + C contents (up to only 23%), which is thought to be the threshold for a viable genome (Bertaccini and Duduk 2009, Weintraub and Jones 2010). They also contain extrachromosomal DNA such as plasmids. Since phytoplasmas cannot be grown in axenic culture, advances in their study are mainly achieved by molecular techniques. Molecular data on sesame phytoplasmas have provided considerable insight into their molecular diversity and genetic interrelationships, which has in turn served as a basis for sesame phytoplasma phylogeny and taxonomy. Classification of phyllody phytoplasma associated with sesame has been attributed to at least three distinct strains worldwide including aster yellows, peanut witches' broom, and clover proliferation group (Al-Sakeiti et al. 2005, Khan et al. 2007).

Based on restriction fragment length polymorphism (RFLP) analysis of polymerase chain reaction–amplified 16S rDNA, sesame phyllody phytoplasma infecting sesame in Myanmar (termed as SP MYAN) belongs to the group 16SrI and subgroup 16SrI-B. Sequence analysis has confirmed that SP-MYAN is a member of *Candidatus Phytoplasma asteris* and it is closely related to that of sesame phyllody phytoplasma from India (DQ 431843) with 99.6% similarity (Khan et al. 2007, Win et al. 2010). RFLP profiling and sequencing reveal that phytoplasma associated with sesame phyllody in Pakistan has the greatest homology to 16SrII-D group phytoplasmas (Akhtar et al. 2009), whereas in a separate study from the same country (i.e., in Pakistan), molecular evidence of the cause of the sesame phyllody has been found to be phytoplasma belonging to subgroup 16SrII and its sequence is essentially reported to be identical to that of the phytoplasma causing sesame phyllody in Oman (Akhtar et al. 2008). Similarly, phytoplasma causing sesame phyllody in Yazd Province of Iran belongs to the 16SrII group, which is peanut witches' broom phytoplasma (Esmailzadeh-Hosseini et al. 2007). Interestingly, in the neighboring Turkey, phytoplasma associated with sesame phyllody belongs to 16S rDNA group closely related to clover proliferation group 16SrVI-A (Sertkaya et al. 2007).

Witches' broom symptom in sesame resembling sesame phyllody in Oman is caused by the phytoplasma strains (SIL, SIF) clustered with Omani Lucerne witches' broom forming a distinct lineage separate from groundnut witches' broom and sesame phyllody (Thailand) phytoplasma strains (Nakashima et al. 1995, 1999, Al-Sakeiti et al. 2005, Khan et al. 2007).

Transmission

The pathogen is transmitted by the leafhopper vectors (order: Homoptera). In India, Thailand, and Upper Volta, sesame phyllody is transmitted by *Orosius orientalis* (Matsumura) (*O. albicinctus*), whereas in Turkey and Iran, sesame phyllody is transmitted by *Circulifer haematoceps* (Mulsant and Rey) (Dehghani et al. 2009). However, Esmailzadeh-Hosseini et al. (2007) first reported transmission of a phytoplasma associated with sesame phyllody in Iran by *O. albicinctus*. The pathogen has also been experimentally transmitted to the cotton plant by the vector *O. cellulosus* (Lindberg). Attempts to transmit the pathogen through sap in Iran and through seed in Thailand have given negative results (Tan 2010).

Epidemiology and Disease Cycle

The pathogen has a wide host range and survives on alternate hosts like *Brassica campestris* var. *toria*, *B. rapa,* and *Cicer arietinum*, which serve as source of inoculum. The pathogen is transmitted by the leafhopper, *O. albicinctus*, in most sesame-growing areas in the world as discussed earlier. Most optimum acquisition period of vector is 3–4 days, and inoculation feeding period is 30 min. The incubation period of the pathogen in leafhoppers may be 15–63 days and 13–61 days in sesame. Nymphs are incapable of transmitting the phytoplasma. Vector population is more during summer and less during cooler months.

There is a significant positive correlation between phyllody incidence with maximum and minimum temperature and negative correlation with maximum relative humidity and rainy days, which could be then consequently related to increase or decrease in vector population in the respective environmental conditions (Choudhary and Prasad 2007).

The incubation period is considerably increased during winter months (October–January) due to low temperature. Among the weather factors, the night temperature (minimum temperature) prevailing from the 30th to the 60th day after sowing is found to have a greater increase of disease incidence. The minimum acquisition feeding period has been observed to be 8 h, while the minimum infection feeding period is 30 min during May and June. Both male and female insects are equally efficient in transmitting the pathogen. The nymphs of the insect are capable of acquiring the pathogen, but they are unable to transmit it, as by the time the incubation period is completed, they reach the adult stage. Once the leafhoppers have picked up the pathogen and become infective, the adult leafhoppers remain so throughout the remainder of their lives without replenishment of the pathogen from infected plants.

Even a single leafhopper may be able to cause infection. It is interesting that leafhoppers show a marked preference for the diseased plants over healthy ones. The diseased plants have been reported to harbor an insect population about two to six times the population on healthy plants—due to higher moisture, higher nitrogen, and lower calcium and potassium contents of the diseased plants. Lower content of calcium and potassium in the diseased plants is suspected to be the factor vulnerable for easy stylet and ovipositor penetration. Higher incidence of phyllody occurs when sesame crop is fertilized with phosphorus without nitrogen (Borkar and Krishna 2000); there also exists a positive correlation between days to maturity of sesame crop and phyllody incidence (Gopal et al. 2005).

Disease Management

Host Plant Resistance

Selections of disease-resistant sesame lines, which would flower within 40–50 days after sowing, appear to be desirable and important from the yield viewpoint under Indian conditions (Kolte 1985, Selvanarayanan and Selvamuthukumaran 2000). From India, a considerable number of genotypes such as RJS 78, RJS 147, KMR 14, KMR 29, Pragati, IC 43063 and IC 43236 (Singh et al. 2007), SVPR-1 (Saravanan and Nadarajan 2005), AVTS-2001-26 (Anandh and Sevanarayanan 2005), Swetta-3, RT-127, No. 171 (Dandnaik et al. 2002), TH-6 (Anwar et al. 2013), and three wild species, that is, *S. alatum*, *S. malabaricum*, and *S. yanaimalaiensis*, are resistant to phyllody with mean incidence below 5%, which can be utilized as donor parents in resistance breeding to phyllody disease (Saravanan and Nadarajan 2005, Singh et al. 2007). A single recessive gene governs resistance in cultivated varieties (KMR 14 and Pragati), whereas wild species possess a single dominant gene conferring resistance to phyllody (Singh et al. 2007). Phyllody resistance in a land race of sesame is reported to be under the control of two dominant genes with complementary (9:7) gene action (Shindhe et al. 2011). Some genotypes in India have not been observed to show phyllody symptoms. Such genotypes are Ny-9, Sirur, Local, NKD-1037, K-50, TC-25, RT-15H, OCP-1827, No. 5, No. 16, No. 17, No. 18, No. 21, No. 23, and No. 24 (Dandnaik et al. 2002). Interspecific hybrids between *S. alatum* and *S. indicum* are, however, moderately resistant to phyllody (Rajeshwari et al. 2010). Advanced phyllody disease–resistant sesame mutant lines with earliness, more capsules, and high harvest index have been developed in Pakistan under the series NS11-2, NS11 P2, NS100 P2, NS 103-1, and NS240 P1 and phyllody disease–resistant sesame. These mutant lines can be of great potential use in breeding for disease resistance (Sarwar and Akhtar 2009).

Some other sesame lines as JT-7, JT-276, and N-32, though not resistant to the disease, have been found useful to escape the disease (Kolte 1985).

Chemical Control

Insect vector management is the method of choice for limiting the outbreaks of phytoplasmas in sesame. At the time of sowing, soil may be treated with Thimet® 10 G at the rate of 10 kg/ha or with Phorate 10 G at the rate of 11 kg/ha or with Temik® 10 G at the rate of 25 kg/ha to get the management of the disease through vector control (Nagaraju and Muniyappa 2005). An effective degree of management is obtained if the aforementioned treatment is combined with spraying of the crop with Metasystox® (0.1%) or with any other effective chemical (Misra 2003, Rajpurohit 2004b).

Tetracycline sprays at 500 ppm concentration at the flower initiation stage have proved to be effective against phyllody, but recovery is temporary.

A possibility of biochemical control by spraying manganese chloride has been indicated. It appears that manganese chloride oxidizes the phenol and protects or inhibits the enzymes, brining the auxin level to normal. Once hyperauxin is oxidized, the plant can gain its normal conditions (Purohit and Arya 1980).

Cultural Control

An appropriate sowing date may be useful in avoiding severe occurrence of the disease. The incidence of the disease is reported to be reduced considerably by sowing the crop in early August under Indian conditions. The reduced population of the vector in the growth period of sesame plants is perhaps important in keeping the disease under check (Mathur and Verma 1972, Nagaraju and Muniyappa 2005).

OTHER SESAME DISEASES

Other diseases of sesame are given in Table 7.1.

TABLE 7.1
Other Diseases of Sesame

Disease	Pathogen/Causal Agent	Geographical Distribution	Disease Description	References
Powdery mildew	*Oidium erysiphoides* (Fr.) *O. sesami* (Sreenivasulu et al.) *Sphaerotheca fuliginea* (Schlecht) Pollacci *Leveillula taurica* (Lav.) Trnaud *Erysiphe cichoracearum* DC	Ethiopia, Australia, India, Iraq, Japan, Tanzania, and Uganda Malawi, India, Sudan India, Sicily, Venezuela China, Thailand	Symptoms of the disease start as small whitish spots on the upper surface of the leaves. The spots coalesce to form a single spot, finally covering the entire leaf surface with dirty white fungal growth. Generally, the mildew is confined to the upper surface of the leaves. The perithecial stage may or may not be observed on sesame leaves. The disease causes a yield loss of 42%. For every 1% increase in disease severity, there is a yield loss of 5.63 kg/ha. There is a significant negative correlation between days to maturity and powdery mildew occurrence and severity. Use of host resistance appears to be the most promising method for the control of the powdery mildews of sesame. Several genotypes from India, namely, BB 3-8, TKG-22, NSKMS-260, G-55, Si 3315/11, hybrid Si 3315/11xSVPR, Co1, hybrids Co1 x S13216 and Co1 x YLM123, are resistant to powdery mildew. Inheritance of resistance/tolerance to powdery mildew reveals that susceptibility is dominant over tolerance and is controlled by two independent recessive genes with complementary epistasis. The crop can be protected from the powdery mildews by spray application of wettable sulfur (0.2%) fungicides or by dusting sulfur dust at the rate of 20 kg/ha or Karathane spray (1%).	Kolte (1985), Rao and Rao (2001), Kumaresan and Nadarajan (2002), Sarvanan and Nadarajan (2004), Gopal et al. (2005), Biju et al. (2011), Sravani et al. (2012), Rao et al. (2011, 2012)
Corynespora blight	*Corynespora cassiicola* (Berk. and Curt.) Wei.	Colombia, India, Tanzania, the United States, Venezuela	Dark, irregularly shaped spots appear on leaves and stems. They enlarge, become brown with light centers, and coalesce forming a blotchy configuration. Extensive defoliation occurs and the affected plants die. Affected stems are bent irregularly on the lesions. Cankers of various sizes also appear on the stem. In mature plants, the infected stem cracks lengthwise and breadthwise. The pathogen perpetuates through plant debris and infected seeds. It is, however, inactivated when infected seeds are stored at 26°C–28°C with 50% relative humidity. Despite having evidence of genetic diversity among isolates of *C. cassiicola*, there are no correlations between the morphological characteristics or rDNA-ITS region sequences and their host or geographical origin, ISSR markers being useful for intraspecies population studies in the pathogen. Genotypes IC-205292, IC-205 561, IC-205633, and Krisna are moderately resistant to the disease. Seed treatment with carbendazim at 0.1% and spraying the crop with mancozeb can be effective in disease management.	Kolte (1985), Verma et al. (2005), Choudhary et al. (2006), Qi et al. (2011), Singh et al. (2000)

(Continued)

TABLE 7.1 (*Continued*)
Other Diseases of Sesame

Disease	Pathogen/Causal Agent	Geographical Distribution	Disease Description	References
Brown angular leaf spot	*Cylindrosporium sesami* Hansford.	Nigeria, Saudi Arabia, Sudan, the United States, Venezuela; epiphytotic occurrence reported in the United States and Venezuela	Spots on the leaves are water soaked, brown, and limited to veinal areas and assume an angular shape. They are 2–20 mm in diameter and may enlarge rapidly to coalesce into extensive necrotic areas. In the ease of severe infection, defoliation occurs. The upper surface of the spot on the leaves shows the presence of dark subepidermal fungal acervuli. The affected leaves frequently show the association of spots caused by *A. sesami* and/or *C. sesami.* The fungus survives through the seed. Selection V-16 is reported to be resistant to the disease. Seed treatment with commonly used fungicides (1–3 g/kg seed) can be effective for the control of the disease.	Kolte (1985), Verma et al. (2005)
Angular leaf spot	*Cercospora sesamicola* Mohanty	Nigeria, Nicaragua, India, Panama	Leaf spots are angularly limited by leaf veinlets measuring 1–8 mm in size. Initially, the spots are minute and become visible as chlorotic lesions on the upper surface of the leaves; later, when the affected tissues become necrotic, the color of the spots changes to dark brown, whereas on the corresponding lower surface of the leaves, the color of the spots remains olivaceous brown. The fruiting bodies of the fungus might become visible on both surfaces of the leaves but chiefly on the lower surface. *C. sesamicola* perpetuates only through viable sclerotia in crop debris and possibly through infected seeds.	Kolte (1985), Verma et al. (2005)
Aerial stem rot	*Helminthosporium sesami* Miyake	Philippines, China, Japan, the United States	Leaf lesions vary from small brown spots 1 mm in diameter to large elongated lesions of about 2–20 mm. Lesions of the stem range from small flecks 1 mm in diameter to large, sunken, dark-brown spots 10 × 40 mm in size. High humidity favors spread of the disease, and young plants are much more susceptible than mature ones. Plants less than 21 days old are more susceptible than mature ones. Nitrogen increases the susceptibility. Phosphorus or potash alone or phosphorus and calcium show decrease in the severity of infection.	Kolte (1985)
Stem blight	*Alternaria alternata* (Fr.) Keissier	India	Blighting of stem is the major symptom. Seed treatment with captan at 3 g/kg seed and foliar spray with copper oxychloride (0.3%) at 20, 40, and 60 days after sowing.	Rao and Rao (2002)

(*Continued*)

TABLE 7.1 (*Continued*)
Other Diseases of Sesame

Disease	Pathogen/Causal Agent	Geographical Distribution	Disease Description	References
Bacterial blight	*Xanthomonas campestris* (Pammel) Dowson pv. *sesami* (Sabet and Dowson) dye	Sudan, India, Pakistan, the United States, Venezuela	Small, water-soaked, light-brown lesions develop on the margin of the cotyledonary leaf about 10–12 days after sowing. The lesions may spread, rapidly covering the entire cotyledons, which consequently become dry. About 4% mortality due to the disease in 4–6-week-old seedlings has been reported. If the seedling survives, dark brown, water-soaked spots also appear on the true leaves. In a severe infection, the lesions extend to the stem through the petiole, leading to the formation of brown discoloration, resulting in systemic invasion and death of the plant. The disease is known in Sudan as *Marad ed Dum*, meaning thereby the blood disease, due to the red color of infected plant tissue. Seed can carry the pathogen up to a period of 16 months. A weed plant, *Acanthospermum hispidum*, is reported to be susceptible to *X. campestris* pv. *sesami*. This host acts as a source of survival of the bacterium in its dried leaves from year to year. The bacterium does not survive in sesame plant debris in soil from year to year. The bacterium enters the host primarily through stomata and quickly becomes vascular. The secondary spread is by spattering rains. High temperature and humidity favor the disease. Seedling infection of sesame is most severe at soil temperature of 20°C. Infection does not take place when soil temperature is 40°C. The disease also becomes severe when the soil moisture is 30%–40% and relative humidity is 75%–87%. Seedling infection can be used as a valid test for determining the resistance of sesame to this disease. Three sesame genotypes SG-34, SG-22, and Sg-55 are resistant to bacterial blight. Chemical or antibiotic seed treatment or hot-water treatment of seed and antibiotic sprays to check secondary spread are the same as described for bacterial leaf spot. Streptocycline seed treatment for 2 h followed by three sprays at 10-day interval of streptocycline plus copper oxychloride effectively controls the disease.	Kolte (1985), Rai and Srivastava (2003), Samina et al. (2007), Isakeit et al. (2012), Naqvi et al. (2012)
Bacterial wilt	*Ralstonia solanacearum* biovar III	Andaman and Nicobar Islands and Assam State of India, Iran, Japan, South China	Soil treatment with combinations of bleaching powder, streptocycline, and mustard cake significantly controls the disease. The variety Pb Til No1 is less affected by this bacterium.	Dubey et al. (1996), Hazarika and Das (1999), Hua et al. (2012)

(*Continued*)

TABLE 7.1 (*Continued*)
Other Diseases of Sesame

Disease	Pathogen/Causal Agent	Geographical Distribution	Disease Description	References
Leaf curl	Leaf curl virus	India, Nigeria, Pakistan, Sierra Leone, Tanzania, Uganda, Zaire	Symptoms of the disease are characterized by curling of the leaves and marked thickening of the veins on the underside of the leaf, combined with a reduction in leaf size. Leaves may also become leathery, possessing dark-green color. Severely affected plants remain stunted and bear few flowers and capsules. The disease is considered to be a serious one causing considerable reduction in yield, especially when the infection takes place at the early stage of crop growth. The incidence of the disease in certain years is reported to be to the extent of 60% in India. The sesamum lines NP-6, T-13-3-2, 65-1/11, 67-13-1/2-1, Entebbe ex Uganda, and sesame mutant lines NS 11-P2, NS100-P2, NS103-1, and NS 240 P1 (developed in Pakistan) have shown a high degree of resistance to leaf curl. Inheritance of resistance to leaf curl from hybridization between resistant *S. radiatum* and *S. indicum* reveals that resistance to leaf curl and lodging is controlled by two independently assorting genes with both dominant alleles F and *L*, producing plants that are resistant to leaf curl and lodging. Two foliar sprays of the insecticide methyl demeton at 1 ml/L reduce leaf curl incidence.	Kolte (1985), Sarwar and Akhtar (2009), Rajpurohit (2004b), Falusi and Salako (2003)
Mosaic disease	Sesame mosaic virus	Czechoslovakia, India	Conspicuous chlorotic areas of irregular shape appear on the leaf lamina. The basal portion of the leaf may remain green, while the top portion and margins of the leaves become yellow, interveinal areas are usually yellow, and the young leaves may become completely yellow. As the disease progresses, top leaves become gradually smaller in size, and plants may not grow further. The disease is caused by the virus that is transmitted through sap and grafting.	Kolte (1985)
Cowpea aphid–borne mosaic virus disease	Cowpea aphid–borne mosaic virus	Paraguay	Yellowing and curling down of sesame leaves and shortening of the internodes.	Gonzalez-Segnana et al. (2011)

REFERENCES

Abd-El-Moneem, K.M.H. 1996. Effect of micronutrients on incidence of sesame char coal rot and wilt disease complex. *Assiut J. Agric. Sci.* 27: 181–195.

Abdel-Salam, M.S., M.M.A. El-Halim, and Q.I.M. El-Hamshary. 2007. Enhancement of *Enterobacter cloacae* antagonistic effects against the plant pathogen *Fusarium oxysporium*. *J. Appl. Sci. Res.* 3: 848–852.

Abdou, E., H.M. Abd-Alla, and A.A. Galal. 2001. Survey of sesame root rot/wilt disease in Minia and their possible control by ascorbic and salicylic acids. *Assiut J. Agric. Sci.* 1(23): 135–152.

Agarwal, P.C., U. Dev, B. Singh, I. Rani, and R.K. Khetarpal. 2006. Seed-borne fungi detected in germplasm of *Sesamum indicum* L. introduced into India during last three decades. *Indian J. Microbiol.* 46: 161–164.

Ahmed, Q., A. Mohammad, Q. Ahmad, and A. Mohammad. 1989. Screening of fungicides against sesame wilt. *Pesticides* 23: 28–29.

Ahuja, D.B., T.S. Rajpurohit, M. Singh et al. 2009. Development of integrated pest management technology for sesame (*Sesamum indicum*) and its evaluation in farmer participatory mode. *Indian J. Agric. Sci.* 79: 808–812.

Akhtar, K.P., M. Dickinson, G. Sarwar, F.F. Jamil, and M.A. Haq. 2008. First report on the association of a 16SrII phytoplasma with sesame phyllody in Pakistan. *Plant Pathol.* 57: 771.

Akhtar, K.P., G. Sarwar, and H.M.I. Arshad. 2011. Temperature response, pathogenicity, seed infection and mutant evaluation against *Macrophomina phaseolina* causing charcoal rot disease of sesame. *Arch. Phytopathol. Plant Prot.* 44: 320–330.

Akhtar, K.P., G. Sarwar, M. Dickinson, M. Ahmad, M.A. Haq, S. Hameed, and M.J. Iqbal. 2009. Sesame phyllody disease: Symptomatology, etiology and transmission in Pakistan. *Turk. J. Agric. For.* 33: 477–486.

Al-Ahmad, M. and A. Saidaai. 1998. *Macrophomina* (Char coal) rot of sesame in Syria. *Arali J. Plant Prot.* 6: 88–93.

Ali, S. and R.B. Singh. 2003. Management of insect-pests and diseases of sesamum through integrated application of insecticides and fungi*cides. Crop Res.* (*Hisar*) 26: 275–279.

Al-Rawi, A.F., F.K. Al-Fadhil, and R.A. Al-Ani. 2001. Histological methods for detection and monitoring distribution of phytoplasmas in some crops and weed plants in central Iraq. *Arab J. Plant Prot.* 19: 3–11.

Al-Sakeiti, M.A., A.M. Al-Subhi, N.A. Al-Saady, and M.L. Deadman. 2005. First report of witches'-broom disease of sesame (*Sesamum indicum*) in Oman. *Plant Dis.* 89(5): 530.

Ammar, S.E., M.S. El-Shazly, M.A. El-Ashry, M.A. Abd El-Sattar, and M.A.S. El Bramawy. 2004. Inheritance of resistance to *Fusarium* wilt disease in some sesame lines. *Egypt. J. Appl. Sci.* 19: 36–55.

Anandh, G.V. and V. Selvanarayanan. 2005. Reaction of sesame germplasm against shoot webber *Antigastra catalaunalis* Duponchel and phyllody. *Indian J. Plant Prot.* 33: 35–38.

Anwar, M., E. Hasan, T. Bibi, H.S.B. Mustafa, T. Mahmood, and M. Ali. 2013. TH-6: A high yielding cultivar of sesame released for general cultivation in Punjab *Adv. Life Sci.* 1(1): 44–57.

Armstrong, J.K. and G.M. Armstrong. 1950. A *Fusarium* wilt of sesame in United States. *Phytopathology* 40: 785.

Ashi, A. 2006. Sesame (*Sesamum indicum*). In: *Genetic Resources, Chromosome Engineering and Crop Improvement*, ed., R.J. Singh. CRC Press, Boca Raton, FL, pp. 231–280.

Badri, J., N.A. Ansari, V. Yepuri et al. 2011. Assessment of resistance to *Fusarium* wilt disease in sesame (*Sesamum indicum* L.) germplasm. *Aust. Plant Pathol.* 40: 471–475.

Bakheit, B.R., A.A. Ismail, A.A. El-Shiemy, and F.S. Sedek. 2000. Triple test cross analysis in four sesame crosses (*Sesamum indicum* L.) 2. Yield, yield components and wilt infection. *Acta Agron. Hung.* 48: 363–371.

Bakhshi, A., H.R. Etebarian, H.A. Aminian, and M. Ebrahimi. 2010. A study on genetic diversity of some isolates of *Macrophomina phaseolina* using molecular marker; PCR-RFLP and RAPD. *Iranian J. Plant Prot. Sci.* 41: 103–112.

Basavaraj, M.K., H. Ravindra, G.K. Girijesh, C. Karegowda, and B. Shivayogeshwara. 2007. Evaluation of sesame genotypes for resistance to leaf blight caused by *Alternaria sesami*. *Karnataka J. Agric. Sci.* 20: 864.

Basirnia, T. and Z. Banihashemi. 2005. Vegetative compatibility grouping in *Fusarium oxysporum* f. sp. *sesami* the causal agent of sesame yellows and wilt in Fars province. *Iranian J. Plant Pathol.* 41: 243–255.

Basirnia, T. and Z. Banihashemi. 2006. Seed transmission of in *Sesamum indicum* in Fars Province. *Iranian J. Plant Pathol.* 42: 117–123.

Bayoumi, T.Y. and M.A.S. El-Bramawy. 2007. Genetic analyses of some quantitative characters and *Fusarium* wilt disease resistance in sesame. In: *Eighth African Crop Science Society Conference*, El-Minia, Egypt, October 27–31, 2007, pp. 2198–2204.

Bayounis, A.A. and M.A. Al-Sunaidi. 2008a. Efficacy of some plant powders in protecting Sesame seeds against *Macrophomina phaseolina* in greenhouse. *Univ. Aden J. Nat. Appl. Sci.* 12: 233–243.

Bayounis, A.A. and M.A. Al-Sunaidi. 2008b. Effect of some plant extracts on growth inhibition of *Macrophomina phaseolina*. *Univ. Aden J. Nat. Appl. Sci.* 12: 469–480.

Bertaccni, A. and B. Duduk. 2009. Phytoplasma and phytoplasma diseases: A review of recent research. *Phytopathol. Mediterr.* 48: 355–378.

Biju, C.K., V.B. Hosagoudar, and S. Sreekumar. 2011. Additions to the powdery mildews of Kerala, India. *Biosci. Discov. J.* 2: 38–42.

Borkar, S.G. and A. Krishna. 2000. Effect of fertilizers on development of phyllody in sesamum. *J. Mycol. Plant Pathol.* 30: 422–423.

Butler, E.J. 1918. *Fungi and Disease in Plants*. Thacker Spink and Co, Calcutta, India.

Cardona, R. 2008. Effect of green manure and *Trichoderma harzianum* on sclerotia population and incidence of *Macrophomina phaseolina* on sesame. *Rev. Fac. Agron. Univ. Zulia* 25: 440–454.

Cardona, R. and H. Rodriguez. 2002. Evaluation of *Trichoderma harzianum* for the control *of Macrophomina phaseolina* on sesame. *Fitopatol. Venez.* 15: 21–23.

Cardona, R. and H. Rodriguez. 2006. Effects of *Trichoderma harzianum* fungus on the incidence of the charcoal rot disease on sesame. *Rev. Fac. Agron. Univ. Zulia* 23: 44–50.

Chattopadhyay, C. and R.K. Sastry. 2000. Methods for screening against sesame stem–root rot disease. *Sesame Safflower Newsl.* 15: 68–70.

Chattopadhyay, C. and R.K. Sastry. 2001. Potential of soil solarization in reducing stem–root rot incidence and increasing seed yield of sesame. *J. Mycol. Plant Pathol.* 31: 227–231.

Chattopadhyay, C. and R.K. Sastry. 2002. Combining viable disease control tools for management of sesame stem–root rot caused by *Macrophomina phaseolina* (Tassi) Goid. *Indian J. Plant Prot.* 30: 132–138.

Cho, E.K. and S.H. Choi. 1987. Etiology of a half stem rot in sesame caused by *Fusarium oxysporum*. *Korean J. Plant Prot.* 26: 25–30.

Choi, S.H., E.K. Cho, and Y.A. Chae. 1987. An evaluation method for sesame (*Sesamum indicum* L.) resistance to *Phytophthora nicotianae* var. *parasitica*. *Korean J. Crop Sci.* 32: 173–180.

Choi, S.H., E.K. Cho, and W.T. Cho. 1984. Epidemiology of sesame *Phytophthora* blight in different cultivation types. *Res. Rep. Rural Dev. Korea Republic* 26: 64–68.

Choudhary, C.S. and S.M. Prasad. 2007. Influence of weather parameters and vector population on incidence and development of sesamum phyllody. *Indian Phytopathol.* 60: 198–201.

Choudhary, C.S., S.M. Prasad, and R.B. Sharma. 2006. Role of meteorological factors on development of *Corynespora* blight disease of sesame. *J. Plant Prot. Environ.* 3: 91–94.

Choudhary, C.S., S.M. Prasad, and S.N. Singh. 2004. Effect of sowing date and fungicidal spray on *Macrophomina* stem and root rot and yield of sesame. *J. Appl. Biol.* 14: 51–53.

Chung, B.K. and K.S. Hong. 1991. Biological control wit*h Streptomyces* spp. of *Fusarium oxysporum* f. sp. *vasinfectum* causing sesame wilt and blight. *Korean J. Mycol.* 19: 231–237.

Chung, B.K. and S.O. Ser. 1992. Identification of antagonistic *Streptomyces* sp. on *Phytophthora nicotianae* var. *parasitica* and *Fusarium oxysporum* f. sp. *vasinfectum* causing sesame wilt and blight. *Korean J. Mycol.* 20: 65–71.

Cousin, M.T., K.K. Kartha, and Delattre. 1971. Sur la presence d'organisms de type mycoplasme dans tubes cribles de *Sesamum orientale* L. atteient de phyllodie. *Rev. Plant Pathol.* 50: 430.

Dandnaik, B.P., S.V. Shinde, S.N. More, and N.P. Jangwad. 2002. Reaction of sesame lines against phyllody. *J. Maharashtra Agric. Univ.* 27: 233.

Dehghani, A., M. Salehi, M. Taghizadeh, and Y. Khajezadeh. 2009. Studies on sesame phyllody and effect of sowing date and insecticide spray on disease control in Khuzestan. *Appl. Entomol. Phytopathol.* 77: 23–36.

Deepthi, P., C.S. Shukla, K.P. Verma, and E. Sankar Reddy. 2014. Yield loss assessment and influence of temperature and relative humidity on char coal rot development of sesame (*Sesamum indicum* L.). *BioScan* 9: 193–195.

Dinakaran, D. and S.E.N. Mohammed. 2001. Identification of resistant sources to root rot of sesame caused by *Macrophomina phaseolina* (Tassi.) Goid. *Sesame Safflower Newsl.* 16: 68–71.

Dubey, A.K., S. Gupta, and T. Singh. 2011. Induced vivipary in *Sesamum indicum* L. by seed borne infection of *Phytophythora parasitica* var. *sesame*. *Indian J. Fundam. Appl. Life Sci.* 1: 185–188.

Dubey, L.N., K.K. Das, and D.K. Hazarika. 1996. Evaluation of some chemicals against bacterial wilt of *Sesamum*. *Indian J. Mycol. Plant Pathol.* 26: 94–95.

El-Bramawy, M.A., K. Veverka, S. Vaverka et al. 2001. Evaluation of resistance to *Fusarium oxysporum* f. sp. *sesami* in hybrid lines of sesame (*Sesamum indicum* L.) under greenhouse conditions. *Plant Prot. Sci.* 37: 74–79.

El-Bramawy, M.A.S. 2003. Breeding studies for *Fusarium* wilt resistance in sesame (*Sesamum indicum* L.). PhD dissertation, Suez Canal University, Ismailieh, Egypt.

El-Bramawy, M.A.S. 2006. Inheritance of resistance to *Fusarium* wilt in some sesame crosses under field conditions. *Plant Prot. Sci.* 42: 99–105.

El-Bramawy, M.A.S. 2008. Assessment of resistance to *Macrophomina phaseolina* (Tassi.) Goid in some sesame collections. *Egypt. J. Appl. Sci.* 23(2B): 476–485.

El-Bramay, M.A.S., S.E. El-Hendawy, and W.I. Shaban. 2009. Assessing the suitability of morphological and phenological traits to screen sesame genotypes for *Fusarium* wilt and char coal disease resistance. *Plant Prot. Sci.* 45: 49–58.

El-Bramawy, M.A.S. and W.I. Shaban. 2007. Nature of gene action for yield, yield components and major diseases resistance in sesame (*Sesamum indicum* L.). *Res. J. Agric. Biol. Sci.* 3: 821–826.

El-Bramawy, M.A.S. and W.I. Shaban. 2008. Short communication. Inheritance of yield, yield components and resistance to major diseases in *Sesamum indicum* L. *Spanish J. Agric. Res.* 6: 623–628.

El-Bramawy, M.A.S. and A.O.A. Wahid. 2006. Field resistance of crosses of sesame (*Sesamum indicum* L.) to charcoal root rot caused by *Macrophomina phaseolina* (Tassi.) Goid. *Plant Prot. Sci.* 42: 66–72.

El-Bramawy, M.A.S. and A.O.A. Wahid. 2007. Identification of resistance sources to *Fusarium* wilt, char coal rot and *Rhizoctonia* root rot among sesame (*Sesamum indicum* L.) germplasm in Egypt. In: *The Eighth ACSS Conference*, Mina, Egypt, October 27–31, 2007, Vol. 8, pp. 1893–1990.

El-Bramawy, M.A.E.S., S.E. El-Hendawy, and W.I. Shaban. 2009. Assessing the suitability of morphological and phenotypical traits to screen sesame accessions for resistance to *Fusarium* wilt and charcoal rot diseases. *Plant Prot. Sci.* 45: 49–58.

El-Bramawy, M.A.E.S.A. and H.E. Embaby. 2011. Anti-nutritional factors as screening criteria for some diseases resistance in sesame (*Sesamum indicum* L.) genotypes. *J. Plant Breed. Crop Sci.* 3: 352–366.

El-Bramawy, M.A.S.A. and O.A.A.A. Wahid. 2009. Evaluation of resistance of selected sesame (*Sesamum indicum*) genotypes to *Fusarium* wilt disease caused by *Fusarium oxysporum* f. sp. *sesami*. *Tunisian J. Plant Prot.* 4: 29–39.

El-Deeb, A.A., A.A. Hilal, T.A. Radwan, A.A. Ali, and H.A. Mohamed. 1985. Varietal reaction and wilt diseases of sesame. *Ann. Agric. Sci. Mushtohor* 23: 713–721.

Elewa, I.S., M.H. Mostafa, A.F. Sahab, and E.H. Ziedan. 2011. Direct effect of biocontrol agents on wilt and root-rot diseases of sesame. *Arch. Phytopathol. Plant Prot.* 44: 493–504.

El-Fiki, A.I.I., A.A. El-Deeb, F.G. Mohamed, and M.M.A. Khalifa. 2004b. Controlling sesame charcoal rot incited by *Macrophomina phaseolina* under field conditions by using the resistant cultivars and some seed and soil treatments. *Egypt. J. Phytopathol.* 32: 103–118.

El-Fiki, A.I.I., F.G. Mohamed, A.A. El-Deeb, and M.M.A. Khalifa. 2004a. Some applicable methods for controlling sesame charcoal rot disease (*Macrophomina phaseolina*) under greenhouse conditions. *Egypt. J. Phytopathol.* 32: 87–101.

El-Hamid, M.A.H.S. and A. El-Bramawy. 2010. Genetic analysis of yield components, and disease response in sesame (*Sesamum indicum* L.) using two progenies of diallel crosses. *Res. J. Agron.* 4: 44–46.

El-Kasim, A., M. Kamal, F.G. Fahmy, A. Ali, and M.A. Sellam. 1991. Effect of preceding crop on the incidence of sesame wilt disease. *Assiut Agric. Sci.* 22: 99–113.

El-Shakhess, S.A.M. and M.M.A. Khalifa. 2007. Combining ability and heterosis for yield, yield components, charcoal-rot and *Fusarium* wilt disease in sesame. *Egypt. J. Plant Breed.* 11: 351–371.

El-Shemey, A., M.M. El-Tantay, and K.A. Mohamed. 2005. Effect of water stress on yield and yield components for some sesame varieties and its relation with wilt disease. *Ann. Agric. Sci. Moshtohor* 43: 1045–1055.

El-Wakil, D.A., A.M. Mahdy, and R.Z. El-MenShawy. 2011. Scanning electron microscopy study of sesame seeds infected with *Macrophomina phaseolina*. *J. Agric. Sci. Technol.* A1: 96–99.

Enikuomehin, O.A. 2005. Cercospora leaf spot disease management in sesame (*Sesamum indicum* L.) with plant extracts. *J. Trop. Agric.* 43(1–2): 19–23.

Enikuomehin, O.A. 2010. Seed sorting of sesame (*Sesamum indicum* L.) by salt density and seed-borne fungi control with plant extracts. *Arch. Phytopathol. Plant Prot.* 43: 573–580.

Enikuomehin, O.A., A.M. Aduwo, V.I.O. Olowe, A.R. Popoola, and A. Oduwaye. 2010. Incidence and severity of foliar diseases of sesame (*Sesamum indicum* L.) intercropped with maize (*Zea mays* L.). *Arch. Phytopathol. Plant Prot.* 43: 972–986.

Enikuomehin, O.A., M. Jimoh, V.I.O. Olowe, E.I. Ayo-John, and P.O. Akintokun. 2011. Effect of sesame (*Sesamum indicum* L.) population density in a sesame/maize (*Zea mays* L.) intercrop on the incidence and severity of foliar diseases of sesame. *Arch. Phytopathol. Plant Prot.* 44: 168–178.

Enikuomehin, O.A. and O.T. Peters. 2002. Evaluation of crude extracts from some Nigerian plants for the control of field diseases of sesame (*Sesamum indicum* L.). *Trop. Oilseeds J.* 7: 84–93.

Enikuomehin, O. A., V.I.O Olowe, O.S. Alao and M.O. Atayese. 2002. Assessment of *Cercospora* leaf spot disease of sesame in different planting dates in South-Western Nigeria. *Moor J. Agri. Res.* 3: 76–82.

Erwin, D.C. 1983. *Phytophthora: Its Biology, Taxonomy, Ecology and Pathology*. APS Press/American Phytopathology Society, St Paul, MN, pp. 237–257.

Esmailzadeh-Hosseini, S.A., A. Mirzaie, A. Jafari-Nodooshan, and H. Rahimian. 2007. The first report of transmission of a phytoplasma associated with sesame phyllody by *Orosius albicinctus* in Iran. *Aust. Plant Dis. Notes* 2: 33–34.

Eswarappa, V., K.T.V. Kumar, and H.V. Rani. 2011. Inheritance study and identification of RAPD marker linked to *Alternaria* blight resistance in sesame (*Sesamum indicum* L.). *Curr. Biotica* 4: 453–460.

Fallahpori, A., H. Aminian, N. Sahebani, and S.A. Esmailzade Hosseini. 2010. Investigation of 10 sesame local germplasms to damping off caused by *Fusarium oxysporum* f. sp. *sesami*. In: *Proceedings of the 19th Iranian Plant Protection Congress*, Tehran, Iran, p. 3.

Fallahpori, A., H. Aminian, N. Sahebani, and S.A. Esmailzade Hosseini. 2013. Evaluation of peroxidase activity in two resistant and susceptible sesame germplasm s to *Fusarium* damping-off caused by *Fusarium oxysporum* f. sp. *sesami*. *Iran. J. Plant Pathol.* 49: 35–36.

Falusi, O.A. and E.A. Salako. 2003. Inheritance studies in wild and cultivated 'Sesamum' L. species in Nigeria. *J. Sustain. Agric.* 22: 375–380.

FAOSTAT. 2011. Food and Agricultural Organization of the United Nations—Statistical data base. http://faostat.org. Accessed on September 11, 2014.

Farhan, H.N., B.H. Abdullah, and A.T. Hameed. 2010. The biological activity of bacterial vaccine of *Pseudomonas putida2* and *Pseudomonas fluorescens3* isolates to protect sesame crop (*Sesamum indicum*) from *Fusarium* fungi under field conditions. *Agric. Biol. J. North Am.* 1: 803–811.

Gabr, M.R., O.I. Saleh, N.A. Hussein, and M.A. Khalil. 1998. Physiological studies and cell-wall degrading enzymes of *Fusarium oxysporum* f. sp. *sesami* and *Macrophomina phaseoli*: The causal organisms of wilt and root rot diseases of sesame. *Egypt. J. Microbiol.* 33: 595–610.

Gao, S.G., H.Y. Liu, J.M. Wang, Y.X. Ni, and B.M. Tian. 2011. Isolation and analysis of NBS-LRR resistance gene analogs from sesame. *J. Henan Agric. Sci.* 40: 81–85.

Ghallab, K.H. and A.T. Bakeer. 2001. Selection in sesame under greenhouse and field conditions for improving yield *and Fusarium* wilt resistance. *Ann. Agric. Sci. Moshtohor* 39: 1957–1976.

Gonzalez Segnana, L., M. Ramirez de Lopez, A.P.O.A. Mello, J.A.M. Rezende, and E.W. Kitajima. 2011. First report of Cowpea aphid-borne mosaic virus on sesame in Paraguay. *Plant Dis.* 95: 613.

Gopal, K., R. Jagadeswar, and G.P. Babu. 2005. Evaluation of sesame (*Sesamum indicum*) genotypes for their reactions to powdery mildew and phyllody diseases. *Plant Dis. Res.* (*Ludhiana*) 20: 126–130.

Guleria, S. and A. Kumar. 2006. *Azadirachta indica* leaf extract induces resistance in sesame against *Alternaria* leaf spot disease. *J. Cell Mol. Biol.* 5: 81–86.

Gupta, I.J. and H.S. Cheema. 1990. Effect of microsclerotia of *Macrophomina phaseolina* and seed dressers on germination and vigour of sesamum seed. *Seed Res.* 20: 27–30.

Hafedh, H.Z.A., H.M. Aboud, F.A. Fattah, and S.A. Khlaywi. 2005. Evaluation the antagonistic efficiency of thirty four isolates of *Trichoderma* spp. against *Macrophomina phaseolina* under laboratory and greenhouse conditions. *Arab J. Plant Prot.* 23: 44–50.

Hazarika, D.K. and K.K. Das. 1999. Incidence of bacterial wilt of *Sesamum* in relation to sowing dates and varieties. *Plant Dis. Res.* 14: 130–133.

Hegde, D.M. 2009. Can India achieve self reliance in vegetable oils? Souvenir published during *National Symposium on "Vegetable Oils Scenario: Approaches to Meet the Growing Demands"*, Indian Society of Oilseeds Research, Directorate of Oilseed Research, Hyderabad, India.

Hoque, M.Z., M.D. Hossain, F. Begum, and M.M. Alam. 2009. Efficacy of some fungitoxicants against *Cercospora sesami* causing leaf spot of sesame. *Bangladesh J. Plant Pathol.* 25: 75.

Hua, J.L., B.S. Hu, X.M. Li, R.R. Huang, and G.R. Liu. 2012. Identification of the pathogen causing bacterial wilt of sesame and its biovars. *Acta Phytophylacica Sin.* 39: 39–44.

Hyun, J.W., Y.H. Kim, Y.S. Lee, and W.M. Park. 1999. Isolation and evaluation of protective effect against *Fusarium* wilt of sesame plants of antibiotic substance from *Bacillus polymyxa KB-8*. *Plant Pathol. J.* 15: 152–157.

Ibrahim, M.E. and A.M. Abdel-Azeem. 2007. Towards an integrated control of sesame (*Sesamum indicum* L.) charcoal rot caused by *Macrophomina phaseolina. Arab Univ. J. Agric. Sci.* 15: 473–481.

Isakeit, T., B.T. Hassett, and K.L. Ong. 2012. First report of leaf spot of sesame caused by *Xanthomonas* sp. in the United States. *Plant Dis.* 96: 1222.

Jahagirdar, S., K.N. Pawar, M.R. Ravikumar, S.T. Yenjerappa, and B.G. Prakash. 2003. Field evaluation of sesamum genotypes for multiple disease resistance. *Agric. Sci. Dig.* 23: 61–62.

Jayashree, K., V. Shanmugam, T. Raguchander, A. Ramanathan, and R. Samiyappan. 2000. Evaluation of *Pseudomonas fluorescens* (Pf-1) against blackgram and sesame root-rot diseases. *J. Biol. Control* 14: 255–261.

John, P., N.N. Tripathi, and N. Kumar. 2010. Efficacy of fungicides against charcoal-rot of sesame caused by *Rhizoctonia bataticola* (Taub.) Butler. *Res. Crops* 11: 508–510.

John, P., N.N. Tripathi, and K. Naveen. 2005. Evaluation of sesame germplasm/cultivars for resistance against charcoal rot. *Res. Crops* 6: 152–153.

Jyothi, N., A. Ansari, V. Yepuri et al. 2011. Assessment of resistance to *Fusarium* wilt disease in sesame (*Sesamum indicum* L.) germplasm. *Aust. Plant Pathol.* 40: 471–475.

Kalita, M.K., K. Pathak, and U. Barman. 2000. Yield loss in sesamum due to *Phytophthora* blight in Barak Valley of Assam. *Indian J. Hill Farm.* 13: 42–43.

Kalita, M.K., K. Pathak, and U. Barman. 2002. Yield loss in sesamum due to *Phytophthora* blight in Barak Valley Zone (BVZ) of Assam. *Ann. Biol.* 18: 61–62.

Kang, S.W. and H.K. Kim. 1989. *Gliocladium virens*: A potential biocontrol agent against damping-off and *Fusarium* wilt of sesame (*Sesamum indicum* L.) in problem fields in Korea. *Res. Rep. Rural Dev. Adm. Crop Prot., Korea Republic* 31: 19–26.

Kavak, H. and E. Boydak. 2006. Screening of the resistance levels of 26 sesame breeding lines to *Fusarium* wilt disease. *Plant Pathol. J.* 5(2): 157–160.

Kavak, H. and E. Boydak. 2011. Trends of sudden wilt syndrome in sesame plots irrigated with delayed intervals. *Afr. J. Microbiol. Res.* 5: 1837–1841.

Khan, A.J., K. Bottner, N. Al-Saadi, A.M. Al-Subhi, and I.M. Lee. 2007. Identification of phytoplasma associated with witches' broom and virescence diseases of sesame in Oman. *Bull. Insectol.* 60: 133–134.

Kolte, S.J. 1985. *Diseases of Annual Edible Oilseed Crops*, Vol. II: *Rapeseed-Mustard and Sesame Diseases*. CRC Press, Boca Raton, FL.

Kolte, S.J. 1997. Annual oilseed crops. In: *Soilborne Diseases of Tropical Crops*, eds., R.J. Hillocks and J.M. Waller. CAB International, Wallingford, U.K., pp. 253–276.

Kolte, S.J. and P.A. Shinde. 1973. Influence of plant extracts of certain hosts on growth and sclerotial formation of *Macrophomina phaseoli* (Mabl.) Ashby. *Indian Phytopathol.* 26: 351–352.

Kumar, P. and U.S. Mishra. 1992. Diseases of *Sesamum indicum* in Rohilkhand: Intensity and yield loss. *Indian Phytopathol.* 45: 121–122.

Kumaresan, D. and N. Nadarajan. 2002. Combining ability and heterosis for powdery mildew resistance in sesame. *Sesame Safflower Newsl.* 17: 1–4.

Kvashnina, E.S. 1928. Preliminary report of the survey of diseases of medicinal and industrial plants in North Caucasus. *Bull. North Caucasian Plant Prot. Stn.* 4: 30.

Leppik, E.E. and G. Sowell. 1964. *Alternaria sesami*, a serious seed borne pathogen of worldwide distribution. *FAO Plant Prot. Bull.* 12(1): 13.

Li, D.H., L.H. Wang, Y.X. Zhang et al. 2012. Pathogenic variation and molecular characterization of Fusarium species isolated from wilted sesame in China. *Afr. J. Microbiol. Res.* 2(6):149–54

Li, L.L., S.Y. Fang, X.P. Hyang et al. 1991. Identification of *Macrophomina phaseolina* resistant germplasm of sesame of China. *Oil Crops China* 1: 3–6.

Liu, L.M., H.Y. Liu, and B.M. Tian. 2012. Selection of reference genes from sesame infected by *Macrophomina phaseolina. Acta Agron. Sin.* 38: 471–478.

Liu, Z.H., X.Y. Huang, M. Liu, H. Yang, J. Sun, and Z.H. Zhao. 2010. Biological characteristics of *Fusarium oxysporum* of the sesame stem blight causal agent. *J. Shenyang Agric. Univ.* 41: 417–421.

Lubaina, A.S. and K. Murugan. 2013a. *Mikania scandens* (L.) Wild.—A plant based fungicide against *Alternaria* leaf spot disease in *Sesamum orientale* L.: Some observations. *Int. J. Curr. Res. Rev.* 5: 22–31.

Lubaina, A.S. and K. Murugan. 2013b. Induced systemic resistance with aqueous extract of *Mikania scandens* (L.) Wild against *Alternaria sesami* (Kawamura) Mohanty and Behera in *Sesamum orientale* L. *J. Crop Sci. Biotechnol.* 16: 269–276.

Mahdizadeh, V., N. Safaie, and E.M. Goltapeh. 2012. Genetic diversity of sesame isolates of *Macrophomina phaseolina* using RAPD and ISSR markers. *Trakia J. Sci.* 10: 65–74.

Mahdy, E.E., B.R. Bakheit, M.H. Motawea, and I.M. Bedawy. 2005a. Pedigree selection for resistance to *Sclerotium bataticola* in three sesame populations. *Assiut J. Agric. Sci.* 36: 57–72.

Mahdy, E.E., B.R. Bakheit, M.M. Motawea, and I.M. Bedawy. 2005b. Pedigree selection for resistance to *Fusarium oxysporum* in three sesame populations. *Assiut J. Agric. Sci.* 36: 141–157.

Maheshwari, D.K., R.C. Dubey, A. Aeron et al. 2012. Integrated approach for disease management and growth enhancement of *Sesamum indicum* L. utilizing *Azotobacter chroococcum* TRA2 and chemical fertilizer. *World J. Microbiol. Biotechnol.* 28: 3015–3024.

Maiti, S., S.R. Hegde, and S.B. Chattopadhyay. 1988. *Handbook of Annual Oilseed Crops*. Oxford and IBH Publishing Co, Pvt. Ltd., New Delhi, India, pp. 109–137.

Martinez-Hilders, A. and H. Laurentin. 2012. Phenotypic and molecular characterization of *Macrophomina phaseolina* (Tassi) Goid. coming from the sesame production zone in Venezuela. *Bioagro* 24: 187–196.

Martinez-Hilders, A., Y. Mendoza, D. Peraza, and H. Laurentin. 2013. Genetic variability of *Macrophomina phaseolina* affecting sesame: Phenotypic traits, RAPD markers and interaction with crop. *Res. J. Recent Sci.* 2: 110–115.

Marri, N.A., A.M. Lodhi, J. Hajano, G.S. Shah, and S.A. Maitlo. 2012. Response of different sesame (*Sesamum indicum* L.) cultivars to Alternaria leaf spot disease (*Alternaria sesami*) (Kawamura) Mohanty & Behera. *Pak. J. Phytopathol.* 24: 129–132.

Mathur, Y.K. and J.P. Verma. 1972. Relation between date of sowing and incidence of sesamum phyllody and abundance of its cicadellid vector. *Indian J. Entomol.* 34: 74–75.

Melean, A.J. 2003. Resistance of white seeded sesame (*Sesamum indicum* L.) cultivars against charcoal rot (*Macrophomina phaseolina*) in Venezuela. *Sesame Safflower Newsl.* 18: 72–76.

Misra, H.P. 2003. Efficacy of combination insecticides against til leaf webber and pod borer, *Antigastra catalaunalis* (Dupon.), and phyllody. *Ann. Plant Prot. Sci.* 11: 277–280.

Mohanty, N.N. 1958. *Cercospora* leaf spot of sesame. *Indian Phytopathol.* 11: 186.

Mohanty, N.N. and B.C. Behera. 1958. Blight of sesame (*Sesame orientale* L.) caused by *Alternaria sesami* (Kawamura). Comb. *Curr. Sci.* (*India*) 27: 492.

Moi, S. and P. Bhattacharyya. 2008. Influence of biocontrol agents on sesame root rot. *J. Mycopathol. Res.* 46: 97–100.

Mudingotto, P.J., E. Adipala, and S.B. Mathur. 2002. Seed-borne mycoflora of sesame seeds and their control using salt solution and seed dressing with Dithane M-45. *Muarik Bull.* 5: 35–43.

Munoz-Cabanas, R.M., S. Hernandez-Delgado, and N. Mayek-Perez. 2005. Pathogenic and genetic analysis of *Macrophomina phaseolina* (Tassi) Goid. on different hosts. *Rev. Mexi. Fitopatol.* 23: 11–18.

Nagaraju and V. Muniyappa. 2005. Viral and phytoplasma diseases of sesame. In: *Diseases of Oilseed Crops*, eds., G.S. Saharan, N. Mehta, and M.S. Sangwan. Indus Publishing Co, New Delhi, India, pp. 304–317.

Nahunnaro, H. and B.A. Tunwari. 2012a. Natural selection of four sesame resistant cultivars against *Cercospora* leaf spot (CLS) disease (*Cercospora sesami* Zimm.) in the Nigerian Southern and Northern Guinea Savannahs. *World J. Agric. Sci.* 8: 540–546.

Nahunnaro, H. and B.A. Tunwari. 2012b. Field evaluation of some selected plant extracts on the control of *Cercospora* leaf spot (CLS) disease (*Cercospora sesami* Linn.) and their effects on the seed yield of sesame (*Sesamum indicum* L.). *Int. J. Agric. Res. Rev.* 2: 1045–1050.

Naik, M.K., A.S. Savitha, R. Lokesha, P.S. Prasad, and K. Raju. 2007. Perpetuation of *Alternaria sesami* causing blight of sesame in seeds and plant debris. *Indian Phytopathol.* 60: 72–75.

Nair, R., K.D. Thakur, S. Bhasme, S. Vijayan, and S. Tirthakar. 2006. Control of root rot (*Rhizoctonia bataticola*) of sesamum by different bioagents. *J. Plant Dis. Sci.* 1: 236–237.

Nakashima, K., W. Chaleeprom, P. Wongkaew, P. Sirithorn, and S. Kato. 1999. Analysis of phyllody disease of phytoplasma in sesame and Richandia plants. *J. Jpn. Inst. Res. Center Agric. Sci.* 7: 19–27.

Nakashima, K., T. Hayashi, W. Chaleeprom, P. Wongkaew, and P. Sirithorn. 1995. Detection of DNA of Phytoplasmas associated with phyllody disease of sesame in Thailand. *Ann. Phytopathol. Soc. Jpn.* 61: 519–528.

Naqvi, S.F., M. Inam-ul-Haq, M.I. Tahir, and S.M. Mughal. 2012. Screening of sesame germplasm for resistance against the bacterial blight caused by *Xanthomonas campestris* pv. *sesame*. *Pak. J. Agric. Sci.* 49: 131–134.

Narayanaswamy, R. and B. Gokulakumar. 2010. ICP-AES analysis in sesame on the root rot disease incidence. *Arch. Phytopathol. Plant Prot.* 43: 940–948.

Nyanapah, J.O., P.O. Ayiecho, and J.O. Nyabundi. 1995. Evaluation of sesame cultivars to resistance to *Cercospora* leaf spot. *E. Afr. Agric. For. J.* 6: 115–121.

Ojiambo, P.S., P.O. Ayiecho, R.D. Narla, and R.K. Mibey. 2000b. Tolerance level of *Alternaria sesami* and the effect of seed infection on yield of sesame in Kenya. *Exp. Agric.* 36: 335–342.

Ojiambo, P.S., P.O. Ayiecho, and J.O. Nyabund. 2008. Severity of Alternaria leaf spot and seed Infection by *Alternaria sesami* (Kawamura) Mohanty and Behera, as affected by plant age of sesame (*Sesamum indicum* L.). *J. Phytopathol.* 147: 403–407.

Ojiambo, P.S., R.K. Mibey, R.D. Narla, and P.O. Ayiecho. 2003. Field transmission efficiency of *Alternaria sesami* in sesame from infected seed. *Crop Prot.* 22: 1107–1115.

Ojiambo, P.S., R.D. Narla, P.O. Ayiecho, and R.K. Mibey. 2000a. Infection of sesame seed by *Alternaria sesami* (Kawamura) Mohanty and Behera and severity of *Alternaria* leaf spot in Kenya. *Int. J. Pest Manag.* 46: 121–124.

Paik, S.B., E.S. Do., J.S. Yang, and M.J. Han. 1988. Occurrence of wilting disease (*Fusarium* spp.) according to crop rotation and continuous cropping of sesame (*Sesamum indicum* L.). *Korean J. Mycol.* 16: 220–225.

Palakshappa, M.G., S.G. Parameshwarappa, M.S. Lokesh, and D.G. Shinde. 2012. Management of *Cercospora* leaf spot of sesame. *Int. J. Plant Prot.* 5(1): 160–162.

Pathak, D.M., A.M. Parakhia, and L.E. Akbari. 2012. Symptomatology and transmission of sesame phyllody disease caused by phytoplasma. *J. Mycol. Plant Pathol.* 42: 479–484.

Pathak, N., A.K. Rai, R. Kumari, A. Thapa, and K.V. Bhat. 2014. Sesame crop: An underexploited oilseed holds tremendous potential for enhanced food value. *Agric. Sci.* 5: 519–529.

Pathirana, R. 1992. Gamma ray-induced field tolerance to Phytophthora blight in sesame. *Plant Breed.* 108: 314–319.

Patil, S.K., P.K. Dharne, and D.A. Shambharkar. 2001. Response of different promising genotypes of sesame to major insect pests and leaf spot disease. *Sesame Safflower Newsl.* 16: 72–74.

Pineda, J.B. 2001. Evaluation of *Trichoderma harzianum* application methods to the soil for control of *Macrophomina phaseolina* in sesame. *Fitopatol. Venez.* 14: 31–34.

Poswal, M.A.T. and S.M. Misari. 1994. Resistance of sesame cultivars to *Cercospora* leaf spot induced by *Cercopora sesami*. Zimm. *Discov. Innov.* 6: 66–70.

Prasad, N., S.P. Sehgal, and P.D. Gemavat. 1970. Phytophthora blight of sesamum. In: *Plant Disease Problems: Proceedings of the First International Symposium on Plant Pathology*. Indian Phytopathology Society, IARI, New Delhi, India, pp. 331–339.

Purohit, S.D. and H.C. Arya. 1980. Phyllody: An alarming problem for sesamum growers. *Agric. Dig.* 4: 12.

Qi, Y.X., X. Zhang, J.J. Pu et al. 2011. Morphological and molecular analysis of genetic variability within isolates of *Corynespora cassiicola* from different hosts. *Eur. J. Plant Pathol.* 130: 83–95.

Rai, M. and S.S.L. Srivastava. 2003. Field evaluation of some agro-chemicals for management of bacterial blight of sesame. *Farm. Sci. J.* 12: 81–82.

Rajeswari, S., V. Thiruvengadam, and N.M. Ramaswamy. 2010. Production of interspecific hybrids between *Sesamum alatum* Thonn and *Sesamum indicum* L. through ovule culture and screening for phyllody disease resistance. *S. Afr. J. Bot.* 76: 252–258.

Rajpurohit, T.S. 2002. Influence of intercropping and mixed cropping with pearl millet, green gram and mothbean on the incidence of stem and root rot (*Macrophomina phaseolina*) of sesame. *Sesame Safflower Newsl.* 17: 40–41.

Rajpurohit, T.S. 2003. Studies on fungicide, plant product and bioagent as spray in sesame disease management. *J. Eco-Physiol.* 6: 87–88.

Rajpurohit, T.S. 2004a. *Trichoderma viride*: Biocontrol agent effective against *Macrophomina* stem and root rot of sesame. *J. Eco-Physiol.* 7: 187–188.

Rajpurohit, T.S. 2004b. Efficacy of foliar spray on incidence of diseases of sesame. *Sesame Safflower Newsl.* 19: Unpaginated.

Rani, S.U., R. Udhayakumar, and D.J. Christopher. 2009. Efficacy of bioagents and organic amendments against *Macrophomina phaseolina* (Tassi) causing root rot of sesame. *J. Oilseeds Res.* 26: 173–174.

Rao, G.V.N. and M.A.R. Rao. 2001. Assessment of yield losses in sesame (*Sesamum indicum* L.) due to powdery mildew and its management. *Indian J. Plant Prot.* 29: 165–167.

Rao, G.V.N. and M.A.R. Rao. 2002. Efficacy of certain fungicides against stem blight of sesamum caused by *Alternaria alternata* (Fr.) Keissier. *Indian J. Plant Prot.* 30: 86–87.

Rao, N.R. and M. Vijayalakshmi. 2000. Studies on *Alternaria sesami* pathogenic to sesame. *Microbiol. Res.* 155: 129–131.

Rao, P.V.R., V.G. Shankar, J.V.P. Pavani, V. Rajesh, A.V. Reddy, and K.D. Reddy. 2011. Evaluation of sesame genotypes for powdery mildew resistance. *Int. J. Bio-Resour. Stress Manag.* 2: 341–344.

Rao, P.V.R., G. Anuradha, M.J. Prada et al. 2012. Inheritance of powdery mildew tolerance in sesame. *Arch. Phytopathol. Plant Prot.* 45: 404–412.

Rayatpanah, S., S.A. Dalili, and E. Yasari. 2012. Diversity of *Macrophomina phaseolina* (Tassi) Goid based on chlorate phenotypes and pathogenicity. *Int. J. Biol.* 4: 54–63.

Robertson, H.E. 1928. Annual report of the mycologists, Burma for the year ended 30th June, 1928, 10pp. Abstr. *Rev. Appl. Mycol.* 8: 355, 1928.

Sağır, P., A. Sağır, and T. Söğüt. 2009. The effect of charcoal rot disease (*Macrophomina phaseolina*), irrigation and sowing date on oil and protein content of some sesame lines. *J. Turk. Phytopathol.* 38: 33–42.

Sağır, P., A. Sağı, and T. Sogut. 2010. The effect of sowing time and irrigation on Charcoal rot disease (*Macrophomina phaseolina*), yield and yield components of sesame. *Bitki Koruma Bult.* 50: 157–170.

Salehi, M. and K. Izadpanah. 1992. Ecology and transmission of sesame phyllody in Iran. *J. Phytopathol.* 135: 37–47.

Samina, B., M. Irfan-ul-Haque, T. Mukhtar, G. Irshad, and M.A. Hussain. 2007. Pathogenic variation in *Pseudomonas syringae* and *Xanthomonas campestris* pv. *sesami* associated with blight of sesame. *Pak. J. Bot.* 39: 939–943.

Sangle, U.R. and O.M. Bambawale. 2004. New strains of *Trichoderma* spp. strongly antagonistic against *Fusarium oxysporum* f. sp. *sesame. J. Mycol. Plant Pathol.* 34: 107–109.

Saravanan, S. and N. Nadarajan. 2004. Reaction of sesame genotypes to powdery mildew disease. *Res. Crops* 5: 115–117.

Saravanan, S. and N. Nadarajan. 2005. Pathogenicity of mycoplasma like organism of sesame (*Sesamum indicum* L.) and its wild relatives. *Agric. Sci. Dig.* 25: 77–78.

Sarwar, G. and K.P. Akhtar. 2009. Performance of some sesame mutants to phyllody an leaf curl virus disease under natural field conditions. *Pak. J. Phytopathol.* 21: 18–25.

Sarwar, G. and M.A. Haq. 2006. Evaluation of sesame germplasm for genetic parameters and disease resistance. *J. Agric. Res. (Lahore)* 44: 89–96.

Sattar, A.M.H., A.M. Salem, and A.A. Bayounes. 2006. Physical and biological control of charcoal root rot in sesame caused by *Macrophomina phaseolina* (Tassi) Goid. *Arab J. Plant Prot.* 24: 37–40.

Savitha, A.S., M.K. Naik, and K. Ajithkumar. 2011. Eco-friendly management of *Alternaria sesami* incitant blight of sesame. *J. Plant Dis. Sci.* 6: 150–152.

Savitha, A.S., M.K. Naik, and K. Ajithkumar. 2012. *Alternaria sesami*, causing blight of sesame produces toxin and induces the host for systemic resistance. *Asian J. Res. Chem.* 5: 1176–1181.

Sehgal, S.P. and N. Prasad. 1966. Variation in the pathogenicity of single-zoospore isolates of sesamun *Phytophthora*: Studies on the penetration and survival. *Indian Phytopathol.* 19: 154–158.

Sehgal, S.P. and N. Prasad. 1971. Instability of pathogenic characters in the isolates of sesamum *Phytophthora* and effect of host passage on the virulence of isolates. *Indian Phytopathol.* 24: 295–298.

Selvanarayanan, V. and T. Selvamuthukumaran. 2000. Field resistance of sesame cultivars against phyllody disease transmitted by *Orosius albicinctus* Distant. *Sesame Safflower Newsl.* 15: 71–74.

Sertkaya, G., M. Martini, R. Musetti, and R. Osler. 2007. Detection and molecular characterization of phytoplasmas infecting sesame and solanaceous crops in Turkey. *Bull. Insectol.* 60: 141–142.

Shabana, R., A.A. AbdEl-Mohsen, M.M.A. Khalifa, and A.A. Saber. 2014. Quantification of resistance of F6 sesame elite lines against char coal rot and *Fusarium* wilt diseases. *Adv. Agric. Biol.* 3: 144–150.

Shah, R.A., M.A. Pathan, M.M. Jiskani, and M.A. Qureshi. 2005. Evaluation of different fungicides against *Macrophomina phaseolina* causing root rot disease of sesame. *Pak. J. Agric. Eng. Vet. Sci.* 21: 35–38.

Shalaby, I.M.S., R.M.A. El-Ganainy, S.A. Botros, and M.M. El-Gebally. 2001. Efficacy of some natural and synthetic compounds against charcoal rot caused by *Macrophomina phaseolina* of sesame and sunflower plants. *Assiut J. Agric. Sci.* 32: 47–56.

Shalaby, O.Y.M. and A.T. Bakeer. 2000. Effect of agricultural practices on root rot and wilt of sesame in Fayoum. *Ann. Agric. Sci. Moshtohor* 38: 1399–1407.

Shalaby, S.I.M. 1997. Effect of fungicidal treatment of sesame seeds on root rot infection, plant growth and chemical components. *Bull. Fac. Agric. Univ. Cairo* 48: 397–411.

Shekharappa, G. and P.V. Patil. 2001a. Chemical control of leaf blight of sesame caused by *Alternaria sesami*. *Karnataka J. Agric. Sci.* 14: 1100–1102.

Shekharappa, G. and P.V. Patil. 2001b. Field reaction of sesame genotypes to *Alternaria sesami*. *Karnataka J. Agric. Sci.* 14: 1103–1104.

Shim, K.B., D.H. Kim, J.H. Park, S.W. Lee, K.S. Kim, and J.H. Rho. 2012. A new black sesame variety 'Kangheuk' with lodging and phytophthora blight disease resistance, and high yielding. *Korean J. Breed. Sci.* 44: 38.

Shindhe, G.G., R. Lokesha, M.K. Naik, and A.G.R. Ranganath. 2011. Inheritance study on phyllody resistance in sesame (*Sesamum indicum* L.). *Plant Arch.* 11: 775–776.

Shivas, R.G., C.A. Brockway, and V.C. Beilharz. 1996. First record in Australia of *Cercospora sesami* and *Pseudocercospora sesami*, two important foliar pathogens of sesame. *Aust. Plant Pathol.* 25: 212.

Shokalu, O., O.A. Enikuomehin, A.A. Idowu, and A.C. Uwala. 2002. Effects of seed treatment on the control of leaf blight disease of sesame. *Trop. Oilseeds J.* 17: 94–100.

Sidawi, A., G. Abou Ammar, Z. Alkhider, T. Arifi, E. Alsaleh, and S. Alalees. 2010. Control of sesame wilt using medicinal and aromatic plant extracts. *Julius-Kuhn-Archiv.* 428: 117.

Silme, R.S. and M.I. Cagrgan. 2010. Screening for resistance to *Fusarium* wilt in induced mutants and world collection of sesame under intensive management. *Turkish J. Field Crops* 15: 89–93.

Simosa, C.N.C. and M. Delado.1991. Virulence of four isolates of *Macrophomina phaseolina* on four sesame (*Sesamum indicum* L.) cultivars. *Fitopathol. Venez.* 4: 20–23.

Singh, A., S.K. Singh, and Kamal. 2000. Three new species *of Corynespora* from India. *J. Mycol. Plant Pathol.* 30: 44–49.

Singh, P.K., M. Akram, M. Vajpeyi, R.L. Srivastava, K. Kumar, and R. Naresh. 2007. Screening and development of resistant sesame varieties against phytoplasm. *Bull. Insectol.* 60: 303–304.

Soner Slme, R. and M.I. Çagrgan. 2010. Screening for resistance to *Fusarium* wilt in induced mutants and world collection of sesame under intensive management. *Turkish J. Field Crops* 15: 89–93.

Sravani, D., Y. Vijay, D. Bharathi, V.L.N. Reddy, M.V.B. Rao, and G. Anuradha. 2012. Genetics of powdery mildew resistance in Sesamum (*Sesamum indicum* L.). *J. Res. ANGRAU* 40: 73–74.

Su, Y.L., H.M. Miao, L.B. Wei, and H.Y. Zhang. 2012. Study on separation and purification techniques of *Fusarium oxysporum* in sesame (*Sesamum indicum* L.). *J. Henan Agric. Sci.* 41: 92–95.

Subrahmaniyan, K., D. Dinakaran, P. Kalaiselven, and N. Arulmozhi. 2001. Response of root rot resistant cultures of sesame (*Sesamum indicum* L.) to plant density and NPK fertilizer. *Agric. Sci. Dig.* 21: 176–178.

Tan, A.S. 2010. Screening phyllody infected sesame varieties for seed transmission of disease under natural conditions. *Anadolu* 20(1): 26–33.

Thiyagu, K., G. Kandasamy, N. Manivannan, V. Muralidharan, and S.K. Manoranjitham. 2007. Identification of resistant genotypes to root rot disease (*Macrophomina phaseolina*) of sesame (*Sesamum indicum* L.). *Agric. Sci. Dig.* 27: 34–37.

Tripathi, S.K., U.S. Bose, and K.L. Tiwari. 1996. Economics of fungicidal control of *Cercospora* leaf spot of sesamum. *Crop Res.* 12: 71–75.

Tripathi, S.K., P. Tamarkar, and M.S Baghel. 1998a. Influence of dates of sowing and plant distance *on Cercospora* leaf spot of sesamum. *Crop Res.* 15: 270–274.

Tripathi, U.K., S.B. Singh, and P.N. Singh. 1998b. Management of Alternaria leaf spot of sesame by adjustment of sowing dates. *Ann. Plant Prot. Sci.* 6(1): 94–95.

Tunwari, B.A. and H. Nahunnaro 2014. Effects of botanical extracts and a synthetic fungicide on severity of Cercospora leaf spot (*Cercospora sesame* Zimm.) on Sesame (*Sesamum indicum* L.) yield attributes under screen house condition in Ardo-Kola, Taraba State, Nigeria. *Int. J. Sci. Technol. Res.* 3: 17–22.

Uwala, A.C. 1998. Evaluation of Chlorithalonil, Benlate and Agrimycin 500 on the control of leaf spot disease and their effects on the yield of sesame in Southern Guinea Savannah. In: *Proceedings of the First National Workshop on Beniseed*, eds., L.D. Busari, A.A. Idowu, and S.M. Misari. NCRI, Badeggi, Nigeria, March 3–5, 1998, pp. 201–206.

Uzun, B. and M.I. Cagrgan. 2001. Resistance in sesame to *Fusarium oxysporum* f. sp. *sesami. Turkish J. Field Crops* 6: 71–75.

Vasudeva, R.S. and H.S. Sahambi. 1955. Phyllody in sesamum (*Sesamum orientale* L.). *Indian Phytopathol.* 8: 124–129.

Verma, M.L. 2002. Effect of soil types on diseases and yield of sesame (*Sesamum indicum* L.). Abstr. In: *Asian Congress of Mycology and Plant Pathology and Symposium on Plant Health for Food Security*, University of Mysore, Mysore, India, p. 53.

Verma, M.L. and R.P. Bajpai. 2001. Effect of bioinoculants on Phytophthora blight and Macrophomina root/stem rot of sesame (*Sesamum indicum* L.). Abstr. In: *National Symposium on Plant Protection Strategies for Sustainable Horticulture*, Society of Plant Protection Science, SKAUAT, Jammu, India, p. 113.

Verma, M.L., N. Mehta, and M.S. Sangwan. 2005. Fungal and bacterial diseases of sesame. In: *Diseases of Oilseed Crops*, eds., G.S. Saharan, N. Mehta, and M.S. Snwan. Indus Publishing Co., New Delhi, India, pp. 209–303.

Wang, L.H., Y.X. Zhang, D.H. Li et al. 2011. Variations in the isolates of *Macrophomina phaseolina* from sesame in China based on amplified fragment length polymorphism (AFLP) and pathogenicity. *Afr. J. Microbiol. Res.* 5: 5584–5590.

Wei, W., S. Wei, H. Zhang, F. Ding, T. Zhang, and F. Lu. 1999. Breeding of a new sesame variety Yuzhi 11. *J. Henan Agric. Sci.* 14: 3–4.

Wei, W.L., X.O. Qi, L.H. Wang et al. 2011. Characterization of the sesame (*Sesamum indicum* L.) global transcriptome using illumina paired-end sequencing and development of EST_SSR markers. *BMC Genomics* 12: 451.

Weintraub, P.G. and P. Jones. 2010. *Phytoplasmas: Genomes, Plant Hosts and Vectors*. CAB International, Oxfordshire, U.K., p. 147.

Win, N.K.K., C.G. Back, and H.Y. Jung. 2010. Phyllody Phytoplasma infecting sesame (*Sesamum indicum*) in Myanmar. *Trop. Plant Pathol.* 35: 310–313.

Wuike, R.V., V.S. Shingare, and P.N. Dawane. 1998. Biological control of wilt of sesame caused by *Fusarium oxysporum* f. sp. *sesame*. *J. Soil Crops* 8: 103–104.

Yadava, D.K., S. Vasudev, N. Singh, T. Mohapatra, and K.V. Pabhu. 2012. Breeding major oil crops: Present status and future research needs. In: *Technological in Major World Oil Crops*, Vol. I: *Breeding*, ed., S.K. Gupta. Springer Science+Business Media LLC, New York, pp. 17–51. doi:10.1007/978-1-4614-0356-2_2.

Zhao, H., H.M. Miao, H.T. Gao, Y.X. Ni, L.B. Wei, and H.Y. Liu. 2012. Evaluation and identification of sesame germplasm resistance to *Macrophomina phaseolina*. *J. Henan Agric. Sci.* 41: 82–87.

Zhang, X.R., Y. Cheng, S.Y. Liu, and X.Y. Feng. 2001. Evaluation of sesame germplasms resistant to *Macrophomina phaseolina* and *Fusarium oxysporum*. *J. Oil Crop Sci.* 23: 23–27.

Zhang, Y.X., L.H. Wang, D.H. Li, W.L. Wei, Y.Z. Gao, and R. Xiu. 2012. Association mapping of sesame (*Sesamum indicum* L.) resistance to *Macrophomina phaseolina* and identification of resistant accessions. *Sci. Agric. Sin.* 45: 2580–2591.

Ziedan, E.H., I.S. Elewa, M.H. Mostafa, and A.F. Sahab. 2011. Application of mycorrhizae for controlling root diseases of sesame. *J. Plant Prot. Res.* 51: 355–361.

Ziedan, E.H., M.H. Mostafa, and I.S. Elewa. 2012. Effect of bacterial inocula on *Fusarium oxysporum* f. sp. *sesami* and their pathological potential on sesame. *Int. J. Agric. Technol.* 8: 699–709.

Section VI

Safflower

Safflower (*Carthamus tinctorius* L.) is a highly branched, herbaceous, self-compatible, thistlelike annual plant. It varies greatly in height ranging from 0.3 to 1.5 m. It belongs to the family Compositae with 12 pairs ($2n = 24$) of basic chromosome number. Its haploid genome is approximately 1.4 Gb (Garnatje et al. 2006). Li et al. (2011) found that there are at least 236 known micro-RNAs (mRNAs) expressed in the safflower, 100 of which are conserved across the plants. The safflower genome is large and complex and it has not been fully sequenced, and relatively little is known about its encoded genes. As of October 2011, only 567 nucleotide sequences, 41,588 expressed sequences tags (ESTs), 162 proteins, and 0 genes from *C. tinctorius* have been deposited in the National Center for Biotechnology Information (NCBI)'s GenBank database. Interestingly, abundant genomic data for *C. tinctorius* and comprehensive sequence resources for studying the safflower transcriptome datasets have been generated that will serve as an important platform to accelerate studies of the safflower genome (Huang et al. 2012). The plant has many branches each terminating in a flower. The inflorescence is a dense capitulum of numerous regular flowers. The flowers are self-pollinated, but cross-pollination to the extent of 16% may occur under natural conditions. Each branch usually has 1–5 flower heads containing 15–20 seeds per head. The seed is ovate, having a flat top with longitudinal ribs, and represents the Cypsela type of fruit. The cultivated forms of safflower are supposed to have originated either from *Carthamus lanatus* Linn or from *C. oxyacantha* Bieb, evidently in two primary centers of origin—the mountainous regions of Afghanistan and of Ethiopia. Safflower cultivation has now extended over many parts of the world, both in the tropics and in the subtropics in more than 60 countries worldwide commercially producing about 600,000 tons seed yield annually. India (producing over half of the world produce), the United States, and Mexico are the three leading producers, with Ethiopia, Kazakhstan, China, the Arab world, Argentina, and Australia accounting for most of the remainder (Yadava et al. 2012). Safflowers have long taproots that facilitate water uptake in even the driest environments enabling these crops to be grown on marginal lands where moisture would otherwise be limited. Thus, safflower is a drought- and salt-tolerant crop. It can be grown in a range of soil types, but well-drained medium to heavy textured soils are best suited for its growth. Earlier, this crop had been grown for its flowers that can be used for dyes as well as in teas and as food additive. Extracts from the florets have been used to reduce hypertension and blood cholesterol levels. Currently, safflower is preferred for its high-quality seed oil that is rich in polyunsaturated fatty acids. Prof. Paulden Knowles, the *father of California safflower*, has accumulated and documented a large collection of safflower germplasm that is currently being maintained by the United States Department of Agriculture (USDA) for further use in research towards improvement of safflower crop. Such a collection is also being maintained at National Bureau of Plant Genetic Resources (NBPGR), New Delhi, and Indian Institute of Oilseeds (ICAR), Hyderabad, India. Safflower diseases are described in this chapter.

8 Safflower Diseases

Alternaria BLIGHT

SYMPTOMS

Seedlings from severely affected plants show lesions on the hypocotyls and/or cotyledons. Dark necrotic lesions measuring up to 5 mm in diameter may be formed on hypocotyls. Both hypocotyl and cotyledonary symptoms are commonly observed in the same plants. In some instances, the hypocotyl infection results in damping-off of the seedlings. In mature plants, small brown to dark-brown concentric spots of 1–2 cm diameter appear on leaves. The center of the mature spots is usually lighter in color. The spots frequently coalesce into large irregular lesions bearing the spores of the fungus. The fully mature spots tend to develop shot holes, and in severe infections, irregular cracking of the leaf blade occurs. The stems and petioles suffer less severe damage with elongated spots. On flower heads, the fungus first attacks the base of the calyx and later spreads to other parts of the flower. Infected flower buds shrivel without opening, and seeds obtained from a severely infected crop may show a dark sunken lesion on the testa.

GEOGRAPHICAL DISTRIBUTION AND LOSSES

Alternaria leaf blight of safflower was first reported from India by Chowdhury (1944). The disease is now reported from all over the world from safflower-growing countries such as Argentina, Australia, Ethiopia, Israel, Italy, Kenya, Pakistan, Portugal, Russia, Spain Tanzania, the United States, and Zambia. The effect of the disease is reported to be quite serious in the northern Great Plains area of the United States (Bergman and Jacobsen 2005) and in the states of Bihar and Madhya Pradesh of India. Reports of severe damage of experimental safflower crops due to the disease have been made from Kenya and Tanzania in East Africa. Yields of safflower infected by the disease may be reduced in highly susceptible varieties by 50%–90% when a week of humid weather occurs following flowering but before maturity. A significant negative correlation has been established between disease severity and yield (Chattopadhyay 2001). Seeds of affected plants become discolored, showing reduced oil content with significant increase in the level of free fatty acids in the seeds that adversely affect the seed germination.

PATHOGEN

The pathogen is *Alternaria carthami* Chowdhury.

The mycelium is septate inter- and intracellularly with slight constrictions at the septa. It is subhyaline when young but becomes dark colored on maturity. The conidiophores are stout; erect; rigid; unbranched; septate; straight or flexuous, sometimes geniculate; and brown or olivaceous brown, paler near the apex, and arise singly or in clusters through the epidermis or stomata. The conidiophores are sometimes swollen at the base and measure 15–85 μm in length and 6–10 μm in width.

The conidia are borne on conidiophores and are solitary or in very short chains. They are smooth, straight or curved, obclavate, and light brown and translucent in shade and possess a long beak. The conidia sometimes show constrictions at the septa. They measure 36–171 μm (with beak) and 36–99 μm (without beak) in length and 12–28 μm in width. The spores have 3–11 transverse septa and up to 7 longitudinal or oblique septa. The beak of the spores is 25–160 μm long and 4–6 μm thick at the base—tapering to 2–3 μm with up to 5 transverse septa. The beak is almost hyaline at the apex and light brown near the base. Some spores may be seen without beaks. Conidial beaks may form chlamydospores in culture. The optimum temperature at which the fungus grows is in the range of 25°C–30°C. It also tolerates a wide pH range, though maximum growth occurs at pH 6.0.

Epidemiology and Disease Cycle

The pathogen survives through seed as well as viable conidia of *A. carthami* on debris of naturally infected susceptible safflower varieties (Prasad et al. 2009, Gayathri and Madhuri 2014). *A. carthami* is readily isolated from seeds using relatively simple techniques. Isolation methods used in conjunction with the planting of seed to assess seedling health would appear to offer the most reliable means of detecting the presence of *A. carthami* in seed (Awadhiya 2000). The primary infection develops from infested seeds obtained from affected plants. Spines present on the leaf margin are the site of infection by the pathogen (Borkar 1997). An opening diameter of 120 μm at the apex of individual spines is a prerequisite for infection through spines. Spine apex openings vary with the location of the spine on the leaf margin and the relationship between the position of the spines on the leaf margin, and infection is governed by the diameter of the openings at the spine apex. Spores produced on lesions developing in plants grown from infected seeds become secondary sources of inoculum, and the pathogen occurs on the crop throughout the growing season. The macrolide antibiotics brefeldin A (BFA) and 7-dehydrobrefeldin A (7-oxo-BFA) have been characterized as phytotoxins and pathogenicity factors from *A. carthami*; the toxins are known to inhibit the endoplasmic reticulum–Golgi flux and processing (Kneusel 1994, Driouich et al. 1997). Rains coupled with high relative humidity above 80% and temperature in the range of 21°C–32°C under irrigated conditions accompanied by heavy dew or frequent showers, cyclonic storms especially at seedling, and grain formation stages favor the disease (Sastry and Chattopadhyay 2005, Gud et al. 2008, Murumkar et al. 2008a).

Disease Management

Host Plant Resistance

There is a considerable variation in response to *A. carthami* infection by a range of safflower varieties (Muñoz-Valenzuela et al. 2007, Thomas et al. 2008). Some genotypes such as EC 32012, NS 133, CTS-7218, HUS 524, and CTV 251 (Desai 1998); GMV 1175, GMV-1199, and GMV-1585 (Indi et al. 2004); GMV-5097, GMV-5133, and GMV-7017 (Murumkar et al. 2009a); and Ellite Line 21–33 (Pawar et al. 2013) show high degree of tolerance to *A. carthami* under high disease pressure and are identified as the most promising genotypes to be used in breeding program for incorporation of resistance to the disease. Semispiny to nonspiny genotypes of safflower are known to show a variable degree of tolerance to *A. carthami* infection. It is possible to combine high yield with high degree of tolerance to the disease (Mundel and Chang 2003, Harish Babu et al. 2005). Four wild *Carthamus* species, namely, *C. palaestinus*, *C. lanatus*, *C. creticus*, and *C. turkestanicus*, are reported to be immune to *Alternaria* leaf spot in laboratory as well as under field screening. Twenty-four F1s derived from crosses between *C. tinctorius* × *C. creticus*, *C. tinctorius* × *C. oxyacantha*, *C. tinctorius* × *C. turkestanicus*, *C. tinctorius* × *C. lanatus* × *C. palaestinus*, and *C. oxyacantha* × *C. tinctorius* have been screened to show no infection (immunity) by *A. carthami*. These resistant lines would serve as base material in disease resistance breeding to tag the resistant genes at molecular level for marker-assisted selections in the field for *Alternaria* blight resistance (Prasad and Anjani 2008a).

Seedling resistance of safflower genotypes to *A. carthami* is reported to be monogenic recessive, whereas adult plant resistance is under the control of two duplicate loci where at least one locus at homozygous recessive conditions confers the adult plant resistance (Gadekar and Jambhale 2002a).

Production of plants resistant to *A. carthami* via organogenesis and somatic embryogenesis (Kumar et al. 2008) and molecular breeding has paved the way for possible transgenic safflower plants that can be used to breed for *Alternaria* blight resistance in safflower. The cloned esterase gene degrading the BFA (phytotoxin and pathogenic factor) provides the basis for generation of transgenic safflower plants (Kneusel et al. 1994).

Chemical Control

Seed treatment with mefenoxam + thiram or with difenoconazole + mefenoxam is effective in reducing the primary infection (Jacobsen et al. 2008). Secondary infection can be controlled by spraying the crop with any of the foliar fungicides such as mancozeb at 0.25%, difenoconazole at 0.5%, AAF (carbendazim 12% + mancozeb 63%) at 0.2% (Sumitha and Nimbkar 2009), and fosetyl at 0.1% (Bramhankar et al. 2001). For effective and economic management of the disease, the first spray of carbendazim at 0.1% should be given immediately after disease appearance (generally at rosette stage, i.e., 25 days after sowing) followed by need-based second and third spray at 15 days after the first spray and during flowering and seed setting stage, respectively (Murumkar et al. 2008a, 2009a). Fungicide application results in lower pathogen transmission from plants to seeds and from seeds to plants (Bramhankar et al. 2002). Two newer fungicides azoxystrobin (Quadris) and pyraclostrobin (Headline) have been registered as foliar fungicides for the control of *Alternaria* blight of safflower crop in Australia as reported by Bergman and Jacobsen (2005). These fungicides are also reported to be effective for the control of the *Alternaria* blight of safflower in the United States (Wunsch et al. 2013).

Cultural Control

The occurrence of the disease may be prevented by using disease-free seeds for sowing. Such seeds can be obtained from early-sown dry land crops rather than from irrigated areas. Alternatively, the infested seeds may be treated with fungicides as discussed earlier. Crop rotation and strict sanitation of crop debris effectively manage the disease. Basal soil application of KCl at 67 kg/ha significantly reduces the disease severity and increases the safflower seed yield (Chattopadhyay 2001). This practice can be integrated with spray application of effective fungicides and suitable sowing dates for better disease management.

Effect of Plant Extracts

Antifungal activities of extracts of various plants such as *Nerium*, *Datura*, garlic bulb, *Lantana*, *Eucalyptus*, neem, onion bulb, and *Ocimum* species have been demonstrated against *A. carthami*, which can be exploited further for practical disease management (Shinde et al. 2008, Ranaware et al. 2010, Taware et al. 2014).

Fusarium WILT

SYMPTOMS

Symptoms of the disease are manifested at all stages of growth. In the seedling stage, cotyledonary leaves show small brown spots either scattered or arranged in a ring on the inner surface, and they may become shriveled and brittle and sometimes tend to become rolled and curved. The seedlings that survive the fungal attack regain vitality at the early stage of blossoming and again show symptoms of the disease at the time of seed setting. The symptoms become quite distinct when the plants are in the 6th to 10th leaf stage and about 15 cm in height. Four important characteristics of the symptoms may be helpful to identify the disease at this stage. These are (1) unilateral infection on branches and leaves, (2) golden-yellow discoloration of the leaf followed by wilting, (3) epinasty, and (4) vascular browning appearing only on one side of the root and stems of plants with unilateral top symptoms. The symptoms develop in acropetalous succession. The reddish-brown vascular discoloration of the root, stem, and petiole tissue of infected plants will vary considerably in intensity, depending on varietal reaction, severity of infection, and environmental conditions. On older plants, the lateral branches on one side may be killed, while the remainder of the plant apparently remains free from the disease. Such plants may show partial recovery between bud formation and early blossoming, but the symptoms may reappear later. The severely infected plants produce small-sized flower heads that are partially blossomed. A large number of ovaries fail to develop seeds, or they may form blackish, small, distorted, chaffy, and abortive seed.

Geographical Distribution and Losses

Fusarium wilt of safflower was first observed in the Sacramento Valley of California, USA, in 1962 (Klisiewiez and Houston 1962) and in India in 1975 (Singh et al. 1975). The disease is also reported from Egypt (Zayed et al. 1980). Now, it is identified to be the most serious disease in all safflower-growing areas in India (Murumkar and Deshpande 2009). Plants grown from infected seed seldom survive beyond the seedling stage, thereby indicating that losses in stand of the crop may occur when infected seed is sown. The disease incidence in the United States is reported to be 10%–20% in most fields and as high as 50% in some fields. Yield losses may reach to 100% if susceptible varieties are grown in fields with a history of severe *Fusarium* wilt (Sastry and Chattopadhyay 2005). In India, it has appeared as a serious threat to safflower cultivation, destroying up to 25% of plants, amounting to considerable yield loss in the Gangetic valley. Fusarial mycotoxins, namely, diacetoxyscirpenol, T-2 toxin, and 12,13-epoxytrichothecene, have been reported to be produced in sufficient quantities on infested seeds of safflower in storage to be capable of causing mycotoxicosis.

Pathogen

The pathogen is *Fusarium oxysporum* Schlecht. f. sp. *carthami* Klisiewicz and Houston.

The fungus is readily isolated from diseased plant parts on potato dextrose agar (PDA). The mycelium is delicate pink en masse or white usually with a purple tinge, sparse to abundant, branched, and septate. Microconidia are borne on simple phialids arising laterally on the hypha or on short sparsely branched conidiophores, abundant, oval to elliptical, one celled, and slightly curved and measure 5–16 × 2.2–3.5 μm. The macroconidia are hyaline, may be up to 5 septate but are mostly 3 septate, are constricted at septa, are borne in sporodochia, are straight or curved, are often pointed at the tip with rounded base, and measure 10–36 × 3–6 μm mostly 28 × 4–5 μm. Chlamydospores are one celled, smooth, and faintly colored and measure 5–13 × 10 μm in size. They are formed abundantly and are both terminal and intercalary, usually solitary but occasionally could be formed in chains (Sastry and Chattopadhyay 2005).

Epidemiology and Disease Cycle

The fungus perpetuates through seed as well as through soil. Mycelium and spores contaminate the seed surface, but hyphae are also reported to be present in the parenchymatous cells of the seed coat of infected seed. The survival of the fungus through soil is mainly through chlamydospores in plant debris. Penetration by the pathogen, which makes its entry into host cells by mechanical means, is easier when the plants are in the seedling stage and tissues are soft. Shriveling of cortical cells is noted in the case of infected plants. It appears that the infection is facilitated by production of polygalacturonase, pectin methyl esterase, cellulase, and protease enzymes. Mycotoxins—diacetoxyscirpenol and T-2—have been detected in diseased safflower plants. The pathogen is also reported to secrete diacetoxyscirpenol, T-2 toxin, fusaric acid, and lycomarasmin in culture filtrate. But the exact role of either enzymes or toxins, as earlier, is little worked out or not known. It is, however, reported that the virulence of *F. oxysporum* f. sp. *carthami* is directly correlated with the amount of fusaric acid it produces. It is also reported that the virulence is lost if fusaric acid production is prevented. The amount of seed infection per head may be limited by the extent of fungus spread in vascular tissue in the stem or lateral branch and seed head. The fungus apparently invades the seed through the vascular strands that extend into the seed through the pericarp–receptacle junction. Fungus spread in the tissue of pericarp and seed coat is intra- and intercellular. Isolates of *F. oxysporum* f. sp. *carthami* collected from different geographical areas have been found to show variation in morphology, culture characteristics, and pathogenicity (Sastry and Chattopadhyay 2003, Prameela et al. 2005, Murumkar and Deshpande 2009, Raghuwanshi and Dake 2009, Somwanshi et al. 2009). The fungus is specific in its pathogenicity on safflower and six other species of *Carthamus*. Four physiological races have had been distinctly identified by differential reactions of safflower varieties (Gila, Nebraska 6, UC-31 and US Biggs) in the United States to 14 isolates of *F. oxysporum* f. sp. *carthami* (Kilsiewicz 1975). *In India, also four distinct races have been identified based on differential reactions of 54 pathogen isolates of the pathogen using four (96-508-2-90, A1, DSF-4, and DSF-6) differential safflower lines. Molecular analysis of genetic variability using random amplified polymorphic DNA (RAPD), microsatellite, and ITS-RLFP markers has revealed three distinct groups among 54 isolates of F. oxysporum* f. sp. *carthami* (Prasad et al. 2004, 2007). *Thus a variety that exhibits resistance in one area may show susceptibility in another area.* The disease has been found associated with soils having a pH 4.3–5.0. In India, the disease is more prevalent along the Ganges River in acidic soils. The disease is also favored by high nitrogen and warm, moist weather. The wilt is reported to be more severe in fallow land and less severe where paddy or millets are cultivated before safflower. In uplands where the soil is neutral to alkaline and clay in texture, the disease incidence is reported to be low (Kolte 1985). Disease severity is favored by high temperature stress, poor drainage, and soil compaction. Any factor that contributes to reduced rate of root growth increases the plant's susceptibility to *Fusarium* wilt. High plant population also increases plant stress and favors infection. The effect of *F. oxysporum* f. sp. *carthami* is most apparent during flowering when plants and its productivity are more sensitive to stress. The disease severity decreases with lowering of temperature (from 21°C to 15°C) during the end of December to the first week of February, but it can increase as the temperature increases (23.6°C) under Indian conditions. Seedlings show less susceptibility to the disease with increase in age, and it differs with respect to different varieties and inoculum density (Sastry and Chattopadhyay 1999a).

Disease Management

Host Plant Resistance

Water culture technique using pathogen culture filtrate at 3.5% is proved to be very useful in screening for resistance of large number of genotypes of safflower (Shinde and Hallale 2009, Waghmare and Datar 2010). Thus, sources of resistance to *Fusarium* wilt disease in wild and cultivated *Carthamus* species have been identified. Wild safflower species like *C. oxyacantha*, *C. lanatus*, *C. glaucus*, *C. creticus*, and *C. turkestanicus* are immune to wilt. Resistant plants have been obtained by selection and reselection from advanced breeding lines derived from crosses of *C. oxyacantha* × *C. tinctorius* and *C. tinctorius* × *C. turkestanicus*. Some of the most promising safflower genotypes that are highly resistant to wilt are GMU-1533 (Gadekar and Jambhale 2002b); 86-93-36A, 237550, VI-92-4-2, and II-13-2A (Sastry and Chattopadhyay 2003); GMU-1702, GMU-1706, and GMU-1818 (Chavan et al. 2004); 96-508-2-90 (Anjani et al. 2005); HUS 305 (Sastry and Chattopadhyay 2003, Raghuwanshi et al. 2008, Singh et al. 2008b); WR-11-4-6, WR-8-24-12, WR8-19-10, WR-4-6-5, WR-5-20-10, and WR-8-17-9 (Singh et al. 2008b); released hybrids DSH-129, NARI-NH-1, and NARI-H-15; and released cultivars A-1, PBNS-40, and NARI-6 (Murumkar et al. 2008b, 2009b, Prasad and Suresh 2012). Information has been generated on the use of molecular markers for genotyping safflower cultivars (Sehgal and Raina 2005) as well as for characterization of safflower germplasm (Johnson et al. 2007).

C. lanatus ($2n = 22$) and the alloploid, produced after colchicine treatment of seedlings from a cross *C. lanatus* × *C. tinctorius* ($2n = 24$), are highly resistant to the disease. The disease resistance in the alloploid appears to be governed by dominant genes contributed by the

C. lanatus genome. Accumulation of antifungal compound, carthamidin (4. 5, 7. 8-tetrahydoxy flavone), in infected plants has been found responsible for resistance of the plants to infection. Resistance to *F. oxysporum* f. sp. *carthami* in some genotypes is governed by two dominant genes with complementary type of gene action, whereas in others, it is governed by inhibitory type of gene action (Shivani et al. 2011) and in still others seedling resistance is reported to be simple monogenic dominant, whereas adult plant resistance is found to be under the control of epistatic nonallelic interactions (Gadekar and Jambhale 2002b). The development of long-term wilt-resistant varieties may, however, be impeded, if additional races evolve in the natural population of *F. oxysporum* f. sp. *carthami* (Kolte 1985).

Chemical Control

Seed treatment with fungicides such as captan; carboxin, thiram, or a mixture of carboxin + thiram; benomyl; and carbendazim + mancozeb at 0.1% or 0.2% can reduce surface contamination by *F. oxysporum* f. sp. *carthami* and becomes effective in eliminating the pathogen from the seed, but all these are more effective when combined with wilt-tolerant varieties or cultural practices (Sastry and Jayashree 1993, Govindappa et al. 2011b).

Cultural Control

Nonhost crops such as chickpea, lentil, pea, and wheat, usually grown with safflower as a mixed crop or crop succession of safflower with these crops in India, have been found to increase safflower yield with a decrease in the wilt incidence by the secretion of compounds inhibitory to the growth of the pathogen (Kolte 1985, Sastry and Chattopadhyay 1999a, Sastry et al. 1993). Additionally, chickpea and wheat in its rhizosphere increase the population of antagonistic microflora that subsequently check the growth of the pathogen significantly. Exudates and extractives of the roots of *Ruellia tuberosa* L. show significant protective and curative action against the safflower wilt. The root extractive shows the potentiality of a foliar fungicide. The inhibitory effect of *R. tuberosa* on *F. oxysporum* f. sp. *carthami* is attributed to the 2,6-dimethoxy quinone, acacetin, and C16-quinone contents of root exudates and extractives. It is reported that wilt of safflower can be controlled by planting *R. tuberosa* (a local weed found in India) in the safflower field (Kolte 1985). Another method to control the disease is topping of the plants at the seedling stage to encourage vegetative growth: safflower produces a chemical imparting resistance to the plant at the flowering stage that increases when there is more vegetative growth. Soil solarization involving the method of covering the deep ploughed and irrigated fields with transparent polyethylene sheets for 6 weeks during the peak summer month is useful in causing the temperature in the soil beneath to rise to 40°C–53°C that sufficiently kills *F. oxysporum* f. sp. *carthami*. Soil solarization also stimulates the population of antagonistic microflora against the pathogen and reduces inoculum density of the pathogen (Sastry and Chattopadhyay 1999b, 2001).

Biological Control

Trichoderma harzianum, *Bacillus subtilis*, and *Pseudomonas fluorescens* (Gaikwad and Behere 2011b, Govindappa et al. 2011b), *Trichoderma viride* (Patibanda and Prasad 2004, Singh Saroj et al. 2006), and *Aspergillus fumigatus* (Gaikwad and Behere 2001) have been found to be antagonistic to the growth of *F. oxysporum* f. sp. *carthami* indicating their potential usefulness for the control of the disease. Local isolates of *Trichoderma* species show more promising results (Waghmare and Kurundkar 2011). Integrated disease management using different methods of disease control has been always useful (Sastry et al. 2002). For example, integrating the seed treatment with *T. harzianum* or *T. viride* at 4–10 g/kg seed with moderately susceptible safflower variety A-1 (Prasad and Anjani 2008b) or with the spray application of NSKE at 5% results in significant control of the disease with an increase in safflower yield (Singh Saroj et al. 2006).

Effect of Plant Extracts

Leaf extracts of *Parthenium hysterophorus*, *Leucaena leucocephala*, *Vinca rosea*, *Gliricidia maculata*, *Ocimum basilicum*, *Eucalyptus globulus*, *Azardica indica*, *Datura metel*, and *Bougainvillea spectabilis* have been found to inhibit the mycelial growth of *F. oxysporum* f. sp. *carthami* and also to reduce the percent wilt incidence. All the leaf extracts tested, however, are inferior to Thiram in reducing the percent wilt incidence of safflower (Kolase et al. 2000).

Phytophthora ROOT ROT

Symptoms

Phytophthora root rot can occur from preemergence to near-maturity stages of the safflower crop. On succulent plants of 2–3 weeks of age, the first visible symptom is water soaking and collapse of cortical tissue of the lower stems. The softening of the stem weakens the young plants that they fall over, shrivel, and die. On older plants near the bloom stage, black necrotic lesions encompass the roots and sometimes extend 2–5 cm above the ground on the lower part of the stems. There is a high negative correlation between lesion length on roots and percentage of live seedlings (Nasehi et al. 2013). The cortex of the affected roots ranges in color from dark brown to greenish black. In advanced stages, the vascular tissue and pith also become necrotic and dark colored. Leaves of such plants sometimes turn yellow and the entire plant then wilts. The wilting is the common symptom of *Phytophthora* root rot, and most of the infected plants do not recover from wilting. The taproot and lateral roots of affected plants totally rot.

Dead plants can occur individually or in patches. Irrigated plants killed by *Phytophthora* are most evident 4–5 days after watering and can easily be identified by a bleached-green color. Symptoms develop similarly in susceptible plants regardless of the varieties.

Geographical Distribution and Losses

Phytophthora root rot of safflower was first observed in 1947 in Nebraska, USA (Classen et al. 1949). It is also reported to occur in Afghanistan, Argentina, Australia, Dominican Republic, India, Iran, Mexico, and Venezuela (Kolte 1985, Nasehi et al. 2013).

Most of the irrigated safflower crops in the western parts of the northern Great Plains in California, USA, were reported to have been damaged due to *Phytophthora* root rot in 1950–1951. This subsequently resulted in forceful limitation of the safflower crop in the Imperial Valley of California, and it was limited to dry land and subirrigated land. Average losses in yield due to *Phytophthora* root rot in the United States have been reported to be about 3%. In the Isfahan Province of Iran, the incidence of the disease has been observed to be about 30% during 2005–2007 crop seasons (Nasehi et al. 2013). In a few instances in certain years, 80% of the plants in a crop have been reported to be killed.

Phytophthora root rot has also limited the development of safflower as a major commercial crop in the New South Wales area in Australia. Although the disease occurs in many other countries in which safflower is indigenous, its economic importance has not been realized, probably because of resistance of local varieties to the local races of the pathogen and also because of the fact that in most of those countries, the crop is grown usually on dry land or on the border of irrigated fields. The disease has not attained serious proportions during the recent past since late 1990s in India (Prasad and Suresh 2012).

Pathogen

Multiple species of *Phytophthora* are known to cause infection in safflower including *Phytophthora drechsleri* Tucker (*P. cryptogea* Pethyb. and Laff.). Early literature established that *P. drechsleri* has the potential to be devastating in safflower production (Banithashemi 2004, Banithashemi and Mirtalebi 2008). The main pathogen is, therefore, *P. drechsleri* Tucker (*P. cryptogea* Pethyb. and Laff.).

The mycelium is hyaline, nonseptate, and branched, having uniform width of 4.5 μm. The sporangiophores are narrower than the hyphae, and the sporangia are hyaline to faint color, thin walled, nonpapillate, and pyriform to ovate and measure 24–38 × 15–24 μm. The zoospores measure 10–20 μm in diameter. Approximately 75% germination of washed zoospores occurs during a period of 3 h in water.

The fungus does not form chlamydospores. The oospores develop singly in the oogonia, and fully formed oospores are spherical, smooth, thick walled, yellow to bright brown, and granular and measure 16–45 μm in diameter. Compatible mating strains of *P. drechsleri* produce oospores abundantly in paired cultures on seed extract media and on safflower plant material.

The oospores germinate at temperatures between 15°C and 30°C, the optimum germination being at 24°C. The germ tube of the germinating oospore usually penetrates an oogonial wall or may pass through the oogonial stalk, terminating the formation of a sporangium and release of a variable number of motile zoospores. The germination percentage is increased with the advancing age of oospores (Kolte 1985, Sastry and Chattopadhyay 2005).

Epidemiology and Disease Cycle

Assuming that compatible fungal strains are present enabling the pathogen to form oospores, it is possible that the fungus survives through the oospores. Chlamydospores are not formed by the fungus in culture, but the presence of the chlamydospores of *P. drechsleri* in roots of artificially inoculated plants of common weed species has been reported. But the exact role of the chlamydospore is not known. Saffron thistle (*C. lanatus*) is an important alternative host of *P. drechsleri*. Other weed species may also act as alternative hosts and produce further means for the survival of the pathogen in soil.

Safflower plants exude substances into the soil that stimulate the growth of germinating zoospores that probably influence root infection. Zoospores of *P. drechsleri* encyst and begin to germinate within 1 h, and germination reaches to 100% within 3 h on hypocotyls of susceptible host. The germ tube may show a positive growth response toward the wound in contrast to random growth on uninjured epidermis. The infection hyphae penetrate directly. Intercellular penetration is accomplished at junctions of epidermal cell walls at the onset of spore germination or after further growth of the germ tubes. Penetration is sometimes preceded by a spindle-shaped enlargement of hyphal tips. Intracellular invasion occurs after initial penetration at cell wall junctions is accomplished. Infection sites become visible in stained epidermal cell layers as small pores between cell walls or by deposits are found on the inside of cell walls in direct contact with the invading hyphae. The presence of the deposits at infection sites is evident only in resistant Biggs safflower. Massive intercellular and intracellular spread of the hyphae occurs in tissues of the cortex and vascular regions of susceptible hypocotyls within 48 h and causes cell collapse and disruption of cell wall. The possibility of production of pectolytic enzymes by the pathogen has been speculated in the infection process.

It has been found that resistant varieties in one locality may be susceptible in another indicating prevalence of physiological races of the pathogen. Such pathogenic races have been reported to exist in the United States. Isolates of *P. drechsleri* designated as isolate nos. 201, 5811, and 45116 have been reported as distinct races. Of these races, two are capable of rapid growth on agar media at a temperature of 35°C, and the other shows little, if any, growth at 35°C. *P. drechsleri* isolates show variation in virulence, and certain isolates appear to be more virulent on stem than on roots.

Induction of water stress predisposes safflower plants to infection by *P. drechsleri* (Duniway 1977). The effect of such water stress conditions can actually be seen under natural conditions when drought condition prevails before irrigating the crop, followed by the severe development of the disease after irrigation (Sastry and Chattopadhyay 2005). Wet soil conditions and flooding of the fields are expected to be important factors when production, release, and movement of zoospores are the factors limiting the disease development. Increased flooding time after infection of intact plants results in a greater percentage of affected plants.

Soil temperature is an important factor influencing the pathogenicity of *P. drechsleri* to safflower. The optimum temperature for disease development is 25°C–30°C, correlated closely with the most favorable temperature (30°C) for radial growth of the pathogen in culture. Soil temperature of 17°C is unfavorable for the development of root rot. At least a portion of the temperature effect is directly on the host. The pathogen is favored in vitro by high temperature (27°C–30°C), and it is also pathogenic under higher temperature. Pronounced increase in plant death occurs when the plants are exposed to temperature of 27°C and low light intensity. Reductions in the resistance of safflower by these factors, however, do not appear to be as great or as likely to occur in the field as does the predisposing influence of water stress. Safflower plants show reduced susceptibility to *P. cryptogea* after prior adaptation of roots to hypoxic (low oxygen) condition due to the formation of root aerenchyma and phytoalexin synthesis (Atwell and Heritage 1994).

Disease Management

Host Plant Resistance

Safflower varieties "Gila" and "US 10" (both developed by crossing "Nebraska 10" with "WO 14"), "Frio," "Vte," and "VFR-1" are reported to be moderately resistant against only one pathogenic race, and the resistance in these varieties is governed by a single dominant gene (Sastry and Chattopadhyay 2005). These varieties are suitable to irrigated culture, provided they are subirrigated or grown on beds with furrow-irrigated systems for recommended durations. The VFR-1 possesses more resistance than either Gila or US 10. The resistant reaction of VFR-1 cotyledons to *P. drechsleri* appears to be indicative of its root resistance, which also is conditioned by a single dominant gene. The "VFR-1" hypocotyl is, however, susceptible to *P. drechsleri* indicating that the cotyledonary reaction is not indicative of hypocotyl reaction. The cotyledonary reactions of Gila and "US 10" are not indicative of their root resistance, which appears to be conditioned by a dominant gene.

The moderately resistant varieties, as previous, fail to show resistance to *Phytophthora* root rot under heavy flood–irrigated or water-logged conditions of soils. The highest level of resistance (lower-stem and root rot resistance) is reported in a safflower introduction selected at Biggs, California. This selection was then named Biggs safflower, possessing resistance to all races of *P. drechsleri.* The hypocotyl resistance of Biggs safflower is conditioned by a single recessive factor. The Big safflower is also reported to be resistant under heavy flood–irrigated conditions. Commercial varieties with this level of resistance are not available. The Biggs safflower is not suitable for commercial production because of late maturity and low oil content. However, it can be a very useful source of resistance in a breeding program. Several safflower germplasm lines from all over the world have been screened, and genotypes in UC 150 and UC 164 series have been reported to be resistant (Kolte 1985). The most resistant genotypes under Iran conditions are KW 9, KW 12, and KW 15 (Nasehi et al. 2013). It may further be added that breeding safflower for resistance to *P. drechsleri* is complicated by the existence of different factors conditioning resistance in either the root, hypocotyl, or cotyledons.

It has been made clear that the host-resistant mechanism in this host–pathogen combination is activated upon penetration of the epidermis. Safynol (*trans-trans*-3, 1 l-tridecadiene-5-7. 9-triyne-l, 2-diol) and dehydro safynol (*trans*-1 l-tridecene-3, 5. 7. 9-tetrayne-1, 2 diol) antifungal polyacetylene compounds have been indicated as disease-resistant factors in stem rot of safflower incited by *P. drechsleri* (Johnson 1970, Allen and Thomas 1971).

Chemical Control

Seed treatment with captan and soil drenching at 0.2% is useful in reducing the preemergence mortality due to seedling blight (Prasad and Suresh 2012).

Cultural Control

Recommendations to commercial growers that can assist in reducing damage from this disease include growing safflower in beds, not permitting water to stand in the field after irrigating and not growing the crop successively on the same land. Rotation with nonsusceptible crops may also be desirable (Kolte 1985, Prasad and Suresh 2012).

RUST

Symptoms

Safflower rust has two pathological phases that become visible as (1) root and foot disease in the seedling phase expressing the rust symptoms on cotyledons, hypocotyls, etc., and (2) foliage phase disease, at later stages of plant growth, expressing the rust symptoms on leaves, flowers, fruits, etc.

The rust in the seedling phase mainly develops because of the infection of emerging seedlings due to basidiospores resulting from germination of seed- or soilborne teliospores. Initially, orange-yellow spots representing pycnia appear on cotyledons; this may be accompanied by drooping and wilting of the seedlings. The color of such spots later changes due to the development of uredinoid aecidia called primary uredia. Many such uredia develop as pustules, and adjacent pustules coalesce to form large rust pustules. The presence of the rust pustules has also been reported on the underground part, for example, taproot and lateral roots. According to Schuster and Christiansen (1952), longitudinal cracking of epidermal and cortical tissue of the infected area is seen frequently. Some of the cracking is mainly due to the adventitious roots that are sent out at the points of infection. These roots may provide means of survival for wilted plants. The stem of 8–10-week-old seedlings can be infected, and formation of orange-yellow pycnia can also be noticed on them. On relatively older plants, girdling of the invaded area due to collapse of the tissue is a very characteristic symptom. Such plants remain erect due to the stiff stem, but the leaves of such plants are generally in a wilted condition. Due to windstorm or rains, these plants often break at the girdled area. The foliar phase of the disease is characterized by the appearance of uredial pustules on leaves, flowers, and fruits. The uredia remain scattered, crumpent on leaves, and these have a chestnut-brown color. The teleutospores are formed in the uredopustules when the safflower plant matures, giving a dark-brownish color to the rust-affected plant parts.

Geographical Distribution and Losses

The most important disease of safflower is the rust caused by *Puccinia carthami* Cda. It was first described by Corda attacking *C. tinctorius* L. in Bohemia in 1840 (Arthur and Mains 1922). Occurrence of this disease is reported in all areas of commercial production of safflower and is endemic over wide areas of safflower's natural range (Kolte 1985). It has been more recently reported to occur in Oman (Deadman et al. 2005), on snow lotus (*Saussurea involucrata* (Kar. & Kir.) in China (Zhao et al. 2007), and in Romania–Bulgaria cross-border regional areas (Anonymous 2014). The disease is more serious in countries where the crop is grown year after year. This precludes the monoculture of the safflower. Severe epiphytotics of this rust were reported in Nebraska in 1949 and 1950, after the introduction of safflower crop there (Schuster and Christiansen 1952). Currently, it is rarely a problem in the Great Plains of the United States because it occurs late enough in the season that yields are not affected (Lyon et al. 2007). Under Indian conditions, though the disease appears to have caused severe yield losses before 1990, now the disease has not been recorded to be a significant factor in limiting safflower production for the last 10–15 years (Prasad et al. 2006, Singh and Prasad 2007). However, the seed and seedling infection is further considered to be of economic importance as it provides the source of inoculum for initiating foliage infection. Additionally, severely contaminated seeds will not germinate well if saved for future plantings (Lyon et al. 2007). For foliage rust to cause significant reduction in yield, heavier infection must occur on the upper and lower leaves before the full-bloom period. When near-isogenic varieties, one resistant to rust (WO-14) and the other susceptible (N-8), are grown under conditions of a rust epidemic, the yield of the susceptible variety is reported to be 65% of the resistant, whereas under rust-free conditions, the relationship is found to be 95%. The average annual loss due to safflower rust in the United States has been estimated to be about 5%, costing about one million dollars (Kolte 1985). The major loss from safflower rust is the reduction in stand from planting untreated teliospore-infested seed or in planting where viable soilborne teliospores exist. Using artificially infested seed, the stand loss is recorded to be 98%, but only about 20% stand loss has been reported from the use of naturally rust-infested seed. Field trials with rust-resistant and rust-susceptible safflower varieties have shown that rust-infected but rust-resistant varieties exhibit stand loss of 26%. But the surviving plants of such resistant varieties have growth compensation ability, and loss in yield remains nonsignificant as compared to the stand loss of 55%–97% in susceptible varieties with a significantly reduced yield.

Pathogen

The pathogen is *P. carthami* (Hutz.) Corda. *P. carthami* is an obligate pathogen with an autoecious life cycle on *Carthamus* spp. It is a macrocyclic rust, and since true aeciospores are naturally omitted from the life cycle, the rust is reported to be of the *brachy-form* type. The uredosori are found scattered on both sides of the leaves usually near the pycnia. In some cases, uredia are formed in between two very closely situated pycnia. Uredosori contain numerous globoid or broadly ellipsoid uredospores measuring 21–27 × 21–24 μm in size. The wall of the spore is 1.5–2.0 μm thick. The uredospores have 3–4 equatorial germ pores, and they are light chestnut brown and echinulate. Teleutosori are formed in uredosori. The teliospores are bicelled, ellipsoid, 36–44 × 24–30 μm, slightly or not constricted at septa, chestnut-brown, rounded or somewhat obtuse at both ends, finely verrucose, 2.5–3.5 μm thick at the side with the spores usually depressed from the apical position. The teliospores are hyaline, fragile, and mostly deciduous with 10 lx-long pedicel. Pycnia, usually formed in groups, are subepidermal and flask shaped or spherical and measure 80–100 μm in diameter. A large number of flexuous hyphae are found protruding, and numerous pycniospores are seen oozing out through the ostiole. Normally, as described earlier, the true aeciospores are not formed in *P. carthami*. But in some of the cases, where single sporidial infections are kept undisturbed for 20–30 days after the formation of pycnia, aeciospore-like spores are produced. Such spores are termed primary uredospores or uredinoid aeciospore because of their position in the life cycle of the rust and morphological resemblance to aeciospores. The uredinoid aecia are amphigenous and chestnut brown and measure up to 0.4 μm in diameter, associated with pycnia in clusters.

Epidemiology and Disease Cycle

P. carthami is mainly perpetuated through teleutospores that remain dormant on the seed or on the buried debris of the previous crop throughout the uncropped season. Two types of teliospores have been reported. One of the two types is known to have the ability to germinate shortly after their formation, whereas the other type shows a dormancy period of 5–6 weeks.

Under field conditions, teliospores (showing dormancy) survive for 12 months, but not for 21 months. Affected safflower straw stored at 5°C has been reported to contain viable teliospores even after 45 months of storage at such conditions. Under natural conditions, the uredospores do not survive. However, they have been reported to remain viable for over a year under dry storage conditions at 8°C–10°C. At room temperature, the uredospores lose their viability within 3 weeks. On infected plants, the uredospores remain viable for 3 weeks at 30°C–31°C and for 3 days at 52°C–55°C. It is interesting that above 40°C, the rust tends to form teliospores directly (Kolte 1985). Some of the wild *Carthamus* species act as collateral hosts in the survival of *P. carthami* (Sastry and Chattopadhyay 2005). In India, this rust is commonly seen on wild safflower *C. oxyacantha*, and it appears that this host gets infected a month earlier than the cultivated safflower. Besides, viable teliospores have been observed on this wild safflower during the off-season, suggesting a potential source of survival of the pathogen. Other *Carthamus* species, for example, *C. glaucus* M B, *C. lanatus* L., *C. syriacus* (Boiss) Dinsm., and *C. tenuis* (Boiss) Bornm., also appear to act as collateral hosts to *P. carthami*. Out of the two types of teliospores as reported by Prasada and Chothia (1950), it is the resting teliospore that oversummers and remains viable to bring about the fresh primary infection in the following season. However, the primary infection in the safflower crop may also be initiated by teliospores formed on wild safflower species, especially by those teliospores that do not require a dormant period after their formation. These may infect the safflower crop directly or may attack the wild species first, and the uredospores then formed on wild species may be blown to initiate infection in the cultivated safflower. It is reported that the volatile substances, especially the polyacetylenes, from safflower crop debris stimulate the germination of teliospores. The optimum temperature for the germination of teliospore is 12°C–18°C. The teliospores germinate normally by producing four-celled promycelium with a cell bearing short sterigma and a kidney-shaped sporidium. Such a gametophytic generation, as it becomes visible through formation of sporidia, causes root and foot infection by direct penetration of epidermis or cortex of seedlings while it is underground during the seed germination process and before plant emergence. A higher percentage of seedlings showing the root and foot phase of the disease is favored by a lower temperature range of 5°C–15°C, whereas such an infection is hindered by a temperature of 30°C and 35°C. Soil moisture variation in the range of 35%–80% of water-holding capacity has not been found to influence the seedling infection due to rust. One of the important characteristics of seedling infection due to *P. carthami* is the elongation and hypertrophy of the affected seedlings. A week after primary infection by sporidia, orange spots consisting of spermogonia appear on cotyledons, and after 2 or 3 days, primary uredosori develop around them. These infect the first leaves, thus setting up the first foci of infection. Late in the crop season, secondary uredospores, the sporophytic generation of the fungus, cause foliage infection. The uredospores germinate by giving rise to a germ tube over a temperature range of 8°C–35°C, but the optimum temperature is between 18°C and 20°C. The germ tube forms the appressorium in the substomatal vesicle, facilitating the penetration of leaf tissues through stomata. Cool temperature and high relative humidity favor the infection. The incubation period is reported to be 10–14 days depending on the temperature. At optimum 18°C–20°C temperature, the incubation period is 10 days, whereas at 35°C, the rust uredospores germinate only in traces, and the infection may not occur. Further, at 40°C, the uredospores do not germinate at all. Artificial inoculation of leaves with uredospores has been found to give only uredo- and teliospores of *P. carthami*, and uredospore inoculation on seed does not cause seedling infection. Since *P. carthami* is an autoecious macrocyclic rust that rapidly completes the sexual cycle, maximum opportunity exists for the recombination. Different races of the rust have been identified in the United States. Different rust differential hosts to identify the races have been established in the United States (Kolte 1985).

Disease Management

Host Plant Resistance

Reaction of infected hypocotyls to rust has been used as a measure of resistance of safflower to *P. carthami*. The highly susceptible seedlings show abundant sporulation on the hypocotyl, and they do not survive, but the resistant seedlings do not show hypocotyl elongation; sporulation occurs only on cotyledons and the seedlings are not killed. Seedling rust resistance appears in most cases to be both physiologically and genetically related to foliage rust resistance. Lines with resistance to the foliage phase are also resistant to the seedling phase of the disease. Seedlings with a high level of resistance to the foliage phase exhibit less than 5% seedling death due to the seedling phase of rust. A close correlation has been found with the seedling rust resistance test as an efficient method for screening for foliage rust resistance. It is, therefore, concluded that foliage rust resistance may be effectively screened by the seedling test. The microliter drop (with a known number of teliospores suspended in a 1 mL) method has potential usefulness in host range and screening for resistance of large number of genotypes (Bruckart 1999). Reaction of several safflower introduction and selections for resistance to rust have been studied (Zimmer and Leininger 1965, Kalafat et al. 2009). Some safflower lines have been reported to be resistant to foliage as well as seedling phases of the disease. These are PI 170274-100, 193764-66, 199882-82, 220647-98, 220647-55, 250601-109, 250721-93, 253759-62, 253911-25, 253912-9, 253913-5-72, 253914-5-108, 253914-7-9, and 257291-68. Other genotypes such as No. 30 and No. 26 in Turkey (Kalafat et al. 2009) and No. 1 and Tayan No. 1 in China are resistant to rust (Liu et al. 2009). The safflower line N-l-1-5 is moderately susceptible to the foliage phase of the disease, but it has been found outstandingly resistant to the seedling phase of the rust. Other such lines possessing a high degree of seedling resistance are PCA, PI 195895, and 6458-5. The seedling resistance of N-l-1-5 is governed by a single dominant gene (N). In certain situations, this source should be given a prime consideration in breeding for seedling rust resistance. Theoretically, utilization of seedling rust resistance may have the same influence on the development of races of *P. carthami* as elimination of alternate host would have on the development of races in a heteroecious species of the same genus. Because the major source of primary infection for the foliage phase is seedling infection, the utilization of seedling rust-resistant varieties would reduce the amount of primary inoculum and would consequently reduce the opportunity for new pathogenic strains to arise from the vegetative recombination.

Noncultivated species with chromosome numbers $2n = 20$, 40, or 64 have been found resistant to the foliage phase of the rust. Resistance available in *C. oxyacantha* is governed by single dominant gene (OYOY), and this has been successfully transferred to the cultivated safflower. Safflower lines with A, M, or N genes have sufficient resistance to rust. The rust resistance present in certain varieties is not linked or suppressed by the gene controlling an economically important thin-hull character of safflower. Virulence of *P. carthami* is reported to be inherited in a recessive manner.

Several commercial varieties of safflower have been developed utilizing the different sources of resistance in a breeding program. However, the development of the new races, as seen earlier, has rendered the resistance ineffective depending upon the prevalence of races in a particular area. Induced resistance in safflower by exogenous chemicals such as salicylic acid (SA), oxalic acid, and vitamin K3 and by spray application of certain nutrients at certain concentrations is typically a systemic acquired resistance (SAR) characterized by systemicity and durability (Dordas 2008, Chen 2009). When the first and second leaves are sprayed with 4 mmol/L of SA, the activity of the defense enzymes (polyphenol oxidase, peroxidases, phenylalanine ammonia lyase [PAL], etc.) increases on the third and fourth leaves with decrease in rust disease index (Chen 2009).

Chemical Control

Seed dressing with fungicides such as maneb, mancozeb, captafol, and thiram (each at 0.2%–0.3% concentration) has been reported to check the seedling infection of safflower rust. The use of systemic fungicides such as oxycarboxin seed treatment has been found most effective in inhibiting spore germination and in the management of the disease when 24–48 oz of the fungicide is used for 100 lb of seeds. Two sprays of systemic fungicides like calixin at 0.05% at an interval of 15 days are useful in the management of the foliar phase of the rust on safflower (Prasad and Suresh 2012, Varaprasad 2012).

Cultural Control

Cultural practices such as avoiding growing safflower in low-lying areas, avoiding monocrop culture of safflower, and avoiding delay of irrigation until the crop exhibits moisture stress symptoms are effective in the management of safflower rust (Varaprasad 2012).

BROWN LEAF SPOT OR FALSE MILDEW

Symptoms

Grayish-chestnut to brown spots of 2–10 mm in diameter appear on the lower leaves. The undersurface of the spot may show the presence of white growth of the fungus, owing to the emergence of tufts of conidiophores bearing conidia (Minz et al. 1961, Rathaiah and Pavgi 1977). The disease is sometimes termed as *false mildew*. The spots may coalesce to cause withering of large area of the leaf. The capitulum may also be affected. The primary symptomatic differences between *Alternaria* and *Ramularia* are that *Alternaria* spots have a shotgun pattern with different colors to the leaf as the disease progresses while *Ramularia* spots are uniformly brown and the underside of the leaf has a white appearance due to the presence of fungal bodies on the underside of the leaf. The differences are apparent with training and experience but are otherwise difficult to distinguish to a casual observer.

Geographical Distribution and Losses

Brown leaf spot of safflower was first observed in 1924 in Siberia in the former Soviet Union. The disease was then reported to occur in several other countries, for example, Ethiopia, France, India, Israel, and Pakistan (Kolte 1985). The most important disease problem of safflower in northwest Mexico (particularly in the Yaqui Valley in the state of Sonora) is reported to be false mildew since the 2000 and has been common most years since then in that country causing losses in crop yield in the range of 6%–90% (Montoya 2005, 2008, Muñoz-Valenzuela et al. 2007). The disease has also been reported to occur in California, USA, and in Argentina, South America (Hostert et al. 2006). The disease is reported to cause a sharp decrease in yield and quality of the seed in the former Soviet Union and adversely affect growth of the plant. Epiphytotic occurrence of the disease was reported at Phaltan in the Maharashtra state of India in the 1981–1982 and in the 1988–1989 crop seasons (Sastry and Chatopadhyay 2005).

Pathogen

The pathogen is *Ramularia carthami* Zaprometov.

The hyphae are hyaline and septate and measure 2–3 μm in diameter. Prior to formation of sclerotia, the hyphae become dark brown, thick walled, and closely septate, increasing to 7 μm in diameter. The sclerotia are formed by continued multiplication of cells of a single hyphal branch just below the epidermis. The mature sclerotia are chestnut brown, spherical to globose, and markedly raised above the level of epidermis. They measure 40–80 × 50–70 μm in size.

The conidiophores are hyaline and unbranched and measure 15–81 × 3–5 μm. The conidia are one or two celled, rarely three celled, hyaline, and cylindrical with rounded apices and measure 14–25 × 4.5–6 μm. The spermogonia may develop within the old conidial stomata. Mature spermogonia are ovate to globose; dark brown to black; and at first embedded subepidermally in the leaf tissue, later becoming erumpent and ostiolate and measure 45–110 × 40–150 μm. The spermatia measure 3.5–4 μm. The possibility of *Mycosphaerella* Johanson as the perfect stage of *Ramularia carthami* is suspected.

Epidemiology and Disease Cycle

The conidial germ tubes enter the leaf by penetration of the stomata. On entering the substomatal chamber, the hyphae begin to spread intercellularly. The hyphae are never seen to penetrate the living host cell, but they penetrate after collapse and death of host cells. The pathogen is mainly airborne, spreading by means of conidia. The infection develops successfully at a temperature of >28°C coupled with high humidity (Patil and Hegde 1988). The disease becomes severe under irrigated conditions favoring the epiphytotic occurrence of the disease. The disease does not occur on a rainfed safflower crop.

Disease Management

Host Plant Resistance

The reaction of wild *Carthamus* sp. has been studied. *C. oxyacantha* and *C. flavescens* are resistant to *R. carthami*. In India, the safflower lines NS 133, HOE, 999, and 1021 (Kolte 1985) and IG FRI-116 (Kumar and Joshi 1995) are reported to be moderately resistant to brown leaf spot. Resistance to brown spot can be found in selections from the original brown spot–resistant GPB4 selection and from directed crosses of GPB4 onto varieties or experimental lines of safflower. Such improved brown spot–resistant safflower lines/varieties are S-746, S-334, S-336, and S-736 (Weisker and Musa 2013) and the two most resistant lines being 04-787 and 04-765, which should be used as a source of resistance to breed improved varieties of safflower (Muñoz-Valenzuela et al. 2007). A new linoleic variety "CLANO-LIN" tolerant to false mildew has been released in Mexico (Borbon-Gracia et al. 2011).

Chemical Control

Spraying the crop with copper oxychloride (0.3%) or mancozeb (0.25%) has been found to manage the disease. Aureofungin has also been found effective in the management of the disease. Three sprays of mancozeb (0.2%) or carbendazim (0.05%) at 15-day intervals starting at 55 days after sowing are also effective in the management of the disease (Patil and Hegde 1989, Prasad and Suresh 2012).

Cercospora LEAF SPOT

Symptoms

Safflower plants are affected a few weeks after planting or when plants are in the flowering stage. Symptoms on leaves are characterized by the formation of circular to irregular brown sunken spots measuring 3–10 mm in diameter. The lower leaves show the symptoms first, and gradually, the middle and upper leaves are also affected. The spots have a yellowish tinge at the border and they are sometimes zonate. As the disease progresses, the leaves turn brown and show internal necrosis, and the entire leaves may be distorted. Under moist conditions, the spots have a velvety grayish-white appearance caused by sporulation of the fungus. Minute black fructification of the pathogen may be seen on both upper and lower sides of the spots of affected leaves. Stems and nodes may also be affected. In case the disease becomes quite severe, the bracts are also affected and show the presence of reddish-brown spots. Affected flower buds turn brown and die. The entire capitulum may also be affected without formation of seeds.

Geographical Distribution and Losses

The *Cercospora* leaf spot of safflower is worldwide in occurrence, particularly when safflower is grown in a large area as a pure crop. It is reported to occur in Ethiopia, India, Iran, Israel, Kenya, the Philippines, the former Soviet Union, and the western Great Plains and Northern Plains area in the United States (Mündel and Huang 2003). Epiphytotic occurrence of the disease was reported in the Coimbatore area in the southern part of India in 1921, 1924, and 1925. However, information on estimates of losses caused by the disease is not available. Observations made in 2006–2007 in Montana, USA, have demonstrated that safflower is an additional host for the sugar beet pathogen, *Cercospora beticola*. This creates new potential disease problems for both crops if grown within 4 years of each other (Lyon et al. 2007). This provides further evidence that safflower is an alternative host of *C. beticola*. This is of significant importance since irrigated safflower is increasingly being evaluated for rotation with sugar beet in Montana, USA, and two crops are occasionally grown adjacent to each other (Lartey et al. 2005, 2007).

Pathogen

The pathogen is *Cercospora carthami* (H. and P. sydow) Sundararaman and Ramakrishnan.

The mycelium is hyaline, smoky brown, septate, and branched and collects in the stomatal areas where stromata are formed. The conidiophores emerge separately or in fascicles (tufts of 12–20 conidiophores) on both leaf surfaces. Under wet conditions, they emerge directly from the epidermis (Kolte 1985). The conidiophores are simple, septate, occasionally branched, erect, and variable, measuring 104.74–209.6 × 4.6 μm in size. The conidia are hyaline, linear, with 2–20 septate, and borne on the conidiophores acrogeneously. They are broad at the base and taper toward the end in a whiplike manner, measuring 2.5–5 × 50–300 μm. The length of the conidia and number of septa vary according to prevalent environmental conditions. The conidia germinate readily in water, giving germ tubes from both ends as well as from the sides. Each cell is capable of giving out a germ tube (Sastry and Chattopadhyay 2005).

Epidemiology and Disease Cycle

C. carthami is reported to have a restricted host range, and it does not infect other plants except *Carthamus* sp. The pathogen perpetuates through a vegetative saprobic mycelium and through viable stromata embedded in crop debris. Stromata of the pathogen appear as small black dots in concentric rings on diseased leaves. The disease cycle is initiated when wind-blown or water-splashed conidia land on safflower and germinate in the presence of free moisture. The fungus infects plant parts through natural openings or wounds or through direct penetration. Heavy and continuous early morning dew or other free moisture is essential for infection, and the disease is most severe during warm, moist weather. The *Cercospora* leaf spot pathogen is disseminated by wind, water splashing, and movement of infested plant material (Lyon et al. 2007).

Disease Management

Though high degree of host plant resistance sources are known, five genotypes, namely, 8-12-1, SSF-650, 2-10-2, 4-13-1, and 2-11-2, are tolerant to both *Cercospora* leaf spot and aphid attack (Akashe et al. 2004). The disease can be managed by spray application of 1% Bordeaux mixture. Dithiocarbamate fungicides (0.25%) or copper oxychloride (0.3%) might also be effective in the management of the disease (Prasad and Suresh 2012).

Seed treatment with thiram 3 g/kg and spraying of mancozeb 2.5 g or carbendazim 1 g/L of water may be useful in the disease management. Four strains of rhizobacteria (GBO-3, INR937a, INR937b, and IPC11), when micromobilized with the safflower seed, have been found to be inducers of systemic resistance in safflower preventing infection caused by *C. carthami* (Govindappa et al. 2013). Few specific cultural control strategies have been developed for *Cercospora* leaf spot. Crop rotations of 3 years or longer to nonhosts (small grains or corn), through incorporation of crop debris, and avoidance of overhead and excessive irrigation will likely reduce the incidence and severity of *Cercospora* leaf spot (Lyon et al. 2007).

Macrophomina (*Rhizoctonia*) ROOT ROT

Symptoms

Initially, dark-brown to black lesions are formed on the roots. Later, infected plants may show a characteristic silvery discoloration of the epidermal and subepidermal layer of the stem base and the root (ashy stem and root). The fungus spreads up to the vascular and pith tissues of the stem, finally forming numerous small sclerotia, like finely powdered charcoal (charcoal rot) giving the infected tissues a grayish-black color. Sclerotia are found along the vascular elements and bordering the pith cavity. Affected plants are stunted and ripen prematurely. A new type of distinct stem-split symptom is reported more recently to occur on 30-day-old safflower plants as minute cracks 2–3 cm above the soil surface that extend to both upward and downward directions resulting in the formation of wide split. The split portion becomes hollow and brown with white to gray mycelia mat of the fungus on the inner surface (Govindappa et al. 2005). Such plants fail to withstand.

Geographical Distribution and Losses

In general, the disease is considered to be of less importance in normal crop–growing season in winter months, but because of changing climate in recent years, the disease has assumed wide prevalence in warm temperate and tropical regions of the world. It causes serious yield losses especially in dry seasons in Iran (Mahdizadeh et al. 2011, Lotfalinezhad et al. 2013). The *Rhizoctonia* phase of the disease is sporadic that regularly causes 1%–10% yield losses all over India (Prasad and Suresh 2012). The incidence of the disease is negatively correlated with yield and height of the crop (Chattopadhyay et al. 2003).

Pathogen

Macrophomina phaseolina (Tassi) Goid is the pathogen, which is the pycnidial stage of *Rhizoctonia bataticola* (Taub) Butler. The details of the characteristics of the pathogen and disease cycle have been described under peanut and sunflower diseases. Genetic diversity analysis using RAPD markers and UPGMA cluster analysis could distinguish isolates prevalent in safflower-growing areas into two major groups. Dendrograph generated by cluster analysis reveals varied levels of genetic similarity, and it ranges from 50% to 55% (Prasad et al. 2011, Navgire et al. 2014).

Disease Management

Host Plant Resistance

Seed germination using towel paper and infested soil cup techniques has been developed at the Indian Institute of Oilseeds, Hyderabad, India, for screening safflower germplasm lines for resistance to the disease (Prasad and Navneetha 2010). However, resistance sources have not been detected either in cultivated or in wild safflower. Diameter of lower stem (DLS) of safflower has been found to have positive and significant correlation with length and width of the necrotic lesion on the stem of safflower; hence, DLS trait should be used as an index for indirect selection of resistant genotypes in safflower (Pahlavani et al. 2007). Some of the disease-tolerant genotypes are IUT-k 115, GUA-va 16, CW-74, AC-Stirling (Pahlavani et al. 2007), AKS-152 and AKS-68 (Ingle et al. 2004), and NARI-6, SSF-658, A-2, PBNS 12, and PBNS 40 (Prasad and Suresh 2012). Four genotypes, namely, GMU-3259, GMU-3262, GMU-3306, and GMU-3316, are identified to be highly resistant with no seedling infection, whereas three genotypes GMU-3265, GMU-3285, and GMU-3297 are found to be resistant with only up to 1%–10% seedling mortality (Salunkhe 2014). These can be used in breeding program to improve resistance in safflower to charcoal rot and root rot caused by *M. phaseolina.*

Chemical Control

No practically useful economic chemical method is recommended for the control of the disease. However, the seed-borne inoculum of the pathogen can be minimized by treating the safflower seed with thiram or carbendazim (Subeej25 DS) at 2 g/kg seed for the control of the disease and for better plant stand establishment in the field (Prashanti et al. 2000a, Prasad and Suresh 2012).

Cultural Control

The use of clean seed, the application of organic matter, long rotations with nonhost crops, and avoidance of excessively dense plant populations and sanitation, including the burial of debris by hand or by plough during summer, have each been suggested as cultural practices (Prasad and Suresh 2012) in the disease management.

Biological Control

Biocontrol agents such as *Trichoderma harzianum*, fluorescent Pseudomonads (*P. fluorescens*), and *Bacillus subtilis* obtained from the rhizosphere soil of safflower and finally prepared as talc-based formulations are used as seed treatment; these biocontrol agents at 10 g/kg prove to be effective in the control of the disease and in triggering defense-related enzymes involved in phenylpropanoid pathways and phenols that induce systemic resistance. High activity of peroxidase, PAL, chitinase, polyphenol oxidase, and beta-1,3-glucanase could be observed in *P. fluorescens*– and *T. harzianum*–treated safflower plants after challenge inoculation with *M. phaseolina* (Prashanti et al. 2000b, Kaswate et al. 2003, Singh et al. 2008a, Govindappa et al. 2010, 2011a). Soil amendment with saw dust + soil in the ratio of 1:10 when combined with seed treatment with *T. harzianum* at 4 g/kg seed shows lowest preemergence mortality due to *M. phaseolina* (Deshmukh et al. 2003).

OTHER DISEASES OF SAFFLOWER

The other diseases of safflower are given in Table 8.1.

TABLE 8.1
Other Diseases of Safflower

Disease	Pathogen/Causal Agent	Geographical Distribution	Disease Description	References
Powdery mildew	*Erysiphe cichoracearum* f. sp. *carthami* (Milovtzova)	Afghanistan, France, India, Israel, the former Soviet Union, and the United States	Symptoms are characterized by the presence of white growth of the fungus on the safflower leaves. The spineless genotypes are more susceptible to the disease. The disease can be controlled by one or two sprays of wettable sulfur (0.2%) or Karathane (0.1%).	Kolte (1985), Prasad and Suresh (2012)
Powdery mildew	*Sphaerotheca fuliginea* (Schlecht.) Pollacci	South Korea	Powdery mildew on leaves.	Kwon et al. (2000)
Powdery mildew	*Leveillula taurica* (Lev) f. sp. *carthami* Arn.	Israel, the former Soviet Union, Sudan	Powdery mildew on leaves.	Sharifnabi and Saeidi (2004)
Fusarium root rot	*Fusarium solani* (Martius) Sacc.	Iran	Root rotting. There is sufficient genetic variation for resistance, and selection can be effective for the development of resistant germplasm to *F. solani*.	Nasehi et al. (2009)
Rhizoctonia blight	*Rhizoctonia solani* Kuhn, (= *Thanatephorus cucumeris* (Frank) Donk.)	Morocco, United States	Small circular, yellow lesions develop on the hypocotyls at the soil surface. The lesions rapidly increase in size and girdle the hypocotyls. Lateral root development is greatly reduced on severely infected seedlings.	Kolte (1985)
Sclerotinia wilt/rot	*Sclerotinia sclerotiorum* (Lib) de Bary	Argentina, Australia, Canada, India Israel, Russia, Turkey, the United States	Plants become yellowish, turn brown, wilt, and die. Large black sclerotia are formed on the crown, inside the stem, and on adjoining roots though roots themselves are not generally. Shredding of the cortical tissue of the lower stem takes place. Flower heads can fall from the affected plants, leaving the outer involucral bracts in situ. The severity of the disease is increased with increase in rainfall during the crop season. The details of morphology of the fungus, infection process, disease management, etc., are described under sunflower and peanut diseases. Interestingly, soil treatment with biocontrol fungal agent such as *Coniothyrium minitans* is the most effective method under the canopy of safflower for reducing carpogenic germination of sclerotia resulting into reduced production of ascospores, which is further useful in the reduction of the disease incidence.	Kolte (1985), Huang and Erickson (2004)

(*Continued*)

TABLE 8.1 (*Continued*)
Other Diseases of Safflower

Disease	Pathogen/Causal Agent	Geographical Distribution	Disease Description	References
Gray mold	*Botrytis cinerea* Persoon ex Fries		Unseasonal rains during the late stage or in-bloom stage of the crop favor flower infection. Infection of flowers is followed by invasion of the seed head, which causes the latter to become readily detached from the plant. The lower part of the plant provides substances that stimulate infection by *B. cinerea*. The spores of the pathogen become wind borne, and flower heads may be affected at any time from budding to postflowering. The disease is more abundantly seen in the coastal areas, which are subject to fog and in Sacramento Valley of the United States.	Kolte (1985)
Pythium root rots	*Pythium* species: *P. aphanidermatum*, *P. debaryanum*, *P. splendens*, *P. ultimum*	Argentina, Austria, Australia, Canada, Iran, India, the United States	Seedling of all ages shows mortality in patches. Infection begins at the collar region that becomes black and necrotic. Roots of affected plants are severely rotted. The disease becomes severe under irrigation but not under dry conditions. Seed treatment with thiram at 0.2%–0.3% or captan at 0.2%–0.3% is effective in the control of the disease. Resistance to *P. ultimum* is known in genotypes 34040, Arak 281, and Isfahan in Iran, and the resistance is under additive and dominance genetic effects. In Canada, microbial seed treatment with strains of *P. fluorescens*, *Bacillus cereus*, *B. megaterium*, and *Pantoea agglomerans* has been found to give effective control of the root rots/damping-off of safflower.	Kolte (1985), Bardin et al. (2003), Pahlevani et al. (2010), Ghaderi et al. (2011)
Verticillium wilt	*Verticillium albo-atrum* Reinke and Berth (*V. dahliae* Kleb.)	Australia, Iran, India, Morocco, Pakistan, the United States	Interveinal yellowing of the lower leaves with progressive upward discoloration, initially, may be on one side and later may extend to the entire plant. Plants are affected at any stage of the growth during cool weather, but are seldom killed unless the infection is severe. The pathogen survives through soil as well as through seeds. Microsclerotia are found on external and internal tissues of the pericarp and on the testa of some infected seed at harvest. Microsclerotia on the seed remain viable for 2 years. Clean crop cultivation including use of healthy microsclerotia-free seed and crop rotation with nonsusceptible crops should be preferred. Safflower should not follow cotton, sorghum, and peanut. Interrow or strip cropping with these crops should be avoided. The possibility of searching safflower lines possessing high degree of resistance exists in germplasm collection.	Kolte (1985), Koike et al. (2012)

(*Continued*)

TABLE 8.1 (*Continued*)
Other Diseases of Safflower

Disease	Pathogen/Causal Agent	Geographical Distribution	Disease Description	References
Anthracnose	*Colletotrichum simmondsii*, *C. carthami*, *C. gloeosporioides*	Austria, Czech Republic, Portugal	Development of circular spots on leaves and stem blight appears to be anthracnose-like symptoms. Affected plants show light brown elongated lesions on stem, dieback symptoms, and premature death. The date of sowing affects the disease development and degree of resistance of the genotypes to the disease.	Beldan (2012), Carneiro et al. (2012), Park et al. (2005), Uematsu et al. (2012), Vichova et al. (2011)
Downy mildew	*Bremia lactucae Regel* f. sp. *carthami*	Cyprus, Iran, Israel, the former Soviet Union	Mildew growth characterized by downy growth.	Kolte (1985)
Leaf spot	*Septoria carthami* Mourashkinsky	Australia, Bulgaria, Morocco, the former Soviet Union, Turkey	Spots are produced on leaves. The pathogen is likely to be seed borne and assume to be of plant quarantine significance.	Kolte (1985)
Leaf spot	*Ascochyta carthami* Kvashnina	Morocco, the former Soviet Union	Spots on the upper side of leaves show the presence of pycnidia measuring 76–224 μm in diameter.	Kolte (1985)
Bacterial blight or leaf spot	*Pseudomonas syringae* van Hall	India, Pakistan, United States	Necrotic spots and streaks with translucent center encircled with dark-brown to black margins appear on leaves and petioles. Later, the petioles rot, which extends through the stem into the roots killing the plants. The disease becomes severe under conditions of higher temperature, humidity, rains, sprinkler irrigation, and injury due to factors like frost.	Kolte (1985), Sastry and Chattopadhyay (2005)
Necrotic yellow disease	*Spiroplasma citri*	Fars Province of Iran	First report of *Spiroplasma* as the cause of diseases in safflower. Affected plants show stunting, yellowing, phloem discoloration, and local or general necrosis.	Khanchezar et al. (2012)
Safflower mosaic disease	Cucumber mosaic virus	India, Israel, Iran, Morocco, the United States	Safflower leaves show mosaic symptoms characterized by the development of irregular flecks of light green color interspersed with dark green areas. Some leaves become blistered and distorted. The virus is transmitted through sap and the aphids, *Myzus persicae* and *Uroleucon compositae*, but it is not transmitted through seed. Young plants are more susceptible and exhibit severe mosaic symptoms.	Kolte (1985), Kulkarni and Byadgi (2004a,b,c,d, 2005)

(*Continued*)

TABLE 8.1 (*Continued*)
Other Diseases of Safflower

Disease	Pathogen/Causal Agent	Geographical Distribution	Disease Description	References
Stem necrosis disease	Tobacco streak virus	Maharashtra (India)	The first report of the virus affecting safflower. Veinal and leaf necrosis, necrotic streaks on the stem, necrosis of the terminal bud, and plant death.	Rao et al. (2003)
Phyllody	Candidatus Phytoplasma trifolii–related strain (that shares close homology with 16SrVI group)	Afghanistan, Iraq, Iran, Israel	Affected plants show floral virescences, phyllody, proliferation of axillary buds along the stem and little leaf symptoms. The causal agent is closely related to *Phytoplasma* that causes brinjal little leaf and periwinkle little leaf.	Kolte (1985), Salehi et al. (2008, 2009)
Root knot	*Meloidogyne hapla, M. incognita, M. javanica*	India, South Africa, Southeast Asia	Root knot as such is least likely to be the major cause for lower yields of safflower crop. But growing safflower on root knot nematode–infested land may build up the population of nematodes, resulting in economic damage to safflower particularly under high soil temperature at 30°C.	Sastry and Chattopadhyay (2005)

REFERENCES

Akashe, V.B., A.J. Patil, D.V. Indi, and D.R. Mummkar. 2004. Response of different genotypes of safflower to major insect pests and leaf spot diseases. *Sesame Safflower Newsl.* 19: unpaginated.

Allen, E.H. and C.A. Thomas. 1971. Time of course of safynol accumulation in resistant and susceptible safflower infected with *Phytophthora drechsleri*. *Physiol. Plant Pathol.* 1: 235.

Anjani, K., H. Singh, and R.D. Prasad. 2005. Performance of multiple resistant line of safflower (*Carthamus tinctorius*). *Indian J. Agric. Sci.* 75: 178–179.

Anonymous. 2014. Romania-Bulgaria, Safflower (*Carthamus tinctorius* L.). Cross-Border Cooperation Program-European Regional Development Fund, Bucharest, Romania.

Arthur, J.C. and E.B. Mains. 1922. Bullaria. *N. Am. Flora* 7: 482.

Atwell, B.J. and A.D. Heritage. 1994. Reduced susceptibility of roots of safflower to *Phytophthora cryptogea* after prior adaptation of roots to hypoxic conditions. *Aust. J. Bot.* 42: 29–36.

Awadhiya, G.K. 2000. Study of detection methods and location of *Alternaria carthami* Chowdhury in safflower seed. *Gujarat Agric. Univ. Res. J.* 26: 67–68.

Banihashemi, Z. 2004. Light-dependent oospore germination of *Phytophthora cactorum* in the presence of susceptible host plant. *J. Phytopathol.* 152: 683–686.

Banihashemi, Z. and M. Mirtalebi. 2008. Safflower seedling a selective host to discriminate *Phytophthora melonis* from *Phytophthora drechsleri*. *J. Phytopathol.* 156: 499–501.

Bardin, S.D., H.C. Huang, L. Liu, and L.J. Yanke. 2003. Control, by microbial seed treatment, of damping-off caused by *Pythium* sp. on canola, safflower, dry pea, and sugar beet. *Can. J. Plant Pathol.* 25: 268–275.

Bedlan, G. 2012. *Colletotrichum carthami* comb. nov. on safflower (*Carthamus tinctorius*). *J. Kulturpflan* 64: 309–313.

Bergman, J.W. and B.J. Jacobsen. 2005. Control of *Alternaria* blight (*Alternaria carthami* Choud.) of safflower (*Carthamus tinctorius* L.) in the United States Northern Great Plains region. In: *Proceedings of the Sixth International Safflower Conference—SAFFLOWER: A Unique Crop for Oil Spices and Health Consequently, a Better Life for You*, Istanbul, Turkey, June 6–10, 2005, pp. 215–221.

Borbon-Gracia, A., O. Espinnoza, C.L. Montoya et al. 2011. CIANO-LIN: A new cultivar of linoleic safflower. *Rev. Mex. Cienc. Agric.* 2: 5.

Borkar, S.G. 1997. Spines, a new avenue for infection by *Alternaria carthami* on safflower. *J. Mycol. Plant Pathol.* 27: 311–316.

Bramhankar, S.B., M.N. Asalmol, V.P. Pardey, and Y.V. Ingle. 2002. Effect of foliar spray of different fungicides on transmission of *Alternaria carthami* (Chowdhury) from plant to seed and seed to plant. *Ann. Plant Physiol.* 16: 112–114.

Bramhankar, S.B., M.N. Asalmol, V.P. Pardey, Y.V. Ingle, and E.B. Burgoni. 2001. Efficacy of fungicides against *Alternaria* leaf blight of safflower. *PKV Res. J.* 25: 99–101.

Bruckart, W.L. 1999. A simple quantitative procedure for inoculation of safflower with teliospores of the rust fungus, *Puccinia carthami*. *Plant Dis.* 83: 181.

Carneiro, S.M. de T.P.G., M.R. Silva, E.B.L. da Romano et al. 2012 Occurrence of *Colletotrichum gloeosporioides* (Penz.) Sacc. in *Carthamus tinctorius* L., in the Parana state. *Summa Phytopathol.* 38: 163–165.

Chander Rao, S., R.D.V.J. Prasada Rao, V.M. Kumar, D.S. Raman, M.A. Raoof, and R.D. Prasad. 2003. First report of tobacco streak virus infecting safflower (*Carthamus tinctorius*) in Maharashtra, India. *Plant Dis.* 87: 1396.

Chattopadhyay, C. 2001. Yield losses attributable to *Alternaria* blight of safflower (*Carthamus tinctorius* L.) and some effective control measures. *J. Mycol. Plant Pathol.* 31: 298–302.

Chattopadhyay, C., P.D. Meena, R.K. Sastryand, and R.L. Meena. 2003. Relationship among pathological and agronomic attributes for soilborne diseases of three oilseed crops. *Indian J. Plant Prot.* 31: 127–128.

Chavan, R.A., A.A. Bharose, R.B. Pawar, and V.D. Patil. 2004. Screening of safflower germplasm lines and penetration mechanism of *T. viride* against *Fusarium* wilt. *J. Soils Crops* 14: 79–82.

Chen, G. 2009. Induced resistance in safflower against safflower rust disease by exogenous chemicas. Master's dissertation, Sichuan Agricultural University, Ya'an, China.

Chowdhury, S. 1944. An *Alternaria* disease of safflower. *J. Indian Bot. Soc.* 23: 59–65.

Classen, C.E., M.L. Schuster, and W.W. Ray. 1949. New disease observed in Nebraska on safflower. *Plant Dis. Rep.* 33: 73–76.

Deadman, M.L., A.M. Al-Sadi, and S. Al-Jahdhami. 2005. First report of rust caused by *Puccinia carthami* on safflower in Oman. *Plant Dis.* 89: 208.

Desai, S.A. 1998. A note on the preliminary screening of safflower genotypes against *Alternaria* leaf spot in northern Karnataka. *Karnataka J. Agric. Sci.* 11: 824.

Deshmukh, V.V., N.S. Chore, and D.J. Jiotode. 2003. Effect of soil amendments with saw dust and viability of *Trichoderma harzianum* in different carriers. *Crop Res.* (Hisar) 26: 508–511.

Dordas, C. 2008. Role of nutrients in controlling plant diseases in sustainable agriculture: A review. *Agron. Sustain. Dev.* 28: 33–46. doi:10.1051/agro.2007051.

Driouich, A., A. Jauneau, and L.A. Staehelin. 1997. 7-Dehydrobrefelin A, a naturally occurring brefeldin A derivative, inhibits secretion and causes a cis-to-trans breakdown of Golgi Stocks in plant cells. *Plant Physiol.* 113: 487–492.

Duniway, J.M. 1977. Predisposing effect of water stress on severity of Phytophthora root rot in safflower. *Phytopathology* 67: 1434–1443.

Gadekar, D.A. and N.D Jambhale. 2002a. Inheritance of seedling and adult plant resistance to *Alternaria* leaf spot in safflower (*Carthamus tinctorius* L.). *Indian J. Genet. Plant Breed.* 62: 238–239.

Gadekar, D.A. and N.D. Jambhale. 2002b. Genetics of seedlings and adult plants resistant to wilt caused by *Fusarium oxysporum* f. sp. *carthami* in safflower. *Sesame Safflower Newsl.* 17: 81–84.

Gaikwad, S.J. and G.T. Behere. 2001. Biocontrol of wilt of safflower caused by *Fusarium oxysporum* f. sp. *carthami*. In: *Proceedings of the Fifth International Safflower Conference—Safflower: A Multipurpose Species with Unexploited Potential and World Adaptability*, Williston, ND, July 23–27, 2001, pp. 63–66.

Gayathri, D.A. and V. Madhuri. 2014. Seed mycoflora of safflower and its control by using otanicals, bioagents and fungicides—A review. *Int. J. Appl. Biol. Pharm. Technol.* 5: 209–215.

Ghaderi, M., M. Pahlevani, and S.E. Razavi. 2011. Inheritance of resistance to *Pythium ultimum* in safflower determined by generation means analysis. *Aust. J. Crop Sci.* 5(4): 439–446.

Govindappa, M., S. Lokesh, and V.R. Rai. 2005. A new stem-splitting symptom in safflower caused by *Macrophomina phaseolina*. *J. Phytopathol.* 153: 560–561.

Govindappa, M., S. Lokesh, V.R. Rai, V.R. Naik, and S.G. Raju. 2010. Induction of systemic resistance and management of safflower *Macrophomina phaseolina* root-rot disease by biocontrol agents. *Arch. Phytopathol. Plant Prot.* 43: 26–40.

Govindappa, M., V.R. Rai, and S. Lokesh. 2011a. Screening of *Pseudomonas fluorescens* isolates for biological control of *Macrophomina phaseolina* root-rot of safflower. *Afr. J. Agric. Res.* 6: 6256–6266.

Govindappa, M., V.R. Rai, and S. Lokesh. 2011b. In vitro and in vivo responses of different treating agents against wilt disease of safflower. *J. Cereals Oilseeds* 2: 16–25.

Govindappa, M., V.R. Rai, and S. Lokesh 2013. Induction of resistance against *Cercospora* leaf spot in safflower by seed treatment with plant growth-promoting rhizobacteria. *Arch. Phytopathol. Plant Prot.* 36: 1–14. doi:10.1080/03235408.2014.880574.

Garnatje, T., S. Garcia, J. Valles et al. 2006. Genome size variation in the genus *Carthamus* (*Asteraceae*, *Cardueae*): Systematic implications and additive changes during allopolyploidization. *Ann. Bot.* 97: 461–467.

Gud, M.A., D.R. Murumkar, S.K. Shinde, and J.R. Kadam. 2008. Correlation of weather parameters with development of leaf spot of safflower caused by *Alternaria carthami*. In: *Seventh International Safflower Conference—Safflower: Unexploited Potential and World Adaptability*, Wagga Wagga, New South Wales, Australia, November 3–6, 2008, pp. 1–5.

Harish Babu, B.N., V.R. Naik, L. Hanumantharaya, S.G. Raju, and S.D. Yaragoppa. 2005. Evaluation of 6 promising breeding lines of safflower for *Alternaria* tolerance, seed yield and its components. *Karnataka J. Agric. Sci.* 186: 803–806.

Hostert, N.D., C.L. Blomquist, S.L. Thomas, D.G. Fogle, and R.M. Davis. 2006. First report of *Ramularia carthami*, the causal agent of *Ramularia* leaf spot in California. *Plant Dis.* 90: 1260.

Huang, H.C. and R.S. Erickson. 2004. Effect of soil treatment of fungal agents on control of apothecia of *Sclerotinia sclerotiorum* in canola and safflower fields. *Plant Pathol. Bull.* 13: 1–6.

Huang, L., X. Yang, P. Sun, W. Tong, and S. Hu. 2012. The first illumina-based de novo transcriptone sequencing and analysis of safflower flowers. *PLoS One* doi:1371/journal/pone.0038653.

Indi, D.V., D.R. Murumkar, A.J. Patil, and V.B. Akashe. 2004. Screening of safflower germplasm against *Alternaria* leaf spot under field conditions. *J. Maharashtra Agric. Univ.* 29: 344–346.

Ingle, V.N., V.V. Deshmukh, D.J. Jiotode, N.S. Chore, and C.S. Dhawad. 2004. Screening of safflower varieties against root rot (*Rhizoctonia bataticola*). *Res. Crops* 5: 113–114.

Jacobsen, B.J., J.W. Bergman, and C.R. Flynn. 2008. Comparison of safflower fungicide seed treatments. In: *Seventh International Safflower Conference—Safflower: Unexploited Potential and World Adaptability*, Wagga Wagga, New South Wales, Australia, November 3–6, 2008, pp. 1–4.

Johnson, L.B. 1970. Influence of infection by *Phytophthora drechsleri* on inhibitory materials in resistant and susceptible safflower hypocotyls. *Phytopathology* 60: 1000–1004.

Johnson, R.C., T.J. Kisha, and M.A. Evans. 2007. Characterizing safflower germplasm with AFLP molecular markers. *Crop Sci.* 47: 1728–1736.

Kalafat, S., A. Karakaya, M.D. Kaya, and S. Bayramin. 2009. A preliminary study on the reactions of some safflower genotypes to rust disease. *Bitki Koruma Bulteni* 49: 183–187.

Kaswate, N.S., S.S. Shinde, and R.R. Rathod. 2003. Effect of biological agents against different isolates of *Rhizoctonia bataticola* (Taub.) Butler in vitro. *Ann. Plant Physiol.* 17: 167–168.

Khanchezar, A., K. Izadpanah, and M. Salehi. 2012. Partial characterization of *Spiroplasma citri* isolates associated with necrotic yellows disease of safflower in Iran. *J. Phytopathol.* 160: 331–336.

Klisiewicz, J.M. 1975. Race 4 of *Fusarium oxysporum* f. sp. *carthami*. *Plant Dis Rep.* 59: 712.

Klisiewicz, J.M. and B.R. Houston. 1962. *Fusarium* wilt of safflower. *Plant Dis. Rep.* 46: 748–749.

Kneusel, R.E., E. Schiltzg, and U. Maternhl. 1994. Molecular characterization and cloning of an esterase which inactivates the macrolide toxin brefeldin A*. *J. Biol. Chem.* 269: 3449–3456.

Koike, S.T., A. Baameur, K. Maruthachalam, and K.V. Subbarao. 2012. *Verticillium* wilt of spineless safflower caused by *Verticillium dahliae* in California. *Plant Dis.* 96(9): 1383.

Kolase, S.V., C.D. Deokar, and D.M. Sawant. 2000. Effect of plant extracts on the incidence of safflower wilt. *Sesame Safflower Newsl.* 15: 92–94.

Kolte, S.J. 1985. *Diseases of Annual Edible Oilseed Crops*, Vol. III: *Sunflower, Safflower and Nigerseed Diseases*. CRC Press, Boca Raton, FL.

Kulkarni, V.R. and A.S. Byadgi. 2004a. Effect of mosaic disease on yield and oil content of safflower. *Karnataka J. Agric. Sci.* 17: 611–612.

Kulkarni, V.R. and A.S. Byadgi. 2004b. Incidence of safflower mosaic disease in North Karnataka. *Karnataka J. Agric. Sci.* 17: 836–837.

Kulkarni, V.R. and A.S. Byadgi. 2004c. Evaluation of plant extracts against safflower mosaic virus disease. *Karnataka J. Agric. Sci.* 17: 838–840.

Kulkarni, V.R. and A.S. Byadgi. 2004d. Studies on transmission of safflower mosaic disease. *Karnataka J. Agric. Sci.* 17: 841–842.

Kulkarni, V.R. and A.S. Byadgi. 2005. Studies on host range of safflower mosaic disease. *Karnataka J. Agric. Sci.* 18: 169–171.

Kumar, A. and H.K. Joshi. 1995. Development of leaf spot caused by *Ramularia carthami* and reaction of safflower cultivars. *J. Agric. Sci.* (Cambridge) 125: 223–225.

Kumar, J.V., B.D.R. Kumari, G. Sujatha, and E. Castano. 2008. Production of plants resistant to *Alternaria carthami* via organogenesis and somatic embryogenesis of safflower cv. NARI-6 treated with fungal culture filtrates. *Plant Cell Tissue Organ Cult.* 93: 85–96.

Kwon, J.H., S.W. Kang, H.S. Lee et al. 2000. Occurrence of powdery mildew on safflower caused by *Sphaerotheca fuliginea* in Korea. *Microbiology* 28: 51–53.

Lartey, R.T., T.C. Caesar-TonThat, A.J. Caesar, W.L. Shelver, N.I. Sol, and J.W. Bergman. 2005. Safflower: A new host of *Cercospora beticola*. *Plant Dis.* 89: 797–801.

Lartey, R.T., A.W. Lenssen, R.G. Evans, and S. Ghoshroy. 2007. Comparative structural study of leaf spot disease of safflower and sugar beet by *Cercospora beticola*. *Plant Pathol. J.* (Faisalabad) 6: 37–43.

Lotfalinezhad, E., Z. Mehri, and S.J. Sanei. 2013. Temperature response of *Macrophomina phaseolina* isolates from different climatic in Iran. *Annu. Rev. Res. Biol.* 3(4): 724–734.

Li., H., Y. Dong, Y. Sun et al. 2011. Investigation of the micro RNAs in safflower seed, leaf, petal by high-throughput sequencing. *Planta* 233: 611–619.

Liu, T.T., F. Gao, Y.Z. Ren, G.Y. Li, and Y. Tan. 2009. Identification of safflower varieties rust-resistance. *Xinjiang Agric. Sci.* 46: 719–723.

Lyon, D., P. Burgener, R. Harveson, G. Hein, and G. Hergert. 2007. Growing safflower in Nebraska. NebGiude. University of Nebraska-Lincoln, Extension Institute of Agriculture and Natural Resources, Lincoln, NE.

Mahdizadeh, V., N. Safaie, and E.M. Goltapeh. 2011. Diversity of *Macrophomina phaseolina* based on morphological and genotypic characteristics in Iran. *Plant Pathol. J.* 27: 128–137.

Minz, G., M. Chorin, and Z. Solel. 1961. Ramularia leaf spot of safflower in Israel. *Bull. Res. Council Israel, Sect. D* 10: 207.

Montoya, C.L. 2005. High quality and yield grain safflower varieties development in southern Sonora, México. In: *Annual Field Day 2005*, CIANO-INIFAP, Yaqui Valley Experimental Station, Ciudad Obregón, Sonora, Mexico.

Montoya, C.L. 2008. Mexican safflower varieties with high tolerance to *Ramularia carthami*. In: *Seventh International Safflower Conference*, Wagga Wagga, New South Wales, Australia, November 3–6, 2008.

Mündel, H.-H. and H.C. Huang. 2003. Control of major diseases of safflower by breeding for resistance and using cultural practices. In: *Advances in Plant Disease Management*, eds., H.C. Huang and S.N. Acharya. Research Signpost, Trivandrum, India, pp. 293–310.

Muñoz-Valenzuela, S., G.L.C. Musa, L. Montoya-Coronado, and V.M. Rivera-Rojas. 2007. Evaluation of safflower genotypes in Northwest México. In: *New Crops and New Uses*, eds., J. Janick and A. Whipkey. ASHS Press, Alexandria, VA, pp. 193–195.

Murumkar, D.R. and A.N. Deshpande. 2009. Variability among isolates of *Fusarium oxysporum f.sp. carthami. J.Maharashtra Agric. Univ.* 34:178–180.

Murumkar, D.R., M.A. Gud, and A.N. Deshpande. 2008b. Screening of released varieties and hybrids of safflower against wilt (*Fusarium oxysporum* f. sp. *carthami*). *J. Plant Dis. Sci.* 3: 114–115.

Murumkar, D.R., D.V. Indi, V.B. Akashe, A.J. Patil, and M.A. Gud. 2009b. Multiple resistance sources against major diseases and pests of safflower, *Carthamus tinctorius* L. *J. Oilseeds Res.* 26: 175–176.

Murumkar, D.R., D.V. Indi, M.A. Gud, and S.K. Shinde. 2008a. Field evaluation of some newer fungicides against leaf spot of safflower caused by *Alternaria carthami.* In: *Seventh International Safflower Conference—Safflower: Unexploited Potential and World Adaptability*, Wagga Wagga, New South Wales, Australia, November 3–6, 2008.

Murumkar, D.R., D.V. Indi, M.A. Gud, S.K. Shinde, and A.N. Deshpande. 2009a. Fungicidal management of leaf spot of safflower caused by *Alternaria carthami. J. Maharashtra Agric. Univ.* 34: 54–56.

Mundel, H.H. and H.H. Chang. 2003. Control of major diseases of safflower by breeding for resistance and using cultural practices. *Adv. Plant Dis. Manag.* 2003: 293–310.

Nasehi, A., J.B. Kadir, M.N. Esfahani et al. 2013. Screen of safflower (*Carthamus tinctorius* L.) genotypes against *Phytophthora drechsleri* and *Fusarium solani*, causal agents of root rot disease. *Arch. Phytopathol. Plant Prot.* 46: 2025–2034.

Nasehi, A., S. Shafizadeh, S. Rezaie, and M.R. Shahsavari. 2009. Evaluation of relative resistance of safflower (*Carthamus tintorius* L.) genotypes to *fusarium* root rot disease in Isfahan Province. *Seed Plant Improv. J.* 25(1): 623–634.

Navgire, K.D., R.D. Prasad, and A.V. Wadhave. 2014. Genetic diversity of *Macrophomina phaseolina* causing root rot of safflower. In: "*Abstract-cum-Souvenir*" *National Symposium on Plant Pathology in Genomic Era*, Indian Phytopathological Society, New Delhi, India, May 26–28, 2014, p. 45.

Ochoa-Espinoza, X.M. 2012. Evaluation of elite lines of safflower (*Carthamus tinctorius* L.) oleic tolerant to *Ramularia carthami* in the Northwest of México. In: *Proceedings of the Eighth International Safflower Conference*, Hyderabad, India.

Pahlevani, M., A. Ahmadi, and S. Razavi. 2010. Assessment of safflower for susceptibility to *Pythium ultimum*, the causal agent of damping-off. *Plant Breed. Seed Sci.* 62: 17–29.

Pahlavani, M.H., S.E. Razavi, I. Mirizadeh, and S. Vakili. 2007. Field screening of safflower genotypes for resistance to charcoal rot disease. *Int. J. Plant Prod.* 1: 45–52.

Park, S.D., K.S. Park, K.J. Kim, J.C. Kim, J.T. Yoon, and Z. Khan. 2005. Effect of sowing time on development of safflower anthracnose disease and degree of resistance in various cultivars. *J. Phytopathol.* 153: 48–51.

Patibanda, A.K. and R.D. Prasad. 2004. Screening *Trichoderma* isolates against wilt pathogens of safflower, *Carthamus tinctorius* L. *J. Biol. Control* 18: 103–105.

Patil, M.S. and R.K. Hegde. 1988. Epidemiological studies on safflower leaf spot caused by *Ramularia carthami* Zaprometov under irrigated conditions. *J. Oilseeds Res.* 5: 45–48.

Patil, M.S. and R.K. Hegde. 1989. Efficacy of fungicides in control of leaf spot of safflower caused by *Ramularia carthami. J. Oilseeds Res.* 6: 118–122.

Pawar, S.V., Utpal Dey, V.G. Munde, Hulagappa, and A. Nath. 2013. Screening of elite material against major diseases of safflower under field conditions. *Afr. J. Agric. Res.* 8: 230–233.

Prameela, M., B. Rajeswari, R.D. Prasad, and D.R.R. Reddy. 2005. Variability in isolates of *Fusarium oxysporum* f. sp. *carthami* causing wilt of safflower. *Indian J. Plant Prot.* 33: 249–252.

Prasad, R. and H. Chothia. 1950. Studies on safflower rust in India. *Phytopathology* 40: 363.

Prasad, R.D. and K. Anjani. 2008a. Sources of resistance to Alternaria leaf spot among Carthamus wild species. In: *Seventh International Safflower Conference—Safflower: Unexploited Potential and World Adaptability*, Wagga Wagga, New South Wales, Australia, November 3–6, 2008, pp. 1–2.

Prasad, R.D. and K. Anjani. 2008b. Exploiting a combination of host plant resistance and *Trichoderma* species for the management of safflower wilt caused by *Fusaruim oxysporum* f. sp. *carthami* Klisiewicz and Houston. *J. Biol. Control* 22: 449–454.

Prasad, R.D. and T. Navneetha. 2010. Techniques for pathogenicity testing and screening of safflower cultivars against *Macrophomina phaseolina. J. Oilseeds Res.* 27: 199–200.

Prasad, R.D., T. Navaneetha, M.S. Chandra Girish, and C.H. Manasa. 2009. Seed borne nature of *Alternaria carthami* in safflower. *J. Oilseeds Res.* 26: 492.

Prasad, R.D., T. Navaneetha, and N.N. Rao. 2011. Cultural, morphological, pathogenic and molecular diversity in *Macrophomina phaseolina* isolates of safflower from southern India. *Indian Phytopathol.* 64: 247–253.

Prasad, R.D., T.R. Sharma, T.K. Jana, T. Prameela Devi, N.K. Singh, and K.R. Koudal. 2004. Molecular analysis of genetic variability in *Fusarium* species using microsatellite markers. *Indian Phytopathol.* 57: 272–279.

Prasad, R.D., T.R. Sharma, and T. Prameela Devi. 2007. Molecular variability and detection of *Fusarium* species by PCR based RAPD, ISSR and ITS-RFLP analysis. *J. Mol. Plant Pathol.* 37: 311–318.

Prasad, R.D., V. Singh, and H. Singh. 2006. Crop protection. In: *Research Achievements in Safflower*, ed., D.M. Hegde. All India Coordinated Research Project on Safflower, Directorate of Oilseeds Research, Hyderabad, India, pp. 52–77.

Prasad, R.D. and M. Suresh. 2012. Diseases of safflower and their management. In: *Safflower Research and Development in the World: Status and Strategies*, eds., I.Y.L.N. Murthy, H. Basappa, K.S. Varaprasad, and P. Padmavathi. Indian Society of Oilseeds Research, Hyderabad, India, pp. 97–106.

Prashanthi, S.K., S. Kulkarni, and K.H. Anahosur. 2000b. Management of safflower root rot caused by *Rhizoctonia bataticola* by antagonistic microorganisms. *Plant Dis. Res.* 15: 146–150.

Prashanthi, S.K., S. Kulkarani, V.S. Sangam, and M.S. Kulkarni. 2000a. Chemical control of *Rhizoctonia bataticola* (Taub.) Butler, the causal agent of root rot of safflower. *Plant Dis. Res.* 15: 186–190.

Raghuwanshi, K.S. and G.N. Dake. 2009. Fusarium wilt of safflower (*Carthamus tictorius* L.): A review. *J. Plant Dis. Sci.* 4: 110–117.

Raghuwanshi, K.S., G.N. Dake, S.N. Mate, and R.M. Naik. 2008. Pathogenic variations of *Fusarium oxysporum* f. sp. *carthami*. *J. Plant Dis. Sci.* 3: 241–242.

Ranaware, A., V. Singh, and N. Nimbkar. 2010. In vitro antifungal study of the efficacy of some plant extracts for inhibition of *Alternaria carthami* fungus. *Indian J. Nat. Prod. Resour.* 1: 384–386.

Rathaiah, Y. and M.S. Pavgi. 1977. Development of sclerotia and spermogonia in *Cercospora sesamicola* and *Ramularia carthami*. *Sydowia* 30: 148.

Salehi, M., K. Izadpanah, and M. Siampour. 2008. First record of 'Candidatus phytoplasma trifolii'-related strain associated with safflower phyllody disease in Iran. *Plant Dis.* 92: 649.

Salehi, M., K. Izadpanah, M. Siampour, R. Firouz, and E. Salehi. 2009. Molecular characterization and transmission of safflower phyllody phytoplasma in Iran. *J. Plant Pathol.* 91: 453–458.

Salunkhe, V.N. 2014. Screening of safflower germplasm accessions for resistance source against *Macrophomina* root rot. *Bioscan* 9: 689–690.

Sastry, K.R. and C. Chattopadhyay. 1999a. Influence of non-host crops on the survival of *Fusarium oxysporum* f. sp. *carthami*. *J. Mycol. Plant Pathol.* 29: 70–74.

Sastry, K.R. and C. Chattopadhyay. 1999b. Effect of soil solarization on *Fusarium oxysporum* f. sp. *carthami* population in endemic soils. *Indian Phytopathol.* 52: 51–55.

Sastry, K.R. and C. Chattopdhyay. 2001. Effect of soil solarization on wilt incidence and seed yield of safflower (*Carthamus tinctorius* L.). *Indian J. Plant Prot.* 29: 47–52.

Sastry, R.K. and C. Chattopadhyay. 2003. Development of *Fusarium* wilt-resistant genotypes in safflower (*Carthamus tinctorius*). *Eur. J. Plant Pathol.* 109: 147–151.

Sastry, K.R. and C. Chattopadhyay. 2005. Diseases of safflower. In: *Diseases of Oilseed Crops*, eds., G.S. Saharan, N. Mehta, and M.S. Sangwan. Indus Publishing Co., New Delhi, India, pp. 236–267.

Sastry, R.K., C. Chattopadhyay, V. Singh, and D.M. Hegde. 2002. Integrated management of safflower wilt using host resistance, cultural and chemical measures. *J. Mycol. Plant Pathol.* 32: 189–193.

Sastry, K.R. and J. Jayashree. 1993. Eradication of wilt fungus from heavily infected safflower seed. *J. Oilseeds Res.* 10: 227–231.

Sastry, K.R., R.T.J. Rego, and J.R. Burford. 1993. Effects of cropping systems on wilt of safflower. In: *Proceedings of the Third International Safflower Conference*, eds., L. Dajue and H. Yunzhou. Beijing, China, June 14–18, 1993, pp. 636–645.

Schuster, M.L. and D.W. Christiansen. 1952. A foot and root disease of safflower caused by *Puccinia carthami* Cda. *Phytopathology* 42: 211.

Sehgal, D. and S.N. Raina. 2005. Genotyping safflower (*Carthamus tinctorius*) cultivars by DNA fingerprints. *Euphytica* 146: 67–76.

Sharifnabi, B. and G. Saeidi. 2004. Preliminary evaluation of different genotypes of safflower (*Carthamus tinctorius* L.) to *Fusarium* root rot disease. *J. Sci. Technol. Agric. Nat. Resour.* 8: 219–227.

Shinde, A.B. and B.V. Hallale. 2009. Standardization of water culture technique for Fusarium wilt of safflower. *Int. J. Plant Sci.* 4: 248–249.

Shinde, A.B., B.V. Hallale, and A.P. Vaidya. 2008. Antifungal activity of leaf extracts against *Alternaria* blight of safflower. *Natl. J. Life Sci.* 5: 203–206.

Shivani, D., C. Sreelakshmi, and C.V.S. Kumar. 2011. Inheritance of resistance to Fusarium wilt in safflower (*Carthamus tinctorius* L.) genotypes. *J. Res. ANGRAU* 39: 25–27.

Singh, A.K., D.K. Chakrabarti, and K.C. Basuchaudhary. 1975. Two new diseases of safflower from India. *Curr. Sci.* (India) 44: 397.

Singh Saroj, S. Kumar, and D.S. Suryawanshi. 2006. Evaluation of different treatments for the management of safflower insect pests and diseases under rainfed conditions. *Indian J. Entomol.* 68: 286–289.

Singh, V. and R.D. Prasad. 2007. Integrated management of pests and diseases in safflower. Directorate of Oilseeds Research, Hyderabad, India, p. 49.

Singh, V., A.M. Ranaware, and N. Nimbkar. 2008a. Safflower: Unexploited potential and world adaptability. In: *Seventh International Safflower Conference*, Wagga Wagga, New South Wales, Australia, November 3–6, 2008, pp. 1–4.

Singh, V., A.M. Ranaware, and N. Nimbkar. 2008b. Breeding for Fusarium wilt resistance in safflower. In: *Seventh International Safflower Conference—Safflower: Unexploited Potential and World Adaptability*, Wagga Wagga, New South Wales, Australia, November 3–6, 2008, pp. 1–5.

Somwanshi, S.D., G.D. Deshpande, and J.S. Magar. 2009. Differential resistance of safflower genotypes to three isolates of *Fusarium oxysporum* f. sp. *carthami* Klisiewicz and Houston. *Adv. Plant Sci.* 22: 343–346.

Sumitha, R. and N. Nimbkar. 2009. Efficacy of fungicides against Alternaria leaf spot disease in safflower. *Ann. Plant Prot. Sci.* 17: 507–508.

Taware, M.R., V.M. Gholve, and U. Dey. 2014. Bioefficacy of fungicides and plant extracts/botanicals against Alternaria crthami, the causal agent of Alternaria blight of safflower (*Carthamus tinctorius* L.). *Afr. J. Microbiol. Res.* 8: 1400–1412.

Thomas, V., R. Norton, and N. Wachsmann. 2008. Variation in response to *Alternaria carthami* infection by a range of safflower (*Carthamus tinctorius*) varieties. In: *Seventh International Safflower Conference—Safflower: Unexploited Potential and World Adaptability*, Wagga Wagga, New South Wales, Australia, November 3–6, 2008, pp. 1–3.

Uematsu, S., K. Kageyama, J. Moriwaki, and T. Sato. 2012. *Colletotrichum carthami* comb. nov., an anthracnose pathogen of safflower, garland chrysanthemum and pot marigold, revived by molecular phylogeny with authentic herbarium specimens. *J. Gen. Plant Pathol.* 78: 316–330.

Varaprasad, K.S. 2012. Safflower scenario: Status and future strategies. In: *Souvenir, Eighth International Safflower Conference, Safflower Research and Development in the World: Status and Strategies*. Indian Society of Oilseeds Research, Hyderabad, India, January 19–23, 2012, pp. 1–9.

Vichova, J., K. Vejrazka, T. Cholastova, R. Pokorny, and E. Hrudova. 2011. *Colletotrichum simmondsii* causing anthracnose on safflower in the Czech Republic. *Plant Dis.* 95: 79.

Waghmare, S.J. and V.V. Datar, 2010. Studies on seed borne nature of *Fusarium oxysporum* f. sp. *carthami* causing wilt of safflower. *J. Maharashtra Agric. Univ.* 35: 491–492.

Waghmare, S.J. and B.P. Kurundkar. 2011. Efficacy of local isolates of *Trichoderma* spp. against *Fusarium oxysporum* f. sp. *carthami* causing wilt of safflower. *Adv. Plant Sci.* 24(1): 37–38.

Weisker, A.C. and G.C. Musa. 2013. Ramularia leaf spot resistant safflower. Publication No. US8476486 B2. California Oils Corporation, Richmond, CA.

Wunsch, M., M. Schafer, and B. Kraft. 2013. Field evaluation of fungicides for management of rust and *Alternaria* blight on safflower. Carrington Research and Extension Center, North Dakota State University, Carrington, ND.

Yadava, D.K., Sujata Vasudev, N. Singh, T. Mohapatra, and K.V. Prabhu. 2012. Breeding major oil crops: Present status and future research needs. In: *Technological Innovations in Major World Oil Crops*, Vol. I: *Breeding*, ed., S.K. Gupta. Springer Science + Business Media, LLC, New York, pp. 17–51.

Zayed, M.A., A.H. Yehia, M.A. El-Sebaey, and A.M. Gowily. 1980. Studies on the host–parasite relationships of safflower root rot disease caused by *Fusarium oxysporum* SClecht. *Egypt. J. Phytopathol.*12: 63–70.

Zhao, S., G.-L. Xie, H. Li, and C. Li. 2007. Occurrence of rust caused by *Puccinia carthami* on snow lotus in China. *Plant Dis.* 91: 772.

Zimmer, D.E. and L.N. Leininger. 1965. Sources of rust resistance in safflower. *Plant Dis. Rep.* 49: 440.

Section VII

Soybean

The soybean (United States) or soya bean (United Kingdom) (*Glycine max* L. (Merrill) is a species of legumes belonging to the plant family Papilionaceae, which is native to East Asia. The eastern half of North China is believed to be the primary center of origin, and Manchuria, the secondary center of origin. From there, it is believed to have spread to Korea and Japan, where it is widely grown for its edible bean, which has numerous uses (Yadava et al. 2012). Soybeans are now major crops in the United States, Brazil, Argentina, China, and India.

The cultivated soybean plant is an annual plant, generally exhibiting an erect, sparsely branched bush-type growth habit with pinnately trifoliate leaves. Purple or white flowers are borne on short axillary racemes on reduced peduncles. The pods are either straight or slightly curved, usually hirsute. The height of the soybean plant varies from less than 0.2 to 2.0 m. The inconspicuous, self-fertile flowers are borne in the axil of the leaf and are white, pink, or purple. The pods, stems, and leaves are covered with fine brown or gray hairs. There are one to three seeds per pod, and they are usually void to subspherical in shape. The seed coat ranges in color from light yellow to olive green and brown to reddish black. Like most other legumes, soybeans perform nitrogen fixation by establishing a symbiotic relationship with the bacterium, *Bradyrhizobium japonicum* (syn. *Rhizobium japonicum*). For best results, an inoculum of the correct strain of bacteria should be mixed with soybean seed before planting. Modern crop cultivars generally reach a height of around 1 m and take 80–120 days from sowing to harvesting.

The basic chromosome number is $2n = 40$. The cultivated soybean (*G. max*) genome size is estimated to be about 1.12 Gb DNA, and the wild soybean (*Glycine soja*) genome size is about 1.17 Gb DNA (Qi et al. 2014). The first draft sequence and gene models of *G. max* (domesticated soybean) as well as *G. soja* (wild soybean) have been known and available for use in research purposes since 2010 (Kim et al. 2010). The comparison between genome sequences of *G. max* and *G. soja* shows significant differences between genomic compositions of the two. Major traits of agricultural importance including yield and stress tolerance are polygenic, and the presence of these favorable alleles in *G. soja* help breeding program to improve beneficial traits into cultivated soybeans (Kim et al. 2010, Joshi et al. 2013).

In addition to high protein content (40%), the soybean seeds contain 18%–23% oil and thus add to the importance of the species as an edible-oil-yielding crop. Because of the terms of production and international trade and maximum share of about 57% of world's oilseed production, the plant is now classed as an edible oilseed rather than a pulse by the UN Food and Agricultural Organization. Soybean cultivation is successful in climates with hot summers, with optimum growing conditions in mean temperatures of 20°C–30°C. Soybean can be grown in a wide range of soils with optimum growth in moist alluvial soils with a good organic content. The main producers of soybean are the United States (36%), Brazil (36%), Argentina (18%), China (5%), and India (4%). The three largest producers have recorded an average nationwide soybean crop yields of about 3 tons/ha. Analyzing the presently prevailing situation and the amount of available arable land and water resources in Brazil, it is expected to eventually become the number one soybean-producing nation in the world. Already, South America as a continent produces more soybeans than North America (combined U.S. and Canada production). In the past decade, large tracts of fertile land and low labor costs have fueled explosive growth in South America's soybean industry although poor road and rail infrastructure, as well as economic instability and environmental concerns, have been the primary checks to further expansion in South American countries. The introduction of this temperate crop to subtropical climatic conditions made it more vulnerable to problems like seed longevity, poor growth rate due to changed photoperiod, and various biotic and abiotic stresses (Hegde 2009, Yadava et al. 2012). Diseases of soybeans are described in Chapter 9 as follows.

9 Soybean Diseases

SEED ROT AND SEEDLING BLIGHT COMPLEX

Causal Fungi, Symptoms, and Environmental Relations and Economic Importance

Several species of fungi belonging to different genera, namely, *Aspergillus flavus*, *Aspergillus niger*, *Aspergillus fumigatus*, *Mucor mucedo*, *Penicillium chrysogenum*, *Fusarium oxysporum*, *Rhizopus stolonifer*, *Cephalosporium acremonium*, *Rhizopus leguminicola*, *Alternaria alternata*, *Colletotrichum dematium*, *Macrophomina phaseolina*, *Phoma* sp., *Sclerotium rolfsii*, and *Curvularia lunata* can be isolated from the seeds of soybean. A great number of fungi are observed on seed coat and cotyledon, followed by axis. These pathogenic seed-borne and/or soilborne fungi actually penetrate and colonize the seeds. Fungal hyphae are present intercellularly within the host tissues. Maceration, disintegration, and rupture of host cells are observed in infected seeds (Adekunle and Edun 2001, Ellis et al. 2013). *Fusarium solani* and *M. phaseolina* are recorded in the cotyledon and axis (Tariq et al. 2006). Though the frequencies of different microflora are often significantly different between the transgenic and conventional cultivars, the nature of microflora on these has been found to be similar (Villarroel et al. 2004). Germination of seed is directly related to the prevalence of fungi associated with the seed (Shovan et al. 2008) and infested soils (El-Hai et al. 2010).

Pythium species: This may be the first cause of seed rot and damping-off of soybean seedlings in a growing season worldwide causing enormous losses in yield due to lack of plant stand establishment. High-residue fields, and heavy or compacted soils, are at higher risk because of cooler, wetter conditions. Pathogen may attack seeds before or after germination and seeds killed before germination and emergence. On infected plants, the hypocotyls become narrow and are commonly *pinched off* by the disease. Emerged plants may be killed before the first true leaf stage. These plants have a rotted appearance. Diseased plants may easily be pulled from the soil because of rotted roots. The two species *Pythium ultimum* and *Pythium aphanidermatum* cause greater seed rot and damping-off than any other *Pythium* species under Canadian conditions. The interactions between temperature and *Pythium* spp. are more pronounced for *P. aphanidermatum*, which shows an increased percentage of seed rot with an increase in temperature (20°C–28°C), whereas *Pythium irregulare*, *Pythium macrosporum*, and *Pythium sylvaticum* show a decreased percentage of seed rot with an increase in temperature (Wei et al. 2011). Some differential disease responses can be detected between glyphosate-tolerant and glyphosate-sensitive cultivars following the application of certain category of such herbicides. However, glyphosate-tolerant and glyphosate-sensitive soybean cultivars react similarly to most herbicide treatments with respect to root rot and damping-off (Harikrishnan and Yang 2002).

Phytophthora species: This can attack and rot seeds prior to emergence and can cause pre- and postemergence damping-off. It produces tan-brown, soft, rotted tissue. At the primary leaf stage (V1), infected stems appear bruised and soft, secondary roots are rotted, the leaves turn yellow, and plants frequently wilt and die. *Phytophthora sojae* commonly infects at the seedling stage, causing pre- and postemergence damping-off. After emergence, infected plants will be clearly visible in low areas of fields but may also be hidden underneath the canopy of nearby plants within the row. *P. sojae* infection is favored by high soil moisture resulting from excessive rains, poor drainage, and heavy clay soil texture.

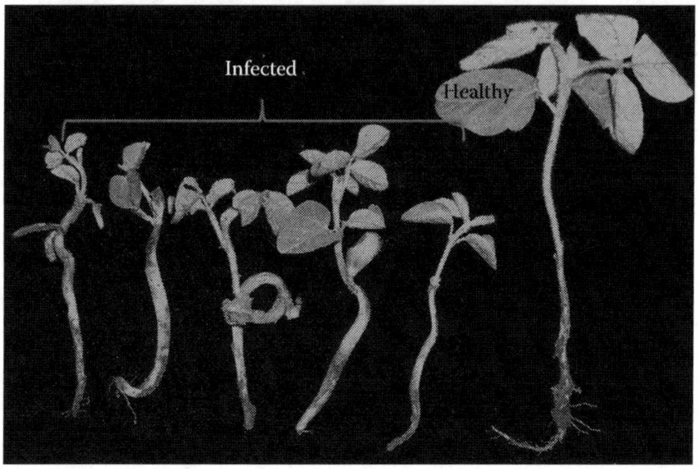

FIGURE 9.1 Seedling blight of soybean caused by *R. solani*. Note the RB lesion on soybean hypocotyls near the soil line. (Courtesy of Dr. Shrishail Navi, Iowa State University, Ames, IA.)

Rhizoctonia solani: This shows the highest ability to infect the seeds and decreases seed germination by 5.26%–15.8% with seedling mortality of 100% (Stephan et al. 2005, El-Hai et al. 2010). Trifluralin enhances the susceptibility of soybean radicles to *R. solani* (Montazeri and Hamdollah-Zadeh 2005). *Rhizoctonia* is more common in wet or moderately wet soils but not in saturated soils and its activity is most in warm soils (over 24°C), where soybean germination is slow or emergence is delayed. Infection may be superficial, causing no noticeable damage, or may girdle the stem and kill or stunt plants. *Reddish-brown (RB) lesion on soybean hypocotyl near the soil line is characteristic of* Rhizoctonia *infection* (Figure 9.1). It normally appears as the weather becomes warm (~27°C), is more often seen in late-planted soybean fields and causes loss of seedlings (damping-off) in small patches or within rows, and is usually restricted to the seedling stage. Stand loss is due to the soft and rotted seed with soil adhering to them. Plants may be killed by *damping-off* before or after *Rhizoctonia* infection.

Fusarium species: Seedling infection in soybean is caused by a complex of different *Fusarium* species that prefer different conditions; some prefer warm and dry soils, while others prefer cool and wet soils. Some species attack corn, wheat, and other host plants. They cause light- to dark-brown lesions on soybean roots that may spread over much of the root system, may attack the tap root and promote adventitious root growth near the soil surface, and may also degrade lateral roots, but usually do not cause seed rot. Less severe infections may degrade without resulting in plant death. Plant stand loss due to the *patchy* nature of *Fusarium* infection occurs in a specific area of the field. *F. oxysporum*, *Fusarium semitectum* (*Fusarium pallidoroseum*), *Fusarium graminearum*, and other fungal species such as *S. rolfsii*, *M. phaseolina*, and *A. flavus* appear to be dominant (Goulart et al. 2000, Goulart 2001, Pant and Mukhopadhyay 2002, Ellis et al. 2011). *Fusarium commune* as the cause of damping-off of soybean is the first report from the United States (Ellis et al. 2013). *F. oxysporum* f. sp. *glycines* reduces seed germination and seedling survivability by 40% and cause preemergence damping-off of soybean seedlings (Begum et al. 2007a).

For soybeans, the soilborne pathogens *P. sojae*, several *Pythium* species, and *R. solani* as described earlier are considered to be the most important seedling pathogens in the North Central states in the United States. A study conducted in Iowa concluded that these three organisms compose 90% of soybean-seedling diseases. Seed-borne fungi such as *Cercospora* species, *Phomopsis longicolla*, or *Fusarium* species can also play a role in seed and seedling disease, particularly if prevailing environmental conditions in the preceding crop season adversely affect the seed production.

Disease Management

Host Plant Resistance

Soybean genotypes showed differences in their reaction to seedling damping-off diseases caused by soilborne fungal pathogens such as *R. solani*, *M. phaseolina*, and *S. rolfsii*; for example, soybean cv Giza 21 shows the least incidence of preemergence damping-off followed by Giza 35 and Giza 83 but only Giza 35 showed the least incidence of postemergence damping-off (Amer 2005). Resistance to damping-off of seedlings caused by *P. aphanidermatum* in soybean cv Archer is governed by a single dominant gene *Rpa1*, which is independent of association of resistance governed by another gene *Rps1K* that confers resistance of Archer to *P. sojae* (Rosso et al. 2008). Many *P. sojae* races are found in South Dakota in the United States. Most of the genes that have been incorporated into soybean for resistance to *Phytophthora* are vulnerable to races found in South Dakota including Rps1k. Producers are required to keep a good history of their fields that are prone to *Phytophthora*, so that they may judge the effectiveness of resistance genes in their varieties. The best strategy would be to plant varieties with Rps1k, Rps3a, Rps6, or a combination thereof.

Resistance in Archer to *Pythium* damping-off and root rot is robust with its efficacy over a number of *Pythium* spp. covering *P. ultimum*, *P. irregulare*, *P. aphanidermatum*, *P. vexans*, and over a range of plant developmental stages of soybean (Bates et al. 2008). Though *Pythium* resistance in soybean cv Archer can withstand the adverse effect of flooding, the disease as such may account for a portion of the negative response of soybean to flooding (Kirkpatrick et al. 2006).

Chemical Control

Identification of seedling disease is essential in fixing the problems as different fungicides are effective in controlling different seedling diseases. Though there are limited choices in fungicide seed treatments for managing the seed rot and seedling diseases, the best seed germination and field emergence of soybean seedlings can be obtained by treating the seeds with thiram at 0.3% (Raj et al. 2002) or with a combination of carbendazim + thiram (Goulart et al. 2000, Sonavane et al. 2011) or with captan or fludioxonil or a combination of fludioxonil + metalaxyl-M (MAXIM XL) (Gally et al. 2004, Ellis et al. 2011) depending on the prevalence of pathogenic fungi. Seeds treated with 0.3% thiram maintain germination above the minimum seed certification standard up to 10 months of storage, after which seed germination can fall below the certification standard. Excellent seed germination with best degree of seedling blight control can still be achieved due to seed treatment with a mixture of thiram + carbendazim + antagonist *Trichoderma harzianum* or *T. hamatum* at 3 + 1 + 4 g/kg seed (El-Sayed et al. 2009, Khodke and Raut 2010). Treatment of soybean seeds with *R. japonicum* (*Rhizobia*) and Thiabendazole (Tecto) induces significant increase in seed germination of soybean in soils infested with *F. solani* and *M. phaseolina* (Al-Ani et al. 2011). Fungicide spray is recommended during the reproductive phase of soybean for disease control and for production of better quality soybean seeds in Brazil (Beal et al. 2007).

Soaking soybean seeds with plant growth chemicals such as ethrel, CCC, or IBA at 200 ppm has been found to be significantly effective in reducing preemergence and postemergence damping-off of soybean seedlings under salinity stress conditions of soils in Pakistan (El-Hai et al. 2010). Systemic resistance in soybean plants can be induced by prior soaking of seeds in a mixture of benzothiadiazole (0.25 g ai/L) and humic acid (4 g ai/L) for the control of damping-off and wilt diseases of soybean caused by *F. oxysporum* in Egypt (Abdel-Monaim et al. 2012, El-Baz et al. 2012).

Cultural Control

Cultural practices are also important. For example, soil drainage, delaying sowing after green manuring, and sowing at temperatures above the pathogen optimum; all precautions may be needed to prevent contamination of tools and irrigation water, in addition to the use of quality seeds. Seed quality affects stand establishment and seedling rot in soybeans, particularly when seedlings are subjected to stress such as excessive moisture and low temperatures. In addition to quality seed, fungicide seed treatments are also highly recommended and often not optional. Fungicide seed treatments benefit stand establishment under adverse conditions such as cool, wet conditions and where pathogens are present. However, a fungicide seed treatment will not turn bad seed into good, and it will only provide a limited benefit under extreme weather and disease conditions. Certain organic soil amendments may have similar effects as an approach for indirect biological control through cultural practices. For example, dried powders of velvet bean and pine bark added to the soil at the rate of 50–100 g/kg of soil can reduce *R. solani*–induced damping-off and root and stem rot disease in soybean (Blum and Rodriguez-Kabana 2006a). Seed treatment with alum recorded maximum seed germination, root length, shoot length, and seedling vigor index (Chandrasekaran et al. 2000a,b).

Biological Control

Biological control has emerged as an alternative and promising means for the management of such type of diseases. Biological control agents like *Gliocladium virens* and *T. harzianum* antagonize pathogens by antibiosis, competition, mycoparasitism, or other forms of direct exploitation (Pant and Mukhopadhyay 2001). Damping-off of seedlings of soybean caused by *S. rolfsii* in the East of Java in Indonesia, Malaysia, Thailand, and the Philippines has been proved to be brought under control by *Actinomycetes* and VAM (Sastrahidayat et al. 2011). Antifungal activity of two bacteria obtained from the soybean rhizosphere as *Pseudomonas fluorescens* BNM296 and *Bacillus amyloliquefaciens* BNM340 has been shown to be antagonistic to *P. ultimum* causing damping-off and is able to increase seedling emergence rate under field conditions (Leon et al. 2009). Coating soybean seeds and roots with spores and mycelia of three antagonists (*Aspergillus sulphureus*, *Penicillium islandicum*, and *Paecilomyces variotii*) gives soybean germinating seeds and seedlings a very good protection from root rot and pre- and postemergence damping-off caused by *P. spinosum*. Applying these biocontrol agents (BCAs) to autoclaved and nonsterilized soil infested with *P. spinosum* provides an excellent way of protection (Al-Sheikh and Abdelzaher 2010).

T. harzianum isolate UPM40 and *Pseudomonas aeruginosa* isolate UPM13B8 have been proved to be the most effective candidates in inhibiting the mycelial growth of *F. oxysporum* f. sp. *glycines*, which causes soybean seed rot (Begum et al. 2007a). Certain bacterial isolates such as the ones belonging to *Bacillus* species (B3, B12, B80) and fluorescent *Pseudomonads* (FLPs) (B43, B51, B63, B64), when obtained from the soybean rhizosphere in Iran, have been found to show strong antagonistic effects against *P. sojae* causing damping-off soybean seedlings indicating their potential use in the management of soybean-seedling diseases (Tehrani et al. 2002, Zebarjad et al. 2006). Soybean seeds coated with a peat-bond formulation of biocontrol bacterial agent *Burkholderia ambifaria* isolate BC-F give significant disease (seedling blight) suppression with significantly greater plant stand over a considerable period of plant growth due to the ability of isolate BC-F to persist for long periods in association with roots of diverse crop plants in different soils and the production of a metabolite(s) with broad-spectrum antibiotic activity (Li et al. 2002).

Effect of Plant Extracts

Seed-borne fungi of soybean can be controlled by using leaf extract of medicinal plant and BCA. Soybean seeds when treated with leaf extracts of *Allium sativum* L. and *Azadirachta indica* A. Juss inhibit the growth of seed-borne pathogenic fungi resulting in control of the seed rot and damping-off of diseases ensuring better establishment of the stand of the plants under field conditions (Rathod and Pawar 2012). The proportion of pathogenic fungi has been found to be the lowest in soybean seeds treated with Biosept 33SL (from grapefruit seed and pulp extract) or Zaprawa Oxafun T (37.5% carboxin + 37.5% thiram) as studied by Patkowska (2006) from Poland.

ANTHRACNOSE

Symptoms

Plants may become infected at any stage of development and as a result exhibit a wide range of symptoms. The soybean is prone to be attacked by *Colletotrichum truncatum* at seed and seedling stages, resulting in pre- and postemergence damping-off (Begum et al. 2010). Seeds colonized with *C. truncatum* produce irregular gray spots with black specks. *C. truncatum* produces compact dark mycelium both intra- and intercellularly in the seed coat, cotyledon, and embryo. Mycelial growth is more abundant in the hourglass layer of the seed coat and hypodermis, where large intercellular spaces are present. Acervuli with setae and abundant hyaline sickle-shaped conidial masses are observed abundantly on the surface of infected seeds. Similar observations are found beneath the inner layers of the seed coat and upper surfaces of embryo and cotyledonary tissues. Brown conidial masses are produced during incubation and liberated in the form of ooze resulting in maceration and disintegration of the parenchyma tissues of the seed coat, cotyledon, and embryo (Begum et al. 2007b).

Infected seedlings that do not die early appear healthy until blossom, but chances of infection tend to increase with maturity and symptoms consist of appearance of brown, irregularly shaped spots on leaves, stem, pods, and petioles (Figure 9.2). The girdling of petioles by large lesions results in premature defoliation. When pods are infected, mycelium may completely fill the cavity and no seeds are produced (pod blanking) or fewer and/or smaller seed form. Seed that does form may appear brown, moldy, and shriveled or may look normal. Dark acervuli develop in lesions on all host tissues. Leaf infections, which generally develop as a result of secondary infection by conidia, may exhibit leaf rolling, necrosis of laminar veins, petiole cankers, and premature defoliation (Figure 9.3). In general, infected plants appear stunted and may have significant yield reduction. This disease is commonly observed at maturity (Galli et al. 2007). Additionally, the presence of pathogens in seeds may lead to significant reductions in seed germination, plant emergence and vigor, duration of seed storage, and crop yield.

Visible and near-infrared reflectance (Vis/NIR) spectroscopy technique has been applied to accurately detect the disease severity of soybean pod anthracnose in China (Feng et al. 2012a). According to the results, Vis/NIR spectroscopy is feasible for the identification of *C. truncatum* on soybean pods. There is a potential to establish an online field application of early plant disease detection based on visible and near-infrared spectroscopy (Feng et al. 2012b). The fungus infects seedlings, stems, petioles, leaves, and pods. Pathogen produces stria-like lesions in the pods of soybean variety "Tai 75" in China. Based on the pathogen isolated, morphological observation, rDNA ITS sequence analyses, and pathogenicity tests, it is demonstrated that the lesion is a type of infection caused by fungal pathogen *C. truncatum*. The same pathogen infects soybean pods resulting in two main types of symptom on various soybean varieties. The stria-like lesions appeared exclusively in the soybean variety "Tai 75," whereas the round blotches could be observed with the other soybean varieties. Pod disease incidence of stria-like lesions in the soybean variety "Tai 75" is observed to be 65.37%, whereas in the other soybean varieties, it could be in the range of 1.02%–12.25%. There is a clear host- and variety-dependent characteristics of infection of fungus *C. truncatum* (Lou et al. 2009).

Geographical Distribution and Losses

Anthracnose is an economically important disease widely distributed in almost all soybean-growing countries of the world. The causal fungus, *C. truncatum* (Schw.) Andrus and Moore, is present in almost all soybean-growing areas of the world (CMI map). It is generally more abundant in subtropical or tropical than temperate zones. It is reported to be the serious disease in Argentina (Daniel Ploper et al. 2001, Ramos et al. 2010), Austria (Zwatz et al. 2000),

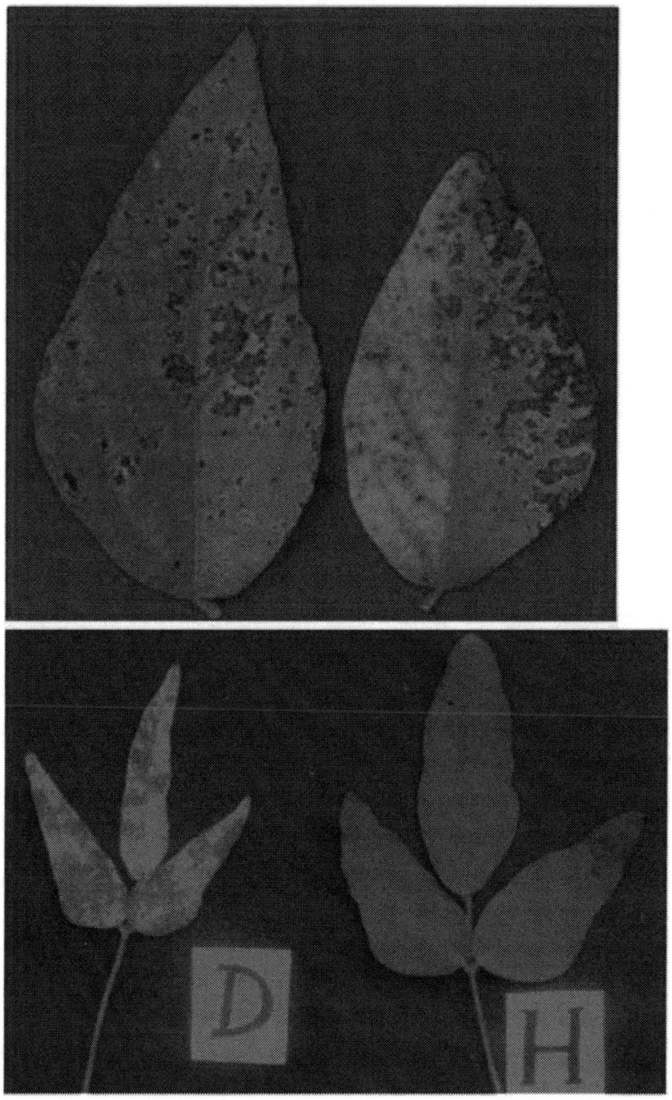

FIGURE 9.2 Anthracnose leaf spots of soybean. (Courtesy of Dr. Anil Kotasthane, IGKV, Raipur, India.)

Brazil (Klingelfuss and Yorinori 2001), China (Feng et al. 2012a), India (Jagtap et al. 2012b), and Zambia (Mayonjo and Kapooria 2003). During the last decade, soybean yield losses have increased as the disease is associated with monocropping, no-till systems, and genetic uniformity of cultivars in the northern Pampeana region of Argentina (Ramos et al. 2010). This disease is severe in these areas, especially when precipitation and relative humidity is high. The combined attack of the disease with frogeye leaf spot (FLS) (*Cercospora sojina*) results in a yield loss of soybean yield in the range of 23.7%–32.5% in India (Mittal 2001). In addition to yield reduction, *C. truncatum* may affect seed quality. Seed-borne nature of the pathogen results in the shift in oil content reduction in the range of 18%–27%, beside the reduced seed germination up to 29.2% and viability by 26.9% coupled with lower seedling vigor (Galli et al. 2005, Nema et al. 2012). Moreover, seed infection can increase the electrolyte leakages compared to healthy seeds (Begum et al. 2008b).

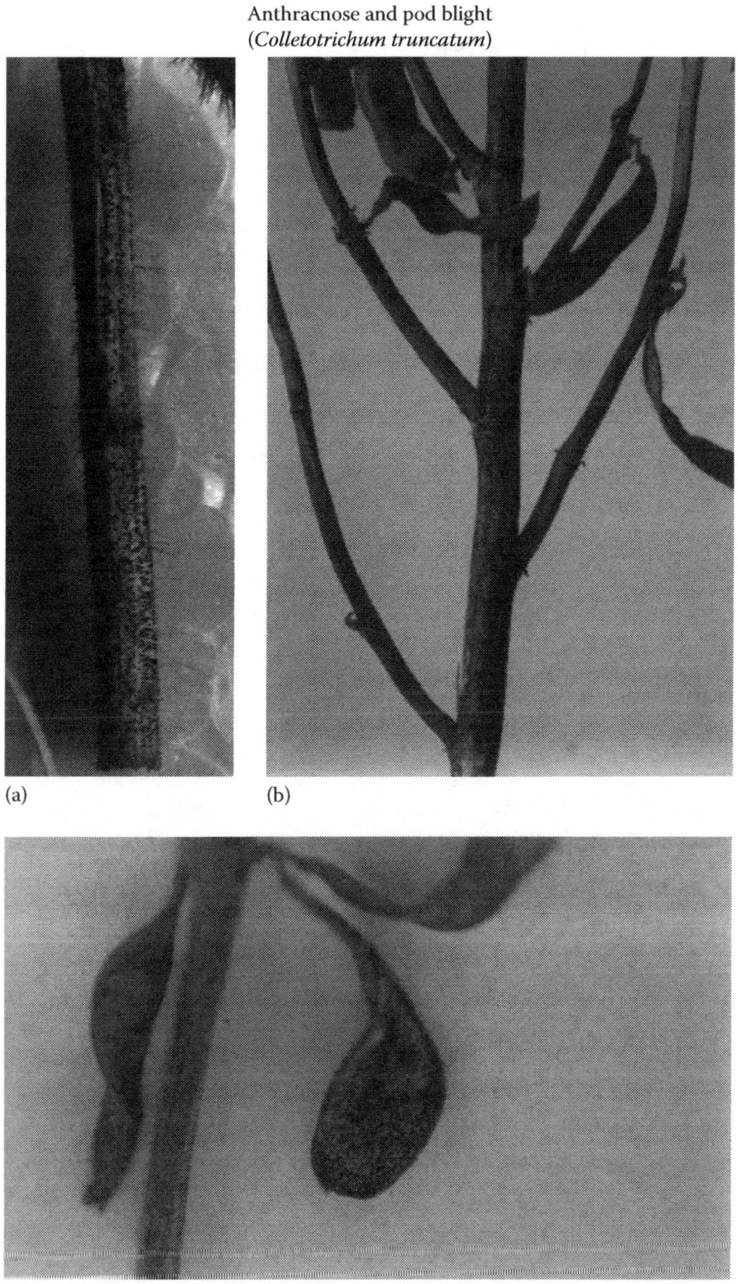

FIGURE 9.3 Anthracnose of soybean. Note the black dot-like acervuli on affected stem and pod of soybean. (a) Infected stem with acervuli of the pathogen. (b) Stem and pod blight symptoms. (c) Infected pods with acervuli of the pathogen. (Courtesy of Dr. G.K. Gupta, ICAR-Directorate of Soybean Research, Indore, India.)

Pathogen

The species most frequently associated with soybean anthracnose is *C. truncatum* (Schw.) Andrus and Moore (= *C. dematium* (Pers. ex Fr.) Grov (= *C. dematium* var. *truncatum* (Schw.)

V. Arx) (teleomorph *Glomerella truncata*). This is the imperfect fungus that belongs to the subdivision *Deuteromycotina* form-class *Deuteromycetes*, form-subclass *Coelomycetidae*, form-order *Melanconiales*, and form-family *Melanconiaceae*.

The morphology of both conidia and setae of *C. truncatum* isolates from soybean has been compared and found to be distinct. Curved conidial shape is more useful than size in isolate determination (Begum et al. 2010). Sucrose is proved to be the best carbon source for growth and sporulation of *C. truncatum* (Jagtap and Sontakke 2009). Growth (mycelial dry weight) is most pronounced at 28°C. Excellent sporulation is observed at 25°C–30°C and at pH 5.5, 6.0 (Singh and Singh 2001). It grows faster on soymilk dextrose agar than potato dextrose agar as soymilk used with agar or used alone as a broth has been found to be the best option for replacing the expensive processed culture media (Xiang et al. 2014). RAPD profiles generated by the random primers exhibit a high degree of variability among different isolates of *C. truncatum*. Infecting soybean and genetic relationships and molecular characterization of *Colletotrichum* species causing soybean anthracnose has been studied using AFLP method (Sharma 2009, Ramos et al. 2013). There appears to be a compatibility through perithecial formation among *Colletotrichum* spp. from chilli and soybean (Guldekar and Potdukhe 2011).

Epidemiology and Disease Cycle

The pathogen is seed borne, although diseased plant debris in the soil may also harbor the pathogen as primary source of inoculum. Percentage of seed-borne infection by *C. truncatum*, however, varies in soybean germplasms. For example, in total, 43 germplasms have been reported to be completely free from seed-borne infection in Bangladesh (Hossain et al. 2001). The production of acervuli of the fungus is obvious on the affected tissues under humid conditions. Consequently, rainy or wet weather facilitates the dispersal of the pathogen to become wind-borne and secondary spread takes place through conidia formed in acrevuli on stems, branches, and pods. Anthracnose disease severity becomes maximum when the average temperatures remain around 28.4°C with average relative humidity 76% and average rainfall 92.5 mm (Singh and Singh 2001). In the lowland situation in the soybean fields under the Chhattisgarh conditions in India, the average percent anthracnose severity is reported to be 76.36%, which is quite higher than the upland situation due to much more favorable environmental temperature (26°C–31.5°C), relative humidity (80%–99%), soil moisture (92%–97%), and soil temperature (23.5°C–28.2°C) in the lowland situation (Shukla et al. 2014). Infection of pods can occur even when they are green. But it remains quiescent until the pods start maturing. *C. truncatum* can thus establish latent infection without showing any visible symptom in all seed components (Begum et al. 2008b). Thus, there appears to be a prevalence of latent infection of *C. truncatum* in soybean at R5.2 growth stage under Brazil conditions (Klingelfuss and Yorinori 2001). Consequently, the development of anthracnose after harvest on apparently healthy pods is from the incipient (latent) infection of the pods in the field.

Virulence of the fungus varies with isolates indicating the existence of distinct strains of the pathogen. Soybean cultivars under natural conditions have been found to be affected by four isolates of *C. truncatum* (*Glomerella tucumanensis*) (Ct 1, Ct 2, Ct 3, and Ct 4). The frequency of isolate Ct 3 is reported to be higher on infected leaves and pods of cv. VLS 1 (Akhtar and Khalid 2008).

Diseases Management

Host Plant Resistance

Seedling test is useful to study host reaction to the pathogen (Costa et al. 2009). The inoculations at stage V1/V2 show differences in the reactions of cultivars when compared to the inoculations made at stage V5/V6. The high resistance at V5/V6 suggests a mechanism of *resistance of adult plant*

(Costa et al. 2006). There is evidence that the activity of phenolics and oxidizing enzymes increases in resistant soybean cultivars (Chandrasekaran et al. 2000d). A number of soybean genotypes such as Ceresia, Essor, Labrador, and Quito (Zwatz et al. 2000); Klaitur, PKV-1, MAUS 13, and Birsa (Gawade et al. 2009a); and Birsasoya-1 and JS (SH) 98-22 (Mahesha et al. 2009) have been shown to possess resistance reaction against the anthracnose disease. Among the aforementioned, Kalitur genotype appears to possess stable degree of resistance to anthracnose (Kumar and Dubey 2007). The most resistant soybean cultivars to damping phase of the disease are as follows: MSOY 8001, Conquista, MSOY 8400, Engopa, and Vencedora (Galli et al. 2007). The resistance in cultivars P30-1-1, Lee, and Himso 333 is governed by single dominant genes that are nonallelic (Kaushal and Sood 2002).

Chemical Control

Fungicide treatments significantly reduce infection on the average by 20%–40%. Folicur (2.5 g tebuconazole/L), with 43% efficacy, can give the best protection against *C. dematium* (Zwatz et al. 2000). In years with particularly high disease pressure, yield increases of up to 20% have been achieved. Considering incremental cost–benefit ratio (ICBR), the most economical treatment that results in giving the highest CBR is the fungicide carbendazim (CBR, 1:14.45) followed by a combination of carbendazim + mancozeb (CBR, 1:8.92) (Jagtap et al. 2012a). Similar results with the use of carbendazim (0.1% spray) have been obtained in a separate study made by Gawade et al. (2009b). Thiophanate methyl should be used as the first choice fungicide to control soybean pod anthracnose, then pyrimethanil, tebuconazole, etc. Applying fungicide two times at the stage of flower beginning, flower flourish, or seed filling of soybean, the control efficiency of soybean pods appears to be significantly higher than applying fungicide one time at the same stage. Applying fungicide one time at the stage of flower beginning and flower flourish or seed filling of soybean, respectively, the control efficiency reaches more than 95% (Wang et al. 2012b). Maximum grain yield (2425 kg/ha) has been obtained with propiconazole 0.05% treatment, which appears to be higher by 28.85% over unsprayed plants, and the differences on seed weight plant-1, 100 seed weight in healthy and infected plants, seem to be significant (Guldekar and Potdukhe 2010). Treatments with carbendazim (double application) and azoxystrobin (single application) have shown the lowest values of premature defoliation due to multiple foliar diseases including anthracnose, and the maximum yield increase could be obtained with single application of azoxystrobin (39.3%) and double application of carbendazim (32.4%) (Ploper et al. 2001). The best management of *C. truncatum* was obtained when the seeds were treated with fludioxonil + mefenoxam and thiabendazole + thiram. These treatments also contributed to improve the physiological performance of the seeds (Pereira et al. 2009). The application of thiophanate methyl resulted in the lowest incidence of pod blight caused by *C. truncatum* (Chaudhary et al. 2005). Propiconazole, Vitavax-200 (carboxin + thiram), at 100–400 ppm (Shovan et al. 2008), and azoxystrobin (Quadris) at 6–9 ounce (Padgett et al. 2003) are also effective in the management of the disease.

Systemic Acquired Resistance

Systemic acquired resistance (SAR) in soybean has been first reported following infection with *C. truncatum* that causes anthracnose disease (Sandhu et al. 2009). Pathogenesis-related (PR) gene GmPR1 is induced following the treatment of soybean plants with the SAR inducer, 2,6-dichloroisonicotinic acid (INA). Soybean GmNPR1-1 and GmNPR1-2 genes show high identities to *Arabidopsis* NPR1. Therefore, SAR pathway in soybean is most likely regulated by GmNPR1 genes (Sandhu et al. 2009). Both total and ortho-dihydroxy phenolics contents are greater in the resistant cultivar JS 89-37 (Chandrasekaran et al. 2000d). Alum at 5% reduced pod blight infection up to 90%. Dipotassium hydrogen phosphate and potassium chloride registered 75% and 65% reduction in pod blight infection over the control at 0.1%, respectively (Chandrasekaran et al. 2000a).

Biological Control

T. harzianum shows significantly variable antagonism ranging from 50.93% to 89.44% reduction of the radial growth of *C. dematium*. Among the promising antagonists, the T3 isolate of *T. harzianum* showed the highest (89.44%) inhibition of *C. dematium* radial growth (Shovan et al. 2008). Introduction of bacterized seeds carrying bacterial isolates with proven growth-promotion capabilities and antagonistic characteristics offer a valid alternative to chemical protectants. FLP strains GRP3, PEn-4, PRS1, and WRS-24 when studied in relation to natural occurrence of anthracnose caused by *C. dematium* also result in significant control of the disease (Tripathi et al. 2006). Soybean seed treatment with the tested bioagents and Rizolex-T also reduce damping-off and increase the survival of plants under field conditions. *Trichoderma lignorum* and *Trichoderma viride* are reported to be more effective than other treatments as they give results as good as Rizolex-T. The aforementioned treatments also increase fresh weight, dry weight, number of pods/plant, number of seeds/plant, and weight of seeds/plant (Saber et al. 2003). *T. harzianum* 5 inhibits the growth of *C. truncatum* (Chandrasekaran et al. 2000b). Bio-priming with *P. aeruginosa* or *T. harzianum* (by increased colony-forming units from 1.2×10^9 to 5.1×10^9 seed-1 after 12 h of bio-priming.) offers an effective biological seed treatment system and an alternative to the fungicide Benlate for the control of damping-off of soybean caused by *C. truncatum* of soybean (Begum et al. 2010). Two fungal BCAs, namely, *Trichoderma virens* isolate UPM23 and *T. harzianum* isolate UPM40, and a bacterial BCA, namely, *P. aeruginosa* isolate UPM13B8, strongly inhibit the growth of *C. truncatum* (Begum et al. 2008), though *T. viride* and *T. harzianum* are reported to be equally effective in reducing the disease (Guldekar and Potdukhe 2010).

Bacillus subtilis strains AP-3 and PRBS-1 isolated from soil samples of Paraná State, Brazil, have been found to be effective in inhibiting soybean seed–pathogenic fungi including *C. truncatum*, indicating their potential usefulness in the biological control of seed-borne infection of *C. truncatum* as well as in promoting soybean growth as the metabolites of AP-3 increase production of root hairs, while the metabolites of PRBS-1 stimulate the outgrowth of lateral roots in soybean (Araujo et al. 2005).

Effect of Plant Extracts

Aqueous leaf extract of garlic, *tulsi* and onion, ginger, and neem leaf extracts at about 20% concentration appears to be the best in inhibiting the radial growth and mycelial dry weight of the pathogen (Shovan et al. 2008, Jagtap et al. 2012a); the 10% leaf extract of *Lawsonia inermis* also reduces anthracnose disease incidence significantly (Chandrasekaran et al. 2000a). Combined application of leaf extract of *L. inermis* (5%) with alum at 1% and 0.1% can give 100% reduction in pod blight infection (Chandrasekaran and Rajappan 2002). Seed treatment with alum (0.1%) + a foliar spray with *L. inermis* leaf extract (1%) + alum (0.1%) reduce leaf anthracnose and pod blight incidences by 7.0% and 4.2%, respectively, with a grain yield of 2191 kg/ha (Chandrasekaran et al. 2000b). Dry hot water extract of *Berberis aristata*, *Boenninghausenia albiflora*, and *Lantana camara* has been shown to be highly potent against *C. truncatum* (Arora and Kaushik 2003). Among the other botanicals tested, minimum percent disease intensity (15.34%) could be recorded in *Trachyspermum ammi* seed extract (5%). The antifungal activity of essential oils from *Hortelã do Campo* (*Hyptis marrubioides*), *alfazema-do-Brasil* (*Aloysia gratissima*), and *erva-baleeira* (*Cordia verbenacea*) has potential as alternatives to synthetic fungicides in the control of anthracnose in soybean seeds (da Silva et al. 2012a).

Overall integrated disease management using all possible options is the best strategy in management of the anthracnose disease of soybean (Chandrasekaran et al. 2000c).

ASIAN SOYBEAN RUST

Symptoms

Two *Phakopsora* species are known to cause soybean rust. The more aggressive species is *Phakopsora pachyrhizi*, known as the Asian (or Australasian) soybean rust. *Phakopsora meibomiae*, the less virulent species, has only been found in areas in the Western hemisphere, and it is not known to cause severe yield losses in soybean. The focus of the disease description in this section is therefore on Asian soybean rust (ASR) caused by *P. pachyrhizi*.

Early symptoms of ASR on the upper side of leaves consist of yellow spots that turn brown then become necrotic, surrounded by wide yellow areas, and chlorosis and brown flecking appear on the lower leaves in the canopy. The key diagnostic features of soybean rust are the cone-shaped angular lesions limited by leaf veins (Figure 9.4). Often, the first lesions appear toward the base of the leaflet near the petiole and leaf veins. This part of the leaflet probably retains dew longer, making conditions more favorable for infection. Lesions enlarge and, 5–8 days after initial infection, rust pustules (uredia, syn. uredinia) become visible. Uredia develop more frequently in lesions on the lower surface of the leaf than on the upper surface (Figure 9.4). Lesions are scattered within yellow areas that appear see-through (translucent) if the affected leaves are held up to the sun. Even though the lesions are small, each lesion often has several pustules. The anamorphic sori (uredo pustules)

Rust (*Phakopsora pachyrhizi*)

(a)

(b)

FIGURE 9.4 Asian soybean rust. Note the angular-shaped lesion-like uredo pustules on the undersurface of soybean leaf. (a) Symptoms of rust on leaf. (b) Urediniosori on lower surface of leaf. (Courtesy of Dr. G.K. Gupta, ICAR-Directorate of Soybean Research, Indore, India.)

are amphigenous, circular, minute (about the size of a pinhead), pulverulent, whitish becoming pale cinnamon brown, scattered or in groups on discolored spots, subepidermal becoming erumpent, and cone like, which can be confused with another disease, bacterial pustule. The uredia open with a round ostiole through which uredospores are released (Goellner et al. 2010). Besides leaves, uredo pustules can also appear on petioles, stems, and even cotyledons, but most rust pustules occur on leaves. Tan lesions on lower leaf surfaces contain small pustules surrounded by a small zone of slightly discolored necrotic tissue. The color of the lesion is dependent on lesion age and interaction with the host genotype. RB lesions with little sporulation indicate a semicompatible reaction, whereas tan lesions with much sporulation indicate a fully compatible reaction. It is used to compare virulence of *P. pachyrhizi* isolates from Asia (Bonde et al. 2006).

The symptoms may be observed at any time during the crop cycle but are more evident at or after flowering. The symptoms progress from lower to upper leaves. The symptoms develop further up the plant until all leaves were infected. As uredo pustules age, they may turn black because of the formation of a layer of teliospores in the pustules, turning pustules from uredinia into telia. Premature defoliation occur in infected plants.

Geographical Distribution and Losses

P. pachyrhizi is widespread in Asia and Oceania (but not in New Zealand). In the neotropics, another soybean rust fungus, *P. meibomiae*, occurs, which was once treated as synonymous with *P. pachyrhizi* but has now been taxonomically segregated (Ono et al. 1992).

The fungus that causes ASR, *P. pachyrhizi*, originally described in Japan in 1902, spread rapidly throughout Southeast Asia in the 1960s. It made a surprise appearance in Hawaii in 1994. It is thus apparent that ASR originated in tropical and subtropical regions of Asia and most likely spread to several African countries via wind currents. It appeared in Uganda in Africa in 1997, and in 2001, it was discovered in South America and moved north above the equator in 2004. Once it moved north of the equator, it moved to North America on wind currents. Now, soybean rust occurs in many countries throughout Africa including South Africa (Jarvie 2009), Asia including India (Hegde et al. 2002, Ramteke et al. 2003), and South and North America covering the United States (Goellner et al. 2010), Canada, and Mexico (Yanez Morales et al. 2009). After reports of its first occurrence in Brazil in 2001 and the continental United States in 2004, research on the disease and its pathogen has greatly increased (Vittal et al. 2012a). This disease destroys photosynthetic tissue and causes premature defoliation and, if untreated prior to the R6 growth stage, can result in severe yield reductions.

Yield losses as high as 20%–80% have been reported, but the amount of loss depends on when the disease begins and how rapidly it progresses. Yield loss equations for the ASR pathosystem using disease intensity at different phenological stages of the crop by manipulating sowing dates have been developed. The variables area under disease progress curve (AUDPC) present high correlation with yield, and variations in the severity of disease between crops affected the relationship AUDPC × yield (Hikishima et al. 2010). In April 2009, a severe rust outbreak in soybean developed at phenological stage R3 of the plants, leading to the complete defoliation (Pérez-Vicente et al. 2010). Yield losses due to the disease have been recorded to be in the range of 10%–90%, depending on the varieties used and local agroclimatic conditions (Sumartini 2010, Hegde and Mesta 2012). Predicting the time of rust appearance in a field is critical to determining the destructive potential of rusts. Mean rust-induced yield reductions have been estimated to be 67% when infection starts at R2 (full bloom) and 37% when infection takes place at R5 (beginning seed) growth stage (Kumudini et al. 2008). Yield loss increases during later sowing periods due to greater inoculum pressure hindering disease management and decreasing grain yield (Cruz et al. 2012, Akamatsu et al. 2013).

Soybean rust has been one of the most important problems in the agribusiness of the most important soybean-producing countries in South America, mainly in Brazil. Since its first detection in Paraguay and in the state of Paraná, Brazil, in 2001, the Asian rust has spread to all parts of Paraguay,

Bolivia, most of Brazil, and parts of Argentina. In the following years (2002 and 2003), it caused an estimated 4.011 million tons grain losses or the equivalent to U.S. $884.425 million (Yorinori 2004, Yorinori and Lazzarotto 2004). It is also reported to be of severe occurrence in Taiwan and Vietnam depending on the conducive environmental conditions for disease development (Tran et al. 2013). When untreated, soybean rust causes yield losses due to premature defoliation, fewer seeds per pod, and decreased number of filled pods per plant. For example, it reduces the weight of grains per plant in susceptible soybean TGx 1950-8F by 94.6% in Nigeria (Ittah et al. 2011).

The disease was reported at epidemic levels in Argentina (Pioli et al. 2005) and in Brazil in 2003/2004 (Nascimento et al. 2012, Roese et al. 2012) and in 2009–2010 (Garces Fiallos and Forcelin 2011). The main phytosanitary problem related to soybean in Brazil is ASR (do Nascimento et al. 2012).

The potential geographical distribution range of soybean rust may include most U.S. soybean production regions and that yield losses would be light in the north but moderate in the south if environmental conditions are conducive (Li and Yang 2009). ASR continues to spread across the southeast and midsouth regions of the United States (Luster et al. 2012). Immunodiagnostic assays using monoclonal antibodies have been developed to detect rust-infected soybeans and ASR spores from sentinel surveillance plots (Luster et al. 2012).

Pathogen

Soybean rust is caused by two species, *P. pachyrhizi* H. Sydow & P. Sydow and, less commonly, *P. meibomiae* (Arthur) Arthur. The latter species (*P. meibomiae*), commonly known as the cause of Latin American rust or legume rust, is found in the Western hemisphere and is not known to cause severe yield losses. In this section, most part of the subject matter is dealt with ASR caused by *P. pachyrhizi*.

Classification

Domain: Eukaryota
Kingdom: Fungi
Phylum: Basidiomycota
Subphylum: Pucciniomycotina
Class: Pucciniomycetes
Order: Pucciniales
Family: Phakopsoraceae
Genus: *Phakopsora*
Species: *pachyrhizi*

P. pachyrhizi is believed to have a microcyclic heteroecious life cycle, producing only uredinia and telia. Stage 0 = pycnial (spermitia) stage and Stage I = aecial stage (aecial spores) have not been found; Stage II = uredinial stage (uredinial spores) is quite common; Stage III = teleuto stage (teleutospores) can be observed but not common; and Stage IV = basidial sage (basidia or sporidia) is not identified.

The uredinia are pustular and open with a round ostiole through which uredospores are released (Goellner et al. 2010). Each pustule contains hundreds of spores. The urediniospores are almost sessile, obovoid to broadly ellipsoid, and 18–34 × 15–24 μm. The spore wall is uniformly ca 1 μm thick, minutely and densely echinulate, and colorless to pale yellowish brown. This coloration is different from many other rust pathogens whose spores are often reddish brown (rust colored). Four to eight (mostly 6, rarely 2–10) germ pores are equatorial or scattered on the equatorial zone or occasionally scattered on and above the equatorial zone of the spore wall. Germination of *P. pachyrhizi* urediniospores occurs through an equatorial (central) pore, producing a germ tube that ends in an appressorium, which the fungus uses to penetrate the host directly or through a stoma.

The pathogen is known almost exclusively by its uredinial stage, and there are only a few records, mainly from Argentina of the occurrence of telial stage. This is thought to be the first report of epidemiological and morphological characterization of ASR in Argentina and the first report of the telial stage of *P. pachyrhizi* on soybean in South America (Carmona et al. 2005). In case the telia are formed, telia are found on infected leaves intermixed with uredinia in old lesions. Teliospores measure 9 × 23.8 μm on average. Telia are hypophyllus, pulvinate and crustose, chestnut brown to chocolate brown, subepidermal in origin, and 2- to 7-spore layered. The teliospores are one celled, irregularly arranged, angularly subglobose, oblong to ellipsoid, and 15–26 × 6–12 μm. The wall is uniformly ca 1 μm thick, sometimes slightly thickened (up to 3 μm) apically in the uppermost spores, and colorless to pale yellowish brown and these have never been shown to germinate (Ono et al. 1992). In fact, the causal agents of soybean rust are two closely related fungi, *P. pachyrhizi* and *P. meibomiae*, which are differentiated based upon morphological characteristics of the telia. *P. pachyrhizi* originated in Asia–Australia, whereas the less aggressive *P. meibomiae* originated in Latin America (Goellner et al. 2010). Twenty-four simple sequence repeat (SSR) markers have been developed for *P. pachyrhizi* (Anderson et al. 2008). The molecular characterization of the pathogen is possible by PCR. Determination of the nucleotide sequence of the internal transcribed spacer (ITS) region reveals greater than 99% nucleotide sequence similarity among isolates of either *P. pachyrhizi* or *P. meibomiae*, but there is only 80% sequence similarity between the two species. Utilizing differences within the ITS region, four sets of polymerase chain reaction (PCR) primers have been designed specifically for *P. pachyrhizi* (Frederick et al. 2002).

Since sporidia, spermogonia, and aecia are not yet known and also if any alternate host is involved, the role of teleuto spores in the life cycle of the pathogen is not completely understood. It seems that urediniospores are the main, if not the only, means of dissemination and spread of the disease.

Considering the lack of a known sexual stage of *P. pachyrhizi*, hyphal anastomosis followed by the parasexual cycle may explain the genetic diversity in virulence among populations of *P. pachyrhizi* (Vittal et al. 2012b). This study establishes a baseline of pathogenic variation of *P. pachyrhizi* in the United States that can be further compared with variation reported in other regions of the world and in future studies that monitor *P. pachyrhizi* virulence with regard to deployment of rust resistance genes (Twizeyimana and Hartman 2012). Detailed information on the taxonomy and molecular biology of the pathogen has been reviewed (Goellner et al. 2010).

Epidemiology and Disease Cycle

Soybean rust pathogen is known to naturally infect 95 species from 42 genera of legumes, inclusive of important weed species like Kudzu vine (*Pueraria lobata*) and major crop species such as common bean (*Phaseolus vulgaris*). Such a broad host range is unusual among rust pathogens that normally have a narrow host range. The significance of the numerous alternative host possibilities for the soybean rust pathogen is that these may serve as an inoculum reservoir or a *green bridge* from one soybean planting season to the next (Jarvie 2009, Goellner et al. 2010). Alternative hosts are not to be confused with alternate host, which is a plant other than the principal host that is needed for a pathogen to complete its life cycle. In frost-free areas, such as South America, Central America, the Caribbean basin, southern Texas, and Florida, the inoculum source could be nearby on volunteer soybean plants, kudzu, or some other alternative host. In areas that experience frost, such as the Midwestern United States, inoculum must be blown in from overwintering sources that may be hundreds of miles away. The climatic and environmental factors are important in determining the development of ASR (Young et al. 2011). Temperature highs common to southeastern states are a factor in the delay or absence of soybean rust in much of the United States. For example, the highest numbers of urediniospores are produced when day temperatures peaked at 21°C or 25°C and night temperatures dipped to 8°C or 12°C (Bonde et al. 2012).

In warm regions, the host species particularly Kudzu vine (*P. lobata*) may harbor the fungus throughout the year or during seasons in which soybeans are not cultivated and may serve as the primary infection source. Kudzu (*Pueraria* spp.) is thus an accessory ASR (caused by *P. pachyrhizi*) that is widespread throughout the southeastern United States (Jordan et al. 2010).

In colder regions where aboveground parts of annual hosts die during winter, no source of new infections in the soybean-growing season has been identified. Soybean rust is sensitive to freezing temperatures and it will not survive anywhere that has adequate cold temperatures to kill off all vegetation. However, low temperature does not seem to be a limiting factor for the survival of *P. pachyrhizi* and that urediniospores could survive on volunteer plants until new soybean plants grow (Formento and de Souza 2006). ASR can only survive for extended periods of time on live host tissue. Therefore, it cannot overwinter anywhere above the freeze line (approximately Tampa Bay, Florida) since its primary hosts, kudzu and soybean, will be dead and defoliated. As such, it will have to blow into such areas each year to cause disease on soybean crops. Each year, inoculum (rust spores) must blow in from infected areas such as South Florida, Mexico, or South America to start the disease over again in the Southeastern United States. Over long distances, *P. pachyrhizi* is mainly spread by wind-borne spores (e.g., in the United States, it is considered that Hurricane Ivan transported it from South America to southern United States). Infections and sporulation by ASR are favored by cooler, wet weather. Hot dry weather will stop the spread of the fungus. *P. pachyrhizi* is unusual in that it penetrates from urediniospores directly through the leaf cuticle without entering stomata. This unusual mode of penetration suggests that disease resistance mechanisms might exist for soybean rust that does not exist for most rust diseases. *P. pachyrhizi* utilizes primarily mechanical force, perhaps with the aid of digestive enzymes, to penetrate the cuticle on the leaf surface. However, the lack of deformation lines in micrographs indicate that digestive enzymes, without mechanical force, are used by the penetration hypha to penetrate the outer and inner epidermal cell walls (Edwards and Bonde 2011). The germination of the uredospores on the soybean leaves occur after 2 h of wetness, with a maximum germination appearing after 4 h of wetness. Wetness interruption affects mainly the spores that initiate the germination (Furtado et al. 2011). Successful infection further is dependent on the availability of moisture on plant surfaces. At least 6 h of free moisture is needed for infection with maximum infections occurring with 10–12 h of free moisture. The development of the disease needs high humidity (>95%) and optimal temperature for infection process, that is, 15°C–28°C. This temperature range commonly occurred in the dry season; therefore, rust disease often attacked soybean in the dry season (Nunkumar et al. 2009, Sumartini 2010, Alves et al. 2011, Mesquini et al. 2011). Urediniospores of the pathogen remain viable during the 11 weeks of storage; the germination of the urediniospores and the severity of ASR are reduced after 3 weeks of storage, and the urediniospores stored at 20°C (plus or minus 2 degrees) for up to 11 weeks are able to cause disease in soybean plants (Beledelli et al. 2012).

This study indicates that extended periods of leaf wetness (18 h) increase disease severity and the rate of spread of the disease in the upper canopy. These results, in combination with spore monitoring, may be used to refine models of pathogen reproduction, prediction, and risk in certain regions (Narvaez et al. 2010).

Spore germination in the dark (40.7%) is found to be statistically different from spore germination in the light (28.5%). The same effect can be observed with appressorium formation, in the dark (24.7%) and in the light (12.8%) (Furtado et al. 2009). The dark incubation period of 8–16 h and light intensity of 600–400 lux (lx) are favorable for the infection of soybean rust urediospores. The infection of soybean rust can be reduced gradually with extended or shortened dark incubation period. The infection rate of the urediospores also decreases gradually with the light intensity increasing or decreasing. Higher light intensity (>3000 lx) or lower light intensity (<200 lx) is disadvantageous for the infection of the urediospores and it is advantageous for the infection of the urediospores with the light intensity changing to the favorable light intensity from the higher or lower light intensity (Mo et al. 2008).

Rain events are the most dominant cause of wetness in the lower canopy. It is revealed that a majority of the wet deposited *P. pachyrhizi* urediniospores would be removed from soybean leaf surfaces by subsequent rainfall, but sufficient percentages of spores (10%–25%) will likely remain on the leaf tissue long enough to germinate and infect during heavy summer rains lasting ≥30 min (Dufault et al. 2010b).

The uredinial stage is the repeating stage. This means that urediniospores can infect the same host on which they are produced (soybean) during the same season. The quantity of urediniospores over the crop fields is positively correlated to the disease severity and incidence as well as to cumulative rainfall and favorable days for *P. pachyrhizi* infection (Nascimento et al. 2012). Epidemics can develop quickly from only a few pustules because spore-producing pustules will develop within a week to 10 days after infection, and hundreds of spores are produced after about 3 weeks. For rust to be damaging, first infections will probably have to occur before the R3 stage of soybean development. Among other environmental factors, sunlight intensity negatively affects *P. pachyrhizi* biology with possible effects on disease epidemiology. Field observations suggest that higher disease severity occurs in shaded environments, such as on soybean leaves in the lower canopy and kudzu leaves under trees, compared with open ground. Soybean rust is more severe in the lower canopy and shaded (20% sunlight) areas, shade duration being at least 2 days (Dias et al. 2011).

Though row spacing and rainfall intensity do not show significant effect on the vertical distribution of uredinia throughout the soybean canopy, approximately half of the urediniospores can be retained within the upper portion of the soybean canopy, and the other half are distributed between the mid- and low-canopy sections (Dufault et al. 2010a).

On average, severe ASR epidemics develop when 18 cloudy days are observed after disease onset, and mild epidemics occur when only 8 cloudy days are observed. In four growing seasons in Brazil and two in the United States, the progress of ASR epidemics does not follow a wavelike pattern, and it results in an exponential distribution of distances to disease locations over time with variable monthly expansion rates. The disease front reach 500 km distance from major inoculum sources after 3 months similarly in both countries. Greater solar radiation intensity is associated with delays in epidemic onset and this knowledge may be useful to improve risk assessments for seasonal ASR epidemics. Variability in disease development across canopy heights in early-planted soybean may be attributed to the effects of solar radiation not only on urediniospore viability but also on plant height, leaf area index, and epicuticular wax, which influence disease development of SBR. These results provide an understanding of the effect solar radiation has on the progression of SBR within the soybean canopy (Young et al. 2012).

The studies made by Ponte et al. (2006) highlight the importance of rainfall in influencing soybean rust epidemics in Brazil, as well as its potential use to provide quantitative risk assessments and seasonal forecasts for soybean rust, especially for regions where temperature is not a limiting factor for disease development. Temperature variables show lower correlation with disease severity compared with rainfall and has minimal predictive value for final disease severity.

Factors that both increase and decrease the risk for ASR epidemics could be prevalent in the United States (Pivonia et al. 2005, Smith 2005), Brazil (Yorinori et al. 2005), Paraguay (Yorinori et al. 2005), and South Africa (Levy 2005). In the United States, soybean rust disease predictions are made on a daily basis for up to 7 days in advance using forecast data from the United States National Weather Service (Tao et al. 2009). Using microsatellite markers, genetic variability in *P. pachyrhizi* spore populations indicates that vertical genetic resistance, provided by single genes, is a risky strategy for soybean breeding programs that aim resistance to ASR (Tschurtschenthaler et al. 2012).

P. pachyrhizi, the soybean rust pathogen, overwinters on kudzu in the southern United States. However, even with severely affected kudzu adjacent to soybean fields, disease symptoms do not occur on soybeans until plants are in midreproductive stages of growth during mid- to late summer. These observations suggest that soybeans are exposed to airborne inoculum of the pathogen long before symptoms occur, and it is hypothesized that these plants may be latently infected (Ward et al. 2012). Soybeans can become infected by the rust pathogen during early stages of plant growth, but symptoms often develop during the midreproductive stages. This extended latent infection period may be an optimum time for fungicide applications.

Surveys of virulence of pathogen population have been carried out in Asia, South America, and the United States for many years, and these studies have identified a wide range of races of *P. pachyrhizi*, by their interaction reaction on a set of differential lines of soybean having five specific genes Rppl, Rpp2, Rpp3, Rpp4, and Rpp5 for rust resistance and two universal susceptible cultivars (Yamaoka et al. 2002, 2011, Pham et al. 2009). ASR resistance genes (Rppl, Rpp2, Rpp3, Rpp4, Rpp5, and Rpp5) are genotyped with five single nucleotide polymorphism (SNP) markers (Monteros et al. 2010). Based on this study, a total of 16 soybean genotypes including cultivars and lines have been selected as a differential set to test the virulence of soybean rust populations from three South American countries, Argentina, Brazil, and Paraguay. Nine differentials are reported to carry resistance to *P. pachyrhizi* (Rpp) genes (Table 9.1).

Of the known Rppl-4 sources of resistance, plant introduction (PI) 459025B (Rpp4) produces RB lesions in response to all of the *P. pachyrhizi* isolates, while PI 230970 (Rpp2) produces RB lesions to all isolates except one from Taiwan, in response to which it produces a susceptible tan (TAN) lesion. PI 200492 (Rppl) and PI 462312 (Rpp3) produce TAN lesions in response to most *P. pachyrhizi* isolates (Pham et al. 2009).

This work will be useful in breeding and management of soybean rust by facilitating the identification of resistant genotypes and targeting cultivars with specific resistance to match prevailing *P. pachyrhizi* pathotypes in a given geographical zone (Twizeyimana et al. 2009).

The regional dynamics of soybean rust, caused by *P. pachyrhizi*, in six southeastern states (Florida, Georgia, Alabama, South Carolina, North Carolina, and Virginia) in 2005 and 2006 could be analyzed based on disease records collected as part of U.S. Department of Agriculture's soybean rust surveillance and monitoring program. Regional spread of soybean rust may be limited by the slow disease progress on kudzu during the first half of the year combined with the short period available for disease establishment on soybean during the vulnerable phase of host reproductive development, although low inoculum availability in 2005 and dry conditions in 2006 also may have reduced epidemic potential (Christiano and Scherm 2007).

TABLE 9.1
Asian Soybean Rust Differential Hosts

Differential		Resistance Genera	Origin
1.	PI 200492	Rpp1	Japan
2.	PI 368039	Rpp1	Taiwan
3.	PI230970	Rpp2	Japan
4.	PI417125	Rpp2	Japan
5.	PI462312	Rpp3	India
6.	PI459025	Rpp4	China
7.	Shiranui	Rpp5	Japan
8.	PI416764	ND	Japan
9.	PI587855	ND	China
10.	PI587880A	Rpp1	China
11.	PI587886	Rpp1	China
12.	PI587905	ND	China
13.	PI594767A	ND	China
14.	BRS 154	ND	Brazil
15.	TKS	ND	Taiwan
16.	Wayne	ND	United States

[a] Rpp1–Rpp5 have been mapped to different loci; ND, not determined.

Disease Management

Host Plant Resistance

High levels of ASR resistance are usually associated with one or a few dominant genes. Six dominant resistance genes (Rpp) as Rpp1 (in genotypes PI 200692, PI 200492, PI 3680390), Rpp2 (in genotypes PI 230970, PI 417125), Rpp3 (in genotype PI 462312 Ankur), Rpp4 (in genotype PI 459025 B), Rpp5 (in genotype Shiranui), and Rpp6 (in genotype PI 567102B) have been identified as capable of conferring ASR resistance in soybean and these have been mapped to different loci (Ivancovich 2008, Meyer et al. 2009, Schneider et al. 2011, Maphosa et al. 2012a, Morales et al. 2012). For example, genotypes PI 200492, PI 561356, PI 587886, and PI 587880A have been analyzed to identify SNP haplotypes within the region on soybean chromosome 18 where the single dominant ASR resistance gene Rpp1 maps, whereas ASR resistance in PI 594538A is governed by Rpp1-b gene (Monteros et al. 2010, Kim et al. 2012). Dominant alleles at three loci conditioning resistance to soybean rust races have been found in Nigeria and the symbols for the three loci controlling resistance to rust in soybean are designated as Rsbr1, Rsbr2, and Rsbr3 (resistance to soybean rust) (Iwo et al. 2012). Differential proteomic analysis of proteins involved in resistance to ASR has been done for understanding the host responses at the molecular level for effective control of the disease (Wang et al. 2012a).

While these dominant genes confer high levels of resistance and are relatively easy to incorporate into new soybean cultivars, they are not effective against all races of *P. pachyrhizi*. Deployment of varieties with new resistance genes is usually followed in a few years by the emergence of races of *P. pachyrhizi* that are virulent on them. This high degree of variability in the soybean rust pathogen is common in many rusts and requires the frequent discovery and incorporation of new sources of resistance. Currently, isolates of *P. pachyrhizi* exist that are virulent on each of the six known genes for resistance.

To select germplasm with levels of resistance to soybean rust, a differentiation must be made between the kinds of lesions on the leaves that are classified into three basic types: the resistant genotypes show RB nonsporulating pustules, whereas the moderately resistant genotypes show rectangular, RB sporulating pustules. The susceptible genotypes exhibit TAN-type gray sporulating pustules of light to medium density on all leaves, and premature defoliation is much common in these genotypes (Paul et al. 2011). The utility of detached-leaf assay for screening large number of genotypes soybean for rust resistance has been demonstrated (Twizeyimana et al. 2007) and that a determination of numbers and sizes of uredinia will detect both major gene and partial resistance to soybean rust (Bonde et al. 2006). Cell wall lignifications are markedly higher in inoculated resistant lines compared with inoculated susceptible lines, indicating a possible protective role of lignin in rust infection development in resistant soybean lines (Lygin et al. 2009, Schneider et al. 2011). Since the resistant genotype forms significantly lower lesion area, the reduced disease severity and the lack of sporulation in the resistant genotype will likely minimize the impact of the disease on canopy photosynthesis and yield (Kumudini et al. 2010).

The Asian Vegetable Research and Development Center (AVRDC) and several national agricultural research and educational institutions in Taiwan have conducted research to incorporate rust (*P. pachyrhizi*) resistance in soybean. Most races of rust in Taiwan produce TAN-type, profusely sporulating lesions and the predominant rust races are complex. Screening of germplasms initially resulted in three resistant lines, namely, PI 200492, PI 200490, and PI 200451. Consequently, PI 200492 has been used to develop three improved rust-resistant cultivars, namely, Tainung 3, Tainung 4, and Kaohsiung 3 in 1967, 1970, and 1971, respectively. Further screening of germplasms by AVRDC showed G 8586 (PI 230970), G 8587 (PI 230971), PI 459024, PI 459025, and *G. soja* (PI 339871) to be resistant. In subsequent years, all the aforementioned germplasms have been observed to be susceptible to rust, apparently due to new races of the pathogen. Recognizing the ineffectiveness of single gene resistance, AVRDC conducted research on rate-reducing and partial resistance and tolerance. A combination of genotypes combining the aforementioned three strategies can withstand rust and give higher yield. The resistant and tolerant materials appear promising in India, Africa, and Latin America (Table 9.2). AVRDC's germplasms are available to any scientist who needs soybean rust-resistant/tolerant materials (Shanmugasundaram et al. 2004).

Soybean lines having resistant reactions to U.S., Brazil, and Paraguay isolates may be important sources for developing elite cultivars with broad resistance to ASR (Li 2009). Soybean major and minor rust-resistant genes showing predominantly additive effects are dispersed among parents and it is possible to select inbred lines superior to the best yielding parent from most crosses (Ribeiro et al. 2007).

The threat posed by soybean rust on soybean production is worsened by resistance breakdown associated with single gene resistance present in most cultivars. The marker gene pyramiding involving gene combination for three independent soybean rust resistance genes, Rpp2, Rpp3, and Rpp4 is feasible and can substantially increase resistance to soybean rust through reduced severity and reduced sporulating lesions (Maphosa et al. 2012a). Soybean genotype UG 5 as parental line has been proved to be the most outstanding one producing the greatest number of resistant populations underscoring the importance of additive gene effects in the control of soybean rust severity and sporulation rate (Maphosa et al. 2012b).

There are differences in virulence among Asian and Brazilian and the Japanese rust populations and should be considered in order to select and use resistant resources. The number of resistant varieties or resistance genes useful in these countries appear limited. Therefore, a resistant cultivar that is universally effective against soybean rust should be developed by pyramiding some major resistance genes and by introducing horizontal resistance (Yamanaka et al. 2010).

An-76, a line carrying two resistance genes (Rpp2 and Rpp4), and *Kinoshita*, a cultivar carrying Rpp5, could contribute differently to resistance to soybean rust and that genetic background plays an important role in Rpp2 activity. All three loci together work additively to increase resistance when they were pyramided in a single genotype indicating that the pyramiding strategy is one good breeding strategy to increase soybean rust resistance (Lemos et al. 2011, Kendrick et al. 2011).

The soybean cultivar Ankur (accession PI462312), which carries the Rpp3 resistance gene, when interacts with avirulent isolate Hawaii 94-1 of *P. pachyrhizi*, elicits hypersensitive cell death that limits the fungal growth on Ankur and results in an incompatible response (Schneider et al. 2011). Some soybean mutant lines obtained through seed irradiation using gamma rays (10–30 kR) and ethyl methanesulfonate (0.4%–0.8%) have been found to show improved rust resistance in India (Basavaraja et al. 2004).

Quantitative PCR (QPCR) assay of fungal DNA (FDNA) screening technique demonstrates its use to distinguish different types of resistance and could be used to facilitate the evaluation of soybean breeding populations, where precise quantification of incomplete and/or partial resistance is needed to identify and map quantitative trait loci (QTL) (Paul et al. 2011).

Molecular Breeding for Rust Resistance

A biotechnological approach may help to broaden resistance of soybean to this fungus. Molecular breeding is considered as a feasible method to improve soybean rust resistance and minimize the adverse effects from overuse fungicides. QPCR assay of FDNA screening technique demonstrates its use to distinguish different types of resistance and could be used to facilitate the evaluation of soybean breeding populations, where precise quantification of incomplete and/or partial resistance is needed to identify and map QTL (Paul et al. 2011).

Molecular markers in a backcross breeding program to introgress the Rpp5 gene of ASR resistance into HL203, an elite Vietnamese soybean variety, have been used (Tran et al. 2013), and a new chitinase-like xylanase inhibitor protein (XIP) from coffee (*Coffea arabica*) (CaclXIP) leaves has been cloned; CaclXIP belongs to a class of naturally inactive chitinases that have evolved to act in plant cell defense as xylanase inhibitors. Its role on inhibiting the germination of fungal spores makes it an eligible candidate gene for the control of Asian rust (Vasconcelos et al. 2011).

Two peptides, Sp2 and Sp39, have been identified that inhibit urediniospore germ tube development when displayed as fusions with the coat protein of M13 phage or as fusions with maize cytokinin oxidase/dehydrogenase (ZmCKX1); when peptides Sp2 and Sp39 in either format are mixed with urediniospores and inoculated to soybean leaves with an 8 h wetness period, rust lesion development is reduced. Peptides Sp2 and Sp39, displayed on ZmCKX1, are found to interact with a 20 kDa protein derived from germinated urediniospores incorporating peptides that inhibit pathogen development and pathogenesis. Such molecular breeding programs may contribute to the development of soybean cultivars with improved, durable rust tolerance (Fang et al. 2010).

Chemical Control

Factors such as recent weather conditions, proximity to sources of ASR, cost of available products, and an estimate of a crops yield potential should be considered when choosing a fungicide program. An immunofluorescence technique in combination with propidium iodide (PI) staining–counterstaining has been developed to specifically detect viable *P. pachyrhizi* urediniospores. The method is rapid and reliable, with a potential for application in forecasting soybean rust based on the detection of viable urediniospores (Vittal et al. 2012b). This system of detection has been touted for use as a potential warning system to recommend early applications of fungicides.

The fungicides used to control ASR include the following: triazoles (cyproconazole, difenoconazole, epoxiconazole, tebuconazole) and strobilurins (azoxystrobin, pyraclostrobin, and trifloxystrobin). The treatments with these fungicides can control the disease showing severity average lower than 2, without difference among them (Soares et al. 2004, Gasparetto et al. 2011, Araujo et al. 2012, Debortoli et al. 2012, Doreto et al. 2012, Pogetto et al. 2012). Applications of triazole and triazole + strobilurin fungicides result in lower rust severity and higher yields compared with other fungicides. The strobilurin fungicides provide the highest yields in many locations; however, severity tends to be higher than that of the triazole fungicide. These fungicides are among the most effective for managing soybean rust and maintaining yield over most locations (Miles et al. 2007, Rezende and Juliatti 2010). The combination of azoxystrobin + cyproconazole or picoxystrobin + cyproconazole is reported to be the most efficient treatment when plants are foliar sprayed with the fungicide mixture at GS R3 and/or GS R5 resulting in the lowest AUDPC values and highest yields with a few exceptions (Mueller et al. 2009, Scherm et al. 2009).

Crop yield increase up to 26.9%, 33.3%, and 38.9% with the application of mancozeb, triforine, and tebuconazole, respectively, under the weekly, 2-weekly, and 3-weekly spray schedules, has been obtained with the highest economic return for mancozeb (Kawuki et al. 2002). Two sprays of triadimefon (Bayleton at 0.1%) are also very effective, as these can completely control the rust infection and increase the yield (32%) over the control. Tilt [propiconazole] when sprayed significantly results in rust control with increase in yield (Khot et al. 2007).

Substances added to the suspension or solutions of fungicides, such as adjuvants (NimbusReg), can influence the fungicide efficacy (Nascimento et al. 2012). The leaf area indices of soybean cultivars influence fungicide drop deposition and fungicide penetration into canopy resulting in the efficiency fungicide application for rust disease control, the fungicide applications being accomplished most successfully in R1 and R4 growth stages (Tormen et al. 2012).

A premix of 60 g azoxystrobin/ha + 24 g cyproconazole/ha when applied at R2 and R5 has been found to be the most efficient treatment in reducing rust severity and AUDPC and increasing yield by 50% (Godoy et al. 2009).

The demethylation inhibitor fungicide myclobutanil can be an effective component of spray programs designed to control the ASR. High degree of xylem systemicity is displayed by myclobutanil in soybean foliage and is a contributory factor toward its commercial effectiveness for the control of ASR (Kemmitt et al. 2008). Fungicides containing protective and curative properties like Silvacur Combi 30 SC and other triazole classes of fungicides could be applied at the first detection of soybean rust symptoms on lower trifoliate leaves. The disease seems to affect soybean after flowering, about 55 days after planting, and mainly during January to March when the weather is cool and moist (Sinha and Reyes 2009).

TABLE 9.2
Soybean Genotypes Resistant (R) or Moderately Resistant (MR) to Asian Soybean Rust (ASBR) as Reported from Different Countries in the World

Genotype	Country	R/MR/Genes	Reference(s)
DS 228 and DS 227	India	R	Khot et al. (2010)
PI 567099A	Paraguay	R (recessive at the Rpp3 locus)	Ray et al. (2011)
Hyuuga, PI 462312 (Ankur) (Rpp3) and PI 506764 (Hyuuga), PI 417089B (Kuro daizu	United States	R (Rpp3), (Rpp5)	Kendrick et al. (2011)
PI 567104B	United States	R (Rpp genes)	Walker et al. (2011)
PI 230970 (Rpp2), PI 459025B (Rpp4), PI 594538A (Rpp1b), PI 561356	United States	R/MR (Rpp2, Rpp4, Rpp1b)	Miles et al. (2011)
PI 594760B, TMG06_0011, TMG06_0012	United States	R	Garcia et al. (2011)
USP 97-08135	Brazil	MR	Araujo and Vello (2010)
PI 398998, PI 437323, and PI 549017, PI 230970 (Rpp2)	Vietnam	R	Pham et al. (2010)
(PI459025B)	United States	R (Rpp4C4)	Meyer et al. (2009)
BR01-18437 inbred line	Brazil	R (as parental line)	Ribeiro et al. (2009)
EC241778 and EC241780	India	R	Ammajamma and Patil (2009)
MNG 10.3 MNG 3.26	Uganda	R	Oloka et al. (2009)
Williams 82 (Rpp1)	United States	R	Paul and Hartman (2009)
GC00138-29, the cross GC00138-29 × Wondersoya	Uganda	R	Kiryowa et al. (2008)
PI 587886 and PI 587880A	United States	R	Ray et al. (2009)
PI 567102B, PI 200492 (Rpp1), PI 230970 (Rpp2), PI 462312 (Rpp3), and PI 459025B (Rpp4)	United States	R	Li (2009)
PI594538A, PI 200492	United States	R	Chakraborty et al. (2009)
EC 241778, EC 241780	India	R	Patil and Ammajamma (2006)
G 33, G 8527, G8586, G 8587, GC 60020-8-7-7-18, GC 87016-11-B-2, GC 87021-26-B-1, SRE-D-14A, SRE-D-14B, and SS 86045-23-2	Uganda	R/MR	Oloka et al. (2008)
PI 200456 and PI 224270	United States	R	Calvo et al. (2008)
PI567102B	United States	R	Li and Young (2009)
Cristalina and *IAC 100*	Brazil	MR (high partial R)	Martins et al. (2007)
Lu Pi Dou and *Hei Dou*	China	Leaf-yellowing prevention characteristic	Yamanaka et al. (2011)
Emgopa 313 and Monsoy 8211	Brazil	R	Azevedo et al. (2007)
Breeding lines: TGx 1835-10E, TGx 1895-50F, and TGx 1903-3F and Accessions (PI 594538A, PI 417089A, and UG-5)	Nigeria, United States, Uganda	R	Twizeyimana et al. (2008)

(*Continued*)

TABLE 9.2 (*Continued*)
Soybean Genotypes Resistant (R) or Moderately Resistant (MR) to Asian Soybean Rust (ASBR) as Reported from Different Countries in the World

Genotype	Country	R/MR/Genes	Reference(s)
PI 379618TC1, PI 417115, PI 423956, and the Shiranui and Kinoshita (PI 200487)	Brazil	R	Costamilan et al. (2008)
BRS 134, BRSMS Bacuri, CS 201, FT-17, FT-2, IACpl1, KIS 601, and OCEPAR 7	Brazil	R (RB-type lesion)	Arias et al. (2008)
PI 506863, PI 567341, and PI 567351B, PI 181456, PI 398288, PI 404134B, and PI 507305, PI 587886, PI 587880A, and PI 587880B, PI 587905 and PI 605779E, PI 594754, PI 605833, PI 576102B, and PI 567104B	Paraguay	R	Miles et al. (2008)
EC 241778 and EC 241780	United States	R	Patil et al. (2004)
MNG 7.13, MNG 8.10, and MNG 1.6	Sub-Saharan Africa and worldwide	R	Tukamuhabwa et al. (2012)
EC 325115, EC 251378, EC 389149, EC 432536, EC 241760, and EC 333917	United States	MR	Patil et al. (2004)
Ankur, PK 1029, TS 98-21, EC 389160, and EC 389165	India	R	Rahangdale and Raut (2004)
JS 19, RPSP-728 and PK 838	India	R	Verma et al. (2004)
Early TGx 1835-10E, late TGx 1838-5E	Uganda	MR	Kawuki et al. (2004)
EC 389160, *EC 393230*, and *TS 98-21*	India	R	Rahangdale and Raut (2003)
PI 567102B, PI 200492 (Rpp1), PI 230970 (Rpp2), PI 462312 (Rpp3), and PI 459025B (Rpp4)	Paraguay	R	Li (2009)
EC241780	India	R	Shivakumar et al. (2011)
TGx 1805-1F, TGx 1951-3F, TGx 1935-3F, and TGx 1972-1F	Nigeria	Highly R	Ittah et al. (2011)
TGx 1949-8F, TGx 1935-5F, TGx 1448-2F, TGx 1965-7F, and TGx 1936-2F	Nigeria	R	Ittah et al. (2011)
EC-241778 and EC-241780	India	R	Parameshwar et al. (2012)
EC 241780, EC 456573(A), EC 456580, EC 427283, EC 481454, EC 457172, EC 481441, and EC 457266(Ku)	India	R	Kurundkar et al. (2011)
TGx1987-62F, TGx1935-3F, TGx1951-3F, TGx1936-2F, TGx1987-10F, TGx1972-1F, and TGx1949-8F	Nigeria	Nigeria (IITA)	Iwo et al. (2012)

Three sprays of hexaconazole alone reduce rust disease severity considerably to higher level and result in significantly higher seed yield (24.79 q/ha), 100 seed weight (14.37), and the inclusion of nimbecidine in the spray schedule not only is more useful in reducing the cost of protection but also gives higher benefits in addition to giving insurance against resistance development by the fungus against hexaconazole (Hegde and Mesta 2012). Combined with the organosilicone adjuvant, Silwet L-77 plus fungicide pyraclostrobin + epoxiconazole contribute to improve soybean rust control increasing the productivity and weight of 1000 grains.

The efficacy of fungicides varies with the cultivars also. For example, three sprays of hexaconazole are sufficient to manage rust and produce high yields in JS-335, while two sprays of the same hexaconazole have been enough to lower disease severity and to obtain high yields for PK-1029 (Hegde et al. 2002). Similarly, the resistant line CB06-953/963 (Rpp4 gene) needs 13.3 days longer than the susceptible cultivar to reach the ETL; late-season fungicide applications reduce rust severity and increase the yield of the resistant cultivar (Koga et al. 2011). The cultivars M-Soy 8199RR and Emgopa 315RR that are less susceptible to disease and a control program termed *monitoring* (in which the appearance of new pustules of the pathogen is monitored to make the decision at each fungicide spray) have been found to be the most effective (da Silva et al. 2011).

The rate of 80 kg/ha K_2O associated with fungicide sprays with azoxystrobin + cyproconazole is promising to reduce the deleterious effects of ASR (Doreto et al. 2012). The glyphosate at rates between 0.84 and 1.68 kg/ha can delay the onset of ASR in soybeans (Feng et al. 2008).

Systemic Acquired Resistance

Silicon (Si) is recognized for its prophylactic role in alleviating diseases when absorbed by plants and has been proposed as a possible solution against soybean rust, caused by *P. pachyrhizi*. Soybean plants supplied with Si show reduction in ASR symptoms (Arsenault-Labrecque et al. 2012, da Cruz et al. 2012). Si can protect soybean plants against soybean rust through mediated resistance. Saccharin (3 mM) applied as a root drench at the second trifoliate (V3) and early reproductive (R1) stages has been found more effective than the foliar spray treatment at inducing SAR (Srivastava et al. 2011). The severity of the soybean rust (area under disease progress curve) is significantly reduced when the soybean plants are fertilized with the combination of 8 and 11 mmol/L of K and Ca, respectively (Pinheiro et al. 2011).

Soil applications of wollastonite ($CaSiO_3$) (Si 0.96–1.92 tons/ha) or foliar applications of potassium silicate (K_2SiO_3) (Si at 500–12,000 mg/kg) may lead to the development of SBR control practices that can benefit organic and conventional soybean production systems (Lemes et al. 2011). Silicon (Si) amendments have been studied as an alternative strategy to control SBR because this element is reported to suppress a number of plant diseases in other host. Potassium silicate (KSi) sprays (40 g/L) could reduce the intensity of soybean rust (Rodrigues et al. 2009). The foliar application of $MnSO_4$ (0.3%) records lower percentage of rust disease index (33.7) compared to the control (89.6) consequently increasing the yield. Considering the effect of $MnSO_4$, in terms of both yield and environmental advantages, it is suggested to replace traditional fungicide application with $MnSO_4$ (Morab et al. 2003).

Cultural Control

There are several cultural practices that may help manage soybean rust. In most areas of the United States where rust must be introduced each year for an epidemic to occur, changing planting and harvest dates may avoid disease. Planting date and soybean cultivar significantly affect disease severity, with severity being higher on soybean crops planted during the wet season than those planted in the dry season. This study suggests that selection of planting date could be a useful cultural practice for reducing soybean rust (Twizeyimana et al. 2011b). For example, early sowing (end of June) of the crop is less damaged (36.15%) when compared with crop sown in mid- and end of July in India (Shukla et al. 2005). For all of the sowing dates, the early-season cultivar, M-SOY 6101, shows a lower risk of being affected by the rust and consequently exhibits less yield loss exhibiting a lower variance in yield, which represents more stability with regard to the inter-annual climate variability, that is, the farmers who use this cultivar will be able to recover more economic benefits (de Avila Rodrigues et al. 2012). Planting dates may also be delayed so that the vulnerable reproductive period occurs during dry conditions that do not favor rust.

In areas where the weather is marginal for rust development, wider row spacing along with lower plant populations may hasten canopy drying, thus reducing the dew period enough to prevent or at least slow disease development. Thus, the row spacing of 60 cm lowers AUDPC values and higher crop productivity (Madalosso et al. 2010). Cultural practices such as the use of reduced seed rates, increased row widths, and row orientation to the sun have been prescribed as environmental modifications that create a microclimate less conducive to foliar disease development. Therefore, it is important to determine the influence of different periods of leaf wetness and respective microenvironments on infection and rust development on soybean plants in a local geographical area in the field (Narvaez et al. 2010). The expression of partial resistance of both cultivars can be influenced due to variation of P and K levels. Lower doses of P and K induce a greater difference in the latent period of the pathogen. The association of genetic cultivar background to mineral nutrition might result in an integrated management disease program, along with evasion and chemical protection strategies (Balardin et al. 2006). The severity of soybean rust is higher in plants under crossed sown lines. The increase in the number of seeds from 15 to 30 per meter in the crossed sown lines reduces the severity of the disease only in some cultivars as in the case of M7211RR cultivar (Lima et al. 2012).

Biological Control

The fungus *Simplicillium lanosoniveum* can colonize *P. pachyrhizi* and significantly lower amounts of DNA of *P. pachyrhizi* and lower rust disease severity when soybean leaves are colonized with *S. lanosoniveum* indicating its potential use in biological control of soybean rust disease (Ward et al. 2012).

Effect of Plant Extracts

The essential oils of *H. marrubioides*, *A. gratissima*, and *C. verbenacea* are fungitoxic by inhibiting 100% of urediniospores of *P. pachyrhizi* and are effective at higher concentration only as preventive treatments in the control of the ASR. But these essential oils at these dosages are not as efficient as the pyraclostrobin + epoxiconazole–based fungicide (da Silva et al. 2012b). It is inferred that the essential oils from *Corymbia citriodora* (*Eucalyptus citriodora*), *Cymbopogon nardus*, *A. indica*, or *Thymus vulgaris* at concentrations of 1.0%, 0.5%, 1.0%, and 0.3% have the potential to reduce infection by *P. pachyrhizi*, agent of the ASR (Medic-Pap et al. 2007).

Integrated Control

An integrated management system must include intensive scouting for ASR during reproductive soybean growth stage; early disease diagnosis; use of moderately rust-resistant cultivars and use of fungicides from groups III and IV (strobilurins, triazoles, and mixture of both); alternative host elimination, including soybean volunteer plants; and early planting dates and diversification in planting dates and may be used in combination with appropriate cultural practices and fungicides when needed. Wider row spacing may also allow better fungicide application and penetration into the canopy, increasing the effectiveness of chemical control (Rupe and Sconyers 2008). Using a spray mixture of cow urine (10%) + plant extract of *Prosopis juliflora* (0.5%) or cow urine (10%) + neem oil (0.5%) has been found to be economically effective in rust disease management of soybean (Jahagirdar et al. 2012).

Spore trapping and aerobiological modeling are useful in maintaining the effectiveness of the Integrated Pest Management (IPM) Pest Information Platform for Extension and Education (ipmPIPE), increasing North American producers' profits by hundreds of millions of dollars each year. In the United States, control practices based on up-to-date maps of soybean rust observations and associated commentary from Extension Specialists delivered by the ipmPIPE may have suppressed the number and strength of inoculum source areas in the southern states and retarded the northward progress of seasonal soybean rust incursions into continental North America (Isard et al. 2011).

SUDDEN DEATH SYNDROME

Symptoms

Sudden death syndrome (SDS), caused by *F. solani* f. sp. *glycines*, is a season-long root rot disease of soybean that results in severe foliar symptoms beginning in late vegetative and early reproductive stages of plant growth. Pattern of symptoms in the field ranges from distinct oval to circular patches to irregularly shaped bands or streaks across the field (Figure 9.5). In severe cases, a majority of field may show the symptoms. The SDS is characterized by root rot followed by the development of foliar symptoms. Root systems may show rotting and discoloration of lateral and tap roots. When split open, internal tissues of taproot and lower stem may show a light-gray to light-brown discoloration. Foliar symptoms begin as scattered yellow blotches in the interveinal leaf tissues. These yellow blotches increase in size and merge to affect larger areas of leaf tissues. Symptoms range from the development of chlorotic spots to severe interveinal chlorosis and necrosis (Figure 9.6). Veins typically stay green. The bright yellow blotches between the green veins give affected leaves a striking appearance. As the interveinal leaf tissue turns brown, it also dries out. Taproots of symptomatic plants are necrotic and stunted and stems exhibit a light tan discoloration, but never the dark-brown discoloration typical for brown stem rot (BSR); the pith of the SDS-affected stem remains white. This is a key symptom to differentiate SDS from BSR, a disease with similar foliar symptoms. The SDS is most recognized by the development of interveinal chlorosis and necrosis on leaves and premature defoliation (Leandro et al. 2012). In severe cases, the leaflets may drop off, leaving the petioles (leaf stalks) attached or they may curl upward and remain attached to the plant (Westphal

FIGURE 9.5 SDS of soybean at the seedling stage. (Courtesy of Dr. Shrishail Navi, Iowa State University, Ames, IA.)

FIGURE 9.6 SDS of soybean at the reproductive stage. (Courtesy of Dr. Shrishail Navi, Iowa State University, Ames, IA.)

et al. 2008). In other diseases that exhibit similar symptoms, the dead leaflets essentially tend to remain attached to the petiole. But these symptoms are not diagnostic by themselves. If the plants are uprooted when soil is moist, small, light-blue patches may be visible on the surface of the taproot near the soil line. These patches are blue spore masses of the fungi that cause SDS. As the root surface dries, the blue color fades, but these blue spore masses, seen in conjunction with the other symptoms mentioned earlier, are strong diagnostic indicators of SD.

Geographical Distribution and Losses

The SDS of soybean caused by *Fusarium virguliforme* was first discovered in Arkansas in 1971 in the United States and in South America in the early 1990s (Roy et al. 1997, Colletto et al. 2008). The disease has spread extensively since then and can be economically devastating depending on disease intensity and timing of disease onset in most soybean-growing regions of the North and South America and the world (Malvick and Bussey 2008, O'Donnell et al. 2010, Leandro et al. 2012, Mbofung et al. 2012). SDS is ranked in the top four on the list of diseases that suppressed soybean yield during 2003–2005 in the United States (Wrather et al. 2003, Aoki et al. 2005, Wrather and Koenning 2006). The extent of yield losses due to SDS depends on the severity and timing of disease expression relative to plant development in regard to yield components. If the disease develops early in the season, flowers and young pods tend to abort. When the disease develops later, the plants produce fewer seeds per pod or smaller seeds. SDS reduces soybean yields in four of the top eight soybean-producing countries in the world, Argentina, Brazil, Canada, and the United States. In the year 2006 alone, yield was reduced by 1.849 million metric tons worldwide (Wrather et al. 2010). Yield suppression of SDS in the United States increased from 3.7 million bushels in 1996 to 34.5 million bushels in 2009 (Wrather and Koenning 2009, Koenning and Wrather 2010). From 1996 to 2007, losses averaged U.S. $190 million a year in the Midwestern U.S. soybean-producing region (Robertson and Leandro 2010). Gibson et al. (1994) estimated yield reduction of 7–34 kg/ha per unit increase in SDS incidence, whereas others have reported total yield decreases of 12%–22% per unit increase in foliar symptom severity. The earlier severe disease develops, the more the yield is reduced. In Argentina, average yield loss is in the range of 192–3770 kg/ha (Mercedes Scandiani et al. 2012). Soybean cyst nematode (*Heterodera glycines*) and *F. virguliforme* causing SDS have a synergistic effect on yield when they occur jointly in the field (Gelin et al. 2006, Xing and Westphal 2009).

Pathogen

SDS is caused by soilborne fungi within a group (clade 2) of the *F. solani* species complex (Aoki et al. 2003, 2005). Phenotypic and multilocus molecular phylogenetic analyses, as well as pathogenicity experiments, have demonstrated that four morphologically and phylogenetically distinct fusaria can induce soybean SDS (O'Donnell et al. 2010). Among the four species of *Fusarium*, *Fusarium brasiliense*, *Fusarium cuneirostrum*, *Fusarium tucumaniae*, and *F. virguliforme*, only two, *F. virguliforme Akoi* (O'Donnell, Homma and Lattanzi) (syn. *F. solani* f. sp. *glycines*) and *F. tucumaniae*, are the main casual fungi in North and South America, respectively (Aoki et al. 2005). The fungus, *F. virguliforme* (syn. *F. solani* f. sp. *glycines*), is semibiotrophic, which grows slowly in culture and is difficult to isolate from diseased plants (Yuan et al. 2008). Once a pure culture is obtained, blue spores and other cultural characteristics distinguish the SDS pathogens from other *Fusarium* species that can infect soybean roots. In North America, the SDS pathogen is considered clonal and has been considered asexual; the pathogen, however, has never been isolated from diseased foliar tissues. Thus, one or more toxins produced by the pathogen have been considered to cause foliar SDS. One such toxin is the *F. virguliforme toxin* (FvTox1) that causes foliar SDS-like symptoms in soybean. This is a low-molecular-weight protein of approximately 13.5 kDa (FvTox1) purified from *F. virguliforme* culture filtrates. FvTox1 induces foliar SDS in soybean, most likely through production of free radicals by interrupting photosynthesis (Brar et al. 2011). Of the four fusaria that have been shown to cause soybean SDS, field surveys indicate that *F. tucumaniae* is the most important and genetically diverse SDS pathogen in Argentina. The first report of sexual reproduction through perithecia formation by a soybean SDS pathogen, that is, *F. tucumaniae* that originated from Argentina, has been made by Scandiani et al. (2010). *F. tucumaniae* life cycle in South America includes a sexual reproductive mode, and thus, this species has greater potential for rapid evolution than the *F. virguliforme* population in the United States, which may be exclusively asexual (Covert et al. 2007). These findings support the hypothesis that the North America SDS pathogen is clonal and *F. virguliforme* in North America and *F. tucumaniae* in South America are the main casual fungi of SDS of soybean (Aoki et al. 2005, Westphal et al. 2008, Scandiani et al. 2011). A new TaqMan real-time PCR assay for the quantification of *F. virguliforme* in soil has been developed. The assay can be used as a diagnostic tool for rapid screens of field and greenhouse soil and for symptomatic and asymptomatic plants (Mbofung et al. 2012).

Epidemiology and Disease Cycle

The SDS pathogen survives between soybean crops as chlamydospores in crop residue or freely in the soil. As soil warms in the spring, chlamydospores near soybean roots are stimulated to germinate and then infect soybean roots (Westphal et al. 2008). Soybean seeds as the primary source of inoculum of SDS pathogen are also evident because the seeds contain the fungus mycelium after 12 months of storage and the fungus is transmissible after 12 months of storage (Balardin et al. 2005). The fungus also can survive in cysts of the soybean cyst nematode (SCN), *H. glycines*. The two pathogens *F. solani* f. sp. *glycines* (syn. *F. virguliforme*) × *H. glycines* act as a complex and the disease development is strongly dependent on high soil moisture (Xing and Westphal 2006).

Evidence for the existence of genetic variation in *F. virguliforme* has been provided and that the minor quantitative traits and environmental interactions are primarily responsible for the variation in aggressiveness found among isolates within the species (Mbofung et al. 2012). Variability of aggressiveness based on measurements of SDS foliar severity, shoot, root, and root lesion lengths; shoot and root dry weights; and total dry weights has been found among isolates (Li et al. 2009). Variability in carbon source utilization among *F. virguliforme* isolates is evident, but it is independent of geographic origin of the isolates (Tang et al. 2010). An international collection of *F. virguliforme* isolates has been established and maintained at the National Soybean Pathogen Collection Center, University of Illinois at Urbana-Champaign in the United States (Li et al. 2009). A real-time QPCR assay to compare the accumulation of genomic DNA among 30 *F. solani* f. sp. *glycines* (FSG) isolates in inoculated soybean roots has been developed. Isolates may differ significantly in their DNA accumulation on a susceptible soybean cultivar when detected and quantified using an FSG-specific probe/primers set derived from the sequences of the nuclear-encoded, mitochondrial small subunit ribosomal RNA gene (Li et al. 2008). Isolates of *F. virguliforme* from corn, wheat, ryegrass, pigweed, lambsquarters, canola, and sugar beet are the asymptomatic hosts of the pathogen (Malvick and Bussey 2008, Kolander et al. 2012). *F. virguliforme* may infect roots of soybean seedlings as early as 1 week after crop emergence. A protoplast-based fungal transformation system for *F. virguliforme* has been developed for the production of a green fluorescent protein (GFP)-expressing fungal transformant. The GFP-expressing fungus can be used to study fungal infection processes including fungal penetration, colonization, and spread, especially at the early stages of disease development (Mansouri et al. 2009). It is apparent that roots are most susceptible to infection during the first days after seed germination and that accelerated root growth in warmer temperatures reduces susceptibility to root infection conducive to foliar symptoms. However, soil temperature may not affect infections that occur as soon as seeds germinate (Gongora-Canul and Leandro 2011b). Cool temperatures are more favorable for root infection by *F. virguliforme* than warmer temperatures. Optimum soil temperature for root rot development is 15°C–17°C with root rot severity being lower at higher temperatures. Interestingly, in contrast to root infection, the expression of foliar symptoms is favored by warmer temperatures of around 22°C–25°C. High soil moisture has been shown to favor SDS. Foliar symptoms are more severe in irrigated fields during wet season. The presence of continuous soil moisture throughout the growing season is most favorable for the development of the SDS. Rate of disease progress increases as inoculum densities increase for both root and foliar disease severities. The incubation period for root and foliar disease severity range from 9 to 18 and 15 to 25 days, respectively (Gongora-Canul et al. 2012).

The pathogen is capable of degrading lignin, which may be important in infection, colonization, and survival of the fungus (Lozovaya et al. 2006), but aboveground symptoms of SDS rarely appear until soybean plants have reached reproductive stages. The fungus produces toxins (FvTox1) in the roots that are translocated to the leaves (Brar and Bhattacharyya 2012). Often, symptoms first appear after heavy rains during reproductive stages; high soil moisture increases the disease severity (Xing and Westphal 2006). SDS is more severe when the SCN (*H. glycines*) is also present in a field and the cultivar is susceptible to both pathogens (Xing and Westphal 2006). The toxin requires light to initiate foliar SDS symptoms. Irrigation treatments during mid- to late reproductive growth stages result in significant increase in SDS foliar symptom development (de Farias Neto et al. 2006).

Both *F. virguliforme* and SCN are widespread. The close association of the pathogens is also apparent in the fact that the SDS pathogen can be isolated from cysts of SCN (Roy et al. 1997). The SCN, *H. glycines*, and the fungus *F. solani* f. sp. *glycines* that causes SDS of soybean frequently co-infest soybean fields. The infection of soybean roots by *H. glycines* does not affect root colonization by the fungus, as determined by real-time PCR. Although both pathogens reduce the growth of soybeans, *H. glycines* does not increase SDS foliar symptoms, and interactions between the two pathogens are seldom significant (Gao et al. 2006).

Although the pathogen may produce spores (macroconidia) on the surface of the taproot during the summer, these spores spread only short distances within a growing season. Over a period of years, flowing water and cultivation practices that move soil can move spores over longer distances within or between fields.

Disease Management

Host Plant Resistance

The use of resistant cultivars is the most effective method for controlling SDS in soybean. Although soybean cultivars that are less susceptible to SDS have been developed, no highly resistant cultivars are available (Njiti et al. 2002). Soybean genotypes with yellow seed coat show a relatively good field response to SDS and a moderate seed yield. Soybean cultivars show differences in their resistance to both the leaf scorch and root rot of SDS. Root susceptibility combined with reduced leaf scorch resistance has been associated with resistance to *H. glycines* (race 14) of the SCN (Kazi et al. 2008). These superior genotypes can be used as potential parents in soybean breeding programs (Gelin et al. 2006, Wen et al. 2014). Providing multiple resistance traits in the same variety is especially important to manage SDS, because both SDS tolerance and SCN resistance are frequently needed in the same variety (Butzen 2010). Research has led to the identification of soybean genotypes with 18 QTLs. However, it is possible that only 11 or 12 loci may contribute to host resistance as some of these loci may have multiple alleles. Some of these QTLs have been shown to be in close proximity to QTL that contribute to resistance to SCN with potential linkage between the two resistance QTLs (Leandro et al. 2012). Multigenic QTL present significant problems to analysis. Resistance to soybean SDS caused by *F. virguliforme* had been partly underlain by QRfs2 that could be clustered with, or pleiotropic to, the multigeneic rhg1 locus providing resistance to SCN (*H. glycines*) (Iqbal et al. 2009).

Soybean genotypes Ripley and PI 567374 both have partial resistance to SDS and the LG D2 QTL should be useful sources of SDS resistance (Farias Neto et al. 2007). The beneficial alleles of the QTL have been shown to be associated with resistance to either foliar disease severity or root rot severity or resistance to both foliar and root rot severity. QTL for resistance to *F. virguliforme* are different from those that confer resistance to *F. tucumaniae*. The report that sexual reproduction occurs in nature in *F. tucumaniae* offers a greater challenge for disease management in regions where this species is found since host resistance to disease be easily overcome. As three other *Fusarium* species, as referred earlier, cause SDS in soybean in South America, it is important to use soybean varieties with broad resistance to the disease in this region (Leandro et al. 2012).

Molecular mechanisms underlying plant resistance and susceptibility to *F. virguliforme* have been studied using *Arabidopsis thaliana*. *A. thaliana* enabled a broad view of the functional relationships and molecular interactions among plant genes involved in *F. virguliforme* resistance. Dissection of the set functional orthologous genes between soybean and *A. thaliana* enabled a broad view of the functional relationships and molecular interactions among plant genes involved in *F. virguliforme* resistance (Yuan et al. 2008). Selection of seedlings in the greenhouse and marker-assisted selection (MAS) are faster and cheaper. DNA markers associated with loci contributing seedling resistance to *F. solani* in the southern and northern U.S. germplasm sources have been established. It is revealed that the SDS resistance can be a pleiotropic effect of shoot and root characters in partially resistant and relatively susceptible genotypes (Njiti and Lightfoot 2006). FvTox1 is an important pathogenicity factor for foliar SDS development, and expression of anti-FvTox1 single-chain variable-fragment (scFv) antibody in transgenic soybean can confer resistance to foliar SDS, and this could be a suitable biotechnological approach for protecting soybean crop plants from toxin-induced pathogen such as *F. virguliforme* (Brar and Bhattacharyya 2012). The fungal genome of *F. virguliforme* has been sequenced by conducting shotgun 454-sequencing. The genome sequence of *F. virguliforme* would become important public resource to a broad community of researchers engaged in developing tools to manage SDS (Srivastava et al. 2014).

Chemical Seed Treatment

Bayer CropScience has developed a chemical ILeVO for soybean seed treatment to provide protection for soybean seedlings from *F. virguliforme*, the fungus that causes SDS. ILeVO-seed treatment protects soybean from early-season infection and reduce late-season chlorosis and necrosis that leads to flower and pod abortion resulting in yield loss. The active ingredient in ILeVO is systemic and moves from the seed into the tissue of both stem and roots of soybean seedlings. The cotyledons and roots act as a sink for ILeVO, enabling the product to stay where it is needed to protect against early-season infection way in advance of SDS visual symptoms appearing in the field (Roden 2014). Since the fungus only colonizes the roots and base of the stem and it does not spread to the leaves and cannot be isolated from foliar portion, the foliar spray of the fungicides is not effective and hence foliar sprays of fungicides are not recommended.

Cultural Control

Soybean roots become less susceptible to xylem colonization and the subsequent development of foliar symptoms as plants mature. Therefore, practices aimed at protecting seed and seedling roots from infection may improve soybean sudden death management (Gongora-Canul and Leandro 2011a). In a regular growing season, an epidemic of SDS is highly correlated with the planting date and the disease tends to be more severe in earlier-planted soybeans in the United States (Navi and Yang 2008). Fields with a history of SDS should be planted later, rather than earlier in the spring. But planting may not be delayed to the point of compromising yield potential. A row spacing × infestation interaction indicated 7% greater yield in narrow rows (38 cm) than wide rows (76 cm) in uninfested plots, with no yield advantage to narrow rows in infested plots. In infested plots with greater SDS symptom expression, the yield advantage of narrow rows may be negated; therefore, cultivar selection is crucial when planting is done in narrow rows to maximize yield (Swoboda et al. 2011). Improving soil drainage, reducing compaction, evaluating tillage systems, and reducing other stresses on the crop if possible, in fields with recurring SDS problems, are useful SDS management practices. For example, a tillage system of disking or ridge till is effective in reducing the incidence of SDS as revealed from the studies done at the University of Missouri in the United States (Wrather et al. 1995).

The likely broad host range limits the efficacy of crop rotation and indicates that crops other than soybean can be damaged by *F. virguliforme* and can maintain or increase inoculum in soil crop rotation to have little impact on SDS incidence and severity (Xing and Westphal 2009, Kolander et al. 2012). Soil suppressiveness against the disease complex of the SCN and SDS of soybean is demonstrated (Westphal and Xing 2011). Chitosan is able to induce the level of chitinase antifungal enzymes to SDS pathogen in soybean resulting in the retardation of SDS development in soybean; it is thus helpful in partially protecting soybeans from *F. solani* f. sp. *glycines* infection (Prapagdee et al. 2007). SDS varies in severity from area to area and from field to field. This requires scouting fields when disease symptoms are present, ideally using GPS tools to map SDS-prone areas and then combination of crop management practices can help minimize the damage from SDS. By taking steps to manage *H. glycines* (resistant cultivars, nematicides), it is possible to help check SDS or at least manage its potential impact if not the disease itself.

CHARCOAL ROT

Symptoms

Symptom expression depends on the soybean plant's growth stage at the time of infection. Infected soybean seedlings show reddish discoloration of the hypocotyls appearing at soil level from root infection. Lesions become dark brown to black and infected seedlings may die under hot dry weather conditions. If wet and cool weather persists, infected seedlings survive but carry the latent infection and symptoms do not develop until plants reach reproductive stages, and only if heat and drought stress the plants. Hence, the disease is also known as *dry weather wilt and summer wilt* (Hartman et al. 1999). If the growing point is killed, a twin-stem plant may develop.

After flowering, the surface tissues (epidermis) of the lower stems of affected plants usually exhibit a light-gray or silvery discoloration and stems often have a shredded appearance. When the epidermis of lower stems and taproots is removed (by scraping with the thumbnail), extremely small, black fungal structures called microsclerotia are found embedded in the diseased tissue, which is the diagnostic feature of charcoal rot. Microsclerotia are tiny black masses of fungal tissue usually so numerous that they resemble charcoal dust, hence the name of the disease (Figure 9.7). The microsclerotia can be best seen with a good hand lens. Positive identification of the microsclerotia distinguishes charcoal rot from other similar diseases. The pycnidial stage is uncommon in soybean in contrast to formation of both microsclerotia and pycnidia on infected stem tissue of other host plants. Splitting the taproot often reveals dark-gray to blue-black streaks within. Later in the season, leaflets turn yellow, then die and shrivel, but remain attached to the plant. And finally, infected plants lose vigor and may die prematurely, and patches of such wilted and died plants are seen in the infested fields.

Geographical Distribution and Losses

Charcoal rot is a disease of economic significance throughout the world. It is widely distributed throughout tropical, subtropical, and warm temperate regions. Its effect is more pronounced in crops under biotic or abiotic stress. Changing global climatic conditions particularly occurrence of frequent drought or drought-like situations are making soybean more vulnerable to this disease. Charcoal rot is endemic in southern states in the United States and is a major problem in the central part of the Midwest, especially in Kansas and parts of Missouri. It is now occurring with greater frequency in the upper Midwest, with outbreaks reported in Illinois, Indiana, Iowa, Minnesota, North Dakota, and Wisconsin.

Root infection by germinating microsclerotia can occur very early in soybean plant development and about 80%–100% incidence of seedling infection can be observed within 3–4 weeks after planting. Based on estimates from 2006 to 2009, charcoal rot is listed as one of the 10 most

Charcoal rot (*Macrophomina phaseolina*)

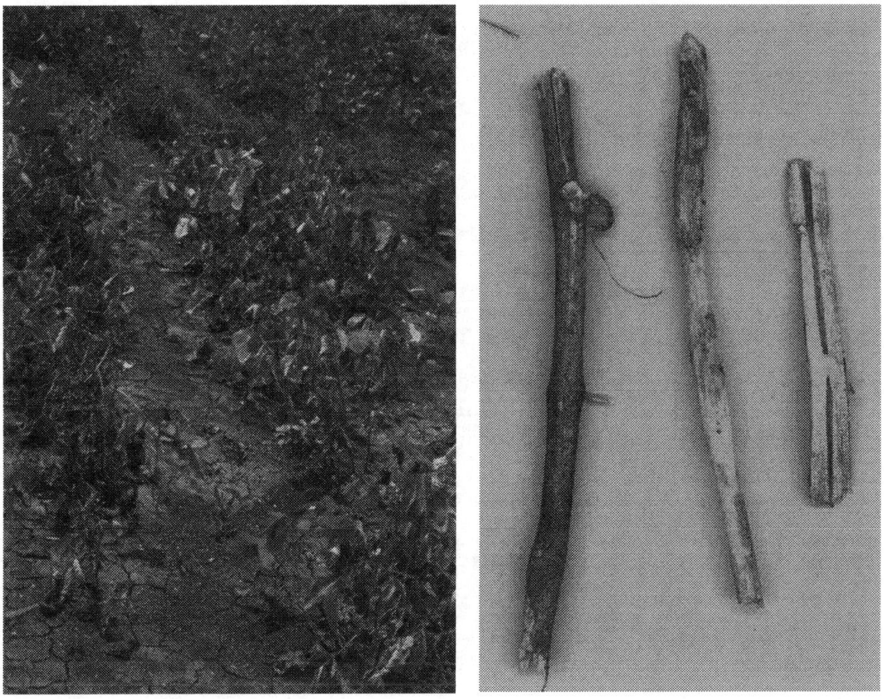

(a) (b)

FIGURE 9.7 Charcoal rot of soybean-seedling mortality under field conditions and infected stems of soybean. (a) Seedling mortality by charcoal rot. (b) Infected lower stem and minute black sclerotia of pathogen in outer corticle and pith region. (Courtesy of Dr. G.K. Gupta, ICAR-Directorate of Soybean Research, Indore, India.)

yield-suppressing diseases in the United States (Koenning and Wrather 2010, Radwan et al. 2013). Charcoal rot ranks second among economically important diseases in the Midsouthern United States next to SCN. Estimated annual loss in soybean in the United States is about seven million bushels, whereas in Brazil, Argentina, and Bolivia, a loss of one million bushels could be attributed to charcoal rot in 1998. Interestingly, yield loss due to charcoal rot in soybean ranges from 6% to 33% even in irrigated environments in the United States (Mengistu et al. 2011). Infection with this pathogen reduces the number of pods per plant, seeds per pod, 100-seed weight, and seed yield. For example, infected plants may yield as low as 67.7 pods/plant, 1.5 seeds/pod, 6.5 g/100-seed and 6.4 g/plant compared to 205.8 pods/plant, 2.2 seeds/pod, 15.2 g/100 seed, and 69.7 g/plant of healthy plants. It is clear that soybean plants infected with charcoal rot have a reduced seed yield representing less than 10% of normal plant seed yield in Iraq (Abbas et al. 2003). Predominantly occurring in most of the soybean-growing states of India, charcoal disease causes 70% or more yield loss in soybean (Ansari 2010). Charcoal rot infection may alter seed composition and nitrogen fixation in soybean. The alteration in seed composition depends on cultivar susceptibility to charcoal rot and irrigation management (Bellaloui et al. 2008).

Pathogen

The disease is caused by the fungus *M. phaseolina* (Tassi) Goid. Its synonyms are *Macrophomina phaseoli* (Maubl.) Ashby, *Rhizoctonia bataticola* (Taub.) Butler, *Sclerotium bataticola* (Taub.) Butler., and *Botryodiplodia phaseoli* (Maubl.) Thir.

Classification
Kingdom: Fungi
Phylum: Ascomycota
Class: Dothideomycetes
Subclass: Incertae sedis
Order: Botryosphaeriales
Family: Botryosphaeriaceae
Genus: *Macrophomina*
Species: *phaseolina*

M. phaseolina is highly variable, differing in size of sclerotia and the presence or absence of pycnidia. *M. phaseolina* has a wide host range and geographic distribution, infecting more than 500 crop and weed species. The fungus is highly variable, with isolates differing in microsclerotial size and the ability to produce pycnidia. Microsclerotial morphology is a key taxonomic characteristic in the identification of this fungus. Cultural and morphological characteristics can vary as a result of continuous subculturing. The optimal temperature for growth in culture ranges from 28° to 35°C. The details of morphological characteristics of *M. phaseolina* have been described earlier under peanut and sunflower diseases chapters. The number of pycnidia that are produced by *M. phaseolina* isolates is dependent on induction medium; however, peanut butter extract–saturated filter paper placed over soy nut butter extract agar (PESEA) allows for greater pycnidia and conidia production than the other media. This conidia inoculum production method can facilitate soybean charcoal rot resistance screening evaluation with different soybean isolates (Ma et al. 2010).

Epidemiology and Disease Cycle

The fungus *M. phaseolina* is a causative agent of charcoal rot diseases in more than 500 plant species. The fungus is primarily soil inhabiting but is also seed borne in many crops including soybean. It survives in the soil mainly as microsclerotia. These are black, spherical to oblong in shape, and typically measure 0.002–0.008 in. in diameter. Microsclerotia produced in host tissues are released into soil as plant tissues decay. Corn, grain sorghum, and cotton generally support lower populations of microsclerotia in soil than does soybean. In dry soils, microsclerotia survive in soil or embedded in host residue for 2 or more years. In wet soils, microsclerotia cannot survive more than 7–8 weeks and mycelia no more than 7 days. Microsclerotia must germinate either on the surface of or in close proximity to roots for infection to occur. Pathogen growth and infection of soybean can occur at emergence and at the cotyledonary stage with 80%–100% of seedlings infected 2–3 weeks after planting. Phytotoxin, botryodiplodin, is suggested to be produced more abundantly by certain isolates of *M. phaseolina* facilitating infection in soybean (Ramezani et al. 2007).

Temperature optima for fungal growth and disease development are high (30°C–37°C). Considerable infection of soybean occurs at these temperatures. Seedling blight of soybean due to *M. phaseolina* is seen in tropical countries only where soil temperatures are at least 30°C at planting. Dry conditions, relatively low moisture and nutrients (NPK), and high temperature ranging from 25°C to 35°C are favorable for the disease at pod formation and filling stage (Ansari 2010). For example, August 2003 was the driest month recorded in Iowa, which may have contributed to the disease outbreaks in that crop season (Yang and Navi 2005). Similarly, in the humid tropics of southwestern Nigeria, areas with high soil moisture levels are unfavorable for the growth and pathogenicity of *M. phaseolina*, while areas with low soil moisture levels favor the growth and pathogenicity of the fungus (Wokocha 2000). Drought stress thus has been proved to increase *M. phaseolina* infections and reduces seedling dry weight in soybeans (Gill-Langarica et al. 2008). Low C:N ratio in the soil and high bulk density as well as high soil moisture content adversely affect the survival of microsclerotia. The scattered literature on these aspects has been reviewed (Gupta et al. 2012). The fungus is seed borne and invariably present in the seed coat of all the infected seeds and moved into the cotyledons (including embryonal axis) of the 40% infected seeds (Tariq et al. 2006, Mengistu et al. 2012). The pathogen can remain viable for 15 months in seeds at room temperature and is transmitted to seedlings during germination by local contact (Kumar and Singh 2000). Although initial infections occur at the seedling stage, they usually remain latent until soybean plant approaches maturity (growth stages R5, R6 and R7). Plants infected after seedling stage generally show no aboveground symptoms until after midseason. There is a significant pathogenic and genetic variability within the soybean isolates of *M. phaseolina* from Iran, India, Italy, and Mexico (Munoz-Cabanas et al. 2005, Jana et al. 2005a,b, Rayatpanah et al. 2012a,b). Genetic variability studies among Brazilian isolates of *M. phaseolina* have revealed that one single root can harbor more than one haplotype. It is significant that *M. phaseolina* isolates from soybean are chlorate-sensitive isolates that grow sparsely with a feathery-like pattern and the isolates characterized by the feathery-like pattern are more virulent on soybean and sunflower (Rayatpanah et al. 2012a). Moreover, cultivation with crop rotation tends to induce less specialization of the pathogen isolates. Knowledge of this variation may be useful in screening soybean genotypes for resistance to charcoal rot (Almeida et al. 2003b). Genetic differentiation of *M. phaseolina* can be altered by crop rotation that *M. phaseolina* is a highly diverse species and also reveals a strong effect of the rotation system on genetic diversity (Almeida et al. 2008).

Cultivation with crop rotation probably tends to induce less diversity of the pathogen isolates (Rayatpanah et al. 2012b). The AFLP analysis has revealed great genetic diversity in *M. phaseolina* since more than 98% of amplified products appear to be polymorphic. But no clear association between geographical origin or host of each isolate and AFLP genotype has been found. A genetic dissimilarity greater than 10% is reported between a group of isolates from Mexico and Italy and isolates from other countries (Munoz-Cabanas et al. 2005).

Single primers of SSRs or microsatellite markers have been used for the characterization of genetic variability of different populations of *M. phaseolina* obtained from soybean and cotton grown in India and the United States. The variability found within closely related isolates of *M. phaseolina* indicated that such microsatellites are useful in population studies and represents a step toward identification of potential isolate diagnostic markers specific to soybean and cotton (Jana et al. 2005b). Universal rice primers (URPs) (primers derived from DNA repeat sequences in the rice genome) using PCR (URP-PCR) are sensitive and technically simple to use for assaying genetic variability in *M. phaseolina* populations (Jana et al. 2005a).

Disease Management

Host Plant Resistance

Strong resistance to charcoal rot does not exist among soybean cultivars. Six genotypes (one genotype in MG III, one in late MG IV, and four in MG V) have been identified as moderately resistant to *M. phaseolina* at levels equal to or greater than the standard DT97-4290, a moderately resistant high-yield potential cultivar (Paris et al. 2006, Ansari 2007). The genotypes identified as having moderate resistance across the 3 years could be useful as sources for developing resistant soybean cultivars (Mengistu et al. 2012). One such first report on soybean genotype with high levels of resistance to charcoal rot is *PI 567562A* and resistance in this genotype is greater than the standard *DT 97-4290* (Mengistu et al. 2012). Mexican lines H86-5030 and H98-1552, as well as Mexican cultivar Suaqui-86, are reported to be moderately resistant to *M. phaseolina* (Gill-Langarica et al. 2008).

Generally, the late maturity groups of soybeans are more tolerant to the disease. Lines B.P-692, J.K-695, and K.S-69035 show the highest tolerance to charcoal rot. Based on the results and the qualitative and quantitative agronomic characteristics, two lines (J.K-695 and B.P-692) have been selected as the suitable cultivars and are introduced as Sari and Telar, respectively, for cultivation in Mazandaran region in Iran (Rayatpanah et al. 2007). Early-maturing cultivars that do not have late reproductive growth stages might coincide with periods of drought stress and high temperatures may help avoid severe damage to the disease during years with hot, dry summer weather conditions. Resistance to this pathogen in some genotypes is associated with drought tolerance. Some drought-tolerant soybean genotypes may resist root colonization by *M. phaseolina*, but this is not true for all drought-tolerant genotypes (Wrather et al. 2008). Ten genotypes (JS 335, G 213, Birsa Sova-1, GS 1, GC 175320, G 9, G-688, NRC 37, DSb 6-1, and RSC 14) have been identified as highly resistant (<1.0% morality) to *M. phaseolina* (Ansari 2007). Soybean cv Rawal is less susceptible to *M. phaseolina* (Ehteshamul-Haque et al. 2007). Planting earlier-maturing varieties in order to shorten the effect of a dry period at the end of the growing season is useful.

The cut-stem inoculation technique, which has several advantages over field tests, successfully distinguishes differences in aggressiveness among *M. phaseolina* isolates, and relative differences among soybean genotypes for resistance to *M. phaseolina* are comparable with results of field tests (Twizeyimana et al. 2012).

Induced Systemic Resistance

Some chemicals may play an important role in controlling the soybean charcoal rot disease, through induction of systemic resistance in soybean plants. The effect of two inducer chemicals, that is, riboflavin (B2) and thiamine (B1), on the induction of systemic resistance in soybean against charcoal rot disease and biochemical changes associated with these treatments in soybean plants have been investigated under greenhouse conditions. Riboflavin (0.1–15 mM) and thiamine (2.5–5 mM) are sufficient for maximum induction of resistance; higher concentration does not increase the effect (Abdel-Monaim 2011).

Plant growth–promoting rhizobacteria (PGPR), such as *B. japonicum* strain USDA 110, *Azotobacter chroococcum*, *Azospirillum brasilense*, *Bacillus megaterium*, *B. cereus*, and *P. fluorescens* when inoculated on soybean plants, result in inducing and enhancing the activity of PR proteins (chitinase and beta-1,3-glucanase), peroxidase, phenylalanine ammonia lyase (PAL), and phenolics and contribute to protect the soybean plants against *M. phaseolina* infection (Attia et al. 2011).

Chemical Control

Since *M. phaseolina* is also seed borne in soybean, seed treatment with effective fungicide can protect the seedlings from infection. Soybean seeds treated with thiophanate methyl applied as 0.1% or 0.2% dry seed treatment or as fungicide slurry with the addition of methyl cellulose result in the highest control of charcoal rot (Lakshmi et al. 2002). Seed treatment with carbendazim (as Bavistin 50 WP) (2.0 g/kg seed) and thiophanate methyl (as Topsin M) (1.0 g/kg seed) is also effective in eliminating the pathogen from infected seeds (Kumar and Singh 2000).

Cultural Control

Several disease management approaches involve the management of populations of microsclerotia using cultural practices to control charcoal rot. It may be a better alternative to suppress charcoal rot by using the no-tillage cropping system in comparison to conventional tillage (CT) system to conserve soil moisture and reduce disease progress (Almeida et al. 2003a, Mengistu et al. 2009a).

Farm practices that increase residue destruction immediately after harvest or those that enhance *Trichoderma* spp. populations may directly or indirectly lower the relative longevity of soilborne pathogens, including *M. phaseolina* (Baird et al. 2003). Water management can limit, but not prevent, colonization of soybean by *M. phaseolina*; excessively dense planting increases drought stress when water becomes limiting. Hence, avoiding excessive seeding rates is practiced so that plants do not compete for moisture, which increases disease risk during a dry season.

Macrophomina infection has been found to be lower in NPK treatment, and the lowest rate of disease development can be observed in the case of the highest NPK combination. By increasing the NK supply, the degree of infection is decreased (Csondes et al. 2008).

Soybean cultivars and other crop species in the host range differ in colonization, and these differences may affect soil densities of the fungus (Kendig et al. 2000). One-year corn–soybean rotation is ineffective in managing charcoal rot since the fungus also causes corn stalk rot. However, the fungus is less damaging to corn than to soybean. Several years of corn or small grain crops rotations are necessary to reduce charcoal rot risk in severely infested fields. Although corn is a host, the microsclerotia numbers are still reduced under this crop. It requires at least 3 years without a soybean crop before microsclerotia levels are low enough to plant soybean again. Once the numbers of microsclerotia are low, a rotation of 1 year of soybean with 1 year of corn may keep microsclerotia numbers at low sustainable level (Kendig et al. 2000).

Microbial communities are more abundant and active in direct seeding (DS) than in CT in response to high nutrient content in soil (Perez-Brandan et al. 2012). Indeed, DS systems present higher soil OM and total N, K, and Ca than CT. Electrical conductivity and aggregate stability (AS) are also improved by DS. Soybean grown in high-quality soil is thus not affected by charcoal rot; however, under CT, disease incidence in soybean appears to have been 54%. These differences are correlated to the higher microbial abundance and activity under DS, the biological component being a key factor determining soil capacity to suppress the soilborne pathogen like *M. phaseolina* (Perez-Brandan et al. 2012).

Biological Control

Application of more than one antagonist of diverse origin is suggested as a reliable means of reducing the variability and increasing the reliability of biological control. *T. harzianum* and plant growth promontory rhizobacteria *P. fluorescens* when tested alone and in combinations for their relative biocontrol potential against *M. phaseolina* causing charcoal rot of soybean result in effective control of the disease (Mishra et al. 2011). *P. aeruginosa* strain Pa5 is a good candidate for use as BCAs against *M. phaseolina* on soybean cv Rawal (Ehteshamul-Haque et al. 2007). *P. fluorescens* isolates Pf-12 and Pf-63 inhibit the mycelial fungal growth of *M. phaseolina* through production of antibiotics as well as volatile metabolites, whereas *B. subtilis* isolates B-13, B-42, B-126, and B-84 do so through volatile and nonvolatile metabolite production. *P. fluorescens* isolates, however, also produce hydrogen cyanide. In greenhouse studies, the *B. subtilis* isolates B-13 and B-126 have been shown to be effective in reducing the intensity of charcoal rot of soybean by 59%–66%. The combinations of isolates B-13 and B-126 are also effective in reducing the intensity of disease (Sharifi-Tehrani et al. 2005). *Bacillus* sp. and *Trichoderma*-inoculated soybeans showed increased plant height, number of pods, vegetative growth, and aerial and radical weights (Cardona Gomez et al. 2000).

One strategy to control charcoal rot is the use of antagonistic, root-colonizing bacteria. *Rhizobacteria* A5F and FPT721 and *Pseudomonas* sp. strain GRP3 are characterized for their plant-growth-promotion activities against the pathogen. *Rhizobacterium* FPT721 exhibits higher antagonistic activity against *M. phaseolina* on dual plate assay compared to strain A5F and GRP3. FPT721 and GRP3 give decreased disease intensity. Lipoxygenase (LOX), PAL, and peroxidase (POD) activities have been detected in extracts of plants grown from seeds treated with rhizobacteria and inoculated with spore suspension of *M. phaseolina* (Choudhary 2011).

Another strategy to control charcoal rot is the use of antagonistic, root colonizing PGPR. Effective biological control by the PGPR isolates indicates the possibility of application of *rhizobacteria* for control of soilborne diseases of soybean including that of charcoal rot in Pakistan and other countries (Inam-Ul-Haq et al. 2012). PGPR, such as *B. japonicum* strain USDA 110, *A. chroococcum*, *A. brasilense*, *B. megaterium*, *B. cereus*, and *P. fluorescens* when inoculated on soybean plants result in inducing and enhancing the activity of PR proteins (chitinase and beta-1,3-glucanase), peroxidase, PAL, and phenolics and contribute to protect the soybean plants against *M. phaseolina* infection (Al-Ani et al. 2011, 2012, Attia et al. 2011).

PGPR as mentioned earlier, phosphate-solubilizing bacteria (*B. megaterium* var. *phosphaticum*), and potassium-solubilizing bacteria (*B. cereus* and *P. fluorescens*) have been proven for their efficacy against *M. phaseolina* on soybean plants and for their influencing effect on percentage of healthy plant and growth. Data suggest the positive impact of PGPR in improving the stand and vigor of soybean plants in *Macrophomina*-infested soil. In the field trial, results have shown that all tested PGPR significantly can decrease root rot and wilt disease incidence. *B. megaterium*–treated plots have been found to be the most effective treatment followed by the combination of *A. chroococcum*, *A. brasilense* and *B. megaterium*. The reduction in disease incidence reflected on plant growth and the apparent bacterial plant growth-promoting and bacterial BCAs could provide a means for reducing the incidence of root rot and wilt disease complex of soybean in addition to avoiding the use of fungicides. Such biocontrol approach should be employed as a part of IPM system (El-Barougy et al. 2009, Attia et al. 2011). For example, seed treatment with *B. japonicum* and *T. viride* and soil application of Zn with B and Fe reduce chaffy pods as well as the disease incidence up to 75%. Seed treatment with *Trichoderma* and irrigation at the time of moisture stress reduce the intensity of disease to about 50% (Ansari 2010).

YELLOW MOSAIC DISEASE

Symptoms

The diseased plants start appearing in the field when the crop is about a month old. Two types of symptoms—yellow mottle and necrotic mottle—are noticeable. The first visible sign of the disease is the appearance of yellow spots scattered on the lamina. They are mostly round in shape. In yellow mottle, the spots diffuse and expand rapidly. The leaves show yellow patches alternating with green areas and also later turn yellow. Such completely yellow leaves gradually change to a whitish shade and ultimately become necrotic. These color changes of affected plants are so conspicuous that the disease can be spotted in the field from a distance (Figure 9.8). In necrotic mottle, the center of yellow spots develops necrosis and the virus becomes systemic in the plant and all newly formed leaves show signs of mottle. There may be a reduction in size of leaves. Number and size of pods per plant and seeds per pod are generally reduced. The pods are deformed and contain shriveled undersized seeds.

FIGURE 9.8 Yellow mosaic of soybean at various stages of crop growth. (Courtesy of Dr. A.K. Tewari, GBPUA&T, Pantnagar, India.)

Geographical Distribution and Losses

At first, it was observed in North India in the early 1970s (Nene 1972) and since then, it has spread at alarming proportions. The disease is now endemic in South Asian countries (India, Pakistan, Bangladesh, Bhutan, Nepal, Sri Lanka). It is also reported to occur in the Philippines and Thailand. In the northern parts of India, the incidence of the disease may range from 20% to 80%. Soybean plants, if infected at prebloom stage, show 16%–73% losses in yield in different cultivars. Yield losses are of lower magnitude with infection of postbloom stage. In India, yield losses of 10%–88% had been reported due to YMD of soybean (Nene 1972, Bhattacharyya et al. 1999).

Pathogen: Mung bean yellow mosaic virus (MYMV) and mung bean yellow mosaic India virus (MYMIV).

Enzyme-linked immunosorbent assay, immunospecific electron microscopy, and whitefly transmission studies reveal that the etiological virus causing YMD in soybean is a begomovirus of the family Geminiviridae. Begomoviruses have characteristic icosahedral geminate particles that encapsidate the genome of circular single-stranded DNA. They infect dicots and are transmitted by the whitefly *Bemisia tabaci* Gennadius. Genomic components of the begomovirus that cause yellow mosaic disease (YMD) in soybean in Delhi, India, when cloned, sequenced, and evaluated for infectivity; nucleotide sequence analysis of the virus isolate revealed more than 89% identity with MYMIV; therefore, it is designated as a soybean isolate of MYMIV (MYMIV-Sb). Total nucleotide and predicted amino acid sequence analysis of MYMIV-Sb with other yellow mosaic virus isolates infecting legumes established dichotomy of the isolates into two species, namely, MYMIV and MYMV. The involvement of at least two distinct viruses in the etiology of soybean YMD in India is established (Usharani et al. 2004).

Yellow mosaic virus infecting soybean in northern India is distinct from the species-infecting soybean in southern and western India (Usharani et al. 2004). Girish and Usharani (2005) further determined the complete nucleotide sequences of two soybean-infecting begomoviruses from the central and southern parts of India, and the sequence analyses show that the isolate from Central India is a strain of MYMIV and the southern Indian isolate is a strain of MYMV. Thus, involvement of at least two distinct viruses in the etiology of soybean YMD in India is reported (Usharani et al. 2004). YMD of soybean is reported to be caused by soybean isolate of MYMIV (MYMIV-sb) (Radhakrishnan et al. 2008, Yadav et al. 2009). MYMIV-sb is similar to cowpea isolate of MYMIV (MYMIV-cp) in its ability to infect cowpea, but differing from blackgram (MYMIV-bg) and mung bean (MYMIV-mg), which do not infect cowpea (Usharani et al. 2005). Genomic analysis of DNA-A and DNA-B components of the MYMIV isolates shows characteristic differences in complete DNA-B nucleotide sequence correlating with host range differences (Usharani et al. 2005). Interestingly, MYMV virulent variant MYMV-Pp1 has been confirmed through nucleic acid spot hybridization using homologous probes to DNA-A and DNA-B of MYMV-Bg to cause infection in soybean (Biswas 2002).

They have a bipartite genome (two components, viz., DNA- 'A' and 'B'), which replicates via rolling circle replication (RCR) model with the help of few viral and several host factors. MYMIV is a representative of the genus *Begomovirus/Begomoviridae*, which is prevalent in the northern part of Indian subcontinent causing YMD. The most affected leguminous crops by MYMIV are *Cajanus cajan*, *G. max*, *Phaseolus aconitifolius*, *Phaseolus aureus*, *P. vulgaris* "French bean," and *Vigna mungo*. MYMIV possesses bipartite ssDNA genomes named as DNA-A and DNA-B, both being ~2.7 kb in size. Both components share a common region (CR) of about 200 bp containing the important *cis*-elements for viral DNA transcription and RCR.

Bipartite geminiviruses possess two movement proteins (NSP and MP), which mediate the intra- and intercellular movement. In order to accomplish the transport process, the MPs interact with viral nucleic acids in a sequence nonspecific manner (Radhakrishnan et al. 2008). Multiple DNA-B components could be detected with the soybean strain of MYMV species. The nucleotide sequence similarity between the DNA-A components of the two isolates is higher (82%) than that between the corresponding DNA-B components (71%) (Girish and Usha 2005).

In bipartite begomoviruses, DNA-A encodes proteins required for replication, transcription, and encapsidation, whereas DNA-B encodes proteins required for movement functions. Phylogenetic analysis of complete DNA-A and amino acid sequence of various protein products of DNA-A clearly indicate the bifurcation of YMV isolates into two different species—MYMIV and MYMV. More number of isolates representing all geographical regions under soybean cultivation are required to be studied to find out if any recombinant between MYMIV and MYMV exists, as begomoviruses are known to show high frequency of recombinations. Phylogenetic study based on comparison of DNA-A nucleotide sequence of YMV isolates with other begomoviruses revealed a unique feature. Members of the genus *Begomovirus* are known to form clusters according to their geographical origin with distinct branches for viruses from America, Africa, and Asia.

Transmission

Female adults of the vector, *B. tabaci*, are more efficient vectors than males. Minimum acquisition feed time is 15 min and the same time is required for inoculation. Increasing feeding period up to 4 h increases transmission ability. Incubation period (latency) in the vector is at least 3 h, optimum being 5–6 h. Preacquisition starvation of the vector increases the efficiency to acquire the virus. In general, the vector is reported to acquire the virus 1–3 days before symptoms appear. A single viruliferous whitefly can transmit the virus but maximum infection is obtained with 10–20 whiteflies per plant. Neither female nor male adults can retain the virus throughout the life span. Normally, the female adults retain infectivity for 10 days and male adults for 3 days.

Epidemiology and Disease Cycle

Disease development is favored when maximum temperature and relative humidity prevail between 29.9°C–36.2°C and 62%–75%, respectively. The earliest YMD appearance of YMD is usually observed at 26–54 days after sowing (DAS). Disease spread becomes evident at 7–32 days after the initial disease appearance. The efficiency of whitefly (*B. tabaci*) as vector is affected by surrounding crops. YMD incidence is lower when soybean is alternated with mung bean. Cross inoculation tests revealed that YMD from mung bean or urd bean (*Vigna mungo*) is not directly transmitted to soybean, but YMD from soybean can be directly transmitted to French bean (*P. vulgaris*), *Alternanthera sessilis*, *Paracalyx scubiosus*, and *Sida rhombifolia* and vice versa. Disease development reaches its peak at 40–60 DAS then decreases thereafter. The distance-wise spread of YMD does not vary among high-, low-, and medium-risk fields. Epidemic development is observed at 60–70 DAS. At 50 DAS, disease development is positively associated with sunshine hours, relative humidity, cloudiness, temperature, and wind velocity (Gupta and Keshwal 2003).

Long-term surveillance study on disease flare-ups revealed that fields near irrigation canals, water points, low-lying areas and foot hills usually show high disease incidence when compared to unirrigated field plains in the state of Madhya Pradesh in India. Soybean cv. JS 81-335 and *Corchorus olitorius* have been found to act as bridge hosts in bringing inoculum of yellow mosaic virus from mung bean to soybean. Plant species such as *P. scubiosus* is found to act as reservoir host of yellow mosaic virus inoculum. In addition, *A. sessilis* (*A. sessilis*) and *S. rhombifolia*, the weed hosts, have been found to help the multiplication and spread of inoculum. The study of weather parameter on yellow mosaic virus and whitefly population revealed that the rate of disease development is high when maximum temperature and relative humidity range between 31.0°C–36.2°C and 62%–75%, respectively (Gupta and Keshwal 2002).

Disease Management

Host Plant Resistance

Highly YMD-resistant soybean cultivars/genotypes such as SL 295, SL 328, SL 525, SL 603, UPSM 534, PK 1029, PK 1024, PK 416, and JS 9305 can be used as parents in crossing programs (Ramteke et al. 2007). Screening under controlled conditions with artificial inoculation with different isolates of the virus and pyramiding genes conferring resistance will help in breeding for durable resistance to MYMV in soybean (Lal et al. 2005, Ramteke and Gupta 2005). Soybean cultivars PK 1042, PK 1046, Pusa 20, and Pusa 40 are resistant to MYMV-Ppl. The resistant cultivars take longer time (16–29 days) to exhibit symptoms compared to susceptible cultivars (8–17 days) after inoculation (Biswas 2002). Soybean cultivar resistant to MYMIV infection induces viral RNA degradation earlier than the susceptible cultivar (Yadav et al. 2009, Yadav and Chattopadhyay 2014). More recently, out of 500 soybean germplasm lines collected from different parts of the world, only 48 genotypes have been detected to be resistant to YMD over 3 years (2007–2009) of consecutive hotspot screening (Kumar et al. 2014).

The inheritance of YMV resistance studied in two highly resistant varieties DS9712 and DS9814 indicated that the resistance is dominant and is controlled by single major gene (Talukdar et al. 2013). Similarly, the YMV resistance in wild accession, *G. soja*, is governed by a single dominant gene (Bhattacharyya et al. 1999) and the segregating populations generated will act as starting materials for developing improved lines with YMV resistance simultaneously paving the way for mapping the gene for YMV resistance with linked molecular marker. It is possible to develop molecular markers linked to MYMIV resistance to facilitate the genotyping of soybean germplasm for MYMIV reaction. Applying linked marker-assisted genotyping, plant breeders can carry out repeated genotyping throughout the growing season in absence of any disease incidence (Maiti et al. 2011).

A construct containing the sequences of Rep gene (566 bp) in antisense orientation has been used to produce MYMIV-resistant soybean plants, and the inheritance of transgene has been found to follow classical Mendelian pattern transgenic lines (Singh et al. 2013).

Vector Control

The management of the disease through prevention of population buildup of the vector can be possible. Spray of 0.1% metasystox, starting when the crop is about a month old or as soon as single diseased plant is seen in the field, can be useful in preventing severe incidence of the disease. However, control of the disease through control of vectors is often not very effective due to the fact that commonly recommended insecticides do not cause instant death of all individual vectors in the vector population and even a very few surviving population is capable of spreading the disease rapidly. Oil sprays can be more effective because they kill the insects within 15 min but they can be phytotoxic. Soil application of granular systemic insecticides at recommended doses can be a much better option for reducing vector population and delaying the appearance of the disease. Some fungal parasites of *B. tabaci* vector have been reported, which are potentially applicable for the development of biological control of the vector.

SOYBEAN CYST NEMATODE

Symptoms

The aboveground visible symptoms and definite signs of cyst nematode attack become detectable in a field only when the cyst content of the soil has gone very high. Foliar symptoms of SCN infection are not unique to SCN infection. In the first few crop seasons, after entry of the nematodes in a field, the disease goes undetected while the population of the cysts continues to rise. The symptoms at this stage could be confused with nutrient deficiency, particularly iron deficiency, stress from drought, herbicide injury, or another disease. The first signs of infection are groups of plants with yellowing of leaves that have stunted growth. High population densities of the SCN can result in large portions of soybean fields with plants that are severely stunted and yellow. Small patches of poorly growing plants may appear in the field. The plants appear as if suffering from poor nutrition. Suspect fields usually have plants of different heights. Temporary wilting of plants occurs during hotter part of the day. Typical aboveground symptoms of heavy soil infestation are stunting and yellowing. Early senescence or maturation of the crop can be an indirect symptom of SCN.

When several crops of soybean are taken in the same field year after year, the patches of sick plants increase in dimension. The nematode feeds on the roots and root stunting, discoloration, and fewer nodules are belowground symptoms of SCN. The pathogen may also be difficult to detect on the roots, since stunted roots are also a common symptom of stress or other plant disease. Signs of root infection are the presence of adult females and white to brown cysts filled with eggs that are attached to root surfaces. Young females are small white and partly buried in the roots, with only part of them protruding on the surface, whereas older females are larger almost completely on the surface of the root and appear yellowish or brown depending on maturity. Once the cysts have matured, they turn brown and fall off the root.

Geographical Distribution and Losses

SCN is thought to be a native of Asia and has been a problem in China and northeastern Asian countries for centuries. The first documented report of damage by the SCN (*H. glycines* Ichinohe) was by S. Hori in Japan in 1915 (Davis and Tylka 2000). SCN was first reported in the United States in 1954 in North Carolina—an area known to import flower bulbs from Japan. It then spread with the expansion of soybean in the soybean belt (Illinois, Indiana, Iowa, Minnesota, Ohio, Missouri, Wisconsin) and adjacent states in the United States. Currently, this nematode causes more than U.S. $1 billion yield losses annually in the States alone, making it the most economically important pathogen on soybean (Liu et al. 2012). A 4-year study (2006–2009) done in the United States revealed that SCN (*H. glycines*) caused annual losses of $1.286 billion (128.6 million bushel). The SCN caused more yield losses than any other disease during 2006–2009 (Koenning and Wrather 2010). Yields may decrease slowly for a number of years as the population of SCN increases in the soil and infection of roots increases. SCN was detected in Colombia, South America, in the early 1980s and was soon thereafter found in Argentina and Brazil—two of the world's important soybean production countries. Yield losses can reach 100% in Brazil (Dias et al. 2009). The mean yield is reported to be 48% greater for the resistant cultivar compared with the susceptible cultivar in Iran (Heydari et al. 2012). SCN has also been reported from Egypt and Italy. In a survey of the top 10 soybean-producing countries in the world, SCN has been found to be the most damaging pathogen of soybean.

The penetration, feeding, and reproduction in soybean roots by the nematode result in direct yield losses and also allow other diseases to invade soybean roots. SCN can reduce soybean yield by more than 30% with no aboveground symptoms. When SCN infestation is severe, plants can become stunted and chlorotic and in some cases die resulting in up to 100% yield losses. In addition to causing yield loss directly, SCN also interacts with other pathogens (*F. virguliforme* and *Phialophora gregata*) making other diseases (SDS and BSR) worse during the same crop season.

Pathogen: *H. glycines* Ichinohe

Classification

Kingdom: Animalia
Phylum: Nematoda
Class: Chromadorea
Order: Tylenchida
Family: Heteroderidae

Subfamily: Heteroderidae
Genus: *Heterodera*
Species: *glycines* Ichinohe

The disease is caused by a microscopic roundworm, the plant-parasitic nematode, that changes shape as it goes through its life cycle, which forms cysts (overwintering structures) on soybean roots. Like all nematodes, the SCN (*H. glycines*) has six life stages—egg, four juvenile stages (J1–J4), and the adult stage. The duration of the SCN life cycle runs from 3 to 4 weeks, but this may be influenced by environmental conditions (mainly adequate temperature and moisture). The first-stage juvenile occurs in the egg; the worm hatches from an egg in the soil to produce the second-stage juvenile, or J2, nematode. The J2 is worm shaped, 375–520 μm long, and about 18 μm in diameter. It is the only life stage that can penetrate roots, and the third and fourth stages occur in the roots. The J2 enters the root moving through the plant cells to the vascular tissue where it feeds. The J2 induces cell division in the root to form specialized feeding sites. As the nematode feeds, in the root, juveniles become males or females and swell. SCN adults are sexually dimorphic, meaning that they are dissimilar in appearance. The females are swollen and sedentary, and the males are vermiform (worm shaped) and motile. The female eventually becomes flask shaped (0.4 mm in length × 0.12–0.17 mm in diam) and swells so much that its posterior end bursts out of the root and it becomes visible to the naked eye. In contrast, the adult male regains a wormlike shape (1.3 mm long × 30–40 μm in diam) and it leaves the root in order to find and fertilize the large females. Higher percentage of males is produced when the nematodes or host plants are under stress. Males do not feed, but they are required for sexual reproduction (copulation) with females that are exposed on the root surface. The male and juvenile stages must be extracted from soil or plant roots to be viewed under a microscope.

The fully developed yellowish-brown lemon-shaped female (0.6–0.8 mm in length × 0.3–0.5 mm in diam) after fertilization continues to feed as it lays 200–400 eggs in a yellow gelatinous matrix, forming an egg sac, which remains inside in its body, but some eggs may be laid in a gelatinous matrix extruded from the posterior (vulva) of the female. The female then dies. Eggs in the gelatinous matrix may hatch immediately, and the emerging second-stage juveniles may cause new infections. Subsequently as the gravid female dies, its cuticle becomes a brown, hardened structure (the cyst) that encases and protects hundreds of viable eggs. Cysts often fall from roots and remain free in the soil.

About 21–24 days is required for the completion of the life cycle of the nematode. Depending upon the environment, several generations of SCN can be completed in a typical soybean-growing season. A significant proportion of eggs that are retained within cysts are in a dormant state—they do not hatch until soybeans are planted for the next growing season. The overall body of the nematode is covered by a flexible, outer *cuticle*. The outside of the cuticle has a series of fine rings (annulations) that allow the cuticle to bend at any point along the nematode's body. The cuticle is composed mainly of the structural protein collagen, and the cuticle is molted four times to allow growth and maturation of the nematode. The *head* of the nematode can be recognized by the presence of a short, dark spear with basal knobs (the *stylet*) just inside the tip of the head. The stylet is hollow (like a hypodermic needle) and protrudes from the head when used by the nematode for feeding from plant cells and penetrating plant tissues. The very outer tip of the nematode head above the stylet (called the *lip* region) is slightly elevated, rounded, and darkened in J2 of SCN. In a relatively clear area just below the stylet, a round, muscular pumping organ called the metacorpus can be seen—the metacorpus pumps substances (i.e., food and secretions) up and down the esophagus of the nematode. Just below the metacorpus is another relatively translucent area that contains three esophageal glands that overlap the nematode's intestine on the ventral (stomach) side of its body. The intestine can be recognized as a fairly long, dark area extending from the esophageal glands to the tail of the nematode. The tail of SCN J2 tapers uniformly to a fine, rounded tip that is hyaline (Davis and Tylka 2000).

Epidemiology and Disease Cycle

Main source of survival of cyst nematode is the cyst. Eggs within the cyst can survive for 10 or more years. As with many plant-parasitic nematodes in soil, SCNs do not move far from the root zone. In most cases, the natural migration of SCN within a field is defined as *contagious*—small patches of infested areas that gradually enlarge to encompass significant areas of disease. The cysts are usually spread along with soil adhering to farm implements or anything that is contaminated with infested soil including seed-size clumps of dried soil within contaminated seed stocks. Surface drainage water, compost, shoes and feet of workers, movement of animals, and wind-borne dry soil are important means of spread. Even waterfowl and other birds feeding in infested fields may ingest cysts and carry them considerable distances. Diseased areas become much more pronounced in sections of soybean fields that are under environmental stress. It is possible that many soil factors may affect SCN reproduction and soybean yield loss, but only two soil factors are commonly associated with SCN damage and population densities—soil texture and soil pH. SCN is capable of infesting soils of all textures, but symptoms and yield loss generally are greater in sandy soils than medium- and fine-textured soils. SCN-infected roots are stunted and lack fine roots and, thus, can explore much less soil for water and nutrients than healthy roots. Also, coarse-textured soils do not hold water and some nutrients as well as medium- or fine-textured soils and SCN seem to cause greater damage to plants stressed by other factors, such as lack of water and/or minerals. The SCN population densities are more strongly related to high-pH soils (Rogovska et al. 2009, Pedersen et al. 2010). Among the microelement treatments, $FeCl_3 \cdot 6H_2O$ is the best one to inhibit J2 survival with the lowest value of LC50 (Zheng et al. 2010). A combination of soil compaction and real-time PCR enables rapid and sensitive quantification of SCN eggs in soil (Goto et al. 2009). SCN does not produce cysts containing eggs at a soil temperature of 33°C, although it does produce eggs at 25°C and 29°C. At soil temperature above 33°C for 200 h or longer, the egg reproduction ratio is significantly suppressed. After cultivation of resistant Peking, the egg number in the soil is significantly suppressed compared to that after Fuki (Uragami et al. 2005).

SCN has been reported to parasitize a broad range of host plants, encompassing nearly 150 legume and nonlegume genera representing 22 plant families. Several SCN host species are common winter annual weeds in U.S. soybean. The influence of winter annual weed management on SCN population densities has received little attention to date and warrants further investigation (Johnson et al. 2008). The SCN shows considerable degree of pathogenic variability all over the soybean-growing countries in the world (Dias et al. 2005, Rocha et al. 2008, Afzal et al. 2012, Asmus et al. 2012, Matsuo et al. 2012). This variability is large in Brazil, where 11 races (1, 2, 3, 4, 4+, 5, 6, 9, 10, 14, and 14+) have been found. Races 4+ and 14+ are found only in Brazil and differ from the classical 4 and 14 races, respectively, for their ability to parasite *Hartwig*, a North American soybean cultivar previously resistant to all races (Dias et al. 2009). Races are characterized by their ability to reproduce on certain soybean varieties. A system of designating races using the four differentials (Pickett, Peking, PI 88788, and PI 90763) has increased the number of potential races to 16 in the United States (Anonymous 2000).

When susceptible crop is planted, some hatching factor from the roots induces release of larvae. Hatching and emigration of larvae take place actively as a result of rise in temperature followed by host penetration and infection. The most rapid development and greatest female production occur between 20°C and 28°C. Male and female ratios do not differ in this range (Melton et al. 1986). However, the male-to-female ratio is the highest at 30°C–35°C (Rocha et al. 2008). Host penetration and infection occurs at a constant temperature of 20°C–22.2°C (Wang et al. 2009). These nematodes invade the root and partially reorganize root cell function to satisfy their nutritional demands for development and reproduction. After SCN hatch from eggs, the infective second-stage larvae penetrate primary roots or apical meristems of secondary roots. The larvae pierce their stylets into and feed off cells of the cortex, the endodermis, or the pericycle, causing the enlargement of these cells. The group of enlarged cells are called syncytia and serve as feeder cells for the nematode.

Syncytia often inhibit secondary growth of both phloem and xylem. Because a short portion of a root may be attacked by many larvae, the large number of syncytia that develop reduces the conductive elements and results in poor growth and yield of soybean plants, especially under stress of moisture (Wang et al. 2000, Alkharouf et al. 2006).

Disease Management

Once established in a field, SCN cannot be eradicated. However, there are various practices that can be implemented individually or in an IPM program to minimize SCN population densities at low to medium levels and maximize soybean yields in infested fields.

Host Plant Resistance

Effective management of this pathogen is contingent on the use of resistant cultivars. The genetic resistance is the most economical and accepted SCN control method by growers. Cultivars resistant to SCN can show greater yields in both high- and low-yielding environments and provide greater yield stability. These data support the selection of new cultivars that yield well at multiple locations and specifically cultivars with resistance to SCN for fields infested with SCN as a method to increase yield and yield stability (de Bruin and Pedersen 2008). However, host resistance must not be the only option because of the high genetic variability of the pathogen. This variability is large in most soybean-growing areas in the world (Dias et al. 2009).

Although more than 100 PIs (exotic varieties) have been identified with resistance to one or more SCN population designations, current resistant varieties trace to only a few PIs such as Pickett, Peking, PI 437654, PI88788, and PI 90763 from the soybean germplasm collection, which is also referred to as Hartwig resistance or the branded CystX® resistance (Anonymous 2008). The most widely used source of resistance is PI 88788. Because of the spread of multiple SCN races in Hokkaido, the Tokachi Agricultural Experiment Station (Japan) has bred soybeans for SCN resistance since 1953 by using two main resistance resources PI84751 (resistant to races 1 and 3) and Gedenshirazu (resistant to race 3) (Suzuki et al. 2012). It is confirmed that race 1 resistance in PI84751 is independently controlled by four genes, two of which are rhg1 and Rhg4. Suzuki et al. (2012) further classified the PI84751-type allele of Rhg1 as rhg1-s and the Gedenshirazu-type allele of Rhg1 as rhg1-g. I.

Having a variety with the correct source of resistance is the first step. Knowing the level of resistance is the second, and equally important, step. The level of resistance is given by the female index or resistance designation. In a general sense, the level of resistance is determined by how many resistance genes the variety has inherited from the original source of resistance. Both the source of resistance and the level of resistance are important for managing SCN in a field.

Resistance is described by no or limited reproduction of an SCN population on a given variety or genotype of soybean. This resistance is due to several (two, three, four, or more) genes being present and interacting in a soybean genotype. Soybean varieties labeled as resistant to SCN vary greatly in yield and in control of SCN. Both are determined by the genetics of the soybean variety and also the genetics of the SCN population in the field. The results of an HG-type test indicate how well a population will be controlled by the various sources of resistance used to develop soybean varieties. It is important to understand that the SCN designation describes the reaction of a population or group of individual nematodes with different genotypes to a source of resistance. Most field populations actually contain individual nematodes that would have different designations, but the population designation describes the average or majority reaction of the individual nematode genotypes in that population (Niblack et al. 2002). SCN-resistant varieties offer significant yield advantages (as much as 50% or more) over susceptible varieties when grown in heavily infested soil. However, variability of the pathogen enables some individuals to reproduce on resistant varieties thus making them less effective. To reduce the possibility of this happening, some researchers recommend that growers alternate the use of the soybean cultivars with different sources of SCN resistance and also that a susceptible cultivar be grown once after all types of available resistance have been rotated. As far as possible, SCN-resistant varieties with other needed defensive traits, such as tolerance to iron deficiency chlorosis or resistance to SDS or *Phytophthora* root rot, should be preferred.

The HG-type test (*HG* represents *H. glycines*, the scientific name for SCN) is designed to give practical information about how well an SCN population in a field can reproduce on the various sources of SCN resistance. The *HG-type* system that has replaced the race system indicates which genetic sources of soybean resistance any given population of SCN can infect.

What population designation (race or HG type) represents the population of SCN individuals in the field is important to know. The most common population designation, for example, in Minnesota, is race 3 (one of 14 HG types). Knowing the population designation in a field is necessary in order to know what source of SCN resistance in the soybean would be most effective for that field (Niblack et al. 2002, Niblack 2005). Since 2003, the HG-type test has been adopted to replace the race test. This new test includes seven sources of resistance (germplasm lines) and the results are shown as a percentage, indicating how much the nematode population from a soil sample increased on each of the seven lines. This test indicates which sources of resistance would be good for the field being tested and which would be poor. Since the genetic sources of resistance are limited in commercially available soybean varieties, it is important to rotate these *sources of resistance* to delay the buildup of a virulent SCN population. Shift in virulence of SCN is associated with use of resistance from PI 88788. Rotation with alternative sources of resistance is recommended as a means to slow the adaptation to PI 88788 (Niblack et al. 2008). To delay SCN populations developing the ability to reproduce on SCN-resistant soybean varieties, producers should grow varieties with different sources of resistance in different years. If it is not possible to obtain the seed of an SCN-resistant variety with a source of SCN resistance different from what had been previously been used, rotate among different SCN-resistant varieties with the common source of SCN resistance, PI 88788 (Anonymous 2008).

The most common strategy applied by soybean genetic breeding programs in Brazil to introduce SCN resistance has been the selection of lines derived from populations resulting from crosses including adapted genotypes and North American cultivars with resistance derived from *Peking* (*Sharkey*, *Centennial*, *Padre*, *Forrest*, *Gordon*, among others) and/or the PIs 88788 (*Bedford*, *Linford*, *Fayette*, *Leflore*, etc.), 90763 (*Cordell*), and 437654 (*Hartwig*). The resistant cultivars are being developed along with the progress of the breeding programs and they, in turn, begin to replace with advantages of the North American resistant sources. Presently, there are about 50 soybean cultivars resistant to SCN in Brazil (Dias et al. 2009). Soybean germplasm lines S01-9364 (Reg. No. GP-350, PI 646156) and S01-9391 (Reg. No. 351, PI 646157) have value as parents in soybean improvement programs because of their broad resistance to SCN (*H. glycines*) populations (Liu et al. 2012).

Molecular Breeding for Resistance to SCN

Molecular mapping of QTL for resistance to SCN and MAS for breeding for SCN resistance have proven useful in order to assist in the development of SCN-resistant soybean cultivars at many major soybean breeding research institutes in the world (Arelli et al. 2010, Delheimer et al. 2010, Carter et al. 2011, Ferreira et al. 2011, Kim et al. 2011, Liu et al. 2011a,b, Mazarei et al. 2011, Vuong et al. 2011, Wu and Duan 2011, Arriagada et al. 2012, Yuan et al. 2012).

Recent advances in *H. glycines* genomics have helped identify putative nematode parasitism genes, which, in turn, will aid in the understanding of nematode pathogenicity and virulence and may provide new targets for engineering nematode resistance (Niblack et al. 2006). Real-time QPCR has been developed for screening for resistant cultivars, which can serve as a prelude to differentiation of resistance levels in soybean cultivars. With the QPCR assay, the time needed to differentiate highly resistant cultivars from the rest is reduced (Lopez-Nicora et al. 2012). This QPCR assay has the potential to replace the traditional female index-based screening and improve precision in determining infection levels.

Methods for MAS for SCN resistance have been identified (Young and Mudge 2002). Yields of the resistant cultivars are greater than those of the susceptible cultivars, except for the Peking source.

Compared with the susceptible cultivars, cultivars with *H. glycines* resistance from PI 88788 give a 13% increase in yield associated with a 15% increase in growth during R1–R5 growth stages. In cultivars with resistance from Hartwig, a 6% increase in yield is associated with a 4% increase in R1–R5 growth stages duration and increased seed-set efficiency. This work demonstrates that yield increases due to resistance to *H. glycines* can be attained by different physiological mechanisms associated with the different resistance sources and probably are controlled by different genes. This opens the possibility of pyramiding genes conferring resistance by different mechanisms (Rotundo et al. 2010). Recent advances in the study of the interaction between soybean and SCN at the genetic and genomic levels have been reviewed (Mitchum and Baum 2008). A total of 17 QTL mapping papers and 62 marker-QTL associations have been reported for resistance to SCN in soybean. SCN-resistant QTLs have been classified into three categories: suggestive, significant, and confirmed. Confirmed QTLs are credible and can be candidates for fine mapping and gene cloning. QTLs on linkage groups (LGs) G, A2, B1, E, and J are classified as confirmed. QTLs on LGs B2, C1, C2, D1a, D2, L, M, and N are classified into suggestive or significant. A relationship between soybean QTLs and SCN races has been reviewed (Guo et al. 2006).

Soybean PI 404198A is one of the newly identified sources that can provide a broad spectrum of resistance to SCN. QTL has been identified to be associated with resistance to SCN races 1, 2, and 5 in PI 404198A. LGs G and A2 are associated with resistance to race 1. Soybean PI 404198A may carry rhg1 on LG G, Rhg4 on LG A2, and a QTL on LG B1 (Guo et al. 2006). A SNP linked to the QTL of SCN resistance has been validated by comparing sequences amplified from *Hartwig*, a broad-based SCN-resistant line, and *Williams 82*, an SCN susceptible line (Gua et al. 2005, 2006).

Chemical Control

Chemical control with nematicides is not normally used because the economic and environmental costs are prohibitive. There are a few nematicides that are labeled for use against SCN, including the fumigant 1,3-dichloropropene (Telone) and the nonfumigants aldicarb (Temik or Bolster) and oxamyl (Vydate). When applied at planting, the effect of the nematicides may last long enough to provide an economic yield benefit (Schmitt et al. 2004). The performance of the nematicide will depend on soil conditions, temperatures, and rainfall. Yield and economic benefits generally are not guaranteed, but the chemicals are suggested to be applied at the soil depths of 5–15 cm, which can last for 1 month after soybean-seedling emergence for the effective management of the nematodes (Wang et al. 2009). Supplementing resistance with chemicals may improve soybean yield and/or nematode management, so a nematicide application, Aldicarb, {aldicarb[2-methyl-2 (methylthio) propionaldehyde *O*-(methylcarbamoyl) oxime]}, when included in the schedule, increases total plant biomass by 9% during R1–R5 soybean growth stages.

Cultural Control

SCN cannot reproduce if host plants are not present. Hence cultural practices, such as crop rotation, are useful as an effective tactic for SCN management. Because SCN is an obligate parasite (requires a living host), a crop rotation involving SCN nonhost plants like corn, alfalfa, small grains, sunflowers, flax, and canola can decrease the population of SCN (Jackson et al. 2005). For example, annual rotation of resistant soybean and corn results in the lowest SCN population density and produces the highest yield of both crops (Chen et al. 2001, 2007, Chen 2007). Similarly, the 2-year corn–soybean rotation generally results in increased soybean yield, decreased winter annual weed growth, and reduced SCN population density in comparison to when soybean is followed by soybean (Mock et al. 2012). In the North Central region of the United States, corn is almost exclusively used as a nonhost rotation crop with soybean (Miller et al. 2006). However, the data suggest that a single year of rotation of soybean with any other crops like sunflowers and flax before planting a susceptible soybean may not be sufficient in managing SCN (Miller et al. 2006). SCN-resistant soybean cultivars often are incorporated into a multiyear cycle of rotations with nonhost crops—this combination of practices is an excellent integrated management strategy

(Kulkarni et al. 2008, Dias et al. 2009). A certain percentage of SCN individuals can reproduce on resistant varieties. If sources of resistance are not rotated, these individuals can produce a SCN race shift. This will reduce the effectiveness of genetic resistance available in commercial soybean varieties.

Plants that have adequate moisture and nutrients are better able to withstand infection by SCN. In land infested with SCN, maintaining proper soil fertility and pH levels and minimizing other plant diseases, insect, and weed pests that weaken the plants are more critical to maximizing soybean yield than when land is noninfested.

The movement of soil can be best managed by following sanitation practices. If only certain fields on a farm are infested, planting and cultivating of infested land should be done only after noninfested fields have been worked. Soil on equipment should be thoroughly removed with high-pressure water or steam, if available, after working in infested fields. Also, seed grown on infested land should not be planted in noninfested fields unless the seed has been properly cleaned; SCN may be spread in the seed-size soil clumps mixed in with the seed (Davis and Tylka 2000, Schmitt et al. 2004, Donald et al. 2009).

Poultry litter at rates of 8 tons/ha when applied to SCN-infested soil results in the highest reduction in the number of SCN females and egg production (Lima et al. 2011). Anaerobically digested swine manure, which is actually the volatile fatty acid (VFA) manure, when applied to the soybean fields every 35 days, gives better results in reducing the SCN counts by 18%–34% (Xiao et al. 2007). Potassium fertilization at 150–600 mg/dm^3 (Pinheiro et al. 2009) and shallow tillage have been found to be advantageous to decrease the SCN population and to promote the suppressive effects of nonhost or trap crops, such as maize, crotalaria, and red clover (Tazawa et al. 2008).

Biological Control

Cysts and eggs of SCN are often found infected with one of several fungi such as *Fusarium*, *Verticillium*, *Neocosmospora*, *Dictyochaeta*, and more recently *Hirsutella minnesotensis* and *Hirsutella rhossiliensis* (Schmitt et al. 2004, Liu and Chen 2005). Biocontrol methods can play an important role in suppressing occurrence and damage of the nematodes (Chen et al. 2011). *H. minnesotensis* and *H. rhossiliensis* are endoparasites of nematodes, and their biological control potential against *H. glycines* is well known (Liu and Chen 2005). In general, percentage reduction of egg population density in the soil is negatively correlated with soil pH and positively correlated with sandiness. There appears to be no or weak correlation between egg reduction and organic matter. Soil pH and/or texture is important in influencing biocontrol effectiveness (Liu and Chen 2009). *Verticillium chlamydosporium* is another fungal BCA of SCN. Zn^{2+} stimulates the hatching of eggs of SCN. Cu^{2+}, Mn^{2+}, and Fe^{2+}, however, decrease hatching and Cu^{2+} could, therefore, be applied as a supplement to the biological control formulation (Xing et al. 2002). *H. minnesotensis* isolates vary in their efficacy in reducing the nematode population (Qian et al. 2011).

H. rhossiliensis controlled *H. glycines* more effectively in J2-infested soil than in egg-infested soil. Monitoring the population dynamics of a BCA in soil can be precisely studied with real-time PCR and bioassay (Zhang et al. 2008). Natural suppression of SCN exists and becomes increasingly attractive; however, ecological mechanisms leading to the suppressive state are rarely understood. Both bacteria and fungi are potentially involved in the soil suppressiveness to SCN: soil disturbance and biocide application may reduce natural soil suppressiveness that could be potentially associated with soil nematode community diversity and microbial enzyme activities (Bao et al. 2011).

Certain species of arbuscular mycorrhizal (AM) fungi could effectively inhibit the infection processes of SCN. It is proved that the tested AM fungi could significantly decrease SCN damage, reduce disease severity, the number of cysts on roots, the number of cysts and the second-stage juveniles (J2) in the rhizospheric soil, and the number of eggs per cyst. Among the AM fungi tested, *Glomus fasciculatum*, *Gigaspora margarita*, and *Glomus intraradices* are much more effective than *Glomus mosseae* and *Glomus versiforme* against the infection process of SCN (Li et al. 2002).

Effect of Plant Extracts

Mortality of SCN female induced by aqueous extract of the neem plant branches, leaves, and seeds is reported to be 99%, 97%, and 99.9%, respectively. The number of females on the root system when determined 30 days after the incorporation of 15 g of whole leaves/kg of soil or 10 g each of ground branches, whole seeds, and ground seeds/kg soil, the number of females recovered after incorporation of whole leaves, ground branches, whole seeds, and ground seeds has been found to be 1, 32, 9.1, and 0.8/root system, respectively, the differences being significantly different (5%). The number of females in the control roots could be 61, indicating the presence of toxic compounds in neem (Rodrigues et al. 2001). Overall, *Lolium multiflorum* is the most effective of all plant species tested for reducing populations of *H. glycines*, by increasing egg hatching of the nematode in the absence of a host, depleting lipid reserves of the juveniles, and inducing the lowest nematode parasitism of all nonhost residues studied (Riga et al. 2001).

OTHER DISEASES OF SOYBEAN

Brown Spot

Brown spot of soybean is caused by *Septoria glycines Hemmi* (teleomorph: *Mycosphaerella uspenskajae Mashkina & Tomilin*) and occurs in most soybean-growing regions in the world particularly in Argentina, Brazil, China, Pakistan, and the United States. Angular RB spots that vary in size from a pinpoint to 1/5 in. may appear on the lower leaves. Infected leaves turn yellow and fall prematurely. In severely infected fields, the lower half of the plant may lose all its leaves. The primary infection source of *S. glycines* is mainly from conidia within pycnidia surviving in plant residues infected in the previous year. The infection of *S. glycines* may be limited by the duration of water retention on the leaf, with a period of at least 24 h required. Warm, moist weather and poor drainage favor the spread of the disease. Management measures include the use of disease-free seed, crop rotation, deep burial of crop residue, and use of strobilurin foliar fungicides at R3 growth stage (Mirza and Ahmed 2002, Mantecon 2008, Carmona et al. 2010, Cruz et al. 2010).

Downy Mildew

The disease is caused by the fungus *Peronospora manshurica (Naumov) Syd.*, which is of quarantine significance (Singh et al. 2003). It is the most widespread disease of soybean in the world. First, symptoms appear as indefinite yellowish-green areas on the upper leaf surface. Later, these areas become light- to dark-brown spots with yellow-green margins (Figure 9.9). In years favorable

FIGURE 9.9 Downy mildew of soybean. Note the initial symptoms on leaves. (Courtesy of Dr. Shrishail Navi, Iowa State University, Ames, IA.)

for the development of the pathogen, yields of susceptible cultivars may be considerably reduced. Disease symptoms may be systemic and local. Most typical symptoms occur on leaves, in the form of chlorotic spots, which necrose and coalesce with time. Conidiophores and conidia grow over the reverse side of the leaf. *P. manshurica* survive through oospores, which reside on seeds and plant residues. In the course of the growing season, *P. manshurica* proliferates by conidia (Vidic and Jasnic 2008a). Management measures include the growing of resistant cultivars. Soybean cultivar *AGS129* is resistant to downy mildew. Marker OPH-021250 has been found to be present in 13 of 16 resistant soybean cultivars so investigated and absent in susceptible cultivars, thus confirming a potential for MAS for breeding for downy mildew resistance (Chowdhury et al. 2002). The use of healthy seed or seed treatment with fungicides based on metalaxyl, oxadyxil, and mancozeb could be useful in preventing the spread of inoculum through seed.

Purple Seed Stain

This disease is caused by the fungus—*Cercospora kikuchii* (Matsumoto & Tomoyasu) M. W. Gardner. It is reported to occur in almost all soybean-growing regions in the world. This disease often appears late in the season and can cause leaf blighting and staining of the seed. Yield losses are often minimal, but a reduction in seed quality can occur due to staining. In most cases, 7%–13% reduction in emergence can occur in the field. Leaves often have red to purple lesions, less than 1 cm in diameter, which become noticeable in August or early September. Infected seed has a distinctive purple discoloration (purple seed stain), varying from violet to pale purple to dark purple over part or all of the seed coat (Figure 9.10). This discoloration is often confined to the upper two layers of the seed coat. Size of the discoloration may vary from a small spot to the entire seed surface. The pathogen attacks other plant parts and overwinters in diseased leaves and stems as well as in infected seed. Premature defoliation may occur when leaves are severely infected. When infected seeds are planted, the fungus grows from seed coats and infects seedlings. This serves as a primary source of inoculum. Wet weather during the growing season favors the development of the disease. RH above 80% and temperature from 20°C to 24°C are more favorable for the germination of conidia and disease development (Kudo et al. 2011). The fungus overwinters in diseased crop residue as well as on infested seed.

There is a high degree of genetic variability and cercosporin production among isolates (Lura et al. 2011), and the population genetic structure of *C. kikuchii* is different between South America and Japan (Imazaki et al. 2006a). The disease management involves using a variety with greater tolerance. Three genotypes, AG5701 (Asgrow), TV59R85 (Terral), and PI80837, are among the more resistant cultivars to the disease (Jackson et al. 2008, Cai et al. 2009). Clean seed and a fungicide (azoxystrobin or carbendazim) seed treatment crop rotation and removal of residue to reduce infection have been potentially useful strategies in disease management (Imazaki et al. 2006b, Prasanth and Patil 2007).

Frogeye Leaf Spot

FLS, caused by *C. sojina* K. Hara, is a common disease of soybean in most soybean-growing countries of the world. Significant yield losses of soybean (10%–60%) have been attributed to FLS under hot and humid growing conditions (Mian et al. 2008). This disease usually appears late in the growing season and the economic impact is usually minimal. The fungus infects leaves, stems, and pods but is most conspicuous on the leaf. Symptoms occur in midseason and then become more severe after flowering. On the leaf, it causes an *eyespot* lesion composed of a gray or tan central area surrounded by a narrow RB margins. Lesions are 1–5 mm in diameter with a tan center and a dark-red/brown border. Older lesions coalesce, and leaves may appear ragged or with a slight slit in the center of the lesion. Badly infected leaves fall prematurely. The fungus is seed borne and also overwinters in residue and causes weak seedlings.

Eight genotypes such as ID, LMD, NLC, DI, PLLA, Cristalina, Davis, and Uberaba are the most resistant. The additive, dominant and epistatic genetic effects are important for the expression

Cercospora blight and purple seed stain (Cercospora kikuchii)

(a) Symptoms of cercospora leaf blight

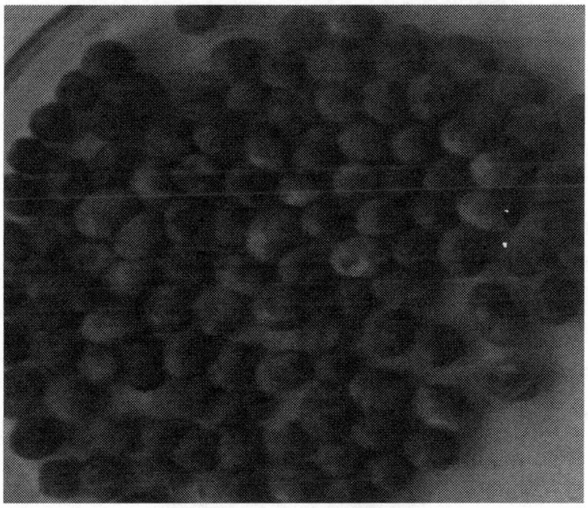

(b) Purple stain on the seeds

FIGURE 9.10 Purple seed stain of soybean. Note the symptoms on leaf (a) and seed stain discoloration (b). (Courtesy of Dr. G.K. Gupta, ICAR-Directorate of Soybean Research, Indore, India.)

of resistance, although the additive genetic effect is reported to be the most important component. These cultivars can be recommended as parents in soybean breeding programs for enhanced resistance to *C. sojina* (Gravina et al. 2004). Advances in research on soybean resistance and inheritance and breeding of resistance against the fungal pathogen *C. sojine* [*C. sojina*] have been reviewed (Cao and Yang 2002). Results demonstrate that the resistance to *C. sojina* is controlled by a dominant gene or a gene block; additive genetic effect and dominance are involved; the effect of the environmental variation is minimum; and the interaction among the genes ranges from the partial to the complete dominance type, depending on the characteristic used in the evaluation of the resistance (Martins Filho et al. 2002). Advances in research on soybean resistance and inheritance and breeding of resistance against the fungal pathogen *C. sojine* [*C. sojina*] have been reviewed (Cao and Yang 2002). *C. sojina* is a dynamic pathogen with extensive virulence or race diversity. Twelve differentials and 11 races of the pathogen have been identified, which should provide the foundation for the identification and comparison of additional soybean resistance genes and new races of *C. sojina* (Mian et al. 2008). Management measures include planting disease-free seed and plowing under crop residue and crop rotation with nonhosts, such as corn or wheat.

Sclerotium Blight (Southern Blight)

Southern blight or southern stem blight is caused by the fungus *S. rolfsii* Sacc. This fungus survives in the soil on organic matter, is favored by hot weather stress, and is recognized by the appearance of white mold on stems at the soil surface causing rotting of stems and roots. Small tan to brown, *mustard seedlike* fruiting bodies (sclerotia) are produced within the white mold growth (Figure 9.11). The disease is most often seen in June, July, and August during very wet periods. Southern blight

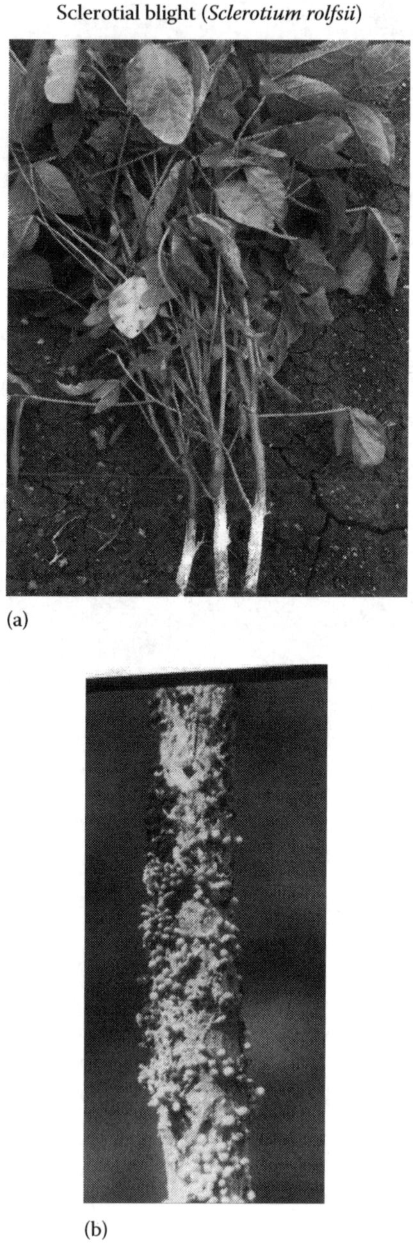

FIGURE 9.11 Sclerotium blight of soybean. Note the presence of fungal growth and mustard seedlike sclerotia on affected plant. (a) White cottony mats of mycelium of pathogen in collar region of seedlings. (b) Reddish-brown sclerotia of pathogen on lower portion of stem of seedling. (Courtesy of Dr. G.K. Gupta, ICAR-Directorate of Soybean Research, Indore, India.)

is very common in fields with moderate to high levels of root-knot nematode (RKN). Occurrence of southern blight in a field is erratic and generally only individual plants are affected. However, in some instances, large numbers of plants may be killed. Plants may be affected at any stage of growth. The first symptom is sudden wilting and subsequent death. The sclerotia are the resting stage of the fungus and will persist in the soil for years. The fungus occurs widely in many soils and is capable of persisting on almost any type of organic matter. On the basis of oxalic acid (OA) production in culture filtrates and pathogenicity on different soybean varieties (cvs. Improved Pelican, Lee, Hardee, and Bragg), the isolates have been grouped into 12 races. Race I is more dominant than other races and highly virulent to all the soybean varieties. There appears to be a positive correlation between OA production and the virulence of the isolates of *S. rolfsii* (Ansari and Agnihotri 2000). It is difficult to manage the disease. Seed treatment with some fungicides such as carboxin and thiram may be effective to limited extent. But integrated approach involving rotation with other crops such as cotton or corn and soil amendments with organic matter can be effective in reducing the inoculum of the pathogen. Dried powders of kudzu (*Pueraria lobata*), velvet bean (*Mucuna deeringiana*), and pine bark (*Pinus taeda*) each at the rate of 25 g/kg stimulate increases in populations of antagonistic microorganism such as *Trichoderma koningii* and *Penicillium citreonigrum* and *Penicillium herquei* and are useful in reducing the incidence of the disease (Blum and Rodriguez-Kabana 2006b). *Bacillus thuringiensis* subsp. *israelensis* has been used to produce chitinase. The addition of chitinase (0.8 U/mg protein) causes increase in seed germination to 90%. *B. thuringiensis* chitinase may contribute to the biological control of *S. rolfsii* and other phytopathogenic fungi in soybean seeds in IPM programs (Reyes-Ramirez et al. 2004).

Diaporthe Pod and Stem Blight/Phomopsis Seed Mold

A complex of soybean diseases is caused by *Diaporthe/Phomopsis* species (D/P complex). D/P complex is grouped into two major taxa: *Diaporthe phaseolorum* var. *sojae* (*Lehman*) *Wehm.* (anamorph = *Phomopsis*) and *Phomopsis longicolla Thomas W. Hobbs.*, which are described as soybean pathogens. The first species includes three varieties: *D. phaseolorum* var. *sojae* (anamorph: *P. sojae*), the causal agent of pod and stem blight, and *D. phaseolorum* var. *caulivora* and *D. phaseolorum* var. *meridionalis*, agents of northern and southern stem canker, respectively. In addition to distinguishing interspecific and intraspecific variability, molecular markers allow the detection of differences among isolates within the same variety (Pioli et al. 2003). *D. phaseolorum* var. *caulivora* (northern stem canker) is the most economically important because it causes wilt and drying of plants during pod development and grain filling. Prematurely wilted plants yield 50%–62% less than healthy plants. *P. longicolla* is the most common and most damaging agent of soybean seed decay. The diseases caused by parasites from this D/P complex genus were first observed and described on soybean in the United States. Presently, they are widespread in most soybean production regions around the world (Li et al. 2004, Santos et al. 2011, Vidic et al. 2011).

Pod and stem blight is caused by the fungus *D. phaseolorum* var. *sojae* (sexual stage) also known as *P. longicolla* (asexual stage). Although plants are infected early in the season, symptoms do not become apparent until after midseason. The disease is identified by the numerous small, black fruiting bodies (pycnidia) appearing on stems and pods of infected plants. The pycnidia are arranged in linear rows on the stems, which is a useful diagnostic criterion to differentiate it from the brown spot and anthracnose diseases symptoms and signs. The pycnidia of the brown spot fungus and acervuli of anthracnose fungus do not occur in rows. RAPD and PCR-RFLP showed that significant variability exists within the population of *D. phaseolorum* var. *sojae*. Infected harvest residues and soybean seeds are the main sources of pathogen inoculum. Humidity and temperature (soil and air) are the main factors that affect the dynamics of fruiting body formation, spore release, establishment of infection, and the development of disease symptoms in soybean. The fungus infects seed and causes them to be shriveled, moldy, and smaller than normal. Seed may be infected but appear normal. Seed infection is the most serious phase of the disease. When infected seeds are planted, the embryo is often killed before emergence or the seedlings are killed at an early stage. Delayed harvest results in an increased incidence of the disease, especially if rain or humid weather and warm conditions prevail. At this period of time, the pathogen is predominantly present in its asexual stage, *P. longicolla*. *P. longicolla* is the primary agent of seed decay and latent infections of seed, although the other members of this genus may cause identical symptoms (Mengistu et al. 2009b). It is characterized by fine cracks that usually develop near the hilum of the infected seed. A white or gray mold may be visible on the seed surface. The yield, grade, viability, and vigor of the seed can be reduced. Yield losses occur because severely infected seeds remain small and light and may be lost during harvest and cleaning operations. Isolation of *P. longicolla* from seed is negatively correlated with percentage of seed germination in irrigated environments but not in the nonirrigated environment (Mengistu et al. 2009b). The fungus overwinters in seed and crop debris. Spores of the fungus are splashed onto developing plants early in the season. Warm, wet, and humid weather during pod fill favors disease development. Varieties that mature late during the cool weather in the growing season or varieties that are short season for an area tend to mature earlier before environmental conditions become warmer and more favorable for seed mold should be preferred. Pod and stem blight can be controlled or reduced by integrating one or more of the control practices such as the use of planting pathogen-free seed, planting later, crop rotation, plowing under soybean debris, and a well-timed harvest.

Target Spot

Target spot is caused by the fungus, *Corynespora cassiicola* (Berk. & M. A. Curtis) C. T. Wei. It is found in most soybean-growing countries, particularly in Brazil and the United States. It is considered to be a disease of limited importance, although its incidence is increasing all over the tropical and subtropical regions. Under favorable climate conditions, it can cause serious damage to soybean. It has become an economically important disease in Brazil in the recent past (Teramoto et al. 2013) and assuming increasing importance in the southeastern United States (Koenning et al. 2006). Symptoms consist of the development of RB leaf lesions that are round to irregular varying from specks to mature spots, which are a centimeter or more in diameter. A dull green or yellowish-green halo commonly surrounds the lesions, which often become concentrically ringed at maturity, hence, the name target spot. Severely affected leaves fall prematurely. Microscopic examination of the lesions can reveal the presence of spores (conidia) typical of *C. cassiicola*. Conidia mostly three to five septate with a central hilum at the base and may range in size from 7 to 22 wide × 39 to 520 μm long. Dark-brown specks to elongated, spindle-shaped lesions form on the petioles and stems. Pod lesions are round and small but may enlarge and merge to cover the entire pod during wet or very humid periods. The fungus may sometimes grow through the pod wall and form small, blackish-brown lesions on the seeds. Large lesions form on the primary roots and growth of the secondary roots may be retarded. The *Corynespora* fungus overseasons in infected soybean debris and seeds and can survive in fallow soil for more than 2 years. The fungus can colonize a wide range of plant residues in soil as well as the cysts of the SCN. Leaf infections occur when free moisture is present on the leaves and the relative humidity is 80% or above. Heavy rainfall associated with hurricanes enhances the disease incidence during September 2004 in the southeastern United States (Koenning et al. 2006). Dry weather inhibits infection and colonization in both leaves and roots. Stems and roots first become infected in the seedling stage. Soil temperatures of 15°C–18°C are optimal for infection and disease development. The pathogen has an extremely wide host range and infects many plant species. There is an abundance of unrecognized genetic diversity within the species and provides evidence for host specialization on certain hosts such as papaya (Dixon et al. 2009). Management practices include the use of disease-resistant cultivars, sanitation involving destruction of crop residues, and avoiding soybean monoculture. Fungicides are rarely justified economically. Among the biological control agents, *T. harzianum* strain RMA-6 and *T. pseudokoningii* strain HMA-3 are reported to be the most effective and can be potentially useful in target spot disease management in soybean (Kaushal 2009).

Sclerotinia Stem Rot

Sclerotinia stem rot (*SSR*) or White mold disease caused by *Sclerotinia sclerotiorum* (Lib.) deBary is prevalent in all soybean-growing regions in the world, being most important in temperate regions and under cool conditions, often at intermediate altitude. Local epidemics outbreaks of the disease have been reported from Argentina, Brazil, Canada, Serbia, and the United States, when weather conditions are favorable for disease development (Vidic and Jasnic 2008b, Alvarez et al. 2012). It is most damaging in years with frequent and abundant rains in the summer. In some fields, more than 50% of plants are infected, causing significant yield reductions (Peltier et al. 2012). Symptoms occur on all aboveground plant parts in the form of thick, white, and soft mycelia and hence the disease is also referred to as *cottony soft rot*. The infected plant parts rot and the infected plants wilt and dry up. *S. sclerotiorum* is polyphagous. The fungus survives by sclerotia, which remain vital in the soil for several years. They germinate and give rise to the mycelium. Alternatively, under favorable weather conditions (humid and cool weather), bowl-shaped fruiting bodies (apothecia) form on the sclerotia. Infection follow colonization of injured or senescent tissue like flowers, cotyledons, or leaves, either by germinated ascospores from asci from the apothecia or directly from mycelium from sclerotia. *S. sclerotiorum* secretes OA and endo-polygalacturonase (endo-PG), which are important pathogenic factors in host plants (Favaron et al. 2004). Symptoms first appear as a watery-soaked lesion followed by cottony growth on the affected plant part with the formation of black irregular-shaped sclerotia. Epidemic development of the disease is favored by temperatures less than 21°C and secondary spread has been shown to occur at 18°C. Continuous moisture on leaves within the canopy or on infected flowers for a period of 48–72 h favors infection by ascospores. There is a potential for field to field dispersal of *S. sclerotiorum* and the majority of ascospores of *S. sclerotiorum* are deposited close to the source (apothecia), where a concentrated area or point source of *S. sclerotiorum* inoculum exists (Wegulo et al. 2000). Forecasting is based on soil moisture, canopy enclosure, senescing leaves, air and soil temperature, and the presence and number of apothecia.

Long-term crop rotation (corn–soybean rotations and compost amendment (Rousseau et al. 2007, Vidic and Jasnic 2008b)) with no soybean tillage (Gracia-Garza et al. 2002) are currently the major methods of controlling this disease. However, 1 year of moldboard plowing will bury sclerotia at least 10 cm in soil and delay the production of apothecia. How this affects SSR development depends on the other factors involved with disease development (Mueller et al. 2002). Fungicides such as thiophanate methyl are another option for the control of SSR but usually recommended in situations where susceptible cultivars must be grown or modification of cultural practices are not disease control options (Muller et al. 2004). The incidence of SSR can be reduced by planting partially resistant cultivars and by implementation of cultural practices that limit pathogen activity. *BCAs* such as *Coniothyrium minitans* CON/M/91-08 (product name: ContansReg. WG), *Streptomyces lydicus* WYEC 108 (ActinovateReg. AG), *T. harzianum* T-22 (PlantShieldReg. HC), *B. subtilis* QST 713 (SerenadeReg. MAX) (Zeng et al. 2012), *Sporidesmium sclerotivorum* [= *Teratosperma sclerotivora*] (Rio et al. 2002), *Clonostachys rosea* BAFC3874 (Rodriguez et al. 2011), and *B. amyloliquefaciens* strains ARP23 and MEP218 (Alvarez et al. 2012) all have been proved to be very effective in reducing the inoculum potential and SSR incidence in soybean.

Few genetic sources of resistance to the pathogen are available to breeders. Therefore, farmers have a continuing demand for new approaches to control the disease. The QTL associated with resistance to *S. sclerotiorum* in soybean genotypes PIs 391589A and 391589B have been identified (Arahana et al. 2001, Guo et al. 2008). SSR markers associated with resistance QTL mapped for SSR resistance may be useful for marker-assisted breeding programs in soybean (Vuong et al. 2008). Biotechnology opens a new avenue to manage this pathogen. Several strategies, including detoxification, defense activation, and fungal inhibition, have potential to engineer *Sclerotinia resistance* (Lu 2003). Enzymes capable of degrading OA have been utilized to produce transgenic resistant plants. Transgenic soybean lines containing the decarboxylase gene (oxdc) isolated from a *Flammulina* sp. have been produced by the biolistic process. Molecular analysis reveals successful incorporation of the gene into the plant genome and shows that the OA decarboxylase (oxdc) gene has been transferred to the progeny plants (Cunha et al. 2010). An oxalate degrading enzyme, oxalate oxidase (OxO), in transgenic soybean has provided white mold resistance equivalent to the best commercial cultivars in a white mold–susceptible background (Donaldson et al. 2001, Cober et al. 2003).

Rhizoctonia Aerial/Foliar Blight

Rhizoctonia aerial/foliar blight of soybean is reported to occur in tropical and subtropical areas worldwide. It is becoming increasingly more important in Brazil (Ciampi et al. 2005, 2008), North Korea (Kim et al. 2005), and the southern United States particularly in Louisiana and North Carolina (Stetina et al. 2006) during prolonged periods of high humidity and high temperatures. This disease of soybean occurs with high disease severity of 50%–75% in tidal swamp land in south Kalimantan in Indonesia (Rahayu 2014). It has been estimated that the disease can cause about 70% losses of foliage and soybean pods. Extensive yield losses (40%–50%) have been reported in soybean when conditions favor disease development. However, Meyer et al. (2006) recorded that *Rhizoctonia* foliar blight of soybean causes higher yield reductions in the range of 60%–70%. Foliar symptoms often occur during late vegetative growth stages on the lower portion of the plant following canopy closure. Initially, leaf symptoms appear as water-soaked, grayish green lesions that turn tan to brown at maturity. The pathogen may infect leaves, pods, and stems in the lower canopy. RB lesions can form on infected petioles, stems, pods, and petiole scars. Long strands of weblike hyphae can spread along affected tissue and small, dark-brown sclerotia form on diseased tissue. Weblike hyphae of *R. solani* spreading along the stem of soybean becomes evident.

The Basidiomycete fungus *R. solani* Kuhn anastomosis group (AG)-1 IA (*Thanatephorus cucumeris* (Frank) Donk) is a major pathogen foliar blight of soybean all over soybean-growing regions in the world. But *R. solani* AG1-IB and AG2-3 are also reported to be the causes of foliar blight of soybean in Japan (Meyer et al. 2006). The pathogen overwinters as sclerotia in soil or plant debris from the preceding crop. During warm, wet weather, mycelium spreads extensively on the surface of plants, forming localized mats of *webbed* foliage. Spread from these localized areas can be rapid when conditions favor disease (high RH and 25°C–32°C). Because this pathogen also causes sheath blight of rice, soybean fields that follow rice with a history of sheath blight are likely to have high incidence of aerial blight. There is little host resistance to *R. solani* in soybean, but some cultivars are less susceptible than others. Planting the least susceptible and best adapted cultivar, rotating it with poor or nonhost crops such as corn or grain sorghum for 2 years, and avoiding narrow row widths and high plant populations are good management practices. When aerial blight is present in highly susceptible cultivars and environmental conditions are favorable for disease, preventive fungicide (strobulirin) applications are the most effective treatments in disease management. Several BCAs including *Trichoderma* species have been reported to be potentially useful in disease management. Nonpathogenic binucleate *Rhizoctonia* spp. (BNR) on the biocontrol of diseases caused by *R. solani* on many crops has been reported in the literature. BNR can induce resistance on soybean against the foliar blight caused by *R. solani* anastomosis group (AG) 1 IA (Basseto et al. 2008).

Powdery Mildew

Powdery mildew caused by the fungus *Microsphaera diffusa* Cooke & Peck (syns. *Erysiphe polygoni DC* and *E. glycines F. L. Tai*) is a minor but common disease of soybeans in many soybean-growing regions of the world particularly in Brazil (Knebel et al. 2006, Araujo et al. 2009) and the United

States (Grau 2006a) and under greenhouse conditions. In its early stages, powdery mildew may be recognized by the presence of small colonies of thin, light-gray or white fungus spreading rapidly on the upper surface of the leaf. Reddening of the underlying leaf tissue sometimes is evident. In time, the whitened areas of fungus enlarge but seldom coalesce to cover all the leaf surface. Many white, powdery patches form on both leaf surfaces and on the stems and pods about midseason. These areas may enlarge to cover much of the aboveground plant parts. Photosynthesis and transpiration are drastically reduced. On very susceptible cultivars, severely affected leaves may turn yellow, wither, and drop prematurely. Heavily infected pods usually contain shriveled, deformed, undeveloped, and flattened seeds. However, the soybean seeds do not become infected. Soybean planted late for a region will lose more yield to powdery mildew than early-planted soybeans. Yield losses ranging up to 14% have been attributed to the disease during certain years when cooler than normal temperatures prevail from flowering to maturity.

Infection occurs when microscopic asexual spores (conidia) land, germinate, and penetrate the epidermal cells. The conidia form several germ tubes, with the first attaching itself to the cells via an anchorage structure (appressorium). A thin filament (infection peg) forms under the appressorium and penetrates the host epidermis. This gives rise to the first feeding structure (haustorium), the only fungus structure found inside the host cells. The rest of the fungus body, or mycelium, grows superficially over the epidermal cells. Conidiophores (asexual fruiting structures) soon develop, giving rise to chains of conidia. Wind-borne conidia start new infections and repeat the disease cycle continuously until soybean plants tissues are available. Cool weather (18°C–24°C) favors disease development, while temperatures above 30°C arrest the growth and reproduction of the fungus. During rainy periods, conidia are washed away, temporarily delaying the secondary spread of the fungus. Speck-sized, black fruiting bodies (cleistothecia) sometimes are produced in mildew colonies late in the fall. Inside the cleistothecia, yellow ascospores (sexual spores) are produced in saclike structures called asci. It is believed that ascospores are released in the spring and serve as primary inoculum.

The only economical management method is to plant resistant soybean varieties. Certain varieties are susceptible in the seedling stage and express resistance about flowering time while others are resistant throughout their lifetime. A single dominant gene has been identified in soybean genotype PI 243540 that provides season-long resistance to powdery mildew, and the powdery mildew dominant resistance gene in PI 243540 has been mapped with PCR-based molecular markers. The map position of the gene is slightly different from previously reported map positions of the only known Rmd locus, which is tentatively called Rmd_PI243540, near the previously known Rmd locus on chromosome 16. The molecular markers flanking the gene will be useful for MAS of this gene (Kang and Mian 2010).

The yield increase due to fungicidal treatments such as thiophanate methyl is usually higher in soybean cultivars that are susceptible to powdery mildew (Hoffmann et al. 2004). Sewage sludge increments elicitation of phytoalexins in soybean and the severity of powdery mildew is reported to be reduced with an increase in the concentration of sludge in the soil and substrate (Araujo and Bettiol 2009).

Phyllosticta Leaf Spot

Phyllosticta leaf spot or leaf blight, caused by *Phyllosticta sojiecola* Massai (syn. *P. glycines*) and teleomorph *Pleosphaerulina sojicola* Miura, is a minor disease of soybean crop, rarely spreading beyond the first few trifoliate leaves. This fungal disease occurs throughout the soybean-growing regions in the United States (Yang 2002). The irregular marginal leaf scorch symptom exhibited on the lower leaves has been identified as *Phyllosticta* leaf spot. The infection starts at the leaf margin and progresses inward, forming an irregular, V-shaped area. Random leaf spots may also accompany the marginal necrosis. Numerous small, black specks (fungus fruiting bodies, or pycnidia) form in older lesions. The fungus may grow from the leaf blades into the petioles and then to the stipules and stem tissues at the leaf scar. Superficial, light-gray, tan, or brownish lesions with a narrow, brown, or purplish border may form on the petioles, stems, and pods. With cool and moist conditions, pods and seeds can be infected, causing seed discoloration. The fungus produces numerous small spores, which can spread to healthy leaves and plants, thereby causing new infection. The fungus can survive on seeds and can be spread with infected seeds. If it is prevalent in seed fields, a seed health test may be conducted before saving the beans for seed. If disease is severe in a production field, consider the use of rotation and tillage to reduce infested residues for the next soybean crop.

Brown Stem Rot (*Phialophora gregata* f. sp. *sojae*)

BSR of soybean caused by *P. gregata* f. sp. *sojae* Kobayashi et al. occurs in many countries including Argentina, Brazil, Canada, Egypt, Japan, Mexico, the United States, and the former Yugoslavia (Gray and Grau 1999, Grau 2006b). Yield losses of 10%–30% are common for susceptible soybean varieties grown in management systems conducive for BSR development. There is no external evidence of the disease in the early reproductive stage and signs of early infection generally go unnoticed unless the stems are cut open and examined or only the mild strain of the pathogen is present. The onset of foliar symptoms typically occurs at growth stages R4 and R5 and foliar symptoms peak at R7. The pathogen causes stem and foliar symptoms that may not always occur together, depending upon pathotype, host genotype, and environmental conditions (Hughes et al. 2002, Malvick et al. 2003). Pathogen pathotype I (genotype A) causes browning of stems as well as foliar symptoms such as interveinal chlorosis, defoliation, and wilting. Symptomatic leaves have a shriveled appearance but remain attached to the stem. Pathotype II (genotype B) causes only browning of stems. Secondary symptoms of BSR are stunting, premature death, decrease in seed number, reduced pod set, and decrease in seed size.

Stem symptoms include brown discoloration of the pith and vascular tissue; foliar symptoms include interveinal necrosis and defoliation (Gray and Grau 1999). While all soybean genotypes may be susceptible to infection by the pathogen, soybean genotypes differ in expressing foliar symptoms. A lack of foliar symptoms is considered as resistance and has been a selection criterion in breeding programs for BSR resistance (Sebastian et al. 1986). Pathogen reproduces asexually by means of conidia inside host plants. Its sexual state has never been found. It is slow growing in culture and is thought to be a poor saprophytic competitor (Adee and Grau 1997). *P. gregata* f. sp. *sojae*, a soilborne vascular pathogen causing BSR of soybean, has been divided into two genotypes, designated as A (pathotype I) and B (pathotype II). These genotypes are differentiated by an insertion or deletion in the intergenic spacer (IGS) region of ribosomal DNA. The two genotypes differ in the type and severity of symptoms they cause and have displayed preferential host colonization. Pathotype I and pathotype II are based on variation in the IGS region of nuclear rDNA marker (Gray 1971, Hughes et al. 2009). The rDNA marker identifies genetically distinct populations. Pathotype I is the defoliating pathotype comprising population A, identifiable by the genotype A rDNA marker (Chen et al. 2000, Hughes et al. 2002), and preferentially infecting susceptible soybean cultivars (Chen et al. 2000, Malvick et al. 2003). Pathotype II is the nondefoliating pathotype comprising population B, identifiable by the genotype B rDNA marker (Chen et al. 2000, Hughes et al. 2002), and preferentially infecting certain resistant soybean cultivars (Chen et al. 2000, Malvick et al. 2003). Field isolation data showed that most isolates obtained from susceptible cultivars belong to population A, whereas most isolates obtained from resistant cultivars belong to population B (Chen et al. 2000, Hughes et al. 2002, Malvick et al. 2003, Malvick and Impullitti 2007). The two populations could be sympatric, residing not only in the same field but also in the same plants under field conditions. The first controlled experimental demonstration that a differential host preference of *P. gregata* f. sp. *sojae* exists toward different cultivars of the same host species has been given by Meng et al. (2005).

The *P. gregata* fungus produces no survival structures, but can overwinter as mycelium in decaying soybean residue previously colonized during the pathogen's parasitic phase. During overwintering, conidia are produced; these conidia are the inoculum for new plants in the spring. Infection occurs through roots by growth stage V3 and progressively colonizes stems. Stem and foliar symptoms are most severe when air temperatures range between 15°C and 26°C during growth stages R4 to R6. Air temperatures in the 32°C range will suppress foliar symptom development. BSR is most severe when optimal soil moisture is present at R1 to R2 followed by dry soil conditions at R5 to R6. The severity of BSR is greater if soils are low in phosphorus and potassium and soil pH is below 6.5. *P. gregata* and *H. glycines*, the SCN, frequently occur together and there is evidence that the severity of BSR is greater in the presence of SCN.

The disease can be managed by the use of disease-resistant cultivars, and commercial soybean varieties have been improved dramatically for resistance to BSR. Most soybean cultivars with SCN resistance tracing to PI 88788 have various degrees of resistance to BSR. However, caution is advised for varieties with SCN resistance derived from Peking or Hartwig, the source of SCN resistance technology, as these two varieties are susceptible to BSR. Successful control of BSR has been obtained through crop rotation, especially if 2–3 years of nonhost crops are spaced between soybean. Soybean is the only known host grown; extended periods of cropping to nonhosts such as corn or small grains or forage legumes effectively lower the inoculum of *P. gregata*. The rate of inoculum decline is directly related to the rate of soybean residue decomposition. Early-maturing soybean cultivars escape the yield-reducing effect of BSR.

Bacterial Blight

Bacterial blight of soybean caused by *Pseudomonas syringae* pv. *glycinea* (Coerper) Youn et al. is the most common bacterial disease of soybeans and it occurs in all soybean-producing regions of the world. Although this disease is of limited importance, it is one of the first leaf spot diseases to appear on young plants (Jagtap et al. 2012b). Bacterial blight has been reported to cause significant yield reductions on susceptible cultivars under heavy disease pressure. In Europe, the disease has not caused great loss, but if the area of soybean production were to increase, losses might be expected on the scale seen in the United States, where in the period 1975–1977, it was by far the most damaging prokaryotic disease, causing an estimated annual average loss of $62 million (Kennedy and Alcorn 1980). In recent years, the disease has become economically important in Serbia in Europe (Ignjatov et al. 2007, 2008).

Bacterial blight is primarily a leaf disease, but symptoms can occur on stems, petioles, and pods. Leaf lesions are at first small, annular, water-soaked, tan-colored spots, which enlarge to 1–2 mm diameter and become dark brown to black with a dark center and a water-soaked margin and surrounded with a narrow yellowish halo. The halo is more noticeable on the upper leaf surface. Bacterial blight is easily confused with Septoria brown spot, a fungal disease that develops first on the lower leaves, whereas the bacterial blight leaf lesions develop first on upper young leaves. A simple test for bacterial blight is to hold infected leaves to the light; bacterial blight spots will be translucent. Lesions can coalesce to produce large irregular areas of dead tissues. The center of all lesions may drop resulting in shot-holed leaves, and the leaf may show ragged and torn appearance. Large black lesions may develop on stem. If the growing point of seedlings is affected, the plant usually dies.

The primary foci in crops derive from seed-borne infection that inhibit germination and on cotyledons cause marginal lesions that enlarge and become dark-brown necrotic. Often, the lesions are covered, particularly on their underside, with a film of a pale grayish bacterial slime that can dry to a thin silvery crust; less commonly, primary foci derive from overwintered infected crop debris. Secondary spread to infect young soybean leaves occurs by means of wind-driven rain and during cultivation or spraying when the foliage is wet at temperatures of 24°C–26°C. Hot, dry weather suppresses its development. There is evidence that the pathogen may be resident epiphyte in buds; this could provide a continued source of inocula. Infection occurs through natural openings on foliage and through wounds that occur commonly on sandy soils by abrasion with sand particles. The bacteria may colonize leaf surfaces without producing symptoms until conditions become favorable. Rain splash and wind-driven aerosols can drive bacteria into natural leaf openings (stomata) causing rapid increases in disease. Psg PG4180 causing bacterial blight of soybean produces the phytotoxin coronatine (COR) in a temperature-dependent manner. COR consists of a polyketide, coronafacic acid (CFA), and an amino acid derivative, coronamic acid, and is produced optimally at 18°C, whereas no detectable synthesis occurs at 28°C. After spray inoculation, PG4180 causes typical bacterial blight symptoms on soybean plants when the bacteria are grown at 18°C prior to inoculation but not when derived from cultures grown at 28°C (Budde and Ullrich 2000).

Host plant resistance genes to bacterial blight in soybean germplasm and a number of physiologic races of *P. syringae* pv. *glycinea* (Psg) have been reported (Fett and Sequeira 1981, Zabala et al. 2006). The interaction of compatible and incompatible races of Psg with different soybean cultivars has been characterized (Cross et al. 1966, Ignjatov et al. 2007). Incompatible interactions lead to a cascade of plant responses (hypersensitive response [HR]) triggered by the action of a resistance gene R and the corresponding avirulent pathogen avr gene (Zabala et al. 2006). The Rpg4 gene may be involved in resistance to Psg in soybean; however, it has another useful function that somehow contributes to soybean productivity in modern agroecosystems, and thus, plant breeders have unintentionally increased its frequency in cultivated germplasm (Farhatullah Stayton et al. 2010). The Rpg4 locus is controlling bacterial blight resistance to *P. syringae* pv. *glycinea* race 4 (Farhatullah Groose et al. 2010). In the incompatible interaction, Psg PG4180 elicits the HR regardless of the bacterial preinoculation temperature (Budde and Ullrich 2000). The complex resistance responses in such incompatible plant–pathogen interactions have been characterized at the molecular level to a larger extent in the model plant *Arabidopsis thaliana* (Quirino and Bent 2003). Durable resistance is difficult to achieve because of many races of Psg.

There is no effective management for bacterial blight when an aggressive race, a susceptible cultivar, and weather favor disease development. However, resistance in susceptible soybean varieties can be induced by salicylic acid (SA), chitosan, beta-aminobutyric acid (BABA), and OA to bacterial blight disease. The concentration of 1000 μg BABA/mL induces the highest resistance among all the inducers. Soybean varieties and application methods all can vary the effect of induced resistance (Liu et al. 2008). Crop rotation to nonhosts such as corn, wheat, and other nonlegume species and tillage will help reduce inoculum. Comparatively narrow rotation should be sufficient to eliminate trash-borne infection from fields since the pathogen does not apparently survive in it for two seasons (Parashar and Leben 1972). Tillage where possible can also help reduce the survival of *P. syringae*–infested debris through burial and rapid breakdown of soybean debris. The epiphyte *P. syringae* pv. *syringae* 22d/93 (Pss22d), isolated from soybean leaves, had been characterized as a promising and species-specific biocontrol strain in vitro and in plant against Psg (Wensing et al. 2010). Seeds produced in heavily infected fields are likely to carry the pathogenic bacteria; hence, it is recommended that seeds should never be saved for sowing from the plants from affected fields. Copper fungicides are labeled for bacterial blight control on soybean, but application needs to be conducted early in the disease cycle to be effective.

Bacterial Pustule Disease

Bacterial pustule disease (BPD) is caused by *Xanthomonas axonopodis* pv. *glycines* (Nakano) Dye (syns. *Xanthomonas campestris* pv. *phaseoli* Smith, Dye) and *Xanthomonas phaseoli* (Smith) Dowson var. *sojensis* (Hedges) Starr and Burkholder. Although this bacterium is widely known as *X. campestris* pv. *glycines*, following DNA–DNA hybridization analysis (Vauterin et al. 1995), the bacterium has been renamed as *X. axonopodis* pv. *glycines* (Xag). It is an economically significant disease in most areas of the world in which the soybean crop is grown. It is widespread in many European countries causing seed losses of up to 28%. It has also become important in India (Khare et al. 2003), Korea (Van et al. 2007), Serbia (Balaz and Acimovic 2008), southern United States, and Thailand (Kaewnum et al. 2005). Yield losses of up to 40% have been reported in certain parts of the world (Prathuangwong and Amnuaykit 1987). However, generally, it is far less economically important than bacterial blight caused by *P. syringae* pv. *glycinea*. The disease occurs typically on soybean foliage and symptoms include small, pale green spots with elevated pustules, which may develop into large necrotic lesions. Pustules are mainly formed by hypertrophy, but hyperplasia may also occur. The appearance of bacterial pustule varies from minute specks to large, irregular, mottled brown areas that arise when smaller lesion coalesces. The spots may enlarge and coalesce, leading to premature defoliation (Narvel et al. 2001). Symptoms on resistant soybean cultivars become visible in the form of small chlorotic spots, but not well-defined pustules or light green chlorosis. Those symptoms are sometimes confused with those of soybean rust. However, pustule lesions are characterized visually by small pustules surrounded by yellowing halos, whereas rust disease forms tan or brown lesion in which uredospores are formed and released through a central pore. The Xag is a motile, gram-negative rod (0.5–0.9 × 1.4–2.3 μm) with a single polar flagellum. Colonies on beef infusion agar are pale yellow, become deep yellow with age, and are small. It is seed borne and can also overwinter in infected host debris on the surface of soil (but less well in buried host debris) or in volunteer plants from infected seed. The pathogen can be transmitted from seed to the seedling. It enters through stomatal openings and proceeds intercellularly. Bulging of epidermal cells initially occurs. Infected cells become yellowish brown and later become deformed and disintegrated. The development of disease requires an optimum temperature of 27°C (+ or −2) with a minimum relative humidity of 83%. Pathogenic variations with regard to aggressiveness on soybean among isolates of Xag have been demonstrated (Ansari 2005, Kaewnum et al. 2005). The isolates differ in their ability to induce an HR on resistant varieties. Xag grows in the xylem vessels of the soybean plants and in the intercellular spaces of the leaves, causing pustule lesions. A major mechanism of this bacterium is the production of an extracellular polysaccharide or xanthan gum that is toxic to the plant tissues and cause them to be necrosis spots. This pathogen is not known to produce any extracellular toxins, degradative enzymes, or plant growth hormones.

The disease can be managed through the use of resistant cultivars. RT-PCR data suggest six candidate genes that might be involved in a necrotic response to Xag in resistant genotype PI 96188 (Van et al. 2007). A nonpathogenic mutant *the M715 mutant* (derived from the pathogenic wild-type strain YR32) shows promise as an effective BCA for BPD in soybeans (Rukayadi et al. 2000). Similarly, *B. subtilis* isolate 210 should be considered as a potential antagonistic agent for BPD of soybean (Salerno and Sagardoy 2003). *B. amyloliquefaciens* strain KPS46 selected as inducer of systemic resistance against Xag can be of potential use. Soybean seeds are treated with KPS46, SA, and harpin (synthetic resistance inducer) prior to sowing and challenged with Xag, the activity of PAL increases, and phenolics are found to accumulate in soybean leaf tissues giving protection to soybean plants against BPD (Buensanteai et al. 2007).

Soybean Mosaic Disease

Soybean mosaic disease (Figure 9.12) is caused by soybean mosaic virus (SMV) and it occurs worldwide causing mild to severe mosaic on susceptible cultivars. SMV infection adversely affects seed quality, oil content, and seed germination. Yield losses due to this disease generally range from 8% to 35%; however, losses as high as 94% have been reported. Infection in the early growth stages has the greatest risk of yield loss and reduced seed quality, compared to infection later in the season. Dual infection with other viruses, that is, mixed infection of SMV with Alfalfa mosaic virus, a common situation, increases the risk of yield loss and reduces seed quality (Malapi-Nelson et al. 2009). Tobacco streak virus and bean pod mottle virus (BPMV) have also been found in multiple infections (Fajolu et al. 2010). SMV and BPMV act synergistically. This means that mixed infections are more severe than single infections. Symptoms on plants with both viruses are very severe and terminal death may occur. Yield losses associated with combined infection can be as high as 66%–86%. The combined infection also increases the level of seed transmission of SMV. Of course, this combination effect is also related to the age of the plant when infection occurs.

FIGURE 9.12 Soybean mosaic symptoms on leaf. (Courtesy of Dr. Shrishail Navi, Iowa State University, Ames, IA.)

Common leaf symptoms of the disease are a mosaic of light and dark green areas that may later become raised or blistered, particularly along the main veins. The youngest and most rapidly growing leaves show the most symptoms, especially at cooler temperatures. Affected plants are stunted with rugose or crinkled leaves that become severely mottled and deformed. The trifoliate leaf blades become slightly narrowed and are puckered along the veins and curled downward. The mottling appears as light and dark green patches on individual leaves. Early-infected plants particularly are stunted with shortened petioles and internodes. Symptoms are more severe when temperatures have been cool (<24°C) and may be masked by warmer (>32°C) conditions. Affected plants produce fewer pods and seeds from infected plants can be mottled black or brown depending on hilum color. Not all infected plants produce mottled seed and seed mottling does not indicate that the virus is present in the seed. Seeds may show reduction in their size in comparison to seeds from healthy plants. Primary leaves of some cultivars may show necrotic local lesions, which merge into veinal necrosis followed by yellowing and leaf abscission.

The virus belongs to the genus *Potyvirus*, group IV (+)ss RNA, under the family Potyviridae. SMV is transmitted by infected seed, and soybean aphid species (*Aphis glycines*) and at least 32 aphid species, belonging to 15 different genera, transmit the SMV in a nonpersistent manner worldwide (Wang et al. 2006). In most varieties, seed transmission is less than 5%, but much greater than seed transmission for BPMV. Spread of the disease can also be done by soybean aphids, which can vector this virus. The timing and incidence of SMV infection depend largely upon the level of primary inoculum and aphid activity. SMV may be introduced into a virus-free region by planting infected seed. The pathogen is spread from plant to plant by aphids. The soybean aphid, *A. glycines*, the most common SMV vector, is the only aphid species that can establish colonies on soybeans. Once an aphid feeds on an infected soybean plant, it only takes a short time (seconds to a few minutes) for the insect to acquire the virus. As the virus-carrying aphids move and feed on healthy plants, the virus will be spread around. In the absence of soybeans, the virus can overwinter on a wide range of other hosts.

SMV is a flexuous rod consisting of positive-sense, single-stranded RNA. Numerous strains of the virus have been identified based upon reactions on a set of differential cultivars. In the United States, SMV has been classified into nine strains using differential reactions on eight soybean cultivars. The strains are currently known at G1 through G7, G7a, and C14. It is probable that additional strains exist, particularly in People's Republic of China and Japan (Zheng et al. 2008). SMV is sap and graft transmissible also. At least 32 aphid species, belonging to 15 different genera, transmit the

SMV in a nonpersistent manner. Virus isolates may show some vector specificity. Infected plants resulting from transmission through seed play an important role in SMV epidemiology. Such plants are primary inoculum sources for SMV. In most cultivars, seed transmission is less than 5%, but no transmission occurs in some cultivars while others can have levels as high as 75%. Infection of soybean plants with SMV has been reported to enhance *Phomopsis* spp. infection, which reduces seed quality. The use of SMV-resistant varieties prevent/reduce SMV and *Phomopsis* spp. seed infection (Koning et al. 2002).

Planting SMV-resistant soybean cultivars is the most economical practice to manage the disease. Several resistance genes have been identified and are effective against some, but not all, virus strains. Based on the differential reactions on a set of soybean cultivars, SMV has been classified into numerous strains. In the United States, nine strains, G1G7, G7a, and C14, are currently recognized. Additional strains have been identified in other countries such as Canada, China, Japan, and South Korea including strains that overcome all known resistance to the virus (Zheng et al. 2005, 2008). At least three independent loci (Rsv1, Rsv3, Rsv4) have been identified for SMV resistance. Multiple resistance alleles have been reported for the Rsv1 and Rsv3 loci (Liao et al. 2002, Zhen et al. 2008). The first dominant resistance gene identified in the soybean line PI 96983 has been designated as Rsv1. Single resistance genes in other cultivars, which confer differential reactions to strains G1 to G7, are found to be alleles at the Rsv1 locus and have been designated as Rsv1y, Rsv1m, Rsv1t, Rsv1k, Rsv1s, and Rsv1n. A new mutation in SMV resulting in overcoming Rsv4 resistance has been reported from Iran (Ahangaran et al. 2013). Some of the most promising soybean genotypes that are resistant to most strains of SMV in the Arkansas state in the United States are Ozark, USG 5002T and USG 5601T. Similarly, SMV-resistant soybean genotypes from India are JS71-05, KHSb2, LSb1, MACS58, MACS124, Punjab1, and VLS2 (Sharma et al. 2014). Soybean genotypes identified with high levels of resistance to SMV from Nigeria with disease incidence of 10% or less are TGx 1440-1E, TGx 1448-E, TGx 1479-1E, TGx 1446-3E, TGx 1371-1E, and TGx 1445-4E, TGx 1440-1E, and TGx.

At present, the use of SMV-free seed 1448-2E (Banwo and Adamu 2000) and avoiding late planting of soybean are the best control measures to preclude loss induced by SMV. Serological seed indexing techniques and/or grow-out tests can be used for virus detection in seed lots. Roguing, in addition to being generally impractical in the field, may not be very effective because of the tendency for symptoms in soybean to be masked above 30°C. Insecticides are not considered effective in reducing transmission of SMV by aphids. Aphids present at spraying are killed, but the field is quickly recolonized by winged aphids and virus transmission can resume. Aphids that contact insecticide residues on the leaf surface are killed but are still capable of virus transmission prior to death.

Bean Pod Mottle Disease (Bean Pod Mottle Virus)

BPMV (genus *Comovirus*, family *Comoviridae*) was first identified in soybean in 1951 in Arkansas (Walters 1958, Ross and Butler 1985) and has caused agronomic problems since this first finding in all production areas in the United States. Worldwide, it is also reported to occur in Brazil (Anjos et al. 1999), Canada (Michelutti et al. 2002), Ecuador (Zettler et al. 1991), Iran (Shahraeen et al. 2005), Nigeria (Ugwuoke 2002), and Peru (Fribourg and Perez 1994). Maximum losses occur when plants are infected at the seedling stage. Yield losses from BPMV alone may reach 2%–20% depending on planting date and geographical area. Yield loss assessments on southern soybean germplasm have revealed reductions ranging from 3% to 52% (Ziems et al. 2007). BPMV produces further financial loss for soybean producers because it causes increased seed coat mottling, which is an irregular pattern or streaking of the hilum associated with BPMV infection. The effects on yield and seed coat mottling are increased when there is a dual infection of BPMV and *SMV*, which causes yield losses >80%. Symptoms on infected soybeans may vary depending on the variety. Foliage symptoms range from mild chlorotic mottling, leaf rugosity in the upper canopy to puckering and severe mosaic in lower leaves, terminal necrosis, and death. A common symptom of BPMV infection is uneven crop maturity or *green stem* in which stems and leaves remain green, even though pods have matured. Young leaves in the upper canopy often have a green-to-yellow mottling that may fade and then redevelop later in the growing season. The green stems are difficult to cut during harvesting. However, it has been shown recently that green stem is independent of BPMV infection when random plants were tested for BPMV at growth stage R6 (Hobbs et al. 2006). In severe cases, malformed leaves and pods may be produced. Infected leaves show reduced turgidity resulting in curling. A reduction in pod set often occurs in infected plants that have undergone moisture stress during dry periods. Infected seed coats, similar to SMV infection, are mottled with brown or black streaks extending from the hilum.

BPMV has a bipartite positive-strand RNA genome consisting of RNA1 (approximately 6.0 kb) and RNA2 (approximately 3.6 kb) that are separately encapsidated in isometric particles 28 nm in diameter (Lomonossoff and Ghabrial 2001). BPMV RNA-1 codes for five mature proteins required for replication, whereas RNA-2 codes for a putative cell-to-cell MP and the two coat proteins (L-CP and S-CP). Separation of segments can be achieved by density gradient centrifugation into three components: top (T), middle (M), and bottom (B). The middle component contains a single RNA1 molecule, whereas the bottom component has RNA2 and the top particle lacks nucleic acid. BPMV is heat stable with a temperature inactivation point of 70°C. Its dilution end point in fresh plant extract is 10,000 and its longevity in vitro is 62 days at 18°C. Unlike SMV, BPMV does not spread very efficiently in seed. The virus is primarily transmitted by the bean leaf beetle (*Cerotoma trifurcata*). The virus has a wide host range among legumes and will be transferred to bean leaf beetles that feed on infected legume plants. The virus can be spread by mechanical injury, especially under wet conditions. The virus has been found in overwintered bean leaf beetle adults that may survive in grass, leaf litter, or even rocks and colonize soybeans as seedlings emerge. Because most flight events of beetles are limited to about 30 m, it is likely that BPMV spread is restricted within and between fields.

The use of soybean cultivars resistant or tolerant to BPMV infection would be the most practical approach. However, currently, no soybean lines have been identified with resistance to BPMV. Soybean cultivars with feeding deterrents against bean leaf beetle may not be sufficient to reduce BPMV incidence in the field. Transgenic soybean lines expressing the BPMV coat protein are resistant to BPMV infection. But this resistance has not been incorporated into commercial soybean cultivars (Reddy et al. 2001). Current management recommendations for reducing BPMV infection include the application of insecticides to manage bean leaf beetle populations to reduce the potential for virus movement. The recommended management of BPMV has been solely based on vector population dynamics, and not on BPMV disease. Delayed soybean planting date has been suggested to manage BPMV. Delayed planting is supposed to help soybean escape the migration period of beetle vectors of the virus.

Root-Knot Disease (*Meloidogyne* species)

Several genera of nematodes parasitize soybean worldwide, and the highest economic impact is attributed to SCNs—*H. glycines* (as described earlier in pages 397–405), root-knot nematodes (*Meloidogyne* species), lesion nematode (*Pratylenchus* species), and reniform nematode (*Rotylenchulus* species).

Root knot nematodes (RKNs) are biotrophic parasites of the genus *Meloidogyne*. The four most common species are *Meloidogyne incognita* (*Kofoid and white*) *Chitwood*, *Meloidogyne hapla Chitwood*, *Meloidogyne javanica* (*Treub*) *Chitwood*, and *Meloidogyne arenaria Chitwood*, but only *M. incognita* and *M. hapla* have been found to be important on soybeans worldwide. *M. hapla*, the northern RKN, is generally considered less damaging on soybean than the southern RKN, the *M. incognita* in the United States (Westphal and Xing 2006). Susceptible soybean plants can be infected at any stage of development. Infected plants are stunted and chlorotic. The interaction effect of the nematode (*M. javanica*) with other soilborne pathogens such as *R. solani* and *M. phaseolina* causes significant decrease in soybean plant height and dried plant weight, especially when nematodes are inoculated 1 week before both fungi and rate of severely damaged plants can reach to about 47.62% and 64.62%, respectively (Stephan et al. 2006). The aboveground symptoms of root-knot disease can be easily confused with other soil-related plant growth–suppressing factors. To confirm RKN infection, it is necessary to excavate root systems and examine them for root galling. Nematode-induced galls consist of globular, irregular deformations within the root system. These swellings are easily distinguished from nodules that are a normal part of soybean root systems. These normal nodules result from infection by beneficial, symbiotic bacteria that fix atmospheric nitrogen for the plant. Beneficial nodules are nearly spherical structures about 1/4 in. in diameter that are attached to the outside of roots. RKN galls, on the other hand, range from 1/8 to 1 in. in diameter and are swellings of the root itself. RKNs are obligate parasites, but they can survive as eggs in the soil for several years. These eggs contain the nematodes in their infective stage: second-stage juveniles. When soil conditions are favorable (when soil temperatures are more than 50°F) and a susceptible host plant is grown, juveniles hatch from the eggs and move through soil in search of host plant roots. When a juvenile finds a suitable root location, generally near the growing tip, it penetrates the root and becomes sedentary. After several molts, a juvenile develops into a mature female, which in turn produces an egg mass containing several hundred new nematode eggs in a gelatinous matrix deposited on the outside of the root. At this point, juveniles either immediately hatch from their eggs or remain dormant within the egg until infection conditions are favorable. The gelatinous matrix is thought to protect the eggs from soil organisms that might otherwise consume the eggs and suppress the nematode's initial inoculum level.

Concerning RKN management, soybean faces the same economic losses and difficulties as other crops. Despite the use of management strategies such as crop rotation with nonhosts and sanitation practices aimed at reducing initial inoculum, sustainable and long-lasting pest management strategies are in high demand. One of the strategies is to deploy novel sources of RKN resistance in soybean breeding programs, for example, using the soybean line PI 595099 (Accession NPGS/GRIN: G93-9223), which is resistant against specific strains and races of nematode species, including *M. javanica*, *M. incognita*, *M. arenaria*, and also the SCN. Another alternative is to introduce genetic modifications in soybean plants to obtain RKN resistance interactions (Bird et al. 2009, Beneventi et al. 2013). Increasing amount of auxin-induced reactive oxygen species (ROS) accumulation in cells in the nematode-inoculated soybean genotype (PI 595099) has an immediate effect on halting pathogenesis. The host coordinate and modulate defenses mostly by the interplay between auxin (Aux), gibberellin (GA), and jasmonate (JA) (Beneventi et al. 2013).

REFERENCES

Abbas, A.I., M. El-Muadhidi, and M.M. Elsahookie. 2003. Dissemination of and economic damage caused by charcoal rot *Macrophomina phaseolina* (Tassi) Goidanich on soybean in Iraq. *Arab J. Plant Prot.* 21: 79–83.

Abdel-Monaim, M.F. 2011. Role of riboflavin and thiamine in induced resistance against charcoal rot disease of soybean. *Afr. J. Biotechnol.* 10: 10842–10845.

Abdel-Monaim, M.F., M.E. Ismail, and K.M. Morsy. 2012. Induction of systemic resistance in soybean plants against *Fusarium* wilts disease by seed treatment with benzothiadiazole and humic acid. *Afr. J. Biotechnol.* 11: 2454–2465.

Adee, E.A. and C.R. Grau. 1997. Population dynamics of *Phialophora gregata* in soybean residue. *Plant Dis.* 81: 199–203.

Adekunle, A.A. and F. Edun. 2001. Histopathology of *Glycine max* (L.) Merril (soybean) seeds and seedlings infected with eight fungal isolates. *J. South Pac. Agric.* 8: 1 12.

Afzal, A.J., A. Srour, N. Saini, N. Hemmati, H.A. El-Shemy, and D.A. Lightfoot. 2012. Recombination suppression at the dominant Rhg1/Rfs2 locus underlying soybean resistance to the cyst nematode. *TAG Theor. Appl. Genet.* 124: 1027–1039.

Ahangaran, A., M.K. Habib, G.H. Mohammadi, S. Winter, and F. Garcia-Arenal. 2013. Analysis of Soybean mosaic virus genetic diversity in Iran allows the characterization of a new mutation resulting in overcoming Rsv4-resistance. *Gen. Virol.* 94: 2557–2568.

Akamatsu, H., N. Yamanaka, Y. Yamaoka et al. 2013. Pathogenic diversity of soybean rust in Argentina, Brazil, and Paraguay. *J. Gen. Plant Pathol.* 79: 28–40.

Akhtar, J. and A. Khalid. 2008. Occurrence and distribution of *Colletotrichum truncation B* causing anthracnose of soybean. *Green Farm.* 1: 44–45.

Al-Ani, R.A., M.A. Adhab, M.H. Mahdi, and H.M. Abood. 2012. *Rhizobium japonicum* as a biocontrol agent of soybean root rot disease caused by *Fusarium solani* and *Macrophomina phaseolina. Plant Prot. Sci.* 48: 149–155.

Al-Ani, R.A., M.H. Mahdi, and H.M. Abood. 2011. Effect of seed treatment with *Rhizobium japonicum* together with thiabendazole in minimizing root rot infection and seedling mortality of soybean caused by *Macrophomina phaseolina* and *Fusarium solani. Arab J. Plant Prot.* 29: 60–67.

Alkharouf, N.W., V.P. Klink, I.B. Chouikha et al. 2006. Time course microarray analyses reveal global changes in gene expression of susceptible *Glycine max* (soybean) roots during infection by *Heterodera glycines* (soybean cyst nematode). *Planta.* 224: 838–852. doi:10.1007/s00425-006-0270-8.

Almeida, A.M.R., R.V. Abdelnoor, C.A.A. Arias et al. 2003b. Genotypic diversity among Brazilian isolates of *Macrophomina phaseolina* revealed by RAPD. *Fitopatol. Bras.* 28: 279–285.

Almeida, A.M.R., L. Amorim, A. Bergamin Filho et al. 2003a. Progress of soybean charcoal rot under tillage and no-tillage systems in Brazil. *Fitopatol. Bras.* 28: 131–135.

Almeida, A.M.R., D.R. Sosa-Gomez, and E. Binneck. 2008. Effect of crop rotation on specialization and genetic diversity of *Macrophomina phaseolina. Trop. Plant Pathol.* 33: 257–264.

Al-Sheikh, H. and H.M.A. Abdelzaher. 2010. Isolation of *Aspergillus sulphureus*, *Penicillium islandicum* and *Paecilomyces variotii* from agricultural soil and their biological activity against *Pythium spinosum*, the damping-off organism of soybean. *J. Biol. Sci.* 10: 178–189.

Alvarez, F., M. Castro, A. Principe et al. 2012. The plant-associated *Bacillus amyloliquefaciens strains* MEP218 and ARP23 capable of producing the cyclic lipopeptides iturin or surfactin and fengycin are effective in biocontrol of *sclerotinia* stem rot disease. *J. Appl. Microbiol.* 112: 159–174.

Alves, M., E.A. Pozza, J.B. Costa, L.G. Carvalho, and L.S. Alves. 2011. Adaptive neuro-fuzzy inference systems for epidemiological analysis of soybean rust. *Environ. Model. Softw.* 26: 1089–1096.

Amer, M.A. 2005. Reaction of selected soybean cultivars to Rhizoctonia root rot and other damping-off disease agents. *Commun. Agric. Appl. Biol. Sci.* 70: 381–390.

Ammajamma, R. and P.V. Patil. 2009. Role of isozymes in rust (*Phakopsora pachyrhizi* Syd.) resistance in soybean. *Int. J. Plant Prot.* 2: 1–3.

Anderson, S.J., C.L. Stone, M.L. Posada-Buitrago et al. 2008. Development of simple sequence repeat markers for the soybean rust fungus, *Phakopsora pachyrhizi. Mol. Ecol. Resour.* 8: 1310–1212.

Anjos, J.R.N., P.S.T. Brioso, and M.J.A. Charchar. 1999. Partial characterization of *Bean pod mottle virus* in soybeans in Brazil. *Fitopatol. Bras.* 24:85–87.

Anonymous. 2000. The soybean cyst nematode problem. Report on plant disease, RPD No. 51, May 2000. Department of Crop Sciences, University of Illinois, Urbana-Champaign, IL, p. 9.

Anonymous. 2008. Soybean cyst nematode management. Field guide. Iowa State University Extension, Iowa Soybean Assoc., Iowa State University of Science and Technology, Ames, IA, p. 55.

Ansari, M.M. 2005. Variability in Indian isolates of *Xanthomonas campestris* pv. *glycines*, the incitant of bacterial pustule of soybean (*Glycine max*). *Indian J. Agric. Sci.* 75: 117–119.

Ansari, M.M. 2007. Evaluation of soybean genotypes against *Macrophomina phaseolina* (*Rhizoctonia bataticola*) causing charcoal rot in soybean. *Soybean Res.* 5: 68–70.

Ansari, M.M. 2010. Integrated management of charcoal rot of soybean caused by *Macrophomina phaseolina* (Tassi) goid. *Soybean Res.* 8: 39–47.

Ansari, M.M. and S.K. Agnihotri. 2000. Morphological, physiological and pathological variations among *Sclerotium rolfsii* isolates of soyabean. *Indian Phytopathol.* 53: 65–67.

Aoki, T., K. O'Donnell, Y. Homma, and A.R. Lattanzi. 2003. Sudden-death syndrome of soybean is caused by two morphologically and phylogenetically distinct species within the *Fusarium solani* species complex-*F. virguliforme* in North America and *F. tucumaniae* in South America. *Mycologia* 95: 660–684.

Aoki, T., K. O'Donnell, and M.M. Scandiani. 2005. Sudden death syndrome of soybean in South America is caused by four species of *Fusarium*: *Fusarium brasiliense* sp. nov., *F. cuneirostrum* sp. nov., *F. tucumaniae*, and *F. virguliforme*. *Mycoscience* 46: 162–183.

Arahana, V.S., G.L. Graef, J.E. Specht, J.R. Steadman, and K.M. Eskridge. 2001. Identification of QTLs for resistance to *Sclerotinia sclerotiorum* in soybean. *Crop Sci.* 41: 180–188.

Araujo, F.F., A.A. Henning, and M. Hungria. 2005. Phytohormones and antibiotics produced by *Bacillus subtilis* and their effects on seed pathogenic fungi and on soybean root development. *World J. Microbiol. Biotechnol.* 21: 1639–1645.

Araujo, F.G., M.R. de Rocha, R.A. da Aguiar, R.A. Garcia, and M.C. da Cunha. 2012. Management of soybean rust with fungicides for seed treatment and foliar application. *Semina: Ciencias Agrarias* (Londrina) 33(Suppl. 1): 2585–2592.

Arelli, P.R., V.C. Concibido, and L.D. Youn. 2010. QTLs associated with resistance in soybean PI567516C to synthetic nematode population infecting cv. Hartwig. *J. Crop Sci. Biotechnol.* 13: 163–167.

Arias, C.A.A., J.F.F. Toledo, L.A. Almeida et al. 2008. Asian rust in Brazil: Varietal resistance. (Facing the challenge of soybean rust in South America.) *JIRCAS Working Rep.* 58: 29–30.

Arora, C. and R.D. Kaushik. 2003. Fungicidal activity of plants extracts from Uttaranchal hills against soybean fungal pathogens. *Allelopathy J.* 11: 217–228.

Arriagada, O., F. Mora, J.C. Dellarossa, M.F.S. Ferreira, G.D.L. Cervigni, and I. Schuster. 2012. Bayesian mapping of quantitative trait loci (QTL) controlling soybean cyst nematode resistant. *Euphytica* 186: 907–917.

Arsenault-Labrecque, G., J.G. Menzies, and R.R. Belanger. 2012. Effect of silicon absorption on soybean resistance to *Phakopsora pachyrhizi* in different cultivars. *Plant Dis.* 96: 37–42.

Asmus, G.L., T.S. Teles, J. Anselmo, and G.T. Rosso. 2012. Race*s of Heterodera glycines* in the Northeast of Mato Grosso do Sul, Brazil. *Trop. Plant Pathol.* 37: 146–148.

Attia, M., N.M. Awad, A.S. Turky, and H.A. Hamed. November 2011. Induction of defense responses in soybean plants against *Macrophomina phaseolina* by some strains of plant growth promoting rhizobacteria. *J. Appl. Sci. Res.* 7(11): 1507–1517.

Baird, R.E., C.E. Watson, and M. Scruggs. 2003. Relative longevity of *Macrophomina phaseolina* and associated mycobiota on residual soybean roots in soil. *Plant Dis.* 87: 563–566.

Balardin, C.R., A.F. Celmer, E.C. Costa, R.C. Meneghetti, and R.S. Balardin. 2005. Possible transmission of *Fusarium solani* f. sp. *glycines*, causal agent of Sudden Death Syndrome, through soybean seed. *Fitopatol. Bras.* 30: 574–581.

Balardin, R.S., L.J. Dallagnol, H.T. Didone, and L. Navarini. 2006. Influence of phosphorus and potassium on severity of soy bean rust, *Phakopsora pachyrhizi. Fitopatol. Bras.* 31: 462–467.

Balaz, J. and S. Acimovic. 2008. Bacterioses of soyabean. *Biljni Lekar* (*Plant Doctor*) 36: 226–235.

Banwo, O.O. and R.S. Adamu. 2000. Evaluation of soyabean (*Glycine max*) cultivars for resistance to soyabean mosaic virus. *Tests Agrochem. Cult.* 21: 41–42.

Bao, Y., D.A. Neher, and S.Y. Chen. 2011. Effect of soil disturbance and biocides on nematode communities and extracellular enzyme activity in soybean cyst nematode suppressive soil. *Nematology* 13: 687–699.

Basavaraja, G.T., P.V. Patil, G.K. Naidu, and P.M. Salimath. 2004. Induced mutations for enhancing resistance to rust (*Phakopsora pachyrhizi*) in soybean (*Glycine max*). *Indian J. Agric. Sci.* 74: 620–622.

Basseto, M.A., W.V. Valerio Filho, E.C. Souza, and P.C. Ceresini. 2008. The role of binucleate *Rhizoctonia* spp. inducing resistance to the soybean foliar blight. *Acta Sci.—Agron.* 30: 183–189.

Bates, G.D., C.S. Rothrock, and J.C. Rupe. 2008. Resistance of the soybean cultivar archer to pythium damping-off and root rot caused by several *Pythium* spp. *Plant Dis.* 92: 763–766.

Beal, I.L., F.A. Villela, and J.C. Possenti. 2007. Application of fungicides in the reproductive phase of soya and physiological quality of the seeds. *Informativo ABRATES* 17: 84–91.

Begum, M.M., M. Sariah, M.A.Z. Abidin, A.B. Puteh, and M.A. Rahman. 2007a. Histopathological studies on soybean seeds infected by *Fusarium oxysporum* f. sp. *glycines* and screening of potential biocontrol agents. *Res. J. Microbiol.* 2: 900–909.

Begum, M.M., M. Sariah, A.B. Puteh, and M.A.Z. Abidin. 2007b. Detection of seed-borne fungi and site of infection by *Colletotrichum truncatum* in naturally-infected soybean seeds. *Int. J. Agric. Res.* 2: 812–819.

Begum, M.M., M. Sariah, A.B. Puteh, and M.A.Z. Abidin. 2008b. Pathogenicity of *Colletotrichum truncatum* and its influence. *Int. J. Agric. Biol.* 10: 393–398.

Begum, M.M., M. Sariah, A.B. Puteh, M.A. Zainal Abidin, M.A. Rahman, and Y. Siddiqui. 2010. Field performance of bio-primed seeds to suppress *Colletotrichum truncatum* causing damping-off and seedling stand of soybean. *Biol. Control* 53: 18–23.

Begum, M.M., M. Sariah, M.A. Zainal Abidin, A.B. Puteh, and M.A. Rahman. 2008a. Antagonistic potential of selected fungal and bacterial biocontrol agents against *Colletotrichum truncatum* of soybean seeds. *Pertanika J. Trop. Agric. Sci.* 31: 45–53.

Beledelli, D., D. Cassetari Neto, L. de S. Cassetari, and A.Q. Machado. 2012. Viability of *Phakopsora pachyrhizi* sidow under host absence. *Biosci. J.* 28: 604–612.

Bellaloui, N., A. Mengistu, and R.L. Paris. 2008. Soybean seed composition in cultivars differing in resistance to charcoal rot (*Macrophomina phaseolina*). *J. Agric. Sci.* 146: 667–675.

Beneventi, M.A., O.B. da Silva, M.E.L. de Sá et al. 2013. Transcription profile of soybean-root-knot nematode interaction reveals a key role of phytohormones in resistance reaction. *BMC Genomics* 14: 322. doi:10.1186/1471-2164-14-322.

Bhattacharyya, P.K., H.H. Ram, and P.C. Kole. 1999. Inheritance of resistance to yellow mosaic virus in interspecific crosses of *soybean*. *Euphytica* 108: 157–159.

Bird, D.M., V.M. Williamson, P. Abad et al. 2009. The genome of root-knot nematodes. *Annu. Rev. Phytopathol.* 47: 333–351.

Biswas, K.K. 2002. Identification of resistance in soybean to a virulent variant of mungbean yellow mosaic geminivirus. *Ann. Plant Prot. Sci.* 10: 80–83.

Blum, L.E.B. and R. Rodriguez-Kabana. 2006a. Dried powders of velvetbean and pine bark added to soil reduce *Rhizoctonia solani*-induced disease on soybean. *Fitopatol. Bras.* 31: 261–269.

Blum, L.E.B. and R. Rodriguez-Kabana. 2006b. Powders of kudzu, velvetbean, and pine bark added to soil increase microbial population and reduce southern blight of soybean. *Fitopatologia Brasileira* 31: 551–556.

Bonde, M.R., S.E. Nester, C.N. Austin et al. 2006. Evaluation of virulence of *Phakopsora pachyrhizi* and *P. meibomiae* isolates. *Plant Dis.* 90: 708–716.

Bonde, M.R., S.E. Nester, and D.K. Berner. 2012. Effects of soybean leaf and plant age on susceptibility to initiation of infection by *Phakopsora pachyrhizi*. Online. *Plant Health Prog.* doi:10.1094/PHP-2012-0227-01-RS.

Brar, H.K. and M.K. Bhattacharyya. 2012. Expression of a single-chain variable-fragment antibody against a *Fusarium virguliforme* toxin peptide enhances tolerance to sudden death syndrome in transgenic soybean plants. *Mol. Plant Microbe Interact.* 25: 817–824.

Brar, H.K., S. Swaminathan, and M.K. Bhattacharyya. 2011. The *Fusarium virguliforme* toxin FvTox1 causes foliar sudden death syndrome-like symptoms in soybean. *Mol. Plant Microbe Interact.* 24: 1179–1188.

Budde, I.P. and M.S. Ullrich. 2000. Interactions of *Pseudomonas syringae* pv. *glycinea* with host and non-host plants in relation to temperature and phytotoxin synthesis. *Mol. Plant Microbe Interact.* 13: 951–961.

Buensanteai, N., D. Athinuwat, and S. Prathuangwong. 2007. *Bacillus amyloliquefaciens* induced systemic resistance against *Xanthomonas axonopodis* pv. *glycines* caused agent soybean bacterial pustule with increased phenolic compounds and phenylalanine ammonia. In: *Proceedings of the 45th Kasetsart University Annual Conference*, Bangkok, Thailand, January 30–February 2, 2007, pp. 364–371.

Butzen, S. 2010. Sudden death syndrome of soybeans. *Crop Insights* (Pioneer) 19: 1–4.

Cai, G.H., R.W. Schneider, and G.B. Padgett. 2009. Assessment of lineages of *Cercospora kikuchii* in Louisiana for aggressiveness and screening soybean cultivars for resistance to *Cercospora* leaf blight. *Plant Dis.* 93: 868–874.

Cao, Y.P. and Q.K. Yang. 2002. Review of study on resistance and inheritance and breeding of resistance to *Cercospora sojine* Hara in soybean. *Soybean Sci.* 21: 285–289.

Calvo, E.S., R.A.S. Kiihl, A. Garcia, A. Harada, and D.M. Hiromoto. 2008. Two major recessive soybean genes conferring soybean rust resistance. *Crop Sci.* 48: 1350–1354.

Cardona Gomez, Y., F. Varon de Agudelo, and O. Agudelo Delgado. 2000. Evaluation of microorganisms as biological control agents of *Macrophomina phaseolina*, causal agent of soyabean charcoal rot. *Fitopatol. Colomb.* 24: 77–81.

Carmona, M.A., M.E. Gally, and S.E. Lopez. 2005. Asian soybean rust: Incidence, severity, and morphological characterization of *Phakopsora pachyrhizi* (Uredinia and Telia) in Argentina. *Plant Dis.* 89: 109.

Carmona, M., R. Moschini, G. Cazenave, and F. Sautua. 2010. Relationship between precipitations registered in reproductive stages of soybean and severity of *Septoria glycines* and *Cercospora kikuchii*. *Trop. Plant Pathol.* 35: 71–78.

Carter, T.E., S.R. Koenning, J.W. Burton et al. 2011. Registration of 'N7003CN' maturity-group-VII soybean with high yield and resistance to race 2 (HG type 1.2.5.7-) soybean cyst nematode. *J. Plant Registrations* 5: 309–317.

Chakraborty, N., J. Curley, R.D. Frederick et al. 2009. Mapping and confirmation of a new allele at Rpp1 from soybean PI 594538A conferring RB lesion-type resistance to soybean rust. *Crop Sci.* 49(3): 783–790.

Chandrasekaran, A. and K. Rajappan. 2002. Effect of plant extracts, antagonists and chemicals (individual and combined) on foliar anthracnose and pod blight of soybean. *J. Mycol. Plant Pathol.* 32: 25–27.

Chandrasekaran, A., V. Narasimhan, and K. Rajappan. 2000a. Effect of defense inducing chemicals on *Colletotrichum truncatum* and its infection in soybean. *Int. J. Trop. Plant Dis.* 18: 125–130.

Chandrasekaran, A., V. Narasimhan, and K. Rajappan. 2000b. Effect of plant extracts, antagonists and chemicals on seedling vigour and anthracnose disease of soybean. *Int. J. Trop. Plant Dis.* 18: 141–146.

Chandrasekaran, A., V. Narasimhan, and K. Rajappan. 2000c. Integrated management of anthracnose and pod blight of soybean. *Ann. Plant Prot. Sci.* 8: 163–165.

Chandrasekaran, A., V. Narasimhan, and K. Rajappan. 2000d. Changes in the activity of phenolics and oxidizing enzymes in soybean cultivars resistant and susceptible to *Colletotrichum truncatum*. *Ann. Plant Prot. Sci.* 8: 176–178.

Chaudhary, H.R., V.P. Gupta, and M. Ali. 2005. Evaluation of compatible combinations of *Bacillus thuringiensis*, insecticide and fungicides for green semilooper and pod blight of soybean (*Glycine max*). *Soybean Res.* 3: 73–75.

Chen, L.J., Y. Zhu, B. Liu, and Y. Duan. 2007. Influence of continuous cropping and rotation on soybean cyst nematode and soil nematode community structure. *Acta Phytophyl. Sin.* 34: 347–352.

Chen, L.J., Y.Y. Wang, X.F. Zhu, and Y.X. Duan. 2011. Review of the biocontrol on soybean cyst nematode (*Heterodera glycines*). *J. Shenyang Agric. Univ.* 42: 393–398.

Chen, S.Y., P.M. Porter, C.D. Reese, and W.C. Stienstra. 2001. Crop sequence effects on soybean cyst nematode and soybean and corn yields. *Crop Sci.* 41: 1843–1849.

Chen, S.Y. 2007. Tillage and crop sequence effects on *Heterodera glycines* and soybean yields. *Agron. J.* 98: 797–907.

Chen, W., C.R. Grau, E.A. Adee, and X.O. Meng. 2000. A molecular marker identifying subspecific populations of the soybean brown stem rot pathogen, *Phialophora gregata*. *Phytopathology* 90: 875–883.

Choudhary, D.K. 2011. Plant growth-promotion (PGP) activities and molecular characterization of rhizobacterial strains isolated from soybean (*Glycine max* L. Merril) plants against charcoal rot pathogen, *Macrophomina phaseolina*. *Biotechnol. Lett.* 33: 2287–2295.

Chowdhury, A.K., P. Srinives, P. Saksoong, and P. Tongpamnak. 2002. RAPD markers linked to resistance to downy mildew disease in soybean. *Euphytica* 128: 55–60.

Christiano, R.S.C. and H. Scherm. 2007. Quantitative aspects of the spread of Asian soybean rust in the southeastern United States, 2005 to 2006. *Phytopathology* 97: 1428–1433.

Ciampi, M.B., E.E. Kuramae, R.C. Fenille, M.C. Meyer, N.L. Souza, and P.C. Ceresini. 2005. Intraspecific evolution of *Rhizoctonia solani* AG-1 IA associated with soybean and rice in Brazil based on polymorphisms at the ITS-5.8S rDNA operon. *Eur. J. Plant Pathol.* 113: 183–196.

Ciampi, M.B., M.C. Meyer, M.J.N. Costa, M. Zala, B.A. McDonald, and P.C. Ceresini. 2008. Genetic structure of populations of *Rhizoctonia solani* anastomosis group-1 IA from soybean in Brazil. *Phytopathology* 98: 932–991.

Cober, E.R., S. Rioux, I. Rajcan, P.A. Donaldson, and D.H. Simmonds. 2003. Partial resistance to white mold in a transgenic soybean line. *Crop Sci.* 43: 92–95.

Colletto, A., A. Luque, G. Salas et al. 2008. Identification of *Fusarium* fungi causing sudden death syndrome of soyabeans in Tucuman and Salta. *Avance Agroindustrial* 29(4): 26–30.

Costa, I.F.D., R.S. da Balardin, L.A. Medeiros, and T.M. Bayer. 2006. Resistance of six soybean cultivars to *Colletotrichum truncatum* (Schwein) in two phenologic stages. *Ciencia Rural* 36: 1684–1688.

Costa, I.F.D., R.S. da Balardin, L.A.M. Medeiros et al. 2009. Reaction of commercial germplasm of soybean to *Colletotrichum truncatum*. *Trop. Plant Pathol.* 34: 47–50.

Costamilan, L.M., C.V. Godoy, R.M. Soares, J.T. Yorinori, and A.M.R. Almeida. 2008. Studies on rust in south Brazil. (Facing the challenge of soybean rust in South America). *JIRCAS Working Rep.* 58: 39–40.

Covert, S.F., T. Aoki, K. O'Donnell et al. 2007. Sexual reproduction in the soybean sudden death syndrome pathogen *Fusarium tucumaniae*. *Fungal Genet. Biol.* 44: 799–807.

Cross, J.E., B.W. Kennedy, J.W. Lambert, and R.L. Cooper. 1966. Pathogenic races of the bacterial pathogen of soybeans, *Pseudomonas glycinea*. *Plant Dis. Rep.* 50: 557–560.

Cruz, C.D., D. Mills, P.A. Paul, and A.E. Dorrance. 2010. Impact of brown spot caused by *Septoria glycines* on soybean in Ohio. *Plant Dis.* 94: 820–826.

Cruz, T.V., C.P. da Peixoto, M.C. Martins et al. 2012. Yield loss caused by rust in soybean cultivars sown in different periods in the western region of Bahia, Brazil. *Trop. Plant Pathol.* 37: 255–265.

Csondes, I., K. Baliko, and A. Degenhardt. 2008. Effect of different nutrient levels on the resistance of soybean to *Macrophomina phaseolina infection* in field experiments. *Acta Agron. Hung.* 56: 357–362.

Cunha, W.G., M.L.P. Tinoco, H.L. Pancoti, R.E. Ribeiro, and F.J.L. Aragao. 2010. High resistance to *Sclerotinia sclerotiorum* in transgenic soybean plants transformed to express an oxalate decarboxylase gene. *Plant Pathol.* 59: 654–660.

da Cruz, M.F.A., L. de F. Silva, F.A. Rodrigues, J.M. Araujo, and E.G. de Barros. 2012. Silicon on the infection process of *Phakopsora pachyrhizi* on soybean leaflets. *Pesq. Agropec. Bras.* 47: 142–145.

Daniel Ploper, L., M. Roberto Galvez, V. Gonzalez, H. Jaldo, M.A. Zamorano, and M. Devani. 2001. Management of diseases affecting the end of the soyabean cultivation cycle. *Avance Agroindustrial* 22: 20–26.

da Rocha, M.R., T.R. Anderson, and T.W. Welacky. 2008. Effect of inoculation temperature and soybean genotype on root penetration and establishment of *Heterodera glycines*. *J. Nematol.* 40: 281–285.
da Silva, A.C., P.E. de Souza, J. da C. Machado, B.M. da Silva, and J.E.B.P. Pinto. 2012a. Effectiveness of essential oils in the treatment of *Colletotrichum truncatum*-infected soybean seeds. *Trop. Plant Pathol.* 37: 305–313.
da Silva, A.C., P.E. da Souza, J.E.B.P. Pinto et al. 2012b. Essential oils for preventative treatment and control of Asian soybean rust. *Eur. J. Plant Pathol.* 134: 865–871.
da Silva, J.V.C., F.C. Juliatti, J.R.V. da Silva, and F.C. Barros. 2011. Soybean cultivar performance in the presence of soybean Asian rust, in relation to chemical control programs. *Eur. J. Plant Pathol.* 131: 409–418.
Davis, E.L. and G.L. Tylka. 2000. Soybean cyst nematode disease. *Plant Health Instructor*. doi:10.1094/PHI-I-2000-0725-01 (updated 2005).
de Araujo, F.F. and W. Bettiol. 2009. Effect of sewage sludge in soil-borne pathogens and powdery mildew severity in soybean. *Summa Phytopathol.* 35: 184–190.
de Avila Rodrigues, R., J.E. Pedrini, C.W. Fraisse et al. 2012. Utilization of the cropgro-soybean model to estimate yield loss caused by Asian rust in cultivars with different cycle. *Bragantia* 71: 308–317.
de Azevedo, L.A.S., F.C. Juliatti, and M. Barreto. 2007. Resistance of soybean genotypes of the cerrado region to rust caused by *Phakopsora pachyrhizi*. *Summa Phytopathol.* 33: 252–257.
Debortoli, M.P., N.R. Tormen, R.S. Balardin et al. 2012. Spray droplet spectrum and control of Asian soybean rust in cultivars with different plant architecture. *Pesq. Agropec. Bras.* 47: 920–927.
de Bruin, J.L. and P. Pedersen. 2008. Yield improvement and stability for soybean cultivars with resistance to *Heterodera glycines* Ichinohe. *Agron. J.* 100: 1354–1359.
de Farias Neto, A.L., G.L. Hartman, W.L. Pedersen, S.X. Li, G.A. Bollero, and B.W. Diers. 2006. Irrigation and inoculation treatments that increase the severity of soybean sudden death syndrome in the field. *Crop Sci.* 46: 2547–2554.
Delheimer, J.C., T. Niblack, M. Schmidt, G. Shannon, and B.W. Diers. 2010. Comparison of the effects in field tests of soybean cyst nematode resistance genes from different resistance sources. *Crop Sci.* 50: 2231–2239.
del Rio, L.E., C.A. Martinson, and X.B. Yang. 2002. Biological control of *Sclerotinia* stem rot of soybean with *Sporidesmium sclerotivorum*. *Plant Dis.* 86: 999–1004.
Dias, A.P.S., X. Li, P.F. Harmon, C.L. Harmon, and X.B. Yang. 2011. Effects of shade intensity and duration on Asian soybean rust caused by *Phakopsora pachyrhizi*. *Plant Dis.* 95: 485–489.
Dias, W.P., V.P. Campos, R.A.S. Kiihl, C.A.A. Arias, and J.F.F. Toledo. 2005. Genetic control in soybean of resistance to soybean cyst nematode race 4+. *Euphytica* 145: 321–329.
Dias, W.P., J.F.V. Silva, G.E.S. Carneiro, A. Garcia, and C.A.A. Arias. 2009. Soybean cyst nematode: Biology and management through genetic resistance. *Nematol. Bras.* 33: 1–16.
Dixon, L.J., R.L. Schlub, K. Pernezny, and L.F. Datnoff. 2009. Host specialization and phylogenetic diversity of *Corynespora cassiicola*. *Phytopathology* 99: 1015–1027.
Donald, P.A., D.D. Tyler, and D.L. Boykin. 2009. Short- and long-term tillage effects on *Heterodera glycines* reproduction in soybean monoculture in west Tennessee. *Soil Tillage Res.* 104: 126–133.
Donaldson, P.A., T. Anderson, B.G. Lane, A.L. Davidson, and D.H. Simmonds. 2001. Soybean plants expressing an active oligomeric oxalate oxidase from the wheat gf-2.8 (germin) gene are resistant to the oxalate-secreting pathogen *Sclerotinia sclerotiorum*. *Physiol. Mol. Plant Pathol.* 59: 297–307.
do Nascimento, J.M., W.L. Gavassoni, L.M.A. Bacchi et al. 2012. Association of adjuvants with picoxystrobin + cyproconazole for Asian soybean rust control. *Summa Phytopathol.* 38: 204–210.
Doreto, R.B.S., W.L. Gavassoni, E.F. da Silva, M.E. Marchetti, L.M.A. Bacchi, and F.F. Stefanello. 2012. Asian rust and soybean yield under potassium fertilization and fungicide, in the 2007/08 crop season. *Semina: Ciencias Agrarias* (Londrina) 33: 941–951.
Dufault, N.S., S.A. Isard, J.J. Marois, and D.L. Wright. 2010a. The influence of rainfall intensity and soybean plant row spacing on the vertical distribution of wet deposited *Phakopsora pachyrhizi* urediniospores. *Can. J. Plant Pathol.* 32: 162–169.
Dufault, N.S., S.A. Isard, J.J. Marois, and D.L. Wright. 2010b. Removal of wet deposited *Phakopsora pachyrhizi* urediniospores from soybean leaves by subsequent rainfall. *Plant Dis.* 94: 1336–1340.
Edwards, H.H. and M.R. Bonde. 2011. Penetration and establishment of *Phakopsora pachyrhizi* in soybean leaves as observed by transmission electron microscopy. *Phytopathology* 101: 894–900.
Ehteshamul-Haque, S., V. Sultana, J. Ara, and M. Athar. 2007. Cultivar response against root-infecting fungi and efficacy of *Pseudomonas aeruginosa* in controlling soybean root rot. *Plant Biosyst.* 141: 51–55.
El-Barougy, E., N.M. Awad, A.S. Turky, and H.A. Hamed. 2009. Antagonistic activity of selected strains of rhizobacteria against *Macrophomina phaseolina* of soybean plants. *American-Eurasian J. Agric. Environ. Sci.* 5: 337–347.

El-Baz, S.M., E.E. Abbas, and R.A.I.A. Mostafa. 2012. Effect of sowing dates and humic acid on productivity and infection with rot diseases of some soybean cultivars cultivated in new reclaimed soil. *Int. J. Agric. Res.* 7: 345–357.

El-Hai, K.M.A., M.A. El-Metwally, and S.M. El-Baz. 2010. Reduction of soybean root and stalk rots by growth substances under salt stress conditions. *Plant Pathol. J.* (Faisalabad) 9: 149–161.

Ellis, M.L., M.M.D. Arias, D.R.C. Jimenez, G.P. Munkvold, and L.F. Leandro. 2013. First report of *Fusarium commune* causing damping-off, seed rot, and seedling root rot on soybean (*Glycine max*) in the United States. *Plant Dis.* 97: 284–285.

Ellis, M.L., K.D. Broders, P.A. Paul, and A.E. Dorrance. 2011. Infection of soybean seed by *Fusarium graminearum* and effect of seed treatments on disease under controlled conditions. *Plant Dis.* 95: 401–407.

El-Sayed, S.A., R.Z. El-Shennawy, and A.F. Tolba. 2009. Efficacy of chemical and biological treatments for controlling soil-borne pathogens of soybean. *Arab Univ. J. Agric. Sci.* 17: 163–173.

Fajolu, O.L., R.H. We, and M.R. Hajimorad. 2010. Occurrence of Alfalfa mosaic virus in soybean in Tennessee. *Plant Dis.* 94: 1505.

Fang, Z.D., J.J. Marois, G. Stacey, J.E. Schoelz, J.T. English, and F.J. Schmidt. 2010. Combinatorially selected peptides for protection of soybean against *Phakopsora pachyrhizi*. *Phytopathology* 100: 1111–1117.

Farhatullah Groose, R.W., M. Raziuddin Akmal, and M. Inayatullah. 2010. Mapping gene for bacterial blight (Rpg4 locus) in soybean. *Pak. J. Bot.* 42: 2145–2149.

Farhatullah Stayton, M.M., R.W. Groose, N.U. Raziuddin Khan, M. Akmal, and M. Inayatullah. 2010. Increase in frequency of the RPG4 gene in soybean cultivars. *Pak. J. Bot.* 42: 997–1002.

Farias Neto, A.L., R. de Hashmi, M. Schmidt et al. 2007. Mapping and confirmation of a new sudden death syndrome resistance QTL on linkage group D2 from the soybean genotypes PI 567374 and 'Ripley'. *Mol. Breed.* 20: 53–62.

Favaron, F., L. Sella, and R. D'Ovidio. 2004. Relationships among endo-polygalacturonase, oxalate, pH, and plant polygalacturonase-inhibiting protein (PGIP) in the interaction between *Sclerotinia sclerotiorum* and soybean. *Mol. Plant Microbe Interact.* 17: 1402–1409.

Feng, L., S.S. Chen, B. Feng, Y. He, and B.G. Lou. 2012a. Spectral detection on disease severity of soybean pod anthracnose. *Trans. Chin. Soc. Agric. Mach.* 43: 175–179.

Feng, L., S.S. Chen, B. Feng, F. Liu, Y. He, and B.G. Lou. 2012b. Early detection of soybean pod anthracnose based on spectrum technology. *Trans. Chin. Soc. Agric. Eng.* 28: 139–144.

Feng, P.C.C., C. Clark, G.C. Andrade, M.C. Balbi, and P. Caldwell. 2008. The control of Asian rust by glyphosate in glyphosate-resistant soybeans. *Pest Manag. Sci.* 64: 353–359.

Ferreira, M.F. da S., G.D.L. Cervigni, A. Ferreira et al. 2011. QTLs for resistance to soybean cyst nematode, races 3, 9, and 14 in cultivar Hartwig. *Pesq. Agropec. Bras.* 46:420–428.

Fett, W.F. and L. Sequeiro. 1981. Further characterization of the physiologic races of *Pseudomonas glycinea*. *Can. J. Bot.* 59: 283–287.

Formento, A.N. and J. de Souza. 2006. Overwinter and survival of Asian soybean rust caused by *Phakopsora pachyrhizi* in volunteer soybean plants in Entre Rios Province, Argentina. *Plant Dis.* 90: 826.

Frederick, R.D., C.L. Snyder, G.L. Peterson, and M.R. Bonde. 2002. Polymerase chain reaction assays for the detection and discrimination of the soybean rust pathogens *Phakopsora pachyrhizi* and *P. meibomiae*. *Phytopathology* 92: 217–227.

Fribourg, C.E. and W. Perez. 1994. *Bean pod mottle virus* affecting *Glycine max* (L.) Merr. in the Peruvian jungle. *Fitopatologia* 29: 207–210.

Furtado, G.Q., S.A.M. Alves, C.V. Godoy, M.L.F. Salatino, and N.S. Massola Jr. 2009. Influence of light and leaf epicuticular wax layer on *Phakopsora pachyrhizi* infection in soybean. *Trop. Plant Pathol.* 34: 306–312.

Furtado, G.Q., S.R.G. Moraes, S.A.M. Alves, L. Amorim, and N.S. Massola Jr. 2011. The infection of soybean leaves by *Phakopsora pachyrhizi* during conditions of discontinuous wetness. *J. Phytopathol.* 159: 165–170.

Galli, J.A., R. de C. Panizzi, S.A. Fessel, F. de Simoni, and M.F. Ito. 2005. Effect of *Colletotrichum dematium var. truncata* and *Cercospora kikuchii* on soybean seed germination. *Rev. Bras. Sementes* 27: 182–187.

Galli, J.A., R. de C. Panizzi, and R.D. Vieira. 2007. Evaluation of resistance of soybean varieties to damping-off caused by *Colletotrichum truncatum*. *Arq. Inst. Biol.* (Sao Paulo) 74: 163–165.

Gally, T., F. Pantuso, and B. Gonzalez. 2004. Seedling emergence of soybean [*Glycine max* (L.) Merrill] after seed treatment with fungicides in three agricultural crop seasons. *Rev. Mexi. Fitopatol.* 22: 377–381.

Gao, X., T.A. Jackson, G.L. Hartman, and T.L. Niblack. 2006. Interactions between the soybean cyst nematode and *Fusarium solani* f. sp. *glycines* based on greenhouse factorial experiments. *Phytopathology* 96: 1409–1415.

Garces Fiallos, F.R. and C.A. Forcelini. 2011. Temporal progress of rust and reduced leaf area and yield components in soybean grain. *Acta Agronomica, Universidad Nacional de Colombia* 60: 147–157.

Gasparetto, R., C.D. Fernandes, C.E. Marchi, and M. de F. Borges. 2011. Efficiency and economic viability of application of fungicides in the control of Asiatic rust of soybeans in Campo Grande, MS, Brazil. *Arq. Inst. Biol.* (Sao Paulo) 78: 251–259.

Gawade, D.B., A.P. Suryawanshi, V.B. Patil, S.N. Zagade, and A.G. Wadje. 2009a. Screening of soybean cultivars against anthracnose caused by *Colletotrichum truncatum*. *J. Plant Dis. Sci.* 4: 124–125.

Gawade, D.B., A.P. Suryawanshi, A.K. Pawar, K.T. Apet, and S.S. Devgire. 2009b. Field evaluation of fungicides, botanicals and bioagents against anthracnose of soybean. *Agric. Sci. Dig.* 29: 174–177.

Gelin, J.R., P.R. Arelli, and G.A. Rojas-Cifuentes. 2006. Using independent culling to screen plant introductions for combined resistance to soybean cyst nematode and sudden death syndrome. *Crop Sci.* 46: 2081–2083.

Gibson, P.T., M.A. Shenaut, V.N. Njiti, R.J. Suttner, and O. Myers Jr. 1994. Soybean varietal response to sudden death syndrome. In: *Proceedings of the 24th Soybean Seed Research Conference*, Chicago, IL, December 6–7, 1994, ed., D. Wilkinson. American Seed Trade Association, Washington, DC, pp. 436–446.

Gill-Langarica, H.R., N. Maldonado-Moreno, V. Pecina-Quintero, and N. Mayek-Perez. 2008. Reaction of improved soybean [*Glycine max* (L.) Merr.] germplasm to *Macrophomina phaseolina* (Tassi) Goidanich and drought stress. *Rev. Mexi. Fitopatol.* 26: 105–113.

Girish, K.R. and R. Usha. 2005. Molecular characterization of two soybean-infecting begomoviruses from India and evidence for recombination among legume-infecting begomoviruses from South-East Asia. *Virus Res.* 108: 167–176.

Godoy, C.V., A.M. Flausino, L.C.M. Santos, and E.M. del Ponte. 2009. Asian soybean rust control efficacy as a function of application timing under epidemic conditions in Londrina, PR. *Trop. Plant Pathol.* 34: 56–61.

Goellner, K., M. Loehrer, C. Langenbach, U. Conrath, E. Koch, and U. Schaffrath. 2010. *Phakopsora pachyrhizi*, the causal agent of Asian soybean rust. *Mol. Plant Pathol.* 11: 169–177.

Gongora-Canul, C.C. and L.F.S. Leandro. 2011a. Plant age affects root infection and development of foliar symptoms of soybean sudden death syndrome. *Plant Dis.* 95: 242–247.

Gongora-Canul, C.C. and L.F.S. Leandro. 2011b. Effect of soil temperature and plant age at time of inoculation on progress of root rot and foliar symptoms of soybean sudden death syndrome. *Plant Dis.* 95: 436–440.

Gongora-Canul, C., F.W. Nutter Jr., and L.F.S. Leandro. 2012. Temporal dynamics of root and foliar severity of soybean sudden death syndrome at different inoculum densities. *Eur. J. Plant Pathol.* 132(1): 71–79.

Goto, K., E. Sato, and K. Toyota. 2009. A novel detection method for the soybean cyst nematode *Heterodera glycines* Ichinohe using soil compaction and real-time PCR. *Nematol. Res.* 39: 1–7.

Goulart, A.C.P. 2001. Incidence and chemical control of fungi on soybean seeds in several counties of Mato Grosso do Sul State, Brazil. *Ciencia e Agrotecnologia* 25: 1467–1473.

Goulart, A.C.P., P.J.M. Andrade, and E.P. Borges. 2000. Control of soybean seedborne pathogens by fungicide treatment and its effects on emergence and yield. *Summa Phytopathol.* 26: 341–346.

Gracia-Garza, J.A., S. Neumann, T.J. Vyn, and G.J. Boland. 2002. Influence of crop rotation and tillage on production of apothecia by *Sclerotinia sclerotiorum*. *Can. J. Plant Pathol.* 24: 137–143.

Grau, C. 2006a. Powdery mildew of soybean. UW Madison Department of Plant Pathology Board of Regents of the University of Wisconsin System, Division of Cooperative Extension of the University of Wisconsin-Extension, Madison, WI, p. 2.

Grau, C. 2006b. Brown stem rot of soybean. UW Madison Department of Plant Pathology and the Division of Cooperative Extension of the University of Wisconsin-Extension, Madison, WI, p. 2.

Gravina, G. de A., C.S. Sediyama, S. Martins Filho, M.A. Moreira, E.G. Barros, and C.D. de Cruz. 2004. Multivariate analysis of combining ability for soybean resistance to *Cercospora sojina* Hara. *Genet. Mol. Biol.* 27: 395–399.

Gray, L.E. 1971. Variation in pathogenicity of *Cephalosporium gregatum* isolates. *Phytopathology* 61: 1410–1411.

Gray, L.E. and C.R. Grau. 1999. Brown stem rot. In: *Compendium of Soybean Diseases*, eds., G.L. Hartman, J.B. Sinclair, and J.C. Rupe. APS Press, St. Paul, MN, pp. 28–29.

Guldekar, D. and S.R. Potdukhe. 2010. Efficacy of different fungicides, botanicals and bioagents against leaf blight of soybean caused by *Colletotrichum truncatum*. *J. Soils Crops* 20: 310–313.

Guldekar, D. and S.R. Potdukhe. 2011. Determination of compatibility through perithecial formation among *Colletotrichum* spp. from chilli and soybean. *J. Soils Crops* 21: 41–43.

Guo, B., D.A. Sleper, P.R. Arelli, J.G. Shannon, and H.T. Nguyen. 2005. Identification of QTLs associated with resistance to soybean cyst nematode races 2, 3 and 5 in soybean PI 90763. *Theor. Appl. Genet.* 111: 965–971.

Guo, B., D.A. Sleper, H.T. Nguyen, P.R. Arelli, and J.G. Shannon. 2006. Quantitative trait loci underlying resistance to three soybean cyst nematode populations in soybean PI 404198A. *Crop Sci.* 46: 224–233.

Guo, X.M., D.C. Wang, S.G. Gordon et al. 2008. Genetic mapping of QTLs underlying partial resistance to *Sclerotinia sclerotiorum* in soybean PI 391589A and PI 391589B. *Crop Sci.* 48: 1129–1139.

Gupta, G.K., S.K. Sharma, and R.K. Ramteke. 2012. Biology, epidemiology and management of the pathogenic fungus *Macrophomina phaseolina* (Tassi) Goid with special reference to charcoal rot of soybean (*Glycine max* (L.) Merrill). *J. Phytopathol.* 160: 167–180.

Gupta, K.N. and R.L. Keshwal. 2002. Studies on the events prior to the development of epidemic in mungbean yellow mosaic virus on soybean. *Ann. Plant Prot. Sci.* 10: 118–120.

Gupta, K.N. and R.L. Keshwal. 2003. Epidemiological studies on yellow mosaic disease of soybean. *Ann. Plant Prot. Sci.* 11: 324–328.

Harikrishnan, R. and X.B. Yang. 2002. Effects of herbicides on root rot and damping-off caused by *Rhizoctonia solani* in glyphosate-tolerant soybean. *Plant Dis.* 86: 1369–1373.

Hartman, G.L., J.B. Sinclair, and J.C. Rupe. 1999. *Compendium of Soybean Diseases*. APS Press, St. Paul, MN.

Hegde, D.M. 2009. Oilseeds production in India: Retrospect and prospect. *Natl. Acad. Agric. Sci. Newsl.* 8: 1–7.

Hegde, G.M., K.H. Anahosur, S. Kulkarni, and M.R. Kachapu. 2002. Assessment of crop loss due to soybean rust pathogen. *Indian J. Plant Prot.* 30: 149–156.

Hegde, G.M. and R.K. Mesta. 2012. Integrated management of soybean rust. *Int. J. Life Sci. Pharm. Res.* 2: 44–48.

Heydari, R., Z.T. Maafi, and E. Pourjam. 2012. Yield loss caused by soybean cyst nematode, *Heterodera glycines*, in Iran. *Nematology* 14: 589–593.

Hikishima, M., M.G. Canteri, C.V. Godoy, L.J. Koga, and A.J. da Silva. 2010. Relationships among disease intensity, canopy reflectance and grain yield in the Asian soybean rust pathosystem. *Trop. Plant Pathol.* 35: 96–103.

Hobbs, H.A., C.B. Hill, C.R. Grau et al. 2006. Green stem disorder of soybean. *Plant Dis.* 90: 513–518.

Hoffmann, L.L., E.M. Reis, C. Forcelini, E. Panisson, C.S. Mendes, and R.T. Casa. 2004. Effects of crop rotation, cultivars and application of fungicides on the yield of grains and leaf diseases in soybean. *Fitopatol. Bras.* 29: 245–251.

Hossain, I., M.R. Islam, and M.M. Hamiduzzaman. 2001. Screening of soybean germplasms for resistance against *Colletotrichum truncatum* infection. *Pak. J. Sci. Indust. Res.* 44: 305–307.

Hughes, T.J., Z.K. Atallah, and C.R. Grau. 2009. Real-time PCR assays for the quantification of *Phialophora gregata* f. sp. *sojae* IGS genotypes A and B. *Phytopathology* 99: 1008–1014.

Hughes, T.J., W. Chen, and C.R. Grau. 2002. Pathogenic characterization of genotypes A and B of *Phialophora gregata* f. sp. *sojae*. *Plant Dis.* 86: 729–735.

Ignjatov, M., J. Balaz, M. Milosevic, and M. Vidic. 2007. *Pseudomonas syringae* pv. *glycinea* economically harmful pathogen of soybean in Vojvodnia. *Biljni Lekar* (*Plant Doctor*) 35: 589–595.

Ignjatov, M., J. Balaz, M. Milosevic, M. Vidic, and T. Popovic. 2008. Studies on plant pathogenic bacterium causal agent of soybean bacterial spots (*Pseudomonas syringae* pv. *glycinea* (Coerper) Youn et al.). In: *Pseudomonas syringae Pathovars and Related Pathogens*, eds., M.B. Fatmi et al. Springers Science+Business Media, the Netherlands, p. 419.

Imazaki, I., Y. Homma, M. Kato et al. 2006a. Genetic relationships between *Cercospora kikuchii* populations from South America and Japan. *Phytopathology* 96: 1000–1008.

Imazaki, I., H. Iizumi, K. Ishikawa, M. Sasahara, N. Yasuda, and S. Koizumi. 2006b. Effects of thiophanate-methyl and azoxystrobin on the composition of *Cercospora kikuchii* populations with thiophanate-methyl-resistant strains. *J. Gen. Plant Pathol.* 72: 292–300.

Inam-Ul-Haq, M., S.M. Rehman, H.M. Zahid Ali, and M.I. Tahir. 2012. Incidence of root rot diseases of soybean in Multan Pakistan and its management by the use of plant growth promoting rhizobacteria. *Pak. J. Bot.* 44: 2077–2080.

Iqbal, M.J., R. Ahsan, A.J. Afzal et al. 2009. Multigeneic QTL: The laccase encoded within the soybean Rfs2/rhg1 locus inferred to underlie part of the dual resistance to cyst nematode and sudden death syndrome. *Curr. Issues Mol. Biol.* 11: 11–19.

Isard, S.A., C.W. Barnes, S. Hambleton et al. 2011. Predicting soybean rust incursions into the North American continental interior using crop monitoring, spore trapping, and aerobiological modeling. *Plant Dis.* 95: 1346–1357.

Ittah, M.A., A. Iwo, E. Osai, and E.E. Aki. 2011. Response of soybean genotypes to soybean rust (*Phakopsora pachyrhizi* (Sydow)) infection in Nigeria. *J. Agric. Biotechnol. Ecol.* 4: 54–69.

Ivancovich, A. 2008. Soybean rust in Argentina. (Facing the challenge of soybean rust in South America.) *JIRCAS Working Rep.* 58: 14–17.

Iwo, G.A., M.A. Ittah, and E.O. Osai. 2012. Sources and genetics of resistance to soybean rust *Phakopsora pachyrhizi* (H. Sydow & Sydow) in Nigeria. *J. Agric. Sci.* (Toronto) 4: 1–6.

Jackson, E.W., C.D. Feng, P. Fenn, and P.G. Chen. 2008. Genetic mapping of resistance to purple seed stain in PI 80837 soybean. *J. Hered.* 99: 319–322.

Jackson, T.A., G.S. Smith, and T.L. Niblack. 2005. *Heterodera glycines* infectivity and egg viability following nonhost crops and during overwintering. *J. Nematol.* 37: 259–264.

Jagtap, G.P., S.B. Dhopte, and U. Dey. 2012b. Survey, surveillance and cultural characteristics of bacterial blight of soybean. *Afr. J. Agric. Res.* 7: 4559–4563.

Jagtap, G.P., D.S. Gavate, and U. Dey. 2012a. Control of *Colletotrichum truncatum* causing anthracnose/pod blight of soybean by aqueous leaf extracts, biocontrol agents and fungicides. *Sci. J. Agric.* 1: 39–52.

Jagtap, G.P. and P.L. Sontakke. 2009. Effect of different carbon sources on growth and sporulation of *Colletotrichum truncatum* isolates causing pod blight disease in soybean. *Curr. Adv. Agric. Sci.* 1: 117–118.

Jahagirdar, S., P.V. Patil, and V.I. Benagi. 2012. Bio-formulations and indigenous technology methods in the management of Asian soybean rust. *Int. J. Plant Prot.* 5: 63–67.

Jana, T., T.R. Sharma, and N.K. Singh. 2005b. SSR-based detection of genetic variability in the charcoal root rot pathogen *Macrophomina phaseolina*. *Mycol. Res.* 109: 81–86.

Jana, T.K., N.K. Singh, K.R. Koundal, and T.R. Sharma. 2005a. Genetic differentiation of charcoal rot pathogen, *Macrophomina phaseolina*, into specific groups using URP-PCR. *Can. J. Microbiol.* 51: 159–164.

Jarvie, J.A. 2009. A review of soybean rust from a South African perspective. *S. Afr. J. Sci.* 105: 103–108.

Johnson, W.G., J.E. Creech, and V.A. Mock. July 2008. Role of winter annual weeds as alternative hosts for soybean cyst nematode. *Crop Manag.* doi:10.1094/CM-2008-0701-01-RV.

Jordan, S.A., D.J. Mailhot, and A.J. Gevens. 2010. Characterization of Kudzu (*Pueraria* spp.) resistance to *Phakopsora pachyrhizi*, the causal agent of soybean rust. *Phytopathology*100: 941–948.

Joshi, T., B. Valliyodan, J.H. Wu, S.H. Lee, D. Xu, and H.T. Nguyen. 2013. Genomic differences between cultivated soybean *G. max* and its wild relative *G. soja*. *BMC Genomics* 14(Suppl. 1): S5. doi:10.1186/1471-2164-14-S1-S5.

Kaewnum, S., S. Prathuangwong, and T.J. Burr. 2005. Aggressiveness of *Xanthomonas axonopodis* pv. *glycines* isolates to soybean and hypersensitivity responses by other plants. *Plant Pathol.* 54: 409–415.

Kang, S.T. and M.A. Mian. 2010. Genetic map of the powdery mildew resistance gene in soybean PI 243540. *Genome* 53: 400–405.

Kaushal, R.P. 2009. Evaluation of fungicides, neem products and bioagents against target leaf spot of soybean caused by *Corynespora cassiicola*. *Soybean Res.* 7: 96–101.

Kaushal, R.P. and O.P. Sood. 2002. Genetics of resistance in soybean to pod blight caused by *Colletotrichum truncatum* and *Diaporthe phaseolorum*. *J. Mycol. Plant Pathol.* 32: 219–221.

Kawuki, R.S., E. Adipala, P. Tukamuhabwa, and B. Mugonola. 2002. Efficacies and profitability of different fungicides and spray regimes for control of soybean rust (*Phakopsora pachyrhizi* Syd.) in Uganda. *Muarik Bull.* 5: 27–34.

Kawuki, R.S., P. Tukamuhabwa, and E. Adipala. 2004. Soybean rust severity, rate of rust development, and tolerance as influenced by maturity period and season. *Crop Prot.* 23: 447–455.

Kazi, S., J. Shultz, J. Afzal, J. Johnson, V.N. Njiti, and D.A. Lightfoot. 2008. Separate loci underlie resistance to root infection and leaf scorch during soybean sudden death syndrome. *TAG Theor. Appl. Genet.* 116: 967–977.

Kemmitt, G.M., G. DeBoer, D. Ouimette, and M. Iamauti. 2008. Systemic properties of myclobutanil in soybean plants, affecting control of Asian soybean rust (*Phakopsora pachyrhizi*). *Pest Manag. Sci.* 64: 1285–1293.

Kendig, S.R., J.C. Rupe, and H.D. Scott. 2000. Effect of irrigation and soil water stress on densities of *Macrophomina phaseolina* in soil and roots of two soyabean cultivars. *Plant Dis.* 84: 895–900.

Kendrick, M.D., D.K. Harris, B.K. Ha et al. 2011. Identification of a second Asian soybean rust resistance gene in Hyuuga soybean. *Phytopathology* 101: 535–543.

Kennedy, B.W. and S.M. Alcorn. 1980. Estimates of U.S. crop losses to procaryote plant pathogens. *Plant Dis.* 64: 674–676.

Khare, U.K., M.N. Khare, and M.M. Ansari. 2003. Bacterial pustule disease of soybean—Present scenario and future strategies. *Soybean Res.* 1: 43–57.

Khodke, S.W. and B.T. Raut. 2010. Management of root rot/collar rot of soybean. *Indian Phytopathol.* 63: 298–301.

Khot, G.G., V.S. Patil, A.M. Tirmali, S.R. Lohate, and V.S. Shinde. 2007. Management of soybean rust caused by *Phakopsora pachyrhizi* Syd. *Karnataka J. Agric. Sci.* 20: 663–664.

Kim, K.S., J.R. Unfried, D.L. Hyten et al. 2012. Molecular mapping of soybean rust resistance in soybean accession PI 561356 and SNP haplotype analysis of the Rpp1 region in diverse germplasm. *TAG Theor. Appl. Genet.* 125: 1339–1352.

Kim, M.S., D.L. Hyten, T.L. Niblack, and B.W. Diers. 2011. Stacking resistance alleles from wild and domestic soybean sources improves soybean cyst nematode resistance. *Crop Sci.* 51: 934–943.

Kim, M.Y., S. Lee, K. Van et al. 2010. Whole genome sequencing and intensive analysis of the undomesticated soybean (*Glycine soja* Sieb and Zucc) genome. *Proc. Natl. Acad. Sci. USA* 107: 22032–22037.

Kim, W.G., S.K. Hong, and S.S. Han. 2005. Occurrence of web blight in soybean caused by *Rhizoctonia solani* AG-1(IA) in Korea. *Plant Pathol. J.* 21: 406–408.

Kirkpatrick, M.T., C.S. Rothrock, J.C. Rupe, and E.E. Gbur. 2006. The effect of *Pythium ultimum* and soil flooding on two soybean cultivars. *Plant Dis.* 90: 597–602.

Kiryowa, M., P. Tukamuhabwa, and E. Adipala. 2008. Genetic analysis of resistance to soybean rust disease. *Afr. Crop Sci. J.* 16(3): 211–217.

Klingelfuss, L.H. and J.T. Yorinori. 2001. Latent infection by *Colletotrichum truncatum and Cercospora kikuchii* in soybean. *Fitopatol. Bras.* 26: 158–164.

Knebel, J.L., V.F. Guimaraes, M. Andreotti, and J.R. Stangarlin. 2006. Influence of row spacing and plant population on late season disease severity, powdery mildew and agronomic characters in soybean Portuguese. *Acta Sci.—Agron.* 28: 385–392.

Koenning, S.R., T.C. Creswell, E.J. Dunphy, E.J. Sikora, and J.D. Mueller. 2006. Increased occurrence of target spot of soybean caused by *Corynespora cassiicola* in the Southeastern United States. *Plant Dis.* 90: 974.

Koenning, S.R. and J.A. Wrather. 2010. Suppression of soybean yield potential in the continental United States by plant diseases from 2006 to 2009. Online. *Plant Health Prog.* doi:10.1094/PHP-2010-1122-01-RS.

Koga, L.J., M.G. Canteri, E.S. Calvo et al. 2011. Chemical control and responses of susceptible and resistant soybean cultivars to the progress of soybean rust. *Trop. Plant Pathol.* 36: 294–302.

Kolander, T.M., J.C. Bienapfl, J.E. Kurle, and D.K. Malvick. 2012. Symptomatic and asymptomatic host range of *Fusarium virguliforme*, the causal agent of soybean sudden death syndrome. *Plant Dis.* 96: 1148–1153.

Koning, G., D.M. TeKrony, S.A. Ghabrial, and T.W. Pfeiffer. 2002. Soybean mosaic virus (SMV) and the SMV resistance gene(Rsv(1)): Influence on *Phomopsis* spp. seed infection in an aphid free environment. *Crop Sci.* 42: 178–185.

Kudo, A.S., L.E.B. Blum, and M.A. Lima. 2011. Aerobiology of *Cercospora kikuchii. Ciencia Rural* 41: 1682–1688.

Kulkarni, S.S., S.G. Bajwa, R. Robbins, T.A. Costello, and L. Kirkpatrick. 2008. Effect of soybean cyst nematode (*Heterodera glycines*) resistance rotation on scn population distribution, soybean canopy reflectance, and grain yield. *Trans. ASABE* 51: 1511–1517.

Kumar, B. and K.S. Dubey. 2007. Assessment of losses in different cultivars of soybean due to anthracnose under field conditions. *J. Plant Dis. Sci.* 2: 110.

Kumar, B., A. Talukdar, K. Verma et al. 2014. Screening of soybean (*Glycine max* (L) Merr) genotypes for yellow mosaic virs (YMV) disease resistance and their molecular characterization using RGA and SSRS markers. *Aust. J. Crop Sci.* 8: 27–34.

Kumar, K. and J. Singh. 2000. Location, survival, transmission and management of seed-borne *Macrophomina phaseolina*, causing charcoal rot in soybean. *Ann. Plant Prot. Sci.* 8: 44–46.

Kumudini, S., C.V. Godoy, J.E. Board, J. Omielan, and M. Tollenaar. 2008. Mechanisms involved in soybean rust-induced yield reduction. *Crop Sci.* 48: 2334–2342.

Kumudini, S., C.V. Godoy, and B. Kennedy. 2010. Role of host-plant resistance and disease development stage on leaf photosynthetic competence of soybean rust infected leaves. *Crop Sci.* 50: 2533–2542.

Kurundkar, B.P., V.S. Shinde, S.B. Mahajan, and M.P. Deshmukh. 2011. Reaction of soybean genotypes against rust. *J. Plant Dis. Sci.* 6: 134–136.

Lakshmi, B.S., J.P. Gupta, and M.S. Prasad. 2002. Influence of methyl cellulose on translocation of thiophanate-methyl in soybean. *Plant Dis. Res.* 17: 278–282.

Lal, S.K., V.K.S. Rana, R.L. Sapra, and K.P. Singh. 2005. Screening and utilization of soybean germplasm for breeding resistance against mungbean yellow mosaic virus. *Soybean Genet. Newsl.* 32: unpaginated. http://soybase.org:8083/articleFiles/45.

Leandro, L.F., N. Tatalovic, and A. Luckew. 2012. Soybean sudden death syndrome—Advances in knowledge and disease management. *CAB Rev.* 7(53): 1–14.

Lemes, E.M., C.L. Mackowiak, and A. Blount. 2011. Effects of silicon applications on soybean rust development under greenhouse and field conditions. *Plant Dis.* 95: 317–324.
Lemos, N.G., A. Lucca e Braccini, R.V. de Abdelnoor et al. 2011. Characterization of genes Rpp2, Rpp4, and Rpp5 for resistance to soybean rust. *Euphytica* 182: 53–64.
Leon, M., P.M. Yaryura, M.S. Montecchia et al. 2009. Antifungal activity of selected indigenous *Pseudomonas* and *Bacillus* from the soybean rhizosphere. *Int. J. Microbiol.* 2009: Article ID 572049.
Levy, C. 2005. Epidemiology and chemical control of soybean rust in Southern Africa. *Plant Dis.* 89: 669–674.
Li, H.Y., R.J. Liu, and H.R. Shu. 2002. A preliminary report on interactions between arbuscular mycorrhizal fungi and soybean cyst nematode. *Acta Phytopathol. Sin.* 32: 356–360.
Li, S., G.L. Hartman, L.L. Domier, and D. Boykin. 2008. Quantification of *Fusarium solani* f. sp. *glycines* isolates in soybean roots by colony-forming unit assays and real-time quantitative PCR. *TAG Theor. Appl. Genet.* 117: 343–352.
Li, S., N.C. Kurtzweil, C.R. Grau, and G.L. Hartman. 2004. Occurrence of soybean stem canker (*Diaporthe phaseolorum* var. *meridionalis*) in Wisconsin. *Plant Dis.* 88: 576.
Li, S. and L.D. Young. 2009. Evaluation of selected genotypes of soybean for resistance to *Phakopsora pachyrhizi. Plant Health Prog.* doi:10.1094/PHP-2009-0615-01-RS.
Li, S.X. 2009. Reaction of soybean rust-resistant lines identified in Paraguay to Mississippi isolates of *Phakopsora pachyrhizi. Crop Sci.* 49(3): 887–894.
Li, W., D.P. Roberts, P.D. Dery et al. 2002. Broad spectrum anti-biotic activity and disease suppression by the potential biocontrol agent *Burkholderia ambifaria* BC-F. *Crop Prot.* 21: 129–135.
Li, X. and X.B. Yang. 2009. Similarity, pattern, and grouping of soybean fungal diseases in the United States: Implications for the risk of soybean rust. *Plant Dis.* 93: 162–169.
Li, Y.H., C. Zhang, Z.S. Gao et al. 2009. Development of SNP markers and haplotype analysis of the candidate gene for rhg1, which confers resistance to soybean cyst nematode in soybean. *Mol. Breed.* 24: 63–76.
Lima, F.B., H.D. Campos, L.M. Ribeiro et al. 2011. Effect of poultry litter on the soybean cyst nematode population reduction. *Nematol. Bras.* 35: 71–77.
Lima, S.F., R. de Alvarez, G. de C.F. Theodoro, M. de F. Bavaresco and K.S.Silva. 2012. Effect of sowing in crossed lines on grain yield and the severity of Asian soybean rust. *Biosci. J.* 28: 954–962.
Liu, D.W., J.J. Chen, and Y.X. Duan. 2011a. Differential proteomic analysis of the resistant soybean infected by soybean cyst nematode, *Heterodera glycines* race 3. *J. Agric. Sci.* (Toronto) 3: 160–167.
Liu, S.F. and S.Y. Chen. 2005. Efficacy of the fungi *Hirsutella minnesotensis* and *H. rhossiliensis* from liquid culture for control of the soybean cyst nematode *Heterodera glycines. Nematology* 7: 149–157.
Liu, S.F. and S.Y. Chen. 2009. Effectiveness of *Hirsutella minnesotensis* and *H. rhossiliensis* in control of the soybean cyst nematode in four soils with various pH, texture, and organic matter. *Biocontrol Sci. Technol.* 19: 595–612.
Liu, S.M., P.K. Kandoth, S.D. Warren et al. 2012. A soybean cyst nematode resistance gene points to a new mechanism of plant resistance to pathogens. *Nature* (London) 492(7428): 256–260.
Liu, X.H., S.M. Liu, A. Jamai et al. 2011b. Soybean cyst nematode resistance in soybean is independent of the Rhg4 locus LRR-RLK gene. *Funct. Integr. Genomics* 11: 539–549.
Liu, Y., G.C. Ma, and L. Feng. 2008. Study on induced-resistances to bacterial blight disease of soybean from chemical application. *Chin. J. Oil Crop Sci.* 30: 116–118.
Lomonossoff, G.P. and S.A. Ghabrial. 2001. Comoviruses. In: *Encyclopedia of Plant Pathology*, Vol. 1, eds., O.C. Maloy and T.D. Murray. John Wiley & Sons, New York, pp. 239–242.
Lopez-Nicora, H.D., J.P. Craig, X.B. Gao, K.N. Lambert, and T.L. Niblack. 2012. Evaluation of cultivar resistance to soybean cyst nematode with a quantitative polymerase chain reaction assay. *Plant Dis.* 96: 1556–1563.
Lou, B.G., W.J. Chen, C. Lin, G.R. Wang, G.M. Xia, and M.Q. Lou. 2009. A new symptom type of soybean pod anthracnose and identification of its pathogen. *Acta Phytophyl. Sin.* 36: 229–233.
Lozovaya, V.V., A.V. Lygin, O.V. Zernova, S. Li, J.M. Widholm, and G.L. Hartman. 2006. Lignin degradation by *Fusarium solani* f. sp. *glycines. Plant Dis.* 90: 77–82.
Lu, G.H. 2003. Engineering *Sclerotinia sclerotiorum* resistance in oilseed crops. *Afr. J. Biotechnol.* 2: 509–516.
Lura, M.C., M.G. Latorre Rapela, M.C. Vaccari et al. 2011. Genetic diversity of *Cercospora kikuchii* isolates from soybean cultured in Argentina as revealed by molecular markers and cercosporin production. *Mycopathologia* 171: 361–371.
Luster, D.G., M.B. McMahon, H.H. Edwards et al. 2012. Novel *Phakopsora pachyrhizi* extracellular proteins are ideal targets for immunological diagnostic assays. *Appl. Environ. Microbiol.*78: 3890–3895.
Lygin, A.V., S.X. Li, R. Vittal, J.M. Widholm, G.L. Hartman, and V.V. Lozovaya. 2009. The importance of phenolic metabolism to limit the growth of *Phakopsora pachyrhizi. Phytopathology* 99: 1412–1420.

Ma, J., C.B. Hill, and G.L. Hartman. 2010. Production of *Macrophomina phaseolina* conidia by multiple soybean isolates in culture. *Plant Dis.* 94: 1088–1092.

Madalosso, M.G., L. da S. Domingues, M.P. Debortoli, G. Lenz, and R.S. Balardin. 2010. Cultivars, row spacing and fungicide application programs on *Phakopsora pachyrhizi* Sydow control in soybean. *Ciencia Rural* 40: 2256–2261.

Mahesha, B., P.V. Patil, and B. Nandini. 2009. Identification of multiple disease resistance sources in soybean. *Crop Res.* (Hisar) 37: 213–216.

Maiti, S., J. Basak, S.S. Kundagrami, A. Kundu, and A. Pal. 2011. Molecular-assisted genotyping of mungbean yellow mosaic India virus resistant germplasms of mungbean and urdbean. *Mol. Biotechnol.* 47: 95–104. doi:10.1007/s 12033-010.9314-1.

Malapi-Nelson, M., R.H. Wen, B.H. Ownley, and M.R. Hajimorad. 2009. Co-infection of soybean with Soybean mosaic virus and Alfalfa mosaic virus results in disease synergism and alteration in accumulation level of both viruses. *Plant Dis.* 93: 1259–1264.

Malvick, D.K. and K.E. Bussey. 2008. Comparative analysis and characterization of the soybean sudden death syndrome pathogen *Fusarium virguliforme* in the northern United States. *Can. J. Plant Pathol.* 30: 467–476.

Malvick, D.K., W. Chen, J.E. Kurle, and C.R. Grau. 2003. Cultivar preference and genotype distribution of brown stem rot pathogen *Phialophora gregata* in the Midwestern United States. *Plant Dis.* 87: 1250–1255.

Malvick, D.K. and A.E. Impullitti. 2007. Detection and quantification of *Phialophora gregata* in soybean and soil samples with a quantitative, real-time PCR assay. *Plant Dis.* 91: 736–742.

Mansouri, S., R. Wijk, M. van Rep, and A.M. Fakhoury. 2009. Transformation of *Fusarium virguliforme*, the causal agent of sudden death syndrome of soybean. *J. Phytopathol.* 157: 319–321.

Mantecon, J.D. 2008. Efficacy of chemical and biological strategies for controlling the soybean brown spot (*Septoria glycines*). *Ciencia e Investigacion Agraria* 35: 173–176.

Maphosa, M., H. Talwana, P. Gibson, and P. Tukamuhabwa. 2012b. Combining ability for resistance to soybean rust in F2 and F3 soybean populations. *Field Crops Res.* 130: 1–7.

Maphosa, M., H. Talwana, and P. Tukamuhabwa. 2012a. Enhancing soybean rust resistance through Rpp2, Rpp3 and Rpp4 pair wise gene pyramiding. *Afr. J. Agric. Res.* 7: 4271–4277.

Martins, J.A.S., F.C. Juliatti, V.A. Santos, A.C. Polizel, and F.C. Juliatti. 2007. Latent period and the use of principal components analysis for partial resistance to soybean rust. *Summa Phytopathol.* 33: 364–371.

Martins Filho, S., C.S. Sediyama, A.J. Regazzi, and L.A. Peternelli. 2002. Genetic analysis of soybean resistance to *Cercospora sojina* Hara. *Crop Breed. Appl. Biotechnol.* 2: 549–555.

Matsuo, E., T. Sediyama, R.D. de L. Oliveira, C.D. Cruz, and R. de C.T. Oliveira. 2012. Characterization of type and genetic diversity among soybean cyst nematode differentiators. *Sci. Agric.* 69: 147–151.

Mayonjo, D.M. and R.G. Kapooria. 2003. Occurrence and variability of *Colletotrichum truncatum* on soybean in Zambia. *Bull. OEPP* 33: 339–341.

Mazarei, M., W. Liu, H. Al-Ahmad, P.R. Arelli, V.R. Pantalone, and C.N. Stewart. 2011. Gene expression profiling of resistant and susceptible soybean lines infected with soybean cyst nematode. *TAG Theor. Appl. Genet.* 123: 1193–1206.

Mbofung, G.Y.C., T.C. Harrington, J.T. Steimel, S.S. Navi, X.B. Yang, and L.F. Leandro. 2012. Genetic structure and variation in aggressiveness in *Fusarium virguliforme* in the Midwest United States. *Can. J. Plant Pathol.* 34: 83–97.

Medic-Pap, S., M. Milosevic, and S. Jasnic. 2007. Soybean seed-borne fungi in the Vojvodina Province. *Phytopathol. Pol.* 45: 55–65.

Melton, T.A., B.J. Jacobsen, and G.R. Noel. 1986. Effects of temperature on development of *Heterodera glycines* on *Glycine max* and *Phaseolus vulgaris*. *J. Nematol.* 18: 468–474.

Meng, X., C.R. Grau, and W. Chen. 2005. Cultivar preference exhibited by two sympatric and genetically distinct populations of the soybean fungal pathogen *Phialophora gregata* f. sp. *soja*. *Phytopathology* 54: 180–188.

Mengistu, A., P.A. Arelli, N. Bellaloui et al. March 2012. Evaluation of soybean genotypes for resistance to three seed-borne diseases. *Plant Health Prog.* doi:10.1094/PHP-2012-0321-02-RS.

Mengistu, A., L. Castlebury, R. Smith, J. Ray, and N. Bellaloui. 2009b. Seasonal progress of *Phomopsis longicolla* infection on soybean plant parts and its relationship to seed quality. *Plant Dis.* 93: 1009–1018.

Mengistu, A., K.N. Reddy, R.M. Zablotowicz, and A.J. Wrather. January 2009a. Propagule densities of *Macrophomina phaseolina* in soybean tissue and soil as affected by tillage, cover crop, and herbicide. *Plant Health Prog.* doi:10.1094/PHP-2009-0130-01-RS.

Mengistu, A., J.R. Smith, J.D. Ray, and N. Bellaloui. 2011. Seasonal progress of charcoal rot and its impact on soybean productivity. *Plant Dis.* 95: 1159–1166.

Mesquini, R.M., R.A. Vieira, K.R.F. Schwan-Estrada, and C.V. Godoy. 2011. Relations of cause and effect among environmental conditions, airborne urediniospores and Asian soybean rust severity. *Biosci. J.* 27: 552–557.

Meyer, J.D.F., D.C.G. Silva, C.L. Yang et al. 2009. Identification and analyses of candidate genes for Rpp4-mediated resistance to Asian soybean rust in soybean. *Plant Physiol.* 150: 295–307.

Meyer, M.C., C.J. Bueno, N.L. de Souza, and J.T. Yorinori. 2006. Effect of doses of fungicides and plant resistance activators on the control of Rhizoctonia foliar blight of soybean, and on *Rhizoctonia solani* AG1-IA in vitro development. *Crop Prot.* 25: 848–854.

Mian, M.A.R., A.M. Missaoui, D.R. Walker, D.V. Phillips, and H.R. Boerma. 2008. Frogeye leaf spot of soybean: A review and proposed race designations for isolates of *Cercospora sojina* Hara. *Crop Sci.* 48: 14–24.

Michelutti, R., J.C. Tu, D.W.A. Hunt, D. Gagnier, T.R. Anderson, and T.W. Welacky. 2002. First report of *Bean pod mottle virus* in soybean in Canada. *Plant Dis.* 86: 330.

Miles, M.R., M.R. Bonde, S.E. Nester, D.K. Berner, R.D. Frederick, and G.L. Hartman. 2011. Characterizing resistance to *Phakopsora pachyrhizi* in soybean. *Plant Dis.* 95: 577–581.

Miles, M.R., C. Levy, W. Morel et al. 2007. International fungicide efficacy trials for the management of soybean rust. *Plant Dis.* 91: 1450–1458.

Miles, M.R., W. Morel, J.D. Ray, J.R. Smith, R.D. Frederick, and G.L. Hartman. 2008. Adult plant evaluation of soybean accessions for resistance to *Phakopsora pachyrhizi* in the field and greenhouse in Paraguay. *Plant Dis.* 92: 96–105.

Miller, D.R., S.Y. Chen, P.M. Porter et al. 2006. Rotation crop evaluation for management of the soybean cyst nematode in Minnesota. *Agron. J.* 98: 569–578.

Mirza, M.S. and Y. Ahmad. 2002. *Septoria glycines*—The cause of brown spots on soybean in Pakistan. *Pak. J. Agric. Res.* 17: 46–48.

Mishra, D.S., A.K. Gupta, C.R. Prajapati, and U.S. Singh. 2011. Combination of fungal and bacterial antagonists for management of root and stem rot disease of soybean. *Pak. J. Bot.* 43: 2569–2574.

Mitchum, M.G. and T.J. Baum. 2008. Genomics of the soybean cyst nematode-soybean interaction. In: *Genetics and Genomics of Soybean*, Vol. 2: Plant Genetics and Genomics: Crops and Models, ed., G. Stacey. Springer, New York, pp. 321–341.

Mittal, R.K. 2001. Yield losses by frog-eye leaf spot and anthracnose diseases in soybean under different sowing dates in the hills. *Indian Phytopathol.* 54: 32–34.

Mo, J.Y., T.X. Guo, X. Li, and X.B. Yang. 2008. Effect of light intensity on infection of soybean rust. *Southwest China J. Agric. Sci.* 21: 997–1001.

Mock, V.A., J.E. Creech, V.R. Ferris et al. 2012. Influence of winter annual weed management and crop rotation on soybean cyst nematode (*Heterodera glycines*) and winter annual weeds: Years four and five. *Weed Sci.* 60: 634–640.

Montazeri, M. and H. Hamdollah-Zadeh. 2005. The effect of trifluralin on *Rhizoctonia solani* (isolate AG4), causal agent of soybean damping off. *Caspian J. Environ. Sci.* 3: 169–172.

Monteros, M.J., B.K. Ha, D.V. Phillips, and H.R. Boerma. 2010. SNP assay to detect the 'Hyuuga' red-brown lesion resistance gene for Asian soybean rust. *TAG Theor. Appl. Genet.* 121: 1023–1032.

Morab, H., R.V. Koti, M.B. Chetti, P.V. Pati, and A.S. Nalini. 2003. Role of nutrients in inducing rust resistance in soybean. *Indian J. Plant Physiol.* 8: 85–88.

Morales, A.M.A.P., A. Borem, M.A. Graham, and R.V. Abdelnoor. 2012. Advances on molecular studies of the interaction soybean—Asian rust. *Crop Breed. Appl. Biotechnol.* 12: 1–7.

Mueller, D.S., C.A. Bradley, C.R. Grau, J.M. Gaska, J.E. Kurle, and W.L. Pedersen. 2004. Application of thiophanate-methyl at different host growth stages for management of *sclerotinia* stem rot in soybean. *Crop Prot.* 23: 983–988.

Mueller, D.S., G.L. Hartman, and W.L. Pedersen. 2002. Effect of crop rotation and tillage system on *sclerotinia* stem rot on soybean. *Can. J. Plant Pathol.* 24: 450–456.

Mueller, T.A., M.R. Miles, W. Morel et al. 2009. Effect of fungicide and timing of application on soybean rust severity and yield. *Plant Dis.* 93: 243–248.

Munoz-Cabanas, R.M., S. Hernandez-Delgado, and N. Mayek-Perez. 2005. Pathogenic and genetic analysis of *Macrophomina phaseolina* (Tassi) Goid. on different hosts. *Rev. Mexi. Fitopatol.* 23: 11–18.

Narvaez, D.F., W.M. Jurick, J.J. Marois II, and D.L. Wright. 2010. Effects of surface wetness periods on development of soybean rust under field conditions. *Plant Dis.* 94: 258–264.

Narvel, J.M., L.R. Jakkula, D.V. Phillips, T. Wang, S.-H. Lee, and H.R. Boerma. 2001. Molecular mapping of Rxp conditioning reaction to bacterial pustule in soybean. *J. Heredity* 92: 267–270.

Nascimento, J.F., J.B. do Vida, D.J. Tessmann, L. Zambolim, R.A. Vieira, and R.R. de Oliveira. 2012. Progress of Asian soybean rust and airborne urediniospores of *Phakopsora pachyrhizi* in southern Brazil. *Summa Phytopathol.* 38: 280–287.

Navi, S.S. and X.B. Yang. February 2008. Foliar symptom expression in association with early infection and xylem colonization by *Fusarium virguliforme* (formerly *F. solani* f. sp. *glycines*), the causal agent of soybean sudden death syndrome. *Plant Health Prog*. doi:10.1094/PHP-2008-0222-01.

Nema, A., M.S. Bhale, and D.C. Garg. 2012. Pod blight of soybean: A prediction model for Central Narmada Valley Zone of Madhya Pradesh, India. Abstract. In: *Third Global Conference on Plant Pathology for Food Security*, Indian Society of Mycology and Pant Pathology, Udaipur, India, January 10–13, 2012, p. 204.

Nene, Y.L. 1972. A survey of the viral diseases of pulse crops in India. GB Pant Univ Agric Tech, Pantnagar. *India J. Res. Bull.* 4: 191.

Niblack, T.L. 2005. Soybean cyst nematode management reconsidered. *Plant Dis*. 89: 1020–1026.

Niblack, T.L., P.R. Arelli, G.R. Noel et al. 2002. A revised classification scheme for genetically diverse populations of *Heterodera glycines*. *J. Nematol.* 43: 279–288.

Niblack, T.L., A.L. Colgrove, K. Colgrove, and J.P. Bond. January 2008. Shift in virulence of soybean cyst nematode is associated with use of resistance from PI 88788. *Plant Health Prog.* doi:10.1094/PHP-2008-0118-01-RS.

Niblack, T.L., K.N. Lambert, and G.L. Tylka. 2006. A model plant pathogen from the Kingdom Animalia: *Heterodera glycines*, the soybean cyst nematode. *Annu. Rev. Phytopathol.* 44: 283–303.

Njiti, V.N. and D.A. Lightfoot. 2006. Genetic analysis infers Dt loci underlie resistance to *Fusarium solani* f. sp. *glycines* in indeterminate soybeans. *Can. J. Plant Sci.* 86: 83–90.

Njiti, V.N., K. Meksem, M.J. Iqbal et al. 2002. Common loci underlie field resistance to soybean sudden death syndrome in Forrest, Pyramid, Essex, and Douglas. *Theor. Appl. Genet.* 104: 294–300. doi:10.1007/s001220100682.

Nunkumar, A., P.M. Caldwell, and Z.A. 2009. Development of *Phakopsora pachyrhizi* on soybean at controlled temperature, relative humidity and moisture periods. *S. Afr. J. Plant Soil* 26: 225–230.

O'Donnell, K., S. Sink, M.M. Scandiani et al. 2010. Soybean sudden death syndrome species diversity within North and South America revealed by multilocus genotyping. *Phytopathology* 100: 58–71.

Oloka, H.K., P. Tukamuhabwa, T. Sengooba, E. Adipala, and P. Kabayi. 2009. Potential for soybean rust tolerance among elite soybean lines in Uganda. *Crop Prot.* 28: 1076–1080.

Oloka, H.K., P. Tukamuhabwa, T. Sengooba, and S. Shanmugasundram. 2008. Reaction of exotic soybean germplasm to *Phakopsora pachyrhizi* in Uganda. *Plant Dis.* 92: 1493–1496.

Ono, Y., P. Buritica, and J.F. Hennen. 1992. Delimitation of *Phakopsora*, *Physopella* and *Cerotellium* and their species on Leguminosae. *Mycol. Res.* 96: 825–850.

Padgett, B., R. Schneider, and K. Whitam. 2003. Foliar-applied fungicides in soybean disease management. *Louisiana Agric.* 46: 7–9.

Pant, R. and A.N. Mukhopadhyay. 2001. Integrated management of seed and seedling rot complex of soybean. *Indian Phytopathol.* 54: 346–350.

Pant, R. and A.N. Mukhopadhyay. 2002. Studies on seed and seedling rot of soybean. *Ann. Plant Prot. Sci.* 10: 401–402.

Parameshwar, G., G.T. Basavaraja, and S. Bhat. 2012. Screening of genotypes against rust and yellow mosaic diseases in soybean (*Glycine max* (L.) Merrill). *Environ. Ecol.* 30: 306–309.

Parashar, R.D. and C. Leben. 1972. Detection of *Pseudomonas glycinea* in soybean seed lots. *Phytopathology* 62: 1075–1077.

Paris, R.L., A. Mengistu, J.M. Tyler, and J.R. Smith. 2006. Registration of soybean germplasm line DT97-4290 with moderate resistance to charcoal rot. *Crop Sci.* 46: 2324–2325.

Patil, P.V., G.T. Basavaraja, and S.M. Husain. 2004. Two genotypes of soybean as promising source of resistance to rust caused by *Phakopsora pachyrhizi* Syd. *Soybean Res.* 2: 46–47.

Patkowska, E. 2006. The use of biopreparations in the control of soybean endangered by pathogenic soil-borne fungi. *Electronic J. Polish Agric. Univ.* 9: 1, Article 19.

Paul, C. and G.L. Hartman. 2009. Sources of soybean rust resistance challenged with single-spored isolates of *Phakopsora pachyrhizi*. *Crop Sci.* 49: 1781–1785.

Paul, C., C.B. Hill, and G.L. Hartman. 2011. Comparisons of visual rust assessments and DNA levels of *Phakopsora pachyrhizi* in soybean genotypes. *Plant Dis.* 95: 1007–1012.

Pedersen, P., G.L. Tylka, A. Mallarino, A.E. MacGuidwin, N.C. Koval, and C.R. Grau. 2010. Correlation between soil pH, *Heterodera glycines* population densities, and soybean yield. *Crop Sci.* 50: 1458–1464.

Peltier, A.J., C.A. Bradley, M.I. Chilvers et al. 2012. Biology, yield loss and control of *Sclerotinia* stem rot of soybean. *J. Integr. Pest Manag.* 3: 1–7.

Pereira, C.E., J.A. Oliveira, M.C.M. Rosa, G.E. Oliveira, and N.J. Costa. 2009. Fungicide treatment of soybean seeds inoculated with *Colletotrichum truncatum. Ciencia Rural* 39: 2390–2395.

Perez-Brandan, C., J.L. Arzeno, J. Huidobro et al. 2012. Long-term effect of tillage systems on soil microbiological, chemical and physical parameters and the incidence of charcoal rot by *Macrophomina phaseolina* (Tassi) Goid in soybean. *Crop Prot.* 40: 73–82.

Pérez-Vicente, L., E. Martínez-de la Parte, M. Pérez-Miranda et al. 2010. First report of Asian rust of soybean caused by *Phakopsora pachyrhizi* in Cuba. *New Dis. Rep.* 20: 32.

Pham, T.A., C.B. Hill, M.R. Miles et al. 2010. Evaluation of soybean for resistance to soybean rust in Vietnam. *Field Crops Res.* 117: 131–138.

Pham, T.A., M.R. Miles, R.D. Frederick, C.B. Hill, and G.L. Hartman. 2009. Differential responses of resistant soybean entries to isolates of *Phakopsora pachyrhizi. Plant Dis.* 93: 224–228.

Pinheiro, J.B., E.A. Pozza, A.A.A. Pozza, A. de S. Moreira, and M.C. Alves. 2011. Effect of potassium and calcium supplied via nutrient solution on the severity of Asian soybean rust. *Rev. Ceres* 58(1): 43–50.

Pinheiro, J.B., E.A. Pozza, A.A.A. Pozza, A.S. Moreira, and V.P. Campos. 2009. Influence of potassium and calcium on the soybean cyst nematode reproduction *Nematol. Bras.* 33: 17–27.

Pioli, R.N., M.V. Cambursano, and E.N. Morandi. 2005. Morphologic and pathometric characterization of the Asian soybean rust (*Phakopsora pachyrhizi*) in Santa Fe Province, Argentina. *Plant Dis.* 89: 684.

Pioli, R.N., E.N. Morandi, M.C. Martinez et al. 2003. Morphologic, molecular, and pathogenic characterization of *Diaporthe phaseolorum* variability in the core soybean-producing area of Argentina. *Phytopathology* 93: 136–146.

Pivonia, S., X.B. Yang, and Z. Pan. 2005. Assessment of epidemic potential of soybean rust in the United States. *Plant Dis.* 89: 678–682.

Ploper, L.D., M. Roberto Galvez, V. Gonzalez, H. Jaldo, and M.R. Devani. 2001. Evaluation of fungicides for control of late season diseases of soybean in Tucuman, Argentina. *Revista Industrial y Agricola de Tucuman* 77: 59–69.

Pogetto, M.H.F., C.G. do A. dal Raetano, R. de S. Christovam et al. 2012. Asian soybean rust control and soybean yield after pyraclostrobin + epoxiconazole spraying in different phenological growth stages. *Summa Phytopathol.* 38: 248–250.

Ponte, E.M., C.V. Godoy, X. Li, and X.B. Yang. 2006. Predicting severity of Asian soybean rust epidemics with empirical rainfall models. *Phytopathology* 96: 797–803.

Prapagdee, B., K. Kotchadat, A. Kumsopa, and N. Visarathanonth. 2007. The role of chitosan in protection of soybean from sudden death syndrome caused by *Fusarium solani* f. sp. *glycines. Bioresour. Technol.* 98: 1353–1358.

Prasanth, P.S. and P.V. Patil. 2007. Purple seed stain of soybean—Its incidence, effect on seed quality and integrated management. *Indian Phytopathol.* 60: 482–488.

Prathuangwong, S. and K. Amnuaykit. 1987. Studies on tolerance and rate reducing bacterialpustule of soybean cultivars/lines. *The Kasetsart J.* (*Nat. Sci.*) 21: 408–420.

Qi, X.P., M.W. Li, M. Xie et al. 2014. Identification of a novel salt tolerance gene in wild soybean by whole-genome sequencing. *Nat. Commun.* 5: Article No. 4340.

Qian, H.L., S.X. Li., Y.L. Xu, C.J. Li, and Y.Q. Sun. 2011. Effects of two *Hirsutella minnesotensis* isolates on soybean cyst nematode (SCN) and soybean growth. *Soybean Sci.* 30: 266–227.

Quirino, B.F. and A.F. Bent. 2003. Deciphering host resistance and pathogen virulence: The *Arabidopsis/ Pseudomonas* interactions as a model. *Mol. Plant Pathol.* 4: 17–30.

Radhakrishnan, G.K., G.A. Splitter, and U. Ramakrishnan. 2008. DNA recognition properties of the cell-to-cell movement protein (MP) of soybean isolate of Mungbean yellow mosaic India virus (MYMIV-Sb). *Virus Res.* 131: 152–159.

Radwan, O., L.V. Rouhana, G.L. Hartman, and S.S. Korban. 2013. Genetic mechanisms of host-pathogen interactions for charcoal rot in soybean. *Plant Mol. Biol. Rep.* 32: 617. doi:10.1007/s11105-013-0686-9.

Rahangdale, S.R. and V.M. Raut. 2003. Evaluation of soybean germplasm lines for rust (*Phakopsora pachyrhizi*) resistance. *Indian J. Agric. Sci.* 73: 120–121.

Rahayu, M. 2014. Identification and pathogenicity of pathogen responsible for aerial blight disease of soybean. *J. Exp. Biol. Agric. Sci.* 2(2s): 279–285.

Raj, R.M., K. Kant, and D.D. Kulshrestha. 2002. Screening soybean cultivars for seed mycoflora and effect of thiram treatment thereon. *Seed Res.* 30: 118–121.

Ramos, A.M., M. Gaily, M.C. Garcia, and L. Levin. 2010. Pectinolytic enzyme production by *Colletotrichum truncation*, causal agent of soybean anthracnose. *Rev. Iberoam. Micol.* 27: 186–190.

Ramos, A.M., L.F. Tadic, I. Cinto, M. Carmona, and M. Gally. 2013. Molecular characterization of *Colletotrichum* species causing soybean anthracnose in Argentina. *Mycotaxon* 123: 457–465.

Ramteke, R. and G.K. Gupta. 2005. Field screening of soybean, *Glycine max* (L.) Merrill. lines for resistance to yellow mosaic virus. *J. Oilseeds Res.* 22: 224–225.

Ramteke, R., G.K. Gupta, B.S. Gill, R.K. Varma, and S.K. Lal. 2007. Development of soybean lines resistant to yellow mosaic virus. *Soybean Res.* 5: 71–74.

Ramteke, R., P.G. Karmakar, G.K. Gupta, R.K. Singh, and I.R. Khan. 2003. Resistance genes for rust and yellow mosaic diseases in soybean—A review. *J. Oilseeds Res.* 20: 195–203.

Ramezani, M., W.T. Shier, H.K. Abbas, J.L. Tonos, R.E. Baird, and G.L. Sciumbato. 2007. Soybean charcoal rot disease fungus *Macrophomina phaseolina* in Mississippi produces the phytotoxin (–)-botryodiplodin but no detectable phaseolinone. *J. Nat. Prod.* 70: 128–129.

Rathod, L.R. and P.V. Pawar. 2012. Antimicrobial activity of medicinal plant to control seed borne pathogen of soybean. *Curr. Bot.* 3: 10–12.

Ray, J.D., W. Morel, J.R. Smith, R.D. Frederick, and M.R. Miles. 2009. Genetics and mapping of adult plant rust resistance in soybean PI 587886 and PI 587880A. *TAG Theor. Appl. Genet.* 119: 271.

Ray, J.D., J.R. Smith, W. Morel, N. Bogado, and D.R. Walker. 2011. Genetic resistance to soybean rust in PI567099A is at or near the Rpp3 locus. *J. Crop Improv.* 25: 219–231.

Rayatpanah, S., S.V. Alavi, and G. Arab. 2007. Reaction of some soybean advanced lines to charcoal rot disease, *Macrophomina phaseolina* (Tassi) Goid, in East Mazandaran. *Seed Plant* 23: 181–189.

Rayatpanah, S., S.A. Dalili, and E. Yasari. 2012a. Diversity of *Macrophomina phaseolina* (*Tassi*) Goid based on chlorate phenotypes and pathogenicity. *Int. J. Biol.* 4: 54–63.

Rayatpanah, S., S.G. Nanagulyan, S.V. Alav, M. Razavi, and A. Ghanbari-Malidarreh. 2012b. Pathogenic and genetic diversity among Iranian isolates of *Macrophomina phaseolina*. *Chilean J. Agric. Res.* 72: 40–44.

Reddy, M.S.S., S.A. Ghabrial, C.T. Redmond, R.D. Dinkins, and G.B. Collins. 2001. Resistance to Bean pod mottle virus in transgenic soybean lines expressing the capsid polyprotein. *Phytopathology* 91: 831–838.

Reyes-Ramirez, A., B.I. Escudero-Abarca, G. Aguilar-Uscanga, P.M. Hayward-Jones, and J.E. Barboza-Corona. 2004. Antifungal activity of *Bacillus thuringiensis* chitinase and its potential for the biocontrol of phytopathogenic fungi in soybean seeds. *J. Food Sci.* 69: 131–134.

Rezende, A.A. and F.C. Juliatti. 2010. Soybean seeds treatment with fluquinconazole in the control of the Asian rust. *Biosci. J.* 26: 84–94.

Ribeiro, A.S., J.U.V. Moreira, P.H.B. Pierozzi et al. 2007. Genetic control of Asian rust in soybean. *Euphytica* 157: 15–25.

Ribeiro, A.S., J.F.F. Toledo, and M.A.P. de Ramalho. 2009. Interference of genotypes × environments interaction in the genetic control of resistance to Asian rust soybean. *Pesq. Agropec. Bras.* 44: 1160–1167.

Riga, E., T. Welacky, J. Potter, T. Anderson, E. Topp, and A. Tenuta. 2001. The impact of plant residues on the soybean cyst nematode, *Heterodera glycines*. *Can. J. Plant Pathol.* 23: 168–173.

Robertson, A. and L. Leandro. 2010. Answers to questions about soybean sudden death syndrome in Iowa 2010. *Integrated Crop Management News*. Iowa State University Extension Services, Ames, IA.

Roden, B. 2014. Bayer crop science submits EPA registration application for ILeVO, first seed treatment to tackle soybean sudden death syndrome. http://www.cropscience.bayer.com/en/Products-and-Innovaion/Brands/SeedGrowh.aspx. Accessed on April 14, 2014.

Rodrigues, A.C., G.N. Jham, and R.D. Oliveira. 2001. Mortality of the soybean cyst nematode in aqueous extracts of neem plant parts. *Nematol. Mediterr.* 29: 173–175.

Rodrigues, F.A., H.S.S. Duarte, G.P. Domiciano, C.A. Souza, G.H. Korndorfer, and L. Zambolim. 2009. Foliar application of potassium silicate reduces the intensity of soybean rust. *Aust. Plant Pathol.* 38: 366–372.

Rodriguez, M.A., G. Cabrera, F.C. Gozzo, M.N. Eberlin, and A. Godeas. 2011. *Clonostachys rosea* BAFC3874 as a *Sclerotinia sclerotiorum* antagonist: Mechanisms involved and potential as a biocontrol agent. *J. Appl. Microbiol.* 110: 1177–1186.

Roese, A.D., C.L.P. de Melo, and A.C.P. Goulart. 2012. Row spacing and Asian soybean rust severity. *Summa Phytopathol.* 38: 300–305.

Rogovska, N.P., A.M. Blackmer, and G.L. Tylka. 2009. Soybean yield and soybean cyst nematode densities related to soil pH, soil carbonate concentrations, and alkalinity stress index. *Agron. J.* 101: 1019–1026.

Ross, J.P. and A.K. Butler. 1985. Distribution of bean pod mottle virus in soybeans in North Carolina. *Plant Dis.* 69: 101–103.

Rosso, M.L., J.C. Rupe, P.Y. Chen, and L.A. Mozzoni. 2008. Inheritance and genetic mapping of resistance to *Pythium* damping-off caused by *Pythium aphanidermatum* in 'Archer' soybean. *Crop Sci.* 48: 2215–2222.

Rotundo, J.L., G.L. Tylka, and P. Pedersen. 2010. Source of resistance affect soybean yield, yield components, and biomass accumulation in *Heterodera glycines*-infested fields. *Crop Sci.* 50: 2565–2574.

Rousseau, G., S. Rioux, and D. Dostaler. 2007. Effect of crop rotation and soil amendments on *Sclerotinia* stem rot on soybean in two soils. *Can. J. Plant Sci.* 87: 605–614.

Roy, K.W. 1997. *Fusarium solani* on soybean roots: Nomenclature of the causal agent of sudden death syndrome and identity and relevance of *F. solani* form B. *Plant Dis.* 81: 259–226.

Roy, K.W., D.E. Hershman, J.C. Rupe and T.S. Abney. 1997. Sudden death syndrome of soybean. *Plant Dis.* 81: 1100–1111.

Rukayadi, Y., A. Suwanto, B. Tjahjono, and R. Harling. 2000. Survival and epiphytic fitness of a nonpathogenic mutant of *Xanthomonas campestris* pv. *glycines*. *Appl. Environ. Microbiol.* 66: 1183–1189.

Rupe, J. and L. Sconyers. 2008. Soybean Rust. American Phytopathological Society The Plant Health Instructor. Doi:10.1094/PHI-1-2008-0401-01.

Saber, M.M., Y.A. Abdou, S.M. El-Gantiry, and S.S. Ahmed. 2003. Biocontrol of anthracnose disease of soybean caused by *Colletotrichum dematium*. *Egypt. J. Phytopathol.* 31: 17–29.

Salerno, C.M. and M.A. Sagardoy. 2003. Antagonistic activity by *Bacillus subtilis* against *Xanthomonas campestris* pv. *glycines* under controlled conditions. *Spanish J. Agric. Res.* 1: 55–58.

Sandhu, D., I.M. Tasma, R. Frasch, and M.K. Bhattacharyya. August 5, 2009. Systemic acquired resistance in soybean is regulated by two proteins, orthologous to *Arabidopsis* NPR1. *BMC Plant Biol.* 9: 105.

Santos, J.M., K. Vrandecic, J. Cosic, T. Duvnjak, and A.J.L. Phillips. 2011. Resolving the *Diaporthe* species occurring on soybean in Croatia. *Persoonia* 27: 9–19.

Sastrahidayat, I.R., S. Djauhari, B. Prasetya, and N. Saleh. 2011. Biocontrol of damping-off disease (*Sclerotium rolfsii* Sacc.) using *Actinomycetes* and VAM fungi on soybean and impact to crop production and microorganism diversity in rhizosfer zone. *Int. J. Acad. Res.* 3: 114–119.

Scandiani, M.M., T. Aoki, A.G. Luque, M.A. Carmona, and K. O'Donnell. 2010. First report of sexual reproduction by the soybean sudden death syndrome pathogen *Fusarium tucumaniae* in nature. *Plant Dis.* 94: 1411–1416.

Scandiani, M.M., M.A. Carmona, A.G. Luque et al. 2012. Isolation, identification and yield losses associated with sudden death syndrome in soybeans in Argentina. *Trop. Plant Pathol.* 37: 358–362.

Scandiani, M.M., D.S. Ruberti, L.M. Giorda et al. 2011. Comparison of inoculation methods for characterizing relative aggressiveness of two soybean sudden-death syndrome pathogens, *Fusarium virguliforme* and *F. tucumaniae*. *Trop. Plant Pathol.* 36: 133–140.

Scherm, H., R.S.C. Christiano, P.D. Esker, E.M. Ponte, and C.V. Godoy. 2009. Quantitative review of fungicide efficacy trials for managing soybean rust in Brazil. *Crop Prot.* 28: 774–782.

Schmitt, D.P., J.A. Wrather, and R.D. Riggs (eds.). 2004. *Biology and Management of the Soybean Cyst Nematode*, 2nd edn. Schmitt & Associates of Marceline, Marceline, MO.

Schneider, K.T., M. van de Mortel, T.J. Bancroft et al. 2011. Biphasic gene expression changes elicited by *Phakopsora pachyrhizi* in soybean correlate with fungal penetration and haustoria formation. *Plant Physiol.* 157: 355–371.

Sebastian, S.A., C.D. Nickell, and L.E. Gray. 1986. Relationship between greenhouse and field ratings for brown stem rot reaction in soybean. *Crop Sci.* 26: 665–667.

Shahraeen, N., T. Ghotbi, M. Salati, and A. Sahandi. 2005. First report of *Bean pod mottle virus* in soybean in Iran. *Plant Dis.* 89: 775.

Shanmugasundaram, S., M.R. Yan, and T.C. Wang. 2004. Breeding for soybean rust resistance in Taiwan. In: *Proceedings VII World Soybean Research Conference, IV International Soybean Processing and Utilization Conference*, Foz do Iguassu, PR, Brazil, February 29–March 5, 2004, pp. 456–462.

Sharma, A.N., G.K. Gupta, R.K. Sharma et al. 2014. *Integrated Pest Management for Soyabean*. Directorate of Plant Protection, Quarantine & Storage, CGO Complex, Faridabad, India, p. 41.

Sharma, R. 2009. Genetic differentiation of host limited forms of *Colletotrichum truncatum* from northwestern Himalayas. *Arch. Phytopathol. Plant Prot.* 42: 960–966.

Sharifi-Tehrani, A., M. Shakiba, M. Okhovat, and Z. Zakeri. 2005. Biological control of *Tiarosporella phaseolina* the causal agent of charcoal rot of soybean. *Commun. Agric. Appl. Biol. Sci.* 70: 189–192.

Shivakumar, M., G.T. Basavaraja, P.M. Salimath, P.V. Patil, and A. Talukdar. 2011. Identification of rust resistant lines and their genetic variability and character association studies in soybean [*Glycine max* (L.) Merr.]. *Indian J. Genet. Plant Breed.* 71: 235–240.

Shovan, L.R., M.K.A. Bhuiyan, N. Sultana, J.A. Begum, and Z. Pervez. 2008. Prevalence of fungi associated with soybean seeds and pathogenicity tests of the major seed-borne pathogens. *Int. J. Sustain. Crop Prod.* 3: 24–33.

Shukla, B.N., P.K. Bhargava, M. Gupta, and R.P. Patel. 2005. Epidemiology and management of soybean rust (*Phakopsora pachyrhizi*). *Indian J. Agric. Sci.* 75: 150–153.

Shukla, C.S., K.P. Verma, P. Deepthi and S. Khalko. 2014. Influence of environmental conditions on the incidence of anthracnose of soybean. Abstr. Paper presented at the National symposium on "Plant Pathology in Genomic Era" organized by Indian Phytopathological Society, New Delhi at Indira Gandhi Krishi Vishwavidyalaya, Raipur, Chattisgarh, India, May 26–28, 2014.

Sinha, A. and H. Reyes. 2009. Evaluation of fungicides for the management of Asian soybean rust. *CARDI Rev.* 8: 14–21.

Singh, B., S. Bhatia, V.C. Chalam et al. 2003. Qualitative processing of transgenic planting material. *Indian J. Agric. Sci.* 73: 97–100.

Singh, B., Q.M.R. Haq, and V.G. Malathi. 2013. Antisense RNA approach targeting Rep gene of Mungbean yellow mosaic India virus to develop resistance in soybean. *Arch. Phytopathol. Plant Prot.* 46: 2191–2207.

Singh, R. and B.K. Singh. 2001. Effect of temperature and pH on growth and sporulation of *Colletotrichum truncatum* causing anthracnose of soybean. *Prog. Agric.* 1: 63–65.

Smith, K.L. 2005. Soybean rust—A threat to agriculture in the United States. *Outlooks Pest Manag.* 16: 110–113.

Soares, R.M., S. de A.L. Rubin, A.P. Wielewicki, and J.G. Ozelame. 2004. Fungicides on the control of soybean rust (*Phakopsora pachyrhizi*) and soybean yield. *Ciencia Rural* 34: 1245–1247.

Sonavane, A.A., B.G. Barhate, and S.J. Bade. 2011. Efficacy of bioagents, botanicals and fungicides on seed mycoflora of soybean. *J. Plant Dis. Sci.* 6: 74–76.

Srivastava, P., S. George, J.J. Marois, D.L. Wright, and D.R. Walker. 2011. Saccharin-induced systemic acquired resistance against rust (*Phakopsora pachyrhizi*) infection in soybean: Effects on growth and development. *Crop Prot.* 30: 726–732.

Srivastava, S.K., X. Huang, H.K. Brar et al. 2014. The genome sequence of the fungal pathogen *Fusarium virguliforme* that causes sudden death syndrome in soybean. *PLoS One* 9(1): e81832. doi:10.1371/journal.pone.0081832.

Stephan, Z.A., H.B. Dawood, and A.R. Nasir. 2006. Effect of root-knot nematode *Meloidogyne javanica* on germination and different ages of soybean seedlings growth and its interaction with the fungi *Rhizoctonia solani* and *Macrophomina phaseolina*. *Arab J. Plant Prot.* 24: 98–101.

Stephan, Z.A., K.S. Juber, and H.B. Dawood. 2005. Isolating fungi from the soybean seeds and plants and their effects on seed germination and seedlings and their biological control. *Arab J. Plant Prot.* 23: 51–56.

Stetina, K.C., S.R. Stetina, and J.S. Russin. 2006. Comparison of severity assessment methods for predicting yield loss to Rhizoctonia foliar blight in soybean. *Plant Dis.* 90: 39–43.

Sumartini. 2010. Rust disease on soybean and its environmentally-friendly control measure. *Jurnal Penelitian dan Pengembangan Pertanian (Indonesia)* 29: 107–112.

Suzuki, C., Y. Tanaka, T. Takeuchi, S. Yumoto, and S. Shirai. 2012. Genetic relationships of soybean cyst nematode resistance originated in Gedenshirazu and PI84751 on Rhg1 and Rhg4 loci. (Special Issue: Soybean breeding in genomic era.) *Breed. Sci.* 61: 602–607.

Swoboda, C.M., P. Pedersen, P.D. Esker, and G.P. Munkvold. 2011. Soybean yield response to plant distribution in *Fusarium virguliforme* infested soils. *Agron. J.* 103: 1712–1716.

Talukdar, A., G.D. Harish, B. Kumar et al. 2013. Genetics of yellow mosaic virus (YMV) resistance in cultivated soybean (*Glycine max* (L.) Merr). *Legume Res.* 36: 263–267.

Tang, E., C.B. Hill, and G.L. Hartman. 2010. Carbon utilization profiles of *Fusarium virguliforme* isolates. *Can. J. Microbiol.* 56: 979–986.

Tao, Z.N., D. Malvick, R. Claybrooke et al. 2009. Predicting the risk of soybean rust in Minnesota based on an integrated atmospheric model. *Int. J. Biometeorol.* 53: 509–521.

Tariq, M., D. Shahnaz, and F.S. Mehdi. 2006. Location of fungi on different parts of soybean seed. *Int. J. Biol. Biotechnol.* 3: 103–105.

Tazawa, J., H. Yamamoto, K. Usuki, and S. Miura. 2008. Effects of tillage method, fertilizer, and crop rotation on population dynamics of soybean cyst nematode (*Heterodera glycines*) and soybean growth. *Jpn. J. Crop Sci.* 77: 33–40.

Tehrani, A.S., A. Zebarjad, G.A. Hedjaroud, and M. Mohammadi. 2002. Biological control of soybean damping-off by antagonistic rhizobacteria. Meded. *Fac. Landbouwkd. Toegep. Biol. Wet. Univ. Gent.* 67: 377–380.

Teramoto, A., T.A. Machado, L.M. Santos, M.R. dos Volf, M.C. Meyer, and M.G. da Cunha. 2013. Reaction of soybean cultivars to *Corynespora cassiicola*. *Trop. Plant Pathol.* 38: 68–71.

Tormen, N.R., F.D.L. da Silva, M.P. Debortoli, J.D. Uebel, D.D. Favera, and R.S. Balardin. 2012. Drop deposition on canopy and chemical control of *Phakopsora pachyrhizi* in soybeans. *Revista Brasileira de Engenharia Agricola e Ambiental* 16: 802–808.

Tran, D.K., Q.A. Truong, T.B. Bui, and D.X. Tran. 2013. Applying molecular breeding to improve soybean rust resistance in Vietnamese elite soybean. *Am. J. Plant Sci.* 4: 1–6.

Tripathi, M., B.N. Johri, and A. Sharma. 2006. Plant growth-promoting *Pseudomonas* sp. strains reduce natural occurrence of anthracnose in soybean (*Glycine max* L.) in Central Himalayan Region. *Curr. Microbiol.* 52(5): 390–394.

Tschurtschenthaler, N.N., E.S.N. Vieira, T. dalla Nora, and I. Schuster. 2012. Genetic variability of *Phakopsora pachyrhizi* accessed by microsatellite markers. *Pesq. Agropec. Bras.* 47: 181–186.

Tukamuhabwa, P., H.K. Oloka, T. Sengooba, and P. Kabayi. 2012. Yield stability of rust-resistant soybean lines at four mid-altitude tropical locations. *Euphytica* 183: 1–10.

Twizeyimana, M. and G.L. Hartman. 2012. Pathogenic variation of *Phakopsora pachyrhizi* isolates on soybean in the United States from 2006 to 2009. *Plant Dis.* 96: 75–81.

Twizeyimana, M., C.B. Hill, M. Pawlowski, C. Paul, and G.L. Hartman. 2012. A cut-stem inoculation technique to evaluate soybean for resistance to *Macrophomina phaseolina*. *Plant Dis.* 96: 121015.

Twizeyimana, M., P.S. Ojiambo, G.L. Hartman, and R. Bandyopadhyay. 2011b. Dynamics of soybean rust epidemics in sequential plantings of soybean cultivars in Nigeria. *Plant Dis.* 95: 43–50.

Twizeyimana, M., P.S. Ojiambo, J.S. Haudenshield et al. 2011a. Genetic structure and diversity of *Phakopsora pachyrhizi* isolates from soyabean. *Plant Pathol.* 60: 719–729.

Twizeyimana, M., P.S. Ojiambo, T. Ikotun, C. Paul, G.L. Hartman, and R. Bandyopadhyay. 2007. Comparison of field, greenhouse, and detached-leaf evaluations of soybean germplasm for resistance to *Phakopsora pachyrhizi*. *Plant Dis.* 91: 1161–1169.

Twizeyimana, M., P.S. Ojiambo, K. Sonder, T. Ikotun, G.L. Hartman, and R. Bandyopadhyay. 2009. Pathogenic variation of *Phakopsora pachyrhizi* infecting soybean in Nigeria. *Phytopathology* 99: 353–361.

Ugwuoke, K.I. 2002. Controlling *Bean pod mottle virus* (BPMV) (Genus *Comovirus*) of soybean with spatial arrangement of maize–soybean in southeastern. *Niger. Agro-Sci.* 2: 27–34.

Uragami, A., M. Morishita, H. Hirokane, F. Sato, S. Tokuda, and H. Higashio. 2005. Control of soybean cyst nematode (*Heterodera glycines*) in vegetable soybean cultivation using a high soil temperature and cultivation of the resistant cultivar, 'Peking'. *Hort. Res.* (Japan) 4: 219–223.

Usharani, K.S., B. Surendranath, Q.M.R. Haq, and V.G. Malathi. 2004. Yellow mosaic virus infecting soybean in northern India is distinct from the species-infecting soybean in southern and western India. *Curr. Sci.* 86: 845–850.

Usharani, K.S., B. Surendranath, Q.M.R. Haq, and V.G. Malathi. 2005. Infectivity analysis of a soybean isolate of Mungbean yellow mosaic India virus by agroinoculation. *J. Gen. Plant Pathol.* 71: 230–237.

Van, K.J., P. Lestari, Y.J. Park et al. 2007. Differential gene expression of soybean [Glycine max (L.) Merr.] in response to *Xanthomonas axonopodis pv. glycines* by using an oligonucleotide macroarray. *J. Crop Sci. Biotechnol.* 10:147–158.

Vasconcelos, E.A.R., C.G. Santana, C.V. Godoy et al. 2011. A new chitinase-like xylanase inhibitor protein (XIP) from coffee (*Coffea arabica*) affects Soybean Asian rust (*Phakopsora pachyrhizi*) spore germination. *BMC Biotechnol.* 11: 14.

Vauterin, L., B. Hoste, K. Kerters, and J. Swings. 1995. Reclassification of *Xanthomonas*. *Int. J. Syst. Bacteriol.* 45: 472–489.

Verma, K.P., M.P. Thakur, K.C. Agrawal, and N. Khare. 2004. Occurrence of soybean rust: Some studies in Chhattisgarh state. *J. Mycol. Plant Pathol.* 34: 24–27.

Vidic, M. and S. Jasnic. 2008a. Downy mildew (*Peronospora manshurica*) of soyabean. *Biljni Lekar* (*Plant Doctor*) 36: 197–201.

Vidic, M. and S. Jasnic. 2008b. White mould—The most harmful disease of soybean. *Biljni Lekar* (*Plant Doctor*) 36: 201–207.

Vidic, M., S. Jasnic, and K. Petrovic. 2011. *Diaporthe/Phomopsis* species on soybean in Serbia. *Pesticidi i Fitomedicina* 26: 301–315.

Villarroel, D.A., R.E. Baird, L.E. Trevathan, C.E. Watson, and M.L. Scruggs. 2004. Pod and seed mycoflora on transgenic and conventional soybean [*Glycine max* (L.) Merrill] cultivars in Mississippi. *Mycopathologia* 157: 207–215.

Vittal, R., J.S. Haudenshield, and G.L. Hartman. 2012a. A multiplexed immunofluorescence method identifies *Phakopsora pachyrhizi* urediniospores and determines their viability. *Phytopathology* 102: 1143–1152.

Vittal, R., H.C. Yang, and G.L. Hartman. 2012b. Anastomosis of germ tubes and migration of nuclei in germ tube networks of the soybean rust pathogen, *Phakopsora pachyrhizi*. *Eur. J. Plant Pathol.* 132: 163–167.

Vuong, T.D., B.W. Diers, and G.L. Hartman. 2008. Identification of QTL for resistance to Sclerotinia stem rot in soybean plant introduction 194639. *Crop Sci.* 48: 2209–2214.

Vuong, T.D., D.A. Sleper, J.G. Shannon, X. Wu, and H.T. Nguyen. 2011. Confirmation of quantitative trait loci for resistance to multiple-HG types of soybean cyst nematode (*Heterodera glycines* Ichinohe). *Euphytica* 181(1): 101–113.

Walker, D.R., H.R. Boerma, D.V. Phillips et al. 2011. Evaluation of USDA soybean germplasm accessions for resistance to soybean rust in the southern United States. *Crop Sci.* 51: 678–693.

Walters, H.J. 1958. A virus disease complex in soybeans in Arkansas (Abstr.). *Phytopathology* 48: 346.

Wang, G.R., Z.F. Sun, W.J. Chen et al. 2012b. Screening of effective fungicides and optimum period for controlling soybean pod anthracnose. *Acta Agric. Zhejiangensis* 24: 258–262.

Wang, J., P.A. Donald, T.L. Niblack et al. 2000. Soybean cyst nematode reproduction in the North Central United States. *Plant Dis.* 84: 77–82.

Wang, R.Y., A. Kritzman, D.E. Hershman, and S.A. Ghabrial. 2006. Aphis glycines as a vector of persistently and nonpersistently transmitted viruses and potential risks for soybean and other crops. *Plant Dis.* 90: 920–926.

Wang, Y., X.Z. Yuan, H. Hu et al. 2012a. Proteomic analysis of differentially expressed proteins in resistant soybean leaves after *Phakopsora pachyrhizi* infection. *J. Phytopathol.* 160: 554–560. doi:10.1111/j.1439-434.2012.01949.x.

Wang, Z.H., L.B. Shi, H.Y. Wu, J. Liu, and X.X. Li. 2009. Distribution and developmental process of *Heterodera glycines* in soybean root. *Sci. Agric. Sin.* 42: 3147–3153.

Ward, N.A., C.L. Robertson, A.K. Chanda, and R.W. Schneider. 2012. Effects of *Simplicillium lanosoniveum* on *Phakopsora pachyrhizi*, the soybean rust pathogen, and its use as a biological control agent. *Phytopathology* 102: 749–760.

Wegulo, S.N., P. Sun, C.A. Martinson, and X.B. Yang. 2000. Spread of *Sclerotinia* stem rot of soybean from area and point sources of apothecial inoculums. *Can. J. Plant Sci.* 80: 389–402.

Wei, L.X., A.G. Cober, E.R. Babcock et al. 2011. Pathogenicity of *Pythium* species causing seed rot and damping-off in soybean under controlled conditions. *Phytoprotection* 91: 3–10.

Wen, Z., R. Tan, J. Yuan et al. 2014. Genome-wide association of mapping of quantitative resistance to sudden death syndrome in soybean. *BMC Genomics* 15: 809. doi:10.1186/1471-2164-15-809.

Wensing, A., S.D. Braun, and P. Buttner. 2010. Impact of siderophore production by *Pseudomonas syringae* pv. *syringae* 22d/93 on epiphytic fitness and biocontrol activity against *Pseudomonas syringae* pv. *glycinea* 1a/96. *Appl. Environ. Microbiol.* 76: 2704–2711.

Westphal, A., T.S. Abney, L.J. Xing, and G.E. Shaner. 2008. Sudden death syndrome of soybean. *Plant Health Instructor*. doi:10.1094/PHI-I-2008-0102-01.

Westphal, A. and L. Xing. 2006. Diseases of soybean: Root knot of soybean. Purdue Extension, Purdue University, West Lafayette, IN. Knowledge to go. BP-130-W.

Westphal, A. and L.J. Xing. 2011. Soil suppressiveness against the disease complex of the soybean cyst nematode and sudden death syndrome of soybean. *Phytopathology* 101: 878–886.

Wokocha, R.C. 2000. Effect of different soil moisture regimes on the development of the charcoal rot disease of soybean by *Macrophomina phaseolina*. *Global J. Pure Appl. Sci.* 6: 599–602.

Wrather, A., G. Shannon, R. Balardin et al. January 2010. Effect of diseases on soybean yield in the top eight producing countries in 2006. *Plant Health Prog.* doi:10.1094/PHP-2010-0125-01-RS.

Wrather, J.A., S.R. Kending, and S.C. Anand. 1995. Effects of tillage, cultivar, and planting date on percentage of soybean leaves with symptoms of sudden death syndrome. *Plant Dis.* 79: 560–562.

Wrather, J.A. and S.R. Koenning. 2006. Estimates of disease effects on soybean yields in the United States 2003 to 2005. *J. Nematol.* 38(2): 173–180.

Wrather, J.A. and S.R. Koenning. 2009. Effects of diseases on soybean yields in the United States 1996–2007. Online. *Plant Health Prog.* doi:10.1094/PHP-2009-0401-01-RS.

Wrather, J.A., S.R. Koenning, and T. Aderson. 2003. Effects of diseases on soybean yields in the United States and Ontario (1999–2002). Online. *Plant Health Prog.* doi:10.1034/PHP// 2003 0325 01RV.

Wrather, J.A., J.G. Shannon, T.E. Carter, J.P. Bond, J.C. Rupe, and A.M.R. Almeida. June 2008. Reaction of drought-tolerant soybean genotypes to *Macrophomina phaseolina*. *Plant Health Prog.* doi:10.1094/PHP-2008-0618-01-RS.

Wu, H.Y. and Y.X. Duan. 2011. Defense response of soybean (*Glycine max*) to soybean cyst nematode (*Heterodera glycines*) race 3 infection. *J. Anim. Plant Sci.* 21: 165–170.

Xiang, Y., T.K. Hermam, and G.L. Hartman. 2014. Using soybean milk to culture soybean pathogens. *Adv. Microbiol.* 4: 126–132.

Xiao, J.L., J. Zhu, S.Y. Chen, W.B. Ruan, and C. Miller. 2007. A novel use of anaerobically digested liquid swine manure to potentially control soybean cyst nematode. *J. Environ. Sci. Health Part B, Pestic. Food Contam. Agric. Wastes* 42: 749–757.

Xing, L.J., D. Chen, W.Z. Liu, and Y. Duan. 2002. Influence of some ions on biocontrol system of soybean cyst nematode. *Chin. J. Biol. Control* 18: 1–5.

Xing, L.J. and A. Westphal. 2006. Interaction of *Fusarium solani* f. sp. *glycines* and *Heterodera glycines* in sudden death syndrome of soybean. *Phytopathology* 96: 763–770.

Xing, L.J., and A. Westphal. 2009. Effects of crop rotation of soybean with corn on severity of sudden death syndrome and population densities of *Heterodera glycines* in naturally infested soil. *Field Crops Res.* 112: 107–117.

Yadava, D.K., Sujata Vasudev, N. Singh, T. Mohapatra, and K.V. Prabhu. 2012. Breeding major oilcrops: Present status and future research needs. In: *Technological Innovations in Major World Oil Crops*, Vol. I: Breeding, ed., S.K. Gupta. Springer Science+Business Media, LLc., New York, pp. 17–51.

Yadav, R.K. and D. Chattopadhyay. 2014. Differential soybean gene expression during early phase of infection with Mungbean yellow mosaic India virus. *Mol. Biol. Rep.* 41: 5123–5134.

Yadav, R.K., R.K. Shukla, and D. Chattopadhyay. 2009. Soybean resistant to Mungbean Yellow Mosaic India Virus infection induces viral RNA degeneration earlier than the susceptible cultivar. *Virus Res.* 144: 89–95. doi:10.1016/s.viruses. 2009.04.0110.

Yamanaka, N., N.G. Lemos, H. Akamatsu et al. 2011. Soybean breeding materials useful for resistance to soybean rust in Brazil. *Jpn. Agric. Res. Q.* 45: 385–395.

Yamanaka, N., Y. Yamaoka, M. Kato et al. 2010. Development of classification criteria for resistance to soybean rust and differences in virulence among Japanese and Brazilian rust populations. *Trop. Plant Pathol.* 35: 153–162.

Yamaoka, Y., Y. Fujiwara, M. Kakishima, K. Katsuya, K. Yamada, and H. Hagiwara. 2002. Pathogenic races of *Phakopsora pachyrhizi* on soybean and wild host plants collected in Japan. *J. Gen. Plant Pathol.* 68: 52–56.

Yanez Morales, M. de J., I. Alanis Martinez, J.M. Soto Rocha et al. 2009. Soybean rust caused by *Phakopsora pachyrhizi* detected in the state of Campeche on the Yucatan Peninsula, Mexico. *Plant Dis.* 93: 847.

Yang, X.B. 2002. Phyllosticta leaf spot on soybean. Iowa State University Extension. *ICM-IC* 488(18): 46.

Yang, X.B. and S.S. Navi. 2005. First report of charcoal rot epidemics caused by *Macrophomina phaseolina* in soybean in Iowa. *Plant Dis.* 89: 526.

Yorinori, J.T. 2004. Soybean rust: General overview. In: *Proceedings VII World Soybean Research Conference, IV International Soybean Processing and Utilization Conference*, Foz do Iguassu, PR, Brazil, February 29–March 5, 2004, pp. 1299–1307.

Yorinori, J.T. and J.J. Lazzarotto. 2004. Current situation of soybean rust in Brazil and in South America. *Documentos—Embrapa Sojja* 236: 27pp.

Yorinori, J.T., W.M. Paiva, R.D. Frederick et al. 2005. Epidemics of soybean rust (*Phakopsora pachyrhizi*) in Brazil and Paraguay from 2001 to 2003. *Plant Dis.* 89: 675–677.

Young, H.M., S. George, D.F. Narvaez et al. 2012. Effect of solar radiation on severity of soybean rust. *Phytopathology* 102: 794–803.

Young, H.M., J.J. Marois, D.L. Wright, D.F. Narvaez, and G.K. O'Brien. 2011. Epidemiology of soybean rust in soybean sentinel plots in Florida. *Plant Dis.* 95: 744–750.

Young, N.D. and J. Mudge. 2002. Marker-assisted selection for soybean cyst nematode resistance. In: *Plant Resistance to Parasitic Nematodes*, eds., J.L. Starr, R. Cook and J.Bridge, CABI, Wallingford, U.K., pp. 241–252.

Yuan, C.P., Y.H. Li, Z.X. Liu, R.X. Guan, R.Z. Chang, and L.J. Qiu. 2012. DNA sequence polymorphism of the Rhg4 candidate gene conferring resistance to soybean cyst nematode in Chinese domesticated and wild soybeans. *Mol. Breed.* 30: 1155–1162.

Yuan, J., M. Zhu, D.A. Lightfoot et al. 2008. In silico comparison of transcript abundances during *Arabidopsis thaliana* and *Glycine max* resistance to *Fusarium virguliforme*. *BMC Genomics* 9: S6.

Zabala, G., J. Zou, J. Tuteja, D.O. Gonzalez, S.J. Clough, and L.O. Vodkin. 2006. Transcriptome changes in the phenylpropanoid pathway of *Glycine max* in response to *Pseudomonas syringae* infection. *BMC Plant Biol.* 6: 26. doi:10.1186/1471-2229-6-26.

Zebarjad, A., A.S. Tehrani, G.A. Hedjarood, and M. Mohammadi. 2006. A study on the effect of several antagonistic bacteria on control of soybean damping-off disease caused by *Phytophthora sojae*. *Iranian J. Agric. Sci.* 37: 671–686.

Zeng, W.T., D.C. Wang, W. Kirk, J.J. Hao. 2012. Use of *Coniothyrium minitans* and other microorganisms for reducing *Sclerotinia sclerotiorum. Biol. Control* 60: 225–232.

Zhang, L.M., E. Yang, M.C. Xiang, X.Z. Liu, and S.Y. Chen. 2008. Population dynamics and biocontrol efficacy of the nematophagous fungus *Hirsutella rhossiliensis* as affected by stage of the soybean cyst nematode. *Biol. Control* 47: 244–249.

Zheng, C., P. Chen, and R.C. Gergerich. 2005. Characterization of resistance to soybean mosaic virus in diverse soybean germplasm. *Crop Sci.* 45: 2503–2509.

Zheng, C., P. Chen, D.X. Li, and R.C. Gregerich. 2008. New sources of resistance to soybean mosaic virus in soybean. *Can. J. Plant Pathol.* 30: 595–603.

Zheng, Y., Y. Duan, S. Chen, J. Sun, and L. Chen. 2010. Responses of soybean cyst nematode *Heterodera glycines* to macroelement and microelement compounds. *Bulgarian J. Agric. Sci.* 16: 172–180.

Ziems, A.D., L.J. Giesler, G.L. Graef et al. 2007. Response of soybean cultivars to *Bean pod mottle virus* infection. *Plant Dis.* 91: 719–726.

Zwatz, B., G. Besenhofer, and R. Zederbauer. 2000. Investigations of damage effects and control methods of two new phytopathogens of soybeans in Austria: *Colletotrichum dematium* (Pers.) Grove f. *truncatum* (Schwein.) Andrus & W. D. Moore and *Diaporthe phaseolorum* (Cooke & Ellis) Sacc. *Pflanzenschutzberichte* 59: 55–66.

Index

G

H

I

K

M

O

P

R

S

T